International Conference on Transport and Infrastructure of the Siberian Region (SibTrans-2019)

IOP Conference Series: Materials Science and Engineering
Volume 760

Moscow, Russia
12 - 15 November 2019

ISBN: 978-1-7138-3500-4
ISSN: 1757-8981

Printed from e-media with permission by:

Curran Associates, Inc.
57 Morehouse Lane
Red Hook, NY 12571

Some format issues inherent in the e-media version may also appear in this print version.

Institute of Physics
Dirac House, Temple Back
Bristol BS1 6BE UK

Phone: 44 1 17 929 7481
Fax: 44 1 17 920 0979

techtracking@iop.org

Additional copies of this publication are available from:

Curran Associates, Inc.
57 Morehouse Lane
Red Hook, NY 12571 USA
Phone: 845-758-0400
Fax: 845-758-2633
Email: curran@proceedings.com
Web: www.proceedings.com

TABLE OF CONTENTS

EVALUATION OF THE STRAIN-STRESS CONDITION OF RAILS 1
A.Y. Abdurashitov, V.P. Sychev

MATHEMATICAL MODEL OF A RADIAL BEARING WITH A LOW-MELTING METAL COATING OF DESIGN MODELS OF HYDRODYNAMIC VISCOELASTIC LUBRICANT FORMED BY MELTING THE SURFACE OF A BEARING BUSH COATED WITH A METALLIC LOW-MELTING COATING 10
K S Akhverdiev, E O Lagunova, M A Mukutadze

PARALLEL OPERATION OF AN INVERTER WITH AN ELECTRICAL AC NETWORK 20
T L Alekseeva, N L Ryabchyonok, L A Astrakhantsev, V A Tikhomirov, N P Astashkov, A L Martusov, M E Alekseev

ANALYSING OIL AND GAS PRODUCTION FACILITIES IMPACT ON ATMOSPHERIC AIR OF POTOCHNOYE FIELD 27
E M Almukhametova, R R Sitdikov, A I Zakirov

SIMULATION OF FRACTIONAL EFFICIENCY OF DIRECT-FLOW CYCLONE WITH INTERMEDIATE DUST COLLECTOR 32
V S Aslamova, A A Aslamov, M I Arschinskii, Y V Zhilycov, E A Rudykh

VERIFICATION OF KNOWLEDGE MODELS BY MEANS OF LOGIC WITH VECTOR SEMANTICS 42
L V Arshinskiy, M S Nitezhuk, M P Dunaev

EXPERT SYSTEM FOR AGGREGATE INDUSTRIAL SAFETY ASSESSMENT AT ENTERPRISES BASED ON KNOWLEDGE TECHNOLOGIES 47
E A Aslamova, L V Arshinsky, V S Aslamova, M V Krivov

PRECISION ADJUSTMENT OF ELECTROMAGNETIC EMITTER OF LARGE-SCALE SYSTEMS SUCH AS GLONASS 57
A V Banshchikov, A V Daneev, V A Rusanov, A A Vetrov

APPLICATION OF GEODETIC DEVICES FOR THE OPERATION OF UNDERGROUND PIPELINES 63
G G Bikbulatova, S V Lyashko

ANALYSIS OF FAILURES OF BEARINGS OF AXLE BOX UNIT WITH POLYAMIDE CAGES AND PROSPECTS OF INCREASING THEIR SERVICE LIFE 68
D V Butorin, N G Filippenko, A V Livshits, S I Popov

DESIGN OF THE TOOLING OF A COMPLEX GEOMETRIC SHAPE FOR ELECTROTHERMAL PROCESSING OF THE POLYMER COVERING PLATE OF THE BOGIE FRICTION WEDGE 78
D V Butorin, A G Larchenko, A A Alexandrov

DETERMINING ELECTROMAGNETIC FIELDS GENERATED BY OVERHEAD POWER TRANSMISSION LINES 86
V P Zakaryukin, A V Kryukov, N V Buyakova

CURRENT TRENDS IN THE DEVELOPMENT OF LOGISTICS PROFESSIONAL DISCOURSE TERMINOLOGY 94
E E Bylina, T A Skopintseva, L V Sludneva, E A Yurkovskaya

DETERMINATION OF ELECTROMAGNETIC EFFECTS OF ELECTRIC TRACTION NETWORKS ON PIPELINES 102
A V Cherepanov, A E Kryukov

STUDY OF THERMAL ZONES IN A HEATING COLLECTOR OF A TRACTION MOTOR NB-418 K6 109
A D Chuiko, S A Christosenkov

USING ONTOLOGICAL AND ARCHITECTURAL APPROACHES FOR THE VEHICLE FLEET MANAGEMENT IN THE ENTERPRISE ENGINEERING CONTEXT 115
V M Kurganov, M V Gryaznov, A N Dorofeev, O B Nastasyak, D V Pervukhin

ABOUT CONTROL OF A NEURON AS AN ELEMENTARY UNIT OF MULTILAYER ARTIFICIAL NEURAL NETWORKS WITH VARIABLE SIGNAL CONDUCTIVITY 121
I A Dovgerd, G A Dovgerd, A V Ignatenkov, A M Olshansky, M G Lysikov, E N Rozenberg

RELIABILITY ANALYSIS OF POWER EQUIPMENT OF TRACTION ROLLING STOCK WITHIN THE EASTERN REGION 127
A M Khudonogov, I A Khudonogov, E Y Dulskiy, P Y Ivanov, I O Lobytsin, A A Khamnaeva

A FUNCTIONAL APPROACH TO THE ANALYSIS OF TECHNOLOGICAL PROCESSES OF USING THERMAL RADIATION IN THE PROCESS OF REPAIRING INSULATION OF ELECTRIC LOCOMOTIVE EQUIPMENT 134
E Y Dulsky, P Y Ivanov, A M Khudonogov, I A Khudonogov

METHODOLOGY FOR EVALUATING OF THE INTERACTION OF WAGONS AND PATH ON THE MOUNTAIN-GORGE AREAS 142
I Yu Ermolenko, V N Zheleznyak

RESEARCH OF HIGH-FREQUENCY INFLUENCE ON POLYMERIC MATERIALS OF ROLLING STOCK PRODUCTS 149
N G Filippenko

SUGGESTIONS FOR IMPROVING THE EFFICIENCY OF REPAIR ACTIVITY 155
A M Gladkih, V Yu Konyuhov, I I Galyautdinov, E I Shchadova

SIMULATION OF AN INTEGRATED PUBLIC TRANSPORT SYSTEM BY THE EXAMPLE OF A COMPACT CITY 161
O Lebedeva, Ju Poltavskaya, V Gozbenko

SEARCH FOR ADDITIONAL CRITERIA TO ESTIMATE THE ELECTRICAL CONNECTION STATE UNDER THE IMPACT OF CURRENTS OF DIFFERENT VALUE 169
I V Ignatenko, S A Vlasenko

EXPERIMENTAL STUDY OF PROFILE CONNECTIONS OF AUXILIARY TOOL MULTI-PURPOSE MACHINES IN CYCLIC LOADING 174
V. A. Ilyinykh

ANALYSIS OF DYNAMIC CHARACTERISTICS OF FREQUENCY-DEPENDENT LINKS 180
V O Kolmakov, O V Kolmakov, E S Ilyin, V S Ratushnyak

CALCULATIONS OF EFFICIENCY IN IMPLEMENTING PROGRESSIVE MOLD FORMING METHODS 189
V Yu Konyuhov, A M Gladkih, I I Galyautdinov, E I Shchadova

STUDY OF THE OPERATION OF HIGH-FREQUENCY ELECTRICAL PLANTS OF RAILWAY CONSUMERS 196
A V Pultyakov, O V Arsentiev, M V Kopanev, V A Alexeenko, G O Arsentiev, E S Shtykin

DETERMINING THE GEOMETRICAL PARAMETERS OF SPRING SUSPENSION OF LOCOMOTIVES 204
M V Kopanev, M O Arsentiev, L Y Kuchera

MAGNETIC VIBRATIONS IN SYNCHRONOUS MACHINES WITH PERMANENT MAGNETS AND FRACTIONAL SLOT CONCENTRATED WINDING 210
D M Toporkov, M A Kovtun, I L Popov

TYPES AND METHODS OF APPLICATION OF INFORMATION TECHNOLOGIES IN THE TRANSPORT INDUSTRY OF SIBERIA 217
O A Krasovskaya, A E Chigir

INFORMATION SYSTEMS OF PASSENGER TRANSPORTATION FORECAST IN TRANSBAIKAL REGION 223
L Y Kriklevskaya, O K Kuklina, E A Mihailova, A S Yahina

REACTIVE POWER SOURCES CONTROL IN RAILROAD POWER SUPPLY SYSTEMS 230
A V Kryukov, A P Kutsyi, A V Cherepanov

SIMULATION OF NON-SINUSOIDAL MODES IN RAILWAY POWER SUPPLY SYSTEMS DURING MOVEMENT OF HIGH-SPEED TRAINS 238
A V Kryukov, A V Cherepanov, A R Shafikov, E S Bezridnyj

DIAGNOSTICS OF HIGH-VOLTAGE INSULATION OF THE RAILWAY TRANSPORT OVERHEAD SYSTEM BY THE METHOD OF SPACED ANTENNAS 246
S M Kutsenko, N N Klimov

DETERMINATION OF THE AREA OF ROBUST STABILITY OF THE SYSTEM WITH A PID CONTROLLER 250
N N Kutsyi, T V Malanova, A P Kutsyi

DIAGNOSTIC OPERATION OF FABRICATIONS OF ROLLING STOCK BY HIGH-FREQUENCY METHOD 258
A G Larchenko, N G Filippenko, A.V. Livshits, S.K. Kargapoltcev

AUTOMATED ASSESSMENT OF THE LOW-RIGID COMPOSITE PARTS INFLUENCE ON THE PRODUCT ASSEMBLABILITY IN THE GEPARD SYSTEM 264
D S Leonovich, D A Zhuravlev, Yu I Karlina, A S Govorkov, A I Karlina

APPLYING INFORMATION TECHNOLOGIES TO A BULLDOZER DESIGN 270
I V Leskovets

GENERATION OF A RATIONAL TRAINING SAMPLE WHEN PREDICTING POWER CONSUMPTION FOR TRAIN TRACTION 278
V.E. Osipova, D.A. Yakovlev

PLANNING OF QUALITATIVE INDEXES OF RAILROAD OPERATIONAL WORK IN POLYGON TECHNOLOGIES 286
E V Malovetskaya, R S Bolshakov, A V Dimov, A A Byshlyago

THE CORRECTION SYSTEM FOR THE DIRECTION OF A LOCOMOTIVE SPOTLIGHT'S LUMINOUS FLUX 292
A S Manikovskij, M G Komogorcev, V V Stepanov

INCREASING EFFICIENCY OF BRAKING CONTROL ALGORITHM FOR A TWO-WHEELED MOTORCYCLE 297
A S Melnikov, V A Kim, A A Melnikov

THE DEVELOPMENT OF A METHOD TO REDUCE WEAR IN THE WHEEL-RAIL SYSTEM 303
E A Milovanova, M A Egorova

THERMODYNAMIC ESTIMINATION OF ADSORPTION ABILITY OF MODIFIED ZEOLITES 309
M V Obuzdina, E A Rush

INVESTIGATION OF SORPTION CHARACTERISTICS OF MODIFIED KLINOPTILOLITES BY THE METHOD OF INFRA-RED ENERGY AND DIFFERENTIAL X-RAY SPECTROSCOPY 314
M V Obuzdina, E A Rush

VOLTAGE REGULATION CHARACTERISTICS OF A TWO-PACK SYNCHRONOUS GENERATOR WITH COMBINED EXCITATION 319
D M Toporkov, I L Popov, A A Blinov, D P Kravchenko, M A Kovtun

ACCURACY ESTIMATE OF THE PAVED SURFACE OF AN AUTOMOBILE ROAD AFTER ITS RECONSTRUCTION 326
L A Pronina, E N Kupreyeva

CALCULATING THE STRESS-STRAIN STATE OF CONNECTING NODES IN PREFABRICATED STRUCTURES FROM COMPOSITE MATERIALS 334
S L Aleskovsky, A A Pykhalov, Van Lam Zyong

THE USE OF ELECTROHYDROPULSE TREATMENT OF WASTE SOLUTIONS IN REFINING PRODUCTION FOR THE RECOVERY OF PLATINUM GROUP METALS 343
D Yu Parfenov, V S Ratushnyak, E S Ilyin, A V Yuryev, V O Kolmakov

METHODS AND ALGORITHMS OF DEVELOPMENT OF A CAD SYSTEM TO DESIGN AUTOMATED PROCESS CONTROL SYSTEMS IN THE TRANSPORT SECTOR 351
D I Sachkov, A V Dimov, R U Upyr

RESEARCH ON ECOLOGICALLY SAFE FLOKULANTS IN CLAY FLUID FLUIDS 357
E V Averkina, E V Shakirova, L A Butakova

ANALYTICAL DESIGN OF CONTROL SYSTEM MATHEMATICAL MODELS FOR MOBILE ROBOTS BASED ON THE METHODS OF INVERSE PROBLEMS OF DYNAMICS AND MODAL PID CONTROLLERS 367
V N Sizykh, S B Antoshkin, R A Daneev, M V Bakanov, A V Livshits, A A Aleksandrov

THE BRAKING SERVICE OF THE LINEAR INDUCTION MOTORS WITH A COMPOUND-EQUALIZED MAGNETIC FLUX FOR MAGNETIC-LEVITATION TRANSPORT 386
A V Solomin

THE EFFECT OF ASYMMETRY OF THE LOADING CYCLE ON HEAT DISSIPATION IN METAL STRUCTURES 395
L Yu Solovyov, A L Solovyov

ADVANCED TECHNOLOGIES OF INTERNATIONAL CARGO CORRESPONDENCE IN RAILWAY TRANSPORT 401
V V Zubkov, N F Sirina

RATIONALIZATION OF PARAMETERS OF HYDRAULIC DAMPERS OF A CARRIAGE 410
A A Tarmaev, G I Petrov, V N Filippov

THE APPLICATION OF RELATIONAL INTERACTIVE LOGIC IN CONTROL OPERATION PROBLEMS BY THE EXAMPLE OF MONITORING THE SERVER EQUIPMENT OF TRANSPORT SYSTEMS 418
D V Zakharov, O V Kuzmin

RECONFIGURATION OF INTELLIGENT ELECTRICAL DISTRIBUTION NETWORKS OF RAILWAYS 426
E A Tretyakov, V T Cheremisin, G E Golovnev

REDUCTION OF ELECTRICAL ENERGY LOSSES OF POWER TRANSFORMERS OF 25 KV TRACTION SUBSTATIONS 435
N P Grigoriev, M S Klykov, V A Tikhomirov, P N Trofimovich

DEVELOPMENT OF METHODS AND MODELS OF COMPUTER-AIDED DESIGN OF SECURITY SYSTEM AGAINST INFORMATION THREATS FOR AVIATION-INSTRUMENT-MAKING 441
V E Trushnikov, N N Baranova, M V Grishin

COORDINATION OF THE MICROWAVE MODULE USED IN THE DOPPLER METER OF THE VELOCITY VECTOR COMPONENTS 447
V E Trushnikov, R R Valitov, M V Grishin

DEVELOPMENT OF MULTIVARIATE EQUIVALENT CYCLIC TESTS OF HALT-BASED AVIATION RADIO EQUIPMENT FOR ASSESSING RELIABILITY 461
V E Trushnikov, A V Komissarov, M V Grishin

ANALYSIS OF STRESSES AND DEFORMATIONS OF ELEMENTS OF AXLE BOXES AND IMPROVEMENT OF THEIR SHAPE 468
L B Tsvik, A A Tarmaev, I S Bocharov, V K Eremeev

ASSESSMENT OF CONTRIBUTION OF HUMAN FACTOR AND FACTORS OF MATERIAL-TECHNICAL SUPPLY TO SAFETY RISKS DUE TO POOR REPAIR AND TECHNOLOGY 477
O I Verevkina

CREATION OF A LEARNING MICROPROCESSOR SYSTEM FOR PROTECTION OF CONTACT NETWORK FEEDERS USING ADAPTIVE PARAMETRIC IDENTIFICATION METHODS 485
M V Vostrikov, K V Menaker, V A Ushakov

Author Index

IOP Conf. Series: Materials Science and Engineering **760** (2020) 011001 doi:10.1088/1757-899X/760/1/011001

PREFACE

The 2019th International Conference on Transport and Infrastructure of the Siberian Region (SibTrans 2019) has successfully taken place at Irkutsk State Transport University on May 21 – May 24, 2019. SibTrans 2019 continues to be a forum mainly for the Commonwealth of Independent States professional and academic community, engaged in locomotive engineering, railway infrastructure development and related areas of engineering.

The major goal of SibTrans 2019 was to bring together scientists and practitioners, engineers and industry researchers to share their experience and novations. The four-day conference provided a great opportunity to initiate discussions on topical issues of instrumentation for testing and maintain railways and locomotives, machine-building technologies and machinery, and discuss the challenges encountered and the solutions adopted.

Scientists and practitioners, mostly from Russia, submitted over ten dozen papers. All the papers were reviewed by the editors and invited experts. The most up-to-date and comprehensive manuscripts were selected for this volume. We sincerely hope that the SibTrans 2019 proceedings will show the readers a broad overview of the latest research results on the field of transportation.

At last, we would like to express our deep appreciation and gratitude to all the members of the SibTrans 2019 scientific committee for their strenuous efforts. Without their assistance and dedication, it would be impossible to hold the discussion at such a high level and to collect quality papers for compiling the conference proceedings. Finally, we want to show appreciation of all the authors, participants and editors of the proceedings for their critical contribution to the work, organization, and holding of the conference and to the preparation of the papers.

With kindest regards,

Dmitriy I. Sachkov,
Nikita V. Martyushev,
Anton P. Kutsiy

Peer review statement

All papers published in this volume of *IOP Conference Series: Materials Science and Engineering* have been peer reviewed through processes administered by the proceedings Editors. Reviews were conducted by expert referees to the professional and scientific standards expected of a proceedings journal published by IOP Publishing.

Evaluation of the strain-stress condition of rails

A.Y Abdurashitov[1], V.P.Sychev[1]

[1]Russian University of Transport (MIIT), 27994, Moscow, 9b9 Obrazcova Street

E-mail: vp@vpm770.ru

Abstract. The paper presents semi-analytical approaches to the method of evaluating strain-stress mode of railway rails when a part of the solution is in the form of ramp functions whose rate of change cannot be adequately considered using the standard method of finite elements. The proposed model of a multipoint boundary value problem allows for effective evaluation of the rail's work against the changing of its profile, i.e. due to its wear.

1. Introduction

Rapid decreasing of functions within the finite element of given dimensions does not allow for using standard methods of finite elements [1]. The paper proposes to resolve the problem in absence of limitations on the length of the investigated section of rail loading. The model of a multipoint boundary value problem allows for effective investigation of rail's work against the changing of its profile, i.e. due to wear [2,3,4,5]. The primary calculated model of the rail is described through a three-dimensional problem of the elasticity theory, specifically isotropic [6,7,8].

2. Materials and methods

Designations: Ω – a domain, described by the system; L – a differential operator, setting conditions within the Ω domain; T – operator, setting differential equations at the domain boundary; $\overline{x} = [\, x_1 \quad x_2 \quad x_3 \,]^T$ – the coordinate vector of the point; $\overline{\sigma} = [\, \sigma_{11} \quad \sigma_{22} \quad \sigma_{33} \quad \sigma_{12} \quad \sigma_{23} \quad \sigma_{13} \,]^T$ – representation of the components of the strain tensor as a vector for the three-dimensional problem of the elasticity theory; $\overline{F} = [\, F_1 \quad F_2 \quad F_3 \,]^T$, $\overline{f} = [\, f_1 \quad f_2 \quad f_3 \,]^T$ – vectors of the composing loads, acting inside the domain and at its boundary (correspondingly); $\overline{u} = [\, u_1 \quad u_2 \quad u_3 \,]^T$ – vector of composing displacements; $\overline{v} = [\, v_1 \quad v_2 \quad v_3 \,]^T$ – vector of the components of the surface normal; $\partial_i = \partial / \partial x_i$, $i = 1,2,3$

The classic statement of the problem of evaluating the strain-stressed rail as a three-dimensional isotropic model of the second elastic boundary value problem is listed as:

$$Lu = \sum_{j=1}^{N} \partial_j \sigma_{ij} = -F_i, \ x \in \Omega; \quad Tu = \sum_{j=1}^{N} v_j \sigma_{ij} = -f_i, \ \ x \in \partial\Omega, \ \ i = 1,\dots,N \tag{1}$$

The following statement of the problem is proposed as the semi-analytical method:

$$\overline{U}' = \tilde{L}\overline{U} + \overline{\tilde{F}}, \text{ where } \overline{U} = \begin{bmatrix} \overline{u} \\ \overline{v} \end{bmatrix}; \ \tilde{L} = \begin{bmatrix} 0 & E \\ L_{vv}^{-1} L_{uu} & L_{vv}^{-1} \tilde{L}_{uv} \end{bmatrix}; \ \overline{\tilde{F}} = -\begin{bmatrix} 0 \\ L_{vv}^{-1} \overline{F} \end{bmatrix} \tag{2}$$

$$\text{where } \overline{F} = [\, F_1 \quad F_2 \quad F_3 \,]^T; \ F_i = \theta F_i + \delta_\Gamma f_i; \ \tilde{L}_{uv} = L_{uv} - L_{uv}^*; \ \tilde{L}_{uv}^* = -\tilde{L}_{uv}; \tag{3}$$

E – identity matrix of the corresponding degree;

$$\theta(x)=\begin{cases}1, & x\in\Omega\\ 0, & x\notin\Omega\end{cases};\ \delta_\Gamma=\partial\theta/\partial\bar{\nu} \tag{4}$$

Characteristic function of the Ω domain and of the delta function of the boundary $\partial\Omega$,

where $\bar{\nu}=[\,\nu_1\ \ \nu_2\ \ \nu_3\,]^T$ – unit vector of the surface normal.

$$\bar{v}=[v_1\ v_2\ v_3]^T=[\partial_3 u_1\ \partial_3 u_2\ \partial_3 u_3]=\partial_3\bar{u}=\bar{u}';\ \bar{v}'=\partial_3\bar{v}; \tag{5}$$

$$L_{vv}=\begin{bmatrix}\bar{\mu} & 0 & 0\\ 0 & \bar{\mu} & 0\\ 0 & 0 & \bar{\lambda}+2\bar{\mu}\end{bmatrix};\ L_{uv}=\begin{bmatrix}0 & 0 & \partial_1^*\bar{\lambda}\\ 0 & 0 & \partial_2^*\bar{\lambda}\\ \partial_1^*\bar{\mu} & \partial_2^*\bar{\mu} & 0\end{bmatrix}; \tag{6}$$

$$L_{uu}=\sum_{j=1}^{2}\partial_j^*\bar{\mu}\partial_j\begin{bmatrix}1 & 0 & 0\\ 0 & 1 & 0\\ 0 & 0 & 1\end{bmatrix}+\begin{bmatrix}\partial_1^*\bar{\mu}\partial_1 & \partial_2^*\bar{\mu}\partial_1 & 0\\ \partial_1^*\bar{\mu}\partial_2 & \partial_2^*\bar{\mu}\partial_2 & 0\\ 0 & 0 & 0\end{bmatrix}+\begin{bmatrix}\partial_1^*\bar{\lambda}\partial_1 & \partial_1^*\bar{\lambda}\partial_2 & 0\\ \partial_2^*\bar{\lambda}\partial_1 & \partial_2^*\bar{\lambda}\partial_2 & 0\\ 0 & 0 & 0\end{bmatrix} \tag{7}$$

where x_3 – coordinate direction of the constancy of physical and geometrical functions of the structure; λ, μ – Lame constants; $\bar{\lambda}=\theta\lambda;\ \bar{\mu}=\theta\mu$.

The same statement of the problem, but as a variational formulation, may be presented as:

$$\Phi(\bar{U})=0.5\cdot(\tilde{\tilde{L}}\,\bar{U},\bar{U})-(\bar{\bar{F}},\bar{U}) \tag{8}$$

or

$$\Phi(\bar{u},\bar{v})=0.5\cdot[(L_{vv}\bar{v},\bar{v})+2(\tilde{L}_{uv}\bar{v},\bar{u})+(L_{uu}\bar{u},\bar{u})]-(\bar{F},\bar{u}) \tag{9}$$

where
$$\tilde{\tilde{L}}=\begin{bmatrix}L_{uu} & \tilde{L}_{uv}\\ \tilde{L}_{uv}^* & L_{vv}\end{bmatrix}=\begin{bmatrix}L_{uu} & \tilde{L}_{uv}\\ \tilde{L}_{vu} & L_{vv}\end{bmatrix};\ \bar{\bar{F}}=\begin{bmatrix}\bar{F}\\ 0\end{bmatrix};\ \tilde{L}_{vu}=\tilde{L}_{uv}^*=-\tilde{L}_{vu} \tag{10}$$

The solution of the problem will be the determination of the extreme point of the functional (9) subject to compliance with the condition (5). The source domain is limited by the dilated domain of the derived form ω, e.g. a parallelepiped, whose approximation is in setting a grid equivalent to the rectangular and best fitting the cross-section of the structure (Fig. 1). The following type of the discrete-continuous model is taken: Network approximation is done in the lateral direction of the rail (axes Ox_1 and Ox_2), while in the longitudinal direction (axis Ox_3) – the problem remains continuous. The domain ω is split into sub-domains – finite continuous element ω_{ij}; $N=N_1\cdot N_2$ (Fig. 2). A different approach is possible, when a local coordinate system Ot_1 and Ot_2 is introduced into a random section of the element, whereas $t_1, t_2\in[0,1]$ and a local renumbering of the nodes is made (Fig.3). For approximation of the unknowns, the displacement components u_1, u_2, u_3 and their derivatives v_1, v_2, v_3 by x_3 are taken as the main nodal unknowns, i.e. for the pq-d node it is u_i^{pq}, v_i^{pq}, i=1,2,3 and, correspondingly, the vectors of the unknowns are:

$$\bar{u}_n^{pq}=\bar{u}_n^{pq}(x_3)=[u_1^{pq}\ \ u_2^{pq}\ \ u_3^{pq}]^T;\ \ \bar{v}_n^{pq}=\bar{v}_n^{pq}(x_3)=[v_1^{pq}\ \ v_2^{pq}\ \ v_3^{pq}]^T. \tag{11}$$

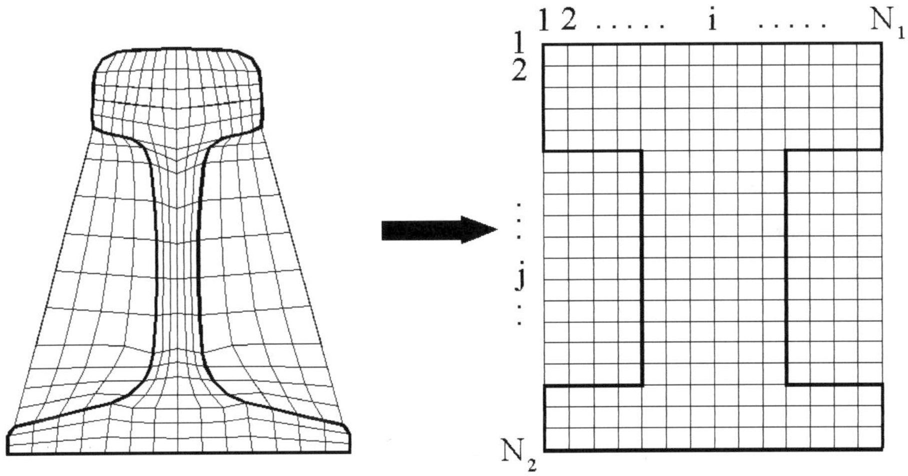

Fig. 1 Example of the approximation network

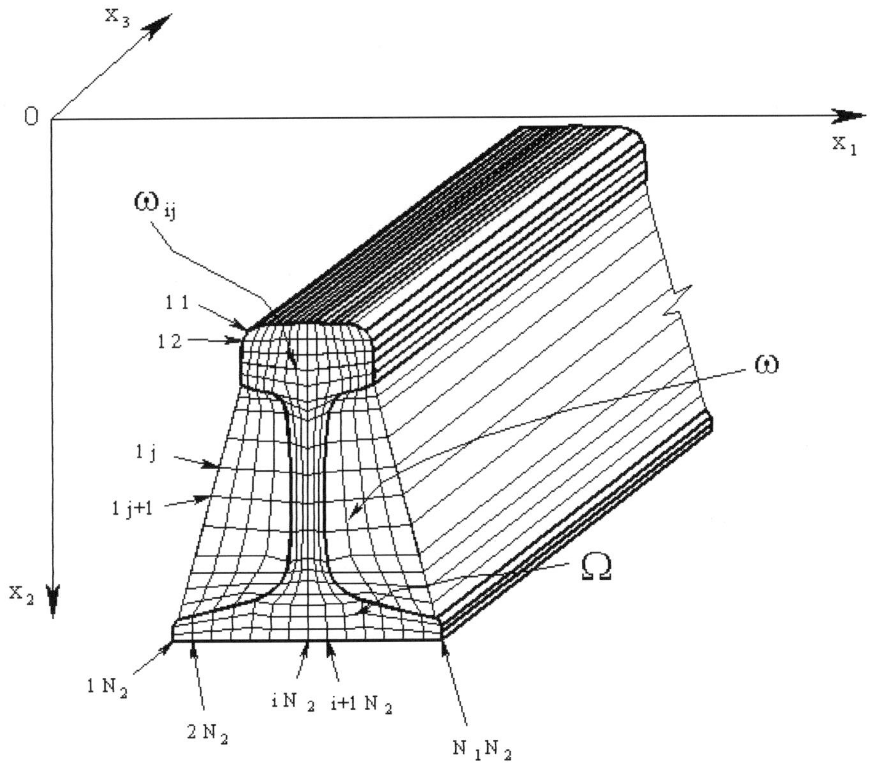

Fig.2 Example of rail discretization

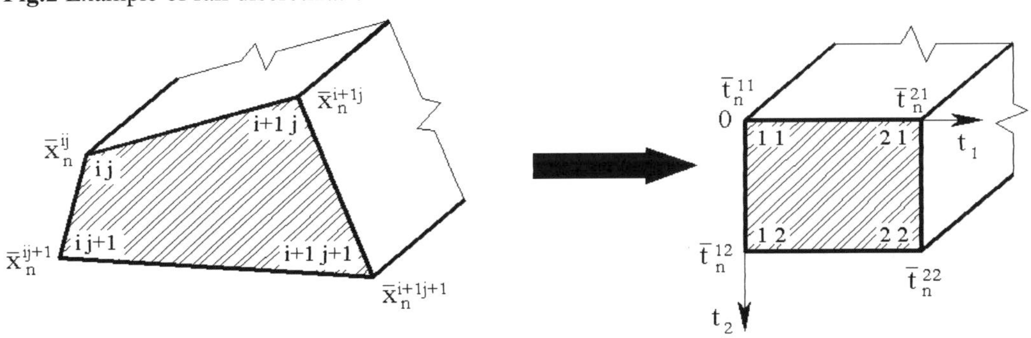

Fig.3 Transition to the local system of coordinates on an element.

Bilinear approximation is adopted for fields u_1, u_2, u_3 и v_1, v_2, v_3 within each finite continuous element:

$$\overline{u} = \overline{u}_n^{11} + t_1\Delta_1\overline{u} + t_2\Delta_2\overline{u} + t_1t_2\Delta_{12}\overline{u}; \quad \overline{v} = \overline{v}_n^{11} + t_1\Delta_1\overline{v} + t_2\Delta_2\overline{v} + t_1t_2\Delta_{12}\overline{v} \tag{12}$$

where $$\Delta_1\overline{u} = \overline{u}_n^{21} - \overline{u}_n^{11}; \quad \Delta_2\overline{u} = \overline{u}_n^{12} - \overline{u}_n^{11}; \quad \Delta_{12}\overline{u} = \overline{u}_n^{22} - \overline{u}_n^{21} - \Delta_2\overline{u} \tag{13}$$

$$\Delta_1\overline{v} = \overline{v}_n^{21} - \overline{v}_n^{11}; \quad \Delta_2\overline{v} = \overline{v}_n^{12} - \overline{v}_n^{11}; \quad \Delta_{12}\overline{v} = \overline{v}_n^{22} - \overline{v}_n^{21} - \Delta_2\overline{v}. \tag{14}$$

The element-wise vector of node loads is formed as follows:

$$\overline{R}^{ij} = \overline{R}^{ij}(x_3) = [\,(\overline{R}_n^{11})^T \ \ (\overline{R}_n^{21})^T \ \ (\overline{R}_n^{12})^T \ \ (\overline{R}_n^{22})^T\,]^T \tag{15}$$

where $$\overline{R}_n^{pq} = \overline{R}_n^{pq}(x_3) = [\,(\overline{R}_{u,n}^{pq})^T \ \ (\overline{R}_{v,n}^{pq})^T\,]^T, \text{ whereas } \overline{R}_{v,n}^{pq} = 0. \tag{16}$$

In order to form the stiffness matrix of the finite continuous element, the initial functional (8) is described as the sum of functional determined on the finite continuous elements:

$$\Phi(\overline{U}) = \sum_{i=1}^{N_1}\sum_{j=1}^{N_2}\Phi_{ij}(\overline{U}^{ij}), \tag{17}$$

where $\Phi_{ij}(\overline{U}^{ij}) = \frac{1}{2}(K^{ij}\overline{U}^{ij}, \overline{U}^{ij}) - (\overline{R}^{ij}, \overline{U}^{ij})$; K^{ij} – stiffness matrix of the finite continuous element of the 24-th order, which is formed using the base variations through the following equation::

$$(K^{ij})_{st} = \widetilde{\Phi}_{ij}(\overline{e}_s + \overline{e}_t) - \widetilde{\Phi}_{ij}(\overline{e}_s) - \widetilde{\Phi}_{ij}(\overline{e}_t) + \widetilde{\Phi}_{ij}(\overline{e}_0), \ s = 1,2,...,24; \ t = 1,2,...,24 \tag{18}$$

where $$(e_j)_k = \delta_{j,k}; \ (e_0)_k = 0, \ k = 1,2,...,24; \ \widetilde{\Phi}_{ij}(\overline{U}^{ij}) = 0.5\cdot(K^{ij}\overline{U}^{ij}, \overline{U}^{ij}); \tag{19}$$

$$K^{ij} = \begin{bmatrix} K_{uu}^{11} & K_{uv}^{11} & K_{uu}^{12} & K_{uv}^{12} & K_{uu}^{13} & K_{uv}^{13} & K_{uu}^{14} & K_{uv}^{14} \\ K_{vu}^{11} & K_{vv}^{11} & K_{vu}^{12} & K_{vv}^{12} & K_{vu}^{13} & K_{vv}^{13} & K_{vu}^{14} & K_{vv}^{14} \\ K_{uu}^{21} & K_{uv}^{21} & K_{uu}^{22} & K_{uv}^{22} & K_{uu}^{23} & K_{uv}^{23} & K_{uu}^{24} & K_{uv}^{24} \\ K_{vu}^{21} & K_{vv}^{21} & K_{vu}^{22} & K_{vv}^{22} & K_{vu}^{23} & K_{vv}^{23} & K_{vu}^{24} & K_{vv}^{24} \\ K_{uu}^{31} & K_{uv}^{31} & K_{uu}^{32} & K_{uv}^{32} & K_{uu}^{33} & K_{uv}^{33} & K_{uu}^{34} & K_{uv}^{34} \\ K_{vu}^{31} & K_{vv}^{31} & K_{vu}^{32} & K_{vv}^{32} & K_{vu}^{33} & K_{vv}^{33} & K_{vu}^{34} & K_{vv}^{34} \\ K_{uu}^{41} & K_{uv}^{41} & K_{uu}^{42} & K_{uv}^{42} & K_{uu}^{43} & K_{uv}^{43} & K_{uu}^{44} & K_{uv}^{44} \\ K_{vu}^{41} & K_{vv}^{41} & K_{vu}^{42} & K_{vv}^{42} & K_{vu}^{43} & K_{vv}^{43} & K_{vu}^{44} & K_{vv}^{44} \end{bmatrix}; \ \overline{U}^{ij} = \overline{U}^{ij}(x_3) = \begin{bmatrix} \overline{U}_n^{11} \\ \overline{U}_n^{21} \\ \overline{U}_n^{12} \\ \overline{U}_n^{22} \end{bmatrix}. \tag{20}$$

Here $K_{uu}^{lm}, K_{uv}^{lm}, K_{vu}^{lm}, K_{vv}^{lm}, \ l, m = 1,2,3,4$ – matrixes of the 3-d order.

Based on the element-wise stiffness matrix (20), the K_{uu}^{ij}, K_{uv}^{ij}, K_{vu}^{ij} and K_{vv}^{ij} matrixes of the 12-th order are formed.

$$K_{uu}^{ij}=\begin{bmatrix} K_{uu}^{11} & K_{uu}^{12} & K_{uu}^{13} & K_{uu}^{14} \\ K_{uu}^{21} & K_{uu}^{22} & K_{uu}^{23} & K_{uu}^{24} \\ K_{uu}^{31} & K_{uu}^{32} & K_{uu}^{33} & K_{uu}^{34} \\ K_{uu}^{41} & K_{uu}^{42} & K_{uu}^{43} & K_{uu}^{44} \end{bmatrix};\quad K_{uv}^{ij}=\begin{bmatrix} K_{uv}^{11} & K_{uv}^{12} & K_{uv}^{13} & K_{uv}^{14} \\ K_{uv}^{21} & K_{uv}^{22} & K_{uv}^{23} & K_{uv}^{24} \\ K_{uv}^{31} & K_{uv}^{32} & K_{uv}^{33} & K_{uv}^{34} \\ K_{uv}^{41} & K_{uv}^{42} & K_{uv}^{43} & K_{uv}^{44} \end{bmatrix}; \tag{21}$$

$$K_{vu}^{ij}=\begin{bmatrix} K_{vu}^{11} & K_{vu}^{12} & K_{vu}^{13} & K_{vu}^{14} \\ K_{vu}^{21} & K_{vu}^{22} & K_{vu}^{23} & K_{vu}^{24} \\ K_{vu}^{31} & K_{vu}^{32} & K_{vu}^{33} & K_{vu}^{34} \\ K_{vu}^{41} & K_{vu}^{42} & K_{vu}^{43} & K_{vu}^{44} \end{bmatrix};\quad K_{vv}^{ij}=\begin{bmatrix} K_{vv}^{11} & K_{vv}^{12} & K_{vv}^{13} & K_{vv}^{14} \\ K_{vv}^{21} & K_{vv}^{22} & K_{vv}^{23} & K_{vv}^{24} \\ K_{vv}^{31} & K_{vv}^{32} & K_{vv}^{33} & K_{vv}^{34} \\ K_{vv}^{41} & K_{vv}^{42} & K_{vv}^{43} & K_{vv}^{44} \end{bmatrix}. \tag{22}$$

The (17) will be as follows:

$$\Phi_{ij}(\overline{u}^{ij},\overline{v}^{ij})=0.5[(K_{uu}^{ij}\overline{u}^{ij},\overline{u}^{ij})+2(K_{uv}^{ij}\overline{v}^{ij},\overline{u}^{ij})+2(K_{vu}^{ij}\overline{u}^{ij},\overline{v}^{ij})+(K_{vv}^{ij}\overline{v}^{ij},\overline{v}^{ij})]-(\overline{R}_{u}^{ij},\overline{u}^{ij}) \tag{23}$$

where $\overline{R}_{u}^{ij}=\overline{R}_{u}^{ij}(x_3)=[\,(\overline{R}_{u,n}^{11})^T\ \ (\overline{R}_{u,n}^{21})^T\ \ (\overline{R}_{u,n}^{12})^T\ \ (\overline{R}_{u,n}^{22})^T\,]^T$; (24)

$$\overline{u}^{ij}=\overline{u}^{ij}(x_3)=[\,(\overline{u}_n^{11})^T\ \ (\overline{u}_n^{21})^T\ \ (\overline{u}_n^{12})^T\ \ (\overline{u}_n^{22})^T\,]^T; \tag{25}$$

$$\overline{v}^{ij}=\overline{v}^{ij}(x_3)=[\,(\overline{v}_n^{11})^T\ \ (\overline{v}_n^{21})^T\ \ (\overline{v}_n^{12})^T\ \ (\overline{v}_n^{22})^T\,]^T. \tag{26}$$

The following differential relation corresponds to the functional (23):

$$-K_{vv}^{ij}(\overline{v}^{ij})'+\widetilde{K}_{uv}^{ij}\overline{v}^{ij}+K_{uu}^{ij}\overline{u}^{ij}=\overline{R}_{u}^{ij}, \tag{27}$$

and reflects the stationary condition for a free element regardless of the system of finite continuous elements, where:

$$(\overline{v}^{ij})'=\partial_3\overline{v}^{ij};\quad \widetilde{K}_{uv}^{ij}=K_{uv}^{ij}-K_{vu}^{ij};\quad K_{vu}^{ij}=(K_{uv}^{ij})^{*}. \tag{28}$$

Or we may have the element-wise system in the following form:

$$\begin{cases} \overline{v}^{ij}=\partial_3\overline{u}^{ij} \\ -K_{vv}^{ij}(\overline{v}^{ij})'+\widetilde{K}_{uv}^{ij}\overline{v}^{ij}+K_{uu}^{ij}\overline{u}^{ij}=\overline{R}_{u}^{ij} \end{cases}. \tag{29}$$

Formation of the corresponding global matrixes K_{uu}, K_{uv}, K_{vu}, K_{vv} of the 3N-d order of the system of finite elements of the whole structure is done in accordance with the standard method of forming the global stiffness matrix (the finite scoring method – summing with accrual). Continual operators and the mentioned matrixes are collated as follows:

$$\begin{matrix} L_{uu}\Rightarrow K_{uu} \\ L_{uv}\Rightarrow K_{uv} \end{matrix};\quad \begin{matrix} L_{vu}\Rightarrow K_{vu} \\ L_{vv}\Rightarrow K_{vv} \end{matrix};\quad \widetilde{\widetilde{L}}\Rightarrow K=\begin{bmatrix} K_{uu} & \widetilde{K}_{uv} \\ \widetilde{K}_{vu} & K_{vv} \end{bmatrix}. \tag{30}$$

Coordinates of terminal cross-sections of the rail are: x_3^k, $k=1,2,\ldots,n_k$, and the terminal conditions in these rails are recorded as:

$$B_k^{-}\overline{U}_n(x_3^k-0)+B_k^{+}\overline{U}_n(x_3^k+0)=\overline{g}_k,\quad k=1,2,\ldots,n_k \tag{31}$$

where B_k^{-}, B_k^{+} – matrixes of terminal conditions, squares of the 6N-d order;

$\overline{g}_k$ – the set 6N-dimensional vector of the right parts of the terminal conditions;

$$\overline{U}_n = \overline{U}_n(x_3) = \begin{bmatrix} \overline{u}_n \\ \overline{v}_n \end{bmatrix}, \text{ where } \begin{array}{l} \overline{u}_n = \overline{u}_n(x_3) = [\,(\overline{u}_n^{11})^T \quad (\overline{u}_n^{12})^T \quad \dots \quad (\overline{u}_n^{N_1N_2})^T\,]^T \\ \overline{v}_n = \overline{v}_n(x_3) = [\,(\overline{v}_n^{11})^T \quad (\overline{v}_n^{12})^T \quad \dots \quad (\overline{v}_n^{N_1N_2})^T\,]^T \end{array}. \quad (32)$$

Then the functional (17) shall read as:

$$\Phi(\overline{u}_n,\overline{v}_n) = 0.5[(K_{uu}\overline{u}_n,\overline{u}_n) + (K_{uv}\overline{v}_n,\overline{u}_n) + (K_{vu}\overline{u}_n,\overline{v}_n) + (K_{vv}\overline{v}_n,\overline{v}_n)] - (\overline{R}_u,\overline{u}_n). \quad (33)$$

With the corresponding global resolving system:

$$\left.\begin{array}{l} \overline{U}'_n(x_3) = A\overline{U}_n(x_3) + \overline{\overline{R}}(x_3) \\ B_k^-\overline{U}_n(x_3^k - 0) + B_k^+\overline{U}_n(x_3^k + 0) = \overline{g}_k, \quad k = 1, 2, \dots, n_k \end{array}\right\} \quad (34)$$

$$\text{where } A = \begin{bmatrix} 0 & E \\ (K_{vv})^{-1}K_{uu} & (K_{vv})^{-1}\widetilde{K}_{uv} \end{bmatrix};\ \overline{\overline{R}} = \overline{\overline{R}}(x_3) = -\begin{bmatrix} 0 \\ (K_{vv})^{-1}\overline{R}_u \end{bmatrix};\ \overline{U}'_n = \partial_3\overline{U}_n. \quad (35)$$

After determining the unknown nodal functions per corresponding formulae of the elasticity theory the strains and deformations can be found.

The classic setting for resolving a multi-point boundary problem is as follows:

$$\overline{y}^{(1)} - A\overline{y} = \overline{f}, \quad x \in \bigcup_{k=1}^{n_k - 1}(x_k, x_{k+1}); \quad (36)$$

$$B_k^-\,\overline{y}(x_k - 0) + B_k^+\,\overline{y}(x_k + 0) = \overline{g}_k, \quad k = 1, \dots, n_k \quad (37)$$

where $\overline{y} = \overline{y}(x) = [\,y_1(x) \quad y_2(x) \quad \dots \quad y_n(x)\,]^T$ – the sought n-dimensional vector function; $\overline{f} = \overline{f}(x) = [\,f_1(x) \quad f_2(x) \quad \dots \quad f_n(x)\,]^T$ – of the right parts; $x_k, r = 1, \dots, n_k$ – coordinates of terminal points; A – matrix of coefficients, square of the n order; B_k^-, B_k^+ – matrixes of terminal conditions, square of the n order; $\overline{g}_k$ – n-dimensional vector of the right sides of terminal conditions.

The process of resolving the problems of calculating the structures of that type is accompanied by: the fringe effect, existence of eigen values in the coefficients matrix, geometrically multiple eigenvalue, high order of the system of differential equations (several thousand) [9,10,11]. Therefore the problem is set by utilizing generalized functions for the whole numeric axis ($x \in (-\infty, +\infty)$):

$$\overline{Y}' = A\overline{Y} + \overline{F}, \ x \in (-\infty; +\infty), \ \text{ where } \overline{F} = \overline{f} + \sum_{k=1}^{n_k}\delta(x - x_k)\Delta\,\overline{y}_k; \quad (38)$$

$$B_k^-\,\overline{Y}(x_k - 0) + B_k^+\,\overline{Y}(x_k + 0) = \overline{g}_k, \quad k = 1, \dots, n_k \quad (39)$$

where $\delta(x)$ – the Dirac delta function; Y – the generalized vector function of the unknowns; $\Delta\,\overline{y}_k = \overline{y}(x_k + 0) - \overline{y}(x_k - 0)$ – the vector of the values of finite discontinuities of the values of the elements of the vector function $\overline{y}(x)$ during its "transition" through the terminal point x_k.

The Jordan decomposition of the A matrix is written as follows:

$$A = T\,J\,T^{-1}, \ \text{ where } J = \{J_1, J_2, \dots, J_u\} \quad (40)$$

where T – a nondegenerate matrix of the n-th order whose columns are own and root vectors of the A matrix; J – quasidiagonal Jordan matrix; J_p – is the Jordan cell, corresponding to the eigen value λ_p of the geometrical multiplicity m_p; $\dim J_p = m_p$. Unfortunately, existence of geometrically multiple eigen values of the coefficient matrix, requiring computation of root vectors, is a setback preventing the practical use of the Jordan decomposition [12]. Therefore it is proposed: partial Jordan decomposition, based on using the right and left eigen vectors of the A matrix to be described as:

$$A = A_1 + A_2, \text{ where } A_1 = T_1 J_1 \widetilde{T}_1,\ A_2 = A - A_2. \tag{41}$$

Here T_1 & $\widetilde{T}_1$ – correspondingly, matrixes containing the right and left eigen vectors, corresponding to the plain non-zero eigen values of the A matrix, distributed into columns and lines (rectangular in the general case); J_1 – the diagonal Jordan matrix, corresponding to the geometrically plain non-zero eigen values; A_2 – part of the A matrix, corresponding to the Jordan cells for geometrically multiple and plain zero eigen values.

Sorting of the eigen values $\lambda_p, p = 1,\dots,u$ (and, correspondingly, transformation of matrixes T_1; $\widetilde{T}_1$; J_1) is done subject to meeting the following conditions:

$$\left.\begin{aligned} &\forall\lambda_p,\ p = 1,\dots,l\ \ \exists m_p = 1 \\ &\forall\lambda_p,\ p = l+1,\dots,u\ \ \exists m_p > 1 \end{aligned}\right\}, \text{ where } l = \sum_{p=1}^{u} \delta_{1,m_p};\ \delta_{i,j} = \begin{cases} 1, & \text{если } i = j \\ 0, & \text{если } i \neq j \end{cases} \tag{42}$$

Let P_1 and P_2 be the matrixes of projecting on the sub-domain of left and right eigen and root vectors corresponding to the non-zero geometrically plain and zero values, correspondingly. They could be determined as follows:

$$P_1 = T_1(\widetilde{T}_1 T_1)^{-1}\widetilde{T}_1;\quad P_2 = E - P_1. \tag{43}$$

After sorting and conducting biorthogonalization of eigen vectors and values we write it down in the form which is convenient for resolving structural analysis problems:

$$\varepsilon(x) = T_1 \widetilde{\varepsilon}_0(x) \widetilde{T}_1 + \chi(x,0)[P_2 + \sum_{k=1}^{m_{max}-1} \frac{x^k}{k!} A_2^k] \tag{44}$$

where
$$\chi(x,\lambda_p) = \begin{cases} \operatorname{sign}(x)\theta(-\operatorname{Re}(\lambda_p)x), & \lambda_p \neq 0 \\ 0.5\operatorname{sign}(x), & \lambda_p = 0 \end{cases}; \tag{45}$$

$$\widetilde{\varepsilon}_0(x) = \operatorname{diag}\{\chi(x,\lambda_1)\exp(\lambda_1 x),\ \dots,\ \chi(x,\lambda_l)\exp(\lambda_l x)\};\ m_{max} = \max_{l \le i \le u} m_i \tag{46}$$

The sum in the right part (44) contains not more than four elements. The total solution of the problem (38), (39) is described by the formula:

$$\overline{Y} = \varepsilon * \overline{F} = \varepsilon * \bar{f} + \sum_{k=1}^{n_k} \varepsilon(x - x_k)\overline{C}_k \tag{47}$$

where $\overline{C}_k = \Delta\bar{y}_k,\ k = 1,\dots,n_k$ – the coefficient vectors.

3. Conclusion

Determination of constant coefficients in the general solution from the terminal conditions (39) is done either through the explicit matrix method or through the method of basic variations similarly to [2]. The diagram of the program implementation of the method with the calculation example is shown in Fig. 4.

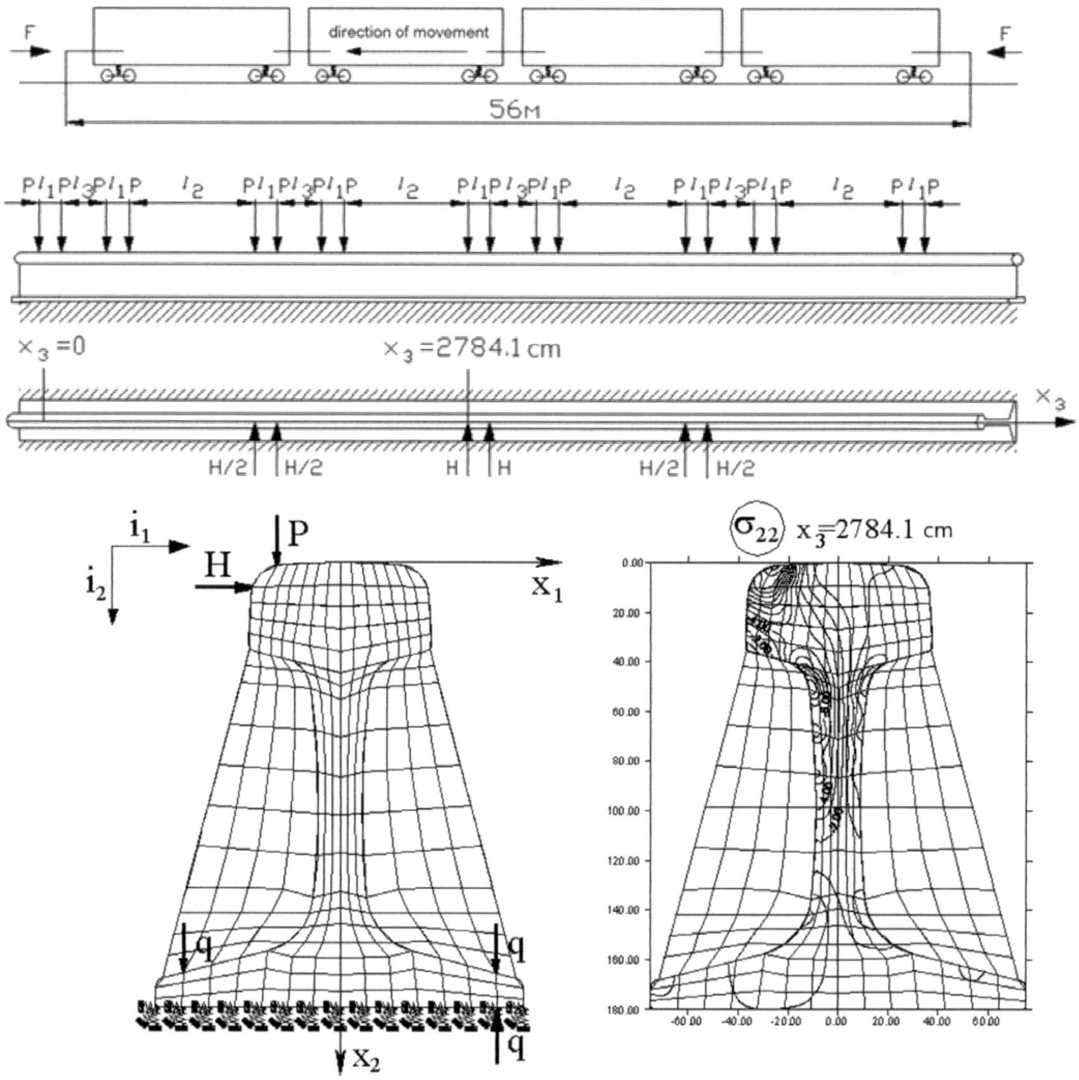

Fig 4. Practical implementation of the method for calculating stress-strain condition of railway rails.

Article I.

Base data for calculations are: rail type P65; geometrical properties of the problem: $l_1 = 185\,\text{cm}$; $l_2 = 680\,\text{cm}$; $l_3 = 342\ \text{cm}$; characteristics of the elastic response of the base: $c_1 = 10\ \text{kN/cm}^2$; $c_1 = 2\,\text{kN/cm}^2$; $c_1 = 2\,\text{kN/cm}^2$; loads from wagons $P = 100\ \text{kN}$; $H = 50\ \text{kN}$; constant distributed load $q = 0.0744\ \text{kN/cm}^2$.

4. Acknowledgments

The results obtained have shown good correlation with theoretic calculations.

References

[1] Zenkevich O S *The method of finite elements in the technique.* (M.: The World) pp 1975-542

[2] Zolotov A B, Larionov A V, Mozgaleva M L, Mskhalaya Zh I 1992 *Statement and approximation of boundary value problems by the extended domain method.* (M: MISI). p 82

[3] Zolotov A B, Abdurashitov A Yu, Akimov P A 2001 Determination of the stress-strain state of a rail by semi-analytical finite element method. *Vestnik VNIIZHT*. **4**. 26-32.

[4] Abdurashitov A Yu, Sychev V P, Abdurashitov Yu A 2016 Selection of models of the stress-strain state of railroad rails for various types of hardening *The introduction of modern structures and advanced technologies in track facilities* **10 10-10 (10)** 64-79

[5] Loktev A, Sychev V, Gridasova E and Stepanov R 2017 Mathematical Modeling of Railway Track Structure under Changing Rigidity Parameters *Nonlinearity. Problems, Solutions and Applications. V.1. Theoretical and Applied Mathematics*. 291-307

[6] Sycheva A V, Sychev V P, Buchkin V A, Bykov Y A 2016 Modeling of railway track operation as a system of quasi-elastic orthotropic layers *Vestnik MGSU* **3** 124-134

[7] Loktev A A, Sychev V P, Ashpiz E S, Lustkii S Ya 2017 Modelling of the Dynamic Contact between a Wheel of a Moving Railway Vehicle and Rails with Evaluation of Defects Emerging upon Their Interaction. Part 2. Contact Problem in the Presence of Defects *Applied Mathematical Sciences* **11 10** 481 – 488

[8] Loktev A, Sychev V, Gluzberg B and Gridasova E Modeling the dynamic behavior of railway track taking into account the occurrence of defects in the system wheel-rail *MATEC Web of Conferences 117, 00108 (2017). XXVI R-S-P Seminar 2017, Theoretical Foundation of Civil Engineering* 1-6

[9] Shur, E A, Borts, A I, Sukhov, A V, Abdurashitov, A Yu, Bazanova L V, Zarizhdek K L 2015 The evolution of rail damage by contact fatigue defects *Vestnik VNIIZHT* **3** 3-9

[10] Sychev P V 2015 Application of the method of the influence of scattered microdamages on the resistance to crack growth in assessing the residual life of the rolling stock of railways *Science and Technology in Transport* **1** 53-58.

[11] Loktev A A, Sychev V P, Vinogradov V V, Buchkin V A 2017 Modelling of the Dynamic Contact between a Wheel of a Moving Railway Vehicle and Rails with Evaluation of Defects Emerging upon Their Interaction. Part 1. The Defects of the Rail and Models of Contact *Applied Mathematical Sciences* **11 10**, 473 – 480

[12] Zolotov A B, Akimov P A, Mozgaleva M L, Shirinsky V I 2000 *Construction of an analytical solution of a multipoint boundary value problem for a system of ordinary differential equations, taking into account the presence of Jordan cells in the matrix of coefficients. «Questions of applied mathematics and computational mechanics». (*M: MGSU) Pp. 61-73.

[13] Maksimenko A, Sychev V, Abdurashitov A 2018 Identification of transverse defects of the top and the bottom of rails *Railway Track and Facilities*. **3**. 2-6

Mathematical model of a radial bearing with a low-melting metal coating of design models of hydrodynamic viscoelastic lubricant formed by melting the surface of a bearing bush coated with a metallic low-melting coating

K S Akhverdiev[1], E O Lagunova[1], M A Mukutadze[1]

[1]Rostov State Transport University, 2, Rostovskogo Strelkovogo Polka Narodnogo Opolcheniya sq., Rostov-on-Don, 344038, Russia

E-mail: lagunova@rambler.ru

Abstract. The work is based on equations of motion of viscoelastic lubricant (Maxwell fluid) for the case of "thin layer", continuity and the formula of the rate of dissipation of mechanical energy. We found asymptotic solution to the degrees of small parameter K, characterizing the rate of dissipation of mechanical energy. This allows determining the molten surface of the bearing sleeve covered with a fusible metal melt, taking into account the dependence of the viscosity of the lubricant and the shear modulus of pressure in the adiabatic process. For zero approximation under adiabatic process conditions, i.e. without taking into account the melt surface, and a solution for the first approximation, i.e., taking into account the melt surface of the bearing sleeve covered with a fusible metal melt, the exact self-similar solution of the problem with incomplete (pre-emergency) filling of the working gap with a lubricant having viscoelastic properties is found.. On the basis of the results obtained for the zero and first approximation, the bearing capacity and friction force are found at incomplete filling of the working gap with a lubricant. In addition, for verification calculations of the obtained theoretical models, a numerical calculation was carried out and as a result, graphs showing a decrease in the friction coefficient by about 20% were constructed.

1. Introduction

Tribosystems of modern machines operate under high load-speed regimes. In these conditions, the use of liquid friction provided by liquid lubricants is the most promising. And in this lubrication mode the absolute advantage belongs to the hydrodynamic process.

It is known [1] that in heavily loaded tribosystems, local centers of melting can occur. This is confirmed by Bowden and Tabor's "welding bridges" [2].

It should be noted that the oxygen of the ambient air very quickly oxidizes the contact surface of mercury and the value of the friction torque for 2 hours of operation increases more than 20 times. In addition, mercury, like any metal melt, is a surface-active liquid that amalgam metals being in contact with it. Therefore, the melt area must be isolated from the environment.

The operation of various valve systems in the primary heat removal circuit of nuclear reactors in the melt of alkali metal is also known.

The wide application of hydrodynamic lubrication in a wide variety of machines and mechanisms required the development of a significant range of new highly effective lubricating liquids [3-9] such as micro-polar, viscoelastic, viscoplastic, electrically conductive and melt fusible metal alloys. The practice of using these liquid lubricants significantly outstrips the development of models and theoretical calculations necessary for their wide application in practice.

Lubrication with liquid metals is used at temperatures at which conventional lubricating media undergo irreversible physical and chemical changes. The advantage of melt lubrication is that the lubricant is formed in the contact area where it is necessary. Melting delivers a sufficient amount of lubricant to the friction zone; there are no mechanical and structural difficulties associated with its supply. Melt lubrication has been studied in many applications, particularly in metal forming and cutting processes. A large number of works are devoted to hydrodynamic calculation of the radial bearing of infinite length in the conditions of incomplete filling of the working gap with lubricant takes into account the dependence of the viscosity of the lubricant on the pressure [10-11]. A significant disadvantage of the friction pair under consideration, working on the melt lubrication, is the low bearing capacity. In addition, the lubrication process is not self-sustaining.

Thus, the development of a design model of sliding bearings operating on lubricants in the form of metal melts, taking into account the above aspects of operation, is of particular importance and scientific interest. An attempt to develop a general methodology for calculating the hydrodynamic lubrication regime for different bearing designs, lubricated with liquids with different physical and mechanical properties, will significantly accelerate their bringing to industrial application.

2. Materials and methods

A mathematical model is proposed on the basis of equations describing the motion of an incompressible viscoelastic lubricant for a "thin layer", continuity and expression of the rate of dissipation of mechanical energy to determine the profile of the molten surface of the bearing sleeve covered with a metal fusible melt also having viscoelastic properties.

3. Results

New multiparametric expressions are developed for the basic performance characteristics of the radial sliding bearing with incomplete filling of the working gap with a viscoelastic lubricant, taking into account the rheological properties, as well as the melt surface of the bearing sleeve coated with a metal fusible coating in the adiabatic process.

Expressions for the load capacity and friction force are obtained taking into account the rheological properties of viscoelastic lubricant and the melt of a fusible metal coating.

The paper summarizes the influence of factors that describe the process of incomplete (pre-emergency) filling of the working gap with a lubricant having viscoelastic properties, taking into account the melt surface of the bearing sleeve coated with a low-melting metal coating, taking into account the dependence of the viscosity of the viscoelastic lubricant and shear modulus on the pressure in the adiabatic process, which significantly complicates the formulation and solution of the problem, but makes its solution interesting and in demand.

3.1. Task setting

The steady flow of an incompressible viscoelastic lubricant in the gap of an infinite radial sliding bearing coated with a low-melting metallic coating is considered in the conditions of the adiabatic process.

The shaft rotates at angular velocity Ω (fig. 1), and the bearing bush is stationary. It is assumed that the space between the eccentrically located shaft and the bearing is not completely filled (pre-fault condition) with a viscoelastic lubricant, and the bearing bush is made of a material with a low melting point.

In the polar coordinate system r, θ, the pole of which is located in the center of the bearing bush, the equation of the shaft contour, the surface of the bearing bush covered with a metal melt, and the molten surface of the bearing bush covered with a low-melting metallic melt will be written as follows:

$$r' = r_0(1+H),\ r' = r_1,\ r' = r_1 + \lambda' f(\theta), \tag{1}$$

where $H = \varepsilon\cos\theta - \frac{1}{2}\varepsilon^2\sin^2\theta + ...$; $\varepsilon = \frac{e}{r_0}$; r_0 is a shaft radius; r_1 is radius of the bearing bush covered with a metal melt; e is eccentricity; ε is relative eccentricity; $\lambda' f(\theta)$ is the bounded function at $\theta \in [0 \div 2\pi]$ subject to definition.

Conditions of the motion of an infinite radial sliding bearing are considered under the following assumptions:

1. All the heat released in the lubricating film goes to the melting surface of the material of the bearing bush.
2. The radial velocity component is slightly less than its circumferential component.
3. The pressure is constant throughout the thickness of the lubricating film.
4. The flow of a Maxwell fluid is given by the equation:

$$\frac{\partial v'_\theta}{\partial r'} = \frac{\tau'}{\mu'} + \frac{\Omega}{G'}\frac{\partial \tau'}{\partial \theta}. \tag{2}$$

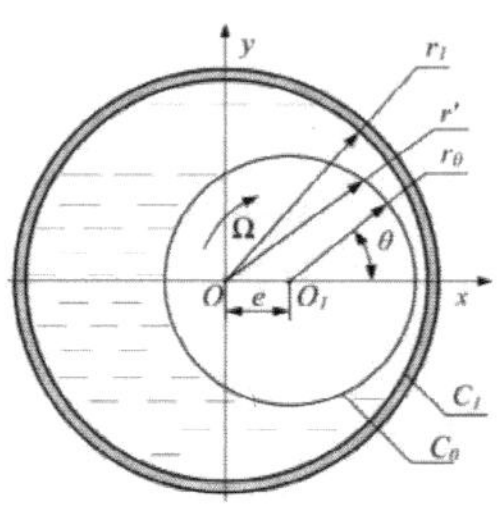

Figure 1. Design scheme.

The dependence of the shear modulus and lubricant viscosity on the pressure is expressed by the dependences:

$$G' = G_0 e^{\alpha' p'},\quad \mu' = \mu_0 e^{\alpha' p'}, \tag{3}$$

where μ_0 is characteristic viscosity, μ' is coefficient of dynamic viscosity of the lubricant, p' is hydrodynamic pressure in the lubricating layer, α' is a constant, G_0 is characteristic value of the shear modulus, G' is shear modulus.

3.2. Initial equations and boundary conditions

As the initial equations, based on the assumptions made, a system of equations of the motion of a lubricant with viscoelastic properties (the Maxwell liquid) for the case of a "thin layer" with allowance for (1), the continuity equation, as well as the energy dissipation rate formula for determining the function due to the molten surface bearing bush covered with a low-melting metal melt, is taken:

$$\frac{\partial^2 v'_\theta}{\partial r'^2} = \frac{1}{\mu'}\frac{1}{r_0}\frac{dp'}{d\theta} + \frac{\Omega}{G'}\frac{1}{r_0}\frac{d^2p'}{d\theta^2},\quad \frac{\partial v'_{r'}}{\partial r'} + \frac{v'_{r'}}{r'} + \frac{1}{r'}\frac{\partial v'_\theta}{\partial \theta} = 0,\quad \frac{d\lambda' f(\theta) r_0}{d\theta}\cdot\Omega L' = 2\mu' \int_{r_1+\lambda' f(\theta)}^{r_0+e\cos\theta} \left(\frac{\partial v'_\theta}{\partial r'}\right)^2 dr', \tag{4}$$

where L' is the specific heat of fusion per unit volume.

The boundary conditions in the case under consideration within terms $O(\varepsilon^2)$ will be written as:

$$v'_\theta = 0,\quad v'_{r'} = 0 \text{ at } r' = r_1 + \lambda' f(\theta),$$
$$v' = r_0\Omega,\quad v'_{r'} = -\Omega e\sin\theta \text{ at } r' = r_0 + e\cos\theta;\quad p'(\theta_1) = p'(\theta_2) = P_g, \tag{5}$$

where θ_1 and θ_2 are respectively, the angular coordinates of the starting and finishing point of the free surface of the lubricant, $\lambda' f(\theta) = \Phi(\theta)$.

When forming an analytical expression for hydrodynamic pressure, we transform the boundary conditions, assuming that the lubricant undergoes a shift at the entrance to the loaded zone. We believe that the lubricant enters the zone of hydrodynamic friction at full relaxation. Then the corresponding boundary conditions for the hydrodynamic pressure can be written in the form:

$$p''(\theta_1) = C'(\theta_1) = 0. \tag{6}$$

Relations between dimensionless and dimensional variables are given in the form:

$$r' = r_0 + \delta r;\ \delta = r_1 - r_0;\ v'_\theta = \Omega r_0 v;\ v'_{r'} = \Omega \delta u;$$

$$p' = p^* p;\ \mu' = \mu_0 \mu;\ G' = G_0 G;\ C' = C^* C;\ C^* = \frac{\mu_0 \Omega r_0}{\delta};\ \alpha' = \frac{\alpha}{p^*};\ p^* = \frac{\mu_0 \Omega r_0^2}{\delta^2}. \tag{7}$$

Substituting (7) into the system of differential equations (4) and the boundary conditions (5) - (6), we obtain the following system of differential equations:

$$\frac{\partial^2 v}{\partial r^2} = e^{-\alpha p}\frac{dp}{d\theta} + \beta e^{-\alpha p}\frac{d^2 p}{d\theta^2},\quad \frac{\partial u}{\partial r} + \frac{\partial v}{\partial \theta} = 0,\quad \frac{d\Phi(\theta)}{d\theta} = K\int_{-\Phi(\theta)}^{h(\theta)} \left(\frac{\partial v}{\partial r}\right)^2 dr, \tag{8}$$

where $\beta = \frac{\mu_0 \Omega}{G_0}$ is the Deborah number; u, v are components of the velocity vector of the lubricating medium; $K = \frac{2\mu_0 \Omega r_0^2}{\delta^2 L'}$ is the parameter characterizing the dissipation rate of mechanical energy; $h(\theta) = 1 - \eta\cos\theta$ the lubricating layer thickness.

And boundary conditions

$$u = 1,\quad v = -\eta\sin\theta \text{ at } r = h(\theta);\quad v = 0,\quad u = 0 \text{ at } r = -\Phi(\theta);$$

$$p(\theta_1) = p(\theta_2) = 0;\ p''(\theta_1) = 0;\ C'(\theta_1) = 0, \tag{9}$$

where $\eta = \frac{e}{\delta};\ \eta_1 = \frac{\lambda'}{\delta};\ \Phi(\theta) = \eta_1 f(\theta)$.

Let us introduce the designations, let $Z = e^{-\alpha p}$. Having differentiated both sides of the equation within terms $O(\alpha\beta)$, the first equation of system (4) takes the form:

$$\beta\frac{d^2 Z}{d\theta^2} + \frac{dZ}{d\theta} = -\alpha\left(\frac{\tilde{C}_1}{h^2(\theta)} + \frac{\tilde{C}_2}{h^3(\theta)}\right). \tag{10}$$

Then the system of equations (8) takes the following form:

$$\beta\frac{d^2 Z}{d\theta^2} + \frac{dZ}{d\theta} = -\alpha\left(\frac{\tilde{C}_1}{h^2(\theta)} + \frac{\tilde{C}_2}{h^3(\theta)}\right);\quad \frac{\partial u}{\partial r} + \frac{\partial v}{\partial \theta} = 0;\quad Z\frac{d\Phi(\theta)}{d\theta} = K\int_0^{1-\eta\cos\theta}\left(\frac{\partial v}{\partial r}\right)^2 dr \tag{11}$$

with the corresponding boundary conditions:

$$u = 1,\ v = -\eta\sin\theta \text{ at } r = h(\theta);\ u = 0,\ v = 0 \text{ at } r = 0 - \Phi(\theta);\ Z(\theta_1) = Z(\theta_2) = 1;\ Z''(0) = 0. \tag{12}$$

Taking K as a small parameter, due to the melt and the rate of energy dissipation, we seek the function $\Phi(\theta)$ as:

$$\Phi(\theta) = -K\Phi_1(\theta) - K^2\Phi_2(\theta) - K^3\Phi_3(\theta) - \ldots = H. \tag{13}$$

Boundary conditions for dimensionless velocity components u and v on the contour $r = -\Phi(\theta)$ can be written as:

$$v(0 - H(\theta)) = v(0) - \left.\left(\frac{\partial v}{\partial r}\right)\right|_{r=0} \cdot H(\theta) - \left.\left(\frac{\partial^2 v}{\partial r^2}\right)\right|_{r=0} \cdot H^2(\theta) - \ldots = 0;$$

$$u(0-H(\theta)) = u(0) - \left(\frac{\partial u}{\partial r}\right)\bigg|_{r=0} \cdot H(\theta) - \left(\frac{\partial^2 u}{\partial r^2}\right)\bigg|_{r=0} \cdot H^2(\theta) - ... = 0. \quad (14)$$

We seek the asymptotic solution of the system of differential equations (11) with allowance for the boundary conditions (12) and (14) in the form of series in powers of the small parameter K:

$$v = v_0(r,\theta) + Kv_1(r,\theta) + K^2 v_2(r,\theta) + ...; \quad u = u_0(r,\theta) + Ku_1(r,\theta) + K^2 u_2(r,\theta) + ...;$$

$$\Phi(\theta) = -K\Phi_1(\theta) - K^2\Phi_2(\theta) - K^3\Phi_3(\theta) - ...; \quad Z = Z_0 + KZ_1(\theta) + K^2 Z_2(\theta) + K^3 Z_3(\theta) - .... \quad (15)$$

Performing the substitution (15) into the system of differential equations (11), taking into account the boundary conditions (12) and (14), we obtain the following equations:

– for zeroth approximation:

$$\frac{\partial^2 v_0}{\partial r^2} = \beta\frac{d^2 Z_0}{d\theta^2} + \frac{dZ_0}{d\theta}, \quad \frac{\partial u_0}{\partial r} + \frac{\partial v_0}{\partial \theta} = 0. \quad (16)$$

With boundary conditions:

$$u_0 = 1, \quad v_0 = -\eta\sin\theta \text{ at } r = 1 - \eta\cos\theta; \quad u_0 = 0, \quad v_0 = 0 \text{ at } r = 0;$$

$$K\Phi_0(0) = Kg_0; \quad Z_0(\theta_1) = Z_0(\theta_2) = 1; \quad Z_0''(\theta_1) = 0. \quad (17)$$

– for the first approximation:

$$\frac{\partial^2 v_1}{\partial r^2} = \beta\frac{d^2 Z_1}{d\theta^2} + \frac{dZ_1}{d\theta}, \quad \frac{\partial u_1}{\partial r} + \frac{\partial v_1}{\partial \theta} = 0; \quad Z_0\frac{d\Phi_1(\theta)}{d\theta} = K\int_0^{h(\theta)}\left(\frac{\partial v_0}{\partial r}\right)^2 dr. \quad (18)$$

With boundary conditions:

$$v_1 = \left(\frac{\partial v_0}{\partial r}\right)\bigg|_{r=0} \cdot \Phi_1(\theta); \quad u_1 = \left(\frac{\partial u_0}{\partial r}\right)\bigg|_{r=0} \cdot \Phi_1(\theta);$$

$$u_1 = 0, \quad v_1 = 0 \text{ at } r = 1\text{-}\eta\cos\theta; \quad Z_1(\theta_1) = Z_1(\theta_2) = 0; \quad Z_1''(\theta_1) = 0,$$

$$K\Phi_1(\theta_1) = K\tilde{\tilde{\alpha}}, \quad \Phi_1(\theta_1) = \Phi_1(\theta_2) = \tilde{\tilde{\alpha}}. \quad (19)$$

The exact self-similar solution of the problem for the zeroth approximation will be sought in the form:

$$v_0 = \frac{\partial \psi_0}{\partial r} + V_0(r,\theta); \quad u_0 = -\frac{\partial \psi_0}{\partial \theta} + U_0(r,\theta); \quad \psi_0(r,\theta) = \tilde{\psi}_0(\xi); \quad \xi = \frac{r}{h(\theta)};$$

$$V_0(r,\theta) = \tilde{v}_0(\xi); \quad U_0(r,\theta) = -\tilde{u}_0(\xi) \cdot h'(\theta). \quad (20)$$

Substituting (20) into the system of differential equations (16), taking into consideration the boundary conditions (17), we get the following system of differential equations:

$$\tilde{\psi}_0'''(\xi) = \tilde{C}_2; \quad \tilde{v}_0''(\xi) = \tilde{C}_1; \quad \tilde{u}_0'(\xi) + \varepsilon\tilde{v}_0'(\xi) = 0; \quad \beta\frac{d^2 Z_0}{d\theta^2} + \frac{dZ_0}{d\theta} = -\alpha\left(\frac{\tilde{C}_1}{h^2(\theta)} + \frac{\tilde{C}_2}{h^3(\theta)}\right); \quad (21)$$

And boundary conditions:

$$\tilde{\psi}_0'(0) = 0, \quad \tilde{\psi}_0'(1) = 0, \quad \tilde{u}_0(1) = -\eta\sin\theta, \quad \tilde{v}_0(1) = 0; \quad Z_0(\theta_1) = Z_0(\theta_2) = 1;$$

$$\tilde{u}_0(0) = 0, \quad \tilde{v}_0(0) = 1, \quad \int_0^1 \tilde{v}_0(\xi)d\xi = 0. \quad (22)$$

By direct integration we obtain:

$$\tilde{\psi}_0'(\xi) = \frac{\tilde{C}_2}{2}(\xi^2 - \xi); \quad \tilde{v}_0(\xi) = \tilde{C}_1\frac{\xi^2}{2} - \left(1 + \frac{\tilde{C}_1}{2}\right)\xi + 1; \quad \tilde{C}_1 = 6. \quad (23)$$

3.3. Determination of hydrodynamic pressure

Solving the equation for hydrodynamic pressure $\beta\frac{d^2Z_0}{d\theta^2}+\frac{dZ_0}{d\theta}=-\alpha\left(\frac{\tilde{C}_1}{h^2(\theta)}+\frac{\tilde{C}_2}{h^3(\theta)}\right)$ taking into account the boundary condition $Z_0(\theta_1)=Z_0(\theta_2)=1;\ Z_0''(0)=0$ we obtain:

$$Z_0=\sup_{\theta\in[0;2\pi]}\left|1-\alpha\left[\left(6+\tilde{C}_2\right)(\theta-\theta_1)-\frac{6\eta\left(4+\tilde{C}_2\right)}{1+\beta^2}\times\left(-\beta\sin\frac{(\theta_1+\theta)}{2}\sin\frac{(\theta-\theta_1)}{2}+\cos\frac{(\theta_1+\theta)}{2}\sin\frac{(\theta_1-\theta)}{2}\right)-\alpha C_2\left(e^{-\frac{\theta_1}{\beta}}-e^{-\frac{\theta}{\beta}}\right)\right]\right|, \tag{24}$$

where

$$C_2=\frac{\alpha e^{\frac{\theta_1+\theta_2}{\beta}}}{e^{\frac{\theta_1}{\beta}}-e^{\frac{\theta_2}{\beta}}}\left[(\theta_2-\theta_1)\left(6-\tilde{C}_2\right)-\frac{6\eta\left(4+\tilde{C}_2\right)}{1+\beta^2}\times\sin\frac{(\theta_2-\theta_1)}{2}\left(\beta\sin\frac{(\theta_1+\theta_2)}{2}+\cos\frac{(\theta_1+\theta_2)}{2}\right)\right],$$

$$\tilde{C}_2=-\frac{\frac{1}{\beta^2}C_2e^{-\frac{\theta_1}{\beta}}}{\frac{3\eta\alpha}{1+\beta^2}(\beta\cos\theta_1-\sin\theta_1)}-4.$$

In order to define $\Phi_1(\theta)$ taking into account equation (23), we arrive at the following equation:

$$\frac{d\Phi_1(\theta)}{d\theta}=\frac{h(\theta)}{Z_0}\int_0^1\left(\frac{\tilde{\psi}_0''(\xi)}{h^2(\theta)}+\frac{\tilde{v}_0'(\xi)}{h(\theta)}\right)^2d\xi. \tag{25}$$

Integrating equation (25), we obtain:

$$\Phi_1(\theta)=\frac{1}{Z_0}\left(\int_{\theta_1}^{\theta}\frac{\Delta_1 d\theta}{h^3(\theta)}+\int_{\theta_1}^{\theta}\frac{\Delta_2 d\theta}{h^2(\theta)}+\int_{\theta_1}^{\theta}\frac{\Delta_3 d\theta}{h(\theta)}\right), \tag{26}$$

where

$$\Delta_1=\int_0^1\left(\tilde{\psi}_0''(\xi)\right)^2d\xi=\frac{\tilde{C}_2^2}{12};\ \Delta_2=\int_0^1 2\tilde{\psi}_0''(\xi)\cdot\tilde{v}_0'(\xi)d\xi=\tilde{C}_2;\ \Delta_3=\int_0^1\left(\tilde{v}_0'(\xi)\right)^2d\xi=4. \tag{27}$$

Solving equations (26) with allowance for (27) and condition $K\Phi_1(\theta_1)=K\tilde{\tilde{\alpha}}$, we obtain:

$$\Phi_1(\theta)=\frac{1}{Z_0}\left(\left(\frac{\tilde{C}_2}{12}+\tilde{C}_2+4\right)(\theta-\theta_1)+\cos\frac{(\theta+\theta_1)}{2}\sin\frac{(\theta-\theta_1)}{2}\left(\frac{9\eta}{2}\tilde{C}_2+8\eta\right)+\tilde{\tilde{\alpha}}\right). \tag{28}$$

Then, for the first approximation, we obtain:

$$v_1=\frac{\partial\psi_1}{\partial r}+V_1(r,\theta);\quad u_1=-\frac{\partial\psi_1}{\partial\theta}+U_1(r,\theta);\quad \psi_1(r,\theta)=\tilde{\psi}_1(\xi);\quad \xi=\frac{r}{h(\theta)};$$
$$V_1(r,\theta)=\tilde{v}(\xi);\quad U_1(r,\theta)=-\tilde{u}_1(\xi)\cdot h'(\theta). \tag{29}$$

Substituting (29) into the system of differential equations (18), taking into consideration the boundary conditions (19), we get the following system of differential equations:

$$\tilde{\psi}_1'''(\xi)=\tilde{\tilde{C}}_2;\quad \tilde{v}_1''(\xi)=\tilde{\tilde{C}}_1;\ \tilde{u}_1'(\xi)+\xi\tilde{v}_1'(\xi)=0;\ \beta\frac{d^2Z_1}{d\theta^2}+\frac{dZ_1}{d\theta}=-\alpha\left(\frac{\tilde{\tilde{C}}_1}{h^2(\theta)}+\frac{\tilde{\tilde{C}}_2}{h^3(\theta)}\right); \tag{30}$$

and boundary conditions:

$$\tilde{\psi}_1'(0)=0,\quad \tilde{\psi}_1'(1)=0,\ \tilde{u}_1(1)=0,\ \tilde{v}_1(1)=0;\ \tilde{u}_1(0)=0;\ \tilde{v}_1(0)=M;\quad \int_0^1\tilde{v}_1(\xi)d\xi=0;$$
$$Z_1(\theta_1)=Z_1(\theta_2)=0;\ Z_1''(\theta_1)=0. \tag{30}$$

From conditions $Z_1(\theta_1)=Z_1(\theta_2)=0,\ Z_1''(\theta_1)=0$, we obtain:

$$Z_1=\alpha\left(6M+\tilde{\tilde{C}}_2\right)(\theta-\theta_1)-\frac{6\eta\alpha\left(4M+\tilde{\tilde{C}}_2\right)}{1+\beta^2}\times\sin\frac{(\theta-\theta_1)}{2}\left(\beta\sin\frac{(\theta_1+\theta)}{2}+\cos\frac{(\theta_1+\theta)}{2}\right), \tag{32}$$

where

$$\tilde{\tilde{C}}_2=\frac{-\frac{1}{\beta^2}C_4e^{-\frac{\theta_1}{\beta}}}{\frac{3\eta\alpha}{1+\beta^2}(\beta\cos\theta_1-\sin\theta_1)}-4,$$

$$M=\sup_{\theta\in[0:2\pi]}\left|\frac{\partial v_0}{\partial r}\bigg|_{r=0}\cdot\Phi_1(\theta)\right|=\sup_{\theta\in[0:2\pi]}\left|\left\{\frac{-\eta\sin\theta}{1-\eta\cos\theta}-\frac{3\eta\beta^2\left(1+\tilde{C}_2\right)}{4\left(1+\beta^2\right)}\times\left[-2\sin\frac{(\theta-\theta_1)}{2}\cos\frac{(\theta_1+\theta)}{2}+\right.\right.\right.$$

$$\left.+2\cos\frac{(\theta_1-\theta)}{2}\cos\frac{(\theta_1+\theta)}{2}\right]-3\left(6+\tilde{C}_2\right)(1-\eta\cos\theta)-\frac{3\eta\beta\left(4+\tilde{C}_2\right)}{2\left(1+\beta^2\right)}\times\left[\cos\frac{(\theta_1-\theta)}{2}\left(\sin\frac{(\theta+\theta_1)}{2}-\cos\frac{(\theta_1+\theta)}{2}\right)+\right.$$

$$\left.\left.\left.+\sin\frac{(\theta-\theta_1)}{2}\left(\cos\frac{(\theta_1+\theta)}{2}-\sin\frac{(\theta+\theta_1)}{2}\right)\right]+\frac{\alpha C_2(1-\eta\cos\theta)}{2\beta}\right\}\times\left\{\left(\frac{\tilde{C}_2}{12}+\tilde{C}_2+4\right)(\theta-\theta_1)+\cos\frac{(\theta_1+\theta)}{2}\sin\frac{(\theta-\theta_1)}{2}\left(\frac{9\eta\tilde{C}_2}{2}+8\eta\right)+\tilde{\tilde{\alpha}}\right\}\right|$$

$$C_4=\left[\alpha\left(6M+\tilde{C}_2\right)(\theta_1-\theta_2)-\frac{6\eta\alpha\left(4M+\tilde{\tilde{C}}_2\right)}{1+\beta^2}\sin\frac{(\theta_2-\theta_1)}{2}\times\left(\beta\sin\frac{(\theta_1+\theta_2)}{2}-\cos\frac{(\theta_1+\theta_2)}{2}\right)\right].$$

Then for $Z=Z_0+KZ_1$ we obtain the following expression:

$$Z=1+\alpha(\theta_1-\theta)\left(6+\tilde{C}_2+K\left(6M+\tilde{\tilde{C}}_2\right)\right)+\frac{3\eta\alpha}{\left(1+\beta^2\right)}\sin\frac{(\theta_1-\theta)}{2}\left(\beta\sin\frac{(\theta+\theta_1)}{2}+\cos\frac{(\theta_1+\theta)}{2}\right)\times$$

$$\times\left(4+\tilde{C}_2+K\left(4M+\tilde{C}_2\right)\right)-\alpha C_2\left(e^{-\frac{\theta_1}{\beta}}-e^{\frac{\theta}{\beta}}\right)$$

or

$$Z=1+\alpha F(\theta), \tag{33}$$

where

$$F(\theta)=(\theta_1-\theta)\left[6+\tilde{C}_2+K\left(6M+\tilde{\tilde{C}}_2\right)\right]+\frac{6\eta}{\left(1+\beta^2\right)}\sin\frac{(\theta_1-\theta)}{2}\left(\beta\sin\frac{(\theta+\theta_1)}{2}+\cos\frac{(\theta_1+\theta)}{2}\right)\times$$

$$\times\left(4+\tilde{C}_2+K\left(4M+\tilde{C}_2\right)\right)-\alpha C_2\left(e^{-\frac{\theta_1}{\beta}}-e^{\frac{\theta}{\beta}}\right).$$

Then applying Taylor series expansions $e^{-\alpha p}$, we obtain:

$$1-\alpha p+\frac{\alpha^2p^2}{2}=1+\alpha F(\theta). \tag{34}$$

Solving equation (34) within terms $O(\alpha^3)$, $F^2(\theta)$ for hydrodynamic pressure, we obtain:

$$p=1+\alpha F(\theta). \tag{35}$$

3.4. Results of study and their discussion

Let us now turn to the determination of the basic operating characteristics of the radial bearing. Taking into account (16), (18) and (35) for the component of the supporting force vector and the frictional force, we obtain:

IOP Conf. Series: Materials Science and Engineering **760** (2020) 012002 doi:10.1088/1757-899X/760/1/012002

$$R_x = \frac{\mu\Omega r_0^3}{\delta^2}\int_{\theta_1}^{\theta_2} p\cos\theta d\theta = \frac{\mu\Omega r_0^3}{\delta^2}\times\left[\sin\theta_2 - \sin\theta_1 + +\alpha\left\{\left[\theta_1\left(\sin\theta_2 - \sin\theta_1\right) - \left(\theta_2\sin\theta_2 - \theta_1\sin\theta_1 + \cos\theta_2 - \cos\theta_1\right)\right]\times\right.\right.$$

$$\times\left[6+\tilde{C}_2 + K\left(6M + \tilde{\tilde{C}}_2\right)\right] + \frac{6\eta}{\left(1+\beta^2\right)}\left(4+\tilde{C}_2 + K\left(4M + \tilde{C}_2\right)\right)\times\left[\sin\theta_1\left(\sin\theta_2 - \sin\theta_1\right) - \frac{1}{2}\left(\sin 2\theta_2 - \sin 2\theta_1\right) -\right.$$

$$\left.\left.-\cos\theta_1\left(\sin\theta_2 - \sin\theta_1\right) + \frac{1}{2}\left(\theta_2 - \theta_1 + \cos 2\theta_2 - \cos 2\theta_1\right)\right] - C_2\left(\sin\theta_2 - \sin\theta_1\right)\left(e^{-\frac{\theta_1}{\beta}} - e^{\frac{\theta}{\beta}}\right)\right\},$$

$$R_y = \frac{\mu\Omega r_0^3}{\delta^2}\int_{\theta_1}^{\theta_2} p\sin\theta d\theta = \frac{\mu\Omega r_0^3}{\delta^2}\times\left[-\cos\theta_2 + \cos\theta_1 + \alpha\left\{\left[\theta_1\left(-\cos\theta_2 + \cos\theta_1\right) + \left(\theta_1\cos\theta_1 + \sin\theta_2 - \sin\theta_1\right)\right]\times\right.\right.$$

$$\times\left[6+\tilde{C}_2 + K\left(6M + \tilde{\tilde{C}}_2\right)\right] + \frac{6\eta}{\left(1+\beta^2\right)}\left(4+\tilde{C}_2 + K\left(4M + \tilde{C}_2\right)\right)\times\left[\sin\theta_1\left(-\cos\theta_2 + \cos\theta_1\right) - \frac{1}{2}\left(\theta_2 - \theta_1 - \cos 2\theta_2 + \cos 2\theta_1\right) +\right.$$

$$\left.\left.+\cos\theta_1\left(\cos\theta_2 + \cos\theta_1\right) + \frac{1}{2}\left(\sin 2\theta_2 - \sin 2\theta_1\right)\right] - C_2\left(-\cos\theta_2 + \cos\theta_1\right)\left(e^{-\frac{\theta_1}{\beta}} - e^{\frac{\theta}{\beta}}\right)\right\},$$

$$L_{\text{тр}} = \frac{\mu\Omega r_0^2 e^{-\alpha p}}{\delta}\int_{\theta_1}^{\theta_2}\left[\left.\frac{\partial v_0}{\partial r}\right|_{r=0} + K\left.\frac{\partial v_1}{\partial r}\right|_{r=0}\right]d\theta = \frac{\mu\Omega r_0^2}{\delta}\left(1+\alpha p - \frac{\alpha^2 p^2}{2}\right)\left[\eta\left(\cos\theta_2 - \cos\theta_1\right) -\right.$$

$$-\frac{3\eta\beta^2\left(4+\tilde{C}_2\right)}{4\left(1+\beta^2\right)}\left[\sin\theta_2 - \sin\theta_1 - \left(\theta_2 - \theta_1\right)\cos\theta_1 - \cos\theta_2 + \cos\theta_1 - \left(\theta_2 - \theta_1\right)\cos\theta_1 - 3\left(6+\tilde{C}_2\right)\times\right.$$

$$\times\left(\theta_2 - \theta_1 - \eta\left(\sin\theta_2 - \sin\theta_1\right)\right) - \frac{6\eta\beta\left(4+\tilde{C}_2\right)}{2\left(1+\beta^2\right)}\times\left[-2\cos\theta_2 + 2\cos\theta_1 - \left(\theta_2 - \theta_1\right)\cos\theta_1 -\right.$$

$$\left.\left.-\left(\sin\theta_2 - \sin\theta_1\right)\right]\times\left(1+KM\right) + \frac{\alpha C_2 e^{\frac{\theta_2}{\beta}}}{2\beta}\left(\theta_2 - \theta_1 - \eta\left(\sin\theta_2 - \sin\theta_1\right)\right)\right]. \qquad (36)$$

For the verification calculations, the following values are used based on the obtained theoretical models:

μ_0 = 0,085 Hs/m^2; η = 0,3…1 m; r_0 = 0,019995…0,0493 m; Ω = 100…1800 s^{-1}; p_g = 2 MPa;

δ = 0,05 · 10^{-3}…0,07 · 10^{-3} m; K = 0,00052…0,0000022; α = 0…1; L' = 35,33…38,1 H/m^2.

Based on the results of numerical calculations, the graphs shown in Fig. 2–3 were made.

a) b)

Figure 2. Dependence of components of the supporting force on parameter α, the lubricant characterizing dependence of viscosity on pressure, and on the parameter $\theta_2 - \theta_1$ characterizing the extent of the loaded area of lubricant.

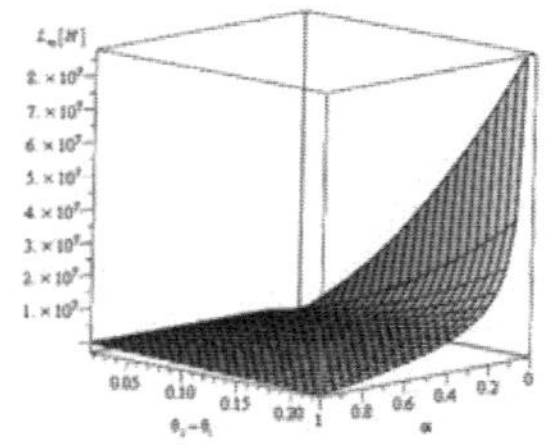

Figure 3. Dependence of a frictional force on parameter α, the lubricant characterizing dependence of viscosity on pressure, and on the parameter characterizing the extent of the loaded area of lubricant.

In the experimental study, a sliding bearing with low-melting metal babbite alloy No. 83 and No. 88 is considered (table 1). According to the results of the experiments, the friction coefficient was determined, which allows to judge the presence of hydrodynamic friction mode both when the bearing is working on a lubricant having viscoelastic properties, and on the melt of a fusible babbite coating. The temperature regime and transition of hydrodynamic friction to boundary friction were also determined. Analysis of experimental studies shows that the melt of a fusion babbite coating affects the coefficient of friction 2-5 times more intense than the rheological properties of the lubricants used. The complex of experimental studies, which confirmed the reliability of the developed theoretical computational models and the data of their numerical analysis, in the considered range of design and operational parameters of tribosystems with low-melting metal melt, as a result of satisfaction convergence of theoretical and experimental results.

Table 1 Results of experimental studies of babbit covering No. 83 ,No. 88

		Theoretical research				Experimental research	
		A truly viscous lubricant		Viscoelastic lubricant		Viscoelastic lubricant	
		low-melting metal coating		low-melting metal coating		low-melting metal coating	
	№	B83	B88	B83	B88	B83	B88
Coefficient of friction	1	0,0021	0,0006	0,0029	0,0014	0,018	0,014
	2	0,0022	0,0008	0,0030	0,0016	0,020	0,016
	3	0,0024	0,0013	0,0032	0,0020	0,022	0,020
	4	0,0028	0,0017	0,0034	0,0024	0,027	0,022
	5	0,0029	0,0023	0,0036	0,0030	0,029	0,023

4. Conclusion

1. A refined design model of a radial sliding bearing operating under conditions of hydrodynamic lubrication with a melt of a low-melting coating is obtained, taking into consideration the dependence of viscosity on pressure under steady conditions of a sliding friction during inexact filling of a working gap with lubricant.

2. A significant contribution of the constructive parameter K due to the melt is shown. With an increase in the design parameter K, the friction coefficient decreases by 60%, and the bearing capacity increases by 16%.

Dependence of the coefficient of friction on the design parameter K, caused by the melt, is close to linear in the range of 0.0014–0.003.

5. Acknowledgments

The work has been carried out under the grant of OJSC RZD No. 2210370 / 22.12.2016 for the development of scientific and pedagogical schools in the field of railway transport.

References

[1] Kragelsky I V, 1968 Friction and wear Mechanical engineering
[2] Bowden, F P, Tabor D 1968 Friction and lubrication of solids. Mechanical engineering
[3] Kropachev, D Y, Grishin, A A, Maslo, A D Methods 2012 For The Rapid Measurement Of The Temperature Of Metal Melt For The Needs Of Machine-Building Enterprises. *Cast and Metallurgy*. **3(66)** 126–127
[4] Wilson 1976 Lubricant with Melt. *Problems of friction and lubrication.* **1** 19
[5] 1991 Physical Quantities. Directory Energoatomizdat
[6] Khavin, V Y 1991 Brief Reference Book of the Chemist Khimiya
[7] Perelman V I 1964 Brief Reference Book of the Chemist Khimiya
[8] Prokopiev V N, Zadorozhnaya, E A, Karavaev, V G, Levanov, I G 2010 Improvement of the Calculation Technique for Complex Loaded Sliding Bearings Lubricated With Non-Newtonian Oils. *Problems of machine building and machine reliability* **1** 63– 67
[9] Beretta G P, Niro A, Silvestri M 1992 Friction bearings lubricated by their own melt or by product of sublimation. *Works of the American Society of Engineers* **1** 86–90
[10] Prokhodko V M, Kotelnitskaya L I 2001 Mathematical Model of Hydrodynamic Lubricant during Melting of Bearing Surface of Radial Bearing. *Friction and wear.* **22** **6** 606–608
[11] Akhverdiev K S, Mukutadze M A, Lagunova E O, Vasilenko V V 2017 Hydrodynamic Calculation of a Radial Bearing Lubricated By a Melt of a Fusible Coating in Presence of a Lubricant. *Bulletin of Rostov State University of Transport.* **2 (66)** 129-135
[12] Lagunova E O 2017 Simulation of Wedge-Shaped Sliding Supports Taking into Account the Rheological Properties of an Electrically Conductive Lubricant. *Bulletin of Southern Ural State University. Series «Machine Building».* **17** **4** 13–25. DOI: 10.14529/engin170402

IOP Conf. Series: Materials Science and Engineering **760** (2020) 012003 doi:10.1088/1757-899X/760/1/012003

Parallel operation of an inverter with an electrical AC network

T L Alekseeva[1], N L Ryabchyonok[1], L A Astrakhantsev[1], V A Tikhomirov[1], N P Astashkov[1], A L Martusov[1] and M E Alekseev[1]

[1]Irkutsk State Transport University, 15, Chernyshevsky St., 664074, Russia

E-mail: talecseeva843@gmail.com

Abstract. The paper studies the matters of the interaction of a source of non-sinusoidal current and voltage with a centralized power supply system for the electric energy recuperation during electric braking of rolling stock on the railroads. Revised energy conservation law in electromagnetic field and spectral analysis of non-sinusoidal voltage and current at the output of the inverter have been used to prove the cause of reduced current of RMS voltage in the overhead system. It has been calculated that controlling the active power of recuperation due to the phase shift of the harmonic current components relative to the similar harmonic voltage components is accompanied by a significant increase in the reactive component of the surge current compared to its active component. Losses of active power in the electric stock recuperating heat exchanger make up 83–87 % of the total power loss in the recuperating heat exchanger and in the overhead system. The reactive component of the equalizing current loads the transformers of the traction substations, therefore the voltage on the buses of the substation switchgear decreases. It is proposed to increase the voltage of the overhead system and the traction substation during the recuperation to maintain the minimum phase shift angle of the harmonic components of the non-sinusoidal current relative to the similar harmonic components in the overhead system.

1. Introduction

The most important factor in reducing the cost of the energy component of material production in the state economy sectors and mitigating the negative impact of energy on the environment is energy saving and power-efficient use of electric energy [1, 2]. During transport operations, significant savings in electric energy consumption for train traction are possible due to the use of potential and kinetic energy of a train. The volume ofregenerative braking in the railway network reaches 2.151 billion kWh, which is about 5 % of the train traction consumption [3, 4]. Currently, due to the potential and kinetic energy of the train, energy recuperation is performed by inverters controlled by an AC network and four-quadrant 4-qS converters, whose power factor decreases to 0.5 and the efficiency goes down to 70 % [5, 6].

The losses of active power with the electric energy recuperation from the AC generators of the electric stock (ES) into the system of external power supply in the electric stock reversible converter comprise 83 – 87 %, and in the overhead system of electric transport power line comprises 13–17 % from the total losses of active power in the traction transformer, reversible converter of electric stock and in the overhead system [7–9]. Due to the decrease in the RMS voltage in the AC overhead system and on the buses of the traction substation during operation of known recuperating heat exchangers, the track capacity of railway sections in mountainous and hilly areas is reduced [10, 11]. The aim of this work is to increase the productivity of railways and to reduce energy consumption for train traction through the use of power-efficient recuperation of electric energy by traction rolling stock.

2. Subject of research and methods

During electrical energy recuperation, limited-power electric stock DC generators interact with traction transformer substations. The shape of the AC voltage curve on their buses is close to sinusoidal. The interaction is carried out through inverters and an overhead network with real and inductive resistance. Reversible converters (RCs) form rectangular alternating current, and 4qS-converters form pulse current, whose envelope is close to sinusoidal form. The power at the output of the inverter is controlled by shifting the rectangular current in phase to the advance side with respect to the voltage in the secondary winding of the ES traction transformer from the minimum advance angle of the inverter thyristors to $\pi/2$. In the overhead system, the reactive component of the rectangular equalizing current increases and significantly exceeds the active component of the reactive current, and energy recuperation stops at $\pi/2$. With a pulsed form of current at the output of the inverter, during the pause between pulses, the electric potential of energy sources is not used for recuperation. For the studies, the refined energy conservation law and new energy characteristics [12] of alternating current energy generators of limited power are used in parallel with the traction transformer substation, which is powered from a centralized energy system. Computer-based mathematical modeling, spectral analysis of non-sinusoidal voltage and current at the inverter output are used to estimate the energy efficiency of parallel operation of the inverter with the AC system.

3. Research method

The power balance at the output of the heat exchanger [13], taking into account the reduction in the duration of the irreversible conversion of electrical energy into a different type of energy by reactive elements and power semiconductors of the recuperating heat exchanger (1):

$$\sqrt{S_G^2 - \Delta S^2} = \sqrt{P^2 + Q^2}, \tag{1}$$

where ΔS is a part of the apparent power S_G , at the output of the recuperative heat exchanger, which is not used for the recuperation of electrical energy and for energy exchange in the electrical circuit;
P is the total active power of the harmonic components of voltage and current at the output of the recuperating heat exchanger;
Q is the total reactive power of the harmonic components of voltage and current at the recuperating heat exchanger output.

The components of the proposed power balance (1) can be calculated and measured using Fourier transform, analytical calculations, mathematical modeling and instrumentation equipment in actual practice.

Let us calculate the power ΔS, which is formed at the output of the recuperating heat exchanger, and which takes into account the electric potential of the energy source while switching the power semiconductors of the inverter:

$$\Delta S = \sqrt{\sum_{k=0}^{n} U_{pk}^2 \cdot I_k^2} = U_P \cdot I_{in}, \tag{2}$$

where U_{pk} is the RMS voltage of the k-th harmonic component at the output of the recuperating heat exchanger while switching the power semiconductors of the inverter;
I_k is the RMS current of the k-th harmonic component at the recuperating heat exchanger output;
k is the number of the harmonic component;
n is the number of the last harmonic taken into account;
where U_P is the RMS voltage at the output of the recuperating heat exchanger while switching the power semiconductors of the inverter;

I_{in} is the RMS current at the recuperating heat exchanger output.

Active power at the recuperating heat exchanger output characterizes a part of electrical energy, which is irreversibly transformed into another type of energy, into thermal energy losses in a power circuit, and is transferred to an external power supply system:

$$P = U_{c0} \cdot I_0 + \sum_{k=1}^{n} U_{ck} \cdot I_k \cdot cos\varphi_k, \quad (3)$$

where U_{c0} is the constant component of the voltage at the recuperative heat exchanger output during the transfer of energy to the AC network;
U_{ck} is the RMS voltage of the k-th harmonic component at the recuperative heat exchanger output during the transfer of energy to the AC network;
where U_0 is the constant component of the current at the recuperative heat exchanger output during the transfer of energy to the AC network;
φ_k is the phase angle of the k-th current harmonic relative to the similar k-th voltage harmonic at the recuperating heat exchanger output.

Reactive power at the recuperative heat exchanger output, which characterizes a part of the electrical energy consumed for energy exchange between the energy source and reactive elements of the electric circuit, is:

$$Q = \pm\sqrt{\sum_{k=1}^{n} U_{ck}^2 \cdot I_k^2 \cdot sin^2\varphi_k}. \quad (4)$$

The angle of apparent power and output electrical resistance φ of the recuperating heat exchanger is:

$$\varphi_\Sigma = arctg\left[\frac{\pm\sqrt{\sum_{k=1}^{n} U_{ck}^2 \cdot I_k^2 \cdot sin^2\varphi_k}}{U_{c0} \cdot I_0 + \sum_{k=1}^{n} U_{ck} \cdot I_k \cdot cos\varphi_k,}\right]. \quad (5)$$

Using expressions (1–5) and mathematical modeling in the Simulink environment of the MATLAB software, it is possible to obtain an estimate of the energy efficiency of the electrical energy recuperation process.

When the reversible converter operates in the mode of 2ES5K electric locomotive inverter in the middle of an inter-substation area with duplicate power supply to the overhead system and the the advance angle of the inverter thyristors increases up to $\beta = 72$ electrical degrees in order to increase the transfer of active power to the external power supply system, a mathematical model of the recuperation process is shown in Figure 1. The mathematical model of a traction transformer substation is presented using an EMF source of 27 500 V with internal resistance and inductance. The overhead system with duplicate power supply 2 has a resistance of 2.5 Ohms and an inductance of 16 mH. Traction transformers of two-section EMF with a rated power of 4350 kVA, rated voltage in the primary winding of 25 kV have 3 sections of consecutively connected secondary windings. The source of DC energy is a DC EMF with real and inductive resistances.

Active power at the recuperating heat exchanger input $P_d = U_{dc} \cdot I_d = 900 \cdot 8375 = 7537{,}5\ kW$ is calculated using readings of the instruments in figure 1 for measuring voltage U_{dc} and the RMS current in the windings of electric machines I_d. The oscillogram of the curves of instantaneous values of voltage and current in the primary and secondary windings of the ES traction transformer is shown in Figure 2. The oscilloscope and powerqui block in Figure 1 allow one to perform calculations of the harmonic components of non-sinusoidal voltage, equalizing current and power balance components at the output of the ES recuperating heat exchanger.

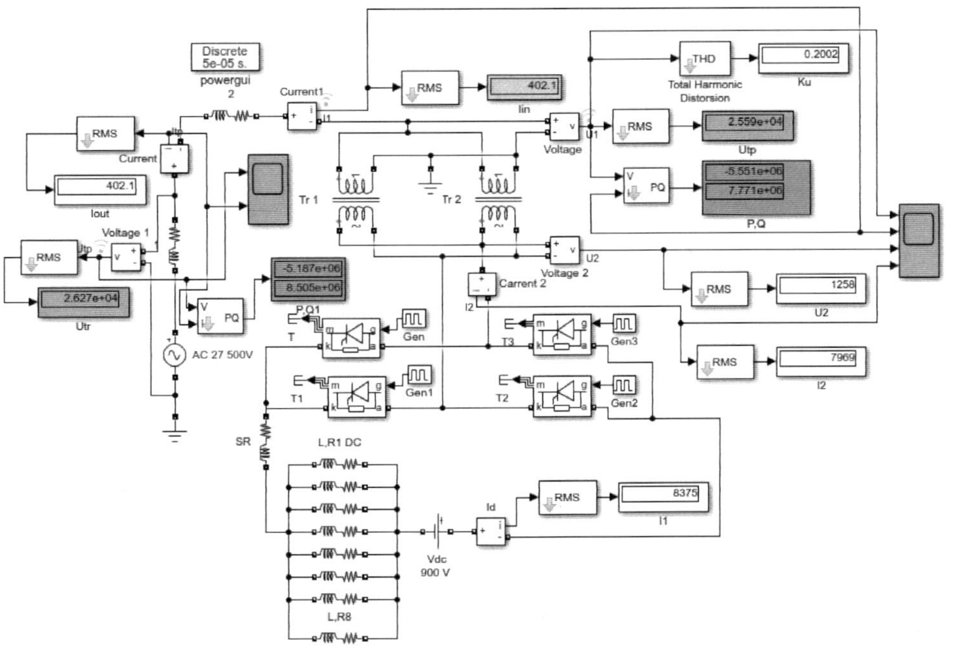

Figure 1. The mathematical model of the process of electrical energy recuperation.

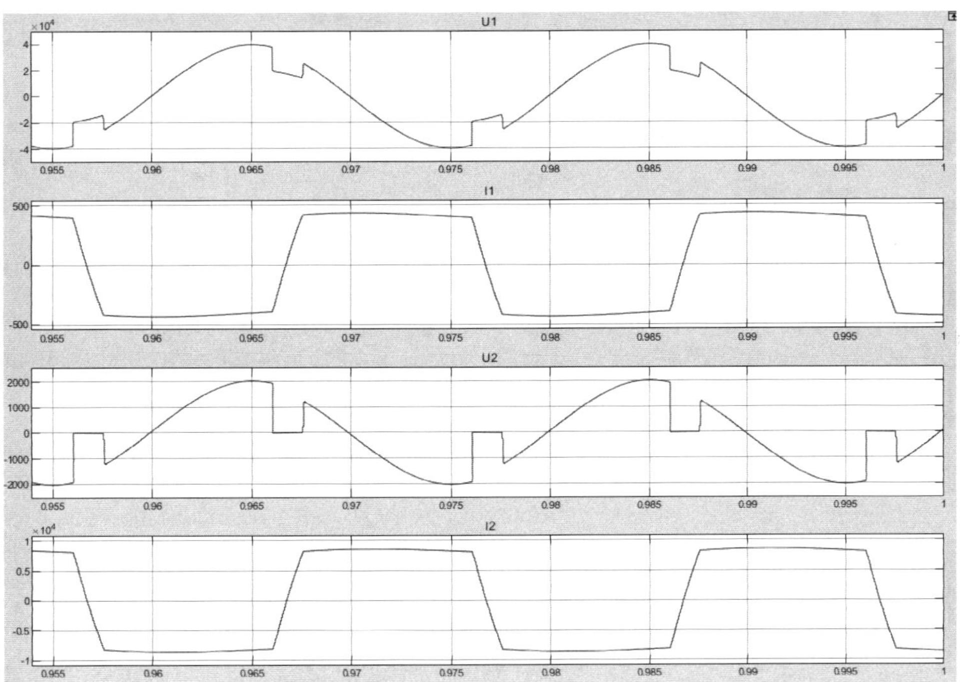

Figure 2. Oscillograms of voltage U_1, current I_1 in the primary and U_2, I_2 in the secondary winding of the traction transformer when the reversible converter is operating in the electric energy recuperation mode.

As a result of the spectral analysis of non-sinusoidal curves U_1 and I_1, the values and initial phases of the similar harmonic components are obtained in Figure 3.

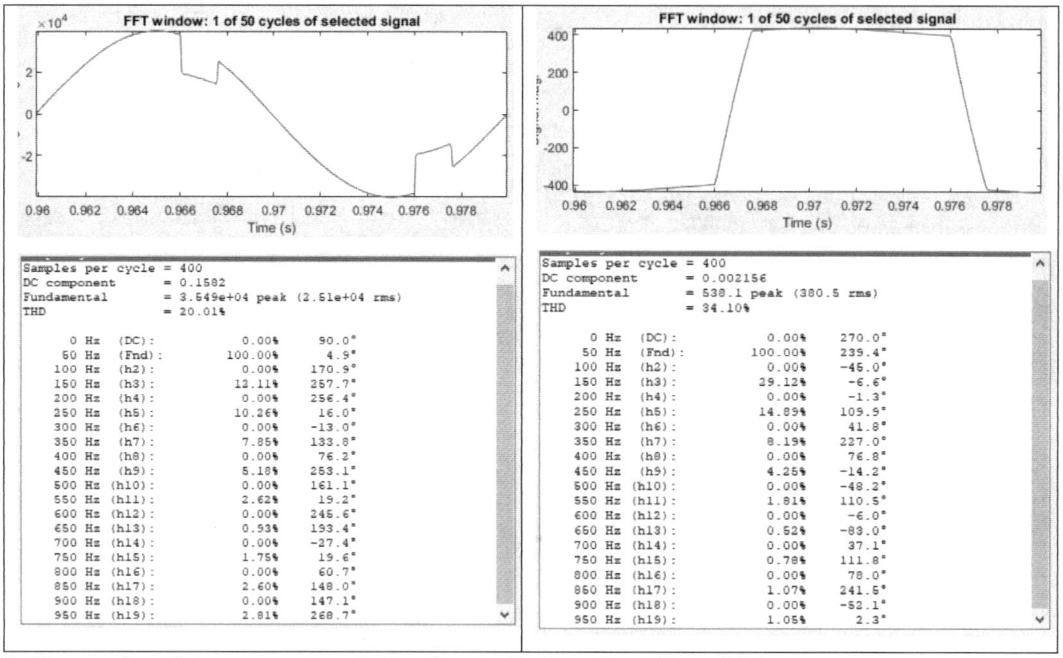

Figure 3. The results of the spectral analysis of periodic functions U_1 and I_1.

The RMS value of the voltage of the first harmonic is $U_{1\,(1)} = 25100\ V, f_1 = 50\ Hz$, the RMS value of the current of the first harmonic is $I_{1\,(1)} = 380.5\ A, f_1 = 50\ Hz$.

According to the formula (3), the active power is calculated taking into account the phase angle of the similar harmonic components with respect to voltage $P_1 = 25100 \cdot 380.5 \cdot \cos(4.9^0\text{-}239.4^0) = -5546.03\ kW$, $P_3 = -33.45\ kW$, $P_5 = -9.92\ kW$, $P_7 = -3.43\ kW$, $P_9 = -0.99\ kW$, $P_{11} = -0.10\ kW$, $P_{13} = +0.05\ kW$, $P_{15} = -0.05\ kW$, $P_{17} = -0.16\ kW$, $P_{19} = -0,\ 18\ kW$ and the total active power of the harmonic components taken into account is $P = -5594.6\ kW$. The error of calculation and mathematical modeling is *0.8 %*. According to the formula (4) the reactive power at the output of the inverter of the electric locomotive is $Q_1 = 25100 \cdot 380.5 \cdot \sin(4.9^0\text{-}239.4^0) = -7775{,}25\ kVAr$, $Q_3 = -335.12\ kVAr$, $Q_5 = -145.65\ kVAr$, $Q_7 = -61.68\ kVAr$, $Q_9 = -21.0\ kVAr$, $Q_{11} = -4.53\ kVAr$, $Q_{13} = -0.46\ kVAr$, $Q_{15} = -1.30\ kVAr$, $Q_{17} = -2.65\ kVAr$, $Q_{19} = -2.81\ kVAr$. The total reactive power of the harmonic components taken into account is $Q = -7784.11\ kVAr$. Using the device P, Q in Figure 1, the reactive power $Q = 7771.0\ kVAr$ of the first harmonic of voltage $U_{1\,(1)}$ and current $I_{1\,(1)}$ was measured.

With the help of formula (5) one can calculate the angle of apparent power, the output electrical resistance of the inverter with electric ES machines $\varphi_{\Sigma} = arctg\,(-7784.11 / -5594.60) = 54.29^o$ which is indicative of the active and real nature of the output electrical resistance of the inverter with electric machines of electric stock. The apparent power at the ES current collector is $S = U_{tp} \cdot I_{tp} = 25590 \cdot 402.1 = 10289.74\ kVA$. The power at the current collector during the switching of the inverter thyristors is $\Delta S = \Delta U \cdot I_{tp} = 9007 \cdot 402.1 = 3621.71\ kVA$.

The power factor of reversible converter in the recovery mode is:

$$K_M = P / S = 5594.6 / 10289.74 = 0.54.$$

The efficiency of reversible converter when operating in the recovery mode is:

$$\eta_{RC} = (P / P_d) \cdot 100\ \% = (5594.6 / 7537.5) \cdot 100\ \% = 74.2\ \%.$$

The power balance (1) is checked at the output of the electric energy recuperating heat exchanger of electric stock and the calculation error is determined as:

$$\sqrt{S_G^2 - \Delta S^2} = \sqrt{P^2 + Q^2};\ \sqrt{10289{,}74^{\,2} - 3621{,}71^2} = \sqrt{5594{,}6^2 + 7784{,}11^2}.$$

$9631.3\ kVA \approx 9586.0\ kVA$, the calculation error is 0.5 %.

Active power losses in the AC overhead system with duplicate power supply with active resistance $R_{ks} = 2.5\ Ohm$ is:

$$\Delta P_{os} = I_{tp}^2 \cdot R_{os} = 402{,}1^2 \cdot 2{,}5 = 404{,}2\ kW.$$

Losses of active power in the overhead system make up 17 %, and in the reversible converter of the electric stock they make up 83 % of the total losses of active power in the overhead system and in the reversible converter of the electric stock.

Active power transmitted to the external power supply system is:

$$P - \Delta P_{os} = 5594.6 - 404.2 = 5190.4\ kW.$$

In the ES current collector during the energy recuperation, the voltage is less than the voltage at the traction substation and is *25590 V*. The voltage on the buses of the traction substation is *26270 V*, that is, deviates by *-0.95 %* from the set voltage when the substation transformer is idle.The sinusoidal distortion of the voltage curve is U_I K_U $= 20.0\ \%$ and in the overhead system is estimated using the instrument Total Harmonic Distorsion (THD) in Figure 1 during the inverter operation.

4. Conclusion

As a result of the study, it was found that the voltage in the overhead system decreased during the recuperation of electrical energy. Nonlinear distortion of the sinusoidal voltage curve in the overhead system reaches 8–20 %. Losses of active power in the traction transformer and reversible converter make up 83–87 %, and losses of active power in the overhead system make up 17 % of the total losses of active power in the traction transformer, reversible converter of electric stock and in the overhead network. Due to the phase shift of the non-sinusoidal voltage at the inverter output by the ES thyristors relative to the alternating voltage on the buses of the traction substation, a non-sinusoidal equalizing current is formed in the electrical circuit.

The reactive component of the equalizing current significantly exceeds the active component of the equalizing current in magnitude with an increase in the inverter thyristors turn-on advance angle to $\beta = \pi/2$. Therefore, the reactive power consumed from the traction substation exceeds the active power at the output of the electric stock inverter. Due to the reactive component of the non-sinusoidal equalizing current, the active power loss in the reversible converter and in the overhead system increases with increasing advance angle of the inverter thyristors. The switching of a high equalizing current by the thyristors of the reversible converters limits the minimum advance angle of the inverter thyristors to $\beta = 36$ electrical degrees due to the increased probability of the inverter triggering. When the advance angle of the inverter thyristors $\beta = \pi / 2$, energy recovery stops, as the reversible converter switches from the inverter operation mode to the rectifier operation mode. When $\beta \geq \pi / 2$, the offset angle of the harmonic components of the equalizing current increases relative to the similar harmonics of the alternating voltage in the overhead system, therefore, electric energy is consumed from the overhead system to create the braking force of the electric stock.

The calculation results confirm the relevance of the development and application of technical solutions that reduce active power losses, increase the RMS voltage in the AC overhead system during the recuperation of electrical energy into the external power supply system.

References

[1] *The energy strategy of the Russian Railways holding for the period up to 2020 and in 2030 perspective. The Order of Russian Railways* 2016 (Moscow), 97 p

[2] Danilov N I and Shchelokov Ya M 2006 *Basics of energy saving* (Ekaterinburg: State-owned vocational college Ural State Technical University USU), 564 p

[3] Shatokhin A P, Baklanov A A and Nezevak V L 2016 On the question of methods of evaluating the regenerative braking effectiveness *Innovative projects and technologies in education, industry and transport: Mater. of scientific conf.* (Omsk: Omsk State Transport University), pp 102–6

[4] Cheremisin V T, Pashkov D V and Ushakov S Yu 2014 Automated monitoring of the energy efficiency of the electric stock of Russian Railways *Trans-Siberian News* **3(19)** (Omsk: Omsk

State Transport University), pp 87–91
[5] Burkov A T 2015 *Electronics and converter equipment* Vol 2 (Moscow: Training and methodological center of railway transport), 307 p
[6] Teigelkotter J and Sprenger D 2000 *Powerful converter on IGBT-transistors for use on railway rolling stock* (Siemens AG, Munich)
[7] Nezevak V L 2019 Modeling the processes of power consumption for traction when changing the parameters of the train schedule on electrified sections of the 3rd and 4th types of the track profile *Modern technologies. System Analysis. Modeling* **61(1)** 156–66. DOI: 10.2673/1813-9108.2019.1(61).156-166
[8] Kryukov A V, Cherepanov A V and Shafikov A R 2018 Simulation of non-sinusoidal modes in railroad power supply systems with asynchronous traction motors for locomotives *Modern technologies. System Analysis. Modeling* **60(4)** 99–108. DOI: 10.26731/1813-9108.2018.4(60).99-108
[9] Vilgel'm A S, Nezevak V L and Shatokhin A P 2013 Comparative efficiency of options for the use of energy regeneration in the railways of direct current *Sci. and education for transport* **1** 243–7
[10] Seo S-S and Han S-W 2019 A Study on control algorithm of thyristor dual converter system for DC electricity supply of urban railway *J. Electrical Engineer. & Technol.* **14(2)** 979–83
[11] Jenella S andRadj Kumar V 2014 Power Electronics and Renewable Energy System *Proceedings of ICEPERES* 225–36
[12] Ryabchenok N L, Alekseeva T L, Yakobchuk K P and Astrakhantsev L A 2015 *Refined energy conservation law* Retrieved from: http://www.rusnauka.com/42_PRNT_2015/Tecnic/5_202603.doc.htm.
[13] Ryabchenok N L, Alekseeva T L, Astrakhantsev L A and Martusov A L 2019 Energy efficiency of electric traction of trains *Modern technology. System Analysis. Modeling* **61(1)** 144–56 DOI: 10.26731/1813-9108.2019.1(61).144-156

Analysing oil and gas production facilities impact on atmospheric air of Potochnoye field

E M Almukhametova[1], R R Sitdikov[1], A I Zakirov[1]

[1]Ufa State Petroleum Technological University 1 Kosmonavtov str. Ufa Republic of Bashkortostan 450062 Russia

E-mail: elikaza@mail.ru

Abstract. Environmental issues are always relevant. Nature is subject to constant human intervention. At the same time, associated institutions are being created to solve the tasks of reducing the level of harmful effects and the full restoration of environment. In this article, we assess the level of harmful emissions and wastes of oil and gas complex facilities during drilling, construction of facilities for field development and their operation, and present basic measures to reduce their impact on the air in Potochnoye field. We have examined the amount of emissions and waste from stationary and mobile sources. During wells drilling and facilities construction, the average annual evidence of pollutant emissions into the atmosphere caused by stationary sources will amount to 141.29 tons per year while the pollutant emissions caused by road and construction equipment and vehicles will be equal to 200.8 tons per year. Every year the amount of emissions into the atmosphere caused by the maintenance of the designed facilities will be 13730.9 tons per year, emissions from vehicles will be 47.1 tons per year. During field development, all work on the field should be carried out in compliance with the standards of the legislation of the Russian Federation. Having analyzed the amount of estimated emissions, a set of measures has been drawn up with the objective to reduce the impact of oil and gas complex on atmospheric air.

1. Introduction

Oil and gas complex expands the territory of field development every year. The number of drilled wells is growing, and the impact on nature is growing as well. In order to equip an oil field, the construction of areal and linear objects (communications) of oil and gas production for various purposes is required. The main task is to bring the field to the planned volumes of oil production. Environmental support on the field aims to regulate the interaction of an enterprise with the environment by means of standardization of impacts [1].

The modern system of field development is distinguished by such stages as hydrocarbon raw materials: exploration and search for deposits, construction (drilling) of exploration and prospecting wells, exploitation and development of explored fields.

Harmful emissions and waste from operations are generated at all stages of field development [2].

IOP Conf. Series: Materials Science and Engineering **760** (2020) 012004 doi:10.1088/1757-899X/760/1/012004

Environmental impact associated with well drilling, arrangement of well clusters and routes of communication corridors to them will be formed during Potochnoye field development. Next, we consider in detail the amount of emissions during field operations and measures to reduce them.

2. Materials and methods

The development of hydrocarbon deposits has a permissive order, specifically, it requires prior permission to operate from relevant state bodies, and is regulated by a package of existing regulatory, methodological and legislative documents [3].

Potential emissions (pollution) into the atmosphere are justified during the environmental impact assessment procedure.

Environmental impact assessment (EIA) is a mandatory part of the design documentation [4]. Such a procedure is organized before the start of the project. An EIA is carried out after assessing the current state of the environment, taking into account the environmental protection measures plan (EPMP). This document contains information on emissions and other negative environmental impacts.

All calculations of quantitative indicators in assessing the environmental impact of facilities were carried out on the basis of consolidated standards calculated for similar projects.

3. Results

This section covers the impact of construction work with the maximum volumes on atmospheric air during the further maintenance of Potochnoye field.

With the further development of Potochnoye field, an additional environmental impact associated with well drilling, arrangement of well clusters and routes of communication corridors to them will be formed.

Potential sources, impact volumes and environmental protection measures of development are considered separately for the components of the natural environment at the stages of well drilling, construction of facilities for field development and their operation [5].

During construction and installation works, the sources of pollutants emission are as follows:

- boiling room operating during drilling operations;
- diesel power plant operating during construction and installation works;
- emissions from car engines and construction equipment operation;
- manipulations with welding equipment;
- molar work;
- pouring work;
- fuel stations;
- mud mixing system [6].

In the operating mode of facilities management, the main sources of emitting harmful substances into the air during oil production are as follows:

- flanges in pipe fittings are loosely connected;
- sealants of movable joints of pumping equipment;
- breather valves of containers;
- air ducts of chemical input unit [7].

When constructing facilities for calculations, the following norms are accepted:

- the norm of pollutant emissions into the atmosphere is per 1 well cluster under construction;
- the norm of waste generation, water consumption, water disposal per one drilled well;
- the amount of rainwater calculated on the average annual runoff from the surface and the area of rainwater harvesting.

When operating the designed facilities, the calculations of the following aspects were carried out:

- atmospheric emissions from processing facilities installed at sites;
- air emissions from mobile vehicles in compliance with the standard for emissions of each pollutant per 1 wellbore;

- production waste in compliance with the norm of industrial waste generation (containers, packaging, etc.) per 1 working production well;
- oily waste in compliance with the standard for the formation of products for oilfield equipment cleaning per 1 ton of produced fluid;
- volumes of produced water in compliance with the indicators of field development.

The mass of harmful substances emitted from stationary and mobile sources during well drilling, construction of facilities and their operation are shown in table 1.

Table 1. The amount of harmful emissions into the atmosphere

Name of the process and sources of emissions	Name of harmful substance	Emission mass, tons per year
1. Well drilling		
1.1. Stationary source	Nitrogen oxides	20.7
	Soot	0.4
	Carbon oxide	14.8
	Fuel oil ash	0.5
	Hydrocarbons	67.7
	Sulphurous anhydride	35.1
	Manganese and its compounds	0.00018
	Iron oxide	0.007
1.2.Vehicles	Carbon oxide	95.2
	Nitrogen oxides	55.6
	Hydrocarbons	20.2
	Soot	7.5
	Lead	0.07
	Sulphurous anhydride	7.5
2. Facilities construction		
2.1. Stationary sources	Nitrogen oxides	0.18
	Carbon oxide	0.88
	Welding spray	1.05
	Solvents	0.02
2.2. Vehicles	Nitrogen oxides	0.4
	Carbon oxide	12.4
	Hydrocarbons	1.7
	Soot	0.07
	Lead	0.02
	Sulphurous anhydride	0.05
3. Facilities operation		
3.1. Stationary sources	Nitrogen oxides	9.4
	Carbon oxide	299.1
	Hydrocarbons	13421.9
	Soot	0.0
	Chemicals (methanol)	0.54
	Benz a pyrene	0.0000001
3.2. Vehicles	Nitrogen oxides	1.8
	Carbon oxide	38.6
	Hydrocarbons	6.1
	Soot	0.17
	Lead	0.036
	Sulphurous anhydride	0.44

4. Discussion

Drilling of wells using drilling rigs with an electric drive was adopted with the objective to calculate emissions.

During the construction of well facilities and drilling facilities, the average annual emissions of polluting substances into the atmosphere caused by stationary sources will amount to 141.29 tons per year and the ones caused by road and construction equipment and vehicles will be 200.8 tons per year.

In the operating mode of project facilities management, pollutants from stationary sources will be released into the atmosphere every year in the amount of 13730.9 tons per year, emissions from vehicles will be 47.1 tons per year.

5. Conclusion

In order to prevent air pollution at an enterprise, the following measures should be taken to reduce the harmful effects:

- complete sealing of equipment: wellheads, systems for receiving and measuring formation fluids coming from testing wells, circulating systems, valves and pipelines, eliminating constant emissions [8];
- selection of equipment taking into account explosiveness, fire hazard and toxicity of products;
- fuels and lubricants must be delivered and stored in sealed containers, it is required to keep a record of spent and consumed fuel and lubricants;
- the process of burning fuel in boiling rooms and heaters is optimized according to regime maps;
- equipping the apparatus (where pressure that exceeds the calculated one may arise) with safety valves in compliance with the requirements of the "Rules for the Construction and Safe Operation of Pressure Vessels";
- pumping oil from the apparatus and drainage during equipment repair, oil and gas discharge from the safety valves into emergency tanks;
- posing control over air pollution [9].

The use of associated and natural petroleum gas should be reflected in projects for a field development [10]. The level of utilizing associated produced gas during the field's operation should be at least 97.5%.

For oil and gas production facilities that have been commissioned at the Potochnoye field, it is necessary to establish standards for harmful emissions into the atmospheric air from each type of equipment located at industrial sites and issue a permit for the emission of pollutants into the atmospheric air in the prescribed manner.

References

[1] Almukhametova E M, Fattakhov D I, Zakirov A I, Faizullin A A 2018 Analysis of enhanced recovery by treatment of bottom-hole zone of Potochnoye field *IOP Conf. Series: Earth and Environmental Science* **194** 082003

[2] Almukhametova E M, Fattakhov D I and Zakirov A I 2018 The efficiency analysis of applying hydraulic fracturing of formation at AB1-2 facility of Potochnoye field. *IOP Conf. Series: Earth and Environmental Science* **194** 082002.

[3] Sadegh Tavallali M and Iftekhar Karimi A 2014 Perspectives on the Design and Planning of Oil Field Infrastructure *Computer Aided Chemical Engineering* ISSN 1570-7946.

[4] Cianciarullo M I 2019 Green construction– reduction in environmental impact through alternative pipeline water crossing installtion, *Journal of Cleaner Production* ISSN 0959-6526.

[5] Khalegh Barati and Xuesong Shen 2017 Optimal Driving Pattern of On-Road Construction Equipment for Emissions Reduction *Procedia Engineering* ISSN 1877-7058.

[6] Faber P, Drewnick F and Borrmann S 2017 Aerosol particle and trace gas emissions from earthworks, road construction, and asphalt paving in Germany: Emission factors and influence on local air quality *Atmospheric Environment* ISSN 1352-2310.

[7] Riddick, Denise S N, Mauzerall L, Kang M, Celia M A, Bressler K, Chu C and Gum C D 2019 Measuring methane emissions from abandoned and active oil and gas wells in West Virginia *Science of The Total Environment*, ISSN 0048-9697.

[8] Zhang L, Cho S, Hashisho Z and Brown C 2019 Quantification of fugitive emissions from an oil sands tailings pond by eddy covariance *Fuel* ISSN 0016-2361.

[9] Anantha Raman L, Deepanraj B, Rajakumar S and Sivasubramanian V 2019 Experimental investigation on performance, combustion and emission analysis of a direct injection diesel engine fuelled with rapeseed oil biodiesel *Fuel* ISSN 0016-2361.

[10] Mukhametshin V V 2017 The need for creation of a unified comprehensive method of geological and field analysis and integration of data on effective influence on the bottom-hole formation zone *Oil Industry* **4** 80-84

Simulation of fractional efficiency of direct-flow cyclone with intermediate dust collector

V S Aslamova[1], **A A Aslamov**[2], **M I Arschinskii**[3], **Y V Zhilycov**[3] **and E A Rudykh**[3]

[1]Irkutsk State Transport University, 15, Chernyshevsky ave., Irkutsk, 664074, Russia
[2]Angarsk State Technical University, 60, Tchaikovsky ave., Angarsk, 665835, Russia
[3]Irkutsk National Research Technical University, 83, Lermontova ave., Irkutsk, 665462, Russia

E-mail: arshmax@mail.ru

Abstract. By means of numerical modeling in the program ANSYS Fluent CFX the evaluation of fractional efficiency in a direct-flow cyclone with intermediate dust collector has been carried out. The boundary conditions were as close as possible to the experimental ones. As a gas phase, the authors take the air under normal conditions, as the solid phase – spherical particles of 30 and 80 microns in size. In order to evaluate the adequacy of the calculation model, a laboratory installation of a DCIC with two separate chambers for the selection of collected particles was made. This design allowed estimating the fractional efficiency of both individual chambers of a direct-flow cyclone DCIC and the apparatus as a whole. The results of the experiments carried out on different types of dust (fly ash, cement) showed a good convergence with the design model

1. Introduction

To date, along with increasing production growth rates, the world is facing pressing environmental issues, in particular the protection of atmospheric air from pollution. The efficiency of new gas treatment plants under development largely depends on the right equipment and the accuracy of engineering calculations in the design of gas treatment equipment. One of the effective elements of a gas processing plant is centrifugal separation in a direct-flow cyclone with an intermediate dust collector (DCIC) [1].

The development of direct-flow cyclones requires a special approach in the calculation of the dust-gas-air mixture motion in separation chambers, which largely determines their efficiency and geometric dimensions [2, 3]. Adequate forecasting of efficiency is very important for the large-scale transition at the stage of design of dust-cleaning equipment, but in the literature on dust collection by now there is no single approach to calculate the efficiency of cleaning in direct-flow cyclones of various designs.

One of the main estimates in calculating the efficiency of gas purification in cyclones is the fractional degree of purification – the ratio of the number of particles collected in the device to the number of incoming dust particles of the same fraction.

The known methods of cyclone efficiency evaluation based on empirical probabilistic functions of fractional efficiency and dust dispersion composition have low accuracy.

For example, the method of calculating the fractional efficiency proposed by the researchers in the works [4–6] assumes a multistage approach, which complicates its use. Each efficiency forecasting method involves certain assumptions and the availability of input data.

Probabilistic and energy method of evaluating the cyclones efficiency [7, 8], assumes subordination of the dust input composition and fractional efficiency of dust collection to the log-normal law of distribution and efficiency evaluation of a gas cleaning apparatus on the probability integral:

$$\eta = \frac{1}{2\pi}\int_{-\infty}^{x} \exp^{\left(-\frac{x^2}{2}\right)} dx = f(x), \tag{1}$$

where $f(x)$ is a function of normal Gaussian distribution and the parameter x is:

$$x = \frac{\lg(\delta_m/d_{50})}{\sqrt{\lg^2\sigma_ч + \lg^2\sigma_\eta}}, \tag{2}$$

where δ_m is the mass-mediated particle diameter, $\lg\sigma_ч$ is the standard deviation of particle diameters logarithms, $\lg\sigma_\eta$ is the value determined by the type of dust collector. Parameter d_{50} is the diameter of particles separated in cyclone by 50 % is based on the result of experimental studies.

According to the NIIOGAZ method [9] the calculation of the cyclones separation efficiency takes into account the influence of the flow velocity w, the dynamic viscosity μ of the stream, the dust density ρ_δ, and the diameter of the cyclone D, in the terms of particle diameter d_{50}, separated by 50 %, working on the technological parameters of the cyclone according to the known operating parameters (T) by the formula:

$$d_{50} = d_{50}^{T}\sqrt{\frac{D}{D_T}\frac{\rho_{\delta T}}{\rho_\delta}\frac{\mu}{\mu_T}\frac{w_T}{w}}. \tag{3}$$

According to this method, the calculation of the d_{50} parameter for any dust collector requires a special experiment on dust with different fractional compositions, as well as the use of such dependencies, which often do not adequately describe the experimental data.

The work [10] proposes to recalculate the cleaning efficiency of η by dust slippage ε from the base cyclone (index 0) to the geometrically similar calculated (index p) at the same average flow rate w in terms of the cyclone, according to the equation:

$$\varepsilon = \frac{(1-\eta_p)}{(1-\eta_0)} = \prod_i \frac{K_{ip}}{K_{i0}}, \tag{4}$$

where K_i is the relative entrainment coefficients.

The method is based on the assumption of independence of the cyclone diameter influence D, mass-mediated particle diameter δ_m and dust flow z on the relative dust slip ε. It is universal, i.e. applicable to any type of cyclones: counter flow, direct flow and counter-rotating flows; does not depend on the dispersion distribution of particles in terms of size and fractional cyclone cleaning parameters.

The work [11] proposes a universal method of evaluating the efficiency of geometrically similar cyclones based on the formula (4). The method takes into account the influence of the average flow rate w, cyclone diameter D, dust concentration z, as well as the joint influence of the mass-mediated particle diameter δ_m and particle density ρ_δ on the relative slip ε – carry-over coefficient K_ρ. The disadvantages of this method are the narrow range of applicability of the cyclone diameter: from 100 to 600 mm. In addition, the average rate of dust flow is not taken into account.

Thus, the literature does not yet provide a generally accepted method of predictive calculation of the separation efficiency of projected cyclones. Since more than 80 % of all collected industrial dust is deposited in cyclones of different types, the development of universal method of separation efficiency predictive calculation is very important.

In order to carry out further studies of fluid dynamics processes in dispersed systems, modern tools of mathematical modeling by the finite element method (FEM) are indispensable.

2. Materials and methods

Experiments to determine the fractional efficiency were carried out on a transparent test device of the DCIC direct-flow cyclone with two separate dust collectors: Intermediate Dust Collector (IDC) and Main Dust Collector (MDC) in Figure 1.

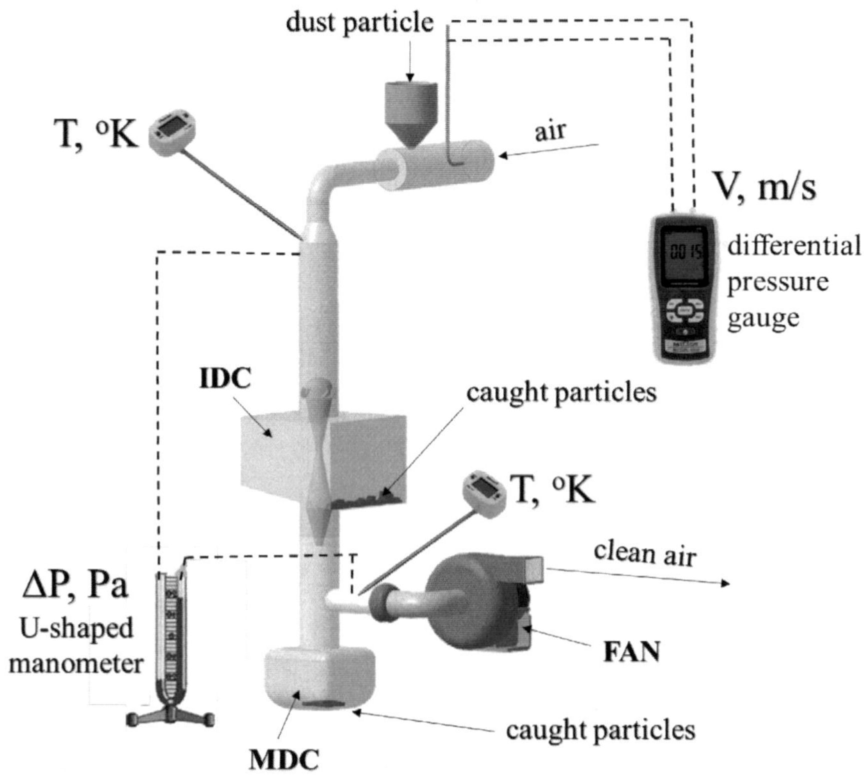

Figure 1. Diagram of test device direct-flow cyclone with intermediate dust collection

Two types of dust were used in the experiment: fly ash CHP (ρ_δ = 1950 kg/m^3) and cement 32.5B (ρ_δ = 3170 kg/m^3) (Figures 2). To determine the exact grain-size composition of the initial and collected dust, we used the method of laser diffraction on the Microtrac S3500 laser diffraction analyzer. The laser analysis system complies with international standards of particle size measurement and is certified in accordance with ISO 13320-1 (International Organization for Standardization). The device is included in the State Register of Measuring Instruments (US.E.27.001.A № 23120), the measuring range – from 0.021 to 2816 microns.

The initial dispersion composition of two dust samples was obtained by laser diffraction. Then the data on the log-normal distribution law were approximated:

$$f(\delta) = \frac{exp\left(-\frac{\left[\frac{ln\delta - M}{\sigma}\right]^2}{2}\right)}{\delta\sigma\sqrt{2\pi}}. \quad (5)$$

The curved lines of the log-normal distribution of particles by size are presented in Figure 2.

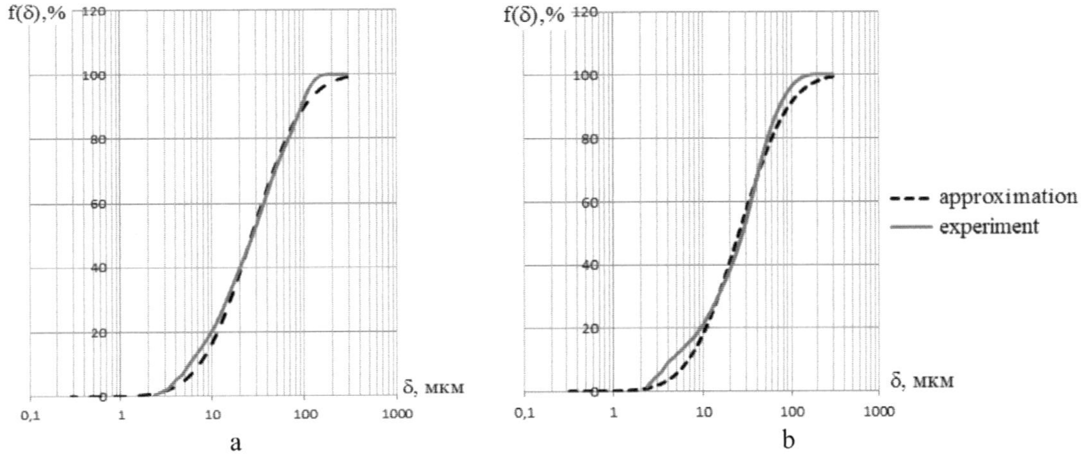

Figure 2. The integral curved lines of particles distribution: a – fly ash; b – cement 32.5B.

The calculated mass-median diameter δ_m and standard deviation σ are presented in table 1.

Table 1. Results of the initial dust dispersal analysis.

	Density ρ_δ, kg/m^3	Mass-median diameter δ_m, μm	RMS logarithm deviation σ
Fly ash	1950	27.27	1.026
Cement 32.5B	3170	25.29	1.019

3. Fractional efficiency studies on the DCIC direct flow cyclone test device

Experiments were carried out on the parameters indicated in Table 2. The efficiency of separation of the intermediate dust collector (η_{idc}), the main dust collector (η_{mdc}) and the entire DCIC direct-flow cyclone was determined:

$$\eta_{idc,mdc} = \frac{m_{caught}}{m_0} \tag{6}$$

$$\eta = \eta_{idc} + \eta_{mdc} \tag{7}$$

where m_0 is the initial mass of dust, m_{caught} is the mass of collected dust.

Table 2. Parameters and results of experiments.

	T, °K	P_{atm}, mm Hg	ΔP, Pa	w_{cp}, m/s	Q, m^3/h	z, g/m^3	m_0, kg/h	η_{idc}, %	η_{mdc}, %	η, %
Fly ash	290	730	2500	7.99	293.7	21.1	6.2	48.06	31.58	79.64
Cement 32.5B	288	730	2450	7.99	293.7	26.1	7.7	47.23	40.16	87.38

Particles collected in the intermediate and main dust collector were examined by laser diffraction. Then, using formula (5), we determined the integral functions of particle distribution and found δ_m and σ (Table 3).

Table 3. Results of the disperse analysis of the collected dust.

	Dust collector	Mass-median diameter δ_m, μm	RMS logarithm deviation σ
Fly ash	IDC	40.75	0.832
	MDC	22.03	0.915
Cement 32.5B	IDC	37.15	0.730
	MDC	23.77	0.947

IOP Conf. Series: Materials Science and Engineering **760** (2020) 012005 doi:10.1088/1757-899X/760/1/012005

On the basis of the obtained integral particle distribution functions, the fractional efficiency in the intermediate and main dust collector and the total fractional efficiency of the DCIC direct-flow cyclone were determined by the formula:

$$\eta(\Delta\delta) = \frac{F_{caught}(\Delta\delta)}{F_0(\Delta\delta)}\eta \quad (8)$$

where $\Delta\delta$ is the difference between the largest and smallest particle sizes nearby, η is the efficiency of dust collectors.

Figure 3 shows the fractional efficiency of the intermediate and main dust collector. Figure 4 shows the total fractional efficiency of the DCIC direct-flow cyclone.

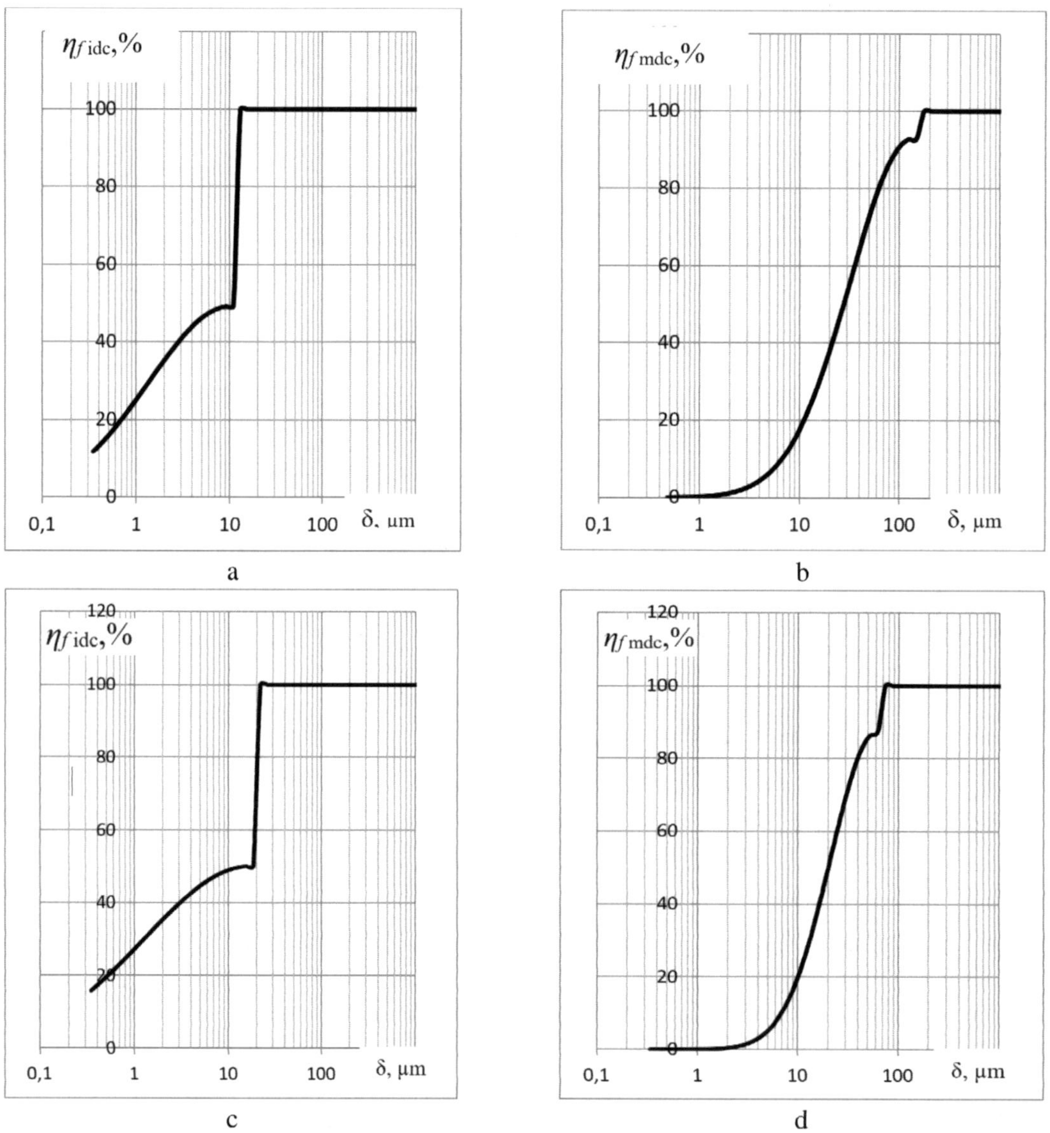

Figure 3. a – fractional efficiency of the intermediate dust collector on ash; b – fractional efficiency of the main dust collector on ash; c – fractional efficiency of the intermediate dust collector on cement; d – fractional efficiency of the main dust collector on cement.

Table 4 shows the experimentally obtained parameters of fractional efficiency d_{50} for the DCIC direct-flow cyclone test device with two separate dust collectors.

Table 4. Parameters of fractional efficiency d_{50} of DCIC direct-flow cyclone.

	Dust collector	d_{50}, μm	RMS logarithmic deviation σ
Fly ash	IDC	11.025	11.025
	MDC	19.72	19.72
	DCIC	6.481	6.481
Cement 32.5B	IDC	15.12	15.12
	MDC	20.18	20.18
	DCIC	6.93	6.93

Table 4 shows that the parameter of fractional efficiency d_{50} varies depending on the density and dispersion composition of the dust.

4. Numerical simulation of the fractional efficiency of DCIC direct flow cyclone by the finite element method (FEM)

The document shows a three-dimensional stationary mathematical model, which allows making the analysis of the motion of particles in a direct-flow cyclone. To solve the problem, we use the CFD calculation software ANSYS CFX, it allows one to model disperse systems. The appearance of the investigated computational model is shown in Figure 5.

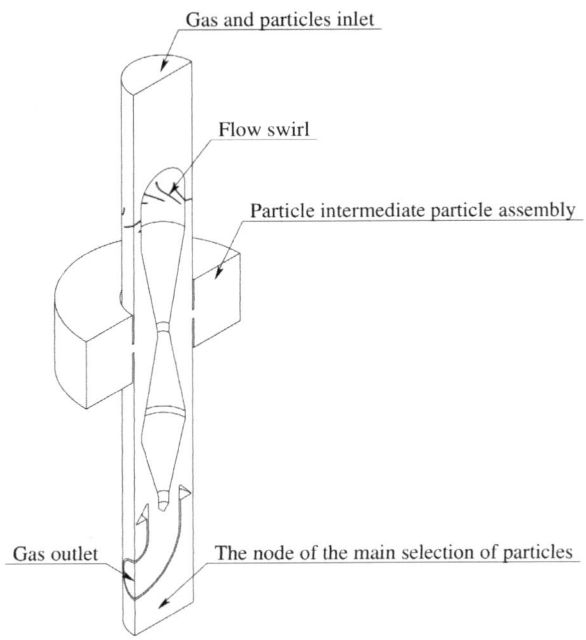

Figure 5. Layout of the studied computational model.

4.1. Setting up the CFD software

In the CFD program ANSYS CFX, the dispersion system modeling is implemented using the models:

- media interaction model;
- model of the exchange of the quantity of motion between the particle and the gas medium;
- model of the motion of the gaseous medium.

4.1.1. Media interaction model. In the computational model, the dispersed system consists of two phases: gas and solid particles. A simulation of the dispersed system in ANSYS CFX is performed using Euler-Euler media interaction model.

Euler-Euler model — in this model, the phases are regarded as interpenetrating continuums; the interaction of media through pressure and interphase exchange coefficients is calculated, taking into account the calculation of Euler variables [12].

4.1.2. Model of the exchange of the quantity of motion between the particle and the gas medium. The choice of other computational models in the cyclone simulation is simpler, but no less important. The Schiller-Naumann model should be used to describe the exchange of the quantity of motion between the particle and the gas medium [13].

According to the model, the drag coefficient of particles is determined in accordance with the following formula:

$$C_D = \frac{24}{Re} \cdot (1 + 0.15 \cdot Re^{0.687}) \tag{9}$$

where *Re* is the particle Reynolds number, which is determined based on the following relation:

$$Re = \frac{\rho_\alpha \cdot |U_\beta - U_\alpha| \cdot d_\beta}{\mu_\alpha} \tag{10}$$

where subscript α denotes that the variable belongs to the main phase (mixture of gases), β belongs to the solid phase.

4.1.3. Model of the motion of the gas medium. The motion of solid particles can be traced taking into account the turbulence of the gas phase.

As a model of turbulence, the RNG (renormalization groups) version of the *k-e* model is used. Of all the CFX turbulence models, it is the most suitable for simulating flow with highly curved particle trajectories [14].

4.1.4. Boundary and initial conditions. The computational geometry with the arrangement of boundary conditions shown in Figure 5 consists of:

- the node of the main selection of particles;
- particle intermediate particle assembly;
- flow swirl;
- entry areas gas and particles inlet;
- purified gas outlet area.

The input parameters for the dispersion medium are shown in Table 5.

Table 5. The input parameters of the dispersion medium.

Parameter description	Value and dimension
Mass flow rate ash solids of 30 micron	4.2 [kg/h]
Mass flow rate ash solids of 80 micron	1.8 [kg/h]
Inlet gas velocity	10 [m/s]

4.1.5. Computational grid. The computational domain is a part of the cyclone design under consideration. On the basis of the preliminary analysis of convergence, a grid mathematical model was constructed in Figure 6, consisting of 7,855,345 tetrahedral elements. The size of the elements in the near-wall layers was set taking into account the criterion y^+, which must satisfy the condition $30<y^+ <300$ as the gas mixture moves in a highly developed turbulent flow.

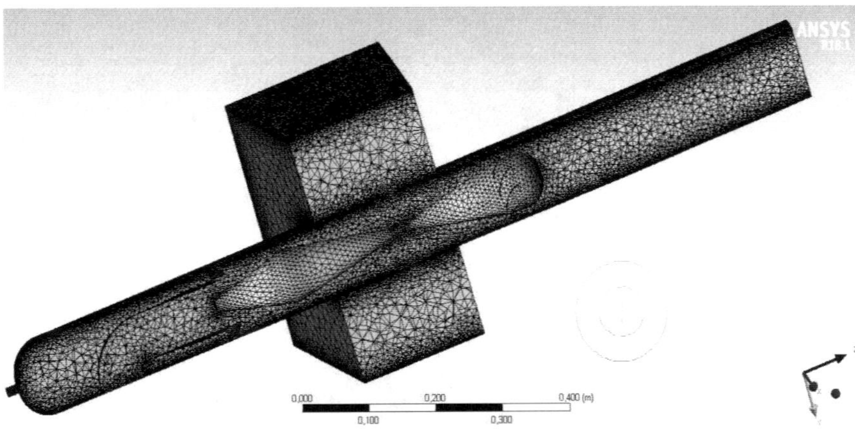

Figure 6. Fragment of the computational grid of the cyclone model.

4.1.6. Conducting a numerical solution. Before starting the calculation, some settings were specified to control the numerical solution scheme:

- First order advection scheme (High Resolution). The advantage of this scheme is its stability and good rate of convergence.
- The time scale was set by a piecewise linear function, which gradually decreases while the iteration number increases (Table 6).

Table 6. Timescale changes.

Iteration number	0-100	100-200	200-300	300-400	400-500
Timescale, sec	0.15	0.10	0.08	0.06	0.04

This technique is caused by a significant difference in the scale of the simulated processes in the various structural elements. The large scale at the beginning of the solution helps to accelerate the convergence of the equations [15].

The system of equations was solved iteratively. The values of the calculated variables in each calculation cell were changed from a certain initial distribution (based on the given initial conditions) to the desired distribution.

In the course of the solution process, two main indicators of the convergence level were monitored:

- the RMS residuals of the calculated variables;
- the global imbalances of the calculated variables.

The solution process was stopped when all the indicators met the convergence criteria (residuals < $10^{-3}...10^{-5}$, imbalances <1 %, the target parameters do not change during the last 100 iterations by more than 1 %).

4.2. Results

As a result of the calculations, the pressure of the gaseous medium and the changes in the volume fractions of solid particles in the computational model were determined. The color map of the pressure distribution of the gas mixture in the cross section of the calculated geometry for a fragment of the calculation model is shown in Figure 7.

As can be seen from the color map of the pressure distribution of the gas mixture for the Euler-Eilor model, ash particles in the region of the location of the intermediate selection unit have a disturbing effect on the gas medium in the extreme regions of the computational model.

Figure 8 shows color maps of changes in the volume fraction of solid particles in the cross section of the calculated geometry.

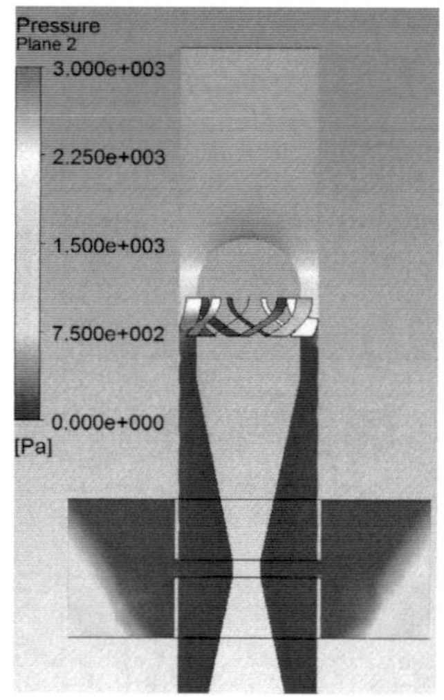

Figure 7. Color maps of the gas mixture pressure distribution

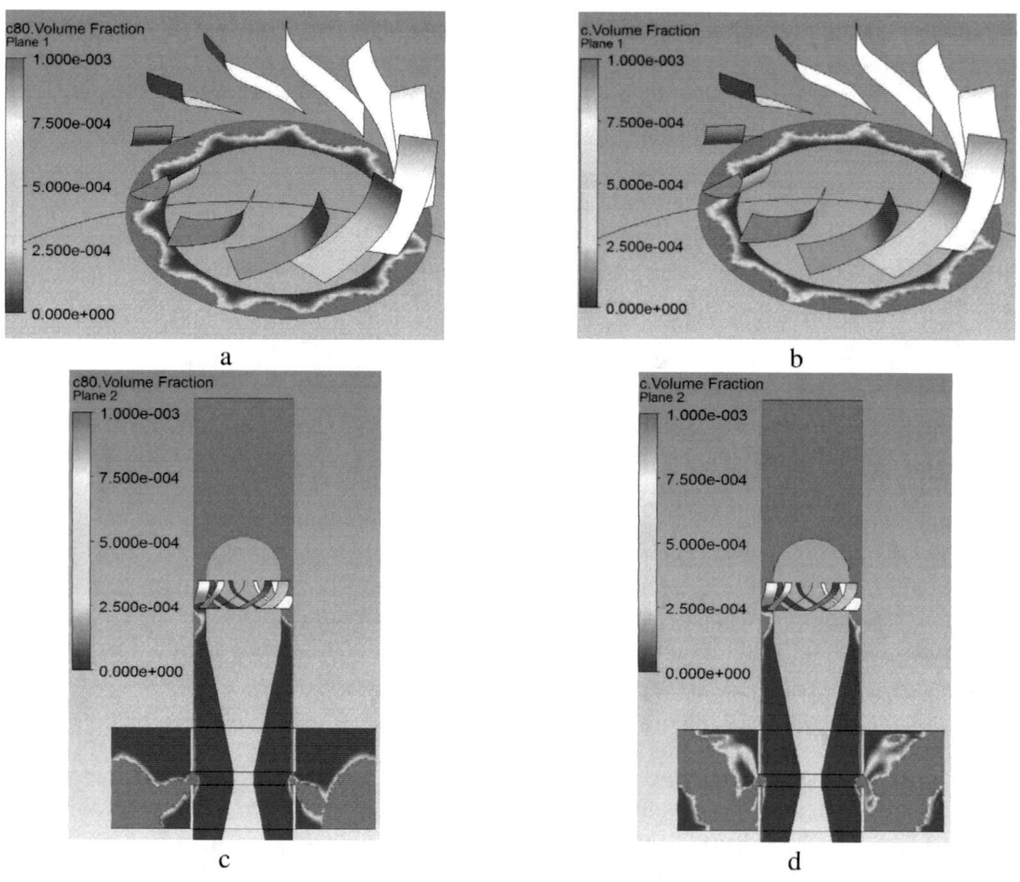

Figure 8. Color maps of the changes in the volume fraction of solid particles: a – distributions of the volume fraction of solid particles 30 microns in size after the swirler; b – distributions of the volume fraction of solid particles 80 microns in size after the swirl; c – volume fraction of solid particles with a size of 30 microns in an intermediate particle extraction unit; d – volume fraction of solid particles with a size of 80 microns in an intermediate particle extraction unit.

The color maps shown in Figure 6 show that the particles after the swirler to the intermediate node for the selection of particles and after it move near the surface of the walls. When it enters the node of the intermediate selection, particles of 80 microns in size move in the direction of the base of this area, and particles of 30 microns in size are evenly distributed throughout the volume, without causing significant disturbances to the gas medium.

5. Conclusion

Numerical modeling of the fractional efficiency of the direct-flow cyclone test device DCIC allowed us to establish that 70 % of the particles 80 microns in size after the swirler fall into the node of the intermediate dust collector, which in turn correlates well with the experimental data.

The developed mathematical model of the flow of dispersed media allows estimating local disturbances associated with the distribution of solid particles in the geometry of a direct-flow cyclone DCIC, and significantly reduces the number and time of experiments at the design and calculation stages using a known method for evaluating the efficiency of dust collectors.

The using of the ANSYS CFX software package will make it possible to further carry out various simulations with different streams of the disperse medium. Thus, specifying in advance the fractional degree of purification and operating parameters of the cleaned dust stream on the test device of a direct-flow cyclone DCIC, and having reliable settings for the model solver in FEM systems, you can accurately calculate the number and diameter of designed direct-flow cyclones with an intermediate dust collector for gas cleaning systems.

References

[1] Aslamova V S, Troshkin O A, and Sherstyuk A N 1991 *Chem. and Petroleum Engineer.* **27(1)** 37–38
[2] Aslamova V S, Aslamov A A and Museva T N 2007 *Bull. of the Tomsk Polytechnic University* **310(1)** 152–156
[3] Aslamova V S, Arshinskii M I, Bragin N A and Zhabei A A 2009 *Chem. and Petroleum Engineer.* **45(5–6)** 301–307
[4] Shilyaev M I 2014 *The methods of calculation for the dust-collecting systems* (FORUM: INFRA-M)
[5] Shilyaev M I, Shilyaev A M and Grischenko E P 2006 *Calculation methods for dust collectors* (Tomsk: TSUAB) 385
[6] Shiljaev A M, Rekunov V S and Kondratyuk A A 2005 *8th Korea-Russia Int.Symposium on Sci. and Technol.* – Proceedings (KORUS 2004) **3** 72–75
[7] Val'dberg A Yu Safonov S G 2005 *Chem. and Petroleum Engineer.* **41(9–10)** 557–559
[8] Val'dberg A Yu Safonov S G 2006 *Chem. and Petroleum Engineer.* **42 (9–10)** 534–537
[9] NIIOGAZ 1970 *Cyclones of the State Research Institute of Industrial and Sanitary Gas Cleaning Guideline Specifications for Design, Fabrication, Assembly, and Operation* 95
[10] Sherstyuk A N and Aslamova V S 1990 *Thermal Engineer.***37(5)** 253–255
[11] Aslamova V S, Aslamov A A, Zhabey A A and Arschinskiy M I 2008 *Rus. J. of Chem. and Chem. Technol.* **51(12)** 80–83
[12] User Guide ANSYS CFX-Solver Modeling Guide 2013 *ANSYS CFX Release 14.5* (Canonsburg: ANSYS, Inc.) 155
[13] Selim M S, Kothari A C and Turian R M *Sedimentation of multisized particles in concentrated sus-pensions. AlChE J.* **29(6)** 55–58
[14] Lyczkowski R W Bouillard J X 2002 *Progress in Energy and Combustion Sci.***28** 543–602
[15] Kondrashov P V 2013 *East Eur. J. of Advanced Technol.* **65 (5–7)** 143–148

IOP Conf. Series: Materials Science and Engineering **760** (2020) 012006 doi:10.1088/1757-899X/760/1/012006

Verification of knowledge models by means of logic with vector semantics

L V Arshinskiy[1], M S Nitezhuk[1] and M P Dunaev[2]

[1]Irkutsk State Transport University, Irkutsk, Russia
[2]Irkutsk National Research Technical University, Irkutsk, Russia

Abstract. The paper highlights the method of verification and identification of contradictions in the production knowledge bases using one class of logic with vector semantics, namely V^{TF}-logic. The general verification algorithm is described. It is noted that the use of such logics, while maintaining the need to iterate over the input facts, offers a procedure of direct inference and analysis of the inconsistency only in the case of conclusions-hypotheses, characterized by the value of truth named "complete contradiction". Tracing back the appropriate chain will identify and resolve the cause of the conflict.

1. Introduction

A knowledge model is one of the modern approaches to the description of subject areas (SA). The difference between knowledge-based models and traditional mathematical models is that the former describe not the subject area itself, but the idea that a person (expert) has about it. That is, a double reflection takes place: the subject area is reflected in the inner world of the expert, and this reflection, in turn, is reflected with certain knowledge models. This makes it possible to use knowledge modeling to describe almost any subject areas, one way or another perceived by man. This significantly distinguishes such models from traditional mathematical models, which work only where mathematical modeling is simple enough to use.

The main consumers of knowledge models today are computer programs called expert systems (ES). Precisely universality and "omnivory" of the knowledge approach to modeling suggest that the software based on it is "seriously and for a long haul", at least until other approaches embrace "the conceivable world" effectively enough to work with appropriate models.

The beginning of knowledge modeling as a separate direction, perhaps, should be counted off from the late '50s, in particular, from the attempt of Newell, Shaw and Simon to create a "General Problem Solver" (GPS) [1]. A distinctive feature of GPS as a software product was its division into two architectural components, which in modern interpretation are known as the knowledge base (KB) and the inference machine. KB contained a model of the subject area, the inference machine used the KB knowledge to obtain information that was not provided by the input (but could be inferred from KB). To some extent, these studies gave rise to a new at that time class of software – expert systems (ES). In Russia (USSR at that time), similar research was carried out in the field of situational control [2].
Today there are several basic ways of representing knowledge – knowledge models. These are semantic networks, frames, models based on logical calculus, primarily propositional calculus and first order predicate calculus, and production model [3]. The rest come down to them one way or another.
The most common model is production due to its external simplicity, modularity, ease of development and support. It is also important that from the algorithmization point of view the system of products

"If..., then..." has an expressive force equivalent to the Turing machine, that is, it has the capabilities of a universal algorithmic model [4].

One of the most important stages of KB creation is verification. This is understandable – development errors reduce the value of the KB as a SA model. The constructed model can be incomplete, contradictory, and unreliable in some cases. The conclusions obtained with its help may contain errors. The problems of development of knowledge models are known as "NON-factors" affecting KB quality [5, 6]. At the same time, many authors consider the possible contradictory of knowledge as the main problem, since the presence of contradictions in the KB reduces the credibility of it as a model of SA.

How to detect contradictions? What kinds of contradictions can occur in the SA knowledge model? Today, these questions are most actively investigated for a productive approach to modeling.

In [7] the following types of contradictions are considered:

1) the internal contradiction of the rules, when elements of production mutually exclude each other, for example $a \to \neg a$ (here "$\to$" and "$\neg$" – symbols of implication and negation);

2) the contradictions between the two rules, which in reasoning leads to mutually exclusive conclusions, like a and $\neg a$;

3) internal inconsistency of chains of reasoning, as a result of which the facts contradicting the known ones can be added to the system;

4) contradictory of chains of reasoning that leads to the fact that the same set of data generates mutually exclusive situations.

The situation is further complicated by the fact that the inconsistency may not appear immediately. It can be hidden in words-synonyms [8, 9], in the features of the subject area [10, 11], etc. [12-14].

In our opinion, majority of traditional knowledge processing methods based on classical logical calculus and truth represented by two-element set {T, F} (or {1, 0})} are poorly adapted for verification of knowledge base contradictory. This makes development of verification methods based on other logical approaches actual.

2. Logics with vector semantics. V^{TF}-logics

In [15] one of the non-classical logical calculus based on the concept of the truth vector was used to work with unreliable and contradictory information. Generally, it refers to an ordered tuple $\langle a_1;...; a_n\rangle$ from the interval [0, 1], each component of which describes the truth of a proposition relative to a certain aspect. This class of logics has received the name of logics with vector semantics [15, 16].

A separate component of a comprehensive assessment of the truth of the object x can be viewed as an aspect. For example, the truth of the statement "car x comfortable" can be evaluated in terms of engine power, ergonomics, suspension, etc. This point of view brings the vector of truth closer to a fuzzy vector of belonging x to a certain set (in this example, a set of comfortable cars). At the same time, categories of truth such as *Truth* or *Falsehood* also may be aspects. In this case, for example, two-dimensional vector logic with aspects ⟨True; False⟩ characterizes the proposition from the point of view of how close it is to Truth and Falsehood separately. The importance of each aspect is determined by its own set of evidences: pros and cons. It is this view and this representation of truth is used in so-called V^{TF}-logics (vector logics with aspects of Truth and Falsehood), which are special cases of logics with vector semantics and a convenient formal apparatus for identifying contradictions in knowledge models. An example of a software system using this approach is presented in [17–19].

3. Detecting contradictions

Different authors may have different interpretations of the contradictory/noncontradictory of the knowledge system. For instance, in [19], a system is considered to be noncontradict, if only admissible states are derived from admissible states, and in order to check the KB for consistency, it is necessary to know all admissible SA states. In [8, 9] it is said that a contradiction arises when KB has semantically incompatible elements. For example, incompatible pairs of properties or concepts. And so on. Despite of the apparent diversity of views, we proceed from the fact that a contradiction is the existence of two mutually exclusive judgments a and $\neg a$ that appear in the system due to any reason.

IOP Conf. Series: Materials Science and Engineering **760** (2020) 012006 doi:10.1088/1757-899X/760/1/012006

Then a check for inconsistency is a test of whether a valid SA state, being entered into the system in the form of a corresponding statement, can produce a situation of the form $a \,\&\, \neg a$ (& is a conjunction symbol). In the light of this, the consistency check by classical means may look as follows:

1) Input of an assertion corresponding to one of the possible SA states (conjunction of facts) into the system.
2) Verification of the statement on admissibility (non-zero truth value).
3) Inference step.
4) Check the updated statement (taking into account the result of the conclusion) for admissibility.
5) If the updated statement is valid and is not final (the next output step is possible), then return to step 3), else
6) If the updated statement is not valid, then the conclusion of the contradiction of the KB and STOP! Further analysis of the contradiction source should be fulfilled either by an expert, or by appropriate automated procedures.
7) If the statement is a goal and other states of SA are possible, then return to step 1, else make the conclusion on the consistency of the KB and STOP.

The process is repeated cyclically until all valid input states are searched through.

It is taken into account that the rules of inference generate only valid (true) judgments from valid (true) judgments. Inadmissibility, in the absence of false productions, is associated with the appearance of the conjunction $a \,\&\, \neg a$ (in one or another form).

We discuss that here the V^{TF}-logic makes its adjustments. The most significant one is that to detect inconsistencies it is not necessary to look for pairs of species a and $\neg a$. The contradictory of judgment a will represent itself by the vector of truth $\langle 1;\, 1\rangle$. To do this, the logical inference procedure should include a stage of combining evidence, characteristic of a plausible, or rather – attached inference [20]. If the inference steps have results:

$$a_1,\, a_1 \to b \vdash b\colon \|b\| = \langle b^+_1;\, b^-_1\rangle;$$
$$a_2,\, a_2 \to b \vdash b\colon \|b\| = \langle b^+_2;\, b^-_2\rangle;$$

and the resulting truth values of the conclusion b is $\|b\| = \langle b^+_1;\, b^-_1\rangle$ and $\|b\| = \langle b^+_2;\, b^-_2\rangle$, the evidence is combined according to the rule:

$$\|b\| = \langle b^+_1 \oplus b^+_2;\, b^-_1 \oplus b^-_2\rangle. \quad (1)$$

Here $\oplus$ – triangular co-norm [21] in infix records. It has the property $x \oplus 1 = 1$. Examples are the well-known expression:

$$x \oplus y = x + y - xy \quad \text{and} \quad x \oplus y = max(x,\, y).$$

As a result, if $\langle b^+_1;\, b^-_1\rangle = \langle 1;\, 0\rangle$ (strict true) and $\langle b^+_2;\, b^-_2\rangle = \langle 0;\, 1\rangle$ (strict false), association rule (1) gives the truth $\langle b^+;\, b^-\rangle = \langle 1;\, 1\rangle$ – a complete contradiction. Having appeared at any step, the contradiction extends to all subsequent consequents. This suggests the possibility of using this inference to automate conflict detection procedures. The algorithm looks the following way:

1) Set the truth value of the input facts in the form of a valid combination of strict truth $\langle 1;\, 0\rangle$ and (if necessary) strict false $\langle 0;\, 1\rangle$.
2) Perform a direct output under the given truth values.
3) Check the truth of terminal facts (hypotheses).
4) If contradictory conclusions are found, find out the cause of the contradiction by reverse tracing the output chain, eliminate the contradiction and return to step 1.
5) Otherwise, if not all possible truth values of the input facts are checked, return to step 1.
6) Otherwise, no contradictions were found. STOP!

Therefore, control of contradictions in a KB using V^{TF}-logic involves setting all valid combinations of truth vectors of input facts, identifying conflicting hypotheses, revealing and resolving contradiction (contradictions) if they are detected. And the task is amenable to automation.

4. Conclusion

Verification of knowledge models is primarily a search for internal contradictions of the knowledge system. To detect a contradiction is to obtain a conclusion of the form $a \& \neg a$ from the admissible set of facts on the model used. This requires a search for all possible facts and all possible conclusions derived from them with constant observation of the facts at each step of the inference, which makes the calculation procedure heavier. The use of the V^{TF}-logics, while preserving the need for too much input facts, provides the procedure for direct display and analysis of the inconsistency only in the case of receiving the findings-hypotheses, characterized by the truth value ⟨1; 1⟩ (complete contradiction). A reverse tracing of the corresponding chain will identify and resolve the cause of the conflict. This procedure looks computationally simpler.

5. References

[1] Newell A, Shaw J C and Simon H A 1959 *Report on a General Problem-Solving Program* (RAND Corporation) p 27
[2] Pospelov D A 1986 *Situational control: theory and practice* (Moscow: Science) p 288
[3] Giarratano J C, Riley G D 2004 *Expert Systems: Principle and Programming* (Course Technology) p 856
[4] Post E L 1943 Formal Reductions of the General Combination Decision Problem *American J. of Mathematics.* **65** 197–215
[5] Narinyani A S 1982 Underdetermined models and operations with underdetermined values *Preprint no 400 of Computing Center of the Siberian Branch of the Academy of Sciencesof the USSR*
[6] Narinyani A S 1986 Underdetermined in the systems of knowledge representation and processing *Bull. of the Academy of Sciences of the USSR. Engineer. Cybernetics* **5** 3–28
[7] Pospelova L Y and Chukanova O V 2009 The search for contradictions in the production knowledge bases *Information and telecommunication systems* **5** (Moscow: National Research Nuclear University) 23–7
[8] Prockuriakov D P 2014 Search for contradictions with the help of production management strategy based on the ontology of the subject area *Proceedings of the XIX Baikal All-Russian conference "Information and Mathematical Technologies in Sci. and Control"* part III (Irkutsk: ISEM SB RAS) p 166–70
[9] Proskuriakov D P 2015 Conflict resolution management in rule-based expert systems *Proceedings of Irkutsk State Technical University* **8(103)** 47–51
[10] Arshinsky V L and Proskuriakov D P 2016 Application of ontologies and case-based reasoning to processing context in event modeling in energy security research *Modern technologies. System analysis. Modeling* **4(52)** 94–100
[11] Dolinyna O N 2010 Classification of errors in the knowledge bases of expert systems *Bull. of SSTU* **4(50)2** 125–30
[12] Antoniou G, van Harmelen F, Plant R and Vanthinen J 1998 Verification and Validation of Knowledge-Based Systems *AI Magazine* **19(3)** 123–26
[13] Zhang D 1993 *Knowledge Base Verification: Issues and Approaches* AAAI Technical Report WS-93-05 pp 148–49
[14] Sun Y, Wu T, Li X and Guizani M 2017 A Rule Verification System for Smart Buildings *IEEE Transactions on Emerging Topics in Computing* **5(3)** 367–79
[15] Arshinskii L V 2007 Substantial and formal deductions in logics with vector semantics *Automation and remote control* **68(1)** 139–48
[16] Arshinskiy L V 2007 Mathematical models for handling incomplete and inconsistent data on the basis of logics with vector semantics *Modern technologies. System analysis. Modeling* **4(16)** p90–9
[17] Arshinskiy L V, Arshinskiy V L, Khomenko A P, Kargapol'tcev S K, Korol'kov B P and Aslamova V S 2017 Logical-mathematical modeling of complex subject areas based on the logics vector semantics *Far East J. of Mathematical Sci. (FJMS)* **101(4)** pp 813–23
[18] Arshinskiy L V, Arshinskiy V L, Dunaev M P, Dunaev A M and Aslamova V S 2017 Development of the expert system of electrical equipment setup based on logic with vector semantics *Far East J. of Electronics and Communications (FJEC)* **17(5)** pp 1119–25

[19] Arshinskiy L V 2012 Application of relational tables in the development of knowledge bases to ill-structured subject areas with incomplete and contradictory information *Modern technologies. System analysis. Modeling* **1(33)** 90–9
[20] Tepandi J 1990 Comparison of Expert System Verification Criteria: Redundancy *Proc. ECAI 90 Conf. Stockholm* pp 49–62
[21] *Handbook on the Artificial Intelligence* vol 2 *Models and Methods* 1990 ed. D A Pospelov. (Moscow: Radio and communication) p 304
[22] Klement E P, Mesiar R and Pap E 2000 *Triangular norms. Trends in Logic* vol 8 (Springer, Dordrecht)

Expert system for aggregate industrial safety assessment at enterprises based on knowledge technologies

E A Aslamova[1], L V Arshinsky[1], V S Aslamova[1] and M V Krivov[2]

[1]Irkutsk state University of railway engineering, 15, Chernyshevsky str., Irkutsk, 664074, Russia
[2]Angarsk state technical University, 60, Tchaikovsky str., Angarsk, 665035, Russia

Abstract. The work presents the knowledge base of the expert system of the aggregate assessment of the level of industrial safety at a hazardous production facility, based on the use of the logico-axiological approach. The fundamental hypothesis: the reduction in production safety is due to the presence of "NE-factors": poor implementation of organizational measures, improper working conditions, inadequate personal protective equipment and so on. Failure to take measures to provide protection against industrial accidents leads to a reduction in the level of industrial safety. According to the approach used, the influence of "NOT-factors" on the level of "non-safety" of production is formalized by products of the type: (SIC)A ->(SIC)B.

1. Introduction

In accordance with Federal Law No. 116 “On industrial safety of hazardous production facilities” [1], industrial safety (IS) of hazardous production facilities (HPFs) including railway facilities shall mean the state of safety of individual and public vital interests against accidents at hazardous production facilities and post-accident consequences of the above accidents.

Despite that there are a number of works on quantifying the industrial safety level at industrial facilities this problem remains unsolved [2].

The IS level assessment is mostly performed using the probabilistic statistical methods based on the calculation of the following performance characteristics [3–8]:

- accident rate at HPF ignoring both the accident type and its consequences, which cannot reflect adequately the actual picture because the number of accidents is 100–200 per year while the total number of supervised HPFs is 110,000;
- rate of fatal injuries at HPF, which is a classical approach in the risk theory evaluating the probability of a causal chain resulting in an accident; still, probability is a rather strict concept that needs a statistical justification to be used, which is not always possible in unique or new industries;
- industrial specialization of an enterprise (ratio of the number of injured workers to the product output), e.g. accident rate with respect to the pipeline length in main pipelines; the disadvantage of this approach is a narrow specialization and inefficiency for the activity assessment at small enterprises with a zero injury incident rate;
- damaging consequences of accidents and incidents; the disadvantage is that some damage parameters are impossible or infeasible to determine;
- indicators of supervisory and control measures such as a number of inspections, violations, disciplinary penalties imposed on officials for allowing violations, and measures focused on increasing the industrial safety level; time of reaction to registered violations; and staff qualification; in [9], the

proposed accident and incident classifier and the HPF IS assessment technique for gas distribution and gas consumption systems based on factor analysis of the safety state considering weights of the relevant factors (number of registered IS requirement violations, accidents, incidents, and injury cases); the disadvantage is the subjectiveness of the assessment and its significant dependence on the thoroughness of IS measures.

It is known that industrial emergencies such as injuries and accidents result from the combined action of multiple background conditions. The classical approach in the risk theory is focused of the probability assessment of this or that causal chain considering the damage done by an event. From this viewpoint, the risk event assessment at an enterprise is the probability assessment of factors that cause risk. Unfortunately, the concept of probability is a rather strict concept that needs a statistical justification to be used, which is not always possible in unique industries. A commonly used concept is subjective probability though such assessments are mostly nominal, which actually makes the probability approach similar to other methods that employ the expert assessment technique [10–13].

A promising approach to the integrated IS assessment of organizational and technical systems characterized by uncertainty, which are hard to represent by statistical models, is based on an expert system based on the logical-axiological quality assessment [12]. This method provides the IS assessment of an industrial system based on fuzzy inference using a production knowledge model. A knowledge base is developed as a result of the system analysis of the causal complex providing accident-free operation and occupational safety of an enterprise. Inference is based on the concepts of efficiency (value) and degree of actualization of factors and classes of factors that affect IS of a system.

The present article is concerned with a technique based on the logical-axiological assessment methodology proposed in [14, 15].

2. Stages of Aggregate IS Assessment Technique

The proposed technique is based on the idea that the normal operation of any complex system, including an accident-free operation system of an enterprise and an occupational safety (OS) system, requires the joint operation of all its subsystems composing it, and the damaged state of one subsystem leads to the damaged state of the whole system. This concept can be formalized by a set of fuzzy causal relationships (productions) of the following type: "If Not-A then Not-B", where B is the assessed component (a system or any of its subsystems) and A is one of its direct subcomponents (functional elements or subsystems). The states of each component are represented by numbers in the interval [0,1]. Here, 1 denotes the normal operation of a component and 0 denotes its loss. In fact, it is the declaration of the fuzzy truth of the propositions "Component A functions normally" and "Component B functions normally". The fuzziness also formalizes the degree of influence of the damage of component A on component B that contains it, which is expressed by a number in the interval [0,1], i.e., it is the declaration of the truth of the causal relationship "Damage in the operation of A causes the damage in the operation of B". Truth of this relationship is called value of A for B.

A system providing accident-free operation and occupational safety at an enterprise is described by the corresponding system of productions (an example of such a system is given below). The assessment is represented as a direct fuzzy attached inference from a knowledge base (KB) consisting of such productions and its implementation is based on the expert system technology (knowledge technology). The starting facts are state assessments of the initial functional elements of an industrial safety system and the result is the aggregate state assessment of the whole system [10–12].

The stages of the aggregate IS assessment technique are as follows:

1. System analysis of causes and factors relevant for the occurrence of an emergencies such as industrial injuries, occupational diseases, or accidents. Determination of the major factors and causes.

2. Formation of causal chains, their division into classes of factors. Determination of Not-factors leading to NOT safety of each class: damage to health of workers and accidents.

3. Determination of measures (factors) providing workplace safety and accident-free operation of the technological process.

4. Formulation of the main factors of each class in productions of the following form:

NOT-$a \rightarrow$ Not-b.

5. Development of survey questionnaires for auditors, who assess the truths (degrees of actualization) of the initial factors at an object of study, and experts, who assess the values (degrees of influence) of the main factor for a class of factors and a class of factors for the industrial safety (IS) level. The assessments are assigned on a scale [0,1] or, which is more convenient from the viewpoint of computation, on an integer scale 0÷100.
6. Survey for auditors and experts.
7. Development of algorithms and software for an expert system (ES) with a database for storing classes of factors, main factors of classes, productions, and expert and auditor survey results.
8. Check of competence and consistency of experts and auditors in ES. Development of aggregate assessments of NOT safety of an enterprise and computing the IS level.

3. Knowledge Base Design

According to the technical literature, factors and causes that affect the industrial injury incident rate can be the following: organizational, sanitary and hygienic, technical, and psychophysiological [16].

Another system of background conditions for the occurrence of an accident and injury hazardous situation is shown in Table 1 [17, 18].

One can see that, in both cases, the initiation and development of an unfavorable situation is caused by "NOT-factors": imperfections, incomplete/low-quality implementation of appropriate measures, poor staff training, lack of appropriate working conditions, etc., which are naturally formalized in the proposed approach. Let us note that the concept of value makes it possible to consider the relevance of each factor for a certain subsystem, and the whole approach considers that some factors can be regarded as key factors, which can devalue the whole system.

In Fig. 1, the causal relationship between the workplace damage to workers (occupational diseases and industrial injuries) and the industrial safety is represented as a tree of classes of factors providing IS and OS [11, 12].

The whole causal chain is formed and divided into classes as follows.

NOT safety of organizational class is determined by the following Not-factors:

- low quality or lack of professional training;
- lack of a work project and instructions on OS;
- violation of technological process organization;
- violation of work and rest schedule, etc.

Table 1. Factors of accident and injury incident rate at industrial enterprises

Group	Factors of accident and injury incident rate	%	In group
Technologies of work performance	Inconvenience of technology of preparation and performance of work	3.8	7.8
	Inconvenience of repair and maintenance	2	
	Complexity of algorithm of actions of a human	1.2	
	Necessity of entering a potentially dangerous area	0.8	
Workers (humans)	Poor emergency skills of staff	12.7	50.1
	Inability to estimate the information about the process state	12.3	
	Poor theoretical knowledge of the processes	7.3	
	Technological indiscipline	8	
	Lack of self-control under conditions of stress	5.6	
	Other disadvantages of workers	4.2	
Environmental conditions	Discomfort of environment in terms of physico-chemical parameters	2.8	16.6
	Low quality of information model of environment state	4.8	
	Possibility of dangerous impacts on humans and machines	9	
Equipment (machines)	Low quality of workplace structures	6	18.1
	Ignorance of working capacity of humans	1.5	
	High energy consumption of energy source	0.7	
	Possibility of dangerous failures	8	
	Other factors of equipment (machines)	1.9	
Others	Other factors	7.4	7.4
Total		100	100

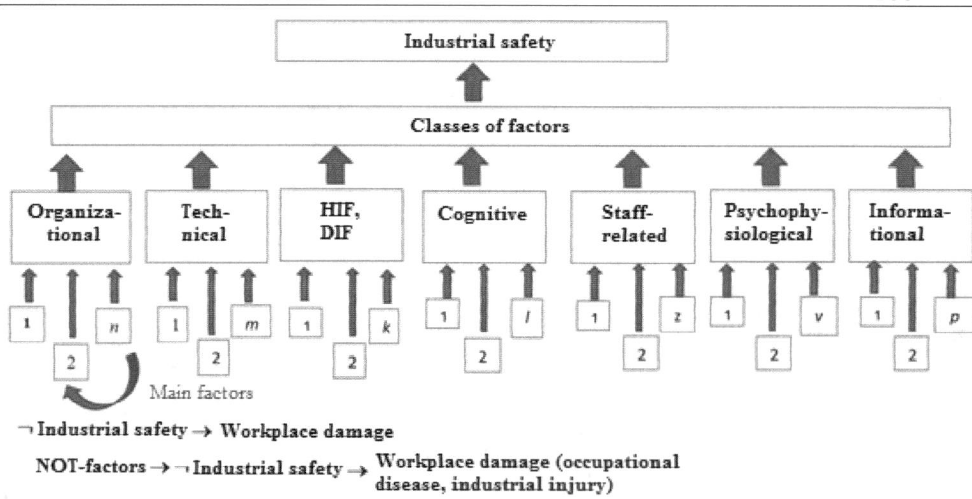

Figure 1. Conceptual model of an industrial safety system

The deficiency in these types of measures leads to "Not-safety" of organizational class.

The deficiency in measures leading to "Not-safety" of the other classes is demonstrated in Table 2.

The knowledge base (KB) is the following system of assessed factor and productions. Causes of health damage to workers are insufficient workplace safety measures.

Table 2. NOT-factors of NOT safety classes

Class	
Technical	insecurity and accident hazard of equipment, devices, collective and personal protective equipment (PPE) non-mechanized and non-automated technological processes non-ergonomic workplaces design imperfection of equipment lack or malfunction of enclosing, protecting, blocking, signaling, and other devices imperfection of technological process non-adherence to preventive repair schedule lack of adequate PPE, etc.
Dangerous and harmful industrial factors (DIF and HIF)	unfavorable industrial microclimate parameters low-quality lighting increased noise and vibration monotonous working conditions high concentration of harmful substances aerosols of predominantly fibrogenic action in air of work area electric and electrostatic fields stressful and intense work process electromagnetic, radioactive radiation, etc.
Cognitive	lack of knowledge and understanding by staff of their role in technological chain lack of knowledge by staff of technical production regulations lack of knowledge by staff of regulatory legal acts on OS and IS lack of knowledge by staff of rules for using PPE lack of knowledge by staff of rules of conduct in emergency situations lack of knowledge and understanding by staff of causes and consequences of accidents, etc.
Psychophy-siological	increased fatigability increased fatigue low concentration of attention slow reaction low stress resistance indiscipline, irresponsibility mismatch between psychological features of a worker to conditions of performed work, etc.
Staff-related	poor professional qualification short experience of work in harmful working conditions inappropriateness of a worker at occupied position, etc.
Informational	lack of warning posters lack of safety signs lack of surface marking lack of barrier for dangerous areas, etc.

The given-below measures of the organizational class of factors provide workplace safety (WS) and IS.

Organizational class of *factors*:

- industrial and occupational safety requirements are met;
- staff training and instruction is timely and of high quality;
- there is a work project and instructions on occupational safety;
- satisfactory organization of workplaces;
- work and rest schedule is adhered to.

In Table 3, measures of the other classes of factors providing WS and IS are listed.

Statements about the state of classes of factors that determine WS and IS (intermediate aggregating facts) are as follows:

- organizational measures are provided;
- technical conditions are appropriate;
- DIF and HIF are appropriate;
- psychophysiological state of workers is appropriate;
- professional qualification of workers is appropriate;
- workers have the necessary knowledge on occupational and industrial safety;
- hazard information signs are available.

The final aggregate inference is as follows: industrial safety and occupational safety at an enterprise is provided and a system for protection of workers from DIF, HIF, and injuries is appropriate.

Using the above facts, they are formulated as corresponding productions.

A production is a structure of the following form: If A→B; in the studied case, NOT-a → NOT-b: NOT appropriateness of the main factor causes NOT appropriateness of the corresponding class of factors.

Some example productions on the main factors of the organizational class are given below:

- NOT(true that)-"staff training and instruction are timely and of high quality" → Not(true that)-"organizational measures are provided";
- NOT-"there is a work project and instructions on occupational safety" → Not-"organizational measures are provided";
- NOT-"workplaces are organized satisfactorily" → Not-"organizational measures are provided";
- NOT-"technological process is adhered to" → Not-"organizational measures are provided";
- NOT-"labor discipline level is high" → Not-"organizational measures are provided";
- NOT-"work and rest schedule is adhered to" → Not-"organizational measures are provided";
- NOT-"there are appropriately used PPE" → Not-"organizational measures are provided".

For the other groups of factors, productions are formulated similarly. In addition, aggregating productions are needed:

- NOT-"organizational measures are provided" → Not-"industrial and occupational safety is provided";
- NOT-"technical conditions are appropriate" → Not-"industrial and occupational safety is provided";
- NOT-"DIF and HIF are appropriate" → Not-"industrial and occupational safety is provided".
- NOT-"psychophysiological state of workers is appropriate" → Not-"industrial and occupational safety is provided".
- NOT-"professional qualification of workers is appropriate" → Not-"industrial and occupational safety is provided";
- NOT-"workers have the necessary knowledge on occupational and industrial safety" → Not-"industrial and occupational safety is provided";
- NOT-"hazard information signs are available" → Not-"industrial and occupational safety is provided".

Other productions for the main factors of the other classes are formulated similarly.

Table 3. Measures of a class of factors providing WS and IS

Class	
Technical	equipment and devices are functioning properly
	there are means of mechanization and automation
	no design imperfection of equipment
	there are properly functioning enclosing, protecting, blocking, and other devices
	technological process of necessary degree of perfection
	adherence to preventive repair schedule
	workplace ergonomics is satisfactory
	technological process is adhered to
	appropriately used, are available and appropriately used
	technical state of equipment is satisfactory
Dangerous and harmful industrial factors (DIF and HIF)	favorable industrial microclimate parameters
	high-quality lighting meeting the standards
	noise and vibration are appropriate
	no monotonous working conditions
	concentration of harmful substances is appropriate
	stressfulness and intensity of the work process are appropriate
	strength of electric and electromagnetic fields is appropriate
	no harmful radiation
Cognitive	knowledge and understanding by staff of their role in the technological chain
	knowledge by staff of technical production regulations
	knowledge by staff regulatory legal acts on occupational and industrial safety
	knowledge of rules and skills for using PPE
	knowledge by staff of rules of conduct in emergency situations
	knowledge and understanding by staff of causes and consequences of accidents, etc.
Psychophy-siological	fatigability of workers is appropriate
	attention is appropriate
	psychological features of workers correspond to conditions or performed work
	reaction rate is appropriate
	stress resistance is appropriate
	discipline and the responsible attitude of staff to performed duties
Staff-related	professional qualification of workers is appropriate
	sufficient experience of work in harmful working conditions
	appropriateness of a worker at the occupied position, etc.
Informational	there are warning posters
	there are safety signs
	there is surface marking
	there is barrier for dangerous areas, etc.

4. Infeence: Aggregate Computation

Truths of the initial facts is assessed on a scale [0,1] or, which is more convenient from the viewpoint of computation, on an integer scale of 0÷100. Truths of intermediate facts (here, they are aggregating facts) are assessed in accordance with the modus ponens rule considering computed truth of the conclusion based on truths of propositions:

$$\neg A,\ \neg A \rightarrow \neg B \vdash \neg B\colon \|\neg B\| = \|\neg A\| \cdot \|\neg A \rightarrow \neg B\|.$$

Here, the $\vdash$ symbol separates propositions from the conclusion; after a colon, the computed truth of the conclusion; $\|\neg A\|$ is the truth of the negation of the fact A, $\|\neg A \rightarrow \neg B\|$ is the truth of the implication "If Not-…, then Not-…" (value of A for B).

IOP Conf. Series: Materials Science and Engineering **760** (2020) 012007 doi:10.1088/1757-899X/760/1/012007

The aggregation is performed using the rule of evidence integration, which, considering the probability nature of events, should be taken in the form:

$||\neg B||=||\neg B||1 + ||\neg B||2 - ||\neg B||1 \cdot ||\neg B||2;$

$||\neg B||1$ and $||\neg B||2$ are the truths of the negation of B computed through different inference chains.

The resulting aggregate industrial safety level is as follows: $||B|| = 1 - ||\neg B||$.

Risk (already not "probable" but certain "fuzzy" risk) can be computed as the product of the final aggregate system assessment by the damage caused by the corresponding event.

Let us consider the formulation of an inference in the following example.

Let there be a fact A_1 = "organizational measures are provided". The auditor's assessment of the degree of influence of organizational class of factors on occupational safety is "good" (0.8).

A_2 = "technical conditions are appropriate", the auditor's assessment of the degree of influence of technical class of factors on occupational safety is "excellent" (0.9).

B = "industrial and occupational safety are provided". According to the expert assessments, value of A_1 for B = 0.4; value of A_2 for B = 0.5.

Inference step 1: $\neg A1, \neg A1 \rightarrow \neg B \vdash \neg B$: $||\neg B|| = 0.2 \cdot 0.4 = 0.08$.

Inference step 2: $\neg A2, \neg A2 \rightarrow \neg B \vdash \neg B$: $||\neg B|| = 0.1 \cdot 0.5 = 0.05$.

Inference step 3: Resulting degree of certainty that the event will occur

$$\neg B: ||\neg B|| = 0.08 + 0.05 - 0.08*0.05 = 0.126.$$

Thus, the deficiency in providing all the industrial safety measures is 12.6 %.

Inference step 4: "industrial and occupational safety are provided, the system of protection of workers from workplace damage" = 1 – 0.126 = 0.876 ≈ 0.9 "excellent".

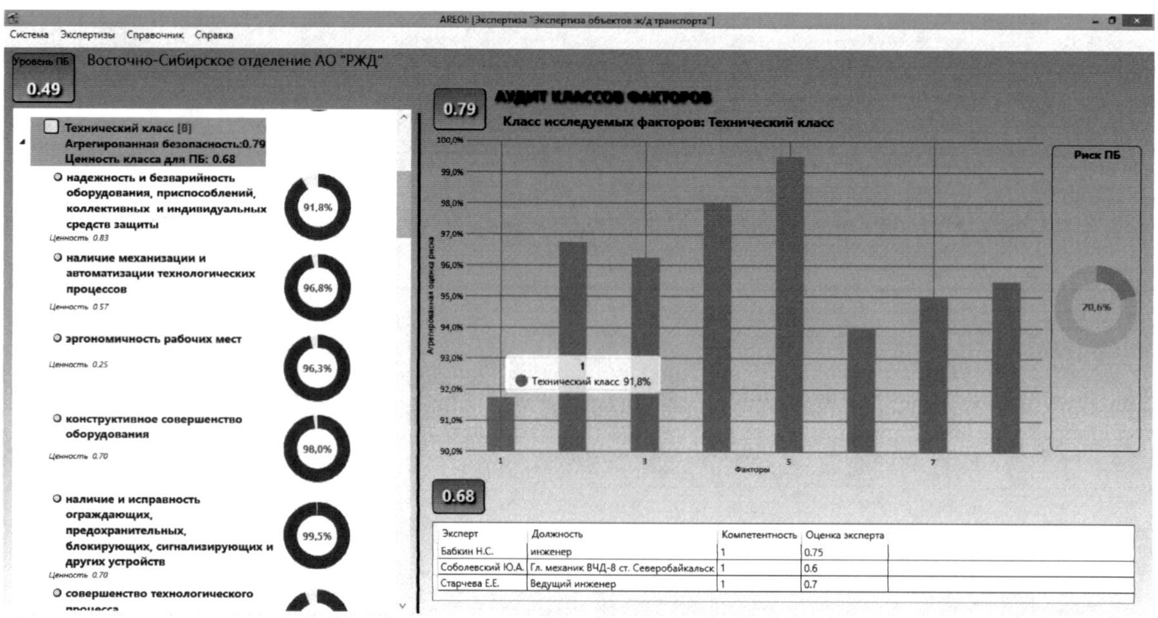

Figure 2. The results of testing the basic factors of the technical class

Hence, the logical-axiological assessment technique provides the aggregate state assessment of the system for providing safety and prevention of injuries at enterprises. The technique is implemented based on the ES technology and relies on Not-factors as causes of a risk situation leading to workplace damage and decrease in the IS level.

The results of testing on real data from objects of the East-Siberian railway are presented in Fig. 2. It can be seen that the level of PB is 0.49, the probability of emergencies is 0.52. The aggregated technical grade safety assessment does not exceed 79 %, although the expert value of this class of factors for industrial safety is the highest (0.68) compared to the values of other classes. Derivatives of histograms of values of basic factors and their execution within the technical class show that the basic conditions such as the availability of adequate protective equipment (95.5 % with a value of 0.90), equipment reliability and reliability, and adaptation (91,8 % with a value of 0.83); compliance with the terms of scheduled maintenance (95.0 % with a value of 0.70). Hence, in front of the decision maker, it is necessary to state the problem of the primary solution of problems in these particular directions, to develop plans of measures for industrial safety of these types.

References

[1] *On industrial safety of hazardous production facilities* 2006 Federal Law of the Russian Federation of March 20, No 35-FZ: adopted by State Duma on June 20, 1997 (as amended of March 4, 2013 No 22-FZ, March 7, 2017 No. 31-FZ)

[2] Kuz'min A A and Chirikov A G 2010 Approach to safety level assessment of hazardous industrial facilities using elements of qualimetry theory *Scientific and Educational Problems of the Civil Defence* **4** 1–9

[3] Buiko K V and Pantyukhova Yu V *Approaches to industrial safety level assessment in organizations operating hazardous industrial facilities* Retrieved from: http://kalugatechnadzor.ru/analitika/371-podxody-k-oczenke-urovnya-promyshlennoj-bezopasnosti-v-organizacziyax-ekspluatiruyushhix-opasnye-proizvodstvennye-obekty (access date: October 24, 2017)

[4] *Unified Standard. Methods for industrial safety level assessment at hazardous industrial facilities*, 2015. URL: https://1cert.ru/stati/metody-otsenki-urovnya-promyshlennoy-bezopasnosti-na-opasnykh-proizvodstvennykh-obektakh (access date: October 25, 2017)

[5] Kuznetsova S N 2009 *Methods of design of mathematical support for natural, technigenic, and environmental risk prediction systems* (St. Petersburg) 20 p

[6] Veltri A, Pagell M, Behm M and Das A A 2007 Data-Based Evaluation of the Relationship between Occupational Safety and Operating Performance *J. of SH&E Res.* **4(1)** 1–22

[7] Ateekh-ur-Rehman L 2012 Safety Management in a Manufacturing Company: Six Sigma Approach *Engineering* **4** 400–7

[8] Anvari A, Zulkifli N and Yusuff R M 2011 Evaluation of Approaches to Safety in Lean Manufacturing and Safety Management Systems and Clarification of the Relationship Between Them *World Applied Sci. J.* **15(1)** 19–26

[9] Pantyukhova Yu V 2011 *Technique for industrial safety level assessment of hazardous industrial facilities of gas distribution and gas consumption systems* (Moscow) 25 p

[10] Arshinsky L V and Aslamova E A 2017 Development of aggregate industrial safety assessment system at an enterprise based on the knowledge technologies *Transport infrastructure of Siberian region: Proce. of the 8th Int. Scientific and Practical Conf.* March 28 –April 1 (Irkutsk: ISTU) pp 401–5

[11] Aslamova E A and Aslamova V S Aggregated Evaluation of Industrial Level Safety of a Dangerous Production Object Based on Knowledge Technologies *Atlantis Press: Part of series: ACSR* **72** 397–400 ISSN: 2352-538X Retrieved from: https://www.atlantis-press.com/proceedings/itsmssm-17/publishing.

[12] Aslamova E A and Arshinsky L V 2017 Technique for aggregate assessment of protection of workers against professional damage based on the knowledge technologies *Informatsionnye i matematicheskie tekhnologii v nauke i upravlenii* **4(8)** 39–51

[13] Korchunov A G 2010 *Methodology for control of product quality attributes in the metalware production technologies based on fuzzy logic models* (Magnitogorsk) 37 p

[14] Arshinsky L V 2015 Technique for aggregate assessment of systems with support of key components *Ontologiya proektirovaniya* **5(2(16))** 223–32

[15] Arshinsky L V 2013 Logical-axiological approach to state assessment of systems *Advanced technologies. Systems analysis. Modeling* **3(39)** 140–6

[16] Filippov A A, Pachurin G V, Shchennikov N I and Kuragina T I 2016 Occupational injuries and trends of their prevention *Advanced high technologies* **1** 45–50
[17] Belov P G 2003 *System analysis and modeling of hazardous processes in technosphere: textbook for students of high schools* (Moscow: Akademiya) 512 p
[18] Belov P G 2014 *Risk management, system analysis, and modeling: textbook and practical work for undergraduate and graduate students*] (Moscow: Yurait) 460 p

IOP Conf. Series: Materials Science and Engineering **760** (2020) 012008 doi:10.1088/1757-899X/760/1/012008

Precision adjustment of electromagnetic emitter of large-scale systems such as GLONASS

A V Banshchikov[1], A V Daneev[2], V A Rusanov[1] and A A Vetrov[1]

[1]Matrosov Institute for System Dynamics and Control Theory of the Siberian Branch of the Russian Academy of Sciences, 130, Lermontov ave., Irkutsk, 664033, Russia
[2]Irkutsk State Transport University, 15, Chernyshevskogo ave., 664074, Irkutsk, Russia

Email: daneev@mail.ru

Abstract. The multidimensional nonlinear regression-tensor model is investigated in the substantiation of necessary and sufficient conditions for the optimal multi-factorial process of precision calibration of the parameters of the electromagnetic radiation source in the geostationary orbit (including the large-scale GLONASS system). A robust-adaptive strategy of a posteriori formation of the target functional of electromagnetic observability of a weighted-distributed signal in an established complex of stationary ground points is proposed on the basis of observations of this signal, made with a tolerance.

1. Introduction

Complex electronic and mechanical processes are the cornerstone of modern orbital precision verification of astroequipment [1–5], which actualizes the issues associated with the development of their mathematical models [6]. In this context, regression models are in demand [7], including an important class formed by regression-tensor systems [8]. These systems, on the one hand, encompass polynomial models, allowing analytic description on the basis of tensor calculus [9], strong differentiability of vector mappings [10], and the theory of extremal problems. On the other hand, these models do not just idly manifest themselves, but actively work [11], acquiring a prominent role in a posteriori multifactor nonlinear mathematical modeling of electron-mechanical [12] and optical-mechanical systems [13], providing (within the optimization of the target quality functional reception of an electromagnetic signal) adaptive tuning of parameters that lower the energy level of the side lobes of electromagnetic emitters [8, 11, 14].

In this connection, the main attention below is paid to the problems posed in the conclusions of the work [14]. Particularly, there is correction of the target functional of the intensity of the observed signal of the source of electromagnetic radiation (SER) in the geostationary orbit in the regression-tensor modeling of the alignment process of the spatial-geometric parameters of the SER. In this case, regression interpretations of multiply connected conditions imposed by complex constraints [15] permitting on the basis of matrix analysis [16] the construction of an optimal mode of adaptive adjustment of SER parameters (in particular, the geometry and orientation of its antenna) in terms of the mathematical input-output model are determined. The predictive model of the SER signal is constructed from the experimental data of orbital verification tests based on the two-criteria identification by the least-squares method (LSM) of covariant tensors [9] of the nonlinear SER equation, as a multidimensional tensor regression with a minimum coordinate-matrix norm [17].

2. Motivation, Terminology and Formulation of Problems

Let R be the field of real numbers, R^n is n-dimensional vector space over R with the Euclidean norm $\|\cdot\|_{R^n}$, $\mathrm{col}(y_1,\ldots,y_n)\in R^n$ be a column vector with elements $y_1,\ldots,y_n\in R$ and let $M_{n,m}(R)$ be the space of all $n\times m$-matrices with elements from R. In addition, we assume that T_m^k is the space of all covariant tensors of k-valence, that is, real multi-linear forms $f^{k,m}:R_1^m\times\ldots\times R_k^m\to R$ with the norm $\left\|f^{k,m}\right\|_{T_m^k}:=\left(\sum t_{\ldots j\ldots}^2\right)^{1/2}$, where $\{t_{\ldots j\ldots}\}$ is the "matrix of coordinates" [9, p. 246] of the tensor $f^{k,m}$ with respect to the canonical basis in R^m.

Let, in the process of orbital alignment, $\{b_i\}_{i=\overline{1,n}}$ be a set of stationary points for ground-based reception of the signal of the SER located on a geostationary satellite (for example, GLONASS systems, i.e., radius vectors, connecting the satellite and points b_i, are constant), some fixed (reference) vector of space-angular parameters of the antenna of the SER [11–15]. v is a purposeful variation of the physical and geometric predictors [7] in the process of precision adjustment. $w(\omega+v)\in R^n$ is the vector of the predictive intensity of the calibration signal of the SER measured at the probing points.

In this formulation, for a compact description of the nonlinear adaptive process of parametric orbital adjustment of the SER, we select for consideration a multidimensional input-output prognostic system described by the vector-tensor k-valent equation of a multi-factorial regression of the form:

$$w(\omega+v)=\mathrm{col}\left(\sum_{j=0,\ldots,k} f_1^{j,m}(v,\ldots,v),\ldots,\sum_{j=0,\ldots,k} f_n^{j,m}(v,\ldots,v)\right)+\varepsilon(\omega,v). \tag{1}$$

Here $\varepsilon(\omega,\cdot):R^m\to R^n$ is not a parametrizable vector function of the class:

$$\|\varepsilon(\omega,v)\|_{R^n}=o\left(\left(v_1^2+\ldots+v_m^2\right)^{k/2}\right), \tag{2}$$

$v=\mathrm{col}(v_1,\ldots,v_m)$, $f_i^{0,m}$ is a "rank 0" tensor representing the intensity of the signal w_i, $i=\overline{1,n}$ in the mode of the orbital electronic geometric tuning of the transmitting antenna of the SER.

The problem of nonlinear regression-tensor modeling of the multifactor optimal process of orbital calibration of the SER is presented and studied in detail in [11, 14] for the 2-valent model (1). With that, analytical solutions of three methodological positions of this problem were obtained:

(I) for a fixed vector-predictor $\omega\in R^m$ and its open neighborhood $V\subset R^m$, the analytic conditions are defined for which the vector function $w(\cdot):V\to R^n$ of the indices of the intensity of the calibration signal of the SER at the probing points satisfies the regression system (1)–(2);

(II) a direct algorithm for identifying the coordinates of the tensors $f_i^{j,m}$, $i=\overline{1,n}$, $j=\overline{0,2}$ in a 2-valent regression-tensor model (1) is constructed on the basis of the solution of a two-criterion LSM-Problem of the form:

$$\begin{cases}\min\left(\sum\limits_{l=1,\ldots,q}\left(\left\|w_{(l)}-\mathrm{col}\left(\sum\limits_{j=0,\ldots,k} f_1^{j,m}(v_{(l)},\ldots,v_{(l)}),\ldots,\sum\limits_{j=0,\ldots,k} f_n^{j,m}(v_{(l)},\ldots,v_{(l)})\right)\right\|_{R^n}^2\right)\right)^{1/2}, \\ \min\left(\sum\limits_{i=1,\ldots,n}\sum\limits_{j=0,\ldots,k}\left\|f_i^{j,m}\right\|_{T_m^j}^2\right)^{1/2},\end{cases} \tag{3}$$

where $w_{(l)} \in R^n$, $v_{(l)} \in R^m$, $l = \overline{1,q}$, respectively, are the vectors of the experimental factor-predictors of the SER, i.e. $w_{(l)}$ is a posteriori "reaction" to the target "variation" relative to the reference vector ω coordinates under the condition $\|v_{(l)}\|_{R^m} < 1$. This inequality is methodologically dictated by the condition (2). q is the number of orbital experiments performed (determined by the representativeness of the model (1)), taking into account the dynamic characteristics of the SER [18, 19];

(III) for the 2-valent regression-tensor model (1), with the given predictor $\omega \in R^m$ and the nominal condition $\varepsilon(\omega,\cdot) \equiv 0$, an analytical solution of the orbital calibration problem is obtained, as a non-linear "v-optimizing" of the variable (relative to the ω vector) predictors of the adjustable physical and geometric parameters of the SER:

$$\max_{v \in R^m} F(v) := r_1 w_1(\omega + v) + \ldots + r_n w_n(\omega + v)\,, \tag{4}$$

where the vector function $v \mapsto w(\omega + v) = \mathrm{col}\big(w_1(\omega+v), \ldots, w_n(\omega+v)\big)$ has a coordinate representation according to the LSM-identified model (1)–(3), $r_i > 0$ is weights reflecting the priority of the ground-based probing points of the SER signal.

Statement of problems (according to the conclusions of [14]):

(i) to determine the necessary and sufficient conditions for the solvability of the optimization task (4) for the 3-valent model (1);

(ii) to construct an algorithm for correcting sufficient conditions for the extremum of the stationary point of task (i) on the basis of a r-parametric adjustment $r \mapsto r^{\mathrm{T}} w(\omega + v)$ of the functional:

$$v \mapsto F(v) = r^{\mathrm{T}} w(\omega + v)\,. \tag{5}$$

3. Optimization of the Process of Adjusting the SER-Parameters

Let us consider the task (i) – optimization of the process of adjustment of the parameters of the SER for the model (1) with $k = 3$. We note that the solution of the concomitant identification Problem (II) with $k = 3$ is a modification of assertion 2 of [14] (see also [20]).

In such a mathematical formulation, the non-linear prognostic equation (1) can be presented in the following vector-matrix-tensor form:

$$w(\omega + v) = c + Av + \mathrm{col}\big(v^{\mathrm{T}} B_1 v + f_1^{3,m}(v,v,v), \ldots, v^{\mathrm{T}} B_n v + f_n^{3,m}(v,v,v)\big) + \varepsilon(\omega, v)\,, \tag{6}$$

where $c \in R^n$, $A \in M_{n,m}(R)$, $B_i \in M_{m,m}(R)$, $i = \overline{1,n}$. Without loss of generality, we assume that each matrix B_i has an upper triangular structure; this simplifies the implementation of the LSM-algorithm (3).

According to (1) with $k = 3$, the functional of electromagnetic observability (5) is twice continuously differentiable, which guarantees the equality of mixed derivatives:

$$\partial^2 F(v_1, \ldots, v_m) / \partial v_g \partial v_p = \partial^2 F(v_1, \ldots, v_m) / \partial v_p \partial v_g \quad \forall g, p = \overline{1,m}\,. \tag{7}$$

Therefore, in the solution of the optimization problem (4) for the 3-valent model (6), we may assume the following Proposition 1 below being the main result, according to (7), Theorem 3 [10, p. 505] and Theorem 7.2.5 of [16]. But first let us preliminary assume that $B_i^* := (B_i + B_i^{\mathrm{T}}) \in M_{m,m}(R)$, $i = \overline{1,n}$, where each B_i is the matrix of the system (6) (the matrix of the tensor $f_i^{2,m}$ in the formulation, when it is not considered symmetric in the system (1)). Moreover, we consider the vector function:

$$v \mapsto \Phi(v) := (r_1 B_1^* + \ldots + r_n B_n^*)^{-1} (A^{\mathrm{T}} + [\nabla_v f_1^{3,m}(v,v,v), \ldots, \nabla_v f_n^{3,m}(v,v,v)])\, r\,,$$

where $\nabla_v f_i^{3,m}(v,v,v)$ is the gradient of the functional $v \mapsto f_i^{3,m}(v,v,v)$.

Proposition 1. *The stationary points* $v^* \in R^m$ *of task* (i) *are the solutions of equation*

$$v^* + \Phi(v^*) = 0. \tag{8}$$

A sufficient condition $F(v^*) = \max\left\{F(v) : v \in R^m\right\}$ *is the requirement that* v^*, *as a stationary point of the functional* (5), *be of elliptic type. In other words, at the point* v^* *for the Hessian* $G(v,r)$ *of the functional* (5), *inequalities:*

$$\det[b_{ij}]_p < 0, \ \ p = \overline{1,m}, \tag{9}$$

must be performed, where $[b_{ij}]_p \in M_{p,p}(R)$, $p = \overline{1,m}$ *are the main Hessian sub-matrices*

$$G(v^*,r) = r_1\left(B_1^* + [\partial^2 f_1^{3,m}(v,v,v)/\partial v_g \partial v_p\big|_{v^*}]\right) + \ldots$$
$$+ r_n\left(B_n^* + [\partial^2 f_n^{3,m}(v,v,v)/\partial v_g \partial v_p\big|_{v^*}]\right) \in M_{m,m}(R),$$

which is equivalent – the characteristic numbers $\lambda_p(v^*,r)$ *of the matrix* $G(v^*,r)$ *satisfy:*

$$\lambda_p(v^*,r) < 0, \ \ p = \overline{1,m}. \tag{10}$$

Corollary 1. *For* $k = 2$, *the Hessian of the functional* (5) *and the conditions* (9), (10) *are invariant to the position of the stationary point* v^*, *while the Hessian is equal to* $G(r) = r_1 B_1^* + \ldots + r_n B_n^*$, *which results in a linear dependence of the numbers* $\lambda_p(r)$, $p = \overline{1,m}$ *on the normalization of the vector* r.

If $\operatorname{rank} G(r) = m$, *then the solution of equation* (8) *is unique and has the form* $v^* = -G^{-1}(r)A^{\mathrm{T}}r$, *which makes the position of the point* v^* *invariant to the normalization of the vector* r.

One of the factors, affecting the geometry of the stationary point v^* of the Proposition 1, is the digital adaptive parametric adjustment $r \mapsto G(v^*,r)$ that results in the performance of elliptical conditions (9) or (10), which is the research subject of the next section.

4. Adaptation of the Target Functional on the *r*-Parameter Family of its Hessians

Let us consider the formulation (ii): for a stationary point of the optimization task (i), construct a numerical procedure for correcting the weight coefficients $r \in R^n$, starting from the fulfillment of the spectral conditions (10); ensuring the elliptic nature of the stationary point v^* of Proposition 1. This formulation is relevant in the problem of orbital calibration of the parameters of the SER, when at some ground points b_j it is necessary to weaken (i.e. $r_j < 0$) the reception of the SER signal.

Let an initial vector $r_0 \in R^n$ of weight coefficients from the formulation (ii) be given. For example, a heuristic choice r_0 can be made based on the equality of its coordinates r_{0i}, $i = \overline{1,n}$ to the values of some functions $\Psi_i : R \to R$ that depend on the values of the functionals $J_i(v) := w_i(\omega + v)$, $i = \overline{1,n}$ from the auxiliary problems of optimal prediction of the quality of orbital alignment of the SER for individual target indicators w_i. We denote by $v^0 \in R^m$ some stationary point of the functional (5) in case when the r-priority of the probing points is equal to r_0. Correspondingly, by $G_0 \in M_{m,m}(R)$ we denote the Hessian of the given functional calculated for the pair (r_0, v^0) and let $G_i := B_i^* + [\partial^2 f_i^{3,m}(v,v,v)/\partial v_g \partial v_p\big|_{v^0}]$, $i = \overline{1,n}$.

Then, for an admissible linear variation Δr of the coordinates of the vector $r_0 = \operatorname{col}(r_{01},\ldots,r_{0n})$ given (by virtue of the comments to formula (4)), the region of this variation $W \subset R^n$ of the form:

$$\Delta r := \operatorname{col}(\Delta r_1,\ldots,\Delta r_n) \in W, \ r_i = r_{0i} + \Delta r_i > 0, \ i = \overline{1,n},$$

Δr-parametric family of linear variations of the Hessian $G(v^0, r_0 + \Delta r)$ is determined by a matrix $m \times m$-manifold of the form:

$$G_0 + \sum_{i=1,\dots,n} \Delta r_i G_i, \ \Delta r \in W. \tag{11}$$

Proposition 2. *Let* $r = r_0 + \Delta r$, $\{(\lambda_p(r_0), x_p), \ p = \overline{1,m}\} \subset R \times R^m$ *be the set of eigen-pairs of Hessian* G_0, *i.e.* $\lambda_p(r_0) x_p = G_0 x_p$, $p = \overline{1,m}$, *and let, starting from the implementation of the manifold* (11), *be given the numbers* $g_{pi} = x_p^{\mathrm{T}} G_i x_p / x_p^{\mathrm{T}} x_p$, *then the eigenvalues* $\lambda_p(v^0, r_0 + \Delta r)$, $p = \overline{1,m}$ *of the Hessian* $G(v^0, r_0 + \Delta r)$ *have the form:*

$$\lambda_1(v^0, r_0 + \Delta r) = \lambda_1(r_0) + \sum_{i=1,\dots,n} g_{1i} \Delta r_i + o\left(\|\Delta r\|_{R^n}\right),$$

$$\dots\dots\dots\dots\dots\dots\dots\dots\dots$$

$$\lambda_m(v^0, r_0 + \Delta r) = \lambda_m(r_0) + \sum_{i=1,\dots,n} g_{mi} \Delta r_i + o\left(\|\Delta r\|_{R^n}\right).$$

Corollary 2. *Let* $k = 2$, $n = m$, $\Lambda(r_0) := \mathrm{col}\left(\lambda_1(r_0), \dots, \lambda_m(r_0)\right)$ *be the vector of characteristic numbers of the matrix* $(r_{01} B_1^* + \dots + r_{0m} B_m^*)$ *and* $\{x_p\}_{p=\overline{1,m}}$ *be the eigenvectors corresponding to them. In addition, let* $\Lambda^* := \mathrm{col}\left(\lambda_1^*, \dots, \lambda_m^*\right)$ *be a certain vector of characteristic numbers, "standard/exemplary" by criterion* (10), *and* $B := [b_{pi}]$ *is the* $m \times m$ *matrix with elements* $b_{pi} = x_p^{\mathrm{T}} B_i^* x_p / x_p^{\mathrm{T}} x_p$.

Then for $r_0 + \Delta r$, *where the variation vector has a representation* $\Delta r = B^{-1}(\Lambda^* - \Lambda(r_0))$, *the eigenvalues of the Hessian* $G(r_0 + \Delta r)$ *will be* $o\left(\|\Delta r\|_{R^n}\right)$ *– close to the reference values* $\{\lambda_p^*\}_{p=\overline{1,m}}$.

5. Acknowledgments

This work was funded by the Russian Foundation for Basic Research (project 19-01-00301).

References

[1] Aleshin I N, Baturin V V, Molodenkov A V et al 2002 Controlling the motion of the space platform complex. V. Algorithms for the alignment of the complex. *Proceedings of the Russian Academy of Sciences. Theory and control systems* **3** 132–9

[2] Tanaka H 2011Surface error estimation and correction of a satellite antenna based on antenna gain-analyses *Acta Astronautica* **68(7)** 1062–9

[3] Lebedev D V and Tkachenko A I 2012 Parametric adjustment of the "camera and stellar sensor" complex installed on a low-orbit spacecraft. *Izvestiya RAN. Theory and control systems* **2** 153–65

[4] Mitrokhin V N and Mozharov E O 2015 The radio-holographic method for controlling the profile of parabolic mirror antennas by the electromagnetic field in the near field. *Bull. of Bauman MSTU. Ser. Instrument making* **4** 81–95

[5] Tkachenko A I 2017 Improvement of the methodology of flight geometric calibration using unknown landmarks *Problems of control and informatics* **2** 112–21

[6] Banshchikov A V, Burlakova L A, Irtegov V D and Titorenko T N 2014 Symbolic computations in modeling and qualitative analysis of dynamic systems *Computational technologies* **19(6)** 3–18

[7] Draper N R and Smith G 2007 *Applied regression analysis* (Moscow: Williams Publishing House) 912 p

[8] Kozyrev V A, Kumenko A E, Rudykh A G and Rusanov V A 2010 Nonlinear regression-tensor analysis of the optimal installation of an electromagnetic radiation source for unauthorized

scanning of its electromagnetic field. *Proc. of higher educational institutions. Instrument making* **53(10)** 10–7

[9] Glazman I M and Lyubich Yu I 1969 *Finite-dimensional linear analysis* (Moscow: Nauka Publ.) 476 p

[10] Kolmogorov A N and Fomin S V 1976 *Elements of the theory of functions and functional analysis* (Moscow: Nauka Publ.) 544 p

[11] Rusanov V A and Daneev R A 2014 On the adaptive tuning of the parameters of the source of electromagnetic radiation in the geostationary orbit *Control systems and machines* **6** 12–7

[12] Karshakov E V 2012 The task of calibrating the electromagnetic system of relative positioning *Controlling large systems* **37** 250–68

[13] Demin A V and Mendeleeva L M 2014 Algorithm for alignment of composite mirrors of high-temperature telescopes *Proc. of higher educational institutions. Instrument making* **57(1)** 51–6

[14] Rusanov V A, Banshchikov A V, Daneev A V, Kumenko A E and Vetrov A A 2016 On the adaptive adjustment of parameters for the source of electromagnetic radiation on a geostationary orbit *Far East Journal of Electronics and Communications* **16(3)** 685–701

[15] Gryanik M V and Loman V I 1987 *Developable mirror antennas of the umbrella type* (Moscow: Radio and Communication) 72 p

[16] Horn R and Johnson C 1989 *Matrix analysis* (Moscow: Mir Publ.) 656 p

[17] Kostrikin A I and Manin Yu I 1986 *Linear Algebra and Geometry* (Moscow: Nauka Publ.) 304 p

[18] Banshchikov A V 2009 Analysis of the dynamics of large-scale mechanical systems by means of computer algebra *Siberian J. of Industrial Mathematics* **12(3)** 15–27

[19] Rusanov V A, Banshchikov A V, Daneev A V, Vetrov A A and Voronov V A 2017 A posteriori simulation of the dynamic model of the elastic element of satellite-gyrostat *Far East J. of Mathematical Sci.* **101(9)** 2079–94

[20] Statnikov R B and Matusov I B 2012 On the solution of problems of multi-criteria identification and development of prototypes *Problems of mechanical engineering and reliability of machines* **5** 20–9

IOP Conf. Series: Materials Science and Engineering **760** (2020) 012009 doi:10.1088/1757-899X/760/1/012009

Application of geodetic devices for the operation of underground pipelines

G G Bikbulatova, S V Lyashko

Department of geodesy and remote sensing, Omsk state UNIVERSITY, 4 Sibakovskaya St., Omsk, 644008, Russia

E-mail: gg.bikbulatova@omgau.org, sv.lyashko@omgau.org

Abstract: The article deals with geodetic devices used for the operation of underground pipelines under the conditions of the northern territories of Russia. The electronic total station Trimble M3, the GPS-device Javad GPS Maxor and the pipeline locators Leica Digicat 650i and Ridgid SeekTech SR-20 are taken into consideration, and their characteristics and accuracy are analyzed.

1. Introduction

Geodetic devices are used at all stages of geodetic support in design planning, construction, reconstruction, and operation of engineering structures. Since such works are carried out everywhere, new territories are currently under development, and new utility lines are being installed, issues regarding the application of different geodetic devices are of crucial and practical importance. By using the devices under new conditions in underdeveloped territories, we get the opportunity to gather new data on the use of different technologies and tools. Underground utilities (networks) include linear structures located below the surface of the earth. Depending on the type of liquids and gases, pipelines are divided into heating pipelines, gas pipelines, water supply, water discharge and drainage systems, and other linear structures [1].

For the positioning of underground networks, the underground space of a human settlement along its streets is used. The density of engineering networks is determined by the size of the territory of the city (settlement), the height of the buildings, and the level of development of the municipal services in the city (settlement).

2. Research

At present time, the most common surveying method in case of linear surveys for the operation of underground utilities is the tacheometric method using electronic total stations.

During the surveying process, the electronic total station is installed at surveying stations, while range rods with reflectors are installed at alignment stations. When pointing at the reflectors, horizontal angles and vertical angles are automatically determined, as well as distances to adjacent points. The results of these measurements are stored in the device memory, thereafter the information is sent to a computer with a special program software for further processing.

There is a wide range of electronic devices that provide high accuracy, for example, the devices Leica, Sokkia.

For the present research, the geodetic support process for the operation of a section of the main gas pipeline in the Beryozovsky District of the Khanty-Mansi Autonomous Okrug was taken into consideration.

The Beryozovsky District is characterized by a variety of landscapes, different types of soil and vegetation. The moistening degree is quite high. The climate of this territory is harsh continental with cold and long winters, and hot but short summers. The average annual air temperature is -3.3 °C, the average air temperature in January is -22.7 °C, while in the hottest month, July, it is +16.1 °C. The absolute minimum temperature is registered in December with -55 °C, and the absolute maximum temperature reaches +35 °C in June. Such temperature drops influence the choice of geodetic instruments for work. The territory of the district is located in the zone of seasonal permafrost, which must be taken into account when conducting various types of engineering work and during the subsequent operation of utilities, especially as regards underground ones.

Precipitation in the district is consistent, especially during the warm period from April to October with 376 mm, while in the cold period from November to March it reaches 127 mm, whereas the annual precipitation is equal to 503 mm [2]. Westward winds are prevalent during the year. In January winds are mainly southeastward, while in July – northward and eastward. In the warm part of the year, northeastward winds prevail, in the cold period – southeastward. The average annual wind speed is 2.4 m/s, the average in January amounts to 2.1 m/s, while in July to 2.4 m/s.

For the geodetic support of works concerning the operation of the utility at study, the electronic total station Trimble M3 was used.

The Trimble M3 device is reliable, quite simple to manage, and has a small weight. Useful options such as the presence of a pointer to the target, an alignment indicator, as well as a display in case of “angle to the right”, should be noticed. Despite the presence of more advanced models, the device provided the required accuracy during the work (Table 1). To transfer data to other devices, the total station possesses RS-232C and USB ports, and a Bluetooth connection option [3].

This device has an integrated Trimble Access field program software. The program is relatively simple, and at the same time has significant data processing capabilities. The program allows the visualization of surveying results and preliminary calculations. Using the program, DXF file can be uploaded in the form of an active map, staking can be made, and the results of executive surveying can be compared with the design drawing. In addition, there is a large set of functions for geodetic calculations.

Table 1. Main characteristics of the device Trimble M3.

Parameter	Value
Linear measurement accuracy with reflectors	±2 mm+2 mm/km
Linear measurement accuracy without reflectors	±3 mm+2 mm/km
Measuring range with reflectors	up to 3000 m
Angle accuracy	2”

Alongside classical methods of large-scale topographic surveys, satellite technology is used.

The essence of a topographic survey using satellite receivers consists in the determination of the planimetric coordinates and heights of terrain points in real time with an error of 2-3 cm. In this case, at least two receivers are used: a basic receiver (at a point with known coordinates) and a movable one (at the point to be determined) [4].

The main advantage of GPS-surveying is that during field measurements the surveyor, controlling the GPS receiver, is personally present at the points to be determined and examines the characteristics of the working area as well as the object. The preliminary processing of measurement data is usually performed directly on the field, then the results are transferred to a computer with a special software for the rendering of topographic plans. During field rendering, the performer can identify errors and correct them, identify areas with insufficient data and survey them in more detail.

In the present case, the device used for GPS-surveying was the GPS-receiver Maxor by the Javad Navigation Systems Maxor company. The Javad GPS Maxor device is a geodetic receiver of high-accuracy class, and it stands out for its reliability and the possibility of high-speed measurements.

The device allows to process GPS and GLONASS signals. The GPS Maxor receiver, depending on the installed chip, can operate with a frequency of GPS L1 and GPS L1+L2 [5]. At the same time, the accuracy in the determination of the baseline length in static mode and fast static mode is 0.3 cm + 1.0 ppm in plan view and 0.5 cm + 1.5 ppm in height. In kinematic mode and on-line mode, the accuracy of distance determination is 1.0 cm + 1 ppm in plan view and 1.5 cm + 1 ppm in height. In offline mode, it is possible to determine a location with an accuracy of two meters [5].

The components of the device are protected by a single durable and compact aluminium alloy casing, which also contains a Li-Ion high-capacity battery (up to 25 hours of continuous operation) and an integrated charger.

One advantage of this device is the possibility of its application under the cold temperatures of the northern territories of Russia.

Of particular note is a very important technology developed for GNSS by Javad Maxor, that is the RAIM technology (Receiver Autonomous Integrity Monitoring), designed to assess and maintain the integrity of the GPS signal reception. The technology is unique in that it eliminates the reception of an excessive number of coordinates and the false determination of the position of an object, which occurs in cases when more satellites than necessary are fixed. This is especially true when simultaneously receiving signals from different systems [5].

For an effective work in the reconstruction, operation, and installation of new pipelines, as well as the production of geodetic works related to the maintenance of these kind of utilities, special devices for the research of underground pipelines have been developed.

To determine the direction of the lines of underground utilities, electronic search devices such as pipeline locators and pipe-and-wire detectors are used, and where these devices cannot be used, trial pitting is carried out [6].

The mechanism and operation principle of pipeline locators is based on the determination of the level of electromagnetic induction, namely, the locator consists of a generator and a receiver which determine the presence of an electromagnetic field. The generator is necessary for tracing the so-called passive lines through which electric current does not flow, that is usually a different kind of pipe, for the search for which electric current of small force is supplied through a special connection [6].

While surveying pipelines, an acoustic method is used to diagnose the state of underground pipelines, which is based on the search for noise from a liquid or gas moving along an engineering network [7].

State-of-art pipeline locators are resistant to adverse weather conditions, these devices are usually compact and easy to transport, equipped with powerful batteries with a large capacity for long-term operation under difficult conditions without charging. Therefore, pipeline locators are widely used in the search and diagnosis of various defects of underground structures [6]. The range of different types of pipeline locators is quite wide [8].

Devices by the Leica company can be included among the most well-known and frequently used geodetic devices, which can guarantee high precision in engineering and geodetic surveys. The pipeline locator Leica Digicat 650i guarantees the necessary precision, being a universal detector that can be used both in passive and active scanning modes. The device allows for the determination of the location and depth of underground electrical conduit networks according to their proper or induced electromagnetic emission.

The device Leica Digicat 650i can be used for the preliminary survey of land plots, which increases safety in the execution of ground works with machine-based or manual excavation. This is especially important taking into account the damp and frozen soil of the northern regions, where the territory of the Khanty-Mansi Autonomous Okrug – Yugra is located. The reliability and efficiency of ground works is increased thanks to the preservation of the integrity of the pipelines. In addition, the pipeline locator Leica Digicat 650i simplifies the research, identification and tracking of linear facilities, in particular if maintenance or technical support of pipelines and traces are needed. In order to trace non-metallic pipelines and to search for damages or obstructions, optional gadgets included in the Leica Digisystem surveying set are used, allowing for a diminishment in the ground work volume.

The pipeline locator Digicat 650i can work on different frequencies and register alternating magnetic fields (Table 2).

Table 2. Main characteristics of the pipeline locator Leica Digicat 650i.

Passive Modes	Signal fixation	Depth
Power Mode	50-60 Hz	Up to 3 m
Radio Mode	15-60 kHz	Up to 2 m
Auto Mode	Combined scanning	2-3 m

When introducing the sonde in non-metallic pipelines, the pipeline locator Leica Digicat 650i in the active Generator Mode can detect signals with a frequency of 8, 33 or 8+33 kHz.

The obtained results are displayed on an LCD with automatic backlighting; in addition, the detector Leica Digicat 650i includes the function of audio notification through headphones or speaker drivers.

Important characteristics of the device Leica Digicat 650i are the presence of a 32-Mb integrated memory and the possibility of wireless interface connection Bluetooth. This features offer the possibility to save data obtained during surveying up to 999 records over 80 hours of work, to control the conditions of the pipeline locator and to determine the depth of pipeline networks from remote. For this purpose tablets, netbooks and other computer systems with the software Logicat installed on can be used. Also data are compatible with the CSV format, making it easier to export and elaborate them using other softwares.

Using GPS-surveying devices allows for the determination of the coordinates of detected networks in real time while showing the depth at which they are running [9].

At our surveying site, a pipeline locator Ridgid SeekTech SR-20 was used to identify defects in underground utilities. The device operates in three different modes: active search, passive search, and sensing option [9].

In the passive search mode, the device receives the signals of all the pipelines installed on the site. After switching to the broadband mode, the device will detect all the objects located close to each other and recognize each of their types. All data on the objects, their condition, type, defects, direction, location, and distance (depth) are displayed on the device screen.

The body of the Ridgid SeekTech SR-20 device is made of high-strength plastic, and a special case is available for transportation. The temperature range for operation stretches from -20 °C to +50 °C. The device is one of the best instruments for the search for underground pipelines at a depth of up to 10 meters [9]. At the same time, if compared with other devices, SeekTech provides the maximum accuracy and reliability of the results, as well as the highest tracing speed. This pipeline locator is the only device that uses multi-directional antennas, directional arrows, and a display that allows to quickly and easily read the obtained information [9].

At present time, the most recent device of this type in the equipment array by SeekTech is the pipeline locator Ridgid Sr-24. It is widely used for accurate detection of underground engineering networks, cables and metal pipes, as well as other types of pipelines. The antennas of the SR-24 point in different directions, thereby, receive the signal field at its full and simplify signal acquisition. The receiver displays the location and direction of engineering networks on the screen. Acoustic and visual alerting for the operator in the event of a disturbed signal field form or other errors is present.

For other types of identification of underground utilities pipe-and-wire detectors are used. In those cases and places where the determination of underground utilities with the help of pipe-and-wire detectors is not possible, the trial pitting method is applied [6]. Trial pitting can be also carried out in order to control the data obtained by electrical methods, to clarify and supplement the available accounting materials, and to check their quality. The trial pitting method is labor consuming and expensive, therefore it is applied only in extreme cases when it is impossible to recur to other methods.

3. Conclusions

Different devices for the geodetic support of the operation of linear underground utilities under the conditions of the northern districts of Russia were analyzed. These devices have proven a high level of reliability and efficiency in the execution of engineering and geodetic surveys, guarantee the necessary precision of works and are thus recommended for use on the territory of cold regions.

Acknowledgments

The study was carried out under the theme "Development of new technologies for geodetic monitoring of engineering structures and natural objects using geodesy and remote sensing methods" state registration number AAAA-A18-118070290063-7

Reference

[1] RSN 1988 *RSN 72-88 Inzhenernye izyskaniya dlya stroitel'stva. Tekhnicheskie trebovaniya k proizvodstvu s"emok podzemnykh (nazemnykh) kommunikatsii* [Republican Building Norms 72-88 Engineering surveys for construction. Technical requirements for the production of surveying of underground (aboveground) pipelines] (online) available at: http://docs.cntd.ru/document/1200001286

[2] Bikbulatova G G 2011 *Issledovanie raspredeleniya atmosfernogo uvlazhneniya na territorii Sibiri i Dal'nego Vostoka* [Research on atmospheric moistening distribution on the territory of Siberia and the Far East] *Omskiy nauchnyy vestnik* **1(104)** 232–235

[3] GEO-otziv (online) 2019 available at: https://www.geo-otziv.ru/products/1732

[4] GKINP 2002 *GKINP (ONTA)-02-262-02 Instruktsiya po razvitiyu s"emochnogo obosnovaniya i s"emke situatsii i rel'efa s primeneniem global'nykh navigatsionnykh sistem GLONASS i GPS* [Geographical and Cartographical Instructions, Norms and Regulations (Basic Normative and Technical Act)-02-262-02 Instruction for the development of surveying control and the surveying of planimetry and relief with the application of the global navigation systems GLONASS and GPS] (online) available at: http://docs.cntd.ru/document/1200030413

[5] GEO-otziv (online) 2019 available at: https://www.geo-otziv.ru/products/970

[6] MDS 2009 *Metodika MDS 11-21.2009* [The technique of MDS 11-21.2009] (online) available at: https://meganorm.ru/Data2/1/4293825/4293825651.htm

[7] Radiodetection (online) 2019 available at: http://www.radiodetection.ru/

[8] NDTcomplekt (online) 2019 available at: https://ndtcomplekt.ru/category_71.html

[9] Trassoiskatel' Leica Digicat 650i [Pipeline locator Leica Digicat 650i] (online) 2019 available at: https://www.rusgeocom.ru/products/trassoiskatel-leica-digicat-650i

[10] Trassoiskatel' Ridgid-SeekTech [Pipeline locator Ridgid-SeekTech] (online) 2019 available at: https://www.soges.ru/trassoiskatel-ridgid-seektech-sr-20

Analysis of failures of bearings of axle box unit with polyamide cages and prospects of increasing their service life

D V Butorin[1], **N G Filippenko**[1], **A V Livshits**[1], **S I Popov**[1]

[1] Irkutsk State Transport University, 15, Chernyshevsky Str., Irkutsk, 664074, Russia

E-mail: den_butorin@mail.ru

Abstract. The article analyzes failures of bearings of the axle box unit with polyamide cages and investigates the influence of climatic conditions on operational characteristics of cages. It is determined that the greatest number of defects during the operation of polyamide cages occurs in the following periods: summer-autumn, autumn, winter, winter-spring, spring. The revealed dependence of the operational properties of polyamide samples on weather and climatic conditions in different seasons of the year allows predicting the change in the operational characteristics of the cage and taking measures to restore and improve them. The authors proposed a method of restoration and improvement of operational properties of cages by processing them using a technique of high-frequency electrothermy. Full-scale tests showed the high efficiency of the proposed method, expressed in reducing the weight of the samples as a result of moisture removal, as well as in an ambiguous change in strength properties. Their ambiguity lies in the fact that cages differently restore their original hardness indicators after processing. And in the periods corresponding to winter and winter-spring, it was possible to restore strength indicators to a level even above the original one. For high-frequency processing of polyamide cages, a device has been developed that provides the required coverage of the workpiece electrodes on the upper, lower and inner side surfaces of the partitions. The test of the developed device made is possible to achieve the desired results, expressed in a significant decrease in the relative humidity of the cages and an increase in their strength characteristics by 7.6% of the initial level. All this confirms the practical importance and progressiveness of the developments presented in the work.

1. Introduction

Ensuring the safety of train traffic, high passenger and cargo turnover at a minimum cost is an important task of the railway transport company. Its solution in the conditions of instability of the domestic economy is based both on the introduction of new technical means and technologies, and on the improvement of existing ones. In this sense, it is important to raise the requirements for the reliability of components and parts of rolling stock, in particular, the increase in the durability of wagons, the service life of which has a great impact on the performance of the rolling stock and, in general, on the economic efficiency of rail transport. Hence it appears that one of the primary plans in the strategy for the development of the railway industry is to equip the rolling stock with modern, new and durable technical means to ensure high reliability [1].

2. The statistics of accidents in railway transport and analysis of failures of bearings of the axle box unit with polyamide cages

Accident statistics in recent years reflects positive achievements in the established strategy. So, as of 2015, in comparison with the previous year, the number of accidents in railway transport increased by 41% [2]. And in the first half of 2016 the number of equipment failures in the infrastructure of JSC "Russian Railways" decreased by 20% [3]. In 2017, in order to increase transportation efficiency, the management of JSC "Russian Railways" set the goal for the Administration office of the rolling stock of the Central Directorate of Infrastructure to reduce equipment failures by at least 15% for JSC "Russian Railways". As a result, for 10 months of 2017, this target indicator was exceeded (-26%) [4]. All this is indicative of the choice of the right strategy.

One of the examples of modern achievements is the renewal of the fleet of Russian wagons and equipping them with axle boxes with cassette type bearings. Historically, axle boxes of wagons "survived" two generations of bearings, and now due to the transition to the third generation, significant results have already been achieved. So, as of 2012, according to the Central Directorate for the Repair of Freight Wagons in the Russian Railways network, the number of rejects due to malfunctions in axle boxes was an absolute majority (95.2%) in the total number of detected rejects (Figure 1) [5].

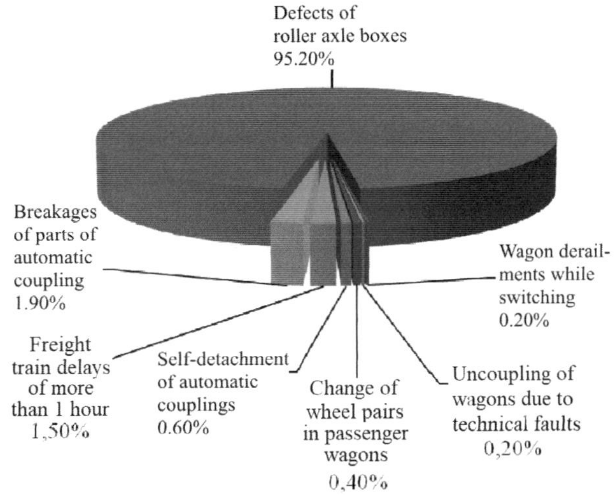

Figure 1. Number of cases of rejects in JSC "Russian Railways" as of 2012 and its types.

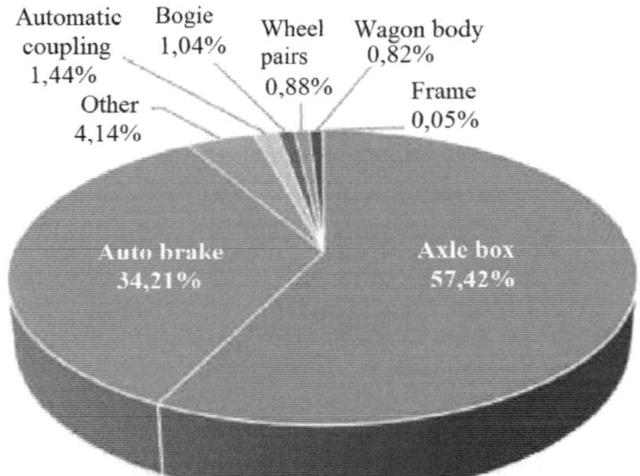

Figure 2. The number of equipment failures in the wagon system according to CAS ARE for 2015-2016.

For 2015 and 2016, according to the comprehensive automated system of accounting, control and analysis of the reliability of equipment failures (CAS ARE), the number of failures of the axle boxes was 57.42% of the total number of equipment failures in the wagon system (Figure 2). But along with the positive dynamics, the largest percentage of defects in the axle box can be attributed to its bearings (Figure 3).

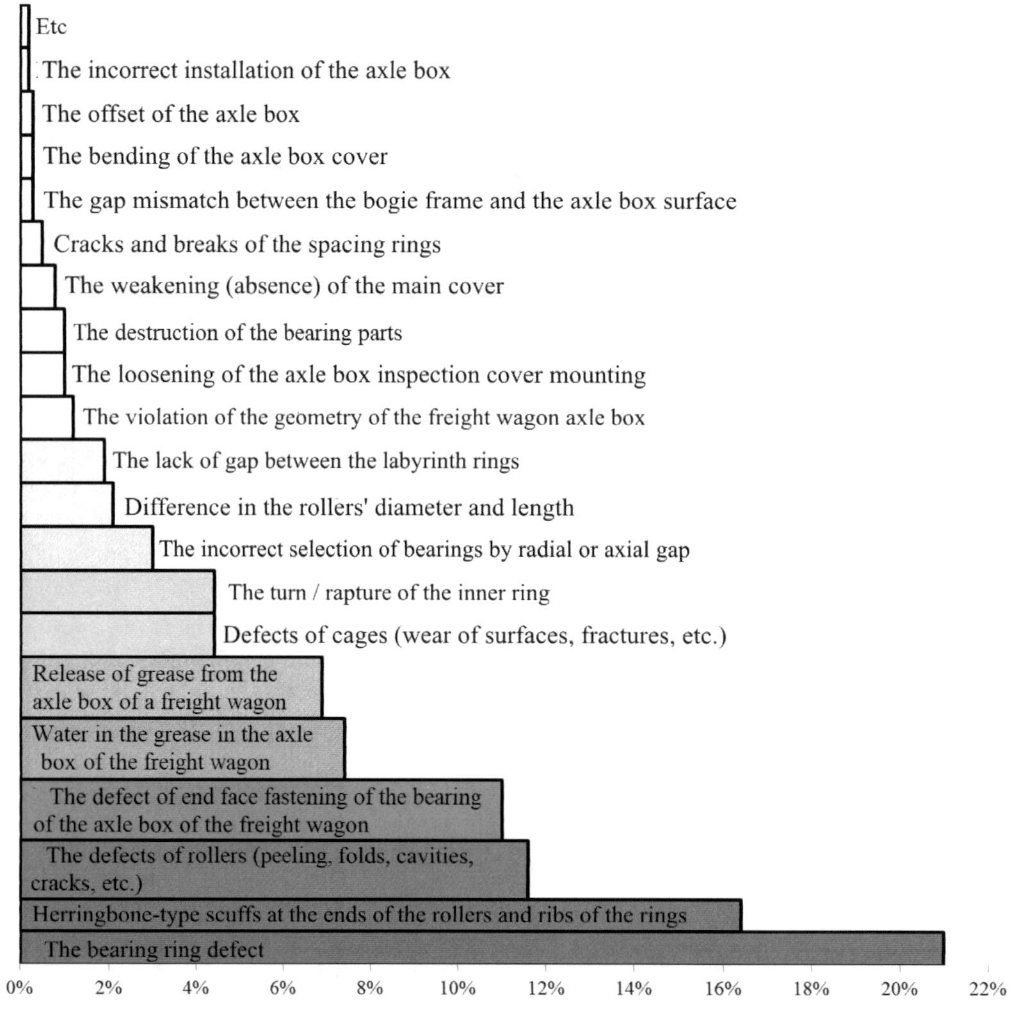

Figure 3. The failure distribution of axle unit for 2015-2016.

Today JSC "RZD" has introduced axle boxes with cassette-type bearings of the following companies: Brenco, Timken and SKF [6-8]. Manufacturers guarantee trouble-free operation of the cassette bearing for eight years of operation or 800 thousand kilometers, as well as provide service within the warranty period and thus exclude the intervention of repair units of JSC "Russian Railways" in their service. This results in an increase in the cost of introducing and maintaining new axle boxes. However, as noted by Sergey Goncharov, the chief engineer of JSC "PGK", "the overall economic effect of the introduction of new axle boxes can be estimated only in the long term" [9]. But today, to reduce the maintenance costs of wagons and reduce downtime of unprocurable rolling stock, work is underway to create service centers that work with cassette bearings. The plans also include the licensing of railway enterprises for a complete overhaul of axle boxes with cassette-type bearings, which includes assembly or disassembly, fault detection, repair or replacement of defective parts.

According to the diagram of the distribution of failures of the axle boxes (Figure 3), the majority of faults occur in bearings, which leads to a failure of the axle box with the ensuing consequences. The most important functional component of the bearing is the cage. Since 1995, Russia has been producing roller bearings with a polyamide cage (Figure 4a). In cassette-type bearings, to keep the tapered rollers at equal distances from each other and maintain their uniform rolling around the entire bearing circumference, a polyamide cage is used as well, but with an improved design (figure 4b) [10].

a

b

Figure 4. The axle box bearings with polyamide cages: a – with cylindrical rollers; b – with tapered rollers.

The operating experience of bearings with polyamide cages in railway transport in the construction of running gears, both wagons and locomotives, has been fully justified for the past two decades and confirmed the prospects of using polymers instead of hard-to-find non-ferrous materials (brass cages). However, along with the advantages, the operating experience of such cages revealed a number of shortcomings that result in a reduction in the period of overhaul operation of the axle boxes. In work [5] it is mentioned that the most common defects of the polyamide cage are cracks (67% of the total number of reasons for rejection of cages, as of 2012), and the level of rejection of cages by "cracks" is not reduced, and in some cases there is an increase in their number. The reasons for the formation of such a defect include:

– the violation of the cage manufacturing technology, which entails the formation of casting defects;
– exceeding operating loads;
– the influence of climatic conditions on the operation of the rolling stock, in particular, on the operation of the axle box, including both water and grease contamination, and changes in the structure and properties of the polyamide cage.

3. Materials and methods

The absorption of moisture by the cage during operation should not exceed the permissible rate (0.4% of the mass fraction), but this is possible when operating the axle box only at temperatures close to 40 °C. For a detailed study of the influence of climatic conditions on the performance properties of the cage, samples were made (width is 4 ± 0,2 mm; height is 6 ± 0,2 mm; length is 50 ± 2 mm) in accordance with GOST 26277-84. Here it should be mentioned that during their manufacture in the batch of polyamide cages, a factory defect was found (Figure 5) formed by casting. It should also be noted that the manufacturers do not even control the occurrence of such defects, and products with this type of rejects are supplied to the operational and repair enterprises of JSC "Russian Railways".

The manufactured samples were kept in simulated weather and climatic conditions corresponding to the operational seasons of the year, as well as weighed on the analytical balance of the model HTR–

120CE (accuracy is up to 0.0001 g) before and after exposure. Moreover, the duration of each experiment was chosen in accordance with GOST 12423-66 and was at least 24 hours. The study itself was conducted in winter, and Figure 6a presents a graphical display of its results.

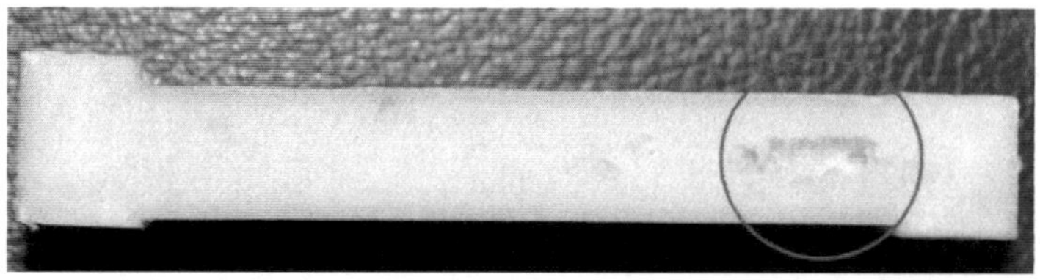

Figure 5. A factory casting defect in the polyamide cage of the axle box bearing

The next stage of this study was to determine the effect of different weather conditions on the strength characteristics of the samples from the cage. To this end, the samples were exposed to such simulated weather and climatic conditions that have the greatest impact on wear, cracking and damage. In accordance with the work [5], seasonal periods of operation of polyamide cages with the greatest number of rejects are: summer-autumn, autumn, winter, winter-spring, spring. After having been kept in simulated conditions, the samples were subjected to hardness and impact strength tests. Hardness measurement of each sample was made at least ten times at different points of its surface using the device TEMP-3. A graphical display of the average hardness values for the specified periods is shown in Figure 6b. Impact strength tests were performed by use of a pendulum impact testing machine 2083 KM-0.4 with an axe 2J by the Charpy method in accordance with GOST 4647-80.

The conducted researches make it possible to reveal dependence of physical and mechanical properties of samples from polyamide on weather and climatic conditions in different seasons of the year. The resulting picture (Figure 6) allows one to predict changes in the operating characteristics of the cage and take measures to restore and improve its operational properties..

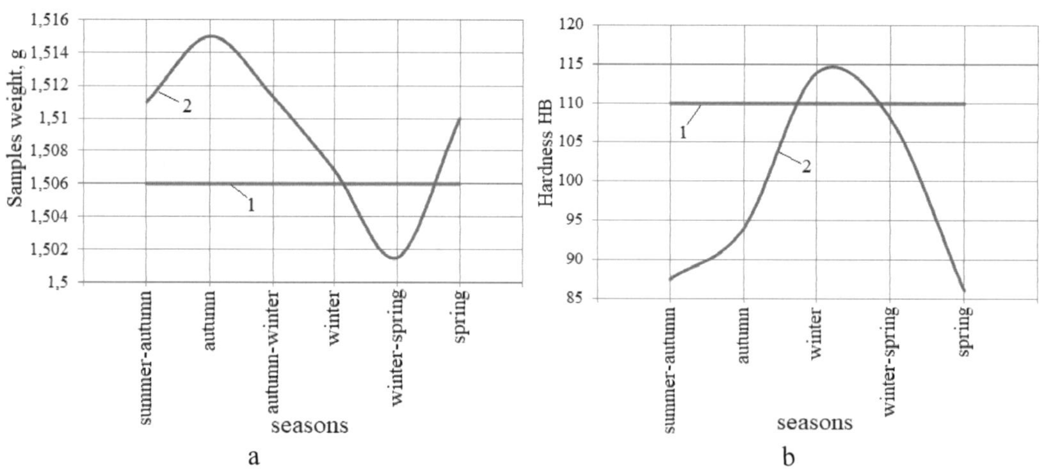

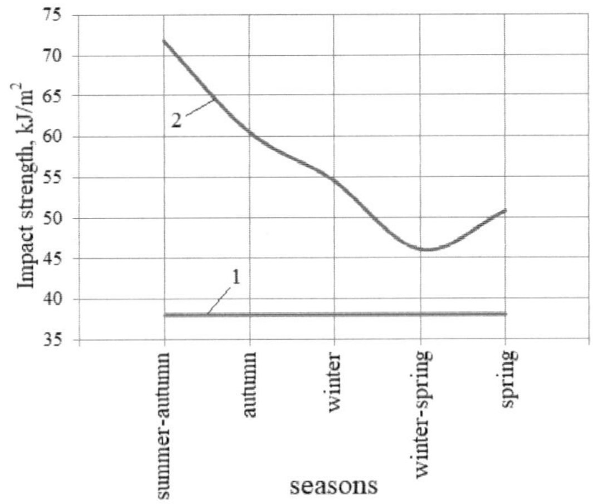

c

Figure 6. Dependence of the influence of weather and climatic conditions by seasons on: a – changes in the weight of polyamide samples; b – changes in the hardness of polyamide samples; c – changes in the hardness of polyamide samples; 1 – before the modeling of weather and climatic conditions; 2 – after the modeling of weather and climatic conditions

4. Method of recovery and increase of operational properties of cages

Within the framework of these studies, it was determined that it is possible to restore and improve the performance properties of a polyamide cage by treating it in the field of high-frequency currents (HFC), in other words, high-frequency (HF) electrothermy [5, 12-14, 20]. This method of processing has proved to be the most progressive, resource-and energy-saving method [11, 13, 15-20]. Its essence consists in placing a polyamide product (and any other polar thermoplastic polymer) between the capacitor plates. Then, high-voltage HF energy is supplied to the capacitor. The alternating electric field formed between the capacitor plates causes dipole polarization in the sample and the emergence of bias current, which leads to the heating of the polymer. It follows therefrom that the source of heat in this processing method is the polymer itself. At the same time, the heat is released evenly and with high intensity. Due to this, the process is practically non-inertial. The process flow diagram of high-frequency processing is shown in Figure 7.

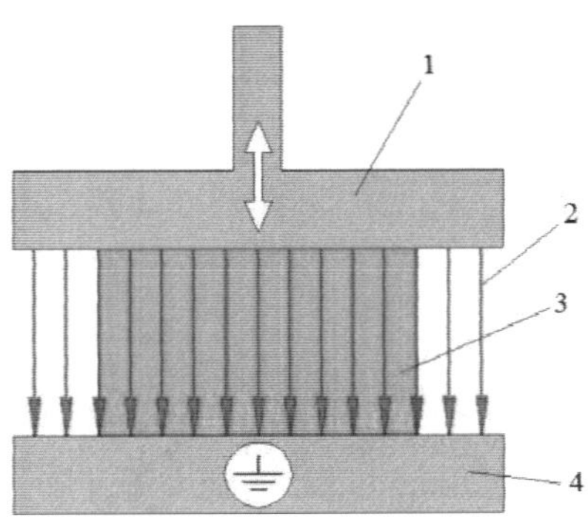

Figure 7. The process flow diagram of HF processing: 1 – the mobile high-potential plate of the working capacitor; 2 – the electromagnetic radiation flux; 3 – the heated polymer; 4 – the low-potential plate of the working capacitor.

To confirm the possibilities of restoring and improving the operational properties of the cage, an additional series of experiments with samples of the same dimensions was carried out (4 ± 0.2×6 ± 0.2×50 ± 2 mm). The experiment included the drying of polyamide samples with the high-frequency equipment of the UZP 2500 model [21] after modeling the above-described weather conditions. This allowed to achieve the results presented in Figure 8.

Analysis of the graphs (Figure 8) showed high efficiency of the HF drying process, expressed in reducing the weight of the samples as a result of moisture removal. As for the strength properties, it should be noted that the samples from the cage sufficiently restore their initial hardness values after HF treatment not in all seasonal periods. The hardness of samples used in periods corresponding to the winter and winter-spring was restored with a quality even higher than the initial one.

A device was developed for the high-frequency processing of polyamide cages (Figure 9). It consists of the upper high-potential and lower low-potential plates on which the upper and lower contact groups are installed. The complexity of the device is determined by the structural features of the cage, consisting of partitions, the upper rim and the lower rim. The resulting device provides the required coverage of the work piece by electrodes on the upper, lower and inner side surfaces of the partitions. The electrodes are simultaneously connected to a high-frequency generator. This design of the device under HF exposure forms a potentials difference between adjacent electrodes, between the upper high-potential plate with electrodes connected through the lower contact group to the lower grounded plate, and the lower grounded plate with electrodes connected through the upper contact group to the upper high-potential plate.

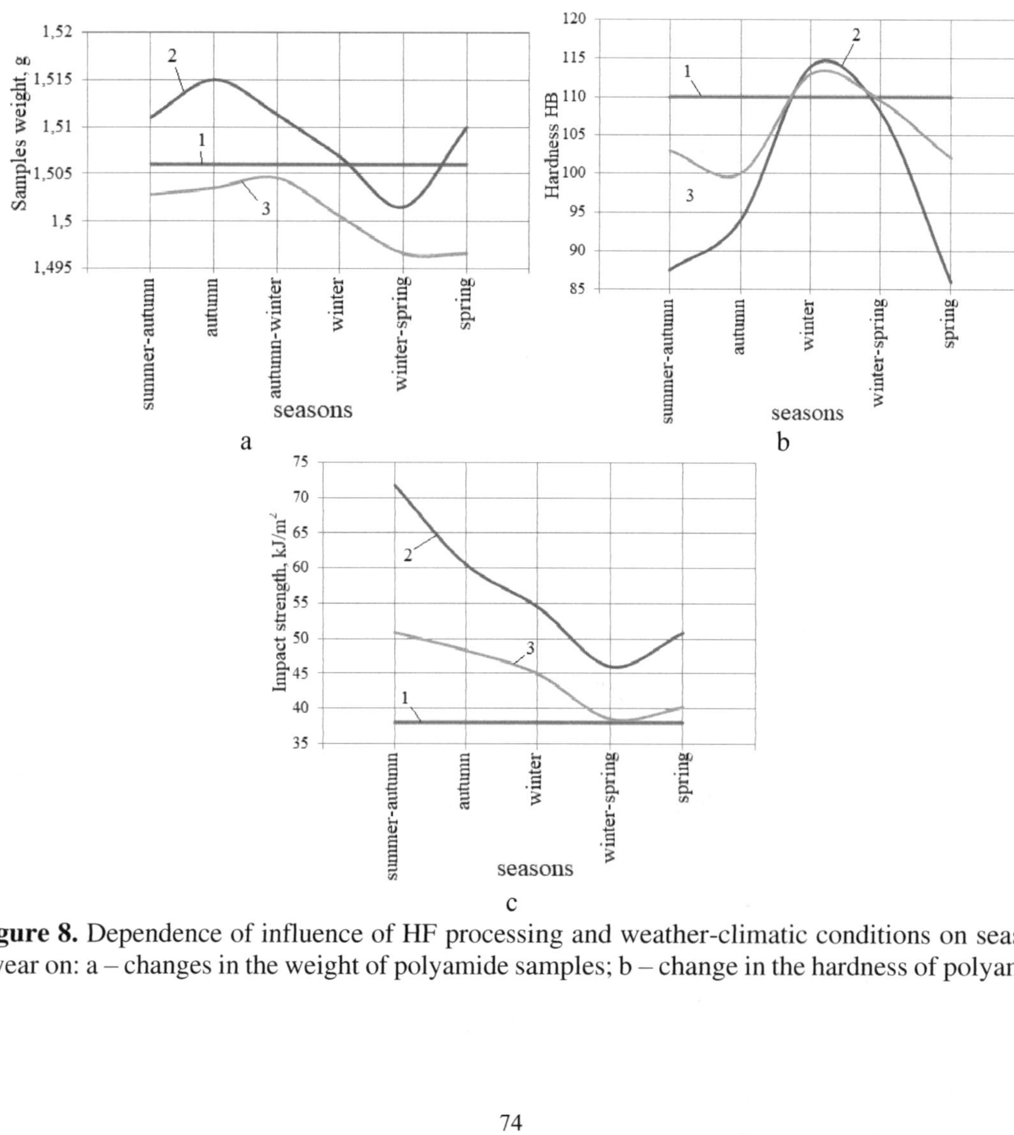

Figure 8. Dependence of influence of HF processing and weather-climatic conditions on seasons of year on: a – changes in the weight of polyamide samples; b – change in the hardness of polyamide

samples; c – change in the impact viscosity of polyamide samples; 1 – before the modeling of weather-climatic conditions; 2 – after the modeling of weather-climatic conditions; 3 – after HF drying.

Using the developed device on high-frequency equipment of the model UZP 2500, polymer cages made of Armamid PA PA 30-1ETM were dried. Table 1 presents the results of high-frequency drying.

Table 1. The results of drying cages on the UZP 2500 installation

Drying time, sec.	Relative humidity,%	Breaking strength, MPa
0	5,6	136,7
60	2,05	137,2
80	1,6	139,3
100	1,15	141,4
120	0,7	144,3

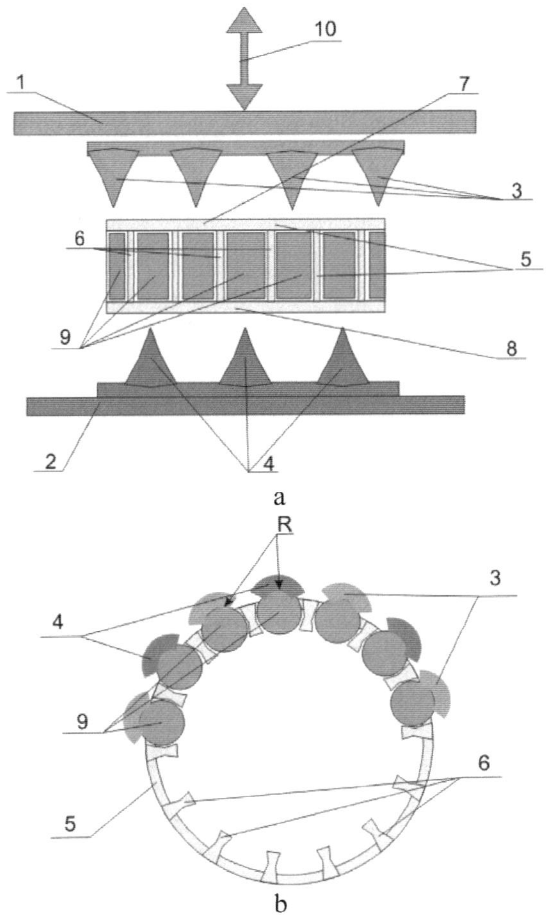

Figure 9. The diagrams of a device for high frequency treatment of parts: a – main view; b – top view; 1 – the top high-potential plate; 2 – the lower low-potential (grounded) plate; 3 – the upper contact block; 4 – the lower contact group; 5 – the cage; 6 – the plates of the cage; 7 – the upper rim of the cage; 8 – the lower rim of the cage; 9 – electrodes (rollers); 10 – the press; R – the radius of curvature.

IOP Conf. Series: Materials Science and Engineering **760** (2020) 012010 doi:10.1088/1757-899X/760/1/012010

From the above data it can be seen that the relative humidity at the initial rates of 5.6% after processing decreased to 0.7%, and the strength characteristics increased by 7.6%, which leads to the conclusion about the practical significance of this work.

5. Conclusion

The results obtained in the work fully confirm the prospects of using high-frequency electrothermy to restore and, moreover, to improve the strength performance properties of polyamide cages, as well as any parts of the rolling stock made of polar polymer materials (for example, polyurethane friction wedge plate in the bogie of a freight wagon). In addition, the high-frequency electrothermy in parallel with the drying also makes it possible to perform diagnosis of polymer parts for the waste casting and non-metallic inclusions [22].

References

[1] Yagodina E I, Suchkov R V 2018, Innovative solutions to improve the reliability of freight wagons Scientific community of students of the XXI century. Technical Sciences: collection of articles on the materials of LXV international student scientific-practical conference (5(64)) Dmitrieva N V (Novosibirsk: ASE Sibac), 355-360

[2] Actual problems in the field of repair of freight wagons, available at: http://www.rzd-partner.ru/zhd-transport/news/aktualnye-problemy-v-sfere-remonta-gruzovykh-vagonov-412278/

[3] In the first half of 2016, the number of failures of technical means on the infrastructure of JSC "Russian Railways" decreased by 20%, available at: http://www.rzd-partner.ru/zhd-transport/news/v-i-polugodii-2016-goda-kolichestvo-otkazov-tekhnicheskikh-sredstv-na-infrastrukture-oao--rzhd--sniz-414667/

[4] The number of failures of technical means is reduced, available at: http://www.gudok.ru/newspaper/?ID=1395012

[5] Popov S I 2013 *Automation of control of technological processes of restoration of operational properties of polymers* (Irkutsk: ISTU)

[6] LLC "EPK-Brenko Podshipnikovaya Kompaniya", available at: http://epk-brenco.ru/ru/

[7] OOO "SKF", available at: http://www.skf.com/ru/index.html

[8] The Timken Company, available at: https://www.timken.com/ru/

[9] Cassette bearings: an overview of the first operating results, available at: http://www.eav.ru/publ1.php?publid=2017-05a11

[10] Axle box with cassette bearings, available at: http://www.pomogala.ru/konsrukt/konstrukt_11.html

[11] Filippenko N G 2012 *Automation of process control of high-frequency processing of polymeric materials* (Irkutsk: ISTU)

[12] Livshits A V 2016 *Automated control of technological processes of high-frequency electrothermy of polymers* (Irkutsk: ISTU)

[13] Butorin D V, Filippenko N G, Livshits A V 2016 Complexed method of automated control high phase transformations in polymeric materials *Instruments and Systems: Monitoring, Control, and Diagnostics* **10** 10-18

[14] Livshits A V, Filippenko N G, Homenko A P, Kargapoltsev S K, Gozbenko V E, Dambaev Z G 2017 Mathematical modelling of the processes of the high-frequency heating of thermoplastics and quality improvement of welded polymeric items *JP JOURNAL OF HEAT AND MASS TRANSFER* **2(14)** 219-226

[15] Bader H 1996 High frequency drying of porous materials *Drying Technology* **7(14)** 1499-1523

[16] Hou A, Wang X, Wu L 2008 Effect of microwave irradiation on the physical properties and morphological structures of cotton cellulose *Carbonate Polymers* **74** 934-937

[17] Lyons P W, Hatcher J D, Sunderland J E 1972 Drying of porous medium with internal heat

generation *Int. J. Heat and Mass Transfer* **5(15)** 897-905
[18] Metaxas A C, Meredith R J 1983 Industrial Microwave Heating *Peter Peregrinus* 111-150
[19] Ohe T, Yoshimura Y 2009 Microwave Irradiation onto Nylon 6 Fibers in Organic Solvent *Sen'i Gakkaishi* **1(65)** 64-70
[20] Butorin D 2018 Automated control system to monitor dielectric losses in polymers *MATEC WEB OF CONFERENCES* (EDP Sciences) 02003
[21] Plant "Industrial electronics Gabrovo» 1987 *Passport UZP 2500A: installation for welding plastics* 60
[22] Larchenko A G 2014 *Automated control system for high-frequency diagnostics in the production and operation of polymer materials* (Irkutsk: ISTU)

IOP Conf. Series: Materials Science and Engineering **760** (2020) 012011 doi:10.1088/1757-899X/760/1/012011

Design of the tooling of a complex geometric shape for electrothermal processing of the polymer covering plate of the bogie friction wedge

D V Butorin[1], **A G Larchenko**[1], **A A Alexandrov**[1]

[1] Irkutsk State Transport University, 15, Chernyshevsky Str., Irkutsk, 664074, Russia

E-mail: den_butorin@mail.ru

Abstract. The article analyzes the operation of the bogie friction wedge. It considers the question of the necessity and possibility of enhancing the reliability and operational resource of wedge-type shock absorbers of railway bogies by increasing the operational properties of a polymer covering plate of a bogie friction wedge. A method for increasing the operational properties of the polymer covering plate, consisting in its processing by high-frequency currents, is proposed. To implement the high-frequency processing of the considered polymeric covering plate and achieve its uniform warming-up throughout the volume, a specialized tooling is developed and constructed according to the technique established earlier. The developed tooling made it possible to implement a uniform non-inertial self-regulating high-frequency heating throughout the entire volume of the polymer covering plate in laboratory conditions with high intensity. As a result of full-scale tests of the developed tooling and the proposed method of high-frequency processing, deviations of temperature values in different parts of the heated polymer plate became apparent. All of this shows that the heating is not completely uniform. The greatest deviation value was 7.9%. As a result of the work, it was possible to establish the reasons for the lack of full uniformity and to determine the direction of further research.

1. Introduction

In the design of a number of modern railway freight bogies (models 18-578, 18-194, 18-194-1) to absorb the vibrations of the wagon arising under the influence of irregularities of the railway track, friction wedges made of heat-strengthened high-grade cast iron VCh 120 with polyurethane covering plates are used (Figure 1) [1-4]. At the same time, the inter-repair run of freight bogie friction wedges, in accordance with the design documentation (CD), should be at least 500 thousand km, and for new-generation freight bogies it should be at least 800 thousand km [5-7]. But despite such guarantees, the friction wedge breaks down much sooner due to the wide range of reasons, including the numerous operational malfunctions of the covering plate (wear, delamination, scuffs, deformation, ruptures, side and corner chipping, cracks, production (factory) rejects) (Figure 2) [8, 9] installed on the inclined surface of the wedge. All this makes it possible to state the relevance of the work on the diagnosis of defects and improvement of the performance of polyurethane covering plates by any suitable means [10].

Along with this, in works [11, 13-17] there is a high progressiveness and efficiency of high-frequency (HF) electrothermy in the processing and diagnosis of polymer products in order to restore and improve their performance properties, which is very appropriate for the problem considered in this paper.

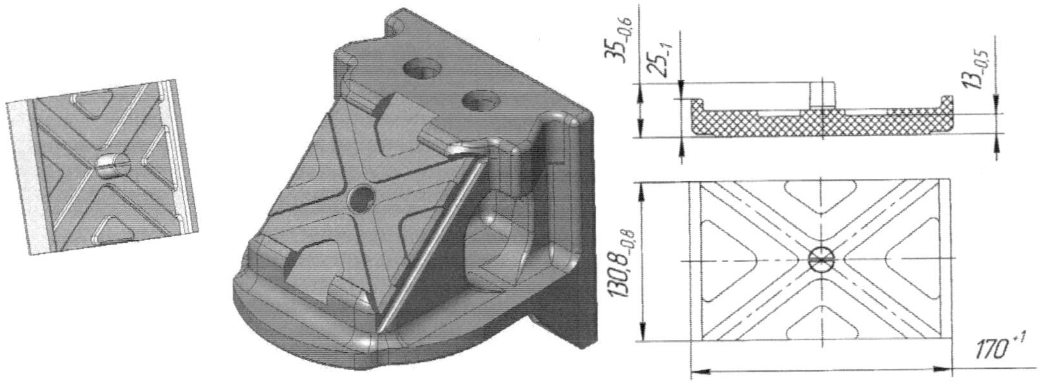

Figure 1. The friction wedge of the bogie model 18-194-1 with a polyurethane covering plate.

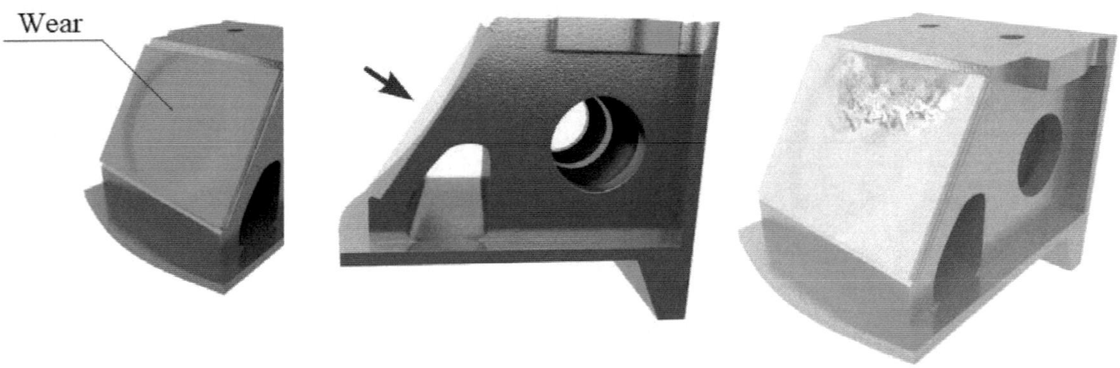

Figure 2. Some types of wear of the polyurethane covering plate by the example of the bogie model 18-194: a, b – wear and delamination; c – scuffs on the work surface

In high-frequency processing, the processed material is placed between the capacitor plates (high-potential and low-potential electrodes), to which a high-frequency current is supplied. When changing the charge on the plates, the polar dipoles of the material move in opposite directions, resulting in molecular friction with the release of heat. The energy of the electric field used to overcome these frictions will be converted into heat.

After analyzing the HF equipment and electrothermy process, as well as taking into account the geometric complexity (different thicknesses) of the polyurethane covering plate and the recommendations stated in [12], it was concluded that it is impossible to ensure the accuracy and efficiency of the HF processing without designing the appropriate tooling (complex profile electrodes) to achieve uniform heating of the covering plate throughout the volume. Based on the above, the purpose of this work was: to develop a tooling to implement the high-frequency processing of products of complex configuration by the example of the polymer covering plate of the freight wagon friction wedge in industrial conditions.

2. Tooling design procedure

Proceeeding from the basic equation of electrothermy (1), it follows that the power released in the unit volume of the polymer in the form of heat quadratically reacts to the change in the interelectrode distance d and the voltage on the plates of the working capacitor U.

$$P_{unit} = 5{,}53 \cdot 10^{-11} \cdot E^2 \cdot f \cdot \varepsilon' \cdot tg\delta , \tag{1}$$

$$E = \frac{U}{d}, \tag{2}$$

where P_{unit} is the specific power of dielectric losses, W/cm^3; E is electric field strength, V/m; f is field frequency, Hz; ε' is the relative permittivity of the treated polymer; tgδ is the dielectric loss tangent of the polymer.

And the voltage of a capacitor depends on its capacitance C (3):

$$U = \frac{q}{C} = \frac{qd}{\varepsilon_o \varepsilon' S}, \tag{3}$$

where q is capacitor charge, C; S is the area of the capacitor plates, m^2; ε_o = 8,854 · 10^{-12} F/m is dielectric constant.

Then, to solve the problem of the planned process of high-frequency processing, which consists in achieving a uniform heating of the polymer covering plate throughout the volume, it is necessary to ensure equal values of the capacity in each section of the part by introducing an air gap. To do this, it is required to make a conditional partition of the polyurethane covering plate into a number of individual elementary surfaces (Figure 3) with a set of equal in area, differentially small capacitors and to calculate the capacity of each capacitor obtained, which in turn will determine the value of the required air gap.

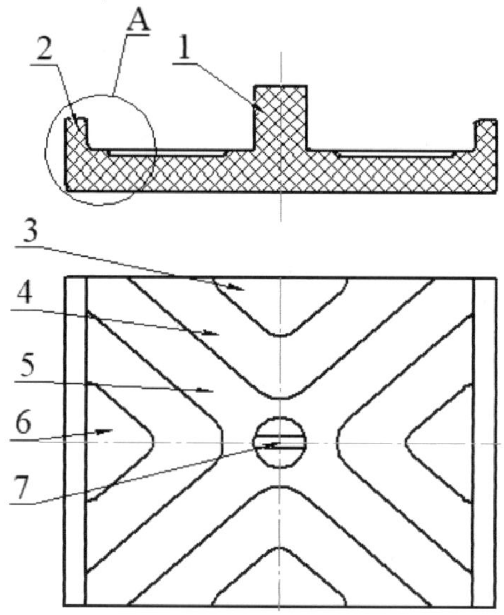

Figure 3. The partitioning scheme of the covering plate of the friction wedge of the bogie model 18-194-1 into elementary surfaces, where 1 is the cylindrical part; 2 is the lateral surface; 3, 5, 6 are lugs; 4 are depressions

First of all, we calculate the capacitance of the conditional surface capacitors with the largest thickness (Figure 3, position 1). The distance between the electrodes is 35 mm. Since all the elementary surfaces obtained by separation are plane-parallel, it is sufficient to divide each surface into the i-th number of capacitors with an area of 1 c.u.^{2}. Then the total capacitance of the capacitor is:

$$C = \sum_{i=1}^{n} \frac{\varepsilon_o \varepsilon' S_i}{d}, \tag{4}$$

where n is the number of differentially small capacitors.

Based on the presence in the cylindrical part of the covering plate of the slot, the thickness of which is less than the total thickness of this part, it is necessary to calculate the capacitance of the capacitors consisting of the covering plate capacity and the air gap capacity. In this case, in accordance with Figure 4, we have a serial connection of the capacitor capacitance, which is calculated by the formula (5):

$$\frac{1}{C_1} = \frac{1}{C_2} + \frac{1}{C_{air}}, \quad (5)$$

where C_{air} is the capacitance of the air capacitor (gap), which is equal to:

$$C_{air} = \frac{C_1 C_2}{C_2 - C_1} = \frac{S_1 S_2}{d_1 S_2 - d_2 S_1}, \quad (6)$$

then the value of the air gap is:

$$d_{air} = \frac{\varepsilon_o \varepsilon'_{air}(d_1 S_2 - d_2 S_1)}{S_1}, \quad (7)$$

where ε'_{air} is the relative permittivity of air.

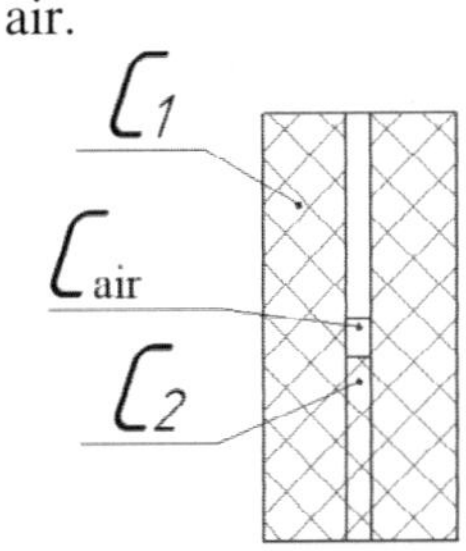

Figure 4. The scheme of the cylindrical part of the covering plate.

The next step is to calculate the capacitance of differentially small capacitors of the remaining surfaces in accordance with the scheme shown in Figure 5. The calculation of the capacitance of the other surfaces are produced similarly in accordance with equations (4)-(7). For the polyurethane covering plate of the friction wedge freight truck of model 18-194 (Figure 6) the tooling designing technique is similar.

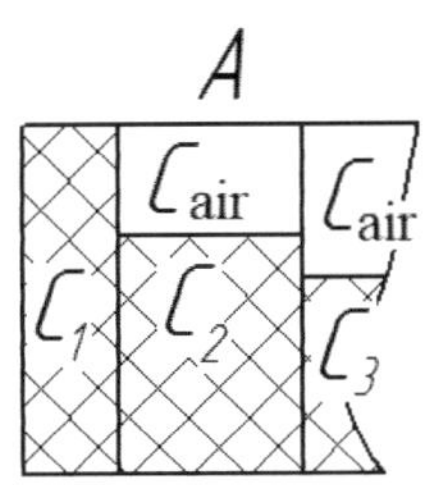

Figure 5. The scheme of condensers, surfaces 2-6.

In accordance with the described technique, the values of air gaps necessary to achieve uniform heating of the polymer covering plate throughout the volume during HF processing were determined, which made it possible to start designing the tooling.

3. Tooling design

In the design of tooling was analyzed recommendations for materials for the manufacture of electrodes. The most suitable material was aluminum alloy AD1. Since this type of alloy has a high electrical conductivity, it is not subject to corrosion and has good technological properties. Based on the configuration of the product in question, it is most advisable to process the covering plate in two cycles, breaking the surface of the lining into flat and cylindrical parts. The general view of the obtained technological system and the tooling developed within this work with division into processing cycles is presented in Figure 7.

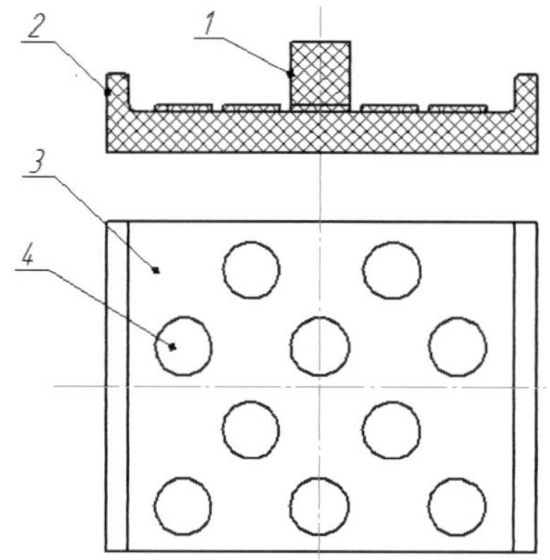

Figure 6. The scheme of partitioning the covering plate of the friction wedge of the bogie model 18-194 into elementary surfaces, where 1 is the cylindrical part; 2 is the side surface; 3 is the deepening; 4 is the ridge.

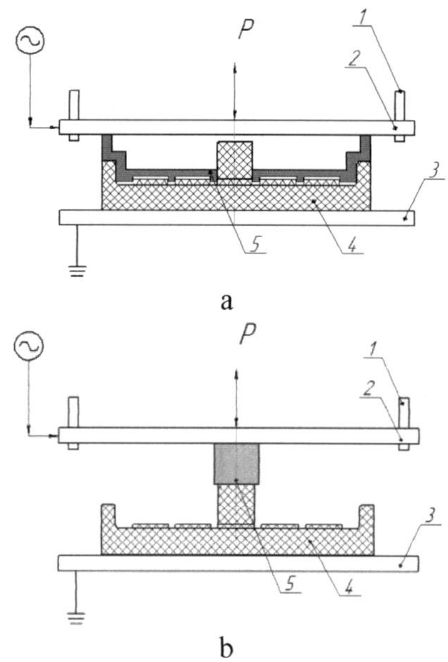

Figure 7. The general view of the technological system of high-frequency processing of the polymer covering plate of the friction wedge of the bogie model 18-194 when heating the flat part (a) and the cylindrical part (b): 1 is the clamp; 2 is the high-potential electrode; 3 is the low-potential electrode; 4 is the polymer covering plate; 5 is the tooling.

4. Experimental part

Approbation of the described technique and the developed tooling for electrothermal treatment of the studied polyurethane covering plate of the friction wedge was carried out on the industrial installation of model UZP 2500 (Figure 8) with an operating frequency of 27.12 MHz [16].

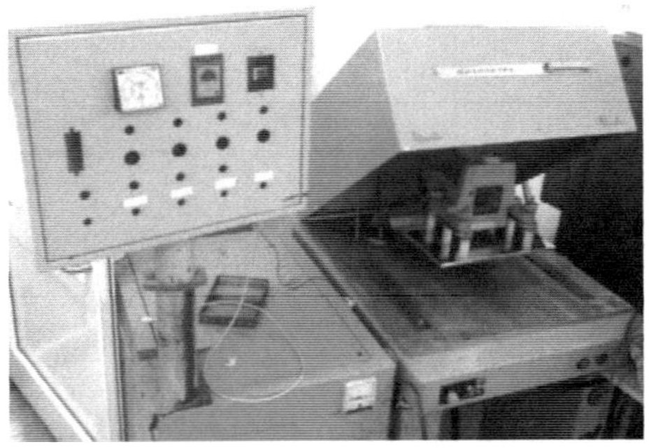

Figure 8. The installation of HF heating of model UZP 2500.

The investigated bogie friction wedge covering plate of model 18-194-1 and the tooling developed for it were installed by holdfasting the covering plate, while simultaneously applying pressure, with a low-potential electrode and of the tooling contact connected to a high-potential electrode of the working capacitor connected to a high-frequency generator (Figure 9).

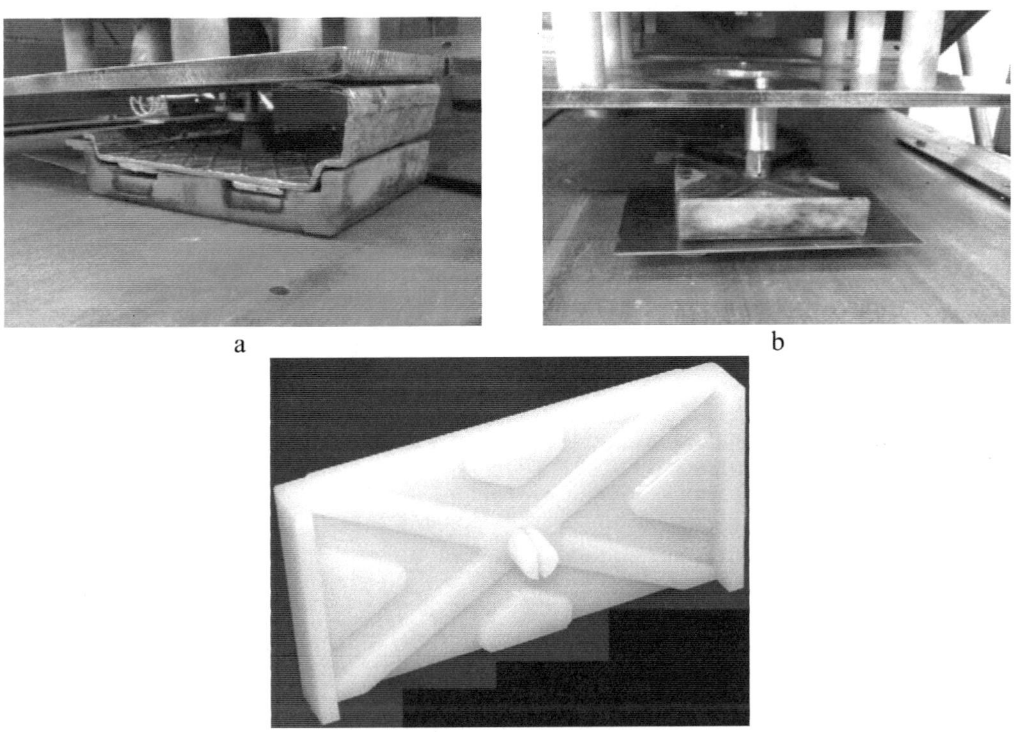

a b

c

Figure 9. The general view of the technological device for HF processing of the flat part (a) and the cylindrical part (b) of the polyurethane covering plate (c).

The processing cycle begins when the operator presses the "HEATING" button of the high-frequency generator located on the remote control. The moment of the end of the processing cycle is also recorded by the operator. For approbation of the works described above, the polyurethane covering plate was heated with a voltage of the tuning capacitor equal to $U = 4000$V for 70 seconds.

Then a series of measurements of the covering plate temperature in different areas was made. As it was impossible to control direct temperature during the high-frequency processing, both on the surface and inside the polymer product, temperature measurements were carried out contactless on the side surface of the covering plate (Figure 10) by the infrared pyrometer ADA TemPro 700 with a measurement accuracy of ±1,5 °C after 70 seconds of the HF heating.

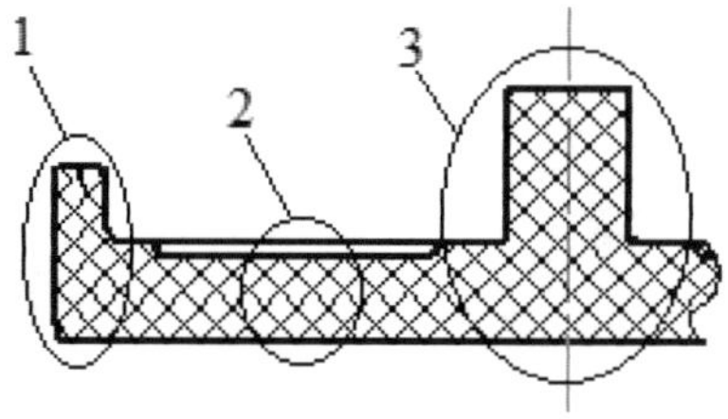

Figure 10. Measuring the temperature of the side surface of the covering plate in the areas of different thickness.

5. Results and Discussion

The following average heating values temperatures were obtained experimentally for each section: section 1 – 137.4 °C; section 2 – 132.8 °C; section 3 – 149.2 °C. The obtained values reflect some deviations of the temperature values, the largest value of which is 7.9% of the experimentally obtained temperature value of section 3. The obtained deviations of the experimental data are due to: firstly, the accuracy of the measuring device; secondly, not taking into account the boundary effects (skin effect) of the electric capacitor, since the classical approach to determining the electrical parameters of the capacitor involves neglecting the overall dimensions of the capacitor plates and considers only a plane-parallel capacitor with electrodes of infinite length [19-21]. Therefore, the purpose of further research will be the development of the above methods, taking into account the skin effects of an electric capacitor.

Nevertheless, the obtained results allow us to state the efficiency of the proposed technique and move on to the development of high-frequency technology for diagnosing and improving the strength properties of polymer products of different configurations by HF energy.

6. Conclusion

The proposed method of calculation and design of tooling to implement the high-frequency processing of polymer products of complex configuration makes it possible to evenly warm the processed products throughout the volume.

References

[1] Velikanov A V, Borshch B V, 2007, Friction wedges of high-strength cast iron, *Bulletin of VNIIZHT,* **2,** 19-22

[2] Zhevnov P M, Belousov V N, 2005, Improvement of bogies of freight cars on results of their tests and operation, *Problems and prospects of development of freight car building: materials of scientific and practical conference,* Kobishchanov V V, (Bryansk: BSTU), 128-131

[3] Velikanov A V, Borshch B V, Fedin V M, Borts A I, Yuryeva E I, 2007, Production technology, quality and efficiency of hardened friction wedges made of high-strength cast iron, *Bulletin of VNIIZHT,* **5,** 19-24

[4] Gabets A V, Levkin I V, Saitov M V, Semenov A V, 2015, Evaluation of the wear of the friction wedge damping of a cargo railcar bogie, *Polzunovskii Herald,* **1(4),** 16-18

[5] Boronenko Yu P, 2013, Strategic tasks of car builders in the development of heavy traffic, *Transport of the Russian Federation,* **5(48),** 3-6

[6] Boronenko Yu P, 2014, The road map of creation of innovative cars of new generation, *Rolling stock of the XXI century: Innovations in freight car building: materials of the International scientific and technical conference,* (St. Petersburg: PGUPS), 17-19

[7] Butorin S M, Efimov P V 2017 The friction wedges of freight bogies with a high between-repairs resource, *Proceedings of Petersburg transport University,* **4,** 605-614

[8] Instructions for repair and operation of freight cars equipped with a node friction vibration damper, manufactured by technology "A. Staki company", available at: https://vunivere.ru/work7617/page2

[9] Faults trucks freight cars, available at: http://www.вагонник.рф/2016/05/blog-post_17.html

[10] Chrysostomou A, Hashemi S, 1998, *Journal of Materials Science,* **33,** 11-65

[11] Butorin D, 2018, Automated control system to monitor dielectric losses in polymers *MATEC, WEB OF CONFERENCES,* (EDP Sciences) 02003

[12] Larchenko A G, 2014, *Automated control system for high-frequency diagnostics in the production and operation of polymer materials,* (Irkutsk: ISTU)

[13] Livshits A V, 2016, *Automated control of technological processes of high-frequency electrothermy of polymers,* (Irkutsk: ISTU)

[14] Filippenko N G, Butorin D V, Livshits A V, 2017, Determination of phase and relaxation transitions in polymer materials, *Automation. Modern technology,* **4(71),** 171-175

[15] Livshits A V, Kargapoltsev S K, Filippenko N G, Butorin D V, 2018, Energy-efficient control of high-frequency processing of polymeric materials, *Bulletin of Irkutsk state technical University,* **12(143),** 86-95

[16] Popov S I, 2013, *Automation of control of technological processes of restoration of operational properties of polymers,* (Irkutsk: ISTU)

[17] Butorin D V, Filippenko N G, Livshits A V, 2016, Complexed method of automated control high phase transformations in polymeric materials, *Instruments and Systems: Monitoring, Control, and Diagnostics,* **10,** 10-18.

[18] Plant "Industrial electronics Gabrovo", 1987, *Passport UZP 2500A: installation for welding plastics* 60

[19] Winders J J, 2002, *Power Transformers: Principles and Applications* p 286

[20] Khainovskii V I, Kozyrev E A, Kopylova O S, Nikitin P V, 2013, Determination of dielectric properties of seeds of agricultural crops taking into account the effects of the lateral electric field of the measuring capacitor, *Scientific journal of Kubsau,* **85(01)**

[21] Shelyuto V A, 1991, A simple formula for the capacitance of the ring capacitor, taking into account the boundary, effects *Journal of technical physics,* **61(2)**

IOP Conf. Series: Materials Science and Engineering **760** (2020) 012012 doi:10.1088/1757-899X/760/1/012012

Determining electromagnetic fields generated by overhead power transmission lines

V P Zakaryukin[1], A V Kryukov[1,2] and N V Buyakova[3]

[1]Transport Electric Power Engineering department, Irkutsk State Transport University, 15, Chernyshevsky Str., Irkutsk, 664074, Russia
[2]Power Supply and Electrical Equipment department, Irkutsk National Research Technical University, 83, Lermontov Str., Irkutsk, 664074, Russia
[3]Power Supply of the Industrial Enterprises department, Angarsk State Technical University, 60, Tchaikovsky Str., Angarsk, 665835, Russia

E-mails: zakar49@mail.ru, and_kryukov@mail.ru, bn_900@mail.ru

Abstract. Under actual operating conditions of electrical networks, it is hard to obtain experimental data that correspond to the maximum levels of the electromagnetic field strengths generated by overhead power transmission lines. Therefore, the analysis of the electromagnetic environment in electric power systems, both at the operated and newly created facilities, is recommended to be performed on the basis of mathematical modeling. The modeling methods and tools developed in IrGUPS and the Fazonord software package implemented on their basis make it possible, after determining the power system mode, to calculate the electromagnetic field strengths generated by multi-wire power transmission lines. In this case, the analyzed lines are considered in close connection with a complex electric power system. To check the adequacy of the simulation, measurements of electric and magnetic field strengths were performed, including comparative calculations for a 110 kV double-circuit line with a ground-wire located within the city. The comparison carried out in rather complex conditions is indicative of the correctness of the method of calculating a multi-wire line electromagnetic field developed in IrGUPS.

1. Introduction

Electromagnetic fields (EMF) generated by high-voltage or high-current transmission lines are a dangerous form of environmental pollution [1–4]. Interference resulting from exposure to EMF can cause disruptions in the operation of electrical and electronic systems, can cause inflammation of flammable substances, etc. [5–6]. Electromagnetic fields have a negative impact on humans [5]. Therefore, the task of determining the strengths of electromagnetic fields generated by power transmission lines is of particular relevance in modern conditions. Under actual operating conditions of electrical networks, it is difficult to obtain experimental data that correspond to the maximum levels of the electromagnetic field strengths generated by overhead power transmission lines. Therefore, the analysis of the electromagnetic environment in electric power supply systems, both at the operated and newly created facilities, is recommended to be performed on the basis of mathematical modeling.

The methods and tools for simulating sinusoidal modes [9, 10] developed in IrGUPS in phase coordinates make it possible to simultaneously calculate the electric power system mode to determine the EMF strength of multi-wire power lines [11–17]. In this case, the analyzed lines are considered in close connection with a complex electric energy system (EES).

When calculating the EMF generated by multi-wire lines, it is necessary to consider the following factors that influence the level of strengths:

- irregularities of the underlying surface;
- grounded metal objects (fencing of territories, buildings with grounded elements inside walls, ceilings and roofs, large metal objects);
- adjacent lines capable of creating significant electromagnetic fields.

The Fazonord software package (SP) makes it possible to use up to one thousand wires in models. This results in the possibility of modeling irregularities of the underlying surface, extended metal objects, adjacent power lines or electric traction networks of AC railways. The use of computational methods requires checking the adequacy of the means used and the correct consideration of the environment that can affect the electromagnetic field.

2. The simulation technique

After determining the EES mode according to the method described in [9], one can calculate the strengths of the electromagnetic field generated by any of the multi-wire power transmission lines belonging to the simulated system. The components of the electric field strength of a system of N wires at a point with coordinates (x, y) are calculated using the following formulas:

$$\dot{E}_Y = -\frac{1}{\pi\varepsilon_0}\sum_{i=1}^{N}\dot{\tau}_i \frac{y_i[(x-x_i)^2 - y^2 + y_i^2]}{[(x-x_i)^2 + (y+y_i)^2][(x-x_i)^2 + (y-y_i)^2]};$$

$$\dot{E}_X = \frac{2}{\pi\varepsilon_0}\sum_{i=1}^{N}\dot{\tau}_i \frac{(x-x_i)yy_i}{[(x-x_i)^2 + (y+y_i)^2][(x-x_i)^2 + (y-y_i)^2]},$$

where $\dot{\tau}_i$ is the wire charge i per unit length, determined from the first group of Maxwell formulas:

$$\dot{\mathbf{T}} = \mathbf{A}^{-1}\cdot\dot{\mathbf{U}}.$$

Here $\dot{\mathbf{U}} = [\dot{U}_1 \;\; \dots \;\; \dot{U}_N]^{\mathrm{T}}$ is the column vector of the voltage of the wires relative to the earth; $\dot{\mathbf{T}} = [\dot{\tau}_1 \;\; \dots \;\; \dot{\tau}_N]^{\mathrm{T}}$ is the column vector of wire charges per unit length; $\mathbf{A}$ is a symmetric matrix of potential coefficients, in which:

$$\alpha_{ii} = \frac{1}{2\pi\varepsilon_0}\ln\frac{2y_i}{r_i}, \quad \alpha_{ij} = \frac{1}{2\pi\varepsilon_0}\ln\frac{\sqrt{(x_i-x_j)^2 + (y_i+y_j)^2}}{\sqrt{(x_i-x_j)^2 + (y_i-y_j)^2}},$$

x_i, y_i are the coordinates of the location of the wire i of radius r_i above the ground (axis Y is directed vertically upward; $y = 0$ corresponds to the surface of the flat earth, axis X is perpendicular to the line, axis Z is directed against the chosen positive current direction in the wires of the line); ε_0 is the electric constant.

After the transition from the complex RMS values of the components $\dot{E}_X$ and $\dot{E}_Y$ to the time dependencies, one can obtain the parametric hodograph equations of the electric field strength vector:

$$E_x(t) = \sqrt{2}\,E_X\sin(\omega t + \varphi_X);\; E_y(t) = \sqrt{2}\,E_Y\sin(\omega t + \varphi_Y),$$

where the factor $\sqrt{2}$ is required since the strengths are calculated according to the RMS values, $\omega = 314$ rad/s.

Field strength reaches the maximum value E_{MAX} at times determined by the following equation:

$$t_{\max} = \frac{1}{2\omega}\mathrm{Arctg}\left(\frac{E_X^2\sin 2\varphi_X + E_Y^2\sin 2\varphi_Y}{E_X^2\cos 2\varphi_X + E_Y^2\cos 2\varphi_Y}\right).$$

The choice of one of the two values of the arctangent is made according to the condition of the negative value of the second derivative:

$$E_X{}^2 \cos 2(\omega\, t_{\max} + \varphi_X) + E_Y{}^2 \cos 2(\omega\, t_{\max} + \varphi_Y) < 0\,.$$

The RMS value of the field strength in a certain direction Ψ, measured from the positive direction of the axis X, is equal to:

$$E_\psi = \sqrt{\frac{1}{2\pi}\int_0^{2\pi} 2[E_X \cos\psi \sin(\omega t + \varphi_X) + E_Y \sin\psi \sin(\omega t + \varphi_Y)]^2\, d(\omega t)} =$$

$$= \sqrt{E_X{}^2 \cos^2\psi + E_Y{}^2 \sin^2\psi + 2 E_X E_Y \sin\psi \cos\psi \cos(\varphi_X - \varphi_Y)}\,.$$

Extreme values of strength are calculated by the formula:

$$E_{\psi E} = \left[\frac{E_X^2 + E_Y^2}{2} \pm \frac{\sqrt{\left(E_X^2 + E_Y^2\right)^2 - 4E_X^2 E_Y^2 \sin^2(\varphi_X - \varphi_Y)}}{2}\right]^{\frac{1}{2}},$$

The plus sign corresponds to the maximum, and the minus sign corresponds to the minimum value.

This formula is given in paper [18], the authors of which indicate that when calculating the field near the surface of the earth, the error of a simple quadratic summation $E = \sqrt{E_X{}^2 + E_Y{}^2}$ usually does not exceed 10 % in the direction of overstating the maximum of the RMS value.

The vertical and horizontal components of the magnetic field strength are calculated using the following formulas:

$$\dot{H}_X = \frac{1}{2\pi}\sum_{i=1}^{N} \dot{I}_i \frac{y - y_i}{(x_i - x)^2 + (y_i - y)^2}\,; \quad \dot{H}_Y = -\frac{1}{2\pi}\sum_{i=1}^{N} \dot{I}_i \frac{x - x_i}{(x_i - x)^2 + (y_i - y)^2}\,.$$

To determine the strengths of the electric and magnetic fields, the computational scheme mode is calculated, the charges and currents of the wires, including the grounded ones, are determined, and the components of $\dot{E}_X$, $\dot{E}_Y$, $\dot{H}_X$, $\dot{H}_Y$ are found.

The described technique is implemented in the "Fazonord" software package [9]. In this case, the strengths of the electromagnetic field can be determined both for a particular mode and for their combination, on the basis of which one can obtain the dynamics of changes in the EMF intensity over time.

The method allows significantly simplifying the calculation of EMF strengths. Indeed, in the traditional formulation, this problem requires the solution of differential equations in partial derivatives. Its solution is much more complicated in the presence of heterogeneities of the underlying surface, as well as the need to take into account extended conductive objects (metal fences, underground pipelines, etc.).

Using a set of grounded wires to simulate inhomogeneity of the earth's surface (with the transmission line route passing along) and conductive objects allows the proposed method to be applied to determine EMF with proper consideration of environmental conditions without additional complications and modifications.

Calculations of EMF strengths by the proposed method can be referred to as integral methods with charge distribution on fictitious grounded conductors located on the surface of a non-flat earth or conductive boundary of the structure. Unlike the applied types of integral methods, the charges on grounded conductors are determined by calculating modes in phase coordinates. After calculating the mode of the system, one can determine the charges of wires per unit length.

3. Experimental conditions

Strengths of the electric and magnetic fields at a frequency of 50 Hz with comparative calculations were measured for a 110 kV double-circuit line with a ground wire located within the city. The measurements were carried out in the interval between the two supports at the intersection of the electric power transmission lines with the railway electrified by the 1 × 25 kV system. At this point the railway is located in a sufficiently deep ditch cut, which did not require the construction of high supports. The measurement site was located on a small hill, at a distance of about 70 m from the railway. The height of the wire suspension above the measurement site was approximately 19 m (ground wire), 16 m, 13 m, 10 m (phase wires). A single-circuit 110 kV dead-end feeder is connected to the line.

At a distance of 10 m from the electric power transmission line with a small lowering of the earth surface there is a metal fence of a garage cooperative, which should have a significant impact on the electric field. A sufficiently large distance to the electrified railroad made it possible to ignore its effect on the magnetic field in the calculations. The electric field of the catenary system also had no effect since the railway was located in the groove and the voltage was relatively small, 25 kV.

According to the regime measurements of double- circuit electric power transmission lines, a power flow of about 9–*j*12 MV·A passes through the circuit closest to the dead-end feeder (the minus sign corresponds to the positive direction of the feeder power flow). The flow in the second circuit is 6–*j*9 MV·A; the dead-end feeder consumes 6+ *j*3 MV·A.

4. Description of the computational scheme

To correctly take into consideration a multi-wire system with facilities that affect the EMF, its model was built in the "Fazonord" software package, including 19 wires (Fig. 1).

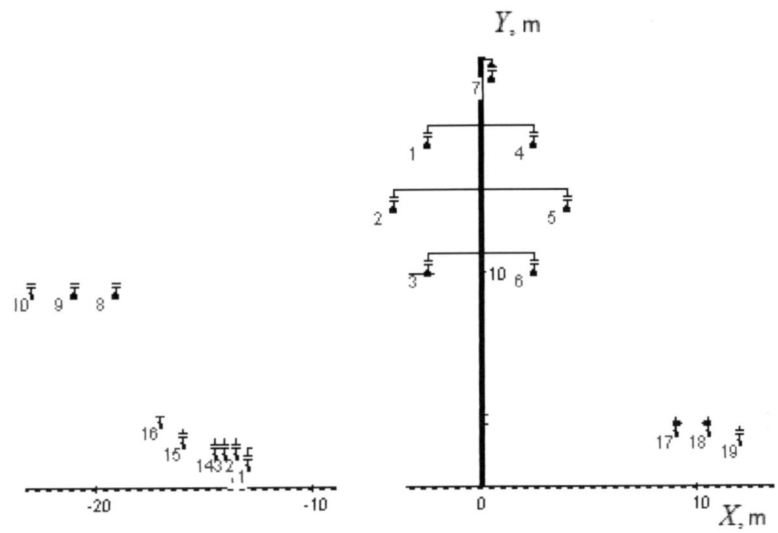

Figure 1. The cross-section of a multi-wire system model

Numbers 1–6 denote phase conductors. Number 7 is assigned to a ground wire connected to the ends of the grounding conductors with a dissipation resistance of 10 Ohms. Numbers 8–10 denote wires of the dead-end feeder. Wires numbered 11–16 are introduced to take into consideration the uneven ground surface. When grounding with a shunt with a resistance of 1 ohm on one side, these wires make it possible to take into account the effect of the elevation of the earth on the electric field, without affecting the magnetic field strength. For the left, more loaded line circuit, the phase voltages were taken symmetrical, 65 kV each, and 66 kV each for the right one. When calculating the mode, the above loads, symmetrically decomposed in phases, resulted in the following currents: 77 A of the left circuit wires, 54 A of the right circuit wires and 34 A of the dead-end feeder. The computational scheme of the Fazonord SP is shown in Fig. 2.

Nodes 2, 4, 6, 7, 8, 9 of the computational scheme are declared balancing; the loads are specified in nodes 16–21. Nodes 10 and 22 correspond to the ground wire. Nodes 11–15 have shunts to earth with a conductivity of 1 S. As a result of the mode calculation, the current of the ground wire amounted to about 0.1 A. Thus, the ground wire, having a significant effect on the electric field, has practically no effect on the magnetic field.

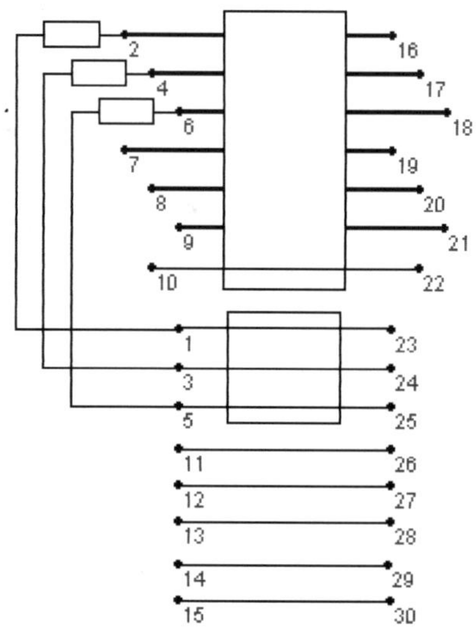

Figure 2. The Fazonord SP computational scheme

5. Results of measurements and calculations of electromagnetic field strengths

Experimental measurements of the electromagnetic field strengths (Tables 1, 2 and Fig. 3, 4) are presented by three groups of measurements taken at different times. During the measurements, the same directions of the axes of coordinates are taken as in the calculation method. The origin of coordinates is located under the center of the electric power transmission lines on the ground surface. The measurements were carried out with a field strength meter P3-50 according to the projections of the strengths to the corresponding coordinate axes at a height of 1.5 m from the ground surface.

Table 1. Maximum values of electric field strength and correlation coefficients of calculated and experimental values at measurement points

Parameter	Calculation	Experiment			Differences, %		
		1	2	3	1	2	3
$E_{\max}$, kW/m	0.53	0.54	0.49	0.59	–1.6	9.6	–17.0
Correlation coefficient		0.94	0.89	0.79	–	–	–

Table 2. Maximum magnitudes of magnetic field strengths and correlation coefficients of calculated and experimental values

Parameter	Calculation	Experiment			Differences, %		
		1	2	3	1	2	3
$H_{\max}$, A / m	0.47	0.54	0.55	0.59	–12.1	-1.8	-6,7
Correlation coefficient		0.92	0.80	0.84	–	–	–

In general, the nature of the dependence of the electric field strength on the coordinate *X*, obtained by calculation, corresponds to similar dependences, built on the basis of experiments. In the first experiment, the maximum values of strengths are almost equal to the design ones. All experiments are characterized by lower values of the field strength below the left chain, according to fig. 1. The calculated values for the right chain lie inside the limits of the scattering region of the experimental values. The simulation adequacy is confirmed by the high values of the correlation coefficients between the experimental and calculated data.

The results of some quantitative discrepancy between the calculated and experimental data on the electric field strengths of the left circuit are attributable to the following factors:

- shielding of the electric field by garages located on the north-east side of the electric power transmission lines at a distance of about 7 m from the measurement site with a coordinate of –12 m, and the garage cooperative buildings on the south side;

- an approximate account of the dead-end feeder;
- shielding of the electric field by the bodies of the crew members carrying out the measurements;
- the presence of errors in the determination of wire heights;
- inaccurate setting of voltage wires and the presence of asymmetry caused by unsymmetrical load.

Tabl. 2 and fig. 4 show the results of measurements and calculations of the magnetic field strength.

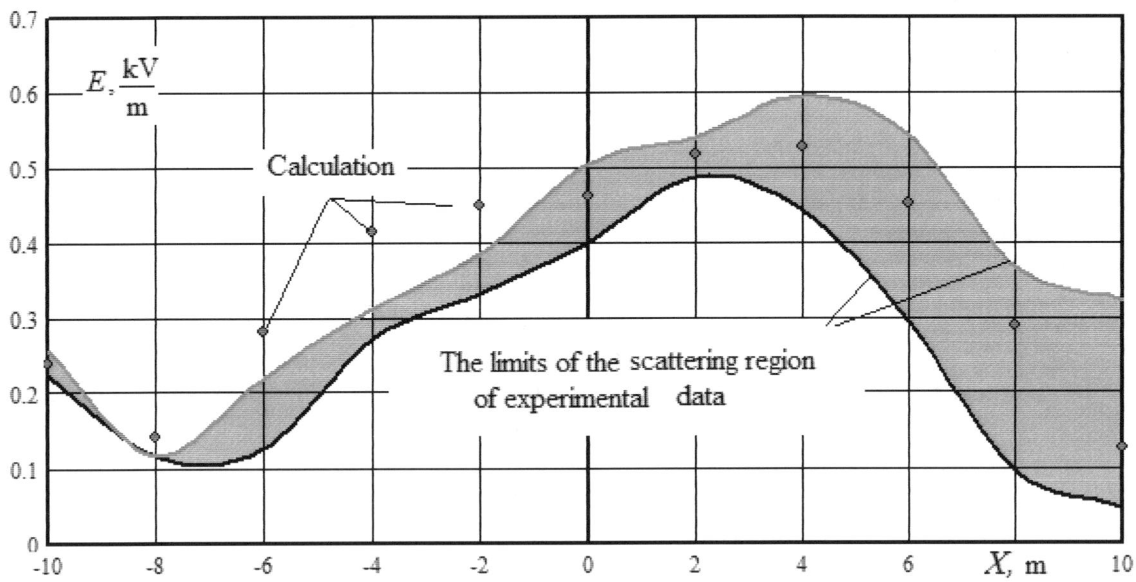

Figure 3. The interval representation of the measurement results of electric field strengths

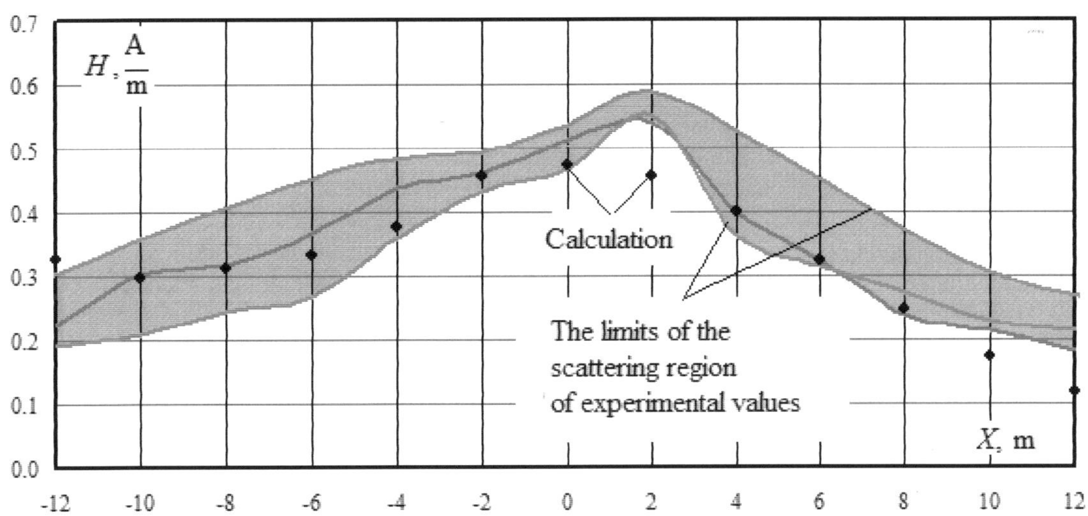

Figure 4. The interval representation of the measurement results of magnetic field strengths

The problem of modeling a magnetic field differs from a similar problem for an electric field by a high level of uncertainty of the operating situation arising due to a large variation in the wire currents, while the voltages of the wires determining the electric field vary in a small interval. However, most of the calculated points of the dependence $H = H(X)$ do not exceed the limits of the scattering region of experimental values.

A more detailed analysis shows that, with a fairly close similarity in the dependences of the H module on the X coordinate for all experiments, there is a difference in the horizontal component of the magnetic field strength obtained by calculation and experimentally. There is a particularly large difference for the third set of experimental data. This is due to the dependence of the magnetic field strength not only on the modules, but also on the phases of the currents in the electric power transmission lines. The latter are highly dependent on the current mode of the EES. Since a double-circuit electric power transmission line is part of a complex network and has two-way power supply, currents can change their phases over wide limits, up to a turn of 180°. This factor greatly affects the horizontal component of strength. The large difference in the current modules of the two circuits results in the emergence of a maximum of the vertical component at the center of the electric power transmission lines.

6. Conclusion

The comparison of the results of experimental measurements of the electric and magnetic fields strengths and the calculated values in fairly complex conditions is indicative of the correctness of the proposed method for calculating the electromagnetic field of a multi-wire line.

References

[1] Apollonsky S M and Bogarinova A N 2006 *Strengths of the air environment on the electrified railways* (St. Petersburg: collection of reports of the ninth Russian scientific and technical conference on electromagnetic compatibility of technical means and electromagnetic safety) pp 579–83

[2] Apollonsky S M 2001 *External electromagnetic fields of electrical equipment and means to reduce them* (St. Petersburg: Bezopasnost' Publ.)

[3] Apollonsky S M and Gorsky A N 2006 *Calculations of electromagnetic fields* (Moscow: Marshrut Publ.).

[4] Blake L B 2007 *Protection against electromagnetic fields. On the impact of household electrical appliances, mobile phones, power lines and other electrical devices on the human body* (Moscow: Ast, Astrel' Publ.)

[5] Sidorov A I and Okrainskaya I S 2008 *Electromagnetic fields near electrical installations of extra high voltage* (Chelyabinsk: South-Ural State Un-ty Publ.)

[6] Tikhonov M N 2011 The mechanism of the influence of natural and man-made electromagnetic fields on life safety *The ecology of industrial production* **4** 24–32

[7] Kusmartseva E V and Yakubovich D M 2016 *The influence of modern sources of electromagnetic fields on human safety in the technical field* (Saratov: Innovations in environmental engineering and protection in emergency situations) pp 46–8

[8] Makhutov N A, Kalmykov V M, Balanovsky V L and Balanovsky L V 2014 *Quality of life and electromagnetic safety* (Quality and life. No. 2) pp 53–7

[9] Zakaryukin V P and Kryukov A V 2005 *Complex non-symmetrical modes in electrical systems* (Irkutsk: Irkutsk university Publ.)

[10] Zakaryukin V P, Kryukov A V 2013 Multifunctional approach to modeling of electric power systems *Modern technologies System analysis Modeling* **4** 100–7

[11] Buyakova N V, Zakaryukin V P and Kryukov A V 2018 *Electromagnetic safety in power supply systems of railways: modeling and control* (Angarsk: Angarsk State Technical University Publ.)

[12] Buyakova N V, Zakaryukin V P and Kryukov A V 2018 Modeling of electrical fields in railway engineering structures *Advances in Engineering Research: International Conference on Aviamechanical Engineering and Transport* vol 158 pp 219–25

[13] Buyakova N, Zakaryukin V, Kryukov A and Nguyen T 2018 Electromagnetic Safety Enhancing

in Railway Electric Supply Systems *E3S, Web of Conf.* **58(01006)** pp 1–6

[14] Buyakova N, Zakarukin V and Kryukov A 2018 Imitative Modelling of Electromagnetic Safety Conditions in Smart Power Supply Systems *Advances in Intelligent Systems Research vol. 158 Vth International workshop "Critical infrastructures: contingency management, intelligent, agent-based, cloud computing and cyber security"* pp 20–5

[15] Buyakova N, Zakaryukin V and Kryukov A 2018 Control of electromagnetic environment in smart traction power supply systems *ES3 Web of Conf. (TPACEE-2018)* **91(01009)** pp 1–11

[16] Kryukov A V, Zakaryukin V P and Buyakova N V 2010 Modeling of the electromagnetic environment on AC railways *Modern technologies, System analysis, Modeling* **2** 169–75

[17] Kryukov A V, Zakaryukin V P and Buyakova N V 2010 Electromagnetic environment control at railway facilities *Modern technologies, System analysis, Modeling* **3** 34–8

[18] Katz R A and Perelman L S 1978 Calculation of the electric field of a three-phase power line *Electricity* **1** 16–9

Current trends in the development of logistics professional discourse terminology

E E Bylina, T A Skopintseva, L V Sludneva and E A Yurkovskaya

Irkutsk State Transport University, Irkutsk, Russia

E-mails: stephanya1@yandex.ru, skopinceva_t@irgups.ru, sludneva54@mail.ru, eayur@mail.ru

Abstract. The article focuses on the study of logistics professional discourse terminology. The study revealed the tendency towards simplification of the morphological and syntactic structure of logistics terms, discovered new productive patterns for abbreviating logistics terms, found out that the existing problem of terminological homonymy of abbreviations could be solved through applying new patterns for creating abbreviations.

1. Introduction

Professional discourse is a specific form of verbal communication in professional settings. This term refers to the language used by professional communities (e.g. logicians, lawyers, doctors, economists) which reflects some shared professional knowledge and use of the same terminology in their intra-community interaction.

Professional discourse has been under close study for the last three decades both by foreign [7] and Russian scholars. There have been numerous attempts at contrasting various terms referring to languages people use in the workplace such as professional discourse [12], workplace discourse [11], specialized discourse [3]. In this study Kong's definition of professional discourse is applied as it blends the three major features of the notion: professional, specialist and workplace. Kong defines it as "the language produced by a professional with specialist training to get something done in the workplace" [12].

In Russian linguistic research a noticeable increase in the interest in linguistic dimensions of professional interaction occurred at the end of the 20th century. The study of professional discourse emerged from the overall discipline of discourse analysis (N.D. Arutyunova, V.Z. Demyankov, V.I. Karasik, Yu.S. Stepanov and others), in which discourse is understood as a form of human communication, including texts and extra linguistic aspects necessary for their generation and interpretation. Professional discourse is interpreted as "a communicative action carried out in the context of a specific communicative situation of professional communication", and based on "knowledge and use of strictly selected linguistic and stylistic means reflecting a specific objective and mental world of communicants" [16: 9]. Strict compliance with all the conventions of a professional discourse is an essential condition of its effectiveness [1].

If compared, both concepts emphasize the idea that professional discourse is a specific language variant but the Russian approach is more theoretically-oriented and presupposes a more detailed study of extra linguistic context in which professional discourses appear. Conversely, the English-language analysis has a more practice-oriented touch.

Logistics is a greatly functional area, the second approach has been therefore followed in this study.

Linguistic research of professional discourse is focused on its verbal representation in the form of oral and written texts. Logistics documentation (including official business correspondence) should be recognized as an integral component of logistics activities and the leading type of verbal representation of logistics professional discourse.

The main lexical unit of professional discourse from the linguistic research perspective is the term [17]. According to the definition by V.M. Leychik, "a term is a lexical unit of a specific language for special purposes, denoting a common – concrete or abstract – theoretical concept of a specific specialist field of knowledge or activity" [15]. So terminology becomes the main object of linguistic analysis of certain professional discourses, both logistics [8, 9, 10, 14] and others such as, for instance, election campaign discourse [20]. In the studies of the logistics terminology system of recent years the leading role of the English-language terminology is recognized. Mastery of logistics terminology is viewed as a means "to get something done", and the easier and quicker it is done by "professionals", the more skilled the "specialist training" is.

2. Methods and Materials

This study was focused on the analysis of logistics terminology with the view of revealing current tendencies of its development. The materials for the study were the logistics terminology selected for "English-Russian Dictionary of Logistics Terms" [5], the logistics abbreviations found on the logistics operators' websites, namely [18, 19] and "List of Abbreviations Used in International Transport Documentation, Commercial Correspondence and Telexes" by D.V. Zerkalov [21].

The resultant data for Part A in the Results Section were obtained by means of:

- secondary data analysis of the 1994-year logistics dictionary by A.K. Kuptsova [14];
- quantitative morpho-syntactical analysis of the contents of the 2008-year logistics dictionary [5] to measure the frequency of the detected types of logistics terms;
- further comparative analysis of the data obtained to find out if there had been any changes in the morphological structure of the present-day logistics terminology.

The resultant data for Parts B-D in the Results Section were obtained by means of:

- qualitative analysis of the abbreviations [18, 21] in order to observe and describe the most productive and innovative patterns of abbreviations within logistics professional discourse;
- interpretation of how certain innovative patterns may become a response to the challenges from terminological homonymy of logistics abbreviations.

3. Results

3.1 Morpho-Syntactic Composition of Logistics Terminology

From a viewpoint of structural composition, a term can be either a word (single-word or one-word terms) or a combination of two or more elements (multi-word terms). Both were detected in "English-Russian Dictionary of Logistics Terms" published in 2008 [5].

The group of single-word terms fall under the morpho-syntactic classification of terms developed by Russian linguist V.S. Grinev-Grinevich [4] who suggests three groups of terms belonging to the category of single words:

1. Simple, e.g. *damage*. They are root words, the stem coincides with the root.
2. Complex, e.g. *palletization*. They consist of the root and some affix(es).
3. Compound, e.g. *europallet*. They consist of several stems, which may occur in the language as free forms.

Multi-word terms, in their turn, are presented by:

1) two-word terms, e.g. *accumulator area*;
2) three-word terms, e.g. *advanced shipping note*;
3) word combinations of four or more words, e.g. *fast moving consumer goods*.

The examples for the classification above are selected from [5].

Further quantitative analysis of the logistics terms measured the frequency of the detected types of terms and found that the multi-word terms (61 %) outnumber the single-word terms (39 %); the complex terms (72 %) outnumber the simple (22 %) and compound ones (6 %)

The multi-word terms represented by two-word, three-word and four-word terminological phrases display the following percentage ratio: the percentage of two-word terms (87 %) exceeds that of the three-word terms (11 %) and four-word terms (2 %).

The results of the analysis are presented in Table 1. The groups of logistics terms were arranged in the descending order from the largest to the smallest ones in percentage points.

Table 1. Morpho-Syntactic Groups of Logistics Terms [a]

Group name	Percentage value (%)
two-word terms	53
complex single-word terms	28.3
simple single-word terms	8.5
three-word terms	6.3
compound single-word terms	2.5
four-word terms	1.4

[a] based on [5]

The table clearly shows that the groups of two-word terms and one-word complex terms (consisting of a root and affixes) are more represented in the logistics terminological system.

In 2015 A.K. Kuptsova analyzed the percentage distribution of the English-language logistics terms according to their morphological and structural features. The study of the terms included into the dictionary "Terminology in Logistics" published in 1994 allowed the scholar to make a conclusion about the prevalence of nominal terms having a two-word structure [14].

In order to reveal current trends in the development of the English-language logistics terminology, the authors of this study conducted a quantitative morpho-syntactic analysis of the contents of the 2008-year "English-Russian Dictionary of Logistics Terms" [5] to measure the frequency of certain types of logistics terms with a view of further comparative analysis of the data of these two studies. The results are shown in Table 2.

Table 2. Comparison of Data

Combined group name	Percentage value (%)	
	1994 [a]	2008 [b]
single-word terms (simple and complex words)	15	36,7
compound terms and multi-word terms	85	63,3
Group name	1994 [a]	2008 [b]
two-word terms	55	83,8
three-word terms	21	10,1
single-word or hyphenated compound terms	18	3,9
word-combinations of four or more words	6	2,2

[a] based on [14]
[b] based on [5]

Thus this study also revealed the prevalence of the compound and multi-word terms over the single-word terms but with a decrease in the percentage (63,3 % in 2008 vs. 85 % in 1994) and the numerical dominance of two-word terms over three- and four-word terms with an increase in the percentage (83,8 % in 2008 vs. 55 % in 1994).

Therefore, there is a significant increase in the number of simple (single-word) terms and two-word terms, which indicates a general tendency to simplify the logistics terminology of the English language, and taking into account its leading role in this professional sphere, logistics terminology as a whole.

3.2. Abbreviations of Logistics Discourse

Another manifestation of the tendency to simplify the logistics terminology is considered to be an extensive use of shortened forms of logistics terms, i.e. logistics abbreviations. It accords with Zipf's law of abbreviation which states that frequently used words tend to be shorter [2]. Logicians abbreviate the terms which are frequently used in logistics documentation.

Logistics is a professional area where speed, accuracy and precision of the execution of a contract, clarity and conciseness of documentation are valued. In trying to save space and time logicians use commonly recognized abbreviations for the most frequently used terms.

The high proportion of abbreviations in the corpus of logistics terminology is evidenced by the data obtained from a study conducted by O.S. Karavaevskaya. According to the author, the share of abbreviations as a percentage of the total number of analyzed terms was 12,9 per cent [9].

In order for logistics operators to respond to the needs of their customers in the most efficient way logistics documents should be precise and easy to understand. In logistics documentation abbreviations perform the function of text compression by shortening a terminological phrase to one word. However, an abbreviated term acquires the status of a word, which is expressed in the presence of lexical meaning, only within a specific professional discourse.

The terminological character of an abbreviation is based on the agreement of discourse community members to use it as a language representation of a certain professional concept, thereby defining the status of the abbreviation as a professional term. So the obligatory requirement imposed on an abbreviated term is its institutional uniformity, which is understood as universality within a certain professional discourse and inclusion into professional terminology lists. It is also required that the abbreviation is easily and unambiguously interpreted by the professional community.

Abbreviated terms are unmotivated, "non-transparent". Unmotivated terms only refer to concepts, not indicating their features [15], therefore, the terminological nature of logistics abbreviations is based on the agreement of the discourse community to use it as a language representation of a certain professional concept.

Such an agreement was reached, for instance, in order to provide internationally accepted definitions and rules of interpretation for the most common commercial terms used in international trade and transportation. Incoterms [6], e.g. CPT for "Carriage Paid to". A professional logistician of any nationality is able to adequately understand these English-language abbreviations on the basis of their generally accepted interpretation and implement them in the most efficient way in their contracts.

Another way to conventionalize the present-day logistics abbreviations is including them into logistics terminology dictionaries, professional glossaries of logistics and lists of terms and abbreviations published by many logistics operators on their websites.

3.3. Some Innovative Patterns for Creating Logistics Professional Discourse Abbreviations

In her study of logistics abbreviations O.S. Karavaiskaya identified the most productive patterns for creating them:

- initials,
  e.g. *MV* for "mother vessel";
  including abbreviations containing numbers,
  e.g. *2PL* for "second party logistics";
  graphic symbols "/" and "-" replacing prepositions,
  e.g. *L / C* for "letter **of** credit", *THC-D* for "terminal handling charges **in** destination";
  and conjunctions,
  e.g. *CIFI & E* for "Cost, Insurance, Freight, Interest **and** Exchange;
- acronyms, e.g. *IATA* for "International Air Transport Association";

- apocopes (clippings),
  e.g. *dim* for "dimension";
- syncopes (contractions),
  e.g. *cnee* for "consignee" and *cnor* for" consignor" [9].

The analysis of contemporary resources of logistics terminology, including lists of abbreviations suggested by logistics operators, revealed some innovative patterns for logistics abbreviations formation:

Pattern 1. Abbreviations including the punctuation mark ".", the Period. This pattern is observed in newly created:

- initialisms written in uppercase letters,
  e.g. *C.Y.* for "container yard";
- acronyms written in lowercase letters,
  e.g. *e.oo.e* for "errors or omissions excepted";
- acronyms written in both upper- and lowercase letters,
  e.g. *Q.w.Q.* for "quantity with quality";
- apocopes (clippings),
  e.g. *occas.* for "occasionally";
- syncopes (contractions),
  e.g. *mfst.* for "manifest/ shipping manifest" [21].

Pattern 2. Acronyms written in lowercase letters,
e.g. *remy* for "referring to my", *retel* for "referring to telegram" [21];

Pattern 3. Abbreviations (initialisms, acronyms, syncopes) starting with a capital letter,
e.g. *Anny* for "annuity", *Qp* for "quality products", *Rly Stn* for "railway station" [21];

Pattern 4. syncopes (contractions) using brackets,
eg. *ql(t)y* for "quality" [21].

3.4. Terminological Ambiguity of Logistics Abbreviations. Synonymy and Homonymy

The logistics terminology system is an area under development. A significant number of websites of logistics operators publish their own lists of frequently-used English-language terms and abbreviations with keys and translation, for instance [18]. As a result, the issue of terminological ambiguity of logistics abbreviations, namely terminological synonymy [10] and homonymy [13], is currently quite pressing.

The terminological synonymy of logistics discourse abbreviations is expressed in the existence of several parallel abbreviations. They can be found in the same and different terminological reference sources. The study detected two types of abbreviation synonyms: spelling synonyms which differ in spelling and structural synonyms which differ in structural composition.

Table 3. Spelling Synonyms

Synonymic Abbreviations	Abbreviated Term / Common Meaning
BL [a]	Bill of Lading
B / L [b]	
B.O.L. [b]	
B / lading [b]	

[a] found in [18]
[b] found in [21]

As Table 3 shows, the abbreviations *BL, B/L, B.O.L.* and *B/lading* are shortened forms of the same term "Bill of Lading" but there are differences in their spelling, so the synonyms of this type can be labelled as spelling synonyms.

IOP Conf. Series: Materials Science and Engineering **760** (2020) 012013 doi:10.1088/1757-899X/760/1/012013

Table 4. Structural Synonyms

Synonymic Abbreviations[a]	Abbreviated Terms	Common Meaning
DTHC	Destination Terminal Handling Charges	charges paid for
DHC	Destination Handling Charges	handling a shipment at
THC-D / THC/D	Terminal Handling Charges in Destination	the port of destination

[a] all the abbreviations are found in [18]

As Table 4 displays, the reason why there may exist variants of abbreviations with the same meaning is that it can be conveyed in various, yet synonymic, terms by means of:

1) omitting words adding no specific information as in "Destination Handling Charges" if opposed to "Destination Terminal Handling Charges";
2) changing the order of the lexical components within a term as in "Terminal Handling Charges in Destination" if opposed to "Destination Terminal Handling Charges";
3) using various graphical symbols as in *THC-D* and *THC/D*. Both the hyphen "-" and the slash "/" are symbols used in abbreviations as substitutes for prepositions.

So in the examples from Table 4 we can observe synonyms of both types: structural and spelling (Line 3). Later research may be done to measure the frequency of these types by means of quantity analytical methods.

Similar to logistics synonymic abbreviations, terminological homonymic abbreviations can also be found both in the same and different terminological reference sources. Homonymic abbreviations have the same spelling but absolutely different meanings as they are shortenings of different terminological phrases.

Table 5. Homonyms

Homonymic Abbreviations	Abbreviated Term / Different Meaning
T/T	Transit Time [a]
	Telegraphic Transfer [b]
D / A	Disbursement Account [c]
	Deposit Account [c]
	Documents Against Acceptance [c]
	Documents Attached [c]

[a] found in [18]
[b] found in [19]
[c] found in [21]

The emergence of new patterns for abbreviations is most likely due to the need to solve the problem of terminological homonymy of abbreviations, as shown in the juxtapositions:

e.g. *a/w, A/W* for "along with" vs. *AW, a.w.* for "actual weight";

n.p. for "notes payable" vs. *n/p* for "nonpayment";

est for "estimated vs. *est.* for "established" [21].

The spelling representation of abbreviations having similar letter composition is a means to contrast the abbreviations in the examples above.

4. Conclusions

The study of current trends in logistics professional discourse terminology revealed a general tendency to simplify logistics terms. The evidence from this study shows that the number of terms having the structure of words and two-word phrases considerably grew in the 2008-year dictionary [5] as compared to the 1994-year dictionary [14].

This tendency is additionally proven by the growing number of logistics terminological abbreviations. Along with existing logistics dictionaries and professional glossaries which include conventional terminology, many present-day logistics operators suggest their own lists of terms, mostly abbreviations. This is supposedly an attempt to conventionalize the terminology they add to their documentation.

One of the most frequent types of terms to appear in logistics operators' terminological lists is a logistics abbreviation. In order to provide opportunity for creating more abbreviations new patterns have been invented. Among the most interesting findings of this study are the new patterns for logistics abbreviations which add the punctuation mark ".", acronyms spelt in lowercase letters.

However, the abundance of logistics abbreviations has led to a considerable terminological ambiguity. It is manifested in existence of synonymic (structural and spelling) and homonymic abbreviations. Nevertheless, the study found out that the problem of terminological homonymy of logistics abbreviations could be resolved through applying new patterns for creating abbreviations.

The present analysis of the English-language terminology of logistics professional discourse is merely another step in linguistic research of logistics terminology, given that logistics is a currently evolving professional area. Further comprehensive study is supposed to follow.

References

[1] Bylina E E 2015 Ineffective promising in modern English discourse *International conference "I.A. Baudouin de Courtenay and Worldwide Linguistics"*: V Baudouin's readings: works and materials in 2 vols, ed by K R Galiullina, E A Gorobec, G A Nikolaeva (Kazan: Kazan university publishing house) vol 2 pp 47–9

[2] Bentz C, Kiela D, Hill F and Buttery P 2014 Zipf's law and the grammar of languages: A quantitative study of Old and Modern English parallel texts *Corpus Linguistics and Linguistic Theory* **10(2)** 175–211

[3] Gotti M 2008 *Investigating Specialized Discourse* (Peter Lang) 249 p

[4] Grinev-Grinevich S V 2008 *Terminology* (Moscow, Publishing house "Academia") 304 p

[5] *English-Russian Dictionary of Logistics Terms, ESR, Aldata Solution* 2008 Retrieved from: http://ocean.mstu.edu.ru/docs/files/20120202_1412-2.pdf.

[6] *Incoterms®Rules* Retrieved from: https://iccwbo.org/resources-for-business/incoterms-rules.

[7] Irimiea S B 2017 Professional Discourse as Social Practice *Eur. J. of Interdisciplinary Studies* **9(1)** 108–19

[8] Kalenteyev S V and Kuzmenko Ju V 2012 On the problems of terminology of modern logistics in the Russian Federation *SUSU Bull., Ser.: Economics and Management* **30** Retrieved from: https://cyberleninka.ru/article/n/o-problemah-terminologii-sovremennoy-logistiki-v-rf,

[9] Karavayskaya O S 2016 Abbreviations in the English terminology of transport logistics *Young Scientist* **28** 1006–8 Retrieved from: https://moluch.ru/archive/132/36686/,

[10] Karavayskaya O S 2017 The phenomenon of synonymy in the English-language terminology of transport logistics *Philological Sciences. Questions of theory and practice* **3–3(69)** Retrieved from: https://cyberleninka.ru/article/n/yavlenie-sinonimii-v-angloyazychnoy-terminosisteme-transportnoy-logistiki

[11] Koester A 2010 *Workplace Discourse* (Bloomsbury Publishing) 208 p

[12] Kong K 2014 *Professional Discourse* (Cambridge University Press) 304 p

[13] Kozhanov A A 2014 Ambiguity: polysemy and homonymy in legal terminology (on the material of the German language) *Bulletin of BSU* **2** Retrieved from: https://cyberleninka.ru/article/n/neodnoznachnost-polisemiya-i-omonimiya-v-yuridicheskoy-terminologii-na-materiale-nemetskogo-yazyka

[14] Kuptsova A K 2015 English as a means of international communication of logisticians *Logistics terminology and SCM* **3(68)** 67–72

[15] Leychik V M 2009 *Terminology: Subject. Methods. Structure* ed 4 (Moscow: Book House "LIBROCOM") 256 p

[16] Minakova LYu 2013 *Teaching foreign-language discourse for students of natural specialties using professionally oriented projects:* Abstract of Diss. of Cand. of Ped. Sci. (Tomsk) 25 p

[17] Shelov S D, Shufung S 2017 Term as a fundamental item of terminology *Questions of*

Linguistics **6** 76–86

[18] *Terms, abbreviations and abbreviations used in international container transport* Retrieved from: http://www.vdnk.ru/site/ru/info-container/port-charges.

[19] *What is T/T?* Retrieved from: Alibaba.com, https://service.alibaba.com/buyer/faq_detail/10609415.htm.

[20] Yurkovskaya E A 2000 Use of idiomatics and language realization of conceptual metaphor ELECTION CAMPAIGN IS RACE *Bulletin of ISLU, Linguistics* **3** 99–107

[21] Zerkalov D V *List of abbreviations used in international transport documentation, commercial correspondence and telexes* Retrieved from: http://www.zerkalov.org.ua/node/1531

IOP Conf. Series: Materials Science and Engineering **760** (2020) 012014 doi:10.1088/1757-899X/760/1/012014

Determination of electromagnetic effects of electric traction networks on pipelines

A V Cherepanov [1] **, A E Kryukov** [1,]
[1] Transport Electric Power Engineering department at Irkutsk State Transport University, 15 Chernyshevsky Str., Irkutsk, 664074

[2] smart_grid@mail.ru, [3] appleforevor@gmail.com

Abstract. The article presents the simulation results of electromagnetic effects of a 25 kV electric traction network on a steel pipeline with a pipe diameter of 250 mm. It is shown that a pipeline with insulating pipe coatings can induce stresses dangerous to operating personnel. At certain points of the pipeline, grounded on both sides, the magnitude of the induced voltage can reach 116 V with an approach width of 100 meters. Therefore, it is required to develop and implement special measures to protect operating personnel. With an increase in the approach width to 400 m, the maximum values of the induced voltages did not exceed the permissible level. In modes of short circuits in the electric traction network, the induced voltages in the pipeline may briefly exceed 500 V. The highest voltages are observed at points located near the short circuit point. In case of short circuits at the terminals of the traction substations, the magnitudes of the induced voltages are reduced to 75 V. To reduce the induced voltages, the following measures can be assumed: reducing the length of parallel passage sections of pipeline and railway, increasing the approach width, installing additional groundings. During the operation of the structure, the magnitude of the "pipe-to-ground" transition resistance may decrease, which will result in a decrease in induced voltages.

1. Introduction

In Russia, there is an extensive network of pipelines for transporting gas, oil, and its refinery products. In some sections of the routes, the pipelines may approach the railways electrified with alternating current [1, 2]. When sections of the pipeline with high-quality insulating coatings are placed parallel to the electromagnetically unbalanced 25 kV electric traction network [1–3], dangerous voltages to ground may occur [4–8]. To elaborate measures to improve electrical safety conditions, it is necessary to develop methods and algorithms of determining induced voltages emerging on pipelines due to the electromagnetic effect of the electric traction network.

The results of determining the induced voltages on the ground-surface pipeline, whose section runs parallel with the railway with a voltage of 25 kV in the overhead system, are described below.

2. The simulation technique

To determine the electromagnetic effects of electric traction networks on pipelines, one can use the methods of simulating modes of traction power supply implemented in the Fazonord software package developed in Irkutsk State Transport University [9]. The principles of using this software for calculating induced voltages are considered in [7, 9–12]. The mutual inductance coupling resistances, as well as the inherent resistances of the wire-to-ground circuits, were calculated using Carson formulas [13], which ensure that the return of currents through the ground is correctly taken into account.

IOP Conf. Series: Materials Science and Engineering **760** (2020) 012014 doi:10.1088/1757-899X/760/1/012014

3. Simulation results for train traffic.

For a considerably ordinary intersubstation railway area of 50 km in length (Fig. 1), electrified with 25 kV alternating current, the operation modes of the traction power supply system were modeled with a pipeline with a pipe diameter of 250 mm included in the multi-wire system. The computational scheme shown in fig. 2 included a model of three traction transformers and two intersubstation areas (ISA). The left ISA was divided into five sections with a length of 10 km. Besides, the scheme presented models of three 220 kV power supply lines. The distance from the pipeline to the axis of the road (approach width) varied from 100 to 400 m.

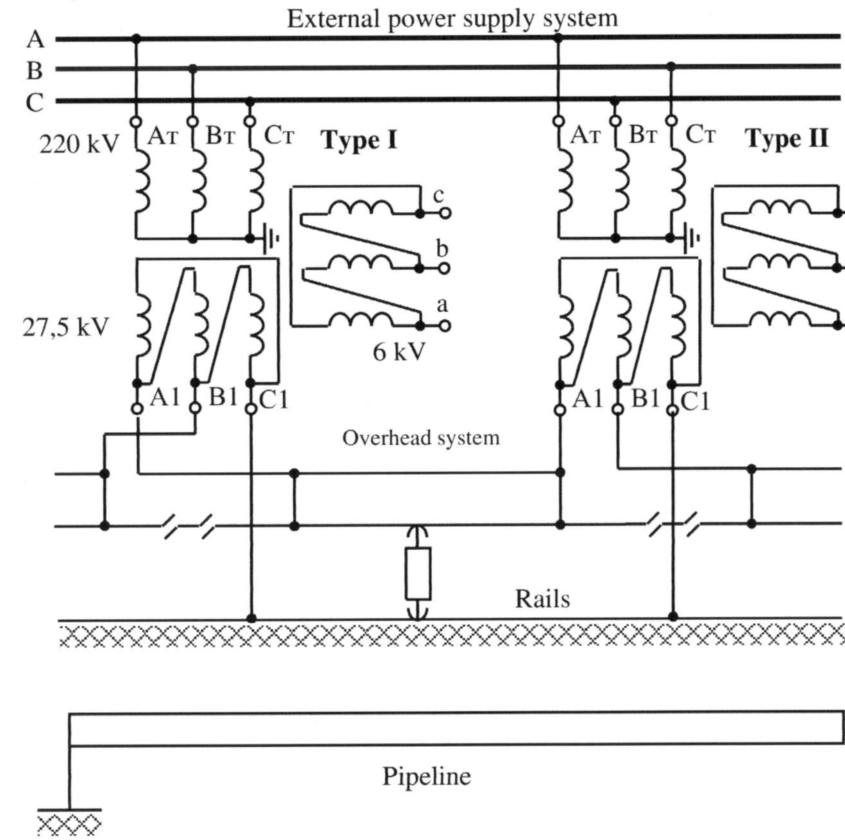

Figure 1. A fragment of the traction power supply system scheme: A, B, C are the phases of supplying power lines

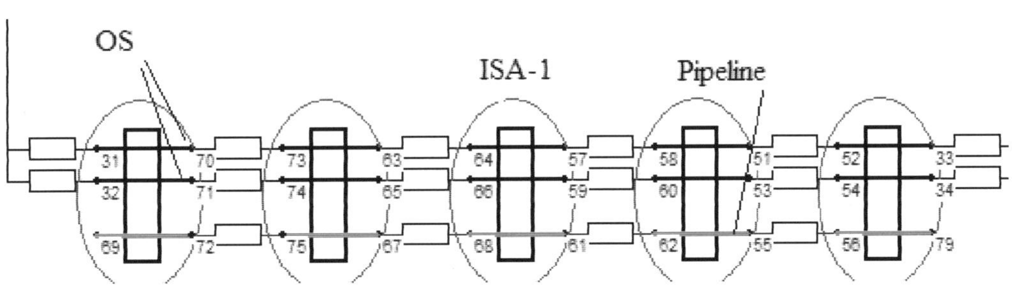

Figure 2. A fragment of the computational model scheme:
ISA is the intersubstation area; OS is the overhead system

The simulation considered the movement of 7 down trains weighing 6300 tons and the same number of up trains weighing 6000 tons. The movement schedule is shown in Fig. 3, and the current profiles of trains are shown in Fig. 4. The simulation took into account the distributed grounding of the pipeline with a resistance of 20 Ohm·km. In addition, presence of stationary groundings with a resistance of 1 Ohm was assumed at the edges of the structure.

The model of sections of the electric traction network with pipeline took into account the distribution of parameters and was formed as a chain-like scheme. In contrast to the results presented in [14], below there are the data on induced voltages that were obtained taking into consideration the influence of higher current harmonics and the electric traction voltage.

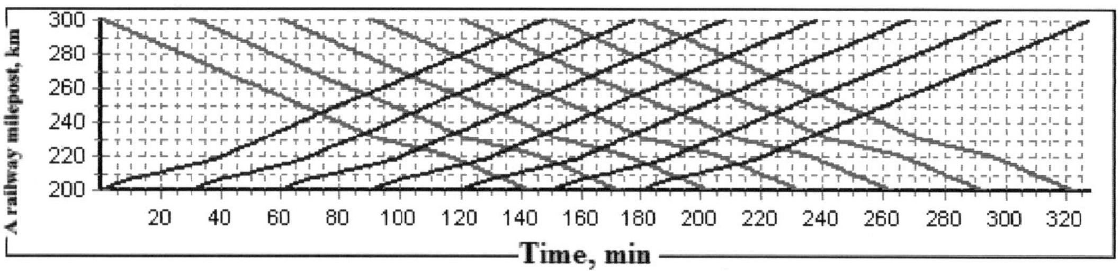

Figure 3. Train movement schedule

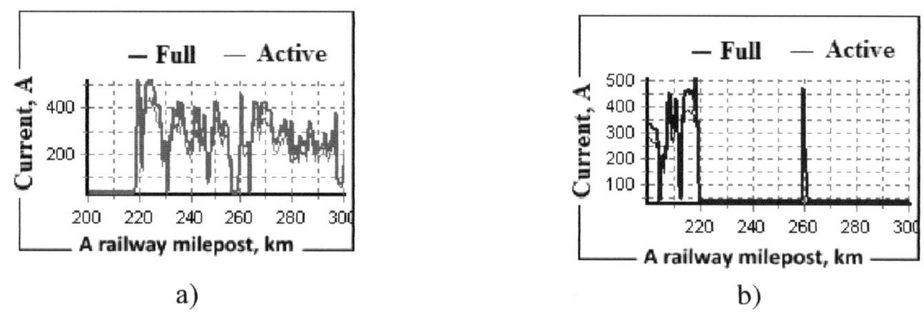

Figure 4. Current profiles of the trains:

a is a 6300 ton down train profile; b is a 6000 ton up train profile

The results of simulation are shown in table 1 and in fig. 5-7.

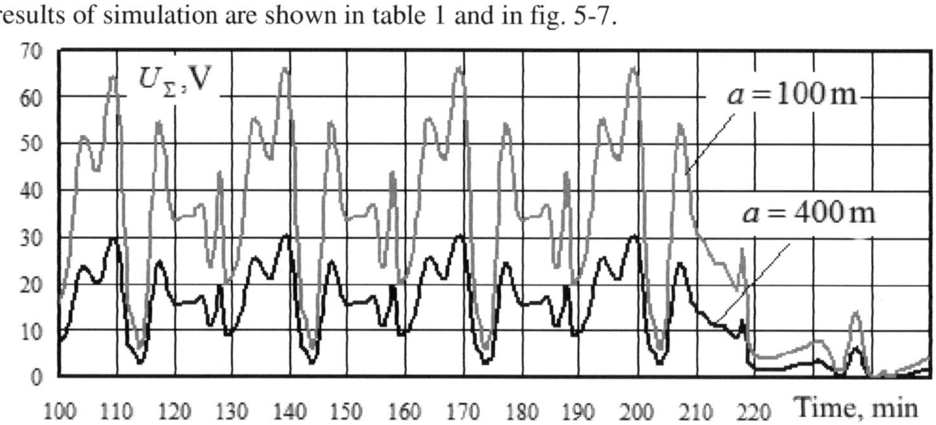

Figure 5. Dependences of induced voltage on time with L = 10 km:

L is the distance from the observation point to the left traction substation;

$U_{\Sigma} = U_1\sqrt{1+k_U^2}$ is the resulting induced voltage; k_U is the total harmonic coefficient; a is the approach width; a is approach width

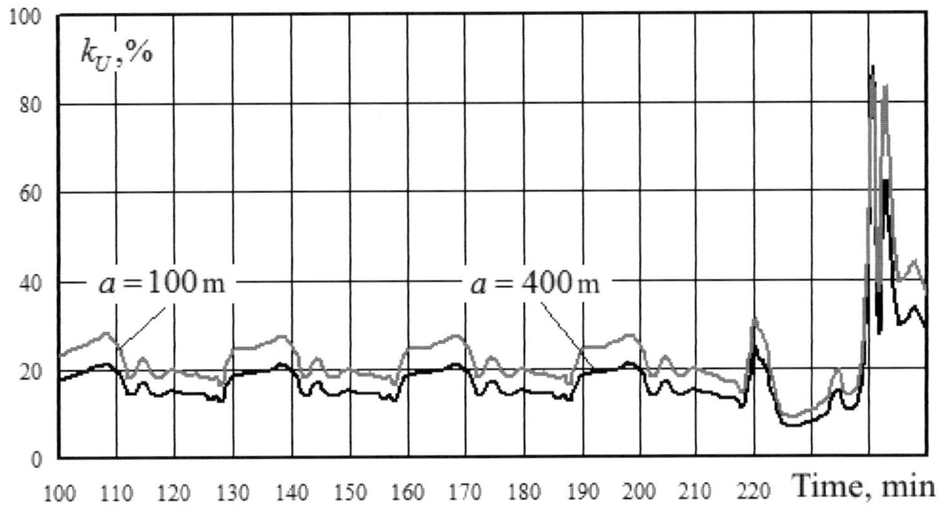

Figure 6. Dependences of the total harmonic coefficient on time with L = 10 km

Table 1. Maximum values of induced voltages and harmonic coefficients

Approach width, m	L, km	U_1, V	U_{hg}, V	U_{Σ}, V	k_U
100	*10*	*64.00*	*17.14*	*66.26*	*86.15*
400	10	29.70	6.11	30.32	**87.62**
100	*20*	*112.80*	*25.24*	***115.59***	*58,74*
400	20	50.90	8.79	51.65	67.27
100	*30*	*63.80*	*10.90*	*64.73*	*28.55*
400	30	28.40	3.76	28.65	21.83
100	*40*	*49.80*	*13.66*	*51.64*	*38.68*
400	40	22.6	4.72	23.09	28.41

Note: U_{hg} is the RMS voltage of higher harmonics

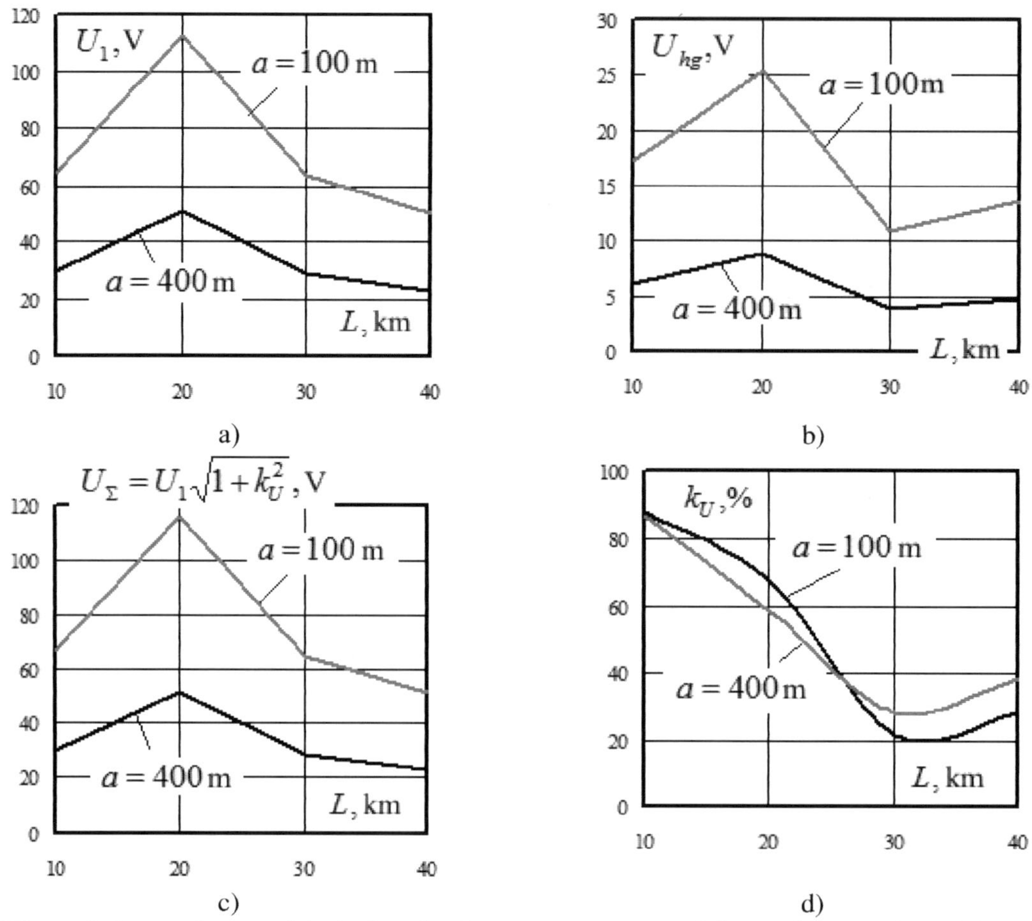

Figure 7. Dependences of the induced voltage components and the total harmonic coefficient k_U on the distance L

4. Simulation results for short circuits (SC) in the electric traction network

The simulation in the Fazonord software package was performed in two versions:

- the short circuit of the overhead line to the rail at the points corresponding to the terminals of the 27.5 kV traction substations (nodes 32 and 34 in Fig. 2);
- the short circuit of the overhead line to the rail in nodes 71, 65, 59 and 53 according to fig. 2;

Electric traction network was modeled on the basis of a chain-like scheme to take into account the distribution of parameters. The results of simulation are shown in fig. 8, 9.

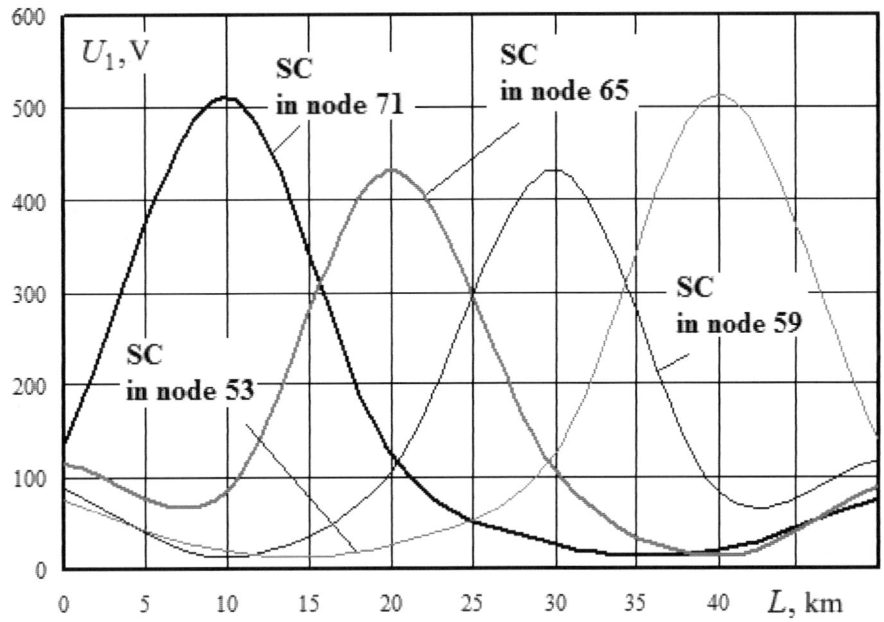

Figure 8. Distribution of induced voltages in short circuit modes of the electric traction network along the pipeline length:

L is the distance from the left traction substation to the observation point

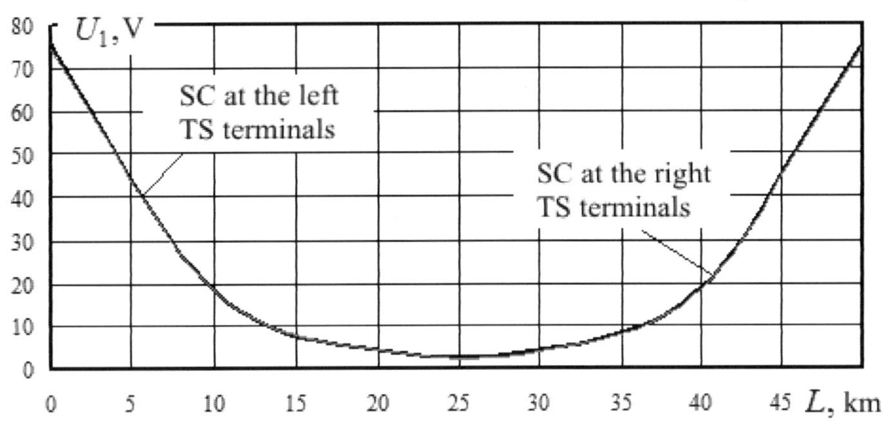

Figure 9. Distribution of induced voltages in short circuit at the terminals of the traction substations along the pipeline length

The simulation results make it possible to formulate the following conclusion: in short circuit modes in the traction network, the induced voltages in the pipeline can briefly exceed 500 V. The highest voltages are observed at points near the short circuit. In case of short circuits at the terminals of the traction substations, the magnitudes of the induced voltages decrease to 75 V.

5. Conclusion

The simulation results make it possible to conclude the following:

1. On the ground-surface pipeline, which has areas of the parallel approach to the AC railway line, there may be dangerous induced voltages to ground. In the calculated example, when the approach width is less than 100 m, the magnitude of the induced voltage reached 116 V. With an increase in the

approach width up to 400 m, the maximum values of the induced voltages did not exceed the permissible level [6].

2. To reduce induced voltages, the following measures can be assumed: reducing the length of parallel passage sections of pipeline and railway, increasing the approach width, installing additional groundings. During the operation of the structure, the magnitude of the "pipe-to-ground" transition resistance may decrease [15], which will result in a decrease in induced voltages.

3. The magnitude of the RMS values of the higher harmonics of the induced voltage in the calculated example reached 25 V. The maximum value of the harmonic coefficient was 88%. However, the values U_1 and U_Σ differed slightly. Nevertheless, in a number of situations that may occur in operational practice, taking into account harmonic distortions while determining the induced voltages on the surface pipeline will make it possible to more properly plan activities for ensuring the personnel safety.

4. In modes of short circuits in the electric traction network, the induced voltages in the pipeline may briefly exceed 500 V. The highest voltages are observed at points located near the short circuit point. In case of short circuits at the terminals of the traction substations, the magnitudes of the induced voltages decrease to 75 V.

References

[1] Strizhevsky I V, Dmitriev V I, 1967 *Theory and calculation of the influence of an electrified railway on underground metal structures* (Moscow: Construction literature Publ.).
[2] Ratner M P, 1966 *Inductive influence of electrified railways on electrical networks and pipelines* (Moscow: Transport Publ.).
[3] Krapivskiy E I., Yabluchanskiy P A, 2013 *Algorithm of the calculation of the electromagnetic influence of the electric AC power line on the underground pipeline* (Mining information and analytical bulletin. No. 2) 213-224.
[4] Kotelnikov A V, Kosarev A B, 1992 *Electromagnetic effect of AC electrical traction networks on metal structures* (Electricity No. 9) 26-34.
[5] Konnova E I, Kosarev A B 1990 *Calculation of the electromagnetic effect of AC networks on metal communications* (Bulletin of Research Institute of Railway Transport. No. 2). 17-19.
[6] Yabluchansky A I, 2008 *Methods of assessing the dangerous effects of high-voltage power lines AC on the projectable gas pipeline* (Moscow: Materials of the industry meeting on corrosion protection) 110–123.
[7] Zakaryukin V P and Kryukov A V 2015 *Determination of the induced voltages when nonparallel power lines are adjacent to one another,* (Power Technology and Engineering vol. 49 no. 4) 304-309.
[8] Borodavkin P P, Berezin V L ,1987*The construction of main pipelines* (Moscow: Nedra Publ).
[9] Zakaryukin V P, Kryukov A V, 2005 *Complex non-symmetrical modes of electrical systems* (Irkutsk: Irkutsk university Publ.).
[10] Zakaryukin V P, Kryukov A V, 2013 *Multifunctional approach to modeling of electric power systems,* (Modern technologies System analysis Modeling No 4). 100-107.
[11] Kryukov A V, Alekseenko V A, 2011*Improving the efficiency of operational management in traction power supply systems* (Modern technologies System analysis Modeling No 4). 158–164.
[12] Zakaryukin V P, Kryukov A V, 2012 *Use of FACTS devices in external power supply systems of railways* (Modern technologies System analysis Modeling No 1),. 267-274.
[13] Carson J R, 1926, *Wave propagation in overhead wires with ground return* (Bell Syst. Tech. J.. 5). 539–554.
[14] Cherepanov A V, Kryukov A E, 2017 *Electromagnetic influence of traction networks of main railways on pipelines* (Chita: Energy in the modern world) 7-12.
[15] Zinevich A M, Glazkov V I, Kotik V G, 1975 *Protection of pipelines and tanks against corrosion* (Moscow: Nedra Publ.).

Study of thermal zones in a heating collector of a traction motor NB-418 K6

A D Chuiko[1], **S A Christosenkov** [1]

[1] Irkutsk state transport university, 11 Magistralnaya str., Chita, 672040, Russia

E-mail: y210672@gmail.com

Abstract. The article provides the data about functional dependencies affecting the heating characteristic of the traction electric motor (faulty soldering in the riser winding). Studies have shown the functional dependence of the temperature drop along the length of the riser (with the coefficient 0.95 %) that allows determining the functional dependence of the riser temperature on its length.

1. Introduction

Currently, there is a large number of different methods for diagnosis and control of DC motors technical state. Literature and patents review shows that the most complete research in the field of development of methods of diagnosis and testing of DC motors has been carried out in various fields of transport, which can be explained by the following circumstances:

- a special significance of engines reliability for traffic safety;
- heavy operating conditions of engines, causing intensive degradation processes in the engine (insulation aging, transfers to the collector, etc.) ;
- long service life of electric locomotives.
- significant economic costs for the restoration of engine performance.

In the field of aviation technology, DC motors are mainly used as actuators of various mechanisms (low-power high-speed motors operating in good conditions at low voltage levels). The most typical failures of such engines are the deterioration of the switching quality, the deterioration of the quality of the current accumulation at the collector node, the defects in bearings, and soldering failure.

In this regard, the development of methods for determining the technical state of the soldering quality is a priority in the development of methods for diagnosis of DC motors in this industry. Nowadays, many methods and techniques have been developed to detect such defects.

The paper presents a study of the temperature drop effect on the functional dependence on the geometry of the riser.

2. Materials and methods

The use of thermal imaging with the approximation of a functional dependency (temperature as a geometry function of the riser winding) and obtaining the equation coefficients a, b, c.

Methods for determining the technical state of various objects depend on the purpose of the study, and they can be divided into methods of diagnosis and control. The tasks of diagnosis and control are formulated in the same way: to determine in which of the pre-established set of states the object is at a determined time interval. The main difference between the diagnosis methods and control is that the

IOP Conf. Series: Materials Science and Engineering **760** (2020) 012015 doi:10.1088/1757-899X/760/1/012015

object of study is considered as a whole. In the diagnosis the object is considered as a whole and as a set of elements. In this case, the state of the main elements influencing the overall state of the object as a whole is considered. Thus, technical diagnostics provides the control of both the object as a whole and its separate elements and determines necessary operations and their sequence to establish the state of the object.

The study of defects in traction motors of electric locomotives, their breakdowns during operation makes it possible to conclude that the traction motor (TM) is considered an unreliable node, the research is necessary to reduce the malfunction.

On the practical side applying the methods of technical diagnosis or control involves the search for defects in a test object, taking into account the following requirements. It is necessary to develop new methods for determining a particular type of defects that rarely occure in the operation of engines and which are economically feasibile.

Thus, the use of diagnostics will allow reducing the number of unplanned repairs and reduce non-production costs.

In case of the faulty soldering, there is an increase in the transient resistance at the junction and, as a consequence, a significant heating described by the Joule – Lenz law [4, 5, 7]:

$$Q = I^2 R \, d T$$

where Q is the amount of heat a measure of the energy of the body, Joule, I is the current in the conductor, Amperes, R is the transient resistance characterizing poor contact, Ohm, dT is the period of time during which heating occurs, seconds.

3. Investigation of the heating effect along the riser length on the defect of faulty soldering of the riser

It allowed identifying the functional dependence represented by Fig. 1.

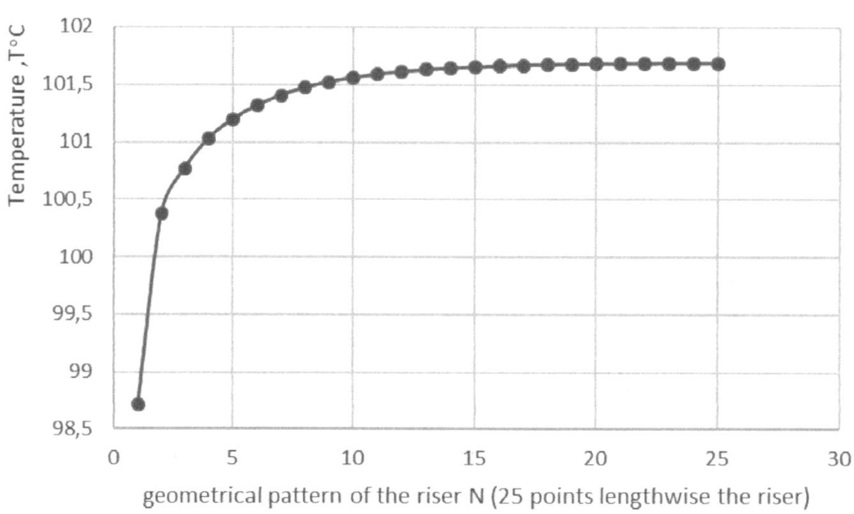

Figure 1. The simulated functional dependence of the riser temperature on the geometrical pattern.

Traction motors have a complex structure as a result of complex physical processes occurring in them; both thermal and electro-magnetic, which generally characterize a traction motor as a complex object for study[8].

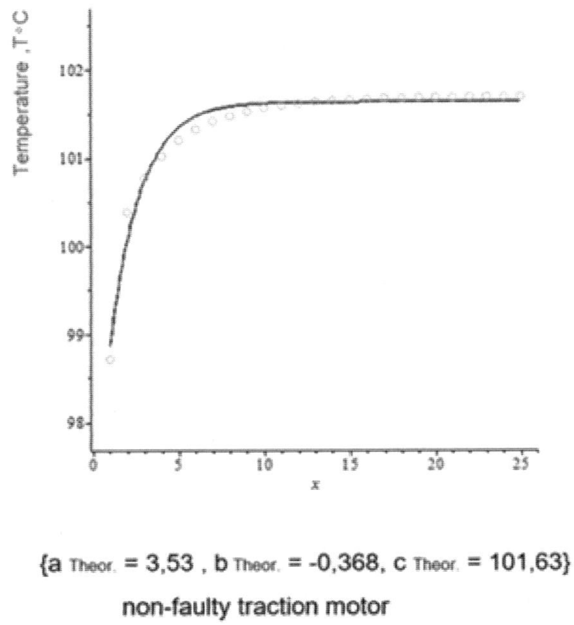

Figure 2. Approximation of the functional dependence.

The research methods of diagnosing traction motors were done at the expense of experts VNIIZHT and different institutes of railway transport, and by experts of locomotive repair plants.

There are various trends in the study of diagnosis methods of traction motors, due to the variety of possible failures and types of different physical damage in a TM.

In this regard, the study of the effect of the TM defects on the working process parameters and the development of methods for determining the type of defect of the traction motor is important.

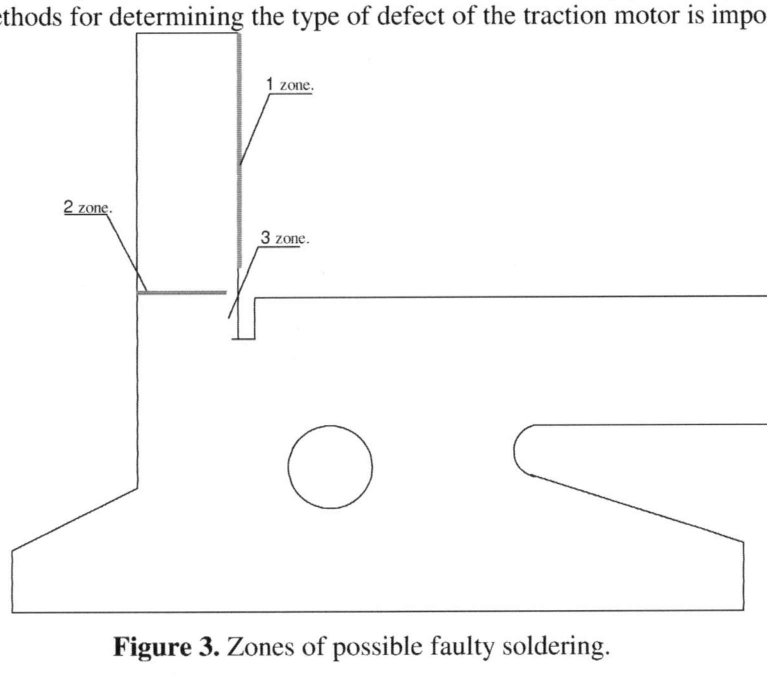

Figure 3. Zones of possible faulty soldering.

IOP Conf. Series: Materials Science and Engineering **760** (2020) 012015 doi:10.1088/1757-899X/760/1/012015

4. Conclusion

Based on the analysis of the damage of the TM on the road network, all major faults that occur in TM can be divided into the following groups. Malfunctions caused by the insulation aging of the motor windings inter-turn short circuits in the windings, the failure of the insulation of the windings on the object. Faults leading to the failure of the normal switching engine of substandard soldering in the riser, deterioration of the current collector in the collector-brush unit of the inaccurate setting of the brush engine in neutral, incorrect setting of the magnetic system additional poles, the asymmetry of the magnetic system of the main poles.

Faults can be caused by defects in the starting motor armature.

The results obtained in the work were used to create a control and measuring complex to determine the type of defect of traction motors NB 418K6.

To detect substandard soldering a riser TM can serve as a temperature distribution in zone 1 (Fig. 3) characterizing the defect as a change in the functional dependence of temperature on the geometry function (the number of points distributed lengthwise the riser). If there is a defect, the functional dependence will change the form and parameters of the mathematical equation.

After studying the defect of faulty soldering of the riser affects the temperature distribution lengthwise the riser.

The obtained data of the experiments with a faulty soldering in the engine riser, allow making the following conclusion. The differences between the graphical dependences of the riser thermal diagrams are evident. Thermal processes are analogues to electrical transients [1, 2, 3, 6]. The possibility of using the least squares method to approximate the mathematical dependence of the experimental data to find the numerical values of the variables of the equation in order to create methods of thermal imaging control of soldered joints.

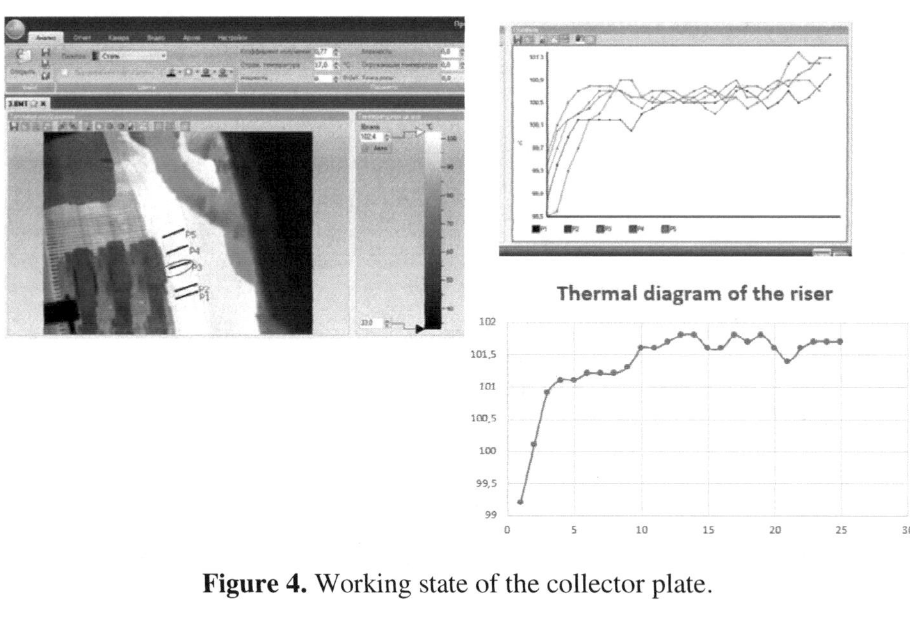

Figure 4. Working state of the collector plate.

IOP Conf. Series: Materials Science and Engineering **760** (2020) 012015 doi:10.1088/1757-899X/760/1/012015

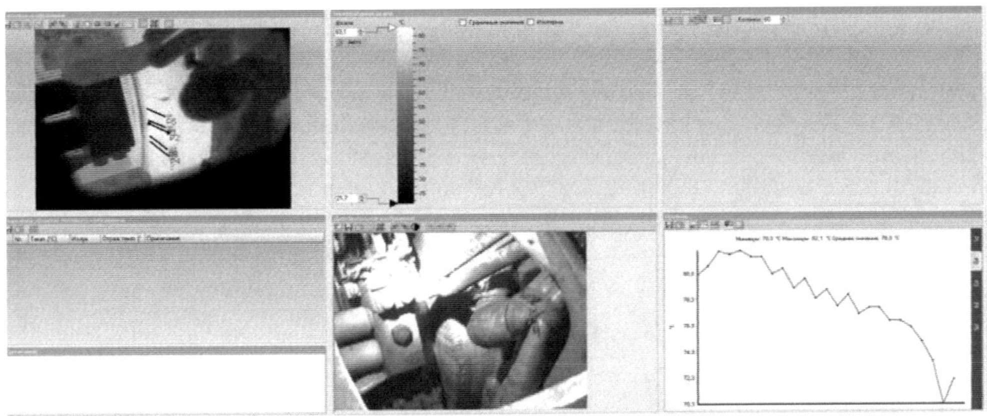

Figure 5. Collector plate malfunction.

Infrared imager position at the test facility

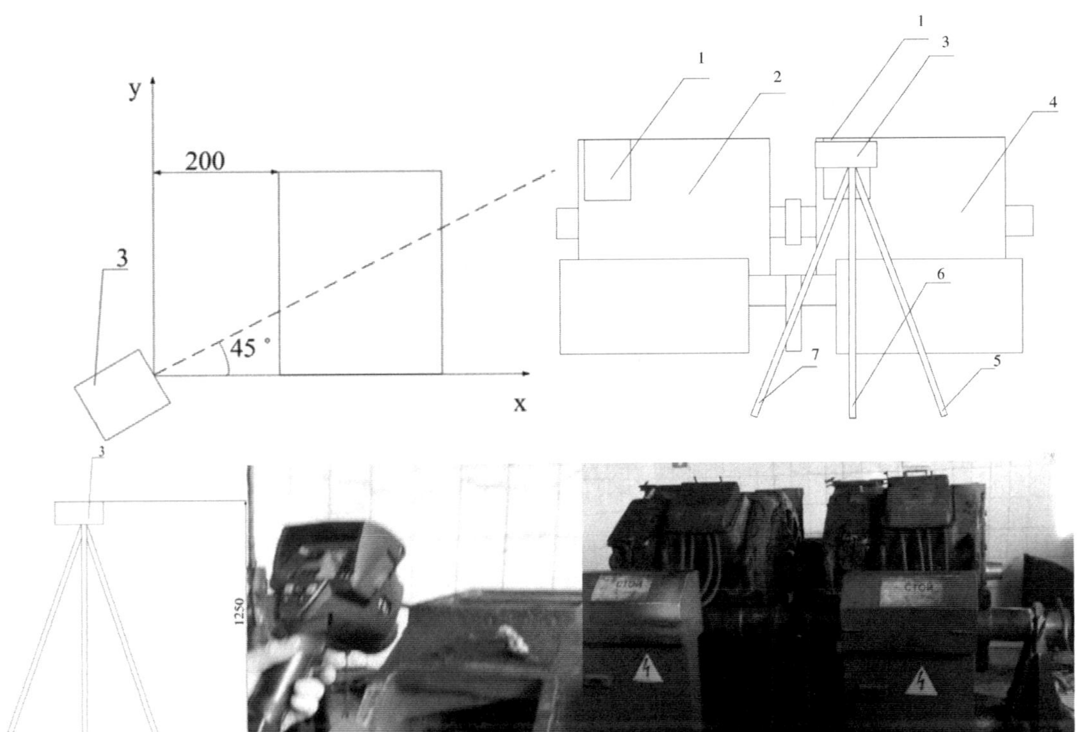

1- commutator inspection hole,
2- traction motor (TM 1),
3- infrared imager,
4 – traction motor (TM 2),
5,6,7 – tripod legs

Figure 6. The technique of thermal imaging for quality control in riser soldering of the collector.

The method of thermal imaging in quality control of the collector risers soldering (Fig. 6) with the test conditions for the NB-418K6 engine is as follows, the current of the motor is Id = 880 A, the engine speed is n = 890 r/pm, the heating time of the collector riser is t initial = 20 Minutes, cooling

IOP Conf. Series: Materials Science and Engineering **760** (2020) 012015 doi:10.1088/1757-899X/760/1/012015

time or TM stopping time = 180 s (3 minutes), the initial temperature of the collector riser is T initial = 21.7°C.

To automate the processing of thermal images in order to identify a defect in the collector plate, it is necessary to use the software based on the conduct of one single direct first collector plate. The presence of a rigid fastening of the thermal imager (Fig. 6) during stress tests allow application the numerical control of subsequent collector plates for the entire traction motor, since the geometry of the motor collector does not change, therefore, the subsequent coordinates of the remaining collector plates are known. The automated process of processing thermal images allows us to reduce the influence of the human factor, and, therefore, has a positive effect on the diagnosis of the engine.

References

[1] Ageev E P, 2019, *Nonequilibrium thermodynamics in questions and answers*. (Moscow: Lenand) 136

[2] Amirkhanov D G, 2014 Technical thermodynamics. *Textbook Amirkhanov D G*, 264 p.

[3] Belov G V, 2016, Thermodynamics in 2 h. *Part 1: Textbook and workshop for academic baccalaureate* 264

[4] Boboshina, S B Physics, 2019, Thermal processes: studies. manual for *2nd ed., ISPR. and additional* 118.

[5] Zakharov A Yu, 2016, Theoretical foundations of physical materials science. *Statistical thermodynamics of model systems: Textbook* 256

[6] Miram A O, 2016, Technical thermodynamics, *Heat and mass transfer: Educational edition* 352

[7] Piralishvili S A, Molecular physics. Thermodynamics. *Condensed States: Textbook sh. A. Piralishvili, E. V. Shalagina et al.* 200

[8] Fedotkin I M, 2015, *Mathematical modeling of technological processes* (Moscow: Lenand) 416

IOP Conf. Series: Materials Science and Engineering **760** (2020) 012016 doi:10.1088/1757-899X/760/1/012016

Using ontological and architectural approaches for the vehicle fleet management in the enterprise engineering context

V M Kurganov[1], M V Gryaznov[2], A N Dorofeev[3], O B Nastasyak[4], D V Pervukhin[5]

[1] Tver State University, 33, Zhelyabova st., Tver, 170100, Russia
[2] Nosov Magnitogorsk State Technical University, 38, Lenin St, Magnitogorsk, 455000, Russia
[3] Financial University under the Government of the Russian Federation, 29, Leningradskiy Ave., Moscow, 125167, Russia, National Research University Higher School of Economics, 33, Kirpichnaya St. Moscow, 105187, Russia
[4] Gazpromavia Aviation Company Ltd, 71/32, Novocheremushkinskaya str., Moscow, 117420, Russia
[5] National Research University Higher School of Economics, 33, Kirpichnaya St. Moscow, 105187, Russia, Financial University under the Government of the Russian Federation, 29, Leningradskiy Ave., Moscow, 125167, Russia

E-mail: adorofeev@hse.ru

Abstract. The transport activities at a large industrial enterprise are usually an auxiliary process. Therefore, its optimization with information technologies is often neglected. As a result, the cost accounting for the transportation of goods for various purposes is often a weighty problem due to inconsistent information flows. The method considered in this article combines the ontological and architectural approaches for the enterprise activities modeling intended to equalize occasionally a somewhat significant gap between business expectations and TMS functionality.

1. Introduction

The practical experience of implementing various information systems shows that this process is often accompanied by considerable difficulties, since IT specialists and business users have knowledge in different areas. This problem has been repeatedly described in literary sources [1, 2, 3, 4] and is particularly typical of specific activities, such as motor road cargo transportation. It is believed that the most complete return on investment into the information system is achieved when business and IT are initially synchronized at the strategic level for implementation of the enterprise mission. For this purpose, various models allowing improving interaction between business and IT in practice have been created in recent years. For example, Strategic Alignment Model (SAM), Model Driven Architecture (MDA), Business and Information Systems MisAlignment Model (BISMAM), etc. acquire currency [5]. However, Transportation Management System (TMS) remains at the periphery of all IT problems in a rapidly changing external environment of vehicle management at industrial enterprises, where cargo transportation is an auxiliary business process solving the issues associated with the flexible adaptation. In this regard, TMS usually solves a narrow range of operational tasks related to the maintenance of mandatory daily document flow without affecting the scope of strategic decision-making, thereby not creating any new value for the business.

Indeed, the automation of industrial enterprises primarily solves the issues of classic MRP/ERP systems related to the planning of production, finance, inventory management and human capital. The tasks of supply chain management are considered primarily in terms of optimizing the cargo transfer, reducing costs and time of transportation. The informational support for the direct operation of vehicles at its best consists of the implementation of GPS/GLONASS monitoring of vehicles and fuel consumption. This, in many cases, limits data collection on the participation of vehicles in the general production processes of a large enterprise. Certainly, fuel consumption by various estimations amounts to 50% of the total cost of vehicle operation. However, expenses on spare parts and vehicles repair should also be considered to ensure trouble-free operation in cargo transportation [6, 7]. It is obvious that a sudden failure of equipment due to low-quality or untimely repair may lead to unplanned failures in the work of the whole enterprise. Thus, the vehicle fleet that often enjoys the least attention during digitization becomes one of the key elements of ensuring sustainable operation of the whole facility [8, 9].

2. Vehicle fleet management as a multidisciplinary system of knowledge and practice

At the same time, the specifics of vehicle fleet digitalization requires profound knowledge in the area of motor vehicles operation from any IT specialist. On the other hand, we could also expect profound knowledge in the information technologies from a transport manager who would be able to create a perfect model of future TMS, or, at least, describe detailed requirements to it. However, as a rule, such questions cause insuperable difficulties in most cases, since transport workers are very poorly aware of the implementation of their business processes in the information system. Thus, the collection of requirements to TMS in practice comes to this: the IT specialist is having hard time perceiving the ideas and proposals of the transport worker, while the transport worker, in turn, sincerely wonders how one doesn't understand things he/she considers to be so simple. The study of paper documents flow does not help a lot in the research of information flows related to the operation of vehicles, as managers of transport departments at different enterprises use different sets of indicators and facts necessary to solve management problems. Thus, it is quite difficult to establish a "reference" model of the document flow. This process is further complicated by the fact that many managers come up with different forms of documents in Excel, which also significantly complicates document formalization process and, subsequently, the information exchange [10].

It should be noted that modern approaches to modeling the activity of the enterprise try to consider its sociocultural and organizational features, where a significant share of overall efficiency may depend on behavior of each individual employee and their personal qualities. For example, vehicle fuel consumption essentially depends on the skills of driver in the transportation process. Accordingly, when modeling the activities of the transport department, you should define the main business actors and factors that motivate them to achieve their business goals, the things they operate and the types of relations between them [11, 12]. It should be noted that the interaction between specialists of different departments, for example, the logistics, marketing or accounting department, often ends up in a mutual misunderstanding, as these specialists use different conceptual apparatus. For example, when considering repairs, the transport manager conceives the structure of the entire vehicle as a whole and conceives a nature of interaction of systems, components and assemblies of the vehicle, their influence on its operation. For an accountant, the main indicator is usually the cost of the part, not its lifetime. Thus, when making decisions on the purchase of spare parts, a transport worker will reason the need to acquire quality parts, though at a higher cost. And the accountant will reason the need to purchase cheaper spare parts. Accordingly, the accountant needs an understandable and accessible model which, on the one hand, would not go too deep into the theory of vehicle operation, and on the other hand, would reflect the essence of impact of part cost on the extent of their wear and, therefore, on their fault tolerance.

Thus, the example shows at least two points of view on the TMS implementation — transport and financial (economic). In practice, they have to constantly face such multidisciplinary tasks when

IOP Conf. Series: Materials Science and Engineering **760** (2020) 012016 doi:10.1088/1757-899X/760/1/012016

introducing an information system for vehicle fleet management at various enterprises. We propose to use a combination of ontological and architectural approaches to model the activities of a transport company. This combination is growing more and more popular among IT specialists as related to the Enterprise Engineering concept, where an organization is also considered as a social system. In this regard, improvement of management efficiency can be achieved, inter alia, by improving interaction between employees. It is TMS that should support the improvement of this interaction. In the context of considered example, you can also add a driver who directly operates this vehicle and may notice a malfunction. However, he can either report it or not report; so, as a result, the vehicle may be laid up for repairs. Consequently, the driver may be out of work for some time. Although, it is obvious that this malfunction can turn into a failure during the operation of the vehicle, which may result in serious consequences.

So, in this example, we see that three different employees, the accountant, the manager of transport department and the driver should agree among themselves and exchange agreements on the possibility of further operation of this vehicle. The accountant should understand that transportation of cargo on this vehicle is possible, but it is necessary to invest in spare parts and repair to prevent failure. The transporter determines the criticality of fault according to the driver's story and diagnostics results. The driver should carefully operate the vehicle and report the malfunction in time. From the point of view of Enterprise Engineering, the decision based on agreements is made at the ontological level.

It should be noted that different organization of business processes and vehicles repair in each particular enterprise affects the document flow and, accordingly, information flows. The repairs can be carried out both by the enterprise's own resources and at third-party vehicle service facilities. In the case of using enterprise's own resources, a diagnostic procedure should be preliminarily performed, then a decision is made on the possible further operation of the vehicle. Then the necessary spare parts are purchased and subsequent repair is performed. However, the organizational structure can significantly differ in the different companies, inter alia, it may be geographically separated. This, respectively, will have a significant impact on the document flow and accounting of these cost items as a whole. Thus, it is necessary to determine all documents, their form and content, which will ensure the exchange of agreements between all interested business actors at the infological level [13]. And actually, the design of the entire TMS as a whole takes place at the data level. Thus, first of all, the organizational structure of the company, business processes, actions of specific performers and their motivation should all be linked in order to effectively use TMS.

3. The study of the structure of the modified lead-tin-base bronze the combination of ontology and architecture of the enterprise to model its activities

Obviously, any transport activities in the enterprise associated with the movement of cargoes, except for internal industrial logistics, are also subject to the regulatory framework regulating it from the point of view of legislation. Thus, decision-making depends on their compliance with the law at the ontological level. At the infological level, the documents must comply with all legal requirements, if any. Accordingly, in addition to economic and technical concepts, it is also necessary to consider the legal concepts and the nature of legal norms influence on the transportation process. So, we see that the semantics of interaction among different entities when performing cargo transportation is described by ontologies of various fields of knowledge.

At present, the lack of a single ontology is a significant problem for the motor transport industry. There have been several attempts to describe separate fragments of the transportation activities. These include the ontology of urban logistics, the ontology of road safety and some others. We have presented the ontology of the business model of a transport company using A. Osterwalder's framework, where the reliability of the transportation process is presented as a value proposition [14]. Thus, the first step to the integration of economic and transport domains was made in this work. In the course of transport ontology development, the main emphasis was made on the improvement of its segments associated with the costs, which also actually affects the reliability of transportation process.

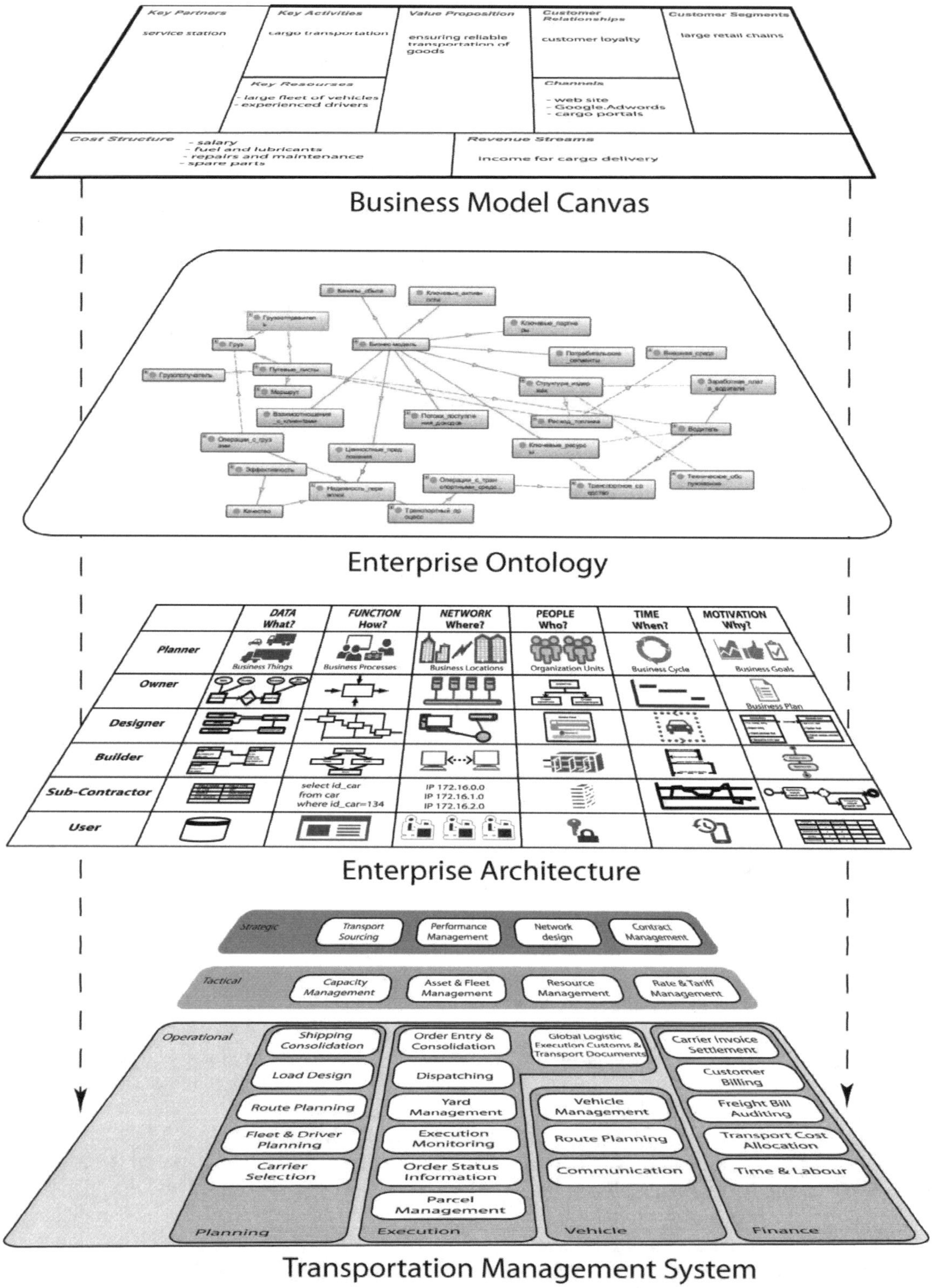

Figure 1. The combination of ontological and architectural approaches in the design of TMS

In this case, as you know, the cost segment is also included in the outline of A. Osterwalder's business model. As a result, an ontology describing the entities influencing fuel consumption, as well as the ontology describing the technical operation of vehicles, including repairs and maintenance, was developed. Thus, with the help of ontologies, we have described the "inner world" of the transport enterprise, the entities involved in the transportation process, the nature of their interaction with each other, as well as the business actors and the nature of their behavior to achieve business goals, and what the enterprise should do to make money [15, 16].

The enterprise architecture concept is designed to answer the questions "Who?", "How?", "When?", "Where?" and "Why?" performs these or those operations for the solution of business tasks [17]. The Enterprise Architecture consists of several layers:

- Business architecture reflecting business activities, business processes that create value for the consumer;
- Application architecture describing the information landscape, i.e. IT services, information systems and information flows;
- Technological architecture that provides integration of business processes, data and applications;
- Infrastructure Architecture which directly reflects the software and hardware configuration;

The work [18] proposes to use the Ontological model as a prototype for Business Architecture. This approach assumes the transfer of semantically related entities of the transport enterprise ontology, including vehicles, documents, drivers, mechanics, etc. into corresponding business objects, business roles that perform certain business processes in a certain location to achieve specific business goals. The development of this approach in relation to transport activities enables to use objects and concepts from different fields of knowledge in the modeling (vehicle management, logistics, human capital management, legislation, economics) by semantically linking them to form the concept of how the enterprise can make money in the best way (Fig.1). By defining the totality of these entities and the nature of their interaction and influence, you can put together the Business Architecture by formalizing business processes and business activities that will be performed by specific business actors according to their business roles [19].

4. Conclusion

The proposed approach in relation to design of transport activities can be used in TMS implementation to ensure flexibility and adaptability, allowing reconstructing a system "on the run". At the same time, various specialists of the enterprise can provide their vision and make their proposals for the project, harmoniously synchronizing them with the strategic goals of the enterprise. Thus, the comprehensive cooperation between all interested persons and subdivisions of the company is ensured. Accordingly, the information flows that seamlessly pass through the operational, tactical and strategic levels of company management can be transformed into management decisions creating new business value. This approach also enables to synthesize value for the company from the informal knowledge of employees and considers their experience in organization model design.

References

[1] Ye F and Wang Z 2013 Effects of information technology alignment and information sharing on supply chain operational performance *Computers & Industrial Engineering* **65** 370–377

[2] Mikalefa P, Patelia A, Batenburgb R. and Weteringb van de R 2013 Investigating the Impact of Procurement Alignment on Supply Chain Management Performance *Procedia Technology* **9**, 310 – 319.

[3] Haes S D and Grembergen W V 2008 Analysing the Relationship between IT Governance and Business/IT Alignment Maturity *The 41st Annual Hawaii International Conference on System Sciences (HICSS 2008)* 428-428

[4] Qrunfleh S and Tarafdar M 2014 Supply chain information systems strategy: Impacts on supply chain performance and firm performance *Int. J. Production Economics* **147** 340–350

[5] Stahl T, Völter M, Bettin J, Haase A and Helsen S 2006 Model-driven software development - technology, engineering, management
[6] Tyszko A, Templin T and Oszczak S 2007 GNSS systems in vehicle fleet management *Transport Problems*, **2** 11–16
[7] Wycoff D F 2009 Implementing an SAP Transportation Management System Solution: a Case Study
[8] Daithankarand J and Pandit T 2014 Transportation Management with SAP TM 9. A Hands-on Guide to Configuring, Implementing, and Optimizing SAP TM
[9] Dorofeev A 2013 Development of Internet-Based Applications for Fleet Management and Logistics *The 15th IEEE International Conference on Business Informatics* 428-433.
[10] Carrese S, Gemma A and La Spada S 2013 Impacts of driving behaviours, slope and vehicle load factor on bus fuel consumption and emissions: a real case study in the city of Rome *Procedia - Social and Behavioral Sciences* **87** 211–221
[11] Dietz J L G 2006 Enterprise Ontology.Theory and Methodology
[12] Janssen T, 2017 Enterprise Engineering. Sustained Improvement of Organizations
[13] Fox M S, Barbuceanu M, Gruninger M, and Lin J 1997 An Organization Ontology for Enterprise Modelling *Simulating organizations* 131-152
[14] Kurganov V, Gryaznov M and Dorofeev A. 2018 Management of transportation process reliability based on an ontological model of an information system *Transportation Research Procedia*, **36**, 392–397
[15] Kang D, Lee J and Kim K 2010 Alignment of Business Enterprise Architectures using fact-based ontologies *Expert Systems with Applications* **37** 3274–3283
[16] Kudryavtsev D and Grigoriev L 2011 The Ontology-based Business Architecture Engineering Framework, New Trends in Software Methodologies, Tools and Techniques. *The 10th International Conference on Intelligent Software Methodologies, Tools and Techniques*, 233-252
[17] Lapalme J, Gerber A, Van der Merwe A., Zachman J, De Vries M. and Hinkelmann K 2016 Exploring the future of enterprise architecture: A Zachman perspective *Computers in Industry* **79**, 103–113
[18] Hinkelmann K, Karagiannis D, Thoenssen B, Woitsch R, Gerber A and Van der Merwe A 2016 A new paradigm for continuous alignment of business and IT: combining enterprise architecture modeling and enterprise ontology *Computers in Industry* **79** 77-86
[19] Kurganov V and Dorofeev A 2016 The Practice of Business and IT Integration in the Transport Company Using Enterprise Architecture Framework *The 7th International Conference on Operations and Supply Chain Management* 377-392

IOP Conf. Series: Materials Science and Engineering **760** (2020) 012017 doi:10.1088/1757-899X/760/1/012017

About control of a neuron as an elementary unit of multilayer artificial neural networks with variable signal conductivity

I A Dovgerd[1], G A Dovgerd[1], A V Ignatenkov[2], A M Olshansky[1], M G Lysikov[1] and E N Rozenberg[1]

[1]Design Bureau JSC "Railway Signalling Institute", 27, b, 1 Nizhegorodskaya St., 109029, Moscow, Russia
[2]JSC "VTB Capital", 10, Presnenskaya Embankment, 123112, Moscow, Russia

E-mail: lexolshans@gmail.com

Abstract. The report is devoted to a possibility of a control of each Hopfield's neuron as an elementary unit of the neural network with variable signal conductivity used for transport scheduling tasks. It is set that direct control of the neuron is impossible in terms of initial data available for initializing the neural network. Also the result of some control algorithms based on statistical computations and signals dynamic is described below.

1. Introduction

Artificial neural networks (ANN) are famous popular algorithms for solving rather difficult and variable tasks at present. Every neural network is a complicated system which could be investigated as a dynamic system. The problem of decreasing the training time of the ANN is very urgent for science and practice and it may be solved using control theory and dynamic representation. It allows us to use computational capacities more efficiently.

At the turn of the XX and XXI centuries in Russia and abroad there were some attempts to apply control theory to ANNs [5, 6, 8]. The characteristic feature of such papers was applying the control theory to multilayer perceptrons with creating optimal time sequences of ANN weight matrices. The sequences were obtained as a result of solving the two-point nonlinear boundary value problem. It led to optimal training rules for ANN weights.

An optimal weight matrix looks like the sum of the products between desired output vectors R^n and concrete output vector calculated permanently during the ANN's training. Finally we get an optimal sequence of weight matrices for ANN.

Paper [7] analyzes the optimality of hyperparameters of ANN using genetic algorithms.

According to the classical rules of the control theory every task of optimal-controlled ANN should consists of the next elements:

1. A model of the controlled object's evolution. The ANN may be described as the set of rules or using differential, integral or discrete equations. The state of ANN as a system can be described as a set of activation functions for each layer or each neuron.

2. Initial and final conditions are being set implicitly via weight functions, values of weights, initial trainspeed, sensitivities to output error etc.

3. Outer impacts the ANN usually can be realized implicitly. For example one kind of outer impacts

is even feed an initial vector on ANN's input.

4. Time period of ANN's functioning can be set directly (it leads to the control problem with fixed ends) or implicitly (control problem with open right end).

5. In majority of cases constraints of ANN's state and constraints of control are not determined for ANN's control problem.

6. In the ANN's control problems a model of the measurements is often absent because the researcher may observe the ANN's out directly.

7. Measurement errors and noises are coupled with some peculiarities of PC computation but sometimes are generated specially using functions of temperature fluctuations.

8. The aims of control are usually formulated as requirements for expected solutions, such as minimal output error, accuracy of classification etc.

9. A control algorithm. At present there are some general algorithms for each kind of ANN [4], such as error back propagation, BFGS, LM, gradient descent, annealing etc. But sometimes ANN requires special methods and algorithms. Its feature is absence of strict requirements which can guarantee us convergence of the ANN.

Hence an ANN is a dynamic system contains all elements of the control task. Applying the control theory methods allows us to create new algorithms of ANN's training and functioning for different tasks, distinctly determine boundaries of using ANNs, investigate ANN's dependence on outer conditions, mode of initialization, ANN's hyperparameters etc.

Optimal control of ANN is not well-known till nowadays especially for non-standard kinds of ANN with other topology. One of them is a neural network with variable signal conductivity and it serves for solving traveler salesman problem, scheduling problem etc. [1].

There were 2 attempts to create optimal control of multilayer ANN with variable signal conductivity based on error signal analysis [1] and total weight analysis using special database. The main result is that this kind of ANN is controllable by output, but there is no solution how to realize this control strategy using weights and layers of the ANN. The ANN with variable signal conductivity has about 10^6-10^7 links and weights; because of this computational complexity is more than 10^{lk}.

The goal of the article is to synthesize a control strategy for an elementary neuron of multilayer ANN with variable signal conductivity and to improve existing solutions.

2. Control of a Hophfield's neuron

Let us describe evolution of an elementary neuron as an electric circuit in terms of J.J. Hopfield's paper [2].

According the topology of the considered ANN output voltage equation can be described as (based on [2]):

$$\frac{dx}{dt} = -\frac{x(t)}{t} + \frac{w}{1+e^{-Power}} + I, \tag{1}$$

where $x(t)$ is output voltage of the neuron, $T = RC$ – the product of neuron's resistance and neuron's capacity, I – the bias value, w – weight of the income link of the neuron, *Power* – the value of neurons' output from previous layer.

For formulation the control problem of the selected neuron let's transform equation (1) as:

$$\frac{dx}{dt} = -x(t) + u(t) + D, \tag{2}$$

where u(t) is desired control function, D is the sum of the second and the third components of the right part in equation (1).

We also have time period $[t_1,t_2]$ and the functional of control's quality (3):

$$I = \int_{t_1}^{t_2} u^2(t)dt \rightarrow min. \tag{3}$$

We arrange that there are no constraints of control.

Initial conditions $x(t_1) = x_1$ are also set.

As we don't need the powers of neuron at the finishing of control $x(t_2)$ (we need only neuron's number to calculate error function, see [3]) so we don't set the value of neuron's output at the final control moment.

So we faced the classical isoperimetric control Lagrange problem. Let's solve it using the Pontryagin's maximum principle.

The Hamiltonian of the system is:

$$H = \Psi(-x(t) + u(t) + D) - u^2(t) = \Psi u(t) - \Psi x(t) + \Psi D - u^2(t). \tag{4}$$

To find the structure of optimal control, let's describe the first derivative of the Hamiltonian by the control function:

$$\frac{dH}{du} = \Psi(t) - 2u(t). \tag{5}$$

The second derivative of it is equal to -2<0, thus equation (5) is the maximum of the control function, and the structure of the optimal control is:

$$u_{opt} = \frac{\Psi(t)}{2}. \tag{6}$$

Let's create the canonical system of equations to find the optimal control:

$$\begin{cases} \frac{dx(t)}{dt} = \frac{dH}{d\Psi} = -x(t) + \frac{\Psi(t)}{2}, \\ \frac{d\Psi(t)}{dt} = -\frac{dH}{dx} = \Psi(t). \end{cases} \tag{7}$$

The condition of transversality is:

$$\delta F\big(t_2, x(t_2)\big) - H(t_2)\delta t_2 + \Psi(t_2)\delta x = 0, t = t_2, \tag{8}$$

where F is a terminal summand of the equation (3).

Let's consider formula (8). As we are solving the Lagrange's problem so the terminal summand is absent and $\delta F(t_2,x(t_2))=0$. The right end of trajectory is fixed thus $\delta t_2=0$ and $H(t_2)\delta t_2=0$. So we see that $\Psi(t_2)\delta x=0$. As the variation δx is rather arbitrary, hence $\Psi(t_2) = 0$.

Thus, $x(t_1) = x_1$ and $\Psi(t_2) = 0$ may be considered as necessary conditions for the system (7).

The general solution of the second equation of the system (7) is:

$$\Psi(t) = C\Psi_0 e^{t-t_1}. \tag{9}$$

Finding the value of C we get the incompatible equation:

$$\Psi_0 e^{t-t_2}. \tag{10}$$

As a result of the task (1)-(3) we may observe that:

1. The optimal control strategy we have found is a monotonic growing function. It is permitted only in the virtually world. For every neuron of the ANN it means the fixation of the control value at the final moment of control.
2. No such control function can be realized because of equation (10).

We may conclude that every neuron of the ANN with variable signal conductivity in the variant of Lagrange problem is impossible and it requires the set of the neuron's output function at t_2.

3. Analysis of ANN phase portraits

Also authors analyze the ANN's behavior using the phase portraits in coordinates ($dE(t)/dt, E(t)$) for the converged and unconverged neural network.

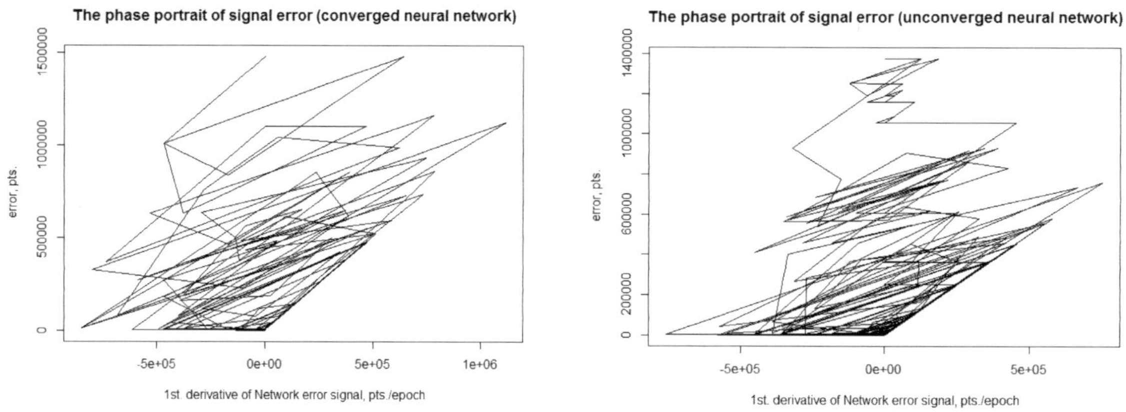

a b

Figure 1. Phase portraits of the ANN: converged – a; unconverged – b.

We may see the similarity. Both of portraits (figure 1) illustrate us sets of quasi-stable cycles and points with a low error level which don't consist of the cycles. Meanwhile the speed of the error is changing during habitation at the cycles with growth of the total error.

Figure 2 presents a phase portrait in coordinates (acceleration of error, speed of error). Quasi-stable cycles are detected too.

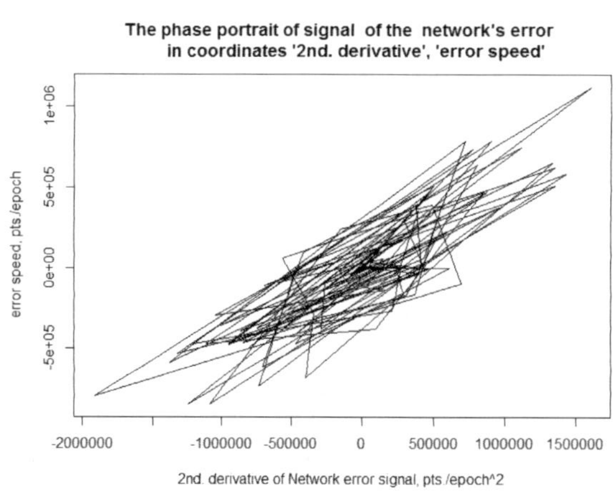

Figure 2. A phase portrait of the 2nd ordered system

Behavior of multilayer ANN with variable signal conductivity [1] can be characterized with sudden jumps of the first and the second derivatives of the error signal. The reason for it is in structure of the network and in the principle of maximal link election during calculation of output.

When an even signal propagates through the connections between neurons, then at the time of activation of the neuron according to the conditions of the problem there is an insensitive zone. An odd signal propagating through the connections finds the maximum odd connection that can lead to the switched-off neuron. In this case the signal propagates through the next link with largest weight. And this connection can be separated from the point of the desired output by a significant distance (network connections are initialized randomly). Thus, we observe a sharp increase in the output error signal.

To mitigate this effect, the model of determining the width of trained links for each neuron was changed according to the formula:

$$s = \left(E(t) + \left(E(t) - E(t-1)\right)\right)^{1/2}, \tag{11}$$

where s is the magnitude of the beam of neuron's links for training, E(t) is the error, t is the epoch number.

Thus, the behavior of the ANN is controlled not only by the error, but also by its change in the previous step. If the ANN tries to increase the error, the width of the link beam increases.

As a result the phase portraits of the error have changed significantly (Figure 3).

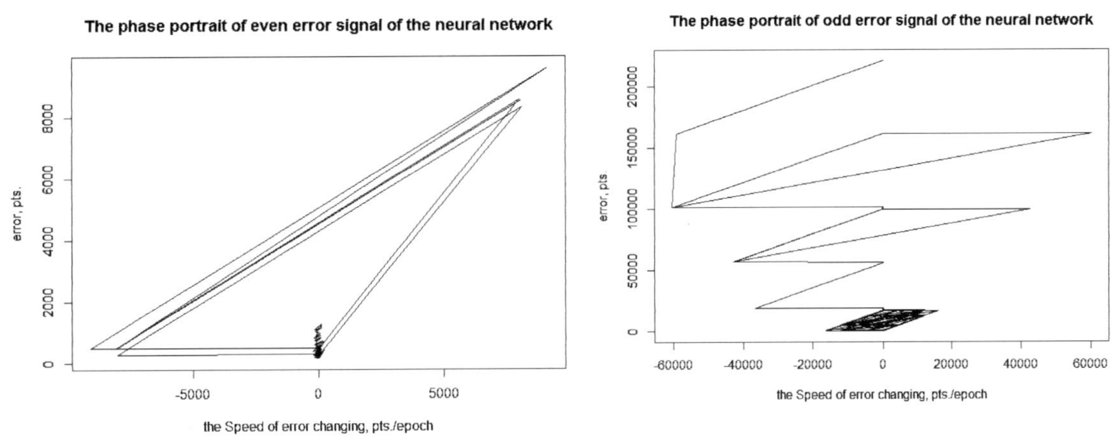

a b

Figure 3. The phase portrait of the ANN considering the current and previous errors (even the error is in a, odd is in b)

Despite the fact that the fully oscillatory behavior of the ANN could not be eliminated, there are significantly lower error values and the ANN is near them for a long time. This allows us to stop the training of the ANN and get a solution for a smaller number of epochs (in 1.7–2 times) than is achieved in the usual network [3].

Another technique to reduce fluctuations in the output of the ANN during the training process is introduction of so-called control errors and post-training.

The essence of post-training is as follows: at the first epoch ANN yields primary errors and writes them as a control. At following epochs the current errors change and are taken to comparison with the results of the control errors of the previous step. If the new error is greater than the previous control error, the procedure of post-training is started; thereby the ANN tries to reduce the error to the minimum value, simultaneously striving to a quasi-stable position. In case when the new value of error is less than control errors of the previous one, the control errors are overwritten. After a certain number of epochs ANN comes to a quasi-stable state, but even there are jumps, on average, once per 275 epochs. Post-training is carried out 25 times, after which the already changed error results are being displayed.

The results are shown in the graphs below (Figure 4).

4. Conclusions

The ANN may be characterized by complex non-linear behavior. As a result of the research authors found two best modes recommended for the control of error signals. For these modes phase portraits are very similar, but post-training gives fewer oscillations within a given area.

There are no classic rigor techniques based on optimal control theory for the considered artificial neural network with variable signal conductivity. These authors suggest a combination of control methods including traditional ANN training techniques and specific approaches.

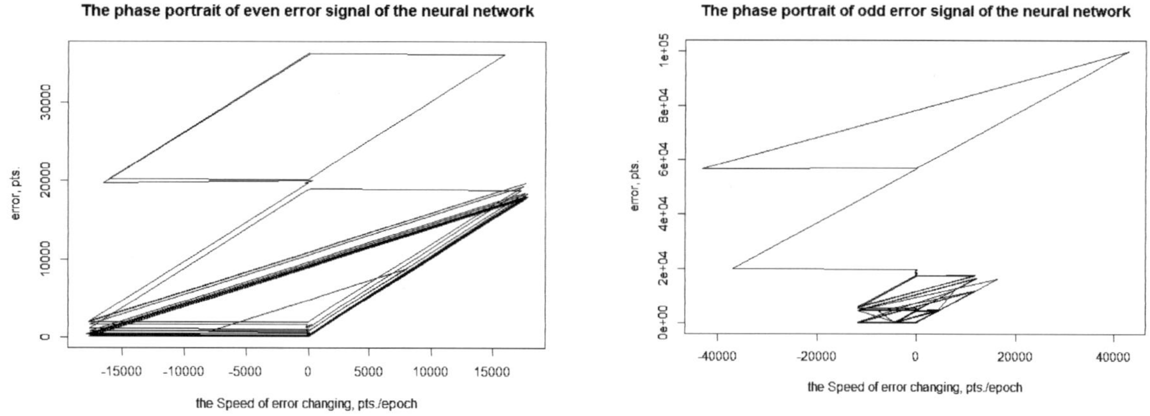

a b

Figure 4. The phase portrait of the post-trained ANN (even error is in a, odd is in b)

Architecture of the post-training method is universal and suitable for every other control technique of the ANN (e.g. PID-controller, neuro controller, etc.). The post-training method provides a solution for a smaller number of epochs (in 1.7–2 times) than others.

Comparison between phase portraits of post-training method and ANN training with previous dynamic (with flexible beam) represents that post-training (in average 1 gap per 200 epochs) reduces oscillations of the error signal to a greater extent than the second method (in average 1 gap per 176 epochs).

5. Acknowledgments

The reported study was funded by the RFBR (project number 17-20-01065 "A theory of railway transport systems' neural network control").

6. References

[1] Olshansky A and Ignatenkov A 2016 About the task of running an artificial neural network with quasiperiodic nature of the behavior of the error signal *Information Technologies* 635–8

[2] Hopfield J J and Tank D W 1985 "Neural" computation of decisions in optimization problems *Biological cybernetics* **52(3)** 141–52

[3] Olshansky A and Ignatenkov A 2016 Development of an artificial neural network for the construction of train schedule *Bull. of the Ryazan state radio engineer. University* **55** 73–80

[4] Nazarov A V and Loskutov A I 2003 *Neural network algorithms of forecasting and optimization of systems* (St. Peterburg: Science and Technology) 384 p ISBN 5-94387-076-8.

[5] Fahotimi O, Dembo A and Kailath T 1989 Neural network weight matrix synthesis using optimal control techniques *Advances in Neural Information Processing Systems-2 (NIPS-2)* (USA, Stanford, Denver, Colorado).

[6] Becerikli Y, Konar A-F and Samad T 2003 Intelligent optimal control with dynamic neural networks *Neural Networks* **16** 251–9 Retrieved from: www.elsevier.com/locate/neunet

[7] Choi Y 2007 *Neuroevolutionary algorithm and software for image processing*: thesis for the degree of candidate of technical Sciences (Tomsk: Tomsk Polytechnic University) 209 p

[8] Andreeva E A and Pustarnakova Yu A 2001 Mathematical model of artificial neural network with Delay *Software products and systems* **3** 6–9

IOP Conf. Series: Materials Science and Engineering **760** (2020) 012018 doi:10.1088/1757-899X/760/1/012018

Reliability analysis of power equipment of traction rolling stock within the Eastern region

A M Khudonogov[1], I A Khudonogov[1], E Y Dulskiy[1], P Y Ivanov[1], I O Lobytsin[1], A A Khamnaeva[1]

[1]Irkutsk State University of Railway Engineering, 15, Chernyshevsky, Irkutsk, 664074, Russia

E-mail: E.Dulskiy@mail.ru

Annotation. One of the most loaded structures, and at the same time, the weakest elements of the power equipment of traction rolling stock, is isolation. The analysis of the operation of power equipment of electric locomotives from the negative impact of climatic factors on it has been made. It was revealed that to ensure reliable and safe operation, a system of maintenance and repair of rolling stock is necessary, taking into account the zonal conditions of operational features. The necessity of creating adaptive systems to maintain optimal values for temperature and humidity conditions for the insulation structures of electric locomotives is shown.

1. Introduction

The reliability of traction rolling stock is one of the main tasks of the railway industry. The economic well-being of the whole country depends on its smooth operation. The weakest link in rolling stock equipment is insulating structures. Insulation is significantly affected by heat and moisture. In the framework of the state task of research and development on the topic "Improving the reliability of power equipment of traction rolling stock" (№AAAA-A19-119010990009-3), an analysis was made of the reliability of power equipment of electric locomotives within the Eastern region.

In the process of operating the power equipment of electric locomotives, from the negative impact of climatic factors on it, natural and artificial influences can be distinguished. Moreover, natural influences depend primarily on external conditions that are characterized by current weather: temperature, humidity, atmospheric pressure, precipitation, wind. In turn, artificial influences are formed in the process of the direct operation of power equipment and other objects located nearby [1, 2]. From considering the influence of climatic factors on the reliability of insulating structures of power equipment, natural climatic influences are of the most significant interest.

To maintain electric locomotives in working conditions and ensure reliable and safe operation, a system of maintenance and repair of rolling stock is necessary, taking into account the zonal conditions of operational features. This study based on a long-term analysis of statistical data on the reliability of extremely loaded electrical equipment of electric locomotives operating on the railways of the Eastern region. Long-term analysis indicated the need for a systematic approach to the analysis of reliability and development of the scientific foundations of the technical maintenance system for electric locomotives, taking into account the zonal features of operation even within the boundaries of the same road or region of circulation [3-5].

IOP Conf. Series: Materials Science and Engineering **760** (2020) 012018 doi:10.1088/1757-899X/760/1/012018

2. The zonal feature of the influence of climatic factors

The objectives of the study include assessing the zonal features of the influence of climatic factors of the Eastern region on the operation of power equipment of electric locomotives. Climate analysis is carried out in the areas of four railways in the region: the Krasnoyarsk Railway, the East Siberian Railway, the Trans-Baikal Railway, the Far Eastern Railway.

The climate of most of the Krasnoyarsk Railway is sharply continental. Moreover, the climate itself is very heterogeneous due to the vast extent of the edge in the meridional direction. Thus, the climate in the northern part of the Krasnoyarsk Railway is exceptionally severe, which is expressed in long winters with significant negative temperatures. In the central and southern parts of the Krasnoyarsk railway, the climate is much milder. These zones belong to the zone of the temperate climatic zone [6]. Moreover, due to the significant remoteness of the entire region from the seas and oceans, the climate here is more severe than in other regions located at similar latitudes.

The climatology of the East Siberian railroad in the central zone is characterized by moderate temperature and a relatively high relative humidity of the surrounding air because of the macroclimatic regulator for this zone of operation of electric locomotives is the largest lake in the world, Lake Baikal. The geographical location of Lake Baikal leaves its mark on the northern direction zone, where during the so-called "Epiphany frosts," the ambient temperature reaches –59 ° C.

The Trans-Baikal Territory occupies a central position; however, its distance from the oceans is not the same. The seas of the Pacific Ocean – the Okhotsk and the Yellow are 850-1000 kilometers away from the Trans-Baikal Territory. The nearest of the seas of the Arctic Ocean – the Laptev Sea is 1700 kilometers away from the region.

The climate in the Far East is monsoon, with characteristic cold winters and humid hot summers. Climatic conditions vary significantly with the nature of the relief, proximity to the sea, and from north to south. For example, the average January temperature in continental regions ranges from −22 °C– in the south, to −40 °C –in the north; on the sea coast – from −18 °C to −24 °C. The average July temperature in the south is +20 °C, in the north +15 °C.

To assess the distribution of climatic features throughout the Eastern region of circulation, we reduce the average temperature values for the years under consideration by one histogram (Figure 1). The analysis of the presented histograms shows that in April and October, all the sections under consideration show different sign temperatures, which can negatively affect the insulation of the power equipment of electric locomotives.

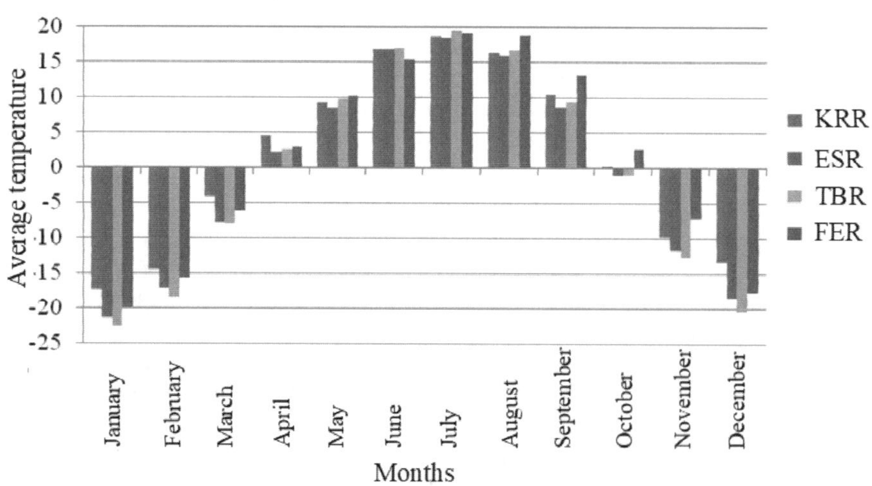

Figure 1. Histogram of average values of air temperature in the Eastern region by climate zones for 2012-2018. (the Krasnoyarsk Railway – KRR, the East Siberian Railway – ESR, the Trans-Baikal Railway - TBR, the Far Eastern Railway road – FER).

Terrain changes are one of the characteristics of the territories of the Eastern region. The influence of the relief affects cloudiness, air temperature, rainfall, and the isolation of mechanisms. Humidity is one of the critical factors from the negative impact on the reliability of the insulation structures of power equipment of electric locomotives. A certain amount of water vapor is almost always present in the air, which is extremely variable. This number may increase or decrease depending primarily on air temperature. This fact causes large fluctuations in the degree of air saturation, depending on the day and year time. Also, humidity significantly depends on the amount of precipitation.

A clear comparison of the distribution of annual precipitation in the Eastern region is presented in the diagrams. Figure 2 shows the total histogram of climatic zones. Figure 3 presents a summary histogram of annual humidity values in the Eastern region by climate zones for 2012-2018.

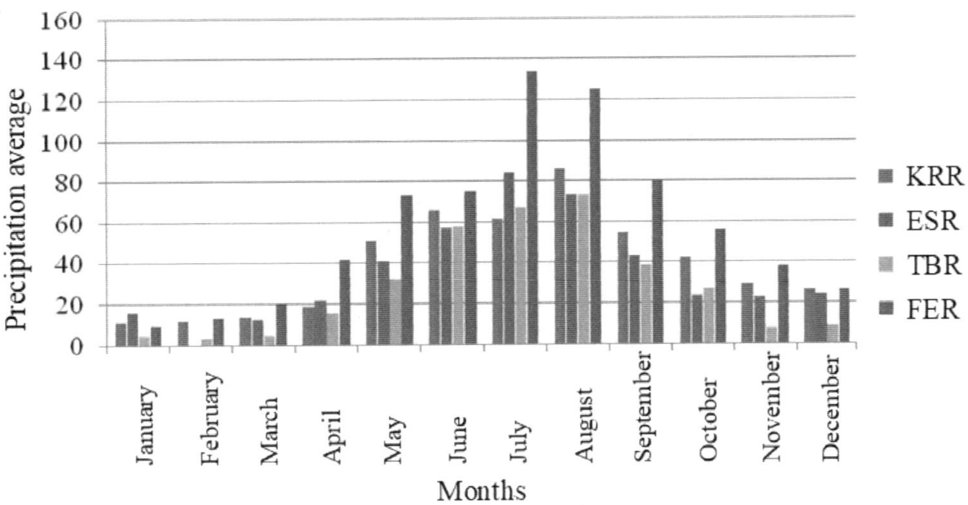

Figure 2. A histogram of average values of total annual precipitation in the Eastern region by climate zones for 2012-2018.

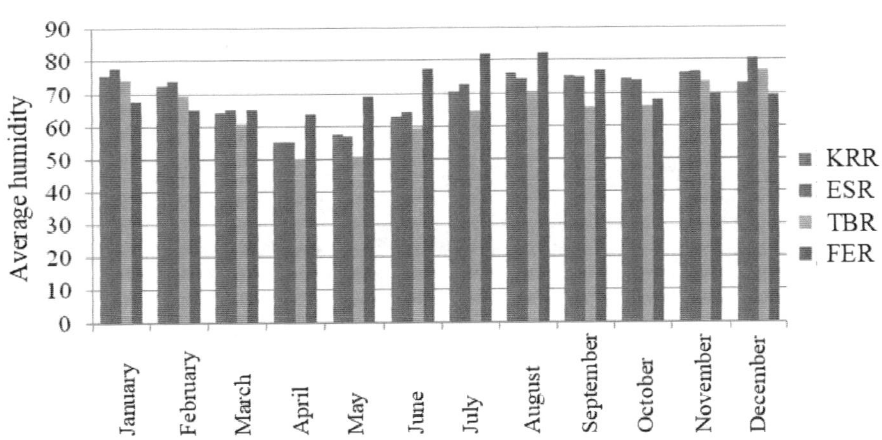

Figure 3. A histogram of average total annual humidity values in the Eastern region by climate zones for 2012-2018.

IOP Conf. Series: Materials Science and Engineering **760** (2020) 012018 doi:10.1088/1757-899X/760/1/012018

Changes in air humidity contribute to the formation of a layer of water on the material. At low humidity, a monomolecular layer of water forms on the surface of the material. At higher humidity, a polymolecular layer begins to form, the thickness of which increases sharply with the relative humidity approaching 90%. At humidity above 90%, the layer of adsorbed water is in a liquid state. The process of adsorbed condensation, i.e., the formation of the thinnest moisture layer associated with the surface by adsorption forces precedes the process of pure physical condensation and may occur at a relative humidity below 100%.

The annual variation in the moisture content of surface air layers depends mainly on local evaporation. In this regard, the first half of the growing season is of the most significant importance. At this time, comparatively little rainfall still falls, and the moisture reserve in the soil after the melting snow cover is small [7]. In this regard, in some years in the central regions, especially along broad river valleys, humidification is insufficient. Atmospheric humidification is closely related to the features of the thermal regime, atmospheric circulation, and the nature of the underlying surface. At an air temperature of 25-40 ° C, with a temperature increase of 2-3 ° C, relative humidity decreases from 100 to 90%, with a temperature increase of 5 ° C – from 100 to 75-85%, with an increase of 10 ° C – to 55-60%.

3. Distribution of failures of insulation of power equipment

Issues of wetting the electrical equipment insulation of electric locomotives always require great attention. Particular attention to the issues of moisture resistance of electrical insulation connects with the development of railway transport. The moisture resistance of equipment is associated with an increase in electrical equipment failures due to the breakdown of insulation.

The impact of extreme negative temperatures on the insulation of power equipment leads to an increase in its resistance, a decrease in leakage currents, and a decrease in return voltages. These residuals contribute to a slower discharge rate. The cyclic effects of negative and positive temperatures lead to irreversible changes in the insulation resistance and a decrease in its breakdown voltage.

Electrical equipment of electric locomotives, by the nature of their work, is in high humidity for a significant part of the year. Moisture penetrates the pores and capillaries of the insulation, worsening its dielectric properties. Even newly manufactured and impregnated insulation absorbs moisture no worse than the old insulation [8].

A significant amount of precipitation, together with the shortcomings of the ventilation systems for cooling the traction engines of electric locomotives, leads to the fact that moisture settles directly on the insulation surface (Fig. 4). Moreover, with a higher degree of probability, a sharp wetting of the insulation can occur, which, in turn, leads to electrical breakdown [9].

Figure 4. The presence of snow on the collector-brush assembly (left) and the magnetic system (right) of the electric traction motor of the locomotive

Atmospheric humidification is closely related to the features of the thermal regime, atmospheric circulation, and the nature of the underlying surface. At an air temperature of 25-40 ° C, with a temperature increase of 2-3 ° C, relative humidity decreases from 100 to 90%, with a temperature increase of 5 ° C – from 100 to 75-85%, with an increase of 10 ° C – to 55-60%.

Studies have shown [10] that moisture penetration into insulation can co-occur at several levels: at the molecular, intermolecular, and capillary levels. The last of them is of the most significant interest since most insulation systems used in the construction of electrical equipment, especially after the long-term operation, pores, microcavities, and through channels, are formed. Their occurrence is mainly due to the shortcomings of the existing technology for repairing insulation using convective electric furnaces. In this case, in the process of surface heating with hot air, a crust forms on the insulation surface, through which fumes of evaporation of volatile organic substances later breakthrough. During operation, under the constant influence of vibrational forces on the polymer insulation, microcavities can pass into microcracks, forming channels connected.

The presence of microcavities is due to the cracks occurrence and the insulation delamination, the material shrinkage during aging. Voids and channels can also appear in the process of applying insulation to the conductors, in particular when using tape materials, which form folds, layers, and loosely join. Significant voids are also observed between the ground conductors of the windings of auxiliary electrical machines. Impregnating varnishes, which contain about 50% of volatile substances, do not fill the formed space. With high humidity, such channels can be excellent conductors of moisture, which can penetrate the depth due to the capillary effect. Under the influence of these forces, the surface of the liquid is covered by a uniformly strained thin film, which tends to give the fluid shape volume with a minimum surface. The surface tension forces develop a molecular pressure in the liquid, typical to its surface, which leads to the penetration of moisture through microcavities.

Figure 5 presents the number of insulation faults of traction electric engines of locomotives in the Eastern region of the distribution of railways. Figure 6 shows the failure rate; that is, the failure rate of the insulation of the power equipment of the traction rolling stock.

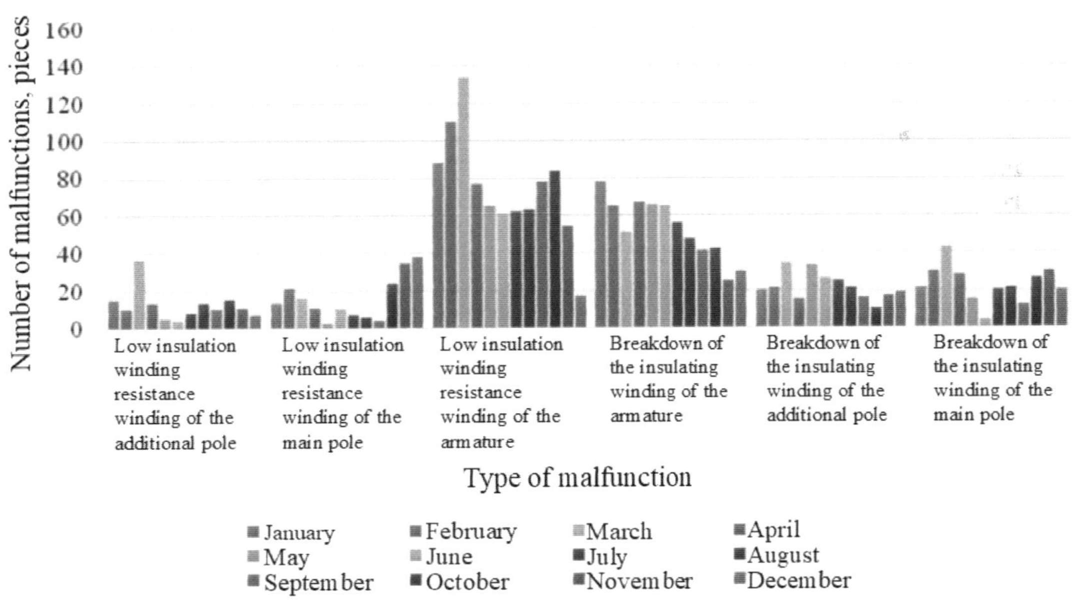

Figure 5. Histogram of failure of insulation of traction electric motors of locomotives in the Eastern region for 2017

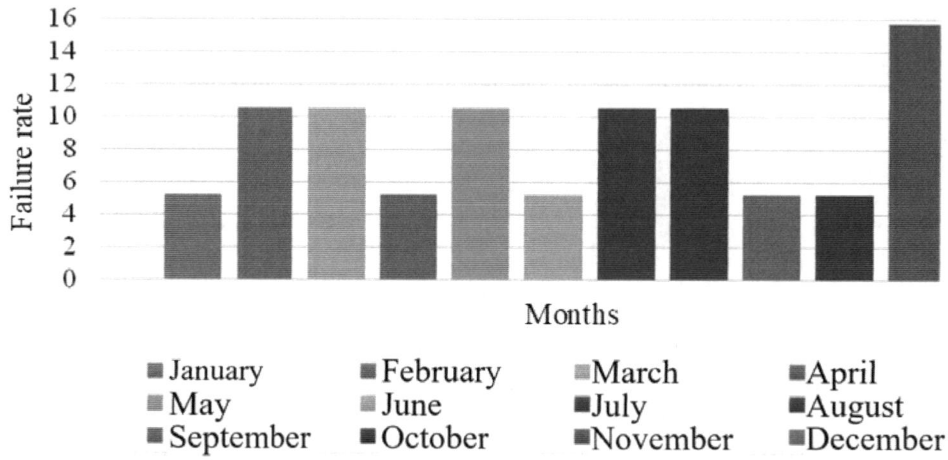

Figure 6. Histogram of the intensity of failure of insulation of power equipment in the Eastern region for 2017

In many cases, premature failure of the insulation of electric transport equipment can be prevented by systematically monitoring the moisture content of the insulation and drying it if necessary [10]. Given the above, it is clear that with the combined effect of all climatic factors, the primary indicator of the reliability of insulation, the dielectric properties, deteriorates to a certain extent. Climatic factors are determined by annual, seasonal, daily fluctuations in temperature and relative humidity, as well as the number of days and the intensity of precipitation in the form of snow, rain, fog. Thus, it is necessary to improve diagnostic methods and tools, as well as maintenance and repair technologies, to reduce the degree of influence of climatic conditions on equipment reliability.

4. Conclusion

As study shows, according to the leading climatic indicators in the Eastern region, the most demanding conditions for the operation of power equipment in the "northern" direction. Long-term monitoring of the operation of the power equipment of electric locomotives showed that the parameter of the flow of their failures in the breakdown of insulation is more significant in some areas of the operation of electric locomotives, in others – less, especially in winter. It is necessary to create adaptive systems to maintain optimal values for the temperature and humidity conditions for the insulation structures of electric locomotives to increase the reliability of the insulation structures of electric locomotives operating in various areas of operation on the railways of the Eastern region.

References

[1] Yao Y, Lu G, Boroyevich D and Khai, Ngo K 2015 Survey of high-temperature polymeric encapsulants for power electronics packaging *IEEE Transactions on Components, Packaging and Manufacturing Technology* **5 (2)** pp 168–181

[2] Meng F, Li W, Zhao A, Cao L, Deng J, Zhang G, Zhi Y and Zhang L 2016 Development of time and frequency domain dielectric response testing device for power apparatus insulation 2016 *International Conference on Condition Monitoring and Diagnosis (CMD)* pp 416–419

[3] Almohaimeed S A and Suryanarayanan S 2017 *A review of air conditioning motor loads stalling on voltage recovery in the saudi electric grid* North American Power Symposium *(NAPS)* pp 1–5

[4] Maheswaran D, Rangaraj V, Kailas K J and Kumar W A 2012 Energy efficiency in electrical systems *IEEE International Conference on Power Electronics, Drives and Energy Systems (PEDES)* pp 1–6

[5] Salama M M, Mansour D A, Abdelmaksoud S M and Abbas A A 2018 Impact of Long-Term Climatic Conditions on the Ageing and Cost Effectiveness of the Oil-Filled Transformer // *Twentieth International Middle East Power Systems Conference (MEPCON)* pp 494–499
[6] Lobytsin I O, Dulskiy E Y and Vasiliev A A 2017 Increasing the reliability of insulating fingers of shield bracket brackets of electric machines of traction rolling stock // *Modern technologies. System analysis. Modeling* **4 (56)** pp 218–224
[7] Stahlhut J, Heydt G T and Selover N J 2008 A Preliminary Assessment of the Impact of Ambient Temperature Rise on Distribution Transformer Loss of Life // *IEEE Transactions on Power Delivery* **4** pp 2000–2007
[8] Das S and Gupta N 2013 Effect of thermal and humidity ageing on space charge accumulation in epoxy resin // *IEEE International Conference on Solid Dielectrics (ICSD)* pp 456–459
[9] García D F, García B, Burgos J C and Villarroel R 2013 A new methodology for determining the moisture diffusion coefficient of transformer solid insulation *IEEE International Conference on Solid Dielectrics (ICSD)* pp 1024–1027
[10] Dulskiy E Y, Ivanov P Y and Khudonogov A M 2017 Basis for Local Methods of Insulation Hardening of Traction Rolling Stock Electrical Machines *Energy Management of Municipal Transportation Facilities and Transport EMMFT 2017: International Scientific Conference Energy Management of Municipal Transportation Facilities and Transport EMMFT* pp 109–119

IOP Conf. Series: Materials Science and Engineering **760** (2020) 012019 doi:10.1088/1757-899X/760/1/012019

A functional approach to the analysis of technological processes of using thermal radiation in the process of repairing insulation of electric locomotive equipment

E Y Dulsky[1], P Y Ivanov[1], A M Khudonogov[1], I A Khudonogov[1]

[1] Irkutsk State University of Railway Engineering, 664074, Russian Federation, Irkutsk, Chernyshevsky st., 15

E-mail: e.dulskiy@mail.ru

Abstract. The article provides an analysis of the interaction of the system "Emitter – Environment – Material" in the drying and hardening of the insulation of electrical equipment of electric locomotives, taking into account the intended purpose of the process. The structural-logical diagram of technological processes of using thermal radiation during repair of insulation materials of windings of power equipment of electric locomotives is presented. This structural and logical diagram indicates the need for an individual approach to the analysis of the system elements' interaction in the process of using heat radiation for the polymerization of impregnating materials for hardening. It is necessary to comply with the parameters of the maximum permissible temperature and heating rate not only for isolation of this class of heat resistance but also for an impregnating material, depending on which losses in the medium depend. A simplified approach to the analysis of the interaction process of the "System" elements in the mathematical model is proposed. The basis of this model is the relationship between the dose of irradiation of polymer insulation with thermal radiation and the expected results on the characteristics and properties of electrical insulating materials.

1. Introduction

The concept of function is one of the main concepts of modern mathematics. This concept can also be encountered both in nature and in technology, studying various phenomena and processes. Any device is considered in a set of interconnected elements. The contents of elements are revealed in the studying process of the system tasks and functions.

2. Structural and logical scheme of technological processes of repair of insulating materials

The technology and technique for hardening the insulation of electrical equipment of electric locomotives, proposed by scientists of Irkutsk State Transport University, can significantly increase the productivity of repair operations with lower costs (electricity) [1, 2, 3].

It is necessary to use mathematical modeling at the design stage to create the most efficient resource-saving industrial plants. As part of the fulfillment of the state task of the research and development work on the topic "Improving the reliability of power equipment of traction rolling stock" (№AAAA-A19-119010990009-3), work is currently underway to compile a set of mathematical models of technological processes for the use of thermal radiation in the repair process of electrical equipment of electric locomotives. In this regard, first of all, it is necessary to analyze the system "Emitter-environment-material" (after this – the "System"). In the case of analyzing the interaction of the system during the re-installation of insulation of the electric equipment of electric locomotives, first of all, it is necessary to take into account the purpose of the technological process of heating the insulating material with heat radiation in general.

In this regard, a structural-logical diagram of technological processes for the use of thermal radiation in the repair of insulation materials of windings of power equipment of electric locomotives was compiled (Figure 1).

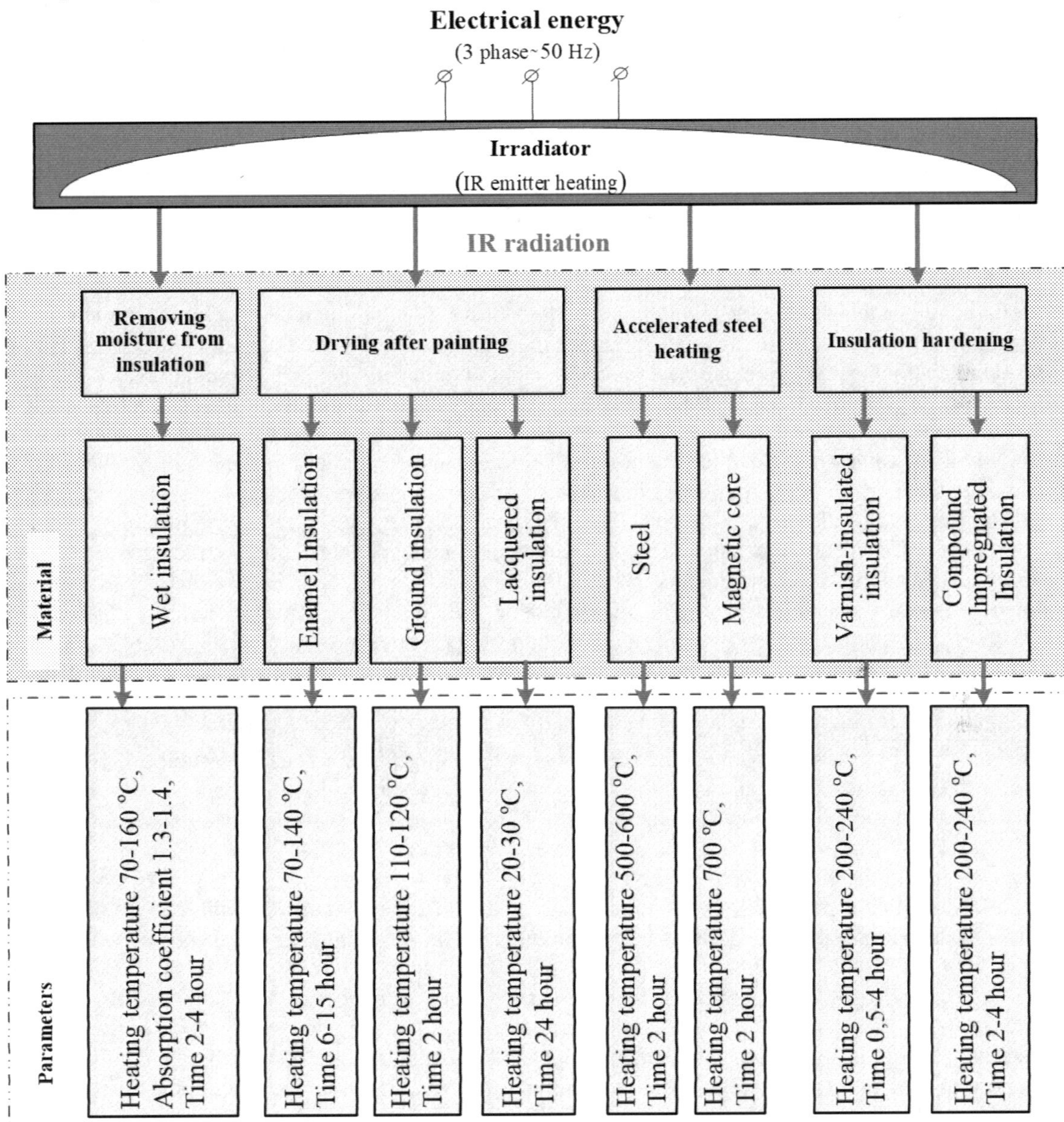

Figure 1. The structural and logical diagram of the technological processes of using thermal radiation in the repair of insulation materials of windings of power equipment of electric locomotives

IOP Conf. Series: Materials Science and Engineering **760** (2020) 012019 doi:10.1088/1757-899X/760/1/012019

Thus, the use of thermal radiation in the repair process of insulating structures can be divided into four technological processes:

– removal of moisture from the insulation;
– drying after painting;
– accelerated heating;
– hardening of insulation.

With the further introduction of thermal radiation in the technology of repairing equipment of electric locomotives, the number of processes for the target feature increase.

The process of heating moistened insulation to dry it is associated not only with the removal of moisture (it is a thermophysical process) but also associated with the need to preserve the structure of polymer insulation as much as possible [4].

Exceeding the maximum permissible temperature and heating rate for a given class of insulation can lead to a violation of the operational characteristics of the equipment. Moreover, this process is technological.

When heating the insulation after painting with enamel, primer or varnish at low temperatures, only a thermophysical process takes place, proceeding with the obligatory observance of the necessary temperature regimes of heating and exposure time.

The use of thermal radiation for accelerated heating of parts of power equipment with high thermal inertia is due to the significant highly concentrated heating of steel parts of power equipment. Heated steel parts may be close to insulating structures. It is necessary to take into account the mutual emission factors of the system surfaces under consideration, as well as the edge effect during the operation of the emitters. In this case, only the thermophysical phenomena of the process are analyzed.

It is entirely different from approaching the analysis of the interaction of the "System" in the process of using heat radiation for the polymerization of impregnating materials to strengthen. In this work, the term "hardening" used in a process in which liquid polymers (varnishes and compounds) are transformed into solid polymers (curing, gelatinization, polymerization) under the influence of thermal radiation in the manufacturing and restoration technologies of insulating structures of electrical equipment for various industries, agriculture and transport [5, 6, 7]. Besides, the term "hardening" laid down the main characteristics and properties of electrical insulating materials: electrical and mechanical strength, heat resistance, cold resistance, moisture and water resistance, flexibility and elasticity, chemical resistance, roughness, cracking resistance and other are critics and properties. As experimental laboratory studies have shown, the complex of these characteristics and properties is directly correlated with the surface roughness of the cured polymer on the insulating structure of electrical equipment — the lower the roughness, the stronger the electrical insulation. However, to ensure these characteristics and properties, it is necessary to remember that the energy characteristics of coherent and incoherent thermal radiation and the physicochemical properties of liquid polymers are the basis for ensuring the modification of polymers with a given roughness functional characteristic in the technological process.

In this case, it is necessary to observe the parameters of the maximum permissible temperature and heating rate not only for this class of insulation but also for the impregnating material. Moreover, the process of hardening is significantly affected by the formation of "gray gas" in the medium [8], in which energy losses occur. Thus, it is necessary to take into account the impregnating material, depending on which losses in the environment depend. So when using varnish-based impregnating materials, losses in the "gray gas" environment are maximal, since a significant part of the solvent evaporates in the form of vapors during the polymerization. And when using impregnating materials based on compounds, these losses are minimal due to different physics of the polymerization process with virtually no use of solvents. The hardening process is also associated with chemical changes in the liquid polymer and subsequent phase transitions as a result of heating by thermal radiation. This process indicates difficulties and difficulties in the analysis of processes for hardening, precisely polymer insulation. Moreover, the variety of modern types of impregnating materials further complicates this task. However, with a simplified approach to the analysis of the process of interaction

IOP Conf. Series: Materials Science and Engineering **760** (2020) 012019 doi:10.1088/1757-899X/760/1/012019

of the elements of the system under consideration, a particular analogy can be drawn in mathematical modeling.

3. Mathematical modeling of the process of polymer insulation hardening by thermal radiation

A mathematical model is necessary to improve the process of hardening of polymer insulation by thermal radiation. A mathematical model establishes the relationship between the dose of exposure of polymer insulation to thermal radiation and the expected final results. This model needs to be tested experimentally for adequacy.

$$\frac{dt}{d\tau} = \upsilon \cdot t \tag{1}$$

t – the heating temperature of polymer insulation, ° C;
v – the specific heating rate of polymer insulation by thermal radiation, ° C.

Formally, this equation is similar to the equation of the rate of a chemical reaction of the first kind. However, in polymer hardening processes, the specific heating rate depends on the power concentration

$$v = f(p) \tag{2}$$

By analogy with the Mono equation

$$v = \frac{v_{max} p}{K_c + p} \tag{3}$$

p – power concentration;
K_c – constant, defining the purpose of the process.

The basis for each considered technological process of insulation repair is its heating by thermal radiation. The insulation material is heated not by sequentially redistributing one type of energy according to the standard scheme "heat" → "heat", as occurs when using contact or convective heat exchange, but by converting electric energy into heat according to the complicated scheme "electric energy"→"heat"→"electromagnetic radiation"→"heat". At the first stages of the process of hardening polymer insulation in the infrared emitter, thermal radiation energy generates due to the influence of electric current on the radiator heating body. The filament most often acts as a heating body. Further, the energy is transferred to the learning surface of the insulation in the electromagnetic waves form and again turns into heat, which subsequently performs one or another technological process, as mentioned above. The indicated stages can be represented in the form of a structural-logical model showing the relationship between input and output parameters.

The structural-logical model shows the energy relationship along the entire path of converting electrical energy in a particular process with its intended purposes. Moreover, it allows you to visually choose mathematical models to justify the technological process of drying and hardening of polymer insulation by thermal radiation. The mathematical model should reproduce the dynamics of the effect of thermal radiation on the insulating structures of the windings of power equipment. In this case, the mathematical model must take into account the spectral characteristics of the emitters, the geometric structural features of the windings, the type of insulation, and the optical properties of the impregnating composition.

In each specific case of mathematical modeling, the purpose of the technological process significantly affects the "System" composition in terms of its expansion and addition of new elements. These entail significant changes in the mathematical model final form is expressed in modifications of the equation in terms of the members' number and the appearance of new coefficients.

IR radiation upon reaching the material is converted into thermal energy as a result of the absorption capacity of the material. In this case, the temperature of the irradiated body reaches a limit

IOP Conf. Series: Materials Science and Engineering **760** (2020) 012019 doi:10.1088/1757-899X/760/1/012019

value, since part of the energy received is returned to the environment. The more significant difference between the material temperature and the ambient temperature, the higher the heat loss. In paragraph 4.7 of this work, experimental studies show determination and analysis are the quantities on which the value of the limiting heating temperature of the irradiated surface depends.

Such parameters are the energy illuminance of the surface of the irradiated material, the absorption and reflectivity of the irradiated surface. It can be expected that these values affect the limiting temperature to the same extent so that with increasing energy illumination and absorption capacity, the limiting temperature also rises. The performed experimental studies prove the correctness of this assumption.

The relationship between these quantities can be expressed by the formula proposed by Tiller and Garber, which has the following form

$$T_{max} = T_0 + \frac{A_\lambda \cdot E}{a \cdot F} \tag{1}$$

T_0 – the ambient temperature;
A_λ – absorption capacity;
α – convection heat transfer coefficient;
F – surface area involved in the heat exchange between raw materials and air.

This formula shows that the limiting temperature can be achieved with lower energy illumination of the infrared emitter if the absorption capacity of the insulating material is sufficiently high. This circumstance can be observed in the example of baking thermosetting synthetic resin varnish.

Thus, the operating surface and insulation temperature depends on the of the entire "System" elements characteristics as a whole, not only on the optical properties of the material from is made. In this connection, it is necessary to analyze the influence of each element of the "system" to identify those elements that have the most significant impact on the hardening process.

Figure 2 shows the structural diagram of the "System". The main characteristics of the "System" structure element evaluate their relationship and their impact on the process as a whole.

As can be seen from the figure, the "System" is conventionally divided into eight elements, each of which is one way or another affects the operability of the subsequent one.

To supply the "System," a single-phase two-wire network with a voltage of 220 V is often used, however, to improve energy efficiency in the development of thermal radiation generators, it is better to use a three-phase system.

To control the operating modes of the emitters in the "System" may be provided with a converter. The converter can be a conventional transformer, as well as a power regulator, which allows control of the heating temperature by changing the spectral composition of the emitters.

Irradiators are combinations of an emitter and a reflector. Currently, there are a massive number of infrared emitters. Their market is continually evolving, and the scope of their use is expanding. This research work has a continuation. Soon, it is necessary to analyze emitters for technological processes of using infrared radiation in the process of repairing insulation, taking into account their spectral characteristics. Ceramic emitters are most widely used in the drying of paints and varnishes. It is in the irradiator that the main flow of thermal radiation is formed. The choice of material and shape of the reflector significantly affects the loss of radiation during operation – often curved aluminum plates with a polished surface used as reflectors. The most common in terms of efficiency and reliability are electric infrared emitters.

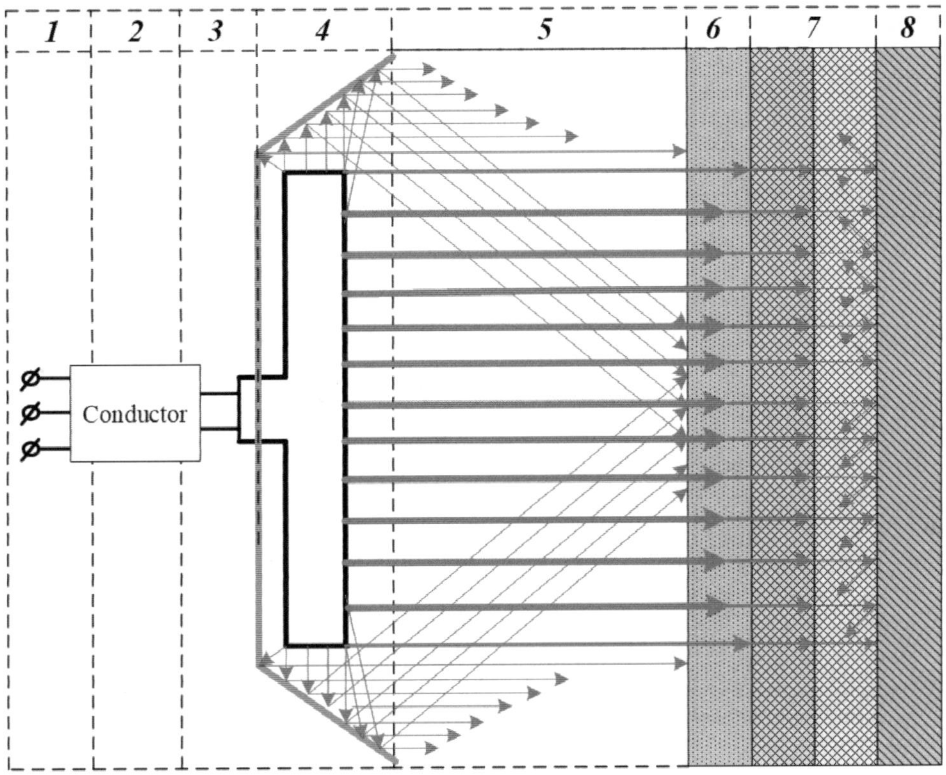

Figure 2. The structural diagram of the "System": 1 – network; 2 – converter; 3 – power line; 4 – irradiator; 5 – environment; 6 – liquidd polymer; 7 – insulation layers; 8 – conductor.

The medium between the irradiator and the heated object is significantly different for each of those processes. In the case of strengthening polymer insulation over the entire period, the medium itself also changes. At the initial time, the medium will mainly represent air. Further, in the process of heating a liquid polymer (polymer material), solvent vapors intensively begin to be released from its surface. In this case, the intensity of the vapor evolution substantially depends on the impregnating material. So for materials based on varnishes, consisting of 90% solvent, the evaporation rate is maximum, and for compounds – the minimum evaporation rate. Thus, the type of liquid polymer significantly affects the chemical composition of the medium, which in turn affects the loss of thermal radiation.

The process of hardening insulation is characterized by the presence of a liquid polymer, which can be varnishes, resins, compounds, or other impregnating materials.

Depending on the type of electrical equipment, various types of insulating materials with different numbers of layers are used. At the same time, the optical characteristics of materials differ significantly, having different indicators of reflection, transmission, and absorption.

4. Conclusion

The main provisions of the analysis of the interaction of the "System" with a functional approach are as follows:

1) The original "System" is considered not as a single whole and indivisible, but as a specific structure consisting of separate parts – subsystems connected and having interconnections.

2) The main structural elements of the "System" are emitters and insulating materials. In this regard, it is necessary to consider the main characteristics of these elements of the system under

consideration. For an insulating material, the main characteristics are geometric, thermophysical, optical, and thermal radiation. For emitters – energy and spectral.

3) The number of factors that insignificantly affect the "System" is limited in this case, which allows us to consider such a system as an object of shallow dimension, reduced to simple mathematical methods.

4) For the mathematical description of the dynamic behavior of the "System," as a first approximation, one can use the differential equation of the energy balance, which also allows one to obtain a dynamic thermal model (chapter 3).

5) The criterion of optimality in the study of the relationship of the elements of the "System" should be considered the maximum allowable temperature, power density, and heating rate for a particular process under study.

The methodological sequence in the analysis of the interaction of the "System" in the case of using thermal radiation in the repair of insulation can be represented by the following points:

1) The structural scheme of the "System" must be clarified for each technological process individually to draw up adequate mathematical models.

2) It is necessary to have the requirements of the process for the maximum allowable temperature or heating rate of the material.

3) When analyzing the stages, it is necessary to take into account the influence of other stages on them as initial and boundary conditions.

As a result, because the mathematical model should adequately reproduce the kinetics and dynamics of heating by thermal radiation of polymers in an appropriate manner [9, 10]. Before developing a mathematical model, it is necessary to carry out a set of experimental laboratory work to study the hardening processes of elements of insulating structures of electrical equipment. Based on these studies, it is necessary to develop a set of mathematical models for calculating the infrared energy supply modes in the manufacturing and repair technologies of insulating structures of electrical equipment of electric locomotives.

References

[1] Dulskiy E Y, Ivanov P Y and Khudonogov A M 2017 Basis for Local Methods of Insulation Hardening of Traction Rolling Stock Electrical Machines *Energy Management of Municipal Transportation Facilities and Transport EMMFT 2017: International Scientific Conference Energy Management of Municipal Transportation Facilities and Transport EMMFT* pp 109–119

[2] Lobytsin I O, Dulskiy E Y and Vasiliev A A 2017 Increasing the reliability of insulating fingers of shield bracket brackets of electric machines of traction rolling stock // *Modern technologies. System analysis. Modeling* **4 (56)** pp 218–224

[3] Lytkina E M 2011 Improving the efficiency of encapsulation of the insulation of the frontal parts of the windings of traction motors of electric locomotives by infrared radiation*: dissertation for the degree of Candidate of tech.* sciences p. 205

[4] Long J, Zhang R, Lu J and Xu F 2019 Heat transfer performance of an integrated solar-air source heat pump evaporator *Energy Conversion and Management* pp 626–635

[5] Shchitov S V, Reshetnik E I and Evdokimov V G 2019 Using infrared radiation in preparation of feed for farm livestock *Indonesian Journal of Electrical Engineering and Computer Science* Vol **15**, pp 148–154

[6] Wu X, Zhang M and Bhandari B 2019 A novel infrared freeze drying (IRFD) technology to lower the energy consumption and keep the quality of Cordyceps militaris *Innovative Food Science and Emerging Technologies* Vol **54** pp 34-42

[7] Dulskiy E Y, Khudonogov A M, Khudonogov I A, Astrakhantsev L A, Ivanov P Y and Tuigunova A 2019 Modeling of the Oscillating Mode of the IR-Energy Supply in the Technology of Restoration of Insulating Fingers of Electric Motors of Locomotives //

International Scientific Conference Energy Management of Municipal Facilities and Sustainable Energy Technologies EMMFT 2018 Vol **1** pp 546-556

[8] Fuqiang W, Hao W and Dayang G 2019 Radiative transfer analysis of semitransparent medium with particles having non-uniform size distribution by differential-integration method *International Journal of Heat and Mass Transfer* Vol **130** pp 342–355

[9] Woodfield J, Alvarez M and Gómez-Vargas B 2019 Stability and finite element approximation of phase change models for natural convection in porous media *Journal of Computational and Applied Mathematics* Vol **360** pp 117–137

[10] Parhizi M and Jain A 2019 Theoretical modeling of a phase change heat transfer problem with a pre-melted or pre-solidified region// 2019 *International Journal of Heat and Mass Transfer* pp 635–643

Methodology for evaluating of the interaction of wagons and path on the mountain-gorge areas

I Yu Ermolenko[1], **V N Zheleznyak**[1]

[1] Irkutsk State Transport University, 15, Chernyshevsky St., Irkutsk, 664074, Russia

E-mail: ermolenko_iy@list.ru

Abstract. In recent years, the number of rolling stock derailments, especially unladen wagons has begun to increase on railways located in hard geographical and climatic conditions in Eastern Siberia. In view of this, we propose the methods for evaluation of the interaction of wagons and railroad track for safety train movement and determination of appropriate schedule for train operation.

1. Introduction

Modern wagons represent a complex statically indefinable mechanical system which has a variety of operational and dynamic loads (vertical, horizontal, transverse and longitudinal). As a rule, these loads are statistical. The movement of trains depends on the profile of the track [1]. Tensile or compressive longitudinal forces on draw-and-buffer gear are caused by the train movement and switching operation. The magnitude and signs of these forces depend on the interaction of wagons with different modes of motion [2-3]. Nowadays, we need to find solutions to the problem of interaction of rolling stock and track on the mountainous sections resulting by the increased weight of the trains (fig. 1). Only in 2018 there were 8 cases of derailment of rolling stock with R = 250-400 m (40%) and R = 800-1200 m (60%).

The particular attention is paid to the problem of maintenance and operation of rolling stock on the areas of the East Siberian Railway (ESSR), located in hard climatic and geographical conditions. The track mileage is 6188 km, the length of federal-aid railway is 3848 km, the length of joint-less track is over 3000 km. The length of curved tracks is 45.1%, including a 25.4% of 250-650 m radius. According to the data about ESSR, the track has about 8,000 right – left curves on such areas. Historically, since 1995, the types of rails such as R65 (88%), R75 (3%) and others (9%) have been laid for the reinforcement of load-carrying capacity. The strength of the railway increased after the transition to the new rail type R65.

More than 20 years have passed. During this period the speed and the weight of trains have changed significantly. The problem is not only in rolling stock, but also in railroad condition, particularly in hard conditions on mountainous sections. This is due to the presence of sharp transitions profile of track, curvature radii, elevations of the outer rail, which have an impact on the ratio of inertial and centrifugal forces and traffic safety, in general.

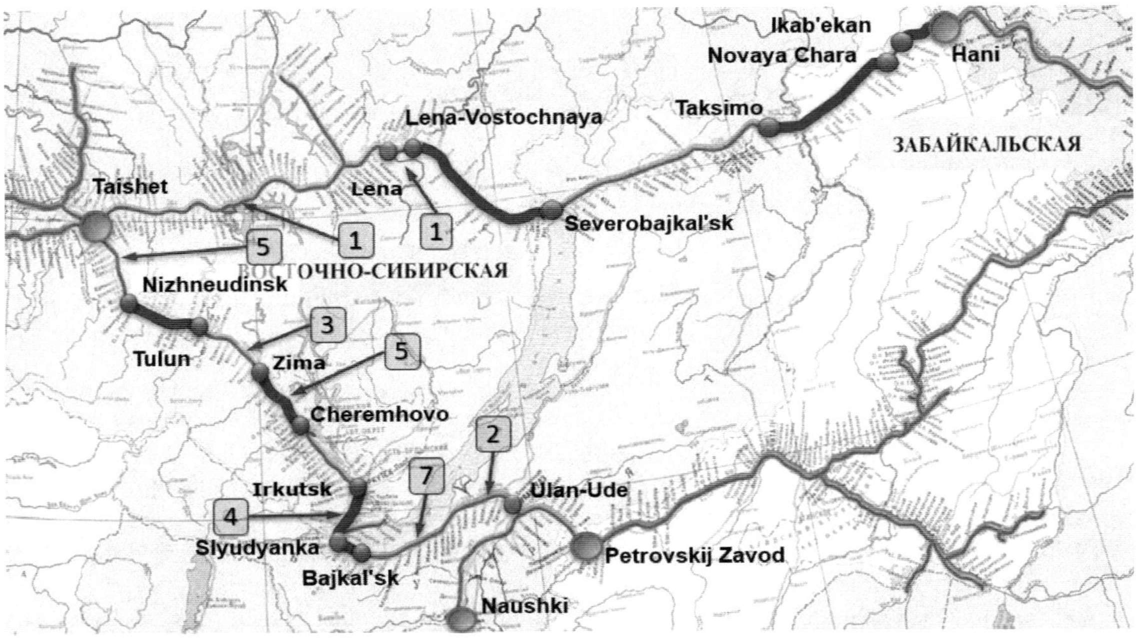

Figure 1. The map of derailments of the rolling stock on network East-Siberian rail roads for 2000-2019.

2. Material and research methods

One of the major challenges in dealing with the problem of safety of rolling stock is to find effective methods and tools that enable likely describe dynamical processes that occur when moving rolling stock on rail way based on actual system states wagon-way [4].

To solve the problem of preventing derailment it's neccesary to explain and change individual rules of maintenance of wagons, regulate the speed of traffic on such areas. It is necessary to investigate various technical condition of cars and tracks [5-7]. For these purposes it's possible and appropriate to use mathematical and computer modelling methods, as well as a real investigation of the reasons of derailment.

Traffic safety and evaluation of dynamic qualities are often associated with the wagon stock wheelset sustainability indicator. Factor of resistance to the rolling into the head rail is an integral indicator of traffic safety. To avoid derailment and ensure sustain ability of reserve boundaries are established, expressed using defined indicators – $K_{stability}$.

$$K_{stability} = \frac{tg(\beta)-\mu}{1+\mu\cdot tg(\beta)} \cdot \frac{P_V}{P_H} \geq \left[K_{stability}\right] \quad (1)$$

For safe driving cars, the value of the coefficient of stock sustainability should not fall below the permissible value of the GOST [8].

In work [9] one of the criteria considered gathering influence of longitudinal forces arising from violations of driving trains. When the cars at an angle to each other (fig. 2) in terms of lateral raises horizontal components of the longitudinal forces; and when the difference in elevation between them-vertical components that create additional loading or unloading the appropriate wheel crew depending on his hesitation.

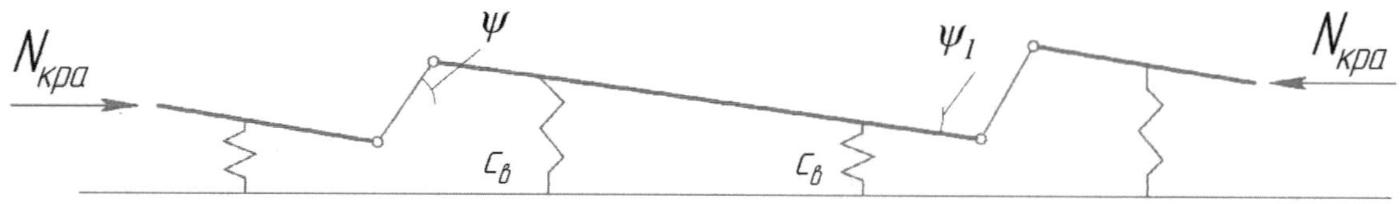

Figure 2. Form of buckling of longitudinal forces squeezing wagon.

3. Research results

To organize the safe movement and determine the optimal modes in operation, we propose the methods of evaluation of the interaction of wagons and tracks on the mountainous sections. There are several stages:

1) the theoretical model of parametric oscillation excitation and galloping body bouncing wagon to carts was developed (fig. 3);

2) the complex tests on the train carriage-laboratories was conducted;

3) the vertical coefficient measurement of dynamics and coefficient of stability was made based on the results of the calculation (fig. 6) (fig. 7);

4) the roller booth was developed (truck model 18-100) to evaluate the dynamic qualities of the wagon within the parameters of the curves plots tracks and various faults of wheels (fig. 8);

5) a set of recommendations for high-speed trains on modes of mountainous sections was developed.

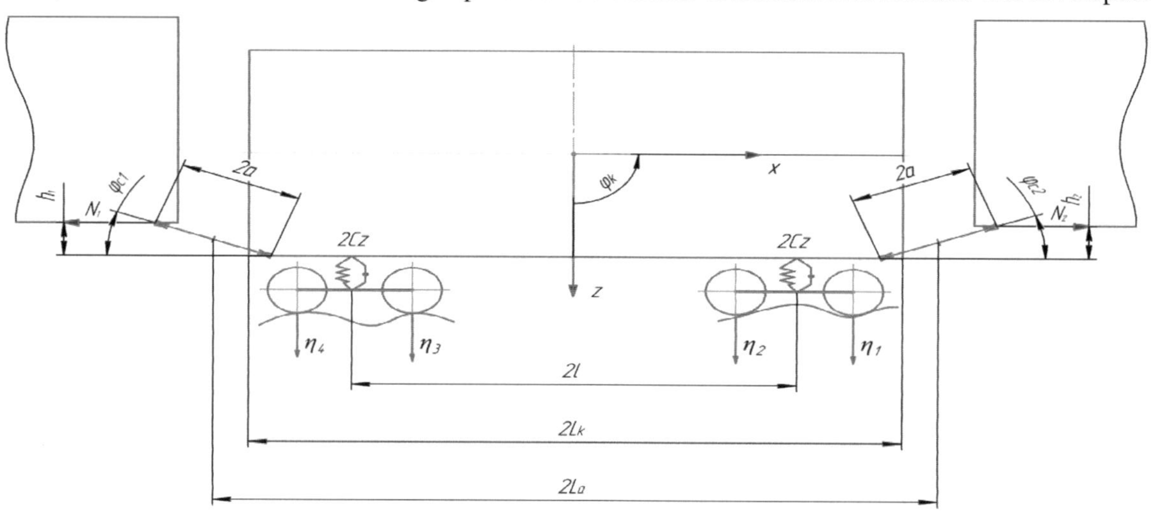

Figure 3. Design of excitation circuit of parametric oscillation and galloping body bouncing wagon to carts.

Stage I. Train brake tests for the wagon. The purpose of the tests is determination of the lateral forces generated, taking into account the longitudinal effort passed from the coupler that affect wagon in the curves of the mountain - a crossover way when performing the prescribed instructions types of braking. To determine the dependence of lateral loads from the longitudinal forces, we executed the experienced travel on caboose, equipped with load cell automatic coupling type SA-3 [10]. According to the results of tests, we built waveform (fig. 4) change of the outstanding longitudinal acceleration when moving the car from outside site path.

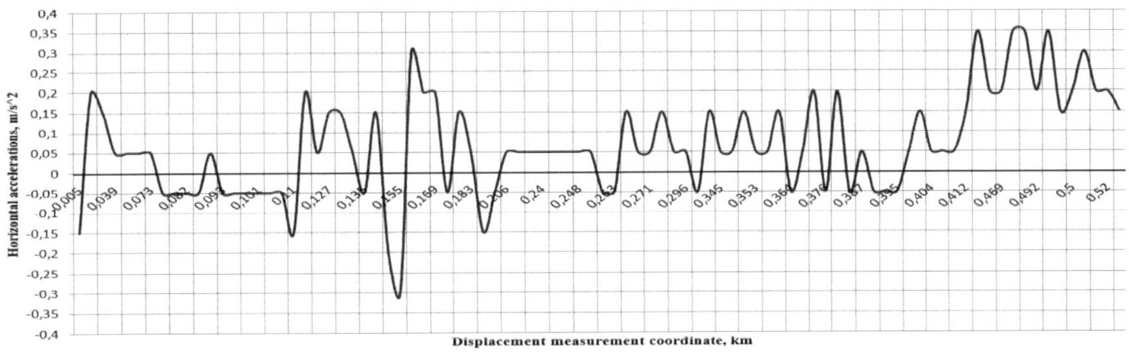

Figure 4. Waveform change of the outstanding longitudinal acceleration.

Train tests on the measuring car. The purpose of the tests is the definition of real characteristics of mining and crossover plots ways to transform coordinates of the unevenness of the track as a function that describes the unevenness of the track under each wheel pair (fig. 5).

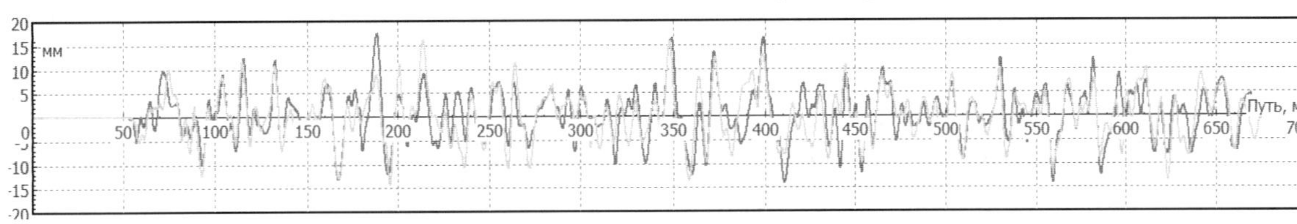

Figure 5. The right and left rails, vertical undulations.

Stage II. Calculation of the coefficient of vertical dynamics of wagon and the coefficient of resistance. The values obtained in tests of train accelerations (longitudinal with coupler) introduced the system of differential equations for parametric oscillation and galloping bouncing wagon. Is used as an optional parameter from the train test and time variable axial force *N(t)* and real roughness parameters path.

The system of differential equations for parametric oscillation, describing the process of bouncing and galloping carriage:

$$\begin{cases} m \cdot \ddot{z} + \left(4Cz + \frac{N(t)}{a}\right) \cdot z = \left(4Cz + \frac{N(t)}{a}\right) \cdot \eta, \\ I_y \cdot \ddot{\phi} + \left(4Cz \cdot l^2 + \frac{N(t)}{a} \cdot l \cdot L\right) \cdot \phi = \left(4Cz \cdot l^2 + \frac{N(t)}{a} \cdot l \cdot L\right) \cdot \eta, \end{cases} \tag{2}$$

where m – tare weight, kg; I_y – the moment of inertia of the bodywork relative to the y axis, kg·m^2; Cz – the stiffness of the spring suspension, N/m; l – car base, m; L – distance between the rear or front rims coupler, m; a – hull length coupler, m; $N(t)$ – time variable axial force; η – real roughness parameters path.

As the object of modelling wagons we selected gatherings over the past 19 years: indoor model 11-2135-01, gondola model 12-132, platform model 13-192-01, tank model 15-150-04, hopper model 19-6870. Wagons were loaded and integrated in an unladen state.

Integrating the resulting differential equations of the second order obtained by the classical method, Runge-Kutta methods of the fourth order helped receive modified coefficient of vertical dynamics in the process of movement on the studied site path (fig. 6). Results of calculation of the vertical dynamics were used in formula (1) and gave real results of calculation of safety factor of sustainability against derailment (fig. 7).

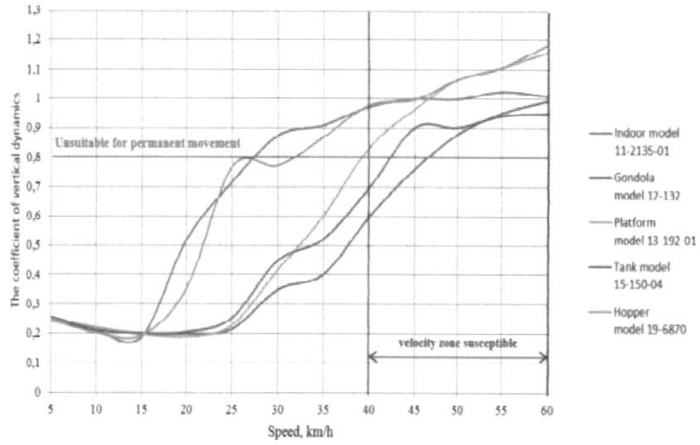

Figure 6. The coefficient of vertical dynamics.

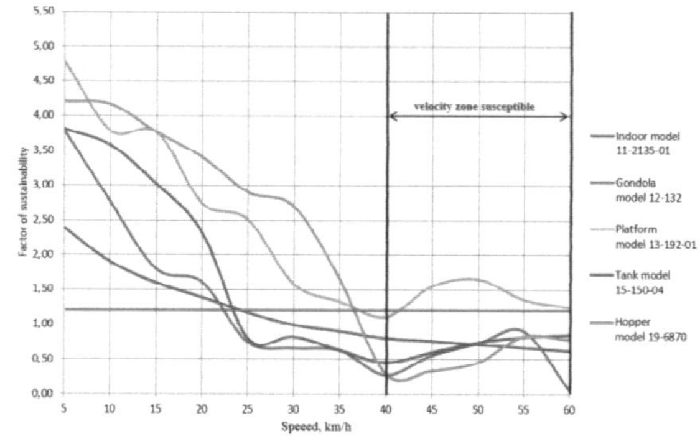

Figure 7. Stability factor.

Analyzing the results (fig. 6-7) we define a range of maximum train speeds on researched mountinous areas.

Stage III. Physical simulation. To evaluate the dynamic qualities of the wagon within the parameters of the curves plots path and various faults wheel (navar) we elaborated a pilot roller booth [11-13]. Stand (fig. 8) allows you to simulate the movement of wagons in straight and curved sections of road with various external rail and eminence to take into account the effect of pressing comb wheels to the lateral surface of the rail with a possibility to change the load along the axis of the wheel.

Figure 8. Roller booth for estimating the dynamic qualities of the wagon.

Measurement of dynamic characteristics arise when driving, made with the help of measuring complex based on MIC-036R. Results are presented in graphical form (fig. 9). Zero envelope chart accelerations on the y-axis correspond to $K_{stability} = 1$. These are measurements limits in accordance with [8]. We see that maximum amplitude of deviations is quite frequent repetitions that has a significant impact on the stability of wheel pair against the derailment.

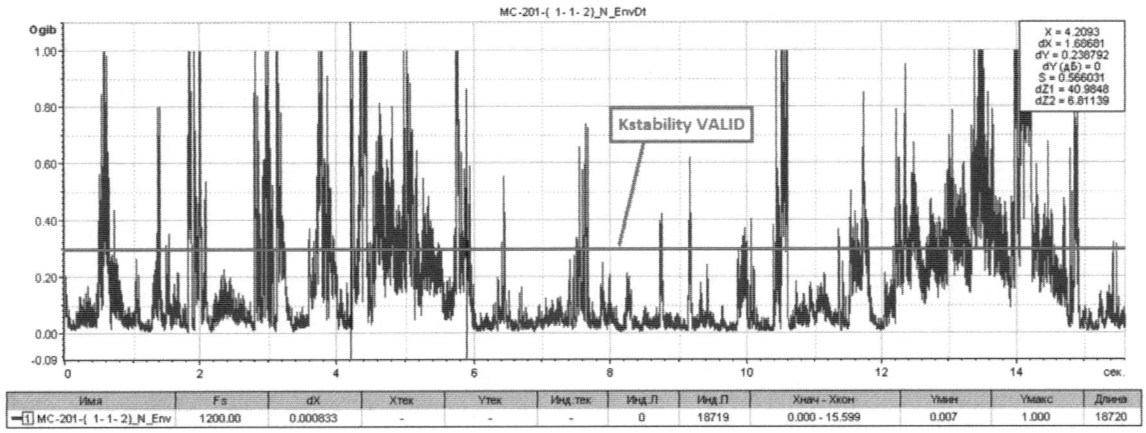

Figure 9. Measurement of the coefficient of stability of wheel pair against the derailment.

Stage IV. Development of technical recommendations for high-speed trains on regimes of mountainous areas. The stages of a complex technical recommendations for reducing power effects on parts of the wagon while driving on the studied sections are due to changes in speed. Knowing features paths (curve radius, elevation of the outer rails) and vertical dynamics, we expect optimum speed in the investigated areas (table 1). Maximum speed limit is determined by the following formulae:

$$V_{\max\ pass} = 0{,}283 \cdot \sqrt{R \cdot (h+115)}\,, \tag{3}$$

$$V_{\max\ gruz} = 0{,}283 \cdot \sqrt{R \cdot (h+50)}\,. \tag{4}$$

where R – curve radius, m; h – elevation of the outer rail, m.

Calculation of the optimal speed of trains for the plot takes into account elevation of the outer rail.

Table 1. Proposals for high-speed modes

Track part	The name of the site, ferrying, station	The proposed path service rate	Speed according to the order of	Research speed	The proposed rolling stock service speed
1	Tayshet - Bayronovka	80	80	from 65 up to 72	up to 70
1	4522-4529 km	60	60	60	up to 60
2	4537 km pk 2	60	60	60	60
2	Bayronovka - Razgon	60	60	from 52 up to 58	up to 60
2	st. Razgon	60	60	60	60
2	Razgon - Oblepiha	60	60	from 52 up to 58	up to 60
2	st. Oblepiha	60	60	60	up to 60

4. Conclusion

Comparative analysis of the data presented shows the calculated speed settings that provide safety for a particular lot of investigated ways. Thus, the application of the proposed method allows a wide range of characteristics and path trains speed modes determining the safest speed range.

References

[1] Ermolenko I Yu 2017 Impact profile path mining area ESRR to power the wheel-rail interaction, taking into account the experimental laboratory car settings *Proceeding of Irkutsk State Technical University* **21(12)** 170-180

[2] Lazaryan V A 1964 *Dynamics of cars* (Moscow: Transport)

[3] Vershinskij S V, Danilov V N, Husidov V D 1991 *Dynamics of cars* (Moscow: Transport)

[4] Lysyk V S 1997 *Causes and mechanism of derailment rail wheels. The problem of wheel and rail wear* (Moscow: Transport)

[5] Ermolenko I Yu, Zheleznyak V N, Kovenkin D A, Voronova Yu V 2016 Limitation of Rolling Stock due to Possible Formations of Surface Defects of Wheel and Rail Riding on Mountain Terrain *The Fifth International Symposium on Innovation and Sustainability of Modern Railway (ISMR'2016)* (China: ECJTU) pp 572-577

[6] Ermolenko I Yu, Zheleznyak V N, Martynenko L V 2018 Force loading of wagon during interaction of truck and bodywork in curves paths *AER-Advances in Engineering Research* **158** 130-134

[7] Ermolenko I Yu, Zheleznyak V N, Martynenko L V 2018 Interrelation of the causes of failure of the axle boxes and the malfunction of the automatic couplings (type SA-3) *Proceedings of the 6th International Symposium on Innovation and Sustainability of Modern Railway (ISMR'2018)* (China: ECJTU) pp 261-264

[8] GOST 33211-2014. Freight wagons. The requirements for durability and dynamic qualities *cars* (Moscow: Standartinform)

[9] Romen Yu S, Mutinshteyn L A, Neverova L I 2013 Influence of longitudinal forces in the trains on the danger of gathering railroad cars depending on their load *Transport of the Russian Federation* **3(46)** pp 64-68

[10] Ermolenko I Yu, Zheleznyak V N 2016 Study of rolling stock dynamics using an experimental laboratory wagon when driving on complex sections of the road *Modern technologies. System analysis. Modeling* **4(52)** pp 199-203

[11] Ermolenko I Yu, Zheleznyak V N, Martynenko L V, Fedyukovich G I 2017 Experimental research stand to assess the dynamic characteristics of the 18-100 truck when driving in straight and curved sections of the road *Aircraft engineering and transport of Siberia IX* (Irkutsk: ISTU) pp 551-514

[12] Zheleznyak V N, Ermolenko I Yu, Fedyukovich G I, Solodov G S, Martynenko L V 2018 Stand to assess the level and nature of the power interaction of the wheel with the rail *patent RF* 2017100776

[13] Ermolenko I Yu, Martynenko L V, Zheleznyak V N 2018 Roller stand to assess the dynamic parameters of the wagon trolley *patent RF* 2017125220

Research of high-frequency influence on polymeric materials of rolling stock products

N G Filippenko[1]

[1]Irkutsk State Transport University, Chernyshevsky, 15, 664074, Russia

E-mail: ifpi@mail.ru

Abstract. The paper presents the main directions of the technological process of high-frequency processing of polymeric materials used in rolling stock products. The necessity of developing a mathematical model of the critical process of exposure to high-frequency energy is substantiated. As a model, an imitational mathematical model of the process of high-frequency processing of polymeric materials that takes into account breakdown phenomena is proposed (based on experimental studies of a large number of the most used polymer materials).

1. Introduction

High-frequency (HF) processing of thermoplastic polymers used in the manufacture of parts and components of rolling stock has proven itself in industrial practice. The main advantages of the method, for example, HF welding, are fast and local (in the weld weld zone) heating of the joining surfaces without penetration of the entire volume of the material. At the same time, the quality of welded joints (strength, tightness), as a rule, exceeds the quality of joints with other welding methods. A self-regulating process, for example, drying polymeric and dielectric materials of large geometrical dimensions, makes HF electrothermal one of the main technological operations in this field [1, 2].

Due to the lack of technical ability to control the temperature in the field of HF exposure, a special role in the electrothermal processing technology is assigned to the task of correctly calculating the process time - the time it takes for the drying temperature or the melting temperature of the thermoplastic to be achieved in the material [1, 3, 4, 5]. But this technique leads to errors associated with the heterogeneity of materials of products, both in chemical composition and in physical condition. Attempts to create mathematical models of the process also do not allow to take into account the properties of polymer products that are changed during transportation, storage and operation. The disadvantages of the well-known mathematical models are also: neglect of changes in the specific power of internal heat sources during the processing cycle, idealization of the design of technological equipment that does not take into account the heterogeneity of the materials being processed, the presence of insulating inserts between the material being processed and the working capacitor electrodes, the critical effect of breakdown phenomena on the product being processed and costly electrodes. Using data from sources [8-10, 11], we can conclude that the defect in the production of the above-described technological preparation reaches 7-12%.

2. Materials and methods

Features of the control process RF processing

Taking into account the peculiarities of the HF process control technique by the parameters of the RF

IOP Conf. Series: Materials Science and Engineering **760** (2020) 012021 doi:10.1088/1757-899X/760/1/012021

generator, namely, the control of the pre-breakdown state and partial discharge dynamics (CR) [4-7, 12], it is obvious that the mathematical model of the process needs to be built based on this particular dependence of the technological system. Moreover, these parameters have a greater impact on the quality of the processed products, and their uncontrolled change leads to emergency stops and significant material losses. The inconstancy of the electrophysical properties of materials ε, tgδ, depending on temperature, the prehistory of the samples, manufacturers and changing technologies for producing polymers, the arrival of new materials in the industry can only be compensated by the results of experimental tests. With such complexity of the task, its solution was found in the compilation of a simulation mathematical model of the process of RF processing of polymeric materials, taking into account the occurrence of a breakdown state.

Experimental tests of polymeric materials significantly different in their physicochemical parameters were carried out. The graphical depiction of experimental data presented in figure 1 shows that the portion of the curves for the onset of the pre-breakdown state, characterizing the relationship between the time of occurrence of the CR and the number of the CR, has a significant view.

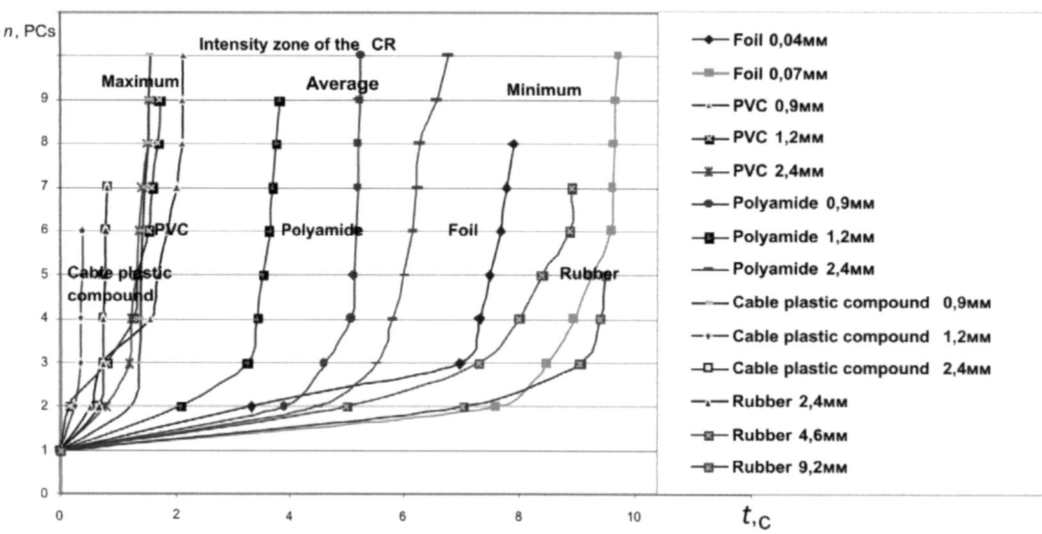

Figure 1. The empirical dependence of the occurrence of CR on the processing time

According to the results of the analysis of the distribution curves of the CR, we can conclude about the exponential development of events, and their characteristic distribution allows us to divide the materials into two groups:

- with the active development of a pre-breakdown state after the first discharge of CR1, when the following condition is fulfilled:

$$\Delta\tau_{CR1} = \tau_{CR2} - \tau_{CR1} < 1{,}2 \text{ c}, \quad (1)$$

where τ_{CR1}, τ_{CR2} - the time of occurrence of the first and second partial discharges, respectively.

- with a long pre-breakdown condition (the time from the first to the subsequent CR is more than 2 s),

$$\Delta\tau_{CR1} = \tau_{CR2} - \tau_{CR1} > 2 \text{ c}. \quad (2)$$

Assuming that the n-th CR, for which

$$\Delta\tau_{CR\,n} > 1{,}2 \text{ c} \quad (3)$$

not the beginning of the development of the pre-breakdown state, we obtain the following family of graphs of the dynamics of the development of the breakdown of various materials, presented in figure

2.

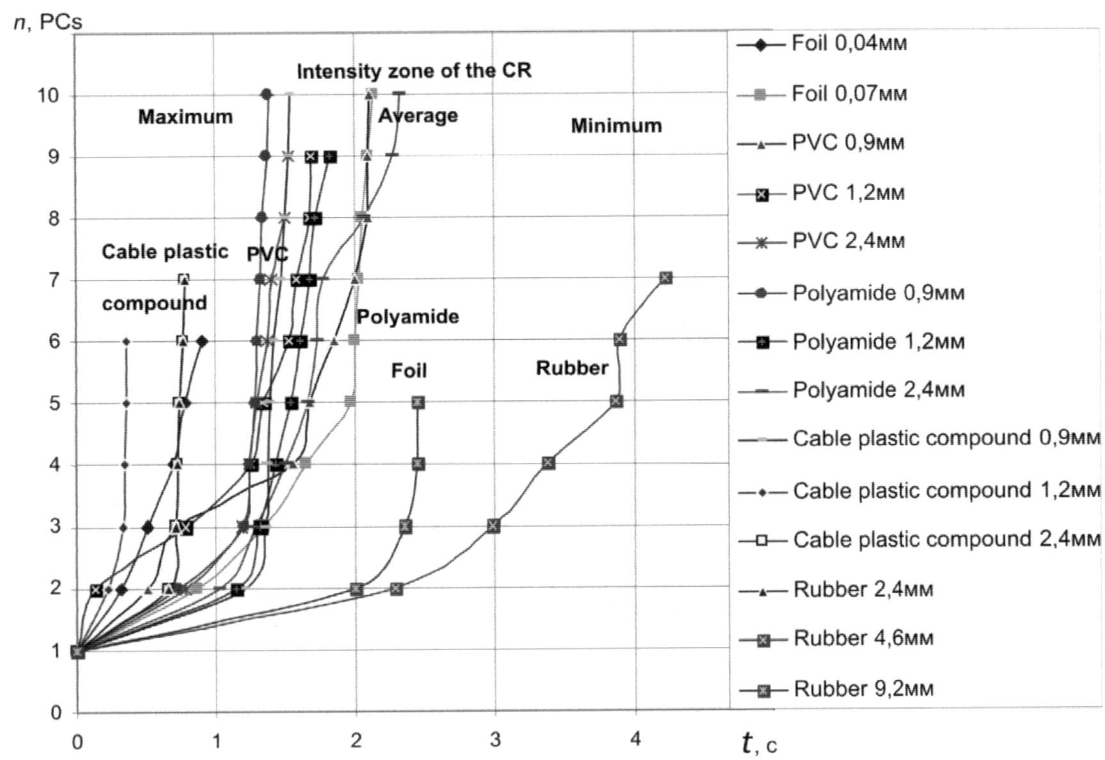

Figure 2. Graph of the dynamics of the development of electrical breakdown

3. Results

Methods of approximation, simplification of initial physical models, reduction of their order, linearization are the main directions, allowing to evaluate the possible dynamic properties of technical systems, the effectiveness of managing their state, to develop special tools to change the dynamic characteristics in the right direction.

Based on this, it was decided to build an imitational mathematical model of the technological process of RF processing of polymeric materials, taking into account the simulation of the development of breakdown by approximation of experimental data using the least squares method.

Having considered the totality of the graphs of occurrence of the CR, we identified a curve that is as far as possible from the others in inter-digit temporal parameters. After conducting a study of its graphical display and approximation by the least squares method, it was determined that this curve is subject to a functional relationship, which has the following form:

$$Y = \exp(A + BX) \tag{4}$$

or in our case:

$$n_{CR} = \exp(A + B\tau_{CR\,pac}) \tag{5}$$

$$\tau_{CR\,pac} = \frac{\ln n_{CR} - A}{B} \tag{6}$$

The obtained approximated curve (4) can be taken as a model of the RF processing process, since it fully reflects the process of occurrence of the CR with the smallest dynamics of the development of the pre-breakdown state.

4. Discussion

In the further presentation of the research materials, the derived dependence (5) will be denoted by the term “partial discharge curve” (CDFR) presented in fig. 4.

A special property of the derived dependence is that it displays the maximum value of empirical data on the time of development of the CR of the polymeric materials under study. It should be noted that the tested samples of polymeric materials (with the aim of choosing homogeneous in composition and quality) underwent conditioning, cleaning, testing for inclusions, roughness, burrs, etc. on physical and chemical composition. At the same time, their humidity, due to the hygroscopicity of polymeric materials, can reach up to 8%. All this significantly reduces the electrophysical parameters of polymers used in industry. Reducing the quality of the material inCReases the likelihood of breakdown. The deterioration of the electrophysical parameters of polymers (always present in real conditions) leads to a shift in the data of its pre-breakdown state to the left area of the curves of our graph, in the area of materials with more active dynamic indicators of the Czech Republic. Therefore, we can say that the CDFR displays the indications of the time of development of the Czech Republic under ideal-laboratory conditions, which are absent in actual production situations. From here, the values of the CR dynamics of polymeric materials, as applied to production conditions, will always be within the zone of the delineated CDFR. Which once again confirms the correctness of the adopted mathematical model of the RF processing process with protection against breakdown phenomena.

Using the free software package Approximator ver. 1.6, we calculated the coefficients A and B, presented in figure 3. The approximation error was satisfactory and amounted to 0.98.

Substituting the coefficients in equation (7, 8), we obtain the equation for the occurrence of a CR in the RF processing for all polymeric materials.

$$n_{CR} = \exp(-1{,}99 + 1{,}8\tau_{CR\,pac}) \quad (7)$$

$$\tau_{CR\,pac} = \frac{\ln n_{CR} + 1{,}99}{1{,}8} \quad (8)$$

The mathematical model of the HF process, which takes into account the occurrence of a pre-breakdown state, corresponds to materials with significant differences in physicochemical parameters and geometrical dimensions (thickness).

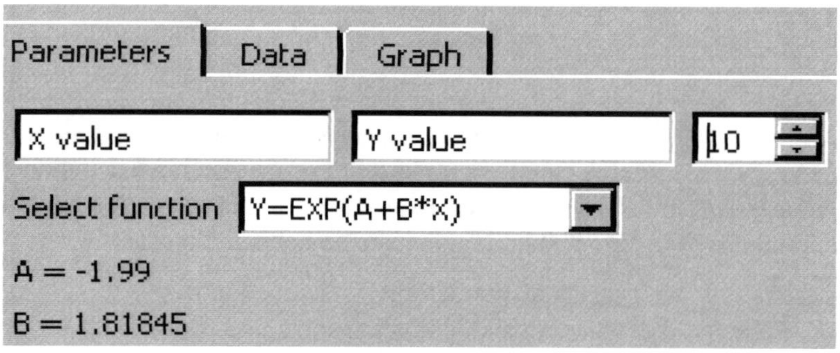

Figure 3. A fragment of the data on the calculation of the CDFR

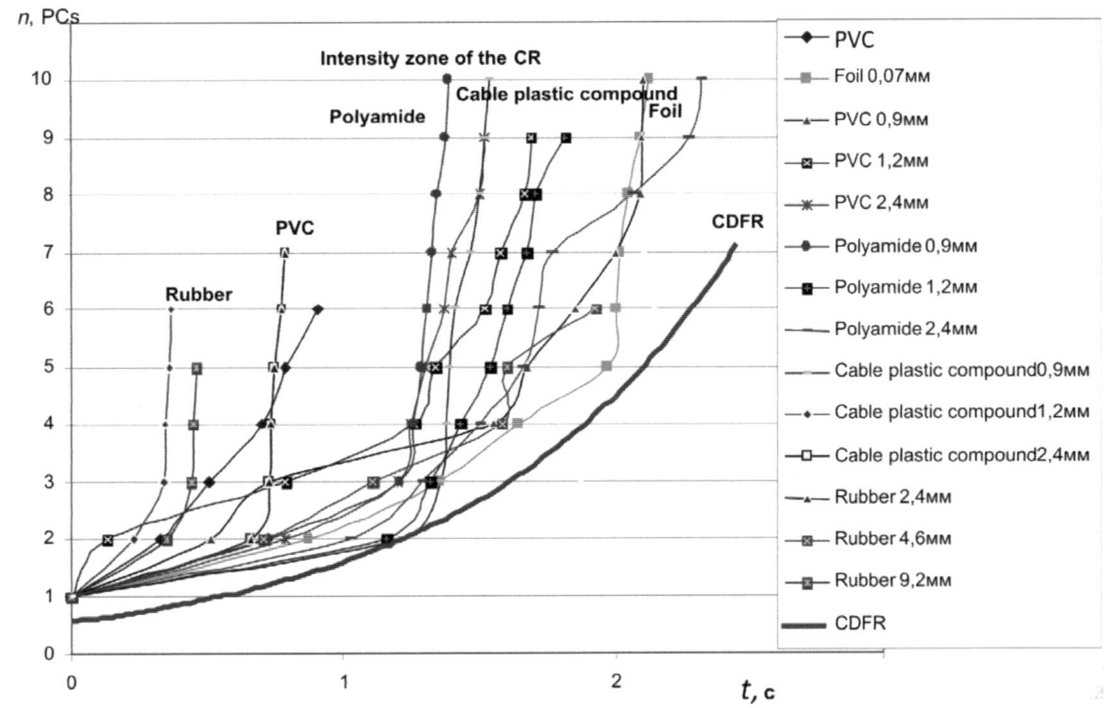

Figure 4. The curve of the dynamics of partial discharges (CDFR)

5. Conclusion

Thus, we can assume that the developed mathematical model of the HF process, which takes into account the occurrence of a pre-breakdown state, corresponds to materials with significant differences in physicochemical parameters and geometrical dimensions (thickness). In this case, it becomes possible, according to a mathematical model of the form (8) determined during the study, to calculate the time of occurrence of a CR event. That, in turn, allows in real time to find the algorithm for controlling the RF exposure process in order to prevent breakdown.

All this is possible and necessary to use for the organization of the control and management of the process of RF processing of polymeric materials, working in a pre-breakdown state, which is the most efficient way of working.

References

[1] Filippenko N G 2012 Matematicheskaya model processa vysokochastotnoj obrabotki polimernyh materialov *Sovremennye tekhnologii. Sistemnyj analiz. Modelirovanie* **1 (33)** 76-79.

[2] Livshits A V 2015 Upravleniye tekhnologicheskimi protsessami vysokochastotnoy elektrotermii polimerov *Problemy mashinostroyeniya i avtomatizatsii* **3** 120-126.

[3] Filippenko N G, Livshits A V and Popov S I Ustrojstvo izmereniya temperatury poverhnosti obekta Russia Patent No. RUS 132549, (19 october 2012)

[4] Aleksandrov A A, Livshits A V and Rudykh A V 2016 Raschet termicheskikh ostatochnykh napryazheniy v zagotovkakh iz alyuminiyevykh splavov *Sovremennyye tekhnologii. Sistemnyy analiz. Modelirovaniye* **1(49)** 52-56.

[5] Livshic A V and Aleksandrov A A 2014 Prognozirovanie temperaturnogo polya dlya opredeleniya ostatochnyh napryazhenij voznikayushchih pri termoobrabotke alyuminievyh splavov *Nauka i obrazovanie: nauchnoe izdanie MGTU im. N.E. Baumana* **7** 36-47.

[6] Livshic A V 2015 Avtomatizirovannaya sistema nauchnyh issledovanij vysokochastotnoj elektrotermii. *Problemy mashinostroeniya i avtomatizacii* **4** 54-60.

IOP Conf. Series: Materials Science and Engineering **760** (2020) 012021 doi:10.1088/1757-899X/760/1/012021

[7] Larchenko A G, Filippenko N G, Livshic A V and Popov S I Ustrojstvo diagnostiki detalej iz poliamidnyh materialov Russia Patent No. RUS 132209, (05 April 2013)

[8] Livshits A V 2014 Issledovaniye protsessa vysokochastotnoy elektrotermii termoplastov na osnove yego matematicheskogo modelirovaniya *Sistemy. Metody. Tekhnologii* **2** **(22)** 90-98.

[9] Livshic A V, Filippenko N G, Popov S I and Larchenko A G 2013 Issledovanie vliyaniya dielektricheskih elementov rabochego kondensatora vysokochastotnoj elektrotermicheskoj ustanovki na process obrabotki polimernyh materialov. *Sovremennye tekhnologii. Sistemnyj analiz. Modelirovanie* **3** **(39)** 270-275.

[10] Fedorova I G and Bezmenov F V 1990 *Vysokochastotnaya svarka plastmass* (L. Mashinostroyeniye)

[11] Netushil A V, Zhukhovitskiy B YA, Kudin V N and Parini Ye N 1959 *Vysokochastotnyy nagrev dielektrikov i poluprovodnikov* (M.-L. Gosenergoizdat)

[12] Spektor S A 1987 *Elektricheskiye izmereniya fizicheskikh velichin: metody izmereniy* (L. Energoatomizdat)

IOP Conf. Series: Materials Science and Engineering **760** (2020) 012022 doi:10.1088/1757-899X/760/1/012022

Suggestions for improving the efficiency of repair activity

A M Gladkih, V Yu Konyuhov, I I Galyautdinov and E I Shchadova

Irkutsk National Research Technical University, 83, Lermontov Street, Irkutsk, 664074, Russia

E-mail: konyuhovvy@ex.istu.edu

Abstract. In order to improve the organization and performance of equipment maintenance, innovative proposals were developed and implemented. The explanations were taken into account when determining the methods for evaluating the maintenance performed for the joint approval of a unified methodology of applying the concept of a standard hour. The planning of performing equipment maintenance by day during the whole month was revised, which, in turn, allowed minimizing significant fluctuations in the scope of work and eliminating peak loads. The works on the maintenance of the mechanical and energy components of a single assembly were revised and combined as much as possible, which ensured the more efficient use of working time. Work has been organized on updating the composition of the operations to be performed, since the scope of maintenance work in the assignment forms is not always representative of the complex of necessary operations. Preventive work with subordinate personnel is organized in the field of improving the industrial standards in the workplace.

1. Introduction
Maintenance is the main preventive measure necessary to ensure reliable operation of equipment between scheduled maintenance and reduction of the total amount of repair work. It provides for supervision of equipment operation, equipment care, maintaining the equipment in good condition, scheduled technical inspections. As a rule, regulated maintenance is performed by the operational and repair personnel of the enterprise.

Recently, in the industrial manufacturing sector, order placement for equipment maintenance services have increased in various measurement systems. Many large enterprises are developing their own information and analytical reference books that represent the costs of performing certain kinds of repair for equipment.

Unfortunately, the existing regulatory framework for the maintenance of the equipment does not provide for the standardizing of each operation that can be carried out as part of the maintenance (the number of the equipment in use reaches tens and sometimes hundreds of thousands of items).

2. Results and discussion
The complexity of the repair work of a particular type is determined based on the number of units of repair complexity and the time norms established for one repair unit. The number of units of repair complexity in the mechanical part of the equipment coincides with the category of complexity. The following technical and economic indicators are used in the analysis and evaluation of the repair service:

- Downtime of equipment under repair per one repair unit. This indicator is determined by dividing the total downtime of equipment under repair by the number of repair units of

IOP Conf. Series: Materials Science and Engineering **760** (2020) 012022 doi:10.1088/1757-899X/760/1/012022

equipment that is being repaired in a given planning period. It is necessary to maximally reduce this time.

- The number of repair units of installed equipment per one repair worker. This number characterizes the labor productivity of repair workers, which should constantly increase.
- The cost of repairing one repair unit, determined by dividing all costs (including delivery notes) for repairs during a certain time (for example, during a year) by the number of repair units of equipment repaired during the same planned period.
- The turnover of the stock of spare parts, which is equal to the ratio of the cost of consumed spare parts to the average balance of them in the storerooms. This indicator should be as big as possible.
- The number of accidents, breakdowns and unscheduled repairs per unit of equipment, which characterizes the efficiency of the planned preventative repair system. It should be minimal.

There is a definite relationship between these indicators. Reduction of equipment repair downtime per one repair unit results in an increase in the number of repair units of installed equipment per repair worker, since the same amount of repair work while reducing time for it can be performed by fewer workers. This leads to a reduction in the cost of repairing one repair unit. The first three indicators are improved through a more efficient organization of repairs and maintenance, which results in an improvement in the fourth indicator. All indicators are analyzed in comparison with the indicators achieved at specialized repair shops [1–6].

The indicator of the cost of normalized working time for rendering services is the normative labor intensity, which is set in standard hours. The number of standard hours required for the manufacture of products / services, is determined by summing up the technically feasible standards of time for all process flow operations. Thus, the customer forms a production program (in standard hours) for the repair company. In this connection, the customer is not able to objectively estimate the scope of work of any unfulfilled (as unnecessary) operation. When conducting an internal audit, the customer, as a rule, checks the availability of human resources (maintenance personnel) at workplaces, thereby determining the number of man-hours worked: the actual cost of working time for rendering services, that is, the actual labor intensity.

As a result, under certain conditions of equipment maintenance, instead of an approved unit of measure for maintenance (a standard hour) the customer uses the number of hours worked by staff, i.e., a kind of "substitution" of indicators occurs when the number of man-hours is deemed to be standard hours.

For clarity, it is proposed to consider the following situation on the conditional example of an enterprise engaged in the maintenance of industrial gearboxes.

For proper maintenance of gearboxes, there is a final list of possible operations to be performed, for which the average standard length of time in hours is established, used only to determine the cost of the contractor for maintenance. The result of the technician's work is a serviceable gearbox, and it is not necessary to perform the entire list of operations to achieve these goals. The list of operations, as well as the time and volume of impact, in each case is determined individually, taking into account the operating conditions, the repair system, and the qualification of the process and maintenance personnel.

Payment for the services of a technician is determined by the price list of his company in accordance with the following entry: maintenance of one gearbox (brand, type and other identification features) is 1,700 rubles regardless of the actual time spent. If the service company carries out maintenance by hundreds of thousands of objects, then their enumeration in the annually approved price list is not appropriate. The price list approves the value of any conditionally constant value, in this case, the cost of a standard hour. A quantity of these values, which measure the result of the maintenance of one unit of equipment, is established by the local reports, which do not require systematic revision.

Estimating effectiveness of measures on increasing the labor productivity is defined as the ratio of actual labor expense to the normative labor expense:

$$K_{ef.} = T_{act.} / T_{norm.} \quad (1)$$

To convert the normative labor input into actual time expenditure, it is adjusted with the help of the labor performance, which directly depends on the qualifications of workers and the engineering and technical level of the enterprise. The actual labor intensity of production is calculated by dividing the working time spent in man-hours by the total volume of production in actual or cost measurements [7–12].

Labor productivity is an indicator of labor resource efficiency. It is measured by the number of products, works in kind or in value terms, produced by one employee for a certain time. One of the methods of measuring labor productivity is *labor method*, based on the calculation of the labor intensity of each product, as well as work. According to this method, labor efficiency is estimated by comparing actual costs with standard costs.

As it follows from the definition of the meaning of the standard hour, this term refers to the section of the economics (labor economics), which studies the processes of reproduction of the labor and the interaction of workers, problems of productivity and efficiency of labor, methods of substantiating the number of employees, etc.

The analysis of publicly available information sources, a summary of the experience of using for the mutual calculations of standard hours by repair companies suggests that the standard hour is a unit of normative labor intensity that differs from the astronomical concept of an hour. This is the value to which the price is linked and the cost of the technical maintenance unit of equipment is obtained. Otherwise, it is necessary to calculate the cost of maintenance of each unit of equipment by making direct calculations, which in turn impedes the task of planning maintenance costs and monitoring their execution.

The standard hour indicator is used in various fields of activity, for example, when evaluating the work of programmers, managers, university professors, etc. Representatives of creative professions (writers, composers, artists) may have a tenfold difference in productivity, since the potential labor productivity of each person is different: the man-hour of different specialists has different returns, which depend on personality type, education, experience, talent and other qualities of a specific person and changes over time [13–17].

The indicator of service in standard hours was widely adopted at car care enterprises to determine the complexity of performing a car repair operation. The total cost of car care services, excluding the cost of spare parts and materials used in the repair, is defined as the product of the number of standard hours and the cost of one standard hour [18–22].

In order to determine the assessment of the effectiveness of equipment maintenance by the repair company, an internal audit was conducted, taking into account the organization of the operation of the maintenance system and the processing of the necessary documentation.

During the period of the audit, the following measures were implemented to improve the efficiency of high-quality maintenance:

- lists of shop equipment of the enterprise were determined and personally assigned with internal orders to the workers of the repair sites via their signature;
- flow charts for maintenance of main and auxiliary process equipment have been developed;
- normative sets of labour saving tools/means for the subteam and team work were developed and formed.

However, the repair organization identified the following deficiencies during the audit:

- insufficient level of qualification of line level engineering and technical workers (foremen) of repair services in the ability to organize the placement of workers in workplaces for equipment maintenance, to monitor the implementation of the shift task, and to make operational decisions when changing the shift task;
- insufficient level of qualification of maintenance personnel: workers have no notion of the required standard necessary to perform certain operations;
- insufficient level of production practices in the workplace.

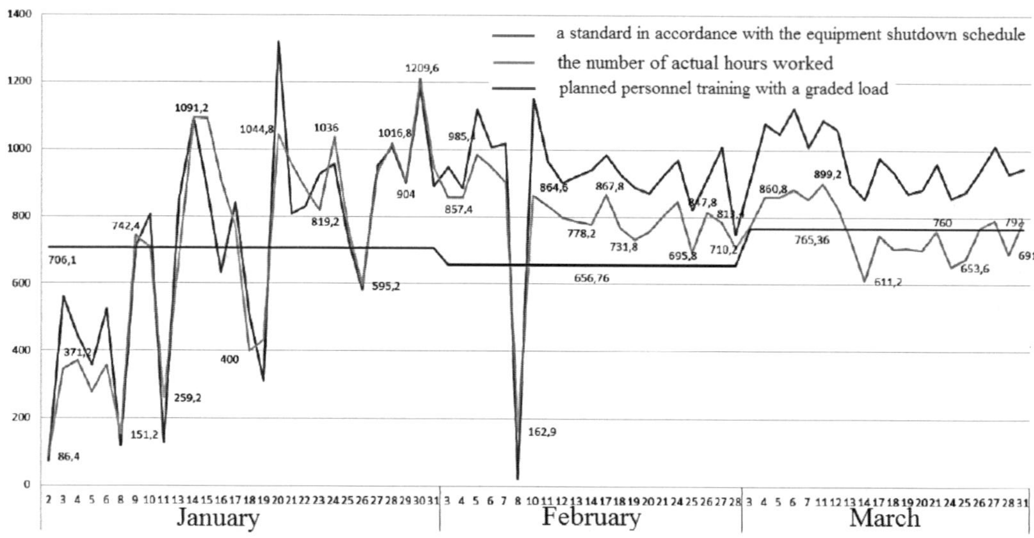

Figure 1. Maintenance of mechanical and processing equipment of a metallurgical plant.

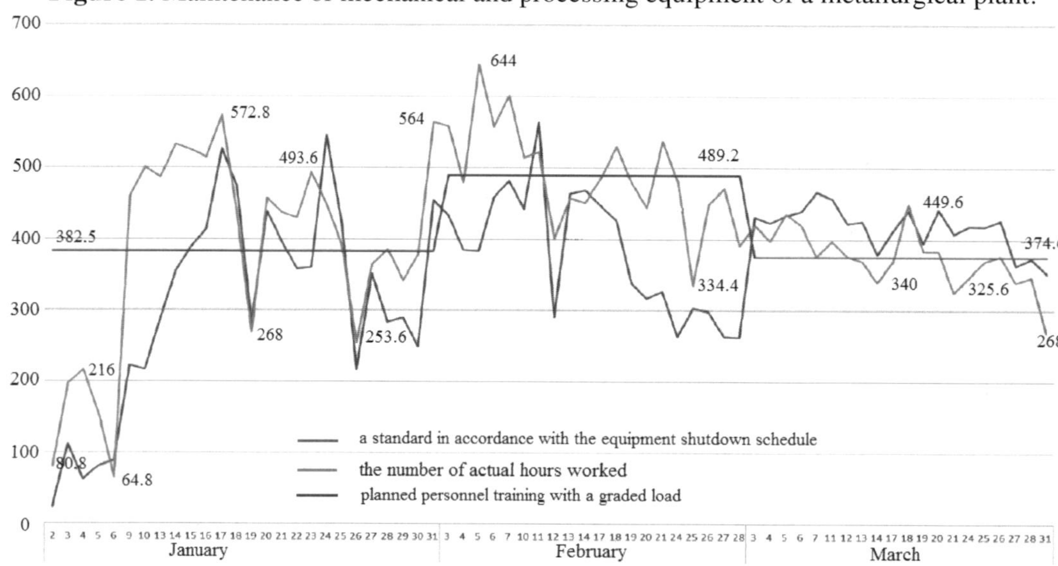

Figure 2. Maintenance of the power equipment of the metallurgical enterprise.

The insufficient level of joint organization of engineering and technical workers in the field, namely:

- sending equipment to maintenance is not systematized by work schedules. The equipment is stopped by the customer service (technologists) to perform maintenance, based on the conditions of the manufacturing program for the production of finished products, which does not allow for high-quality preparation for maintenance and leads to loss of working time when shifting team workers to other production sites;
- there is no systematic approach to the preparation of equipment for maintenance (timely shutdown of equipment, preparation of assignment forms, etc.);
- increased loss of working time for registration of staff permits for work in existing electrical installations;
- in the approved model lists for maintenance of equipment there are items related to their repair, which in fact does not allow performing the entire scope of work specified in the assignment form (the standard only includes the labor costs for maintenance).

All of the above gave evidence of the need to jointly put in order the organization of planning and production of works on the maintenance of fixed assets of the customer [23–27].

From the graphs presented above, you can see that:

- the planned indicator of maintenance performance (in labor hours) is not synchronized with the possible training of personnel involved in the maintenance performance of both mechanical and processing and power equipment of the metallurgical corporate customer;
- in general, the dynamics of the actual personnel training follow the equipment shutdown schedule, while the actual working off exceeds the standard values. The main reasons for deviations are: not shutting down equipment for maintenance, long execution of permits, etc.

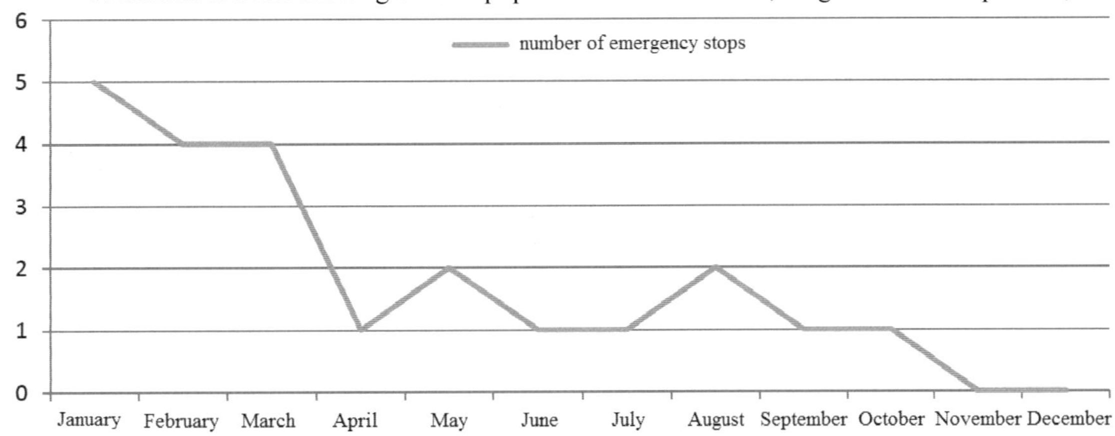

Figure 3. Change in the number of emergency stops during the year after the taken measures.

3. Conclusion

In order to improve the organization and performance of equipment maintenance, the customer and the contractor jointly implemented the following proposals:

- The explanations described above were taken into account when determining the methods for evaluating the maintenance performed for the joint approval of a unified methodology of applying the concept of a standard hour.
- The planning of performing equipment maintenance by day during the whole month was revised, which, in turn, allowed minimizing significant fluctuations in the scope of work and eliminating peak loads.
- The works on the maintenance of the mechanical and energy components of a single assembly were revised and combined as much as possible, which ensured the more efficient use of working time.
- The advanced training courses were organized for maintenance personnel performing maintenance of equipment.
- Work has been organized on updating the composition of the operations to be performed, since the scope of maintenance work in the assignment forms is not always representative of the complex of necessary operations.
- Preventive work with subordinate personnel is organized in the field of improving the industrial standards in the workplace.

Considering the positive dynamics of operational improvements in the organization and maintenance of equipment during the year, the reduction of emergency stops and the increase in equipment reliability, the measures taken were deemed appropriate and will be further implemented.

References

[1] Novitsky N I 2001 *Organization of production in enterprises: a study guide* (Moscow: INFRA-

M Publ.) p 214
[2] Gozbenko V E, Kargapoltsev S K, Kornilov D N, Minaev N V and Karlina A I 2016 *Int. J. of Applied Engineer. Res.* **11(20)** 10367–73
[3] Livshits A V, Filippenko N G, Homenko A P, Kargapoltsev S K, Gozbenko V E and Dambaev Z G 2017 *JP J. of Heat and Mass Transfer* **14(2)** pp 219–26
[4] Kargapoltsev S K, Shastin V I, Gozbenko V E, Livshits AV and Filippenko NG 2017 *Int. J. of Applied Engineer. Res.* **12(17)** 6499–503
[5] Khomenko A P, Gozbenko V E, Kargapoltsev S K, Minaev N V and Karlina A I 2017 *Int. J. of Applied Engineer. Res.* **12(23)** 13773–8
[6] Shastin V I, Kargapoltcev S K, Gozbenko V E, Livshits A V and Filippenko N G 2017 *Int. J.of Applied Engineer. Res.* **12(24)** 15269–72
[7] Bachurin A V 1962 *Economic Encyclopedia. Industry and construction:* in 3 vols (Moscow: Soviet Encyclopedia Publ.)
[8] Gozbenko V E, Kargapoltsev S K and Karlina A I 2019 *IOP Conf. Series: Earth and Environmental Sci.* **229** 012020
[9] Gozbenko V E, Kargapoltsev S K and Karlina A I 2019 *IOP Conf. Ser.: Earth and Environmental Sci.* **229** 012021
[10] Shtayger M G, Balanovsky A E, Kondratev V V, Karlina A I and Govorkov A S 2018 *Advances in Engineer. Res. Confer. Proc.* p360–4
[11] Balanovsky A E, Shtaiger M G, Grechneva M V, Kondratev V V and Karlina A I 2018 *IOP Conference Series: Materials Science and Engineering* **411** 012012
[12] Balanovsky A E, Shtayger M G, Kondratev V V, Huy Vu V and Karlina A I 2018 *IOP Conf. Ser.: Mater. Sci. and Engineer.* **411** 012013
[13] Lvov I V and Pirogov V I 2006 How to measure the productivity of management? *All-Rus. Economic J.* **9**
[14] Gozbenko V E, Khomenko A P, Kargapoltsev S K, Minaev N V and Karlina A I *2017 Int. J. of Applied Engineer. Res.* **12(22)** 12369–72
[15] Nokhrina O I, Rozhikhina I D and Khodosov I E 2016 *Steel in Translation* **46(10)** 686–91
[16] Fedin V M and Borts A I 2009 *Metal Science and Heat Treatment* **51(11–12)** 544–52
[17] Yolkin K S, Yolkin D K, Shtayger M G, Kolosov A D and Ivanov N A 2018 *IOP Confe. Ser.: Mater. Sci. and Engineer.* **411** 012028
[18] Tsekhovoy A 2009 *Autobusiness* **20**
[19] Gozbenko V E, Kargapoltsev S K, Minaev N V and Karlina A I 2016 *Int. J. of Applied Engineer. Res.* 11(**23**) 11132–6
[20] Kargapoltsev S K, Khomenko A P, Gozbenko V E, Minaev N V and Karlina A I 2017 *Int. J. of Applied Engineeri. Res.* **12(22)** 12362–8
[21] Balanovsky A E, Kondratev V V, Shtayger M G, Nebogin S A and Karlina A I 2018 *IOP Conf. Ser.: Mater. Sci. and Engineer.* **411** 012014
[22] Ivanchik N N, Balanovsky A E, Shtayger M G, Sysoev I A and Karlina A I 2018 *IOP Conf. Ser.: Mater. Sci. and Engineer.* **411** 012035
[23] Balanovsky A E, Shtaiger M G, Kondratev V V, Karlina A I and Govorkov A S 2018 *J. of physics: conf. ser.* 012006
[24] Boyarkin D N and Novikov N I 2010 Criteria and indicators of economic efficiency of the repair service, ensuring the efficiency of fixed assets of enterprises *Bull. of Chelyabinsk State University* **14(195)** 127–32
[25] Kondrat'ev V V, Ershov V A, Shakhrai S G, Ivanov N A and Karlina A I 2016 *Metallurgist* **60(7–8)** 877–82
[26] Kondratiev V V, Govorkov A S, Kolosov A D, Gorovoy V O and Karlina A I 2017 *Int. J. of Applied Engineeri. Res.* **12(22)** 12373–7
[27] Iamshchikova I V and Butina N I 2013 *News of Higher Educational Institutions. Investments. Construction. Real estate* **2(5)** 53–58

Simulation of an integrated public transport system by the example of a compact city

O Lebedeva [1], Ju Poltavskaya[1] and V Gozbenko[2]

[1]Angarsk State Technical University, Angarsk, Russia
[2]Irkutsk State Transport University, Irkutsk, Russia

E-mail: kravhome@mail.ru, juliapoltavskaya@mail.ru, vgozbenko@yandex.ru

Abstract. The effectiveness of the transport network depends on the coherence of all message options. Suburban rail transportation and route bus service are travel options often used in medium and large cities in developed and developing countries. Integrating these two travel options reduces total travel time. The study will consider the application of the operational integration model of suburban trains and urban public transport. The model includes two submodels: routing and planning. In the routing submodels, public transport bus routes are generated that pass through the railway station. The planning sub-model will allow developing suburban train schedules with the optimal coordination of timetables for urban bus routes. As an example, the railway station Angarsk and the urban route network are considered. Bus transportation is coordinated with the existing railway timetable for the arrival of local trains. Genetic algorithms are robust optimization techniques and can be used in planning submodels. The result of the study is to determine the routes that require changes in the timetable, taking into account the arriving trains.

1. Introduction

Most cities in developed and developing countries face problems of inefficient operation of the route network for lack of coordination between public transport and objects of aggression. The operation of the transport terminal is rarely taken into account in the operation of the urban network, and each infrastructure object is considered separately. They compete with each other instead of harmoniously interacting. Unhealthy competition leads to a duplication of transport services, an increase in travel time due to the long waiting time for a vehicle at a transfer hub. The efficiency of the transport system can be achieved through overall coordination of the transportation process. Rarely, optimization is carried out on several types of transport at the same time, more often studies are devoted to a specific area [1–4].

Earlier, attempts were made to route and plan the coordinated work of different types of transport in a complex transport system using analytical models. It was assumed that the street-road network is a rectangular area located parallel to one railway line, but its actual configuration was not always taken into account. As an example, let us consider the city of Angarsk, referring to the agglomeration of the Irkutsk region.

2. Expermental studies

The data was collected at the Angarsk passenger railway station, which is a transport terminal where passengers make a transfer to other mode of transport (private, public or taxi). The station is located on the Trans-Siberian Railway, belonging to the East-Siberian railway. All suburban and passenger trains stop there. Neighboring stations are in the west – Kitoy, in the east – Sukhovskaya and Yuzhnaya. That is, there are 4 stations on the line in question. A railway station is a terminal where passengers change mode of transport [5]. At this station a sufficient number of urban public transports pass in different directions, routes no. 3, 7, 8, 9, 27, 28.

Movement surveys were conducted during the morning rush hours, that is, from 7 to 9 am. In the course of road traffic surveys, it was found that the maximum passengers number was from 7 to 8 am, and after 9 am, the suburban traffic decreases. During the survey, the number of passengers leaving the station was counted manually, and was 1304 passengers. The lack of coordination between the bus and train service is observed in the morning and evening rush hours.

The accountants took a selective survey of passengers (328 people), and it was about 25 % of the total. The questionnaire included the following questions: destination, mode of transport and travel time. The passengers who used individual transport or taxis for further travel were asked if they are ready to refuse in favor of public transport if the bus schedule is coordinated with the railway service. A group of passengers who do not need any mode of transport was identified, as the enterprises providing jobs are located within walking distance from the station (less than 5 minutes).

Table 1 shows the potential demand by direction.

Table 1. Potential demand by directions

№	Destination	Potential demand by direction		
		7.00–8.00	8.00–9.00	7.00–9.00
1	Central market	13	23	36
2	Shop “Zarya”	17	18	35
3	Shop “Magistral”	10	15	25
4	Technical College	8	16	24
5	Communication center	20	22	42
6	4th village	21	7	28
7	Gorgaz	10	14	24
8	Firm “Cars”	5	12	17
9	Medical Sanitary Unit-28	9	14	23
10	102nd quarter	8	21	23
11	Orgstroyproekt	4	7	29
12	Microdistrict “Northern”	14	24	11
13	AEMZ	9	14	38

The percentage of passengers who agree to use public transport in the case of optimization route schedule is taken into account in Table 1. It was found 13 destinations, which respondents called more often. Thus, these nodes were identified as priority destinations that are important for the development of the route network.

To optimize the operation of the route network, matrices of potential demand (obtained by interviewing passengers arriving at the station) and consumption of time are required. The following data were used to calculate the matrix of time-consuming: distances to destinations obtained by measuring the route network, an average speed of 15 km/h was set by the time method. The matrices (dimension 14×14) include the station Angarsk and the 13 identified destinations listed in Table 1.

The aim of the study is to develop a model for the operational integration of public transport timetables and railway communications [6–8].

3. Theoretical basis

The general methodology can be presented as follows.

1. Information gathering: data requirements are established; specification of the existing route and transport networks [9] (determination of the length, duration of movement and characteristics of the rolling stock involved in the transportation process); road traffic survey (assessment of the existing distribution of passenger traffic by mode of transport and potential demand by direction).

IOP Conf. Series: Materials Science and Engineering **760** (2020) 012023 doi:10.1088/1757-899X/760/1/012023

2. Routing, which bases its work on the data obtained in the previous stage and includes a realistic assessment of the road network and a matrix of demand distribution. Further, it is possible to use the route network generation algorithm for the purpose of designing.

3. Planning, which includes minimizing the distance of the route network (operating costs) and waiting time, coordinating the work of urban public transport (passenger costs) subject to constraints: load factor, transfer time and established demand.

Routing is performed using a heuristic approach to optimize the network by including additional nodes. When tested on a real network, it provides the best route, taking into account the location of various destinations (nodes), restrictions on communications in the station area and design without further interchanges. The routing algorithm is based on the demand matrix, as it is a fundamental aspect in network design. In addition, the waiting time for rolling stock should be taken into account, since its increase leads to a decrease in the quality of passenger service. To minimize the waiting time, it is proposed to optimize the bus schedule [10].

The proposed heuristic algorithm consists of two stages.

1. Determine the shortest distance from the station to the main destinations.

2. Calculate the deviations of the shortest distances by inserting other nodes into the chain.

A balance must be maintained when satisfying demand by including nodes and increasing the length of the route. The optimal route network is governed by the criterion of maximum demand and the minimum travel time - the distance should not increase more than 1.5 times, therefore, an upper limit is set. In our study, it is 15 km (1 hour) and is determined based on the location of the points of destination specified in the sample.

The limitation of the upper limit of the length is established in the case when the route has a large length. The inclusion of additional nodes in the route network with a small length allows increasing the efficiency of the transport infrastructure [11].

The route network optimization algorithm includes the following operations:

1. Determination of potential demand for stations and preparation of the matrix with node numbers.

2. Compilation of a distance matrix based on the existing route network and determining connectivity of nodes. Convert distance matrix to travel time matrix using average travel speed.

3. Identification of nodes with high demand in order to develop the shortest linking network.

4. Deleting an item from a chain of nodes that is included in the shortest connecting network. The remaining nodes are arranged in descending order relative to demand.

5. Including on the nodes continues until all nodes are set in motion.

6. End nodes are determined using the time minimization criterion.

7. After generation, all routes are checked for inclusion in the network twice [12, 13].

Let us consider a strategy for selecting and inserting nodes into a route. When designing optimal paths, narrow requirements should be established. The strategy of including a node with any priorities in the shortest route consists of 5 stages.

1. We determined by the best possible shortest path for any node to be included in the route. The best way (shortest) path to a concrete node can be determined on the basis of time saving relationships. This ratio is calculated for all shortest routes. The node is inserted into the shortest path that gives the largest value of this ratio.

2. After choosing the shortest path, the best way is indicated.

3. Sometimes, due to the presence of a node or group of nodes at the end of the shortest path, the route is extended. All possible solutions are analyzed and the variant that gives minimal additional delay to passengers is chosen. If this corresponds to their geographical location, to avoid additional delays for sites of higher demand.

4. Due to the presence of many nodes in the immediate vicinity of short routes, return to the same point can not be avoided. A repetitive path can increase the length of the route. To solve this problem – we combine several consecutive nodes and, thus, new routes are being developed.

5. All routes are checked with a stopover at the same node. There is a search for the best option of travel time.

Figure 1 shows the process of adding a node to different routes. The directions with demand above average: j_1, j_2 and j_3. The shortest paths starting from station are "*i*": (i, j_1), (i, j_2) and (i, j_3). Knot to insert is "k_1". Restrictions from the railway station are "*i*" to "*j*": D_{ij}.

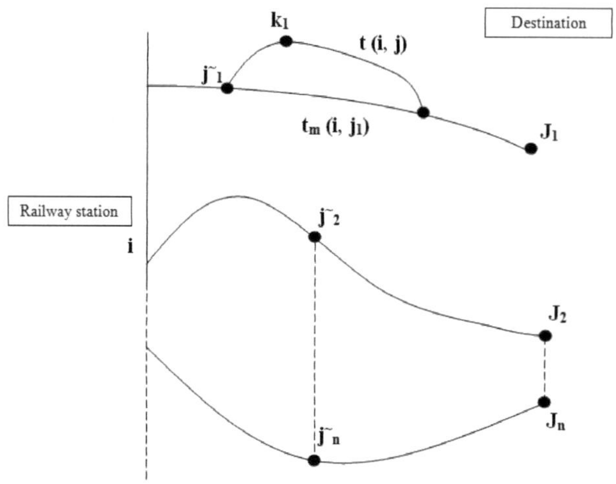

Figure 1. Example of including a node in the route [14].

Restrictions from the railway station "*i*" to "k_1": $D_{i\,k1}$. The nodes on the shortest paths (i, j_1), (i, j_2) and (i, j_3) that are closest to "k_1": $j_1^{\sim}$, $j_2^{\sim}$ and $j_3^{\sim}$ (the shortest paths that are not connected with k_1 are omitted). Travel time along the shortest route (i, j): $t_m\,(i, j)$. Thus, if k_1 is inserted into the route (i, j_1), the travel time will increase from "*i*" to "j_1" due to the deviation of the shortest path. $D_{i\,k1}$ passengers must travel a distance $(j_1^{\sim}\ k_1)$ to reach k_1. Therefore, the travel time from "*i*" to "j_1" through the node k_1 due to the insertion: $t\,(i, j_1)$ increases: $\{t\,(i, j_1) - t_m\,(i, j_1)\}$.

Delay of passengers is $D_{ij1}\ \{t\,(i, j_1) - t_m\,(i, j_1)\}$. An approach time for passengers is from $j_1^{\sim}$ to k_1: $t\,(j_1^{\sim}\ k_1)$. Savings in passenger-minutes thanks to the approach: $D_{i\,k1}\,t\,(j_1^{\sim}\ k_1)$.

Next, the coefficients are calculated for all routes, and the route for which this ratio is maximal is considered. In this case, the choice is given to the option with the maximum demand with less travel time.

The process of optimal planning of the route schedule is rather complicated, attempts were repeatedly made to solve this problem using computer simulation, considering only the transfer time, a combination of optimization models and mathematical procedures. Developing optimal traffic schedules is problematic even for a small transportation network. The task of coordinating a schedule consists of transferring at least between two modes of transport in order to reduce the cost of operating the rolling stock. Restrictions on load factors and transfer time are used to solve transport problems, but the use of a large number of variables and constraints in the objective function create difficulties in solving the problem by traditional optimization methods.

Therefore, genetic algorithms are a reliable optimization method and are suitable for the tasks used in this study. The main difference between the genetic algorithm and the alternative: coding of variables, and stochastic operators instead of deterministic ones are directly used [14–21]. All these functions make the search reliable, allowing one to apply it to a wide range of problems in coordinating traffic schedules. Steps to define a coordinated schedule:

1. Determination of passenger traffic on the developed routes.
2. Decision by applying the objective function and constraints.
3. Use of a genetic algorithm to determine the optimal traffic interval.

The work of public transport should be aimed at meeting the demand of passengers, and minimizing the cost of carriers. For passengers, the level of availability of services, the frequency of rolling stock delivery intervals, and the reduction of waiting times are also important. Carriers are interested in reducing operating costs, minimizing the amount of rolling stock produced on the line during the off-peak period, and increasing profits or achieving break-even levels. Thus, the target function includes the time consuming of passengers, minimizing the time of transfer from the railway to the urban passenger transport. The cost of time consuming is summarized with the operating costs of the rolling stock. Restrictions are connected with minimum and maximum load factors, transfer time and unmet demand [14]. Mathematically, the objective function and constraints can be represented as follows:

$$C_1\{\sum_j \sum_u \sum_l pass_j(bus_j^l - train)\,\delta_j^{u.l}\} + C_2\{\sum_j f_j T l_j\} \tag{1}$$

Limitations:

$$(bus_j^l - train) \leq T_{max} \tag{2}$$

$$(bus_j^l - train) \geq T_{min} \tag{3}$$

$$\frac{Q_{j.max}}{N_j \times CAP} \leq L_{max} \tag{4}$$

$$\frac{Q_{j.max}}{N_j \times CAP} \geq L_{min} \tag{5}$$

$$\sum_j d_{unsat} = 0 \tag{6}$$

where j – number of routes of railway transport;
i – number of public transport routes;
C_1 – cost of time, rub.;
C_2 – cost of operating public transport, rub.;
$pass_j$ – number of passengers arriving at the station;
bus –number of rolling stock departing from a stopping point on the j-th route;
$train$ – number of trains arriving at the station;
$\delta_j^{u.l}$ – coefficient taking into account the possibility of transportation of passengers (depending on the variant, it becomes equal to 0 or 1);
f_j – bus departure frequency on j route (number of trips per hour);
l_j – route j length, km;
T_{max} – maximum allowable transfer time between arriving at the station and departing urban passenger transport;
T_{min} – minimum allowable transfer time between arriving at the station and departing urban passenger transport;
$Q_{j.max}$ – number of passengers at the station on the j-th route for a certain period of time;
N_j – number of bus trips during the entire considered period of time;
CAP – bus capacity;
L_{max} – maximum load factor (assumed to be 1.2);
L_{min} – minimum load factor (assumed to be 1);
d_{unsat} – unsatisfied demand.

The objective function includes travel time between the railway station and vehicle operating costs proportional to the distance of movement. The constants C_1 and C_2 are used to convert the objective function into a currency. The first two limitations are related to the transfer time. Limitation (2) – the transfer time between the arrival of trains and the departure buses on one route must be less than the maximum value. Limitation (3) – transfer time should be minimal. During the survey, the minimum transfer time was 5 minutes. Limitation (4) and (5) ensure that the load factor lies within the maximum and minimum values, and the best level of service and availability are determined in the case of the minimum number of passengers. The maximum load factor is the ratio of the maximum and normal capacity of buses. The last limit is the maximum satisfaction of demand. The choice of constraints is linked to the goals of optimizing operation of public transport in a particular area, since none of the above restrictions is mandatory. These restrictions are set according to their relative importance and magnitude of demand.

IOP Conf. Series: Materials Science and Engineering **760** (2020) 012023 doi:10.1088/1757-899X/760/1/012023

4. Research results

In the process of modeling the real transport network, a genetic algorithm of the objective function and limitations (1)–(6) were used.

The load factors were taken in the range between 1 (minimum load factor) and 1,2 (maximum load factor), the percentage of unmet demand was set at the lower threshold, the waiting time – from 0 to 5 minutes. Demand satisfaction and traffic on different routes are two dominant factors, important for both passengers and carriers. As a result of the survey, it was revealed that the waiting time of transport may take in the range of up to 10 minutes, but when establishing such a high value, it is necessary to take into account the loss of a part of potential passengers for other transportation options (for example, a taxi).

Table 2 shows the schedule for the arrival of suburban trains at the station "Angarsk".

Table 2. Schedule of arrival of suburban trains to Angarsk station

№	Itinerary	Arrival time
1	Malta → Bolshoy Lug	05:57
2	Cheremkhovo → Irkutsk Pass.	06:21
3	Cheremkhovo → Irkutsk Pass.	07:10
4	Irkutsk Pass. → Polovina	07:12
5	Cheremkhovo → Irkutsk Sorting	08:17
6	Irkutsk Sorting → Cheremkhovo	08:38
7	Bolshoy Lug → Cheremkhovo	09:32
8	Polovina → Bolshoy Lug	10:10
9	Zima → Irkutsk Pass.	12:07
10	Zima → Irkutsk Pass.	13:40

Analyzing the time of arrival, it can be noted that in the period from 6 to 8 am, the suburban trains arrive every 30 minutes.

Table 3 shows the beginning of the movement and the interval of the public transport from the train station, which is the final stop.

Table 3. Schedule of public transport from the stopping point "Railway Station"

№	Driving route	Departure time	Interval of movement, min.
1	Route № 3 Railway Station – Shop "Magistral"	06:02	8–9
2	Route № 7 Railway Station – 4th village	06:06	6
3	Route № 8 Railway Station – 12th microdistrict	05:36	5
4	Route № 9 Railway Station – Shop "Mriya"	06:23	7
5	Route № 27 Railway Station – 211st quarter	06:00	10–11
6	Route № 28 Railway Station – AEMZ	06:00	18

When testing the model, the following results were identified: on routes no. 3, 7, 8, the timetable was compiled in such a way that allows passengers arriving at the station to wait for urban passenger transport less than 5 minutes (fig. 1). For routes no. 9, 27, 28 waiting time ranges from 10 to 20 minutes, which adversely affects the comfort of transportation.

As a result of the simulation, several solutions to the problem were obtained – reducing the traffic interval or optimizing the route schedule so that passengers arriving by train did not wait city transport for more than 5 minutes.

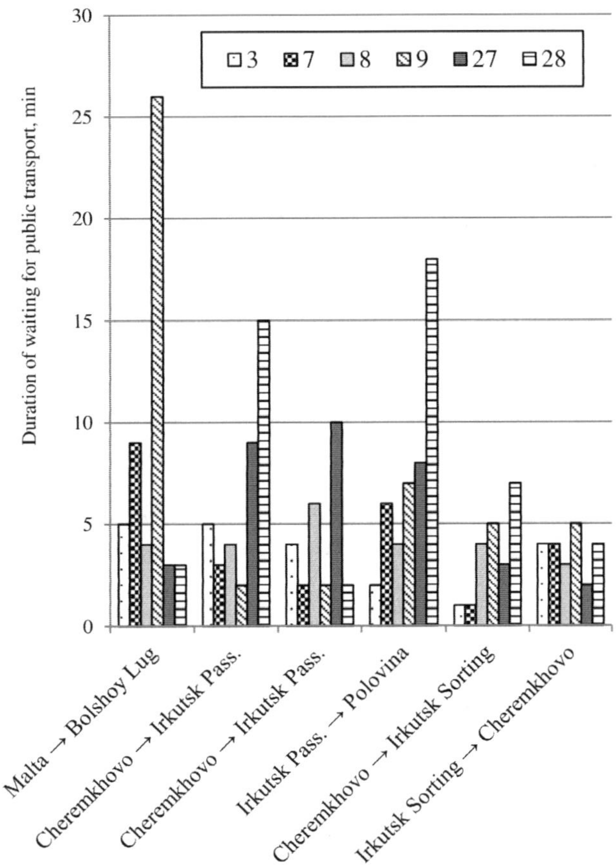

Figure 2. Distribution of public transport waiting times

5. Conclusion

In this study, a model of operational integration of two transportation options was tested: urban public transport (bus) and rail service for one station. Simulation can be carried out at other stations after determining the potential demand for transportation. It will provide an integrated public transport system, in which the railway communication will be considered as the main option of movement, and buses – accompanying. This option will allow the regional transport system to work smoothly, to use routes of optimal length, in order to meet the needs of passengers.

Genetic algorithms are very effective in solving objective, non-linear tasks of coordinating train diagram. The model, tested in the study, allows developing a real transport network, which is important, both for the qualitative satisfaction of passenger demand and for optimizing the haulage contractor. It takes into account such restrictions as: service level (maximum load factor), economy of work (minimum load factor), minimum and maximum transfer times, availability of public transport for all passengers (limit on unmet demand) and maintaining a reasonable balance between load factor and demand satisfaction within an acceptable waiting time. Thus, the model is able to provide satisfactory results (routes and agreed schedules) for passengers and carriers. Therefore, it can be stated that this approach is an important contribution to the modeling of passenger traffic.

References

[1] Lebedeva O, Kripak M and Gozbenko V 2018 Increasing effectiveness of the transportation network by using the automation of a Voronoi diagram *Transportation Res. Proc.* **36** 427–33

[2] Kopylova T, Mikhailov A and Shesterov E 2018 A Level-of-Service concept regarding intermodal hubs of urban public passenger transport *Transportation Res. Proc.* **36** 303–7
[3] Gozbenko V E and Olentsevich V A 2013 Improving the safety of rail transport system based on the automation of placement and fastening of cargo in the car. *News TRANS-Siberian Railway* **1(13)** 110–6
[4] Poltavskaya O Y, Kripak M N and Gozbenko V E 2016 Estimation of traffic flows using GIS technology *Modern technology. System analysis. Modeling* **1(49)** 155–61
[5] Gozbenko V E, Ivankov A N, Kolesnik M N and Pashkova A S *2008 Forecasting Methods and optimization of the transport network taking into account the capacity of passenger and cargo flows deposited manuscript* 330-V2008
[6] Kolesnik M N and Gozbenko V E 2007 Principles of creation of information planning and control system of transportations on motor transport *Modern technology. System analysis. Modeling* **3(15)** 46–52
[7] Kolesnik M N and Gozbenko V E 2007 Algorithm of the automated choice of rolling stock. *Modern technology. System analysis. Modeling* **4(16)** 45–7
[8] Gozbenko V E, Kripak M N and Poltavskaya O Y 2014 M athematical modeling of the transport company *Modern technology. System analysis. Modeling* **2(42)** 120–9
[9] Kripak M N and Lebedeva O A 2016 Assessment of the state of the road network of a large city *Modern technology. System analysis. Modeling* **3(51)** 171–4
[10] Miandoabchi E, Farahani R Z, Dullaert W and Szeto W Y 2012 Hybrid evolutionary metaheuristics for concurrent multi-objective design of urban road and public transit networks *Networks and Spatial Economics* **12(3)** 441–80
[11] Chebbi O and Chaouachi J 2016 Dealing with the Strategic Level of Decisions Related to Automated Transit Networks: A Hybrid Heuristic Approach *Lecture Notes in Computer Sci.* **9668** Springer, Cham, 202–17
[12] Fatnassi E, Chaouachi J, and Klibi W 2015 Planning and operating a shared goods and passengers on-demand rapid transit system for sustainable city-logistics *Transp. Res. Part B Methodol.* **81** 440–60
[13] Birkan G and Kennington J 2009 Design procedures for backbone transport networks with shared protection: optimization-based models, exact and heuristic algorithms, and unavailability computations *Telecommunication Systems* **41(2)** 97–120
[14] Shrivastava P and O'Mahony M 2007 Design of a feeder route network using combined genetic algorithm and specialised repair heuristic *J. of Public Transportation* **10** 99–123
[15] Ash G R and Chang F 1993 Transport network design of integrated networks with real-time dynamic routing *J. of Network and Systems Management* **1(4)** 319–35
[16] Lebedeva O and Kripak M 2018 Modeling of public transport waiting time indicator for the transport network of a large city *MATEC Web Conf., International Conference on Modern Trends in Manufacturing Technologies and Equipment* vol 224
[17] Han L, Guo J and Han J 2015 Research on Interaction Between Traffic Improvement Around the Old Railway Station and Urban Land Utilization –A Case Study in Hohhot Railway Traffic Regulation *Recent Developments in Chinese Urban Planning* pp 247–67
[18] Gozbenko V E, Kargapoltsev S K, Minaev N V and Karlina A I 2016 Simulation of the vibration of the carriage asymmetric parameters in mathcad *Int. J. of Applied Engineer. Res.* **11(23)** 11132–6
[19] Carlina A I, Kargapoltsev S K and Gozbenko V E 2016 Bringing the generalized forces in mathematical models of transportation systems. Modern technology *System analysis. Modeling* **3(51)** 175–9
[20] Revenskiy A A and Gozbenko V E 2011 Modeling the mechanical gear *Modern technology. System analysis. Modeling* **3(31)** 119–23
[21] Akhmadeeva A and Gozbenko V E 2011 Defining the main coordinates of the car with two-stage spring suspension *Modern technology. System analysis. Modeling* **4(32)** 71–6

Search for additional criteria to estimate the electrical connection state under the impact of currents of different value

I V Ignatenko and S A Vlasenko

Far Eastern State Transport University, 47, Seryshev str., Khabarovsk, 680021, Russia

E-mail: vsa_ens@mail.ru

Abstract. The paper reports on search for additional criteria to estimate the electrical connection state in the operating mode characterized by abrupt change in electric current parameters. The results obtained in the course of theoretical studies are based on the use of methods of mathematical and experimental modeling of thermal processes in electrical connections of a traction power supply system. New criteria for estimating the state of the electrical connection for cyclic traction load have been found out. Investigation of the defectiveness ratio in the "heating-cooling" cycle makes it possible to estimate the current state of the electrical connection under service conditions.

1. Introduction

A traction power supply system is an extended object intended to transmit electric power to electric rolling stock. It consists of a set of electrical connections of bolted-type accomplished by clamping the fastened wires with bronze dies. No-failure system-wide operation depends on reliable work of such electrical connections [1]. The peculiar feature of the railway power supply system is that traction load (electric rolling stock) is not uniform within a day, which leads to abrupt change in electric current parameters [2]. The given circumstance complicates a current state estimate of an electrical connection.

2. Problem definition

There are distinguished two states of electrical connection: satisfactory and unsatisfactory. The state depends on the value of transient resistance (R_T), resulting from contiguity of contact surfaces of a wire and clamp dies. Under service conditions the value R_T increases as tightening torque of clamp dies decreases (loosening) and oxide films are formed in places of a wire and clamp contact. For satisfactory state the value of transient resistance should not exceed the value equal to the ohmic resistance of a conductor. It follows, the purpose of work is search for additional criteria to estimate an electrical connection state under service conditions, as these criteria are able to indirectly determine an actual value of R_{T}.

3. Solution

In work [3] it is proved that in the electric current supply mode indirect estimate of value R_T by means of defectiveness coefficient (K_θ), with heating considered, is possible and, as a result, the same holds good for estimate of electrical connection state. However, the diagnostic data on current-carrying clamps obtained with the help of a thermal imaging camera have shown that the lack of the similar data analysis of cooling mode leads to doubtful estimate of current state of bolted-type electrical connection [4].

Therefore, there arises a need to investigate not only heating, but also cooling of an electrical connection, since only joint consideration of both modes can offer profound qualitative evaluation of the existing indicators influencing the electrical connection state estimate.

At present there is no mathematical model of electrical connection estimate within cooling mode. The authors of the given paper offer to transform the universal equation of heat balance of electrical connection [6], without current being taken into account, and receive an expression:

$$(C_j\, m_j + C_w\, m_w)\cdot\frac{\mathrm{d}\theta}{\mathrm{d}t} = -[\,0.01312\cdot\theta^{\,1.49} + \varepsilon\, S\cdot 5.67\cdot 10^{-8}[(T_j+273)^4 - (T_a+273)^4]], \quad (1)$$

where C_w is the specific heat capacity of a wire material (W·s kg^{-1} K^{-1}); C_j is the specific heat capacity of a clamp material (W·s kg^{-1} K^{-1}); m_w is the mass of a wire clamped between dies (kg); m_j is the mass of a clamp (kg); S is the area of an exterior surface of a clamp (m^2); T_a is the air temperature (°C); T_j is the temperature of a clamp (°C); 5,67·10^{-8} (W m^{-2} K^{-4}) is the Stefan-Boltzmann constant; ε is the relative emittance for bronze (ε = 0.6); θ is the excess of temperature of electrical connection over ambient temperature (°C).

To determine temperature of electrical connection we will transform the above equation (1) and receive the dependence defining temperature change on time in the course of cooling [6]:

$$\Delta\theta_C = \left(0.0108Pt\,(C_j\, m_j + C_w\, m_w)^{-1} + \theta_0^{-0.49}\right)^{-2.03}, \quad (2)$$

where P is the perimeter of a wire cross section (m); t is the time period of current flow (s).

The received expression (2) describes temperature variation of clamp dies in the course of time when traction current is not flowing through the electrical connection. It gives a possibility to investigate process of cooling and to estimate its influence on K_θ, which represents the ratio of temperature variations of electrical connection to temperature variations of a wire [7]. As mathematical models of heating and cooling for a wire and an electrical connection are known (figure 1,a), it is possible to investigate K_θ temporal variation in the course of the "heating - cooling" cycle and consider the additional factors influencing the estimate of an electrical connection [8].

In figure 1,b, showing K_θ variation in "heating-cooling" cycle, it is possible to conditionally single out four areas of a curve which describes the thermalphysic processes occurring in the feeding clamp and branching wires. In *the Inertial Component* area when temperature of connection is much lower than that of wire, the function of the defectiveness coefficient, with heating considered, is of the minimum value, as for the defectiveness coefficient value, it depends on connection mass and the material it is made of. A point of an extremum, as regards to thermal characteristics, is explained by inertial properties of a clamp [9].

Then there is a process of determining a coefficient value due to balancing of a gradient of thermal fields of a clamp and the connected wires [10]. The third area represents a steady-set coefficient value in the heating mode. The coefficient value will depend on two components which influence heat balance of the system under discussion. The first component is a ratio of connection resistances and a wire (defectiveness coefficient with resistance considered). The second is a ratio of heat emission coefficients of a connection and a wire.

The fourth area shows sharp increase in K_θ value with a distinctive point of a maximum. Such a behavior of the coefficient is explained by smaller cooling of an electrical connection, than that of a wire. It should be noted that at this moment K_θ value considerably exceeds the rated value and is equal to 1.5 that has testified to unsatisfactory state of an electrical connection. At the same time the value of transient resistance of 25 μΩ adopted in modeling is considered to be rated. Thus, it is proved that the process of cooling cannot be neglected on estimating an electrical connection state; otherwise research results may turn to be erroneous and misleading.

IOP Conf. Series: Materials Science and Engineering **760** (2020) 012024 doi:10.1088/1757-899X/760/1/012024

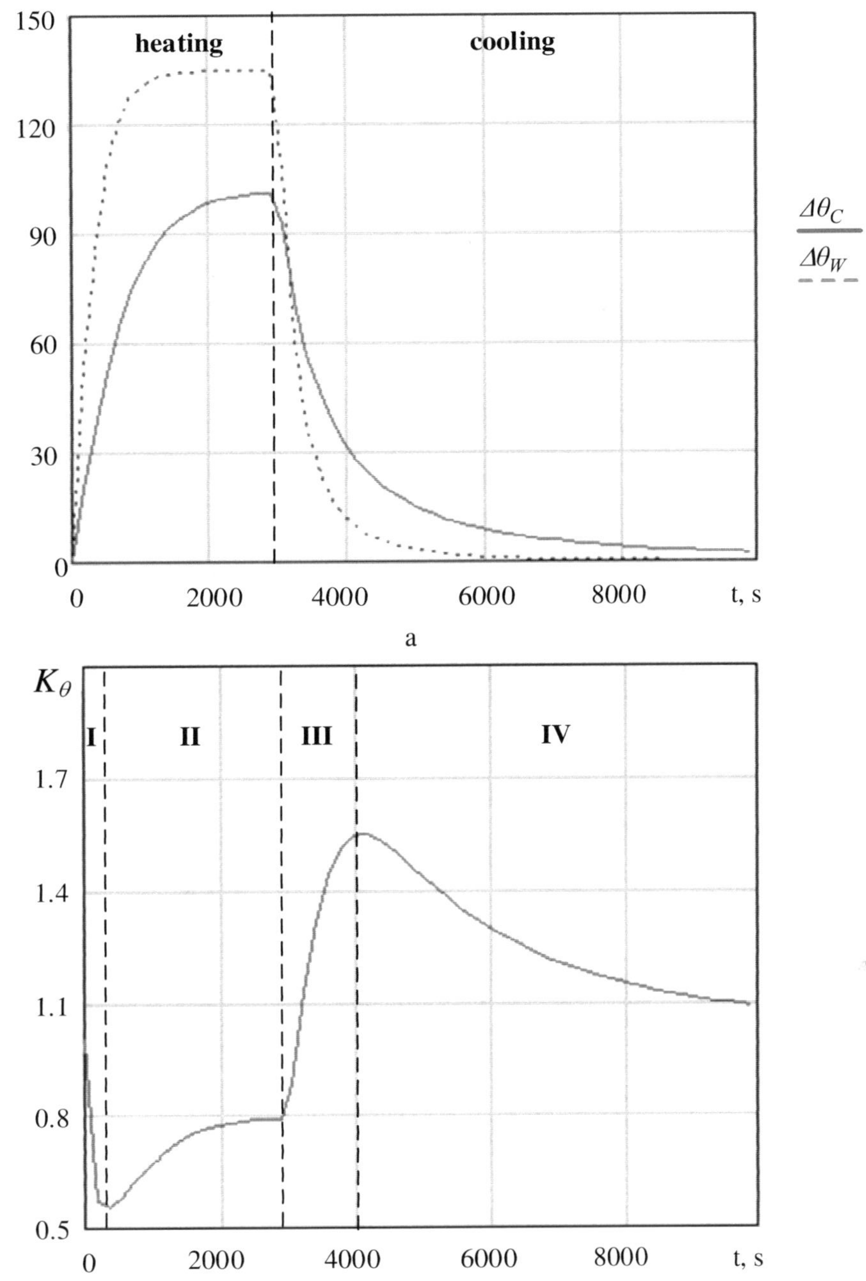

Figure 1. "Heating-cooling" cycle of a clamp: a – temperature variation of a wire $\Delta\theta_W$ and an electrical connection $\Delta\theta_C$ related to the time period of conventional nominal current flow through them I = 600 A, at T_a =24 °C; b – change of defectiveness coefficient K_θ with heating considered at R_T = 25 μΩ.

The maximum value of coefficient being reached, its value decreases up to its steady-state value.

On the basis of the offered mathematical model the Program for calculations in Mathcad computer environment was developed. The Program allows a user not only to consider cooling process but also to change electric current parameters by substituting its value and time period of current flow [11] and [12].

The Program performs calculations of differential equations concerning wire heating and electrical connections [13]. Applying the Program operation algorithm, the value of K_θ is determined.

The Program includes the following data: 6 values of current which correspond to variables $I_1(t)$, $I_2(t)$, $I_3(t)$, $I_4(t)$, $I_5(t)$, $I_6(t)$. The current values are chosen one after another, proceeding from time intervals. The chosen time intervals are designated by a variable – t_m, t_n, t_k, t_d t_c – covering the maximum and minimum values of changes of the graph [14]. The value of transient resistance (r, μΩ) is set as well.

Plotting the graphs of dependence of coefficient K_θ on time, the Program retains the previous data on an electrical connection and a wire and considers the when plots new graphs.

To obtain actual data of the algorithm performance, calculations of transient resistance of different value were carried out. In addition, values of current flowing through an electrical connection during heating and cooling modes were changed too.

Reasoning from the calculations made one can conclude that, in case of an electrical connection cooling, value K_θ, can exceed the rated parameter used to estimate satisfactory and unsatisfactory states of an electric connection.

Being aware of the value of the inclination angle of a change curve K_θ, and knowing that the angle shows the speed at which defectiveness coefficient increases during heating mode, one can predict current-carrying clamp state under the impact of electric load within the whole time interval of [3].

For assessment of adequacy of mathematical model pilot studies of heating of electrical connections were conducted. The experiment was made at a specially designed test bench, which is described in work [6].

The results of an experiment have proved comparability of their values to mathematical model illustrating K_θ value variations in reference to time. The margin of error is no more than 5% (figure 2). Thus, the received mathematical model is adequate.

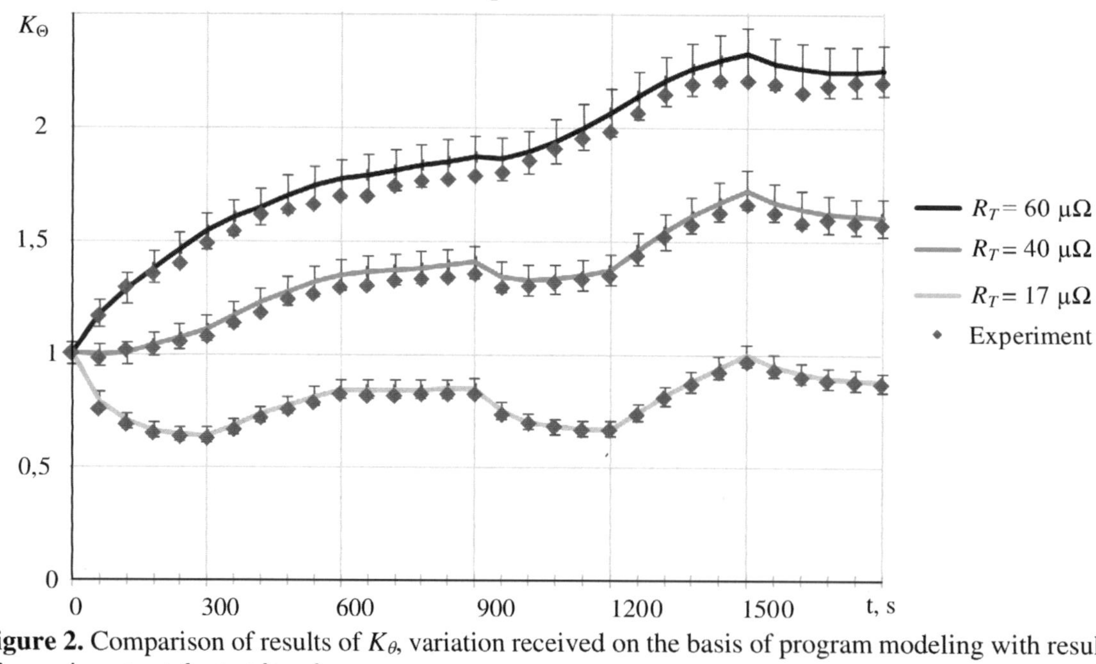

Figure 2. Comparison of results of K_θ, variation received on the basis of program modeling with results of experiments at the test bench

4. Conclusions

There has been received dependence of temperature change of an electrical connection in cooling mode which defines model of K_θ change in reference to time under the impact of currents of different value. The adequacy of model at the experimental test bench has been proved.

Analysis of change of defectiveness coefficient dependence in the "heating-cooling" mode has allowed revealing additional criteria to estimate electrical connection state: steady-state value and speed of its change.

The received results will become a basis for development of the technique allowing estimating current state of an electrical connection under the impact of electric load in actual operating conditions.

References

[1] Grigor'ev N P, Voprikov A V and Parfianovich A P 2016 Increasing the operating life of power transformer winding insulation when upgrading 25-kV alternating current traction substations *Russian Electrical Eng.* **87(2)** 100–3

[2] Li V N and Shnurova N K 2016 On the method of determining optimal electricity consumption from an electric traction network *Russian Electrical Eng.* **87(2)** 97–9

[3] Xin Zhou and Schoepf T 2012 Detection and formation process of overheated electrical joints due to faulty connections *26th International Conference on Electrical Contacts (ICEC 2012)* 12909251

[4] Tumanov A Y, Gumenyuk V I and Tumanov V A 2017 Development of advanced mathematical predictive models for assessing damage avoided accidents on potentially-dangerous sea-based energy facility *IOP Conf. Ser.: Earth Environ. Sci. (EMMFT 2017)* **90** 012027

[5] Grigor'ev N P, Klykov M S, Parfianovich A P and Shestukhina V I 2016 Increasing the power efficiency of a 25-kv alternating current traction power supply system *Russian Electrical Eng.* **87(2)** 113–5

[6] Ignatenko I V and Vlasenko S A 2016 Determination of criteria to find the quality of an electrical connection *Russian Electrical Eng.* **87(2)** 65–7

[7] Wilson C, McIntosh G and Timsit R S 2012 Contact spot temperature and the temperature of external surfaces in an electrical connection *26th International Conference on Electrical Contacts (ICEC 2012)* 12909202

[8] Bukhanov S S, Doykina L A and Tsirkunenko A T 2019 Evaluation of the electromagnetic situation at open switchgears of power stations and substations *IEEE Conference of Russian Young Researchers in Electrical and Electronic Engineering (EIConRus)* 457–61

[9] Ignatenko I V, Konstantinov A M and Demina L S 2016 Determination of the transmission capacity of a 500-kV line equipped with flexible electric power transmission devices *Russian Electrical Eng.* **87(2)** 104–6

[10] Konstantinova Y A, Li V N and Tryapkin E Y 2018 Enhancing Energy Efficiency of 25 kV Traction power system due to balancing current decrease *International Multi-Conference on Industrial Engineering and Modern Technologies (FarEastCon)* 18393188

[11] Bukhanov S S, Zhuravleva L A and Kornienko M N 2018 Improving the reliability of the relay automation system for power facilities *IOP Conf. Ser.: Mater. Sci. Eng.* **450** 072008

[12] Tryapkin E Y, Keino M Y and Protasov F A 2016 Synchronous Phase Measurements in the Automated Monitoring System of Railway Power Supply Facilities *Rus. Electrical Eng.* **87(2)** 110–2

[13] Li V N and Klimenko S V 2016 Russian Electrical Eng. Determination of thermal softening of overhead wire exposed to an electric arc effect **87(2)** 77–9

[14] Galkin A G, Kovalev A A and Shayukhov T T 2018 Integrated approach to the planning of energy consumption by non-traction railway consumers *MATEC Web Conf. Siberian Transport Forum – TransSiberia* **239** 01056

IOP Conf. Series: Materials Science and Engineering **760** (2020) 012025 doi:10.1088/1757-899X/760/1/012025

Experimental study of profile connections of auxiliary tool multi-purpose machines in cyclic loading

V. A. Ilyinykh[1]

[1]Chita, TRANS-Baikal Institute of railway transport, 11, Magistralnaya street, Chita, 672040, Russia

E-mail: ilinykh.viktor5@mail.ru

Abstract. At present wide application in the machine tool industry has a professional connection with equiaxed circuit. These torque transmission connections have a number of operational advantages over traditionally used such as: high accuracy, stiffness, wear resistance, strength and vibration resistance. The practice of using these compounds in the critical nodes of metal-cutting machines in particular multi-purpose CNC machines has shown that these compounds have high flexibility in the requirements for reusable assemblies of the connection when replacing technological sets of tools. Experimental studies of these compounds of models made of optically active material allowed us to study the processes of contact interaction of parts of the compound under cyclic loading.

1. Introduction

The use of the modular principle of the construction of multi-purpose machines (MS), proposed and developed by Professor Averyanov O. I. [1, 2], allows one to solve a number of scientific and practical problems. They are related to ensuring high accuracy and rigidity of individual modules of the machine through the introduction of innovative torque-transfer connections with the profile surface of the coupling parts.

Let us consider the modular system of the auxiliary tool.

According to research works [1, 2, 3, 4] the system includes three subsystems of the auxiliary tool (PVI), namely:

1) PVI for CNC drilling and boring and milling groups;
2) PVI with a cylindrical shank for a CNC turning group;
3) PVI with a basing prism for a CNC lathes group.

The first subsystem of the auxiliary tool, which is used in the MS, is of greatest interest. It should be noted that for connection with the spindle of the machine, tool mandrels with a cone 7:24 with a round cross-section profile are used, as a rule. In some intermediate, the pump is used as a tool joint with a Morse taper, collet and coupling of tools with the cylindrical shank.

The automatic replacement of the toolholder with the taper 7:24 is necessary to ensure the position accuracy of PVI, as well as stiffness in the process of operation of the IPU. Increasing requirements for the quality of the MS accuracy and performance with a significant increase in the number of spindle speeds up to 45000 rpm determine correspondingly high requirements for connections, which often do not meet the traditional and special taper 7:24.

Improving the design of joints based on mandrels SK, HSK and HSK does not always meet the requirements for technological and operational reliability of MS. First of all, this is due to the axial and transverse movements and angular rotations of the PVI during the rotation of the tool mandrels before processing the products, as well as during the manufacture of products [5, 6]. This leads to a decrease in the accuracy of manufacturing products due to the change of bases in the spindle - auxiliary tool - cutting tool. Attempts to improve the accuracy of the quality of the geometry of the shape of the landing surfaces of the mandrels in their manufacture do not allow one to fully ensure the required accuracy, for example:

IOP Conf. Series: Materials Science and Engineering **760** (2020) 012025 doi:10.1088/1757-899X/760/1/012025

the diametrical size of the main bearing holes in the housing parts; deviation from the alignment with the overall axis of the landing surface of the holes for different types of bearings, as well as a number of other technological requirements. There are known works [5, 6, 7] which considered and solved with the replacement of the traditional connections on the profile with the equiaxed contour of the cross section with higher precision and rigidity. At the same time, there are unexplored questions of changes in elastic displacements in the joints of compounds, which is important not only for determining the contact stiffness of the compound, but also for assessing their bearing capacity under cyclic loading.

As is known, the machining of products of the auxiliary tool is under the action of alternating forces. The combined effect of contact pressure and cyclic shear micro-displacements, which are the result of elastic deformations of the conjugated parts of the auxiliary tool, can be the cause of fretting wear.

2. Materials and methods

The article presents the results of experimental studies of models of the progressive moment because connections of details of TK MS occur on the basis of the relevant compounds by the static photoelasticity method.

Polarization-optical methods are widely used in the study of contact problems. Polarization-optical methods, characterized by visibility, allow us to investigate not only the stress state of the interacting bodies throughout the volume and determine the components of the stress tensor, but also visually track how the VAT changes when the position of the interacting bodies relative to each other in the implementation of loading schemes.

The theory of similarity and dimensions were developed in the works of M. M. Frocht, N. So. Prigorovsky, O. K. Slavin, N. D. Tarabasova [8], allow solving a wide class of contact problems and with a sufficient degree of accuracy to assess the stress state in nature and model.

3. Study of stress-strain state of compounds

In order to study the VAT profile connections, we made models of the shaft and hub with the equiaxed contour. An optically active compound based on an epoxy resin cured with maleic anhydride (ED16-M) was used as the material of the models. The thickness of the plates of the material of the models is b = 6 mm. The accuracy of determining the contact angles and the difference of the main stresses was provided by both the technology of making models and the technique of the experiment. Each coupling model was a hub and shaft with an equiaxed contour (figure 1-2).

Thus, the study of RK-3 profile-based connection of models of the polarization-optical method has made it possible to evaluate the influence of radial clearance operating conditions (load cases) and the values of the angles characterizing the position of the parts of the connection, the indirect account of an error of a form to change the values of the contact angle and the maximum shear stresses in the interface zone.

During the experiment, for each case of loading, the positions of the profile connection with respect to the direction of the line of action of the radial force were investigated. It was assumed that the distribution of the load on the contacting surfaces at a particular loading scheme depends on the adopted position, characterized by the position angle ψ.

It is established that the nature of the change in the value of the contact angle and shear stresses for each of the three contact zones is different both in the transmission of torque by the connection and in the account of the joint action of the radial load.

In the first case, this is due solely to the error of the geometric shape of the mating connection models. In the second case, this is explained both by the error of the shape of the mating models and the redistribution of contact stresses in the zones from the action of radial load (figures 1 and 2).

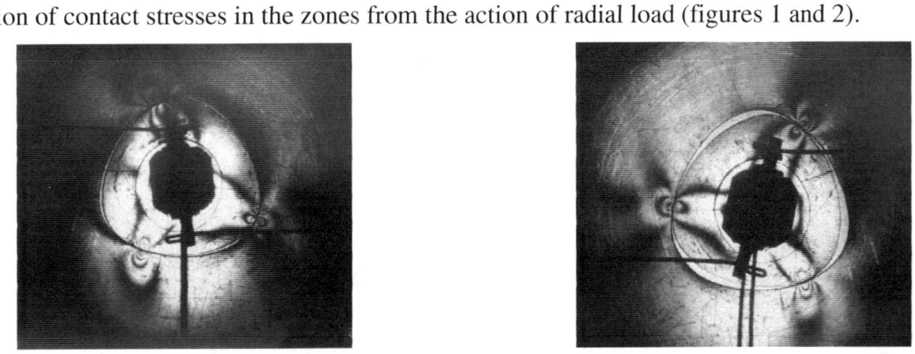

Figure 1 Stress state of models 1, 2 in the scheme of loading by torque

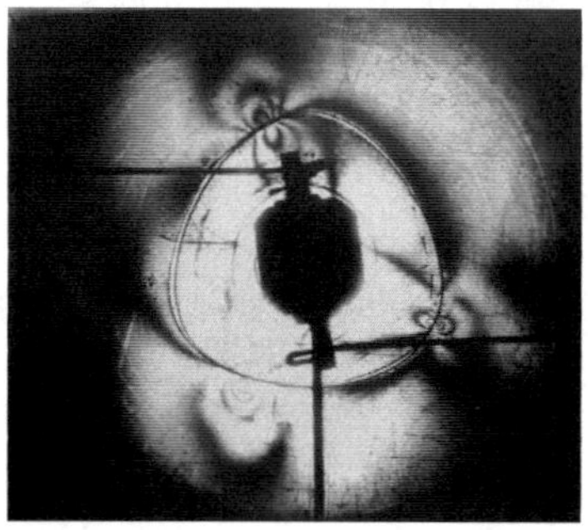

Figure 2 Stress state of models 1, 2 in the scheme of loading by torque and radial force

In the analysis of the graphic dependence of the contact angle on the external load during the transmission of the connection torque and radial force for the corresponding models 1, 2, 3, we set the change in the contact angle from the position of the connection.

The results of the assessment of the values of the contact angles allow one to draw the following conclusions:

1.The magnitude of the contact angles during transmission with the torque connection increases with increasing load, with the nature of the increase being close to linear.

2.The value of the contact angle during transmission by the connection of torque and radial force varies for each contact zone of the connection and depends on the value of the angle of the relative position ψ of the connection.

The estimation of the values of maximum shear stresses is determined using the known dependence of Tymoshenko S. P.:

$$\tau_{max} = \frac{1}{2} \cdot \sigma_0 \cdot n, \tag{1}$$

where τ_{max} – maximum shear stress;

n – the order of bands;

σ_0 – the price band model.

The order of the band is determined by counting the number of bands for each contact zone in the conjugation of the general picture of the stress state, which is fixed in the photo or on the screen BPU – IMASH-KB2.

The price of the strip model is determined by the formula of Freight M. M.:

$$\sigma_0 = \frac{\lambda}{2 \cdot c \cdot t} = \frac{\sigma_0^{0,1}}{t}, \tag{2}$$

where $\sigma_0^{0,1}$ – the price of a strip of material equal to $\frac{\lambda}{2 \cdot c}$, obtained with a model thickness of 1.0 mm.

The price of a strip of material is established experimentally on coordinate-synchronous polarimeter KSP-7 by a compensation method.

According to the values of the maximum tangential stresses, the corresponding graphical dependences for the considered models of coupling, external load schemes and positions are constructed connections (figure 3).

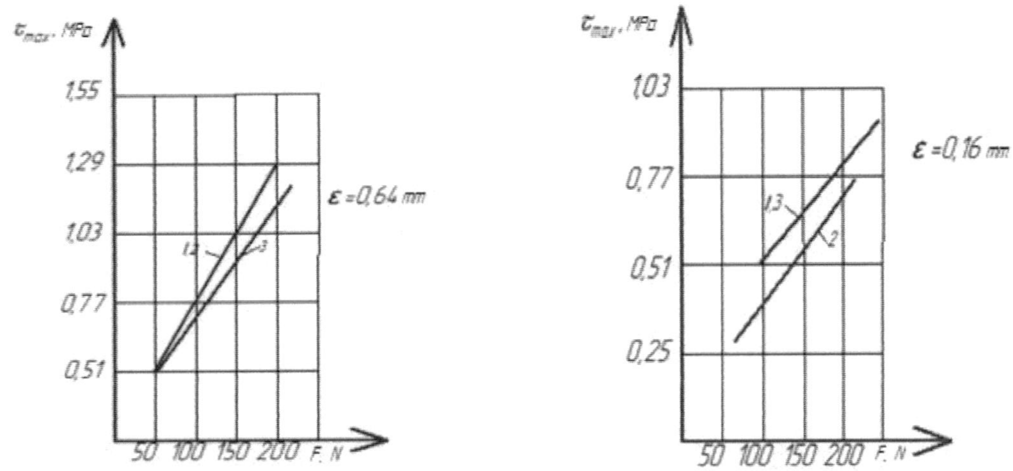

Figure 3 Diagram of the dependence of the maximum shear stresses on the load (for three contact areas)

To determine the influence of the radial load on the stress state of the joints, we made a comparative evaluation of the considered models of coupling 1, 2 for the most loaded contact zones in the transmission of torque, torque and radial force.

Comparative evaluation of contact parameters in different loading schemes and gaps showed a significant effect of the radial component of the complex load F_r (M_{kr}, F_r). Under the influence of radial load, the maximum shear stresses increased by 25%, with an increase in the contact angle by 36.9% for coupling model 1 and for coupling model 2 by 33% and 24.5%, respectively.

According to the research of Undacova N. S. and Suslov A. G. when working with RK-3 profile joints, the influence of the radial component of the complex load is significant from the point of view of fatigue fracture of the contact surfaces of joints, connections and wear.

According to the obtained experimentally values of angles of contact, the calculated values of maximum shear stresses, patterns of interference fringes, it is possible to assess the impact of design parameters (gaps) and operating conditions (load cases, regulations) on the stress state of the joints of the models profile joints.

It can be assumed that the influence of the radial force is noticeable, since it leads to a significant change in the values of the maximum tangential stresses in the contact interaction zone.

Since this preserves the number of contact zones, it can be assumed that this is due to the displacement of the shaft relative to the hub in the direction of the radial force by the number of elastic contact deformations.

In practice, it is proved that after a certain time of operation, under the action of variable shear stresses, the contact surface of this connection is destroyed at voltages below those that are reliably resisted under the action of a static load.

In the calculations of profile connections with the equiaxed contour, the initial data of the maximum shear stresses for the coupling models were taken. The stress state of the contact zones of the coupling models for different angular positions of the joint (at angles from 0° to 360°) was investigated, namely: the difference between the values of the maximum tangential stresses under the action of the torsion moment and the torsion moment and radial force was determined.

We set the limits, change the intervals of values of the difference of the values of the maximum shearing stresses under the action of external loads during the period of connection, which set the cycles variable of the maximum shear stresses for each surface area of the corresponding model (figure 4).

IOP Conf. Series: Materials Science and Engineering **760** (2020) 012025 doi:10.1088/1757-899X/760/1/012025

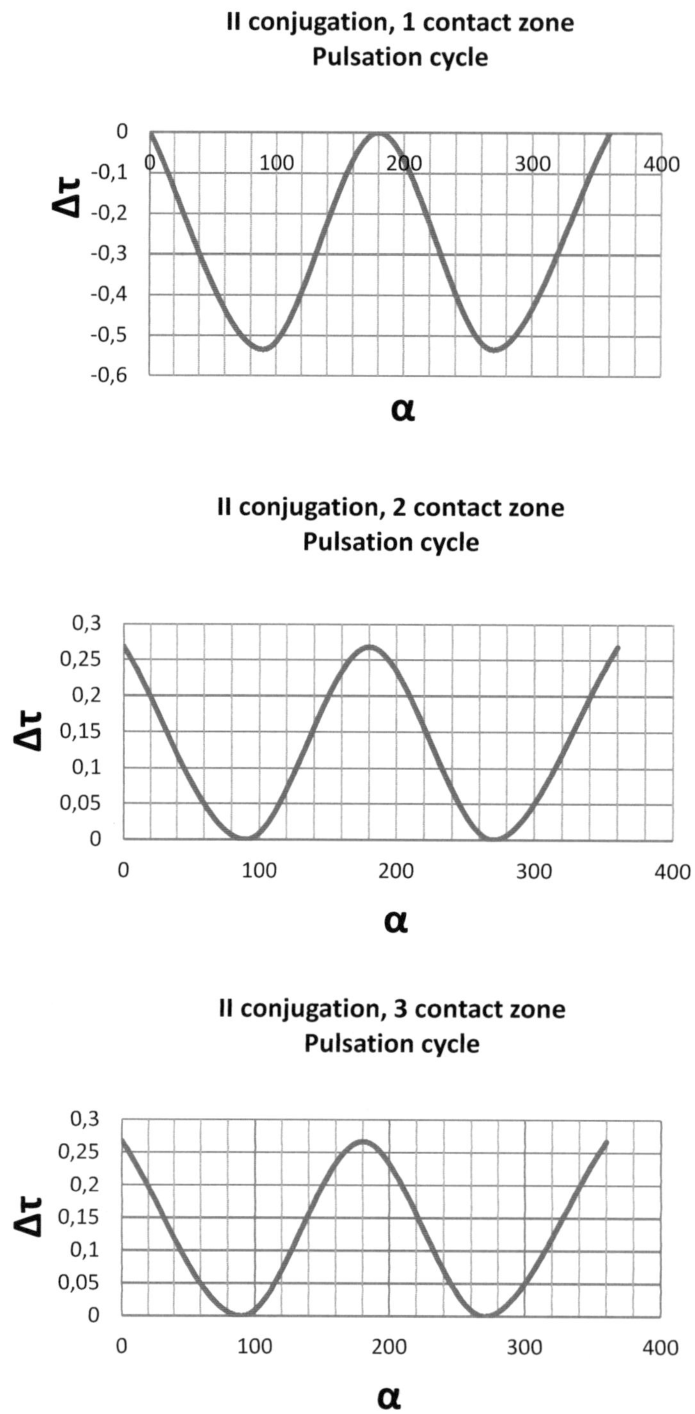

Figure 4 Graphic dependences of change of the tangential stresses, MPa

The analysis of the research results showed that the change in the variables of the maximum shear stresses in the contact zone during the transfer of the complex load by the compound per revolution is subject to a periodic law that is close to sinusoidal. Since the magnitude of the shear stress amplitude also varies for the contact zones during the operation of the connection at the corresponding loading stage, it can be assumed that the joints of the profile connection operate under non-stationary conditions. The work of the profile connection in the conditions of variable shear stresses of small numerical values can be the cause of the process of fretting-wear joints, which reduces its operational reliability, namely: accuracy and contact stiffness.

4. Conclusion

1.As a result of the study of profile connections with an equiaxed circuit, it was found that under the action of radial load the maximum shear stresses increased by 25%, with an increase in the contact angle by 36.9%; for conjugation model 1 and for conjugation model 2 - by 33% and 24.5%, respectively.

2.Calculations have established that the change in the values of the maximum shear stresses during the operation of the profile connection is subject to the pulsation cycle.

References

[1] Averyanov O I, Averyanova I O 2018 *Modular principle of construction of multi-purpose CNC machines: monograph* (Stary Oskol, TNT).

[2] Solomentsev Y M, Kosov M G, Mitrofanov V G 1985 *Modeling of accuracy in the automated design of metal-cutting equipment* (Moscow, NIIMASH).

[3] Ilyinykh V A 2019 Statistical analysis of geometry of profile details of spindle Assembly of multipurpose machines. *Scientific and technical knowledge of SPbSPU* **25(2)** 132-140.

[4] Ilinykh V A, Yarilov V E, Rozhkova E A 2019 Estimation of accuracy and strength of modular system of auxiliary tool on the basis of profile connections *Bulletin of Irkutsk state technical University* **23(3)** 451-462.

[5] Ilyinykh V A 2018 The influence of the conical profile of the compounds on the accuracy and stiffness breadboards multi-task machines. *Scientific-technical Bulletin of SPbSPU* **4** 150-159.

[6] Lineitem V Y, Ilyinykh V A, Arrow V E 2017 Research of the bevel RK-3 profile of compounds on the accuracy and rigidity. *The modern technologies. System analysis. Modeling* **1(53)** 36-39.

[7] Ilyinykh V A, Lineitem V Y, Rozhkova E A 2010 Calculation of the stress-strain state of the hub with the hole of the RC-core by the finite element method. *The modern technologies. System analysis. Modeling.* **3(27)** 69 – 73.

[8] Prigorovsky N I 1983 *Methods and means of determining the fields of deformation and stress* (Moscow, Mechanical Engineering).

IOP Conf. Series: Materials Science and Engineering **760** (2020) 012026 doi:10.1088/1757-899X/760/1/012026

Analysis of dynamic characteristics of frequency-dependent links

V O Kolmakov[1], O V Kolmakov[1], E S Ilyin[1], V S Ratushnyak[1]

[1] Krasnoyarsk Institute of Railway Transport, branch of Irkutsk State Railway Engineering University

E-mail: kolmakov_vo@krsk.irgups.ru

Abstract. The article is devoted to an experimental study of the dynamic characteristics of the four-beam frequency filter under consideration in various power consumption modes. Based on the PSpice program, a typical power supply circuit for LED lamps was modeled. The peaks of the harmonic series reflect the power of the higher harmonics relative to the fundamental harmonic. In the three-phase mode of the lighting systems switching, the filter compensating device according to the four-beam "star connection" scheme is more efficient. The experimental results showed that the content of the 3rd harmonic, when the filter of the compensating device is turned on, exceeds the permissible state standard specification by 5% only in the 6-fold asymmetry mode in the current of the main power supply. Thus, in three-phase networks, using a four-beam passive harmonic filter, it is possible to suppress the higher harmonics of non-sinusoidal voltages and currents effectively. The filtering level of the harmonic components provides a significant reduction in the non-sinusoidality of the current and, as a result, an increase in the quality of electricity, as well as a reduction in power losses in the filter compensating device, a decrease in the overall dimensions of the filter and its cost.

1. 1. Introduction

The power supply of LED lighting which is being actively introduced causes contamination of the power supply network by harmonic components of the load current [1]. The frequency range for the real power consumption of the PSU-40-UA-B40 LED luminaire is shown in Figure 1. The oscillogram of the harmonic spectrum contains the 3, 5, 7 and 9 harmonics [2].

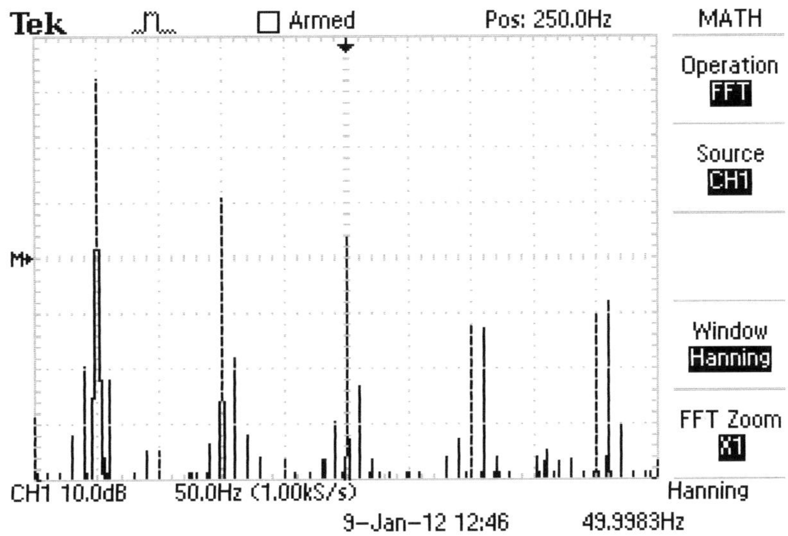

Figure 1. The spectrum of harmonics in the current consumed by the PSU-40-UA-B40 LED lamp.

In practice, the "four-beam" three-phase FCD scheme (Fig. 3) is more preferable than the traditional one (Fig. 2) during the three-phase mode of the lighting system switching [3].

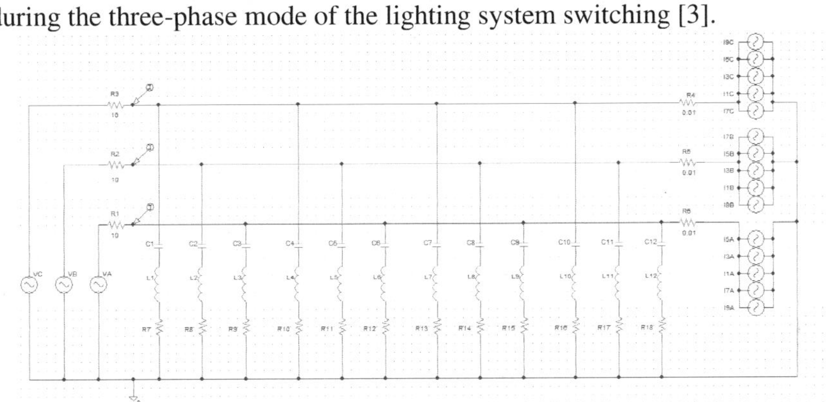

Figure 2. The traditional scheme of the FCD switching-on

The simulation of the three-phase non-linear load with the emission of high-frequency harmonics into the supply network is implemented by IA3, IA5, IA7, IA9 frequency generators for phase A, IB3, IB5, IB7, by IB9 frequency generators for phase B and by IC3, IC5, IC7, IC9 frequency generators for phase C with frequencies and amplitudes corresponding to the harmonic number, which is represented by the network diagram in Figure 3.

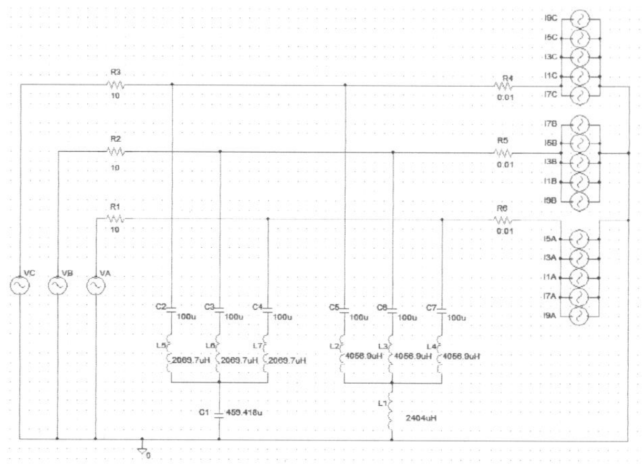

Figure 3. The network diagram with the "four-beam" FCD and non-linear load

The amplitude-frequency characteristics of the voltage and the current of the circuit that simulates the three-phase sinusoidal (VA, VB, VC) power source, non-linear (I1A, I3A, I5A, I7A, I9A), (I1B, I3B, I5B, I7B, I9B), (I1C, I3C, I5C, I7C, I9C) load, and four-beam frequency filters containing R1 - R6, (L1 - L7) and (C1 - C7), have the form shown in Figures 4, a and 4, b.

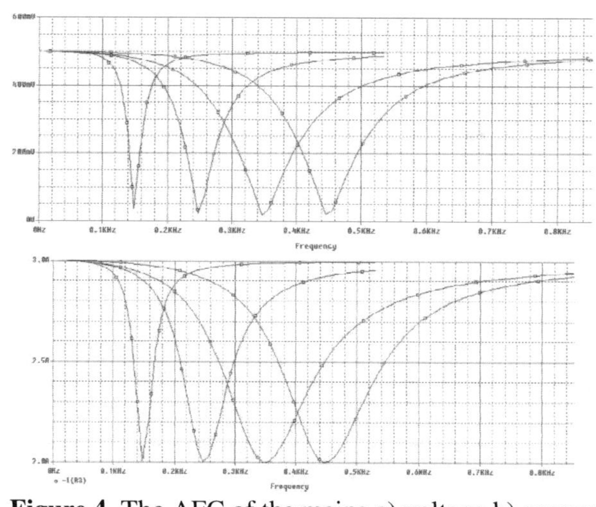

Figure 4. The AFC of the mains a) voltage b) current

To test the filtering properties of the four-beam filter (Fig. 4), the analysis of the non-linear symmetric load mode was performed (Table 1). The current spectrum of the nonlinear symmetric load is presented in Figure 5. The amplitude values of the higher harmonics of the current relative to the main harmonic (Table 1, Figure 5) were obtained by measuring the LED load on the real operating object of "4100" bridge at the Krasnoyarsk Railway (Figure 6).

Table 1. The harmonic values at the symmetrical load

№ Harmonics	Amplitude relative to the main harmonic A	B	C	Frequency, Hz	High frequency generator
1	1	1	1	50	IA1
3	1,2	1,2	1,2	150	IA3
5	0,35	0,35	0,35	250	IA5
7	0,02	0,02	0,02	350	IA7
9	0,18	0,18	0,18	450	IA9

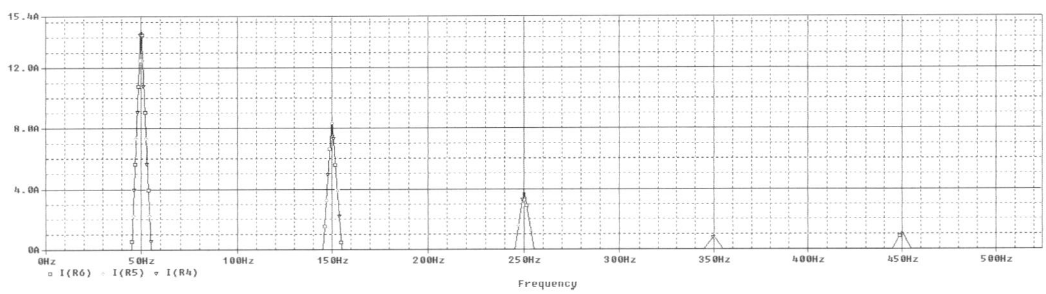

Figure 5. The current spectrum of the nonlinear symmetric load

2. The spectrum of harmonics

Figure 7 shows the current in the supply network (a) and the load current with the symmetric non-linear consumption (b). With the filters in the supply network, the current spectrum is practically free of higher harmonics (Figure 8).

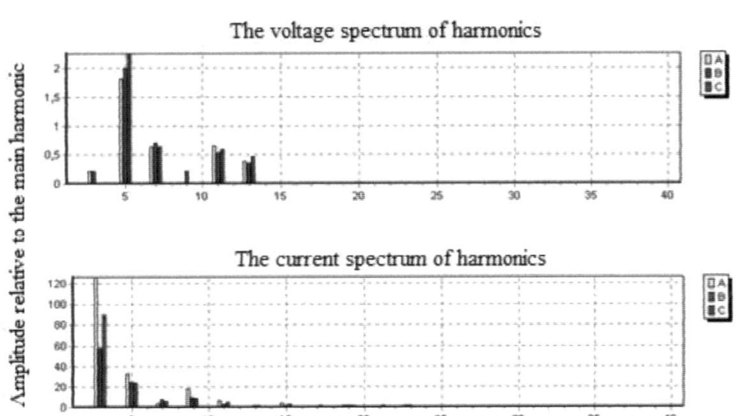

Figure 6. The spectrum of harmonics in the power supply network of LED lamps

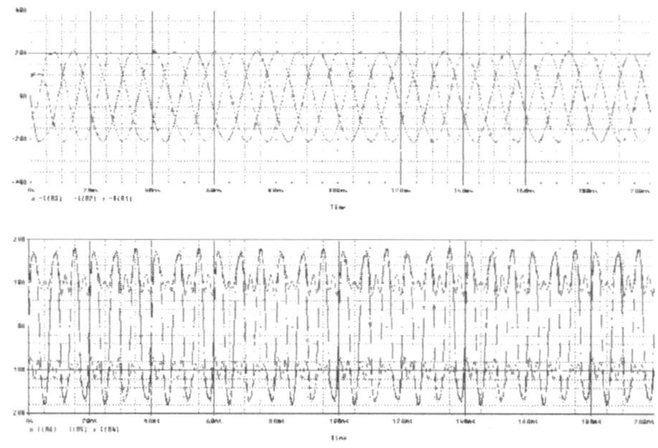

Figure 7. Current: a) in the mains b) in consumption of the symmetric nonlinear load

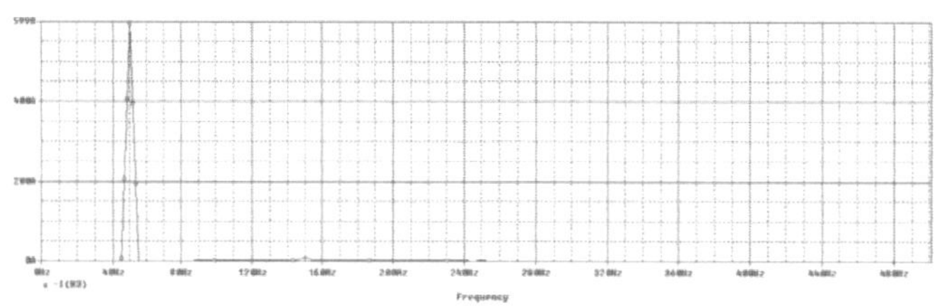

Figure 8. The current spectrum in the mains with the symmetrical non-linear load

In real conditions, filters are switched on in the network under voltage. When the filters are switched on, transients may occur that can cause overvoltage. The nature of transients largely depends on the switching moment. Figures 9, a and 9, b show the transition process with the simultaneous closure of the contacts of the filter switches where one can see the transient for the current almost to be completed in 25 mS, and for the voltage - in 35 mS.

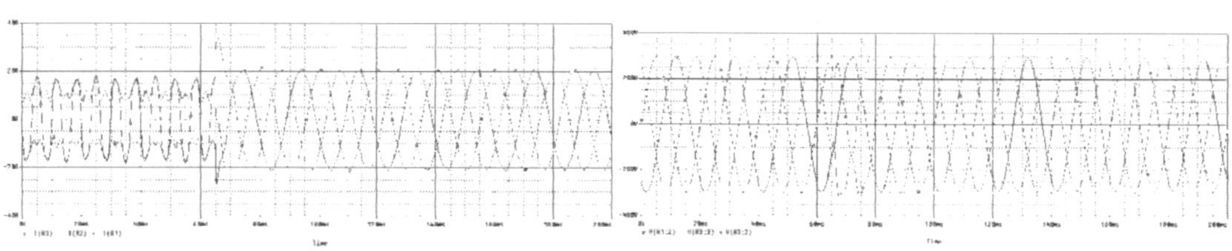

Figure 9. The transient process with the simultaneous switching on the filters and the symmetrical load a) Current change, b) Voltage change.

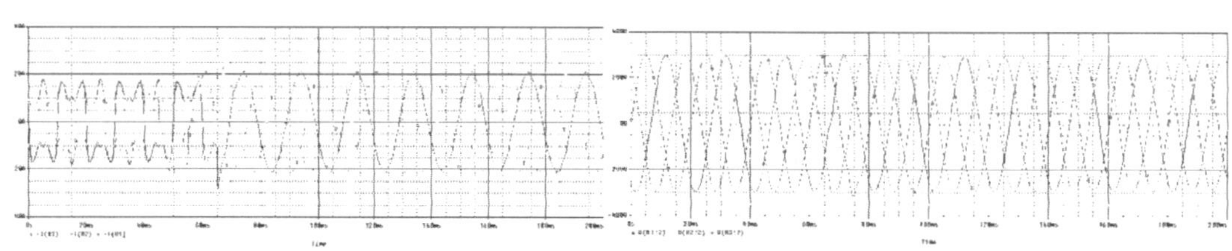

Figure 10. The transient with the non-simultaneous switching on (A: 65ms, B: 55ms, C: 45ms) the filters and the balanced load a) Current change b) Voltage change

IOP Conf. Series: Materials Science and Engineering **760** (2020) 012026 doi:10.1088/1757-899X/760/1/012026

Figures 10, a and 10, b show the transient process for the symmetrical load with the non-simultaneous closure of the switch contacts: phase A-65ms, phase B-55ms, phase C-45ms and one can see the process to stop in all three phases in 40mS – the beginning at 45mS, the ending - at 85mS - after the switching on the first filter.

The considered symmetrical load mode in real electric networks is practically absent. That is why the analysis of the asymmetrical consumption regime is of greater interest. The ratio of currents in the phases for the asymmetrical mode is presented in table 2.

Table 2. The parameters of harmonics under the asymmetric load

№ Harmonics	Amplitude relative to the main harmonic			Frequency, Hz	Frequency generator
	A (15A)	B (15A)	C (30A)		
1	1	1	1	50	IA1
3	1,2	0,58	0,9	150	IA3
5	0,35	0,27	0,25	250	IA5
7	0,02	0,06	0,04	350	IA7
9	0,18	0,08	0,07	450	IA9

Figure 11 shows the frequency range of the current consumption at an unbalanced load. On the spectral plane, the difference in the load of the phases can be seen.

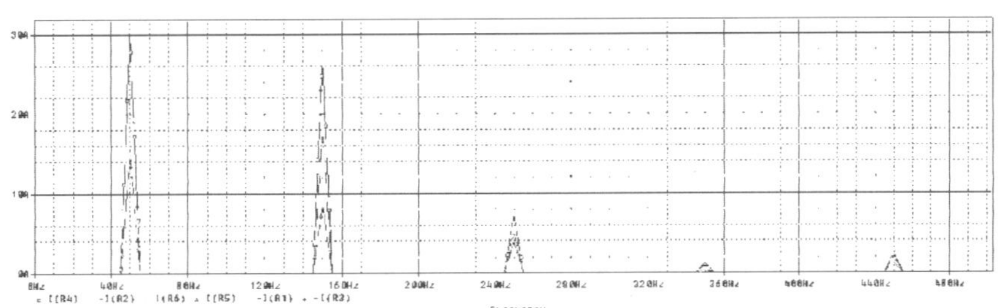

Figure 11. The spectrum of current consumption under the asymmetric load

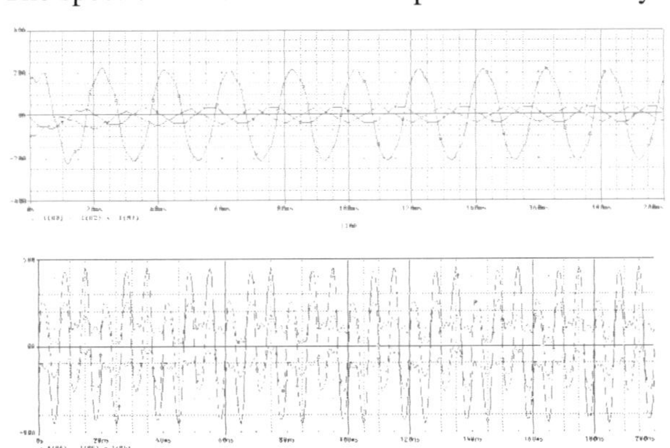

Figure 12. Current a) in the supply network, b) of consumption under the asymmetrical load.

The time characteristic in Figure 12 shows the nature of the current in the power supply network (a) and the current consumption of LED lamps (b).

The type of current curves shows that the network current I (R1), I (R2), I (R3) has the sinusoidal dependence, while the current consumption I (R4), I (R5), I (R6) has the significant distortion due to the influence of the harmonic components of the emission current.

Switching processes under the asymmetric load also need to be analyzed, as well as the factors contributing to overvoltage in the networks.

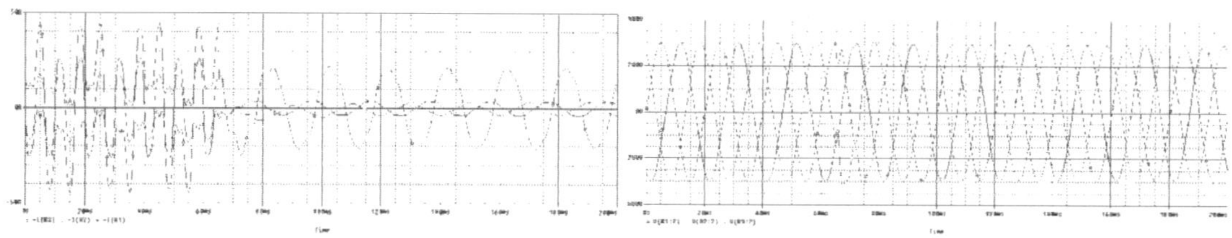

Figure 13. The transient process with the simultaneous switching on the filters and the unbalanced load a) Current change, b) Voltage change.

Figures 13 (a) and 13 (b) show the transition process while simultaneously switching on the filters. The analysis showed that the process of changing the current stopped in 35mS - from 65mC to 100mC, the voltage change stopped in 30mS (65mS-95mS).

Figures 14 (a) and 14 (b) show the transition process when filters are switched non-simultaneously on. The analysis showed that the transient process for current lasted 40mS - (45mC -85mS), and for the voltage of 50mC - (45mS-95mS).

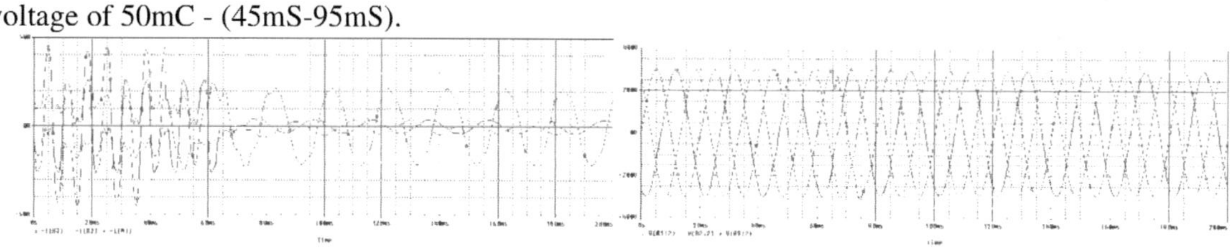

Figure 14. The transient at non-simultaneous (A: 65ms, B: 55ms, C: 45ms) switching on the filters and the asymmetrical load a) current, b) voltage.

Based on this network type, the critical (permissible) multiplicity of asymmetry is determined at which the 3rd harmonic is 5% of the fundamental harmonic [4]. Figure 15 shows the current spectrum at the 2-fold asymmetry (phase A = 15A, B = 15A, C = 30A). The harmonic of the "zero" sequence with frequency of 150 Hz is 1.7% of the main harmonic.

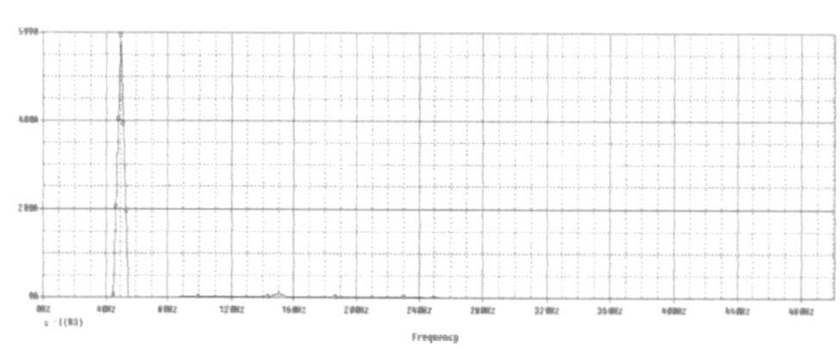

Figure 15. The spectrum of the network current at the 2-fold asymmetry.

Figure 16 shows the current spectrum with the 6-fold asymmetry (phase A = 15A, B = 15A, C = 90A). The harmonic of the "zero" sequence with frequency of 150 Hz is 6% of the main harmonic.

IOP Conf. Series: Materials Science and Engineering **760** (2020) 012026 doi:10.1088/1757-899X/760/1/012026

Figure 17 shows the current consumption of LED lighting of the 4100 bridge of the Krasnoyarsk Railway. The nature of the current is determined by the influence of the 3, 5, 7 and 9 harmonics as the dominant components in the current consumption. When modeling the spectral composition, the relative values of the harmonics were used (Table 3) obtained experimentally.

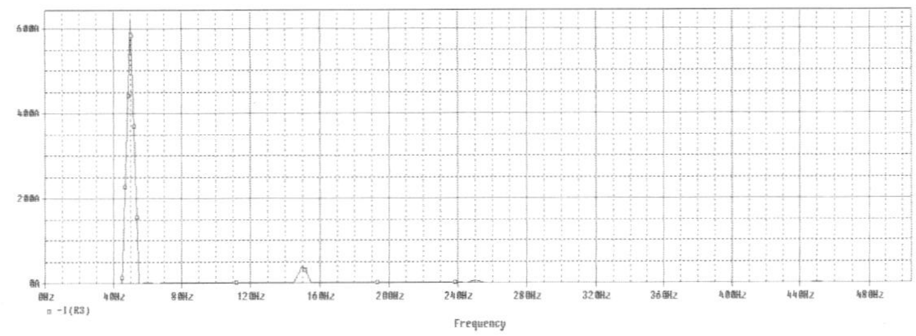

Figure 16. The spectrum of the network current with the 6-fold asymmetry.

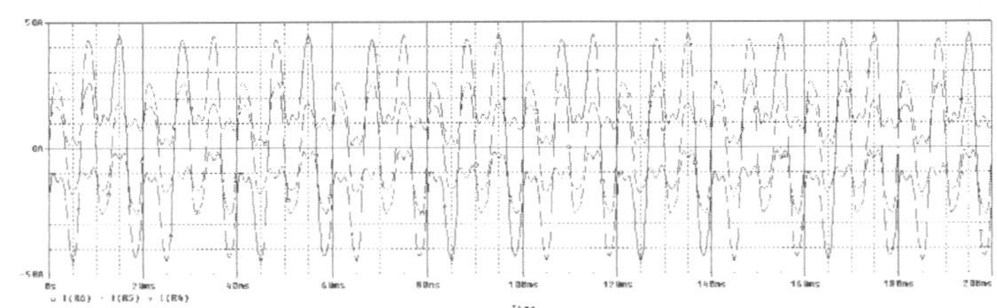

Figure 17. The current consumption in the lighting system of the "4100"bridge.

Table 3. The harmonic parameters for the unbalanced load

№ Harmonics	Amplitude relative to the main harmonic			Frequency, Hz	Frequency generator
	A (15A)				
1	1	1	1	50	IA1
3	1,2	0,58	0,9	150	IA3
5	0,35	0,27	0,25	250	IA5
7	0,02	0,06	0,04	350	IA7
9	0,18	0,08	0,07	450	IA9

When modeling the current consumption (Figure 17) at the "4100" bridge with the use of the "four-beam" filters, the mains current remains sinusoidal (Figure 18) that is confirmed by the frequency diagram (Figure 16).

IOP Conf. Series: Materials Science and Engineering **760** (2020) 012026 doi:10.1088/1757-899X/760/1/012026

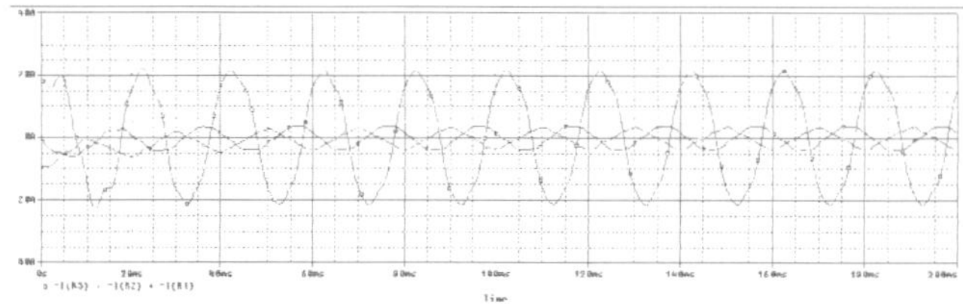

Figure 18. The current in the power supply network of the "4100" railway bridge.

3. Conclusion

The analysis performed shows that the use of the four-beam passive harmonic filter is more efficient in the three-phase networks. The level of filtering of the harmonic components provides a significant reduction in the non-sinusoidality of the supply current and the improvement in the quality of the power supply of networks with the non-linear consumption [5].

References

[1] Klimov V P Problems of higher harmonics in the modern power supply system V. P. Klimov, A. D. Moskalev available at: http://www.tensi.ru/article01.html

[2] IPS available at: www.intas-electro.ru

[3] 2009 Current Harmonics Filtering Systems Topology *IEEE transactions on Power ELECTRONICS* **24 8** 1939–1950.

[4] 2012 State standards P 54149-2010. *Electric Energy. Electromagnetic compatibility of technical equipment. The quality standards for electrical energy in general-purpose power supply systems.* (M.: Standardinform),.

[5] Kolmakov, V O 2014 Selection and calculation of the effective filter for networks with LED lamps *Electrical Energy Quality Management: Proceedings of the Intern. scientific-practical Conf.,* (Moscow: Raduga Printing Services Center LLC) 109–115 ISBN 978-5-905485-9-5.

Calculations of efficiency in implementing progressive mold forming methods

V Yu Konyuhov[1], A M Gladkih[1], I I Galyautdinov[1], E I Shchadova[1]

[1]Irkutsk National Research Technical University, 83 Lermontov Street, Irkutsk, 664074, Russia

E-mail: konyuhovvy@ex.istu.edu

Abstract. The article provides a classical prime cost accounting, outlining steps for calculating the cost at production site. The calculations compare two methods of manufacturing titanium parts for the aviation industry. The first method uses the mold forming by drop forging and the second one uses a modern superplastic manufacturing technology. This comparative calculation will determine the price of production of one set of parts for the SU-30. The considered methods of manufacturing parts will show the prime cost of production and will draw conclusions on further cooperation between Irkutsk Aviation Plant and Irkutsk State Technical University in the field of titanium products mold forming for modern aircraft.

1. Introduction

At any stage of the company's life cycle, there is always the task of accounting, analyzing and managing costs. For example, in contrast to the question of managing a company, where new trends arise from year to year (today it is innovations, yesterday it was mergers and acquisitions, etc.), one question is trivial in the cost accounting: how to calculate them correctly? To solve the principal tasks (how much you need to sell and at what price, so that this product brings profit to the company), one needs to calculate how much it costs to produce one unit of production, that is, to calculate its prime cost. The calculation of the production prime cost is not a new task and is solved in almost every enterprise. The prime cost calculation method is passed “from generation to generation” and has credibility without any evidence.

The classic calculation of the prime cost of production unit implies two steps:

- the first one is the definition of costs, changing in proportion to the volume of production of a particular product (definition of variable costs per unit of production) The calculation is performed by the production of consumption rates of individual elements of the costs and their acquisition cost. The classic representatives of variable costs are raw materials, supplies, components, process energy, piecework wages.
- the second one is the summation of all other expenses incurred during the year, and their division into specific types of products (separation of overhead costs by type of products). The classic representatives of the “other expenses” are the expenses for the maintenance and repair of equipment, buildings, structures, salaries of administrative and management personnel, depreciation charges, administrative expenses. Often these “other” expenses are provided in special documents: statements (estimates) of general shop expenses, general business expenses, general production expenses. As a rule, total costs for individual types of products are divided

in proportion to the wages of the main production workers accrued on these types of products [1-6].

2. Results and discussion

The price of the set is calculated by the formula:

$$C = K_1 \times K_2 \times S \tag{1}$$

where C is the selling price, rub.; K_1 is the working course, rubles/c.u.; K_2 is the trade margin; S is the prime cost, c.u.

In turn, the prime cost, S, is made up of the sum of the cost of raw materials, S_1, c.u.; of manufacturing costs, S_2, c.u.; of general shop expenses, S_3, c.u.:

$$S = S_1 \times S_2 \times S_3. \tag{2}$$

We define each component of the formula (2):

$$S_1 = K_3 \times r \times V \times P_1 \tag{3}$$

where K_3 is the utilization of raw materials; r is the density of the material, kg / m^3; V is the volume of finished products, m^3; P_1 is the cost of raw materials, c.u. / kg;

$$S_2 = K_4 \times \sum_i P_2(i) \times T \tag{4}$$

where K_4 is the total complexity factor; P_2 is the cost of the i-th operation, c.u. / h; T is the time required for the i-th operation, h; i is the operation;

$$T = \frac{F_i}{Z_i} \tag{5}$$

Here F_i are the dimensions of the finished product for this operation; Z_i is the performance of the operation. It is necessary that the units F_i and Z_i be consistent with each other. That is, if Z_i for this operation is determined in m^2/ h, F_i should be measured in m^2.

$$S_3 = (S_1 + S_2) \times K_5 \tag{6}$$

where K_5 is the ratio of general expenses.

Thus, we obtain the final formula for calculating the prime cost [6-15]:

$$C = K_1 \times K_2 \times \left(K_3 \times r \times V \times P_1 + K_4 \times \sum_i P_2(i) \times \frac{F_i}{Z_i} \right) \times (1 + K_5). \tag{7}$$

Using this method of calculating the prime cost, one can compare the processes of manufacturing parts and determine a more profitable production. As an example, let us compare two ways of making a set of parts.

As the consumption of metals grows, the continuously increasing demands of machine building in large forgings and details make it increasingly urgent for metallurgy to create more powerful and efficient equipment for metal forming. For many years, the manufacture of parts with the help of pressure is constantly progressing, and this progress makes it possible to improve the quality of parts and reduce the time of their manufacture. In our region, the strategic enterprise is the Irkutsk Aviation Plant (IAP), where much attention is paid to the development of production technologies, equipment modernization and the development of research and production base. Tens of millions of dollars are annually invested in technical re-equipment. The development of key competencies (technologies) is one of the main tasks facing the enterprise. These are the main technologies that ensure the production of modern aircraft with the required quality and efficiency [2]. Though IAP uses advanced technologies for the manufacture of parts, some workshops still need more in-depth modernization.

It should be noted that the drop forging method of manufacturing titanium parts is already outdated. This mold forming technology no longer allows competing in the global market, and the quality of parts is low compared with those of such competitors as Boeing, Airbus [4, 5, 16-19]. Analogs can be the draw forming of a part into a rigid die, mold forming by an elastic medium [1, 20-22].

The mold forming of parts by drop forging can be replaced by progressive methods of forming in the blank and forging production, namely, forming in the superplasticity mode. Irkutsk National Research Technical University (INRTU) is engaged in this method of mold forming, where the laboratory of "Advanced Forming Methods in Blank and Forging Production" was created. The laboratory equipment, which costs 78 million rubles, is manufactured by special order of the university and is unique for Russia. It is used for superplastic mold forming, combined with diffusion welding of sheet metal alloys. The laboratory equipment allows you to work on the development and testing of new technologies for forming sheet parts from aluminum and titanium alloys to produce a wide range of engineering products, including those for aircraft production (production of parts for the SU-30), to conduct specialized tests of metals and alloys in superplastic deformation modes. To evaluate the economic efficiency of the new method of mold forming using superplasticity, the cost of manufacturing a set of parts was calculated [7, 23-24].

A sheet of titanium with a cross section of 0.8 mm, size 600x1.500 mm at a price of 1525 rubles is used to manufacture one set of parts at IAP (raw materials and materials). The process of manufacturing parts is controlled by an engineer, the work is done by a workshop worker. The manufacturing technology is the drawing of parts by drop forging. The equipment used for the manufacture of parts is shown in table 1.

In INRTU titanium is also used for the manufacture of a set of parts. The parts are molded in the superplastic mode on the presses SPF 60T 2011 (the cost is 3 million rubles).

Comparative calculation of the prime cost of manufacturing titanium parts using various technologies is given in Tables 2-4.

Table 1. Equipment used at IAP for the manufacture of a set of titanium parts.

Designation	Quantity, units	Cost (1,000 rubles).
Hardening furnace RK-180; 29.0 kW	1	800
Hammer ML 1-5 1,5 t; 5.5 kW	1	700
Vibrating shears; 3.0 kW	1	150

Table 2. Calculation of general business expenses, rub.

Expenditure items	IAS		INRTU technology park	
	For year	Per unit of production set	For year	Per unit of production set
The salary of the enterprise managerial staff	0.00	0.00	0.00	0.00
Social charges	0.00	0.00	0.00	0.00
Corporate property tax*	36300.00	3630.00	66000.00	6600.00
Total	36300.00	3630.00	66000.00	6600.00

* are provided by the IAP accounting department and INRTU

Table 3. Calculation of general production (workshop) costs, rub.

Expenditure items	IAS		INRTU technology park	
	For year	Per unit of production set	For year	Per unit of production set
Remuneration of managerial production personnel: engineer*	18000.00	1800.00	15000.00	1500.00
Social charges*	6156.00	615.60	5130.00	513.00
Maintenance of buildings and structures	0.00	0.00	0.00	0.00
Communication services	0.00	0.00	0.00	0.00
Cleaning of production premises	0.00	0.00	0.00	0.00
Social charges	0.00	0.00	0.00	0.00
Depreciation of industrial buildings and structures, workshop equipment*	1212.00	121.20	1212.00	121.20
Labor protection*	8000.00	800.00	8000.00	800.00
Total	33368.00	3336.80	29342.00	2934.20

* are provided by the IAP accounting department and INRTU

Table 4. Calculation of the total prime cost of production, rub.

Expenditure items	IAS		INRTU technology park	
	For year	Per unit of production set	For year	Per unit of production set
Raw materials and supplies*	15250.00	1525.00	15250.00	1525.00
Transportation	0.00	0.00	0.00	0.00
Fuel for process purposes	0.00	0.00	0.00	0.00
Electricity for process purposes*	682.19	68.22	718.574	71.86
Basic and additional wages of production workers*	11000.00	1100.00	26000.00	2600.00
Social charges*	3762.00	376.20	8892.00	889.20
Expenses for production preparation and engineering	0.00	0.00	0.00	0.00
The cost of maintenance and operation of equipment*	31350.00	3135.00	57000.00	5700.00
General production (workshop) expenses	33368.00	3336.80	29342.00	2934.20
Workshop prime cost	95412.19	9541.22	137202.57	13720.26
General business expenses	36300.00	3630.00	66000.00	6600.00
Manufacturing prime cost	131712.19	13171.22	203202.574	20320.26
Non-manufacturing (commercial) expenses	0.00	0.00	0.00	0.00
Total prime cost	131712.19	13171.22	203202.60	20320.26

* are provided by the IAP accounting department and INRTU

Thus, the calculation of the total prime cost is as follows:

- social charges = staff labour remuneration x 0.342;
- overhead (workshop) costs = the salary of managerial and production personnel + social charges + depreciation of industrial buildings and structures, workshop equipment + maintenance of buildings and structures + payment for communication services + cleaning of production premises + labor protection;
- general business expenses = the salary of enterprise managerial staff + social charges + corporate property tax;
- shop cost = raw materials and materials + transportation + fuel for process purposes + electricity for process purposes + basic and additional wages of production workers + social charges + expenses for production preparation and manufacturing + expenses for maintenance and operation of equipment + general production (workshop) expenses;
- production cost = shop prime cost + general business expenses;
- total prime cost = production prime cost + non-manufacturing (commercial) expenses;
- price for one set of parts = total prime cost x 1.4 + total prime cost x 1.4 x VAT [3, 25-29].

Based on the calculations presented in Tabl. 2-4, it can be concluded that the price for one set of parts for the set manufactured at IAS was 21.758.85 rubles, price for the set manufactured at the technology park of INRTU was 33569.07 rubles.

3. Conclusion

As you can see, the production of a set of titanium parts, according to the old technology, is cheaper. However, the use of a new technology of superplasticity makes it possible to obtain parts from brittle heat-resistant alloys, previously made by casting (turbine disks with blades, etc.). This allows us to obtain significantly better properties due to a more uniform microstructure. This improves the quality of parts, prolongs their service life, reduces production time and labor intensity of the process. The use of superplasticity turned out to be more expensive, but, as they say, you have to pay for quality. If we want our modern production to be at its highest and able to compete with Western aviation giants, then it is necessary to switch to modern technologies. At the moment, if there are no potential investors to promote superplasticity in aviation, then it is worth re-qualifying the current trend to private enterprises or in more popular production.

References

[1] Kondakov A I, Vasilyev A S 2007 *Selection of blanks in mechanical engineering: a reference book* (Moscow: Mechanical Engineering Publ.) p 560

[2] Gozbenko V E, Kargapoltsev S K, Kornilov D N, Minaev N V, Karlina A I 2016 Definition of the main coordinates of the car with two-level spring suspension *International Journal of Applied Engineering Research* Vol 11 **20** pp 10367-10373

[3] Livshits A V, Filippenko N G, Homenko A P, Kargapoltsev S K, Gozbenko V E, Dambaev Z G 2017 Mathematical modelling of the processes of the high-frequency heating of thermoplastics and quality improvement of welded polymeric items *JP Journal of Heat and Mass Transfer* v. 14 **2** pp 219-226

[4] Kargapoltsev S K, Shastin V I, Gozbenko V E, Livshits A V, Filippenko N G 2017 Laser alloying of wear surfaces with metal components *International Journal of Applied Engineering Research* vol 12 **17** pp 6499-6503

[5] Khomenko A P, Gozbenko V E, Kargapoltsev S K, Minaev N V, Karlina A I 2017 Comparative analysis of simulation results and test of the dynamics of the wheelset *International Journal of Applied Engineering Research* Vol. 12 **23** pp 13773-13778

[6] Chernilevsky D V 2009 *Machine parts and design basics: a textbook for universities* (Moscow: Mechanical Engineering Publ.) p 656

[7] Gozbenko V E, Kargapoltsev S K, Karlina A I 2019 Environmental benefits of new industrial waste-based lubricant compositions *IOP Conf. Series: Earth and Environmental Science* **229** 012020

[8] Gozbenko V E, Kargapoltsev S K, Karlina A I 2019 Synthesis and structure of sulfur-containing polymers based on polymer industrial waste applied for rail lubrication *IOP Conf. Series: Earth and Environmental Science* **229** 012021
[9] Chernyshenko M S 2011 *Innovation management: a textbook* (Irkutsk: Irkutsk State Technical University Publ.)
[10] Yolkin K S, Yolkin D K, Shtayger M G, Kolosov A D, Ivanov N A 2018 Technologies, which allow to reduce an impact of metal silicon production on the environment *IOP Conference Series: Materials Science and Engineering* **411** 012028
[11] Sobolev A S 2015 Evaluation of economic attractiveness of investment projects at the stage of idea development *Vestnik of ISTU* **10** 249-253
[12] Suceska M 2013 Test Methods for Explosives *Springer Science & Business Media*
[13] Vasina A *Calculation of production costs available at:* http://www.klubok.net/reviews226.html
[14] Shastin V I, Kargapoltcev S K, Gozbenko V E, Livshits A V, Filippenko N G 2017 Results of the complex studies of microstructural, physical and mechanical properties of engineering materials using innovative methods *International Journal of Applied Engineering Research* v. 12 **24** pp 15269-15272
[15] Gozbenko V E, Kargapoltsev S K, Minaev N V, Karlina A I 2016 Simulation of the vibration of the carriage asymmetric parameters in Mathcad *International Journal of Applied Engineering Research* v. 11 **23** pp 11132-11136
[16] Kargapoltsev S K, Khomenko A P, Gozbenko V E, Minaev N V, Karlina A I 2017 Development of new lubricants for reducing the wear of the elements of the path and running parts of rolling stock *International Journal of Applied Engineering Research* v. 12 **22** 12362-12368
[17] Gozbenko V E, Khomenko A P, Kargapoltsev S K, Minaev N V, Karlina A I 2017 Creating of the alternative lubricants and practice of their use *International Journal of Applied Engineering Research* Vol. 12 **22** pp 12369-12372
[18] News of Irkutsk State Technical University available at: http://www.istu.edu/news/9864/
[19] Shchadov I M, Chemezov A V, Konyukhov V Yu, Belyaevskaya T S 2015 Comparative analysis of creation of the enterprise on repair of wheel pairs on the basis on "Norilsktransremont" *Vestnik of ISTU* **5** pp 297-300
[20] Kondratiev V, Govorkov A, Lavrent'eva M, Sysoev I, Karlina A 2016 Description of the heat exchanger unit construction, created in IRNITU *International Journal of Applied Engineering Research* **11 19** 9979-9983
[21] Yolkin K S, Yolkin D K, Shtayger M G, Kolosov A D, Ivanov N A 2018 Technologies, which allow to reduce an impact of metal silicon production on the environment *IOP Conference Series: Materials Science and Engineering* **411** 012028
[22] Balanovsky A E, Kondratev V V, Shtayger M G, Nebogin S A, Karlina A I 2018 Complex metallographic researches of 110G13l steel after heat treatment *IOP Conference Series: Materials Science and Engineering* **411** 012014
[23] Ivanchik N N, Balanovsky A E, Shtayger M G, Sysoev I A, Karlina A I 2018 Capability enhancement of production of activating fluxes for arc welding using ultradispersed products of silicon waste processing *IOP Conference Series: Materials Science and Engineering* **411** 012035
[24] Balanovsky A E, Shtaiger M G, Kondratev V V, Karlina A I, Govorkov A S 2018 Comparative analysis of structural state of welded joints rails using method of Barkhausen effect and ultrasound *Journal of physics: conference series* 012006
[25] Balanovsky A E, Shtaiger M G, Grechneva M V, Kondratev V V, Karlina A I 2018 Comparative metallographic analysis of the structure of st3 steel after being exposed to different ways of work-hardening *IOP Conference Series: Materials Science and Engineering* **411** 012012
[26] Balanovsky A E, Shtayger M G, Kondratev V V, Huy Vu V, Karlina A I 2018 Plasma-arc surface modification of metals in a liquid medium *IOP Conference Series: Materials Science and Engineering* **411** 012013

[27] Nemarov A, Lebedev N, Kondrat'ev V, Kornyakov M, Karlina A 2016 Theoretical and experimental research of parameters of pneumatic aerators and elementary cycle flotation *International Journal of Applied Engineering Research* 11 **20** 10222-10226

[28] Kondratiev V V, Govorkov A S, Kolosov A D, Gorovoy V O, Karlina A I 2017 The development of a test stand for developing technological operation flotation and separation of MD2. The deposition of nanostructures md1 produce nanostructures with desired properties *International Journal of Applied Engineering Research* v. 12 **22** pp 12373-12377

[29] Yaschura A I 2006 *The system of maintenance and repair of general industrial equipment: a guide* (Moscow: Publishing House of Scientific Center ENAS) p 360

IOP Conf. Series: Materials Science and Engineering **760** (2020) 012028 doi:10.1088/1757-899X/760/1/012028

Study of the operation of high-frequency electrical plants of railway consumers

A V Pultyakov[1], O V Arsentiev [2], M V Kopanev[1], V A Alexeenko[1], G O Arsentiev[2] and E S Shtykin[2]

[1]Irkutsk State Transport University, 15, Chernyshevskogo str., Irkutsk, 664074, Russia
[2]Irkutsk National Research Technical University, 83, Lermontova str., Irkutsk, 664074, Russia

E-mail: mvk@irgups.ru

Abstract. The article discusses the issues of improving the quality of electricity in the networks of non-traction railway consumers through the use of energy routers. Model studies of the electricity supply system of railways with high-frequency routers were carried out. Results have been obtained that allow using energy routers to supply electricity to plants of railway consumers with increased requirements for electric power quality. The article presents the results of the calculation of the high-frequency transformer and its experimental characteristics obtained in the study of high-frequency electrical plants of consumers.

1. Introduction

The railway system is the basis of the transport communication of the country and the continent. This is a well-developed infrastructure, including travel facilities, overhead systems, electric power supply, loco depots, repair companies, etc. Like any other complex structure, all elements of which are interconnected in a single technologically process, the system strives to reduce consumption at each production stage, make extensive use of existing internal capabilities and advantages, and improve control over the resources used. In this regard, the power supply of consumers' installations operating in the railway system also has its own specifics, due to the unified power system of railway transport, designed to provide electricity to both traction and other consumers. This feature results in a significant deterioration in the quality of electricity supplied to electrical plants of non-traction consumers.

This is a known issue, and there are a number of technical measures aimed at reducing the impact of traction load on the operation of electrical plants of other consumers, united by a single electric power supply system. However, there is no general solution to the problem. The proposed measures mainly reduce the amount of distortion – they reduce the asymmetry and non-sinusoidal voltage coefficients to the values acceptable by the quality standard (GOST 32144-2013). This does not ensure the complete elimination of distortions in the form of the supply voltage, which, in some cases, is not enough for the normal supply of electrical plants to responsible consumers with increased requirements for electric power quality. Currently, this problem is partially solved by the use of stand-alone generating units, or uninterrupted power sources. These technical solutions cannot be called successful, because, by their nature, they are closer to meeting the requirements for electric power supply of a special group of consumers, and are not designed for long-term use. To solve the problem completely, it is required to develop power plants that allow the consumer to provide high-quality electricity during long-term operation.

Currently, Smart Grid technologies are widely introduced in electric power supply systems. They are based on intelligent generation and consumption of electricity, using distributed generation plants and semiconductor converter devices designed to improve the quality and reliability of generation, power supply and consumer facilities [1–3]. This integrated approach determines the close relationship between all the elements involved in the electric power process. The electric power systems of railway transport are also actively introducing elements of intelligent systems to solve problems related to improving the characteristics of their own electric power supply networks.

One of the ways of introducing intelligent elements into the electric power supply network is to organize a special type of transformer substations, including special transformers and controlled conversion devices, so-called energy routers (ER) [4–6]. These devices are designed to organize controlled energy consumption, connect distributed generation plants, and provide two-way energy flows.

2. Energy Routers

To solve the problem of energy quality in the traction electric power supply networks, when connecting consumer plants with increased requirements to electric power parameters, it is proposed to organize special ERs, which provide almost sinusoidal voltage of stable frequency in combination with the ability to maintain a constant amplitude. Modern elemental base of power electronics allows one to create converter devices that operate reliably in high-voltage networks, capable of switching high currents, and perform high-frequency switching. The power of ER should be comparable with the energy characteristics of power transformers used in electric power supply networks of 10 / 0.4 kV (no more than 2.5–3.2 MVA).

Fig. 1 shows the flow chart of the ER, arranged by the principle of high-frequency transformation [7], with a capacity of 3 MVA.

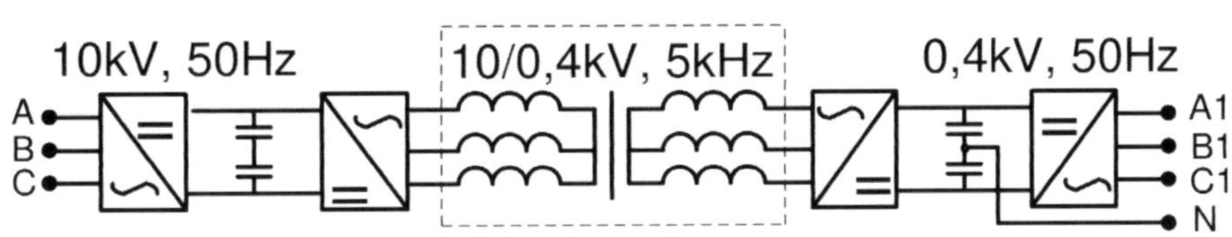

Figure 1. The flow chart of the energy router

The proposed ER scheme consists of an input rectifier (high-voltage, industrial frequency), inverter 1 (it generates 5–15 kHz high-frequency pulses of a rectangular shape), a high-frequency transformer (its magnetic core is made of 0.1 mm electrical steel sheet when used at high frequencies), the output controlled rectifier (low-voltage, high-frequency), inverter 2 (it generates an almost sinusoidal output voltage of the mains frequency, for example, a resonant inverter [8–10]).

In the circuits of small and medium power routers, field Mosfet transistors are used in a high-frequency inverter, which have good dynamic characteristics allowing them to be used in systems with a frequency up to 100 kHz. The operating frequency of such ER is approximately 20 kHz. However, these semiconductor elements have limitations on the magnitude of the transmitted current up to 100 A.

A bipolar IGBT transistor, capable of switching currents above 1 kA, is used in the high-frequency inverter of the proposed ER. The effective use of this transistor is limited by the switching frequency at which the losses are within acceptable limits, ensuring a sufficiently high conversion efficiency (above 90 %). This frequency is approximately equal to the clock frequency of the frequency converter – up to 5 kHz.

2.1 ER model studies

On the basis of the proposed flow chart, an ER model was compiled in the Matlab-Simulink environment, shown in Fig. 2. In accordance with the flow chart, the model has two uncontrolled rectifiers to rectify voltages in the high-voltage and low-voltage parts of the circuit, and two inverters, respectively, to receive pulses of 5 kHz frequency (Invertor_5000Hz) and output 50 Hz voltage (Invertor_50Hz). The ER is designed as a Ehnergorouter subroutine, which can be integrated into any model of the energy system. In this case, the system of traction electric power supply is being investigated (Fig. 3), where the ER is used to supply high-quality electric power to the plants of non-traction consumers.

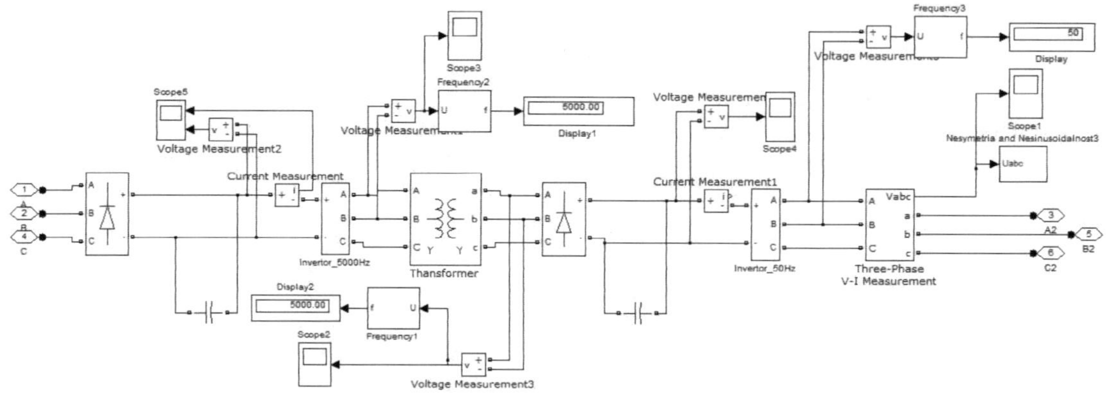

Figure 2. The Energy Router Model

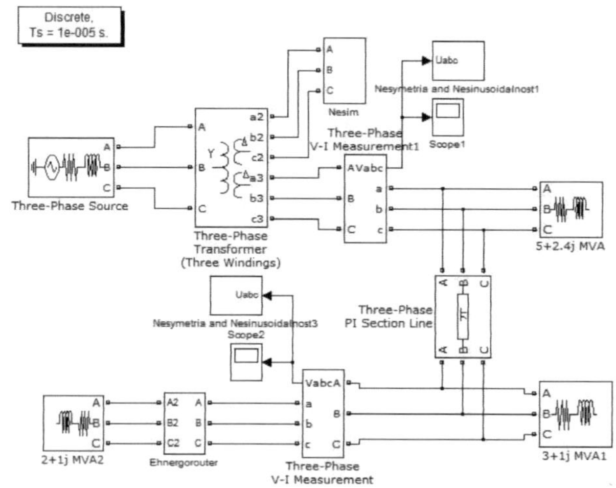

Figure 3. The traction electric power supply system with ER

In the studied model of the traction power supply system, there is a three-winding transformer, the secondary windings of which serve to supply energy to the electric stock and to the plants of non-traction consumers. The load power for the ER is 2.24 MVA, with power factor $cos\varphi = 0.89$. Model studies were conducted with the ER turned on and off.

Fig. 4a shows the simulation results with the ER turned off. Obviously, there is a significant amount of asymmetry (Nesimmetria, 5.265, maximum allowable – 4) and unsinusoidality (Nesinusoidalnost, max. 6.598, maximum allowable – 5) (Fig. 4b) in the form of voltage supplied to the consumer from the traction transformer.

The ER operation is characterized by indicators of digital frequency measurement units (Fig. 3) and voltages on the windings of a high-voltage transformer (Fig. 5a, b).

Fig. 6 presents a diagram of voltage (a) and indicators of quality of electricity (b) supplied to electrical plants of a consumer through the ER.

The type of output characteristics of the ER shows that the voltage supplied to the consumer is almost sinusoidal with a frequency of 50 Hz, there are virtually no deviations in magnitude (0.0565) and in form (max. 0.434).

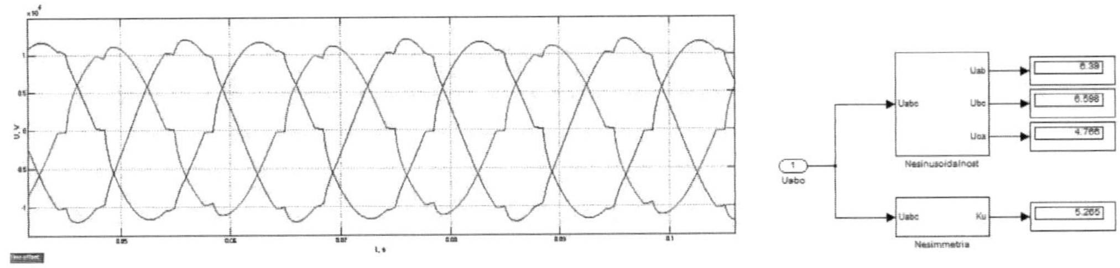

a b

Figure 4. Voltage (a) and electric power quality parameters (b) of a non-traction consumer without an ER

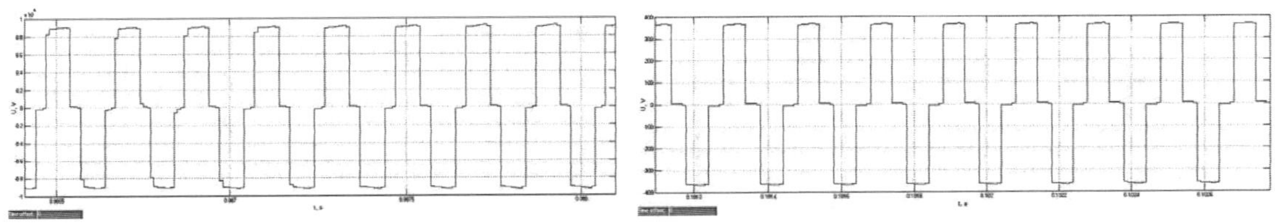

a b

Figure 5. Voltages at the high-frequency transformer, from the side of 10 kV (a), from the side of 0.4 kV (b)

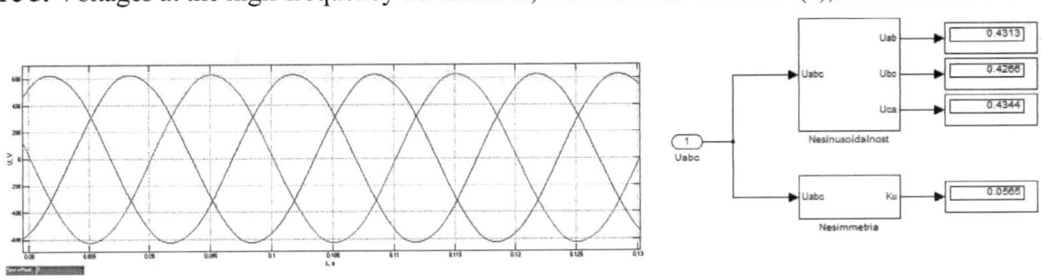

a b

Figure 6. Voltage (a) and power quality parameters (b) when the consumer is powered through the ER

3. Experimental studies of the operation of high-frequency electric plant

The concept of ER and the possibility of technical implementation were experimentally tested on a high-frequency power supply unit of a technological plant developed at Trans-Atom LLC, Russia, by order of the Cuban Railway (Ferrocarriles de Cuba), The Republic of Cuba.

The principle of operation of the converter is based on double conversion of electrical energy. First, the three-phase mains voltage is supplied to the bridge unregulated diode rectifier. Then the rectified voltage, using a bridge single-phase Mosfet transistor inverter, is converted into a rectangular AC voltage with a frequency of up to 15 kHz. The calculated values of the currents in the primary winding are less than 100A, so the use of these high-frequency transistors is justified and expedient. The voltage is further applied to the primary winding of the power transformer. A load is connected to the secondary winding, the permissible value of which depends on the frequency of the voltage pulses.

Regulation of the RMS voltage value at the load is made by changing the interpulse period at the output of the power inverter and, therefore, at the windings of the transformer. The converter operation and protection management is controlled from the control board, which is located in the converter. Fig. 7a shows the arrangement of the high-frequency power supply unit of the process plant.

3.1 Calculation of high-frequency transformer

The resistance of the windings and, as a consequence, their heating is determined by the current density in the wire, which was set when calculating. Based on practical experience, it is recommended to use the value of current density in a copper wire not exceeding 3.2 A/mm^2. The transformer is calculated from the operating conditions of the magnetic system on the linear portion of the magnetization characteristic, since according to technological conditions, the frequency of operation is variable (10–15 kHz), and in order to perform high-precision control, it is better to control the process of tuning the frequency with linear characteristics dependencies. In this case, the determining factor is the choice of the magnitude of the magnetic induction.

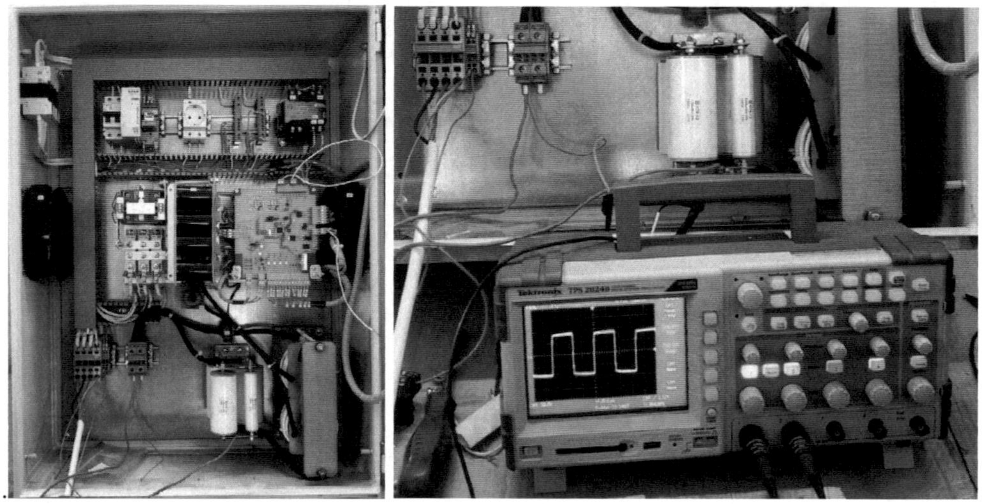

a b

Figure 7. High-frequency power supply unit of process plant: a - arrangement, b - experimental plant

When the basic design conditions are fulfilled, the efficiency of a high-frequency transformer will be at least 95 %. Based on this, an efficiency value of 0.95 pu is used in the calculations.

The determining value for the calculation of the transformer makes it possible to place the windings in the aperture of the core of the magnetic circuit. The recommended value of the fill capacity factor of the core aperture with copper (K_m) for frame shell-type or core-type magnetic circuits is 0.45. With wide frames and a large winding length of one layer, the value of K_m can be as high as 0.5…0.55, as, for example, in magnetic circuits of type B69 and B35. With frameless industrial winding K_m can have values up to 0.6 ... 0,65. In general, the theoretical limit of the value of K_m for layered arrangement of a round wire without insulation in a square aperture is 0.87.

The given practical values of K_m are achievable only if the wire is laid strictly turn to the turn, with thin interlayer and interwinding insulation, and the leads are sealed outside the core aperture (on the lateral winding overhangs). In the manufacture of frame windings in laboratory or pilot production, the value of K_m should not be more than 0.3 ... 0,4.

The power-to-size ratio of the transformer, in watts, is determined by the following expression:

$$P = \frac{4{,}44\,\eta\, S_c\, S_0\, f\, B\, j\, K_m\, K_c}{(1+\eta)100},$$

where η is the transformer efficiency, pu;

S_c and S_o are the cross-sectional areas of the core and the aperture, respectively [square cm];

f is the lower operating frequency of the transformer, Hz;

B is magnetic induction, T;

j is the current density in the wire windings, A / mm^2;

K_m is the fill capacity factor of the core aperture with copper;

K_c is the fill capacity factor of the core section with steel.

With the known voltage in the windings, the number of turns can be calculated by the formula:

$$n_1 = \frac{U_1\,10^4}{4{,}44\,f\,B\,S_c\,K_c},$$

where U, U, ... are the winding voltages in volts, and n, n, n, ... are the number of turns of the windings.

Table 1 shows the main parameters of the high-frequency transformer designed for installation.

Table 1. The results of the calculation of the transformer with ferrite N97 U141 / 78/30 (B67374G0000X197)

S_c	The cross-sectional area of the core, cm^2		13,65	
S_o	Aperture area, cm^2		33,5	
η	Efficiency		0,95	
V	Magnetic induction, T		0,35	
j	Current density in the wire windings, A / mm^2		3,2	
K_m	The fill capacity factor of the core aperture with copper;		0,3	
K_c	The fill capacity factor of the core section with steel.		0,93	
f	Frequency (kHz)	10	12	15
P	Calculated power-to-size ratio, kW	31	37	46
U_1	Primary Voltage, V		535	
U_2	Secondary Voltage, V		100	
I_1	Primary winding current, A		54	
I_2	Secondary winding current, A		290	
n_1	The number of turns in the primary winding	27	23	18
	Wire section in the primary winding		16	
n_2	The number of turns in the secondary winding	5	4	3
	Wire section in the secondary winding		90	
	Weight, kg		6	
	Dimensions, mm (LxWxH)		196x84x150	

According to the calculated data, a high-frequency transformer was made, which is one of the main components of the power supply unit. The use of this transformer has reduced the weight and dimensions of the plant. For example, consider the main indicators of industrial frequency dry power transformers (Table 2).

Table 2. Comparative characteristics of industrial frequency dry transformers

№	Type	Dimensions, mm	Weight, kg
1	TS-25/10	880x520x925	340
2	TS-40/10	940x520x985	420
3	OSU-20 / 0.5	540x420x555	136
4	OSU-40/0,5	740x575x765	283

It is obvious that all the transformers presented in Table 2 are significantly inferior to the developed high-frequency transformer according to the given characteristics (Table 1). Depending on the design, the transformer weights differ by more than 70 (three-phase) and 48 (single-phase) times. In terms of dimensions, the differences range from 3 to 6 times.

3.2 Experimental studies of high-frequency transformer

The high-frequency power supply unit was studied by the oscillography of signals of currents and voltages characterizing the qualitative and quantitative conversion parameters. Fig. 7b shows a general view of the experimental plant, which includes the object of research and the measuring instrument - a Tektronix TPS 2024B oscilloscope. This device makes it possible to obtain accurate frequency values, which is an important indicator in the regulation and fixation of the RMS voltage. To obtain results compatible with computer processing, the Hantek DSO-2090 digital oscilloscope was also used in the studies.

Fig. 8a shows one of the results of experimental studies to determine the characteristics of the high-frequency transformer of the plant — transformation ratio, pulse distortion, phase shift of input and output characteristics, conversion frequency.

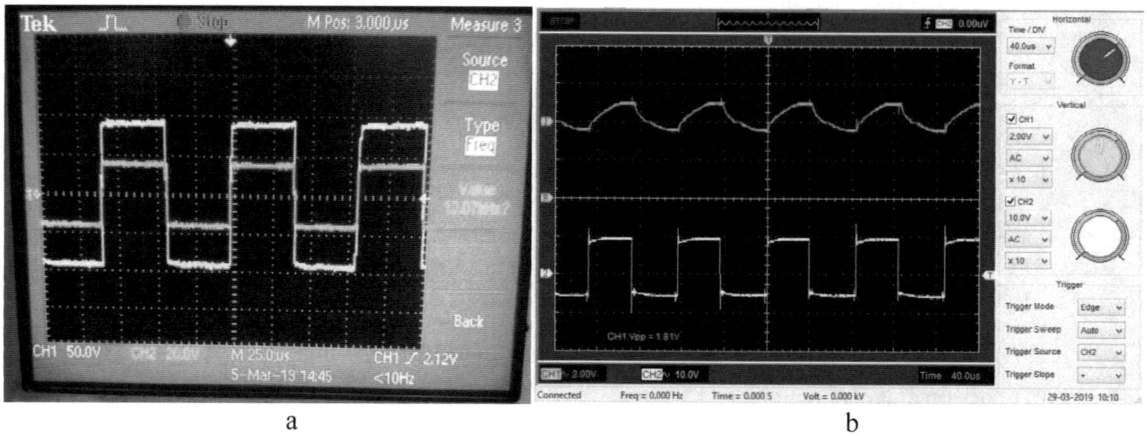

Figure 8. The results of experimental studies: a – characteristics of the high-frequency transformer, b – currents and voltages in the primary winding

The measurements showed that the high-frequency transformer fully complies with the calculated data (table 1): the transformation ratio was 5.4, the voltage pulses have a clearly defined rectangular nature, the signal is transmitted through the magnetic circuit without distortion in shape and angle of phase shift, the frequency was 12.07 kHz.

Fig. 8b shows the combined waveforms of currents and voltages at the high-voltage side of the transformer. The amplitude of the voltage pulses was 550 V at a frequency of 12.1 kHz. The current oscillogram corresponds in time to the voltage pulses, the amplitude value is 100 A. The currents and voltages are practically coincide in phase, that is, after automatic frequency control, reactive power compensation occurs, and the transformer is switched on to a purely active load. Obtaining this operation mode is justified by minimizing the load to obtain more rigid output characteristics of the transformer.

4. Conclusion

It was determined that the inclusion of intelligent elements in the power supply network allows one to solve a number of tasks to improve the quality and reliability of energy supply of railway transport facilities. The use of energy routers as sources of power supply to responsible consumers makes it possible to completely solve the problems of power quality in the networks of non-traction consumers. The results of model studies of the ER with a high-frequency transformer showed that its use completely eliminates the drawbacks of the quality of electricity in traction power supply networks. The developed high-power ER model can be used to analyze the operation of the electric power system, including digital objects. The capabilities of the ER model allow it to be used as an integrating and controlling element in the implementation of intelligent power supply systems.

A design and technical implementation of a high-frequency transformer for a process plant was carried out. The results of experimental studies of the work of high-frequency transformer confirmed its correspondence to the calculated parameters. The effect of the application of these transformers in the power supply units of the electrical devices of consumers very considerably decreases the mass-dimensional indices of power transformer, which substantially increases the consumer properties of these power supply systems.

References

[1] Wang J, Huang A Q, Sung W, Liu Y and Baliga B J 2009 *IEEE Industrial Electronics Magazine* **3(2)** 16–23

IOP Conf. Series: Materials Science and Engineering **760** (2020) 012028 doi:10.1088/1757-899X/760/1/012028

[2] Kuchera L Y, Kopanev M V and Fyodorova N V 2010 *Modern technologies. System analysis. Modeling* **2(26)** 204–8

[3] Kryukov A V, Zakaryukin V P and Alekseenko V A 2012 *Modern technologies. System analysis. Modeling* **2 (34)** 99–105

[4] Xu Y, Zhang J H, Wang W Y, Juneja A and Bhattacharya S 2011 *Proc. 2011 IEEE Int. Conf. on Smart Grid Communication (SmartGridComm)* 31–6

[5] Bulatov Y N, Kryukov A V and Arsentiev G O 2018 *International Ural Conference on Green Energy (UralCon)* 143–8

[6] Bulatov Yu N, Kryukov A V and Arsentyev G O 2018 *Electrical equipment: operation and repair* **4**, 64–9

[7] Bansode Sh G and Joshi P M 2014 *Proceedings of 11th IRF International Conference, 15th June-2014* (Pune, India) 15–21

[8] Pultyakov A V, Dunaev M P and Arsentiev G O 2018 *AER – Advances in Engineering Research*, **158** 109–13

[9] Dunaev M P, Arsentiev O V and Arsentiev G O 2018 *IOP Conf. Ser.: Mater. Sci. Eng.* **327** 052005

[10] Kopanev M V and Arsentyev M O 2018 *AER – Advances in Engineering Research* **158** 190–5

Determining the geometrical parameters of spring suspension of locomotives

M V Kopanev[1], M O Arsentiev[2], L Y Kuchera[1]

[1] Irkutsk State Transport University, 15, Chernyshevskogo str., Irkutsk, 664074, Russia
[2] Irkutsk National Research Technical University, 83, Lermontova str., Irkutsk, 664074, Russia

E-mail: mvk@irgups.ru

Abstract. The paper considers the issues of measuring the geometric parameters of the springs using a laser-optical sensor. The features of an automated rig for measuring the parameters of the springs are presented. The work also investigates the operation of the laser-optical sensor. The results obtained allow using the laser-optical sensor for measuring the geometric parameters of the springs of a rolling stock. The measurements obtained with the laser-optical sensor to determine the parameters of the locomotive suspension springs are presented.

1. Introduction

In accordance with the innovative development program of the Russian Railways Company, the quality of the parts and products used in the railway transport should be high, since the safety of the transport of passengers and goods depends on each mechanical assembly and individual parts.

In this regard, during overhauls of locomotives, all components and all parts included in their composition are inspected, as well as individual components and components are replaced or repaired in accordance with standards and tolerances, tested and adjusted to restore their operational characteristics and guarantee their performance [1-4].

Testing of spring suspension of locomotives includes testing under the action of a test load to determine the value of permanent deformation and under the action of a static load to determine the magnitude of the camber of the spring.

Residual deformation is defined as the difference between the heights of the spring in the free state before and after compression under a test load. The test is performed as follows: the spring is pressed twice with the test load and after removing the load, the spring height is measured in the free state. Then, the spring is again loaded with the test load, completely unloaded, and its height is measured. In this case, the height of the spring in the free state measured first, must remain unchanged.

The springs that have passed the tests for the absence of permanent deformation are tested to determine the actual chamber of the spring under the working static load, which is defined as the difference between the height of the spring in the free state and the height of the spring under the working static load. At the same time, the measured actual chamber of the spring given in the tables should not deviate from the calculated value by more than +12 and -8%.

To date, on the market there are automated spring calibration stands to measure the geometrical parameters of the springs, which are based on two basic non-contact measurement techniques: shadow

IOP Conf. Series: Materials Science and Engineering **760** (2020) 012029 doi:10.1088/1757-899X/760/1/012029

method and triangulation method [5-10].

2. Triangulation measurement method using automated bench

To determine the geometric dimensions of the calibrated spring, a non-contact method is used. For this, an optoNCDT 1700 laser sensor shown in Figure 1 was selected.

This sensor works using the principle of optical triangulation and has high accuracy and resolution.

The measurement results from the digital output of the sensor are sent to an industrial computer. The main advantage of the sensor is the laser system, which allows taking into account the reflective properties of the surface (mirror metal, black rubber or shiny varnish).

Figure 1. OptoNCDT 1700 laser optical sensor

To improve the accuracy of movement of the measuring laser system, profile rail guides are used. The principal design of the rolling guide rails is shown in Figure 2.

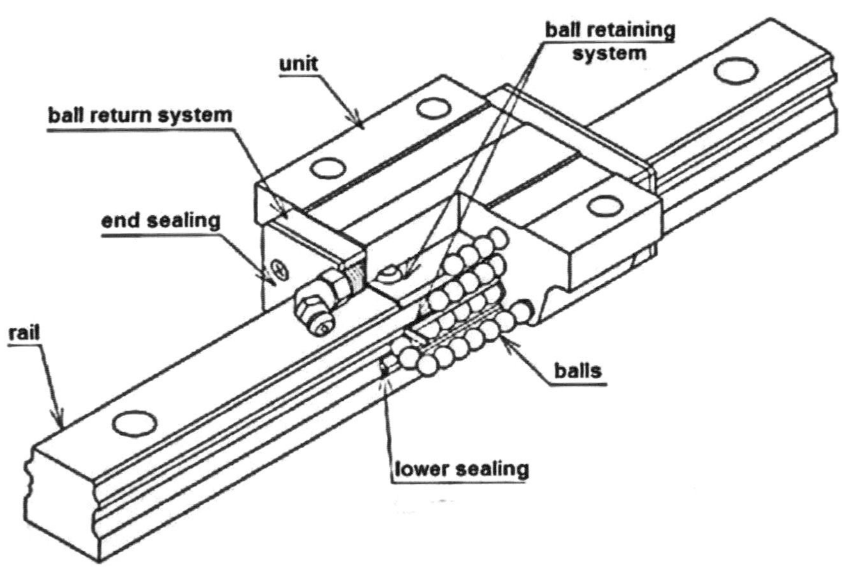

Figure 2. Rail design

A rail on both sides has longitudinal parallel support tracks for rolling elements (balls or rollers). The shape and number of tracks are determined by the type of rolling elements and the operational characteristics of the system. On a rail, a movable carriage is mounted having internal longitudinal support surfaces (their number and shape correspond to the support tracks of the rail) and longitudinal return channels of the rolling bodies. At the ends of the carriage, end plates (caps) are fixed that provide closed movement of the rolling elements from the support tracks of the carriage to the return channels and back. In the carriage, limiters are also mounted that protect the rolling elements from falling out during an emergency displacement of the carriage from the rail.

The profile rail guides have numerous benefits, including low friction, no jamming and slippage, and smooth linear movement, even under heavy loads. Due to their ability to maintain high efficiency and operating parameters over a long period of time, they meet a wide range of tasks solved on general industrial and precision equipment.

The profile rail guides and movable carriage are components of the linear displacement module, driven by an asynchronous motor with a frequency converter.

The linear displacement module is shown in Figure 3.

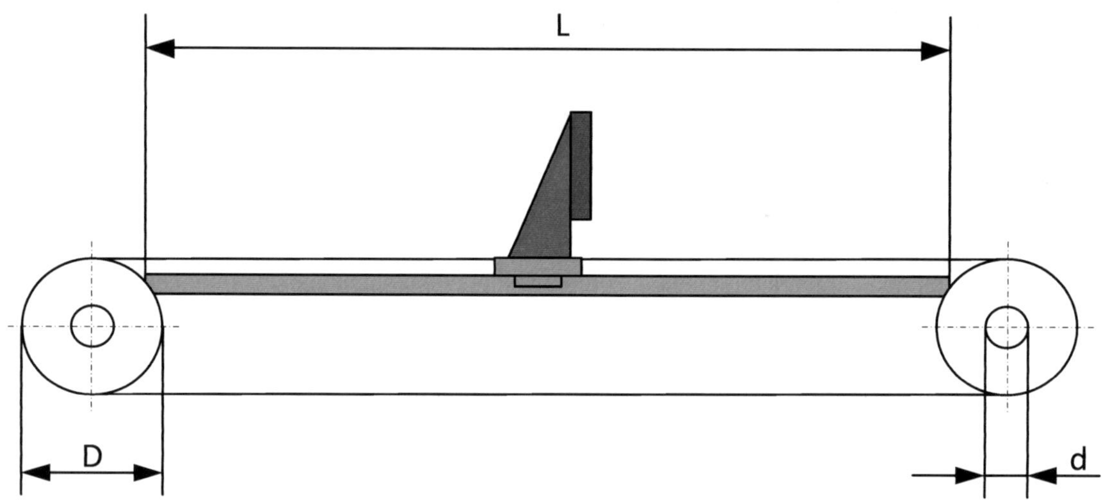

Figure 3. Linear displacement module

The laser sensor is installed on the movable carriage, which measures the distance from the measuring surface of the sensor to the measured spring.

3. Experimental studies of rig measurement system

The geometric dimensions of the spring, as shown in Figure 4, are determined as follows.

The scanning distance obtained between the measurements corresponding to the values L1 and L3, L1'and L3' determines the diameter of the spring rod d1. The distance between the measurements corresponding to the values L1 and L2, L2 and L3, L1'and L2', L2' and L3' determines the radii of the spring bar. In this case, the values of L1, L3, L1', L3' are selected according to the maximum value of the measured distance, while L2 and L2' are selected according to the minimum value of the measured distance to the neighboring spring rods.

The minimum scanned distance to the spring bar closest to the sensor, corresponding to the values of L2 and L2', and the minimum distance to the spring bar, remote with respect to the sensor, determine the outer diameter of the spring D.

The value of the inner diameter of the spring d is determined by the value of the outer diameter D minus the doubled value of the diameter of the bar d1.

The spring pitch is defined as the average scanned distance between the measurements corresponding to the values of L2 and L2'.

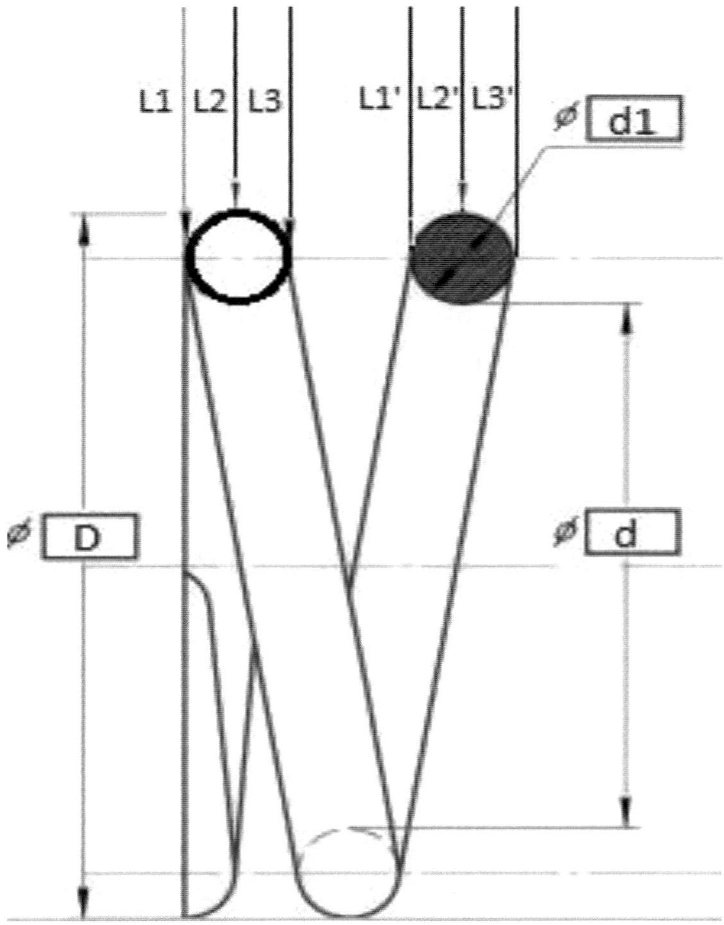

Figure 4. Spring bar diameter measurement

The results of scanning the suspension springs with the optoNCDT 1700 laser-optical sensor are shown in Figures 5 and 6.

In Figure 5, the spring is located on the test rig in a way that its end face is pressed to the right side of the bench.

In Figure 6, the spring is located on the test rig in a free state and is not pressed against the side wall of the bench.

To process the results for the suspension springs scanned by the optoNCDT 1700 laser-optical sensor, a computer program was developed that implements the above-described algorithm for determining geometric dimensions.

The results of processed data obtained during the scanning of the suspension spring by the optoNCDT 1700 laser-optical sensor are shown in Figure 7.

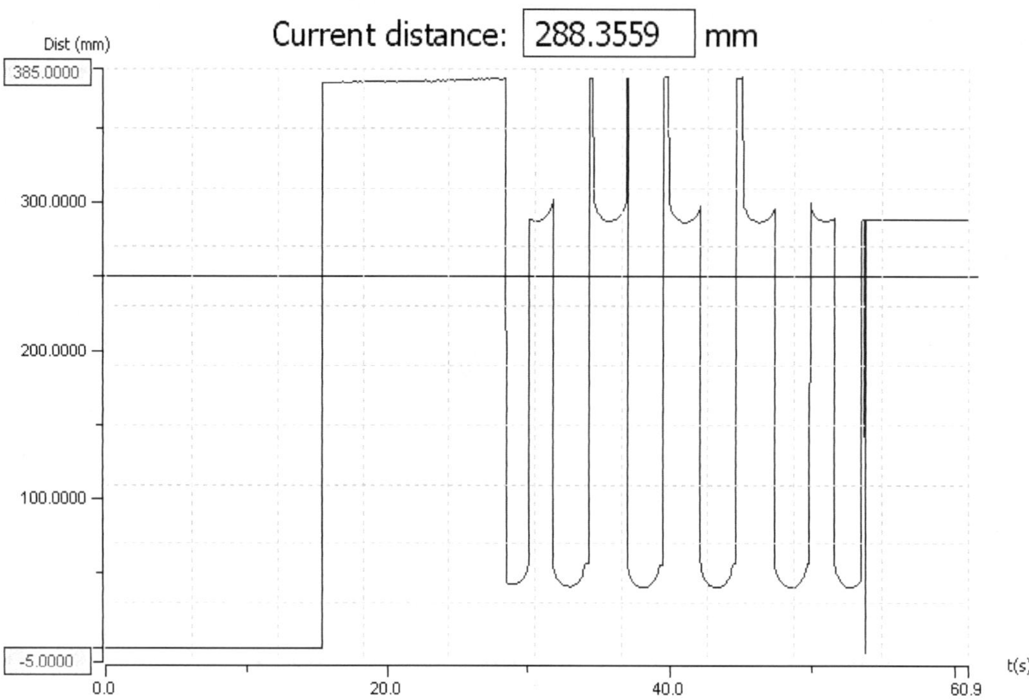

Figure 5. Scanning of spring pressed against the end

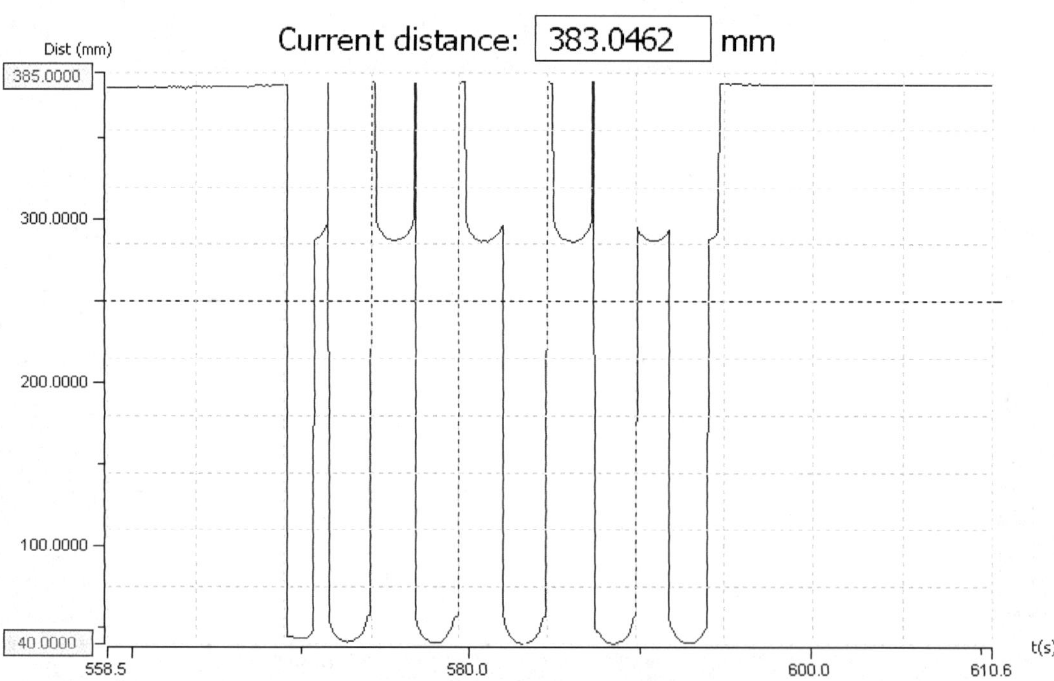

Figure 6. Scanning of spring not pressed against the end

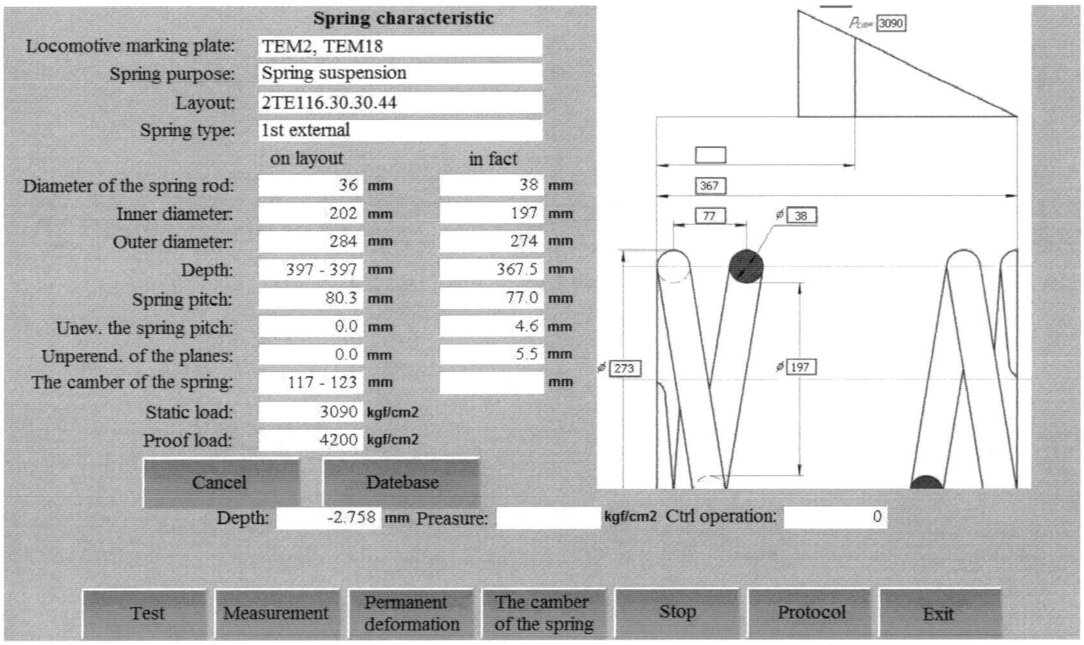

Figure 7. Processed data from the laser-optical sensor

4. Conclusion

By using the optoNCDT 1700 laser-optical sensor, it is possible to measure the geometrical parameters of the spring suspension of locomotives. This ensures high measurement speed, which allows increasing the productivity of spring calibration. The use of a laser-optical sensor also provides high measurement accuracy.

References

[1] Kopanev M V and Kuchera L Y 2015 Novye tehnologii i problemy tehnicheskih nauk, *10th November-2015* 6–11

[2] Kuchera L Y and Kopanev M V 2015 *Organizatsiya bezopasnosti na transporte, Chita* 155–160

[3] Kopanev M V and Kuchera L Y 2018 *Povyshenie effektivnosti proizvodstva i ispolyzovaniya energii v usloviyah Sibiri, 24-28 April- 2018, Irkutsk, Russia,* **1** 147–150

[4] Kopanev M V and Kuchera L Y 2018 *Povyshenie effektivnosti proizvodstva i ispolyzovaniya energii v usloviyah Sibiri, 24-28 April-2018, Irkutsk, Russia* **1**, 150–154

[5] Uzin A N, Arsentiev M O and Gladyshev M A 2017 *Povyshenie effektivnosti proizvodstva i ispolyzovaniya energii v usloviyah Sibiri, 24-28 April-2018, Irkutsk, Russia* **1** 109–113

[6] Wang Z, Lan Q and Wang E 2015/04 *Int. Conf. on Advances in Mechanical Engineering and Industrial Informatics (AMEII 2015),* 2015/04, Atlantis Press 8–15

[7] Zhang B and Tang C 2015 *Int. Conf. on Structural, Mechanical and Material Engineering (ICSMME 2015), 06-08 November – 2015, Dalian, China* 128–132

[8] Lei W, Wang L, Li D and Fang J 2016 *Int. Conf. on Automatic Control and Information Engineering, 22-23 October–2016, Hong Kong* 74-77

[9] Fang S, Su C and Pan Z 2017 *7th Int. Conf. on Manufacturing Science and Engineering (ICMSE 2017), April-2017, Zhuhai, China* 22-27

[10] Dvoynishnikov S, Meledin V, Bakakin G and Rakhmanov V 2016 *Int. Conf. on Mechatronics, Control and Automation Engineering, 24–25 Jul – 2016 Bangkok, Thailand,* https://doi.org/10.2991/mcae-16.2016.42

IOP Conf. Series: Materials Science and Engineering **760** (2020) 012030 doi:10.1088/1757-899X/760/1/012030

Magnetic vibrations in synchronous machines with permanent magnets and fractional slot concentrated winding

D M Toporkov[1], M A Kovtun[1], I L Popov[1]

[1]Novosibirsk State Technical University / Department of Electromechanics, 20, Karla Marksa ave., Novosibirsk, 630073, Russia

E-mail: kovtynaga96@gmail.com

Abstract. Nowadays synchronous machines with permanent magnets are widely used in motor manufacturing due to output parameters. The paper aims to analyse the radial magnetic forces that cause the stator vibration of a synchronous machine with permanent magnets and a fractional slot concentrated winding. The method to assess the vibration parameters of synchronous machine is proposed.

1. Introduction

Vibration is a type of mechanical movement in which each of the points of the body makes a periodically repetitive movement near a relatively fixed position. The scale and nature of the vibration change over time for each case are individual, the analysis of the vibration state of an electric machine is of the greatest interest, the study of which makes it possible to diagnose the technical condition of mechanisms and machines, identify sources of vibration excitation and eliminate many defects that improve the vibration state of an electric machine.

Vibration control solves several problems:

- The prevention of the increased vibration and the intensive wear of the mechanism by vibration
- Diagnosis of dangerous developing defects that can lead to serious damage, and sometimes to the complete destruction of the mechanism.
- Evaluation of the noise level, in order to reduce the impact of the noise level on the person.

Nowadays, permanent-magnet synchronous motors with fractional cog windings with a wide range of advantages are gaining more and more popularity as compared to classical synchronous machines:

- The use of high-coercivity rare-earth magnets based on NdFeB and SmCo with high specific magnetic energy, which makes it possible to manufacture machines with the best mass-dimensional parameters.
- Fraction slot concentrated winding allow in the dimensions of conventional classic low and medium power machines to perform electrical machines with the number of poles reaching 50 or more, resulting in the rejection of the use of the gearbox, cheapening and reducing the weight of the electric drive.
- The use of fractional slot concentrated winding windings reduces copper consumption and axial length of the motor by reducing the length of the end parts, and also simplifies the installation technology, since each coil covers one stator tooth, which reduces the manufacturing costs.

However, in spite of the listed advantages of SMPM (synchronous machines with permanent magnet) with fractional slot concentrated winding, their implementation and widespread use in some areas is limited. This is especially true of high precision drives and those areas where stringent requirements are imposed on vibrations, noise and oscillations of the moment.

Due to the increasing relevance of synchronous motors with fractional slot concentrated winding (SMPM FSCW), due to the advantages described above, there is a need to analyze the vibration state of the electric machine. As the literature review [1-5] showed, the vibration state analysis is provided for machines with classical windings (with the number of slots per pole per phase greater than one), which causes certain difficulties for estimating the vibration level of the PDMS FSCW and designing an electrical machine that satisfies the technical requirements.

2. Problem statement

The literature [2-5], [8] contains a detailed description of the method for analyzing the vibration of electric machines due to the action of mechanical and aerodynamic forces, which is the same for most types of machines. The calculation of the magnetic forces that cause vibration for each type of machine is calculated individually, as a result, the need has arisen to develop a technique for analyzing the vibration state for SMPM FSCW, due to the action of radial magnetic forces.

3. Theory

A. General assessment of the cause of the vibration of a magnetic nature

The cause of vibration is magnetic radial forces acting between the stator and the rotor, which vary in space and time. According to Maxwell, at any point of the air gap a radial force acts, the value of which per unit area can be expressed by the equality:

$$p_r = \frac{B_\delta^2(\alpha,\theta)}{2\cdot\mu_0} \tag{1}$$

Where $B_\delta(\alpha,\theta)$ - the instantaneous value of the magnetic flux density at the point of the air gap, shifted relative to the origin of coordinates by angle α , now of time θ and μ_0 is magnetic permittivity of vacuum.

Magnetic flux density, changing in space and time, corresponds to the radial force arising between the stator and the rotor, which also varies in space and time in a certain way. These forces cause vibrations of individual parts of the machine, and the radial force has the greatest effect on the stator yoke, which is a thin-wall hollow cylinder with the lowest rigidity compared to other parts of the electric machine.

To express the distribution of the radial magnetic forces arising between the stator and the rotor, it is necessary in formula (2) to substitute the magnetic induction in the air gap, which can be found as the product of the sum of the magnetomotive force stator winding and the magnetomotive force of the rotor magnets multiplied by the magnetic permeance of the air gap:

$$B_\delta(\alpha,\theta) = \left[F_S(\alpha,\theta) + F_R(\alpha,\theta)\right]\cdot\Lambda(\alpha,\theta) \tag{2}$$

Therefore, to determine the magnetic flux density in the air gap, and then find the radial magnetic force, it is necessary to determine the magnetomotive force distribution of the stator and rotor fields, as well as the magnetic permeance of the air gap of the electric machine.

Magnetomotive force definition of stator SMPM FSCW described in detail in [6-7], in where the amplitudes and orders of harmonics are calculated for windings with q=$\frac{1}{2}, \frac{2}{5}, \frac{3}{8}, \frac{4}{11}$. Magnetomotive force of each stator harmonics is an infinite series:

$$F_S(\alpha,\theta) = \sum_{\nu=1}^{\infty} F_{m\nu}\sin(\nu\alpha - \theta) \tag{3}$$

The amplitude of the stator magnetomotive force is calculated by the expression:

$$F_{m\nu} = \frac{m\sqrt{2}}{\nu\pi} I_K w k_{y\nu} k_{p\nu} \quad (4)$$

The order of magnetomotive force harmonics is calculated by the ratio:

$$\nu = 3k \pm 1 \quad (5)$$

From the figure in Figure 1 it can be seen that the teeth have a wide range of higher harmonics.

The sum of stator magnetomotive force harmonics creates a resultant magnetomotive force, which determines the occurrence of radial magnetic forces.

The winding coefficient k_o show the reduction of the electromotive force of the distributed winding in comparison with the electromotive force of the winding with the same number of turns. Figure 3 shows a graph of the distribution of a winding coefficient for each harmonic number of magnetomotive force for winding with $q = \frac{3}{8}$.

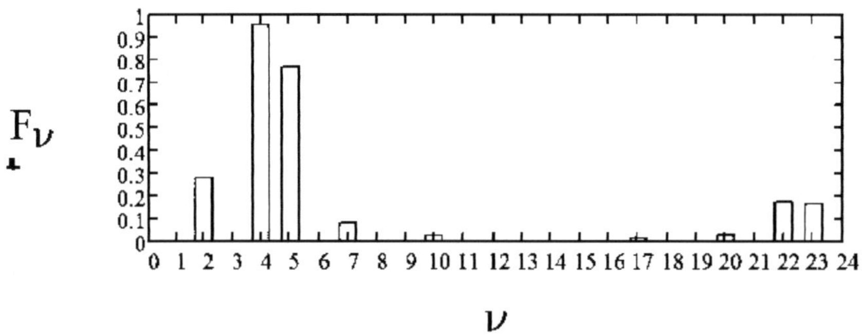

Figure 1. Relative magnitudes of three-phase winding harmonics with $q = \frac{3}{8}$.

Winding scheme with $q = \frac{3}{8}$, shown in Figure 2

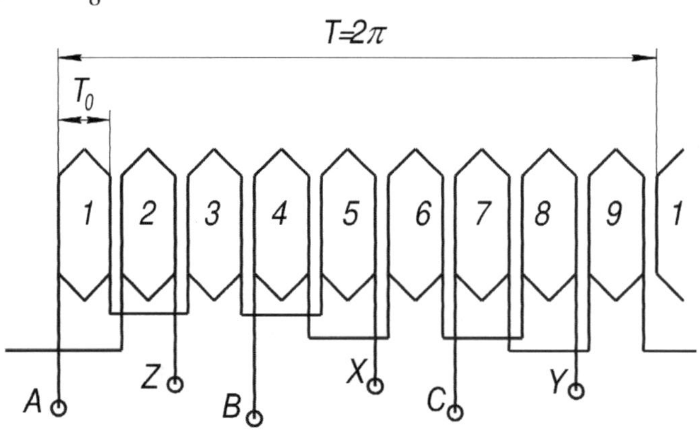

Figure 2. Winding diagram with $q = \frac{3}{8}$.

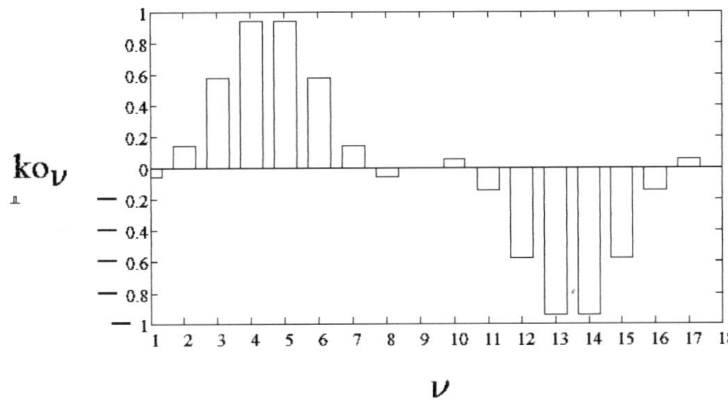

Figure 3. Winding coefficient for winding with $q = \frac{3}{8}$.

B. Determination of the rotor magnetomotive force

The design of the rotor with a surface mounted permanent magnets presented in Figure 3

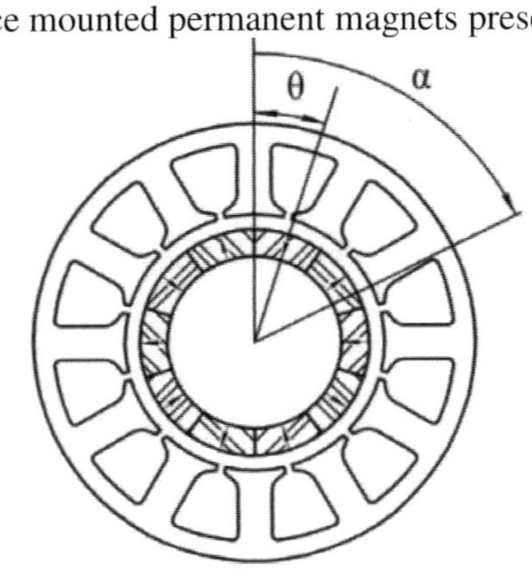

Figure 4. Machine design with radial magnets.

In paper [8], the distribution of rotor magnetomotive force, which in general form represents an infinite harmonic series

$$F_R(\alpha,\theta) = \sum_{\mu=1}^{\infty} F_{m\mu} \sin(\mu\alpha - \theta) \tag{6}$$

Where p - the number of rotor pole pairs

In addition, the distribution of magnetomotive force shown in Figure 4

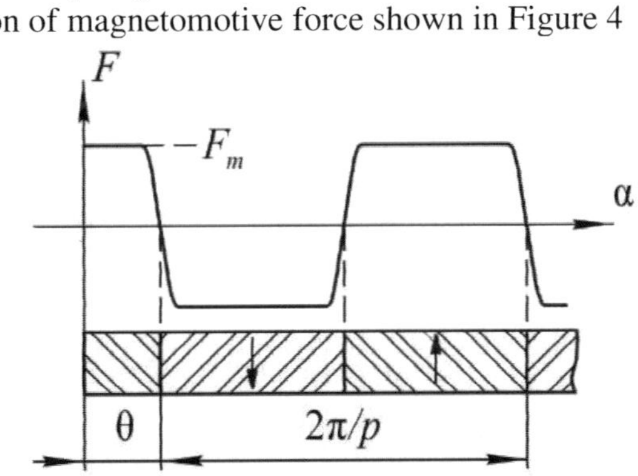

Figure 5. Distribution of magnets magnetomotive force.

C. Determination of air gap magnetic conductivity

If the air gap between the stator and the rotor varies along the inner surface of the stator core of the machine in an arbitrary periodic relationship, then the magnetic conductivity of the air gap, assuming that the rotor around the circumference is smooth, can be represented by an infinite harmonic series:

$$\Lambda(\alpha) = \Lambda_0 + \sum_{v=1}^{\infty} \Lambda_{mv} \cos v z_1 \alpha \tag{7}$$

Where z_1 - number of stator slots

By multiplying the terms of equation (2) and substituting them into equation (1), one can obtain expressions for the harmonics of radial magnetic forces varying in time and space and having different orders. Among which the following can be taken into account from the point of view of creating vibration:

Force waves created by the main field:

$$p_{r1}(\alpha,\theta) = P_1 \cos(2p\alpha - 2\theta) \tag{8}$$

These waves cause vibrations with order

$r=2p$, with frequency $2\frac{\theta}{t}$ and amplitude:

$$p_r = \frac{B_\delta^2(\alpha,\theta)}{4 \cdot \mu_0} \tag{9}$$

Force waves created by the interaction of any pair of higher harmonic fields of the stator and rotor:

$$p_{v\mu} = P_{v\mu} \cos((v \pm \mu)\alpha - (\theta_v \pm \theta_\mu)) \tag{10}$$

With amplitude:

$$P_{v\mu} = \frac{B_v(\alpha,\theta_v) B_\mu(\alpha,\theta_\mu)}{2 \cdot \mu_0} \tag{11}$$

With order $r = v - \mu$ and angular velocity $\dfrac{\theta_v - \theta_\mu}{t}$.

The largest contribution to the occurrence of vibration made by the tooth harmonics of the order $v = Z_1 \pm p$ и $\mu = Z_2 \pm p$, what is described in the paper [8].

4. Experimental results

Vibration measurement was carried out on the motor VDM-80 with,

$$Z = 18, p = 8, q = \frac{3}{8}$$

according to the test program in no-load mode when powered by a frequency converter, without mounting the motor, at the points shown in Figure 5.

The measurement results are shown in Table 1.

Based on the permissible vibration speeds for electric machines, with a height of the axis of rotation up to 132 mm for a category of machines without special vibration requirements, with free fastening, in accordance with the standard with GOST IEC 60034-14, that should not exceed 1.6 mm/s.

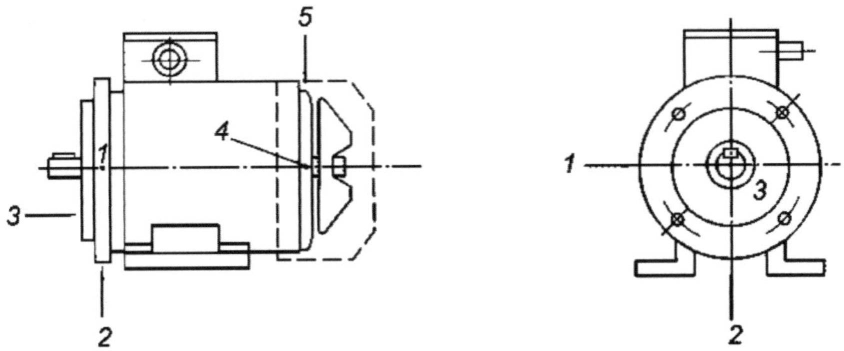

Figure 6. Measuring points of the motor vibration speed.

Table 1. The preset measured values of vibration velocity (mm / s) vdm-80

n (rpm)	1000	1500	3000
Point 1	1,3	0,76	0,91
Point 2	1,0	1,0	1,25
Point 3	0,02	0,07	0,7
Point 4	1,35	0,9	1,7
Point 5	1,0	1,17	2,08

5. Discussion of results

Estimating the vibration state for SDPM SDCs listed in Table 1, one can judge the deviation of experimental results from the permissible values required by GOST standards for vibration velocity values at 3000 rpm. at points 4 and 5. This deviation is unacceptable for the operation of the electric motor and justifies the urgency of applying solutions that reduce the level of vibration velocity to acceptable values.

6. Conclusion

Exploring the expressions for the values of magnetic radial forces for SDPM FSCW, obtained by multiplying the infinite harmonic series in formulas (1) and (2), we can find the amplitudes and frequencies of the forces that cause vibrations, then with the help of constructive methods to neutralize or reduce the magnetic radial force amplitude, and thereby improve the vibration state of the electric machine.

These conclusions will help investigate the influence of vibration due to magnetic forces in SDPM FSCW for more detailed.

References

[1] Balagurov V A, Galteev F F and Larionov A N 1964 Electric machines with permanent magnets (Moscow: Energy)
[2] Goldin A S 1999 Vibration of rotary machines. (Moscow: Mashinostroenie)
[3] Kucher V Ya 2004Vibration and noise in electric machines: written lectures (St. Petersburg: SZTU) 81
[4] Lee S K, White P R 1997 Higher-order time-frequency analysis and its application to fault detection in rotating machinery *Mechanical Systems and Signal Processing* **11 4**. 637-650
[5] Lee S K, White P R 1998 The enhancement of impulsive noise and vibration signals for fault detection in rotating and reciprocating machinery *Journal of Sound and Vibration* **217 3** 485-505
[6] Bukhgolc U G, Komarov A V, Shevchenko A F, Shevchenko L G 1996 Multipolar synchronous machines with fractional slot windings part 2 (Novosibirsk: NGTU)

IOP Conf. Series: Materials Science and Engineering **760** (2020) 012030 doi:10.1088/1757-899X/760/1/012030

[7] Bukhgolc U G, Komarov A V, Shevchenko A F, Shevchenko L G 1996 Multipolar synchronous machines with fractional slot windings part 1 (Novosibirsk: NGTU)
[8] Heller B, Hamata V 1977 Harmonic field effects in induction machines (New York: Elsevier)

IOP Conf. Series: Materials Science and Engineering **760** (2020) 012031 doi:10.1088/1757-899X/760/1/012031

Types and methods of application of information technologies in the transport industry of Siberia

O A Krasovskaya and A E Chigir

Irkutsk National Research Technical University, 83, Lermontov st., 664074, Irkutsk, Russia

E-mail: chigir-1981@mail.ru

Abstract. Transport in Siberia is an integral part of human life. It is one of the important sectors of the national and regional economies. It is necessary to know how transport technologies have evolved. Development of technologies accelerates the economic growth of the region, establishes strong economic ties with developed regions of the country. The regions with a developed transport infrastructure have the most useful conditions for the economic growth. This article aims to describe the development of the transport system in Siberia. It describes the development of information technologies applied in the transport industry of Siberia. Application of the GLONASS navigation system is analyzed by the example of Irkutsk region.

1. Introduction

Currently, information technologies are being modernized and developed. They solve many problems, make the world more comfortable and better. Information technologies are used to optimize information processes, process information which is important in the market conditions.

Modern technologies are used to process a large amount of information. In the transport industry, information technologies are used to manage transport processes. The article describes main technologies used in the transport industry.

The article describes the transport system of Siberia, development of technologies applied in the transport industry of Siberia.

2. Research purpose

The article aims to analyze types and methods of application of transport information technologies, the GLONASS navigation system in Irkutsk region as part of the program "Safety of school transportation". There are different types and methods of application of information technologies in the transport industry. The most widespread ones are GPS9 with three types of sensors, car alarms, the GLONASS.

3. Application of information technologies

The development of transport information technologies is problematic. They are expensive. Therefore, they are rarely used by transport organizations. This causes transport problems. Violation of traffic rules is one of the main transport problems which can be solved using transport technologies. One of the main problems is transportation of schoolchildren. It is difficult to implement the target program of Irkutsk region "Safety of school transportation".

4. Technologies applied in the transport industry

The most widespread technology applied in the transport industry is the GPS.

When using the GPS, sea and road navigation is performed. It also supports the TPMS. It can measure temperature and tire pressure for safe driving. [2]

The GPS has a rearview camera. The camera automatically switches the image from the rearview camera if you turn on the reverse gear. It expands a visibility area and ensures safety.

The GPS has a function of satellite monitoring of transport. The GPS monitors location and speed. To facilitate the control, a parking radar technology is used. It is used to facilitate parking maneuvers, alert the driver about barriers in the blind zone.

Next to the obstacle, it produces a sound and informs about the distance. The sounds are heard at a distance of 1–2 m. At a distance of 0.1–0.4 m, the sound is disturbing. The radar may be activated. The radar is operated along with the rear gear. The radars come into action at a speed of 20 km/h.

The display unit is connected to the sensors through wires installed in the vehicle body. There are wireless sensors [3]. There are three types of radar sensors:

1. The ultrasonic sensor. The principle of its operation is similar to that of an echo sounder: the sensor produces a short ultrasonic jolt (about 40 kHz) and receives the sound reflected by nearby objects. The electronic unit determines the time before radiation and reception of the reflected sound and measures the distance to the object.

Ultrasonic sensors are installed on the bumper, and the concentrated force of emitters creates a sensitivity zone. Initial devices had different emitting and receiving sensors. Only one piezoelectric element transmitted signals. The high price of the material allowed for only mechanical scanning of the rear hemisphere with a single sensor. For blind zones, additional devices are used. Ordinary sensors have only two devices. The most famous sensors use only four devices installed on the rear bumper at a distance of 30-40 cm from each other. This kind of location eliminates "blind zones". In the most complex sensors, two or four devices are located on the front bumper. Special sensors are able to use a large number of devices.

2. The electromagnetic sensor converts electromagnetic waves in the middle of two conductor antennas installed along the body. The RF signal is transmitted to the first conductor, and the second one is connected to an amplitude detector. All changes in the position of objects near the antenna zone influence the level of communication among the conductors and change the amplitude of the signal.

This type of the PDC (DP1) was applied by the Audi company. The main advantage is the lack of external violations of the body design after application of plastic bumpers: antenna sensors are installed inside the bumper.

Electromagnetic Parktronic devices are used in the USA and Europe. In Russia and the CIS, ultrasound Parktronic devices are used.

3. The lidar. The principle of operation is the same as that of the "Side Assist" type systems: after activation of the signal, the indicator light is activated.

Unmanned vehicles and auto driving systems require monitoring of the area around the car. The best way out is scanning laser range finders (lindars). They build a three-dimensional projection near the vehicle at a distance of one hundred meters. Their price is high [6–8].

Although they help drivers, it is not recommended to rely on them. The driver must monitor obstacles. Some objects are not detected by the radar; there are those ones which can cause the radar to be switched on incorrectly [4].

The radar signals when:

- there are ice, snow or other contaminants;
- moving along the unpaved sloping road;
- moving along the rough road;
- there is much noise;
- raining or snowing;
- radio transmitters are switched on;
- towing;

IOP Conf. Series: Materials Science and Engineering **760** (2020) 012031 doi:10.1088/1757-899X/760/1/012031

- parking in cramped conditions.

The radar does not respond to:

- small objects;
- ultrasonic absorbing objects;
- objects reflecting sounds.

The radar does not detect road holes, small and other dangerous objects that are in the blind zone. There are radars that control the zones around the devices allowing the driver not to encounter vehicles [5].

There is one more transport technology. It is car alarms.

The car alarm system prevents thefts of car parts and car thefts. It informs the owner and people near the vehicle with a sound or light, but it does not block a car theft or a theft of spare parts. There are different types of car alarms. The one-way car alarm is the simplest and cheapest system. It performs a limited number of functions: engine locking and door controlling. There is a minimum distance between the vehicle and the key fob for the system to function properly. As for the two-way car alarm system, information about the engine locking is received by the key fob. Figure 1 shows the diagram of the electronic unit of the car alarm which is not expensive. The diagram shows its simplicity and efficiency. When this system is switched to the "Security" mode, the consumption of electric current does not exceed 180 mkA. The SA1 switch can be installed anywhere in the car. The R8 resistor and the C5 capacitor are responsible for the delay time which can be changed by selecting nominal values of these elements.

Car alarms can have or lack a feedback; they can be satellite and GSM ones. They consist of a main unit, a receiver-transmitter, a key fob, a shock sensor, a terminal for turning off the light, a limit switch for the engine compartment of the service button, and an indicator. The alarm system with a feedback informs about the state of the vehicle. It protects against thefts [9].

It does not provide full protection against thefts; however, it reduces their probability [10].

More expensive alarm systems can be connected to GSM / GPRS modules to control the alarm from the phone via SMS and receive alerts. Using this module, one can determine the location of a car (GPS / GLONASS): this information is transmitted to the key fob, telephone and website of the manufacturer. The GSM module can call recorded phone numbers, diagnose messages, notify the movement of vehicles in restricted areas, etc. There are devices used to hack vehicles.

To hack car alarm systems, code grabbers can be used (static, suppression and algorithmic) [13].

A) For static and suppression grabbers, the alarm can intercept the code.

B) Code-grabbers use an algorithm of actions required for the button to be pressed by the owner. They use radio jamming and interception of key fob signals.

C) The algorithmic code grabber identifies the alarm system based on the digital signal of a key fob and becomes a complete copy of the key fob. This method is used by car alarm systems applying KeeLoq and other algorithms to encrypt the signal from the key fob to the main signaling unit.

There are additional security measures.

A special code-security method created by the Magic Systems company is used to identify key fobs used in cryptography.

After receiving the signal, it makes sure that it was sent from its" key fob. After responding to the debut signal, it sends a request in the form of a random number. Then it is processed using an individual algorithm and sent back. The car alarm system processes its signal using the same algorithm. And if they are the same, the order is performed, and the alert is sent to the key fob. An interactive code is used for auxiliary protection against electronic hacking. There is an immobilizer which is an electronic anti-theft device [14].

Immobilizers are used for safe blocking of vehicle modules. They prevent car thefts.

The engine is blocked. If the car is stolen, it prevents it from moving. A good immobilizer will be a significant problem for hijackers.

The car owner is obliged to control the immobilizer. An electronic code key is used; there are models with manual codes. Before starting the engine, the driver inserts a code key into the slot while the immobilizer reads the code and disables the lock. The driver can place a contactless electronic "tag" into the immobilizer area [11]. For manual immobilizers, the code is entered by the owner.

If an immobilizer breaks down, it remains blocked.

Most of the immobilizers automatically switch to the protection mode when the owner takes no actions.

Immobilizers are divided into contact, contactless and regular ones.

Regular immobilizers are used. They do not have relays. This system is closely connected with the engine fuel injection controller. If there is no authorized signal, the controller will not start the engine. The engine is blocked automatically.

The regular immobilizer consists of two main parts:

- the control unit which signals the need for switching on the system;
- the key identified by the control unit. Sometimes it is activated after a password is entered.
- the auxiliary immobilizer has an electromagnetic relay where an electrical circuit breaks down after penetration into the car.

There are immobilizers operating without breaking electrical circuits (CAN immobilizers). The engine is blocked after sending messages by means of interferences. After application of auxiliary devices, it is permissible to block non-electrical systems.

Immobilizers are installed in the electronic zone. They are small which reduces probability of being detected [12]

Ninety-two percent of all Siberian companies use information technologies; over 800 thousand people have already been registered on a single portal of public services.

The navigation system GLONASS is used in the transport industry. It is well known. It was ordered by the Ministry of Defense of the USSR. This system is used for remote control and exact localization of data. The GLONASS differs from the GPS. GLONASS satellites are not synchronous with the Earth's rotation, thereby ensuring stability for themselves. They do not require additional adjustment. The basis of this system is 24 satellites moving over the surface of the Earth in three orbital planes. The GLONASS satellites are located in the medium-altitude circular orbit at an altitude of 19,400 km with an inclination of 64.8° and a period of 11 hours and 15 minutes. This orbit can be used for high latitudes (northern and southern regions), where it is difficult to catch GPS signals. The satellite constellation is deployed in three orbital planes with 8 equally distributed satellites. For the global coverage, 24 satellites are used. To cover Russia, 18 satellites are used. Signals are transmitted at a 38 ° directivity using right-handed circular polarization at a power of 316–500 W (ERP 25–27 dBW). In order to determine the coordinates, the receiver must receive a signal from four satellites and calculate the distance to them. If three satellites are used, it is difficult to determine the coordinates due to errors caused by the inaccuracy of the receiver clock. In 2012, the Ministry of Transport of Russia defined technical requirements for the satellite navigation equipment in order to enhance safety of vehicles transporting passengers and dangerous goods.

The GLONASS and GPS are widely used in Irkutsk region. They are used for implementing a long-term target program "Safety of school transportation". The program is aimed at enhancing safety of schoolchildren transportation. The main threat is violation of traffic rules by school bus drivers. The GLONASS allows the dispatching service to monitor the situation in the bus cabin and prevent emergency situations. The driver has an alarm button on the panel and a microphone. In case of pressing this button, the dispatcher can promptly call emergency services. The GLONASS allows dispatchers to track the route and follow traffic rules. The integrated use of navigation and information technologies enhances safety of children and improves efficiency of school bus monitoring. According to the Ministry of Education of Irkutsk Region, all vehicles transporting children are equipped with the GLONASS system, tachographs and meet all the requirements for children transportation. 461 buses have been purchased. 22,000 schoolchildren from 910 villages are transported to schools.

5. Research results

Information technologies applied in the transport industry were described. It is recommended to develop car alarm systems. The car alarm systems have to be improved in order to reduce the number of car thefts, enhance safety of passenger and cargo transportation.

The GLONASS navigation system used for vehicles transporting schoolchildren is expressive. It is being slowly implemented in Russia. However, the system is important.

6. Conclusion

Information technologies should be used in the transport industry. The correct structure of information flows increases efficiency of routine operations when making management decisions. To process information flows, an information system of an enterprise or a process is needed. The main IT concepts are object selection rules and their quality, principles of interconnectedness between objects and the sequence of structure.

The main technologies used for reading information prepare and present data for making management decisions and performing accounting tasks. The public automated system controlling the transportation of schoolchildren using the GLONASS will solve several important tasks.

Transport information technologies will be developed in close relations with business processes in order to obtain accurate and timely data for operators and users of vehicles. The main functions that stimulate the development of information technologies are as follows: development of transportation logistics, container transportation, delivery of materials, etc.; improvement of the quality of passenger transportation and reduction of transportation costs.

Currently, computer chips used for traffic management systems are being developed. Car owners will be able to exchange information about the road surface or maneuvers. Interlocking devices that do not allow the vehicle to start when the owner is intoxicated are being implemented. A device transmitting signals to the police when an airbag is triggered, an accident has occurred or a vehicle is stolen is being developed.

Information technologies used in the transport industry of Siberia will determine its social and technical development.

References

[1] Yutt V E 1983 *Electrical and electronic equipment of cars* (Moscow: Transport)
[2] Kako N and Yamane Ya 1986 *Sensors and microcomputers* (Leningrad: Energoatomiz dates)
[3] Titze U and Shenk K 1982 *Semiconductor circuitry* (Moscow: Mir)
[4] Horowitz P and Hill W 1984 *The art of circuitry* vol 2 (Moscow: Mir)
[5] *Reference book of the radio amateur designer* 1990 (Moscow: Radio and communication)
[6] Vigleb G 1989 *Sensors: device and application*
[7] Osipovich L A 1979 *Sensors of physical quantities*
[8] Friden J 2005 *Modern sensors. Reference book* (Moscow: Technosphere)
[9] Aleinikov A F, Gridchin V A and Tsapenko M P 2001 *Sensors. Perspective development directions* (NSTU)
[10] Kotyuk A F 2006 *Sensors in modern measurements* (Moscow. Radio and communications)
[11] Pinsky F I, Davtyan R I and Chernyak B Ya 2002 *Microprocessor control systems for automotive internal combustion engines* (Moscow: Legion-Avtokada)
[12] *AvtoVAZ: Technology of repair of bodies and body parts* 1987 ed by B V Prokhorov (Leningrad: Mashinostroenie)
[13] Lavrus V S Series *Informational Edition* iss 1 *Security systems* (Moscow: Science and Technology)
[14] Mortise electromechanical locks controlled by ACS. Types and characteristics of locks *Arm-systems*
[15] Bargagli B, Manes G, Facchini R and Manes A 2012 Acoustic sensor network for vehicle traffic monitoring *Proceedings of the First International Conference on Advances in*

Vehicular Systems, Technologies and Applications (Venice, Italy) 24–29 June

[16] Ahmad F, Basit A, Ahmad H, Mahmud S A, Khan G M and Yousaf F Z 2013 Feasibility of deploying wireless sensor based road side solutions for Intelligent Transportation Systems *Proceedings of the 2013 International Conference on Connected Vehicles and Expo* (Las Vegas, NV, USA) 2–6 December, pp 320–6

[17] Mehrabi A and Kim K 2015 Using a mobile vehicle for road condition surveillance by energy harvesting sensor nodes *Proceedings of the 2015 IEEE 40th Conference on Local Computer Networks* (Clearwater Beach, FL, USA) 26–29 October pp 189–92

Information systems of passenger transportation forecast in Transbaikal region

L Y Kriklevskaya, O K Kuklina, E A Mihailova and A S Yahina

Chita Institute of Baikal State University, 56, Anokhin str., Chita, 672000, Russia

E-mail: lara.kriklevskaya@mail.ru

Abstract. The paper is concerned with the qualitative assessment of socio-economic development of transport industry of the region made with a help of information systems and mathematical methods. The object of this study is a railway transportation system in Transbaikal region. The research was made on the basis of a priori and posteriori approach with the use of recessionary analysis and correlation. The research of basic trends in socio-economic development of Transbaikal region during a long period of time (10 years) comprising the periods of growth and stagnation made it possible to reveal a set of models which affects the volume of transportation. This will help to forecast a strategy of development of Transbaikal region. The article contains the results of empirical study on interconnection between a set of factors and a volume of passenger transportation in Transbaikal region.

1. Introduction

An automated system of operative management of the transportation process (ASOUP) is used on each road of the public limited company “Russian Railway”. Its creation and adoption became a basis of computer network on the rail roads of the country.

Information about all operations made with the models of transportation process is transmitted directly from the place (a station, a locomotive depot, a wagon depot, a loading area, etc.). This information can be generated in the corresponding AWP (automated working place) of the workplace. Further, the information should be transferred to the ASOUP complex by means of the DTN (data transportation network).

The users can obtain information from the ASOUP-2 complex, the databases of which are stored in the DB2 database system that supports SQL queries, through WEB applications developed by the Research and Design Institute of “Russian Railways”, as well as by the ITC staff.

To develop projects in the Chita Information and Computing Center we adopted the technology of JSP-pages for displaying data and the technology of "data loaders" for downloading information into their databases or application tables. The function of the “loaders” is as follows: at a certain point of time, the Main Frame starts a program written by ITC staff, which selects the information of interest and loads it into the appropriate tables, and also clears outdated information after the expiration date. The ‘loaders’ writing languages are Assembler and C.

In this work, we used real data obtained by a request from ASOUP. Then to the data obtained with the Visual Basic programming language, we applied operations previously recorded with macros. The operations were the utilities of MS Excel “Analysis Package”: “Data Analysis” and “Regression”.

Taking into account the experience of state regulation in different fields of economy and in different types of economic activity it is possible to suppose that the strategy of social and economic perspective development should be based on forecasting and modeling method [1, 2]. The creation of some alternative ways of development based on the analysis of retrospective data, as well as the study of endogenous (with weak state intervention) and exogenous (with strong state regulation) within and inter-sectoral relations, allows one to create scenarios of future development.

2. Results

To make a multifactor forecasting of the total passenger transportation in the Transbaikal region a list of the most significant factors was formed.

A dynamic analysis of the change in the total identified factors for the period of 2000-2016 also was done.

Statistical data provided by the Federal State Statistics Service of Transbaikal region were consolidated into one table (table 1); Pr – time period (years) monitoring (1 – 2000, 2 – 2001, …, 17 – 2016).

Table 1. Indicators of multifactorial forecasting of the total passenger traffic in the Transbaikal region for 2000 – 2016

Pr	I_1	I_2	I_3	I_4	I_5	I_6	I_7	I_8
1	104	457.4	74.8	1185.8	67	1405.7	173747	15179.9
2	108.3	423.2	92.6	1178.8	55.3	2011.5	172515	15512.9
3	113.3	503.69	81.4	1165.4	44.6	2993.1	190116	15475.9
4	105.2	440.3	101.8	1152.6	32.3	4017.7	192580	15174.9
5	97.6	453.2	147.1	1142.7	28.8	4808.5	205126	15368.9
6	90.5	516.8	131.1	1132.8	26.2	5907.6	209874	15355.9
7	91.7	505.9	179	1123.3	23.8	7119.2	220909	29560.9
8	87	489.8	232.7	1115.1	23.9	8265.4	232363	27051.7
9	91.1	465.8	258.6	1111.4	20.1	11047.2	264982	26935.9
10	87.8	479.6	269.1	1109.2	19.9	12704.8	283759	27080.7
11	83.2	470.5	274.8	1109	19.3	14204.5	289329	26786.7
12	79.8	478	276.9	1106.2	18.9	15968.8	302968	26414.4
13	70.5	475.6	302.9	1099.4	17.6	17545.6	316405	26996.4
14	71.6	479.4	295.1	1095.2	16.2	19886.1	344076	26750.4
15	78	481.5	352.5	1090.3	18	20520	355349	26762.4
16	71.1	480	235	1087.5	20.4	22980.2	359860	26444.4
17	92.9	477.5	291.7	1083	21.4	22846.4	368297	26519.4

where:
I_1 – the total passenger traffic, mln. people;
I_2 – the employed population, thousand people;
I_3 – the building commissioning, thousand sq.m.;
I_4 – the population size, thou people;
I_5 – the number of the population with the incomes below the subsistence level, thou people;
I_6 – the population's average cash income, ruble;
I_7 – the number of transport involved in the transportation process, el.;
I_8 – the extent of railway line, km.

Dynamic analysis of the indicators states the presence of ambiguous trends in their change. For example, the total of passenger traffic for the period from 2000 to 2016 decreased by 12 million people per year. At the same time, positive trends in this indicator were observed until 2003, from 2004 to 2016 there was a general trend of a decrease in the total value of passenger transportation, however, within the analyzed period, the decline in the level of this indicator is replaced by a slight growth.

The number of vehicles involved in the transportation process during the research period has significantly increased. This trend is possible to explain by sharp reduction in the capacity of transport, i.e. a passenger transportations are carried out by small transport of Gazel type now.

Dynamic analysis of the selected factors for the research revealed ambiguous trends of change over the analyzed period, what in turn can have a negative impact on the formation of the model [3].

Then a matrix of sample values of the pair correlation coefficients was constructed (table 2).

Table 2. The values of paired correlation coefficients

	I_1	I_2	I_3	I_4	I_5	I_6	I_7	I_8
I_1	1.00							
I_2	**-0.30**	1.00						
I_3	-0.84	0.22	1.00					
I_4	0.85	-0.38	-0.93	1.00				
I_5	0.77	-0.41	-0.82	0.91	1.00			
I_6	-0.83	0.22	0.89	-0.93	-0.73	1.00		
I_7	-0.82	0.22	0.90	-0.93	-0.74	1.00	1.00	
I_8	-0.75	0.34	0.87	-0.85	-0.75	0.75	0.75	1.00

The factor I_2 from the whole set of factors (table 2), based on the analysis of the matrix of sample values of the pair correlation coefficients (table 3), is excluded as it slightly influences the factor I_1 (in table 4 the correlation coefficient is in bold).

Let us estimate the parameters of this model using the method of least squares. The values of the estimates of the model coefficients, their standard errors and the actual values of t-criteria are calculated and presented in table 3.

Table 3. Values of estimates of the coefficients of the model

n=17	Parameter values	Standard parameter errors	Actual values of the t -criteria	P-value
b_0	199.45102	532.61475	0.37448	0.71587
I_3	-0.11679	0.08033	-1.45378	0.17666
I_4	-0.19870	0.46193	-0.43015	0.67621
I_5	0.30169	0.50777	0.59415	0.56561
I_6	-0.00843	0.00433	-1.94627	0.08024
I_7	0.00086	0.00049	1.73030	0.11426
I_8	0.00001	0.00077	**0.01000**	0.99222

All the factors in model, except I_6, are statistically insignificant (at a level of significance 5%). However, all factors should not be deleted at the same time. It is possible that their insignificance is due to the influence of the “worst” of the insignificant factors, and in the next step of calculations these factors will turn out to be significant. So, the factor I_8, which is characterized by the smallest value of t-criteria should be deleted. In this case we get the following statistics (table 4).

Table 4. Values of coefficients estimates of the corrected model for the factor “The extent of railway line”

n=17	Parameter values	Standard parameter errors	Actual values of the t -criteria	P-value
b_0	202.39519	423.11676	0.47834	0.64177
I_3	-0.11634	0.06363	-1.82831	0.09472
I_4	-0.20102	0.38111	**-0.52745**	0.60835
I_5	0.30391	0.43555	0.69775	0.49982
I_6	-0.00842	0.00405	-2.07811	0.06190
I_7	0.00085	0.00045	1.91402	0.08198

Further, for the same reasons, we successively delete the factors I4 and I5 from the model. So, get the following statistics (table 5).

Table 5. Values of coefficients estimates of the corrected model for the factors "The population size" and "The number of the population with the incomes below the subsistence level"

n=17	Parameter values	Standard parameter errors	Actual values of the t -criteria	P-value
b_0	-22.06281	60.69390	-0.36351	0.72207
I_3	-0.12204	0.04518	-2.70101	0.01816
I_6	-0.00820	0.00340	-2.41333	0.03129
I_7	0.00088	0.00039	2.22078	0.04475

The model with the estimated parameters has the form:

$$I_1 = -22.06281 - 0.12204 \cdot I_3 - 0.00820 \cdot I_6 + 0.00088 \cdot I_7. \quad (1)$$

The graph of the observed values and predicted values from model (10) of the total of passenger traffic is in figure 1.

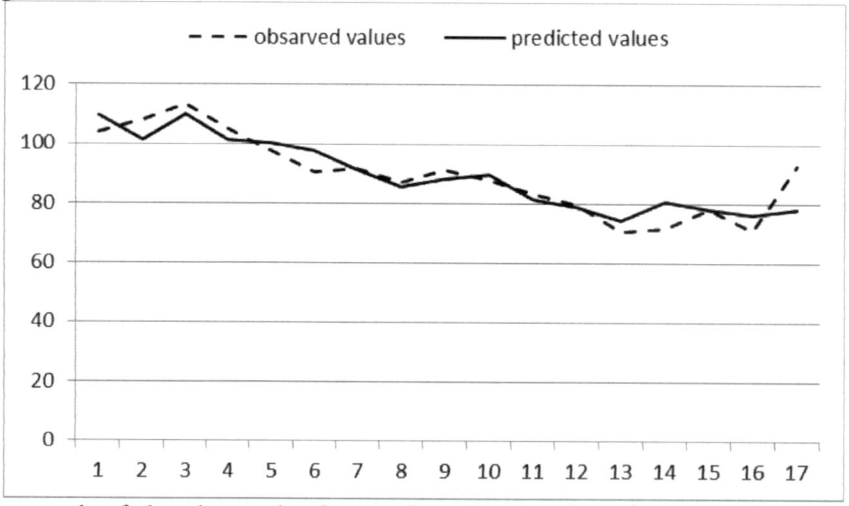

Figure 1. The graph of the observed values and predicted values from model (10) of the total of passenger traffic

The multiple linear regression model (1) approximates the statistical data.

The next stage of the research is the verification of the adequacy of the multiple linear regression model.

IOP Conf. Series: Materials Science and Engineering **760** (2020) 012032 doi:10.1088/1757-899X/760/1/012032

Using Student t-test and Fisher F-criteria to check the statistical reliability of the regression coefficients we get the assumptions of residuals ε_i – they are independent random variables, their mean value is 0, they have the same variance and are subject to a normal distribution.

After constructing the regression equation we check the estimates ε_i of those properties that were supposed.

This is due to the fact that estimates of regression parameters should be unbiased and effective.

The adequacy of the models is based on the following criteria described above:

– peaks criteria;

– criteria based on the coefficient of autocorrelation;

– RS-criteria.

The homoscedasticity of the variance of residuals due to the small sample size is visually estimated according to the graph of the residues represented in the time sequence. If the residuals on graph are in the form of a horizontal strip then we can say that the time effect does not influence the data (table 6).

Table 6. Results of the model adequacy check (1):

Criteria for verification of adequacy			
Peaks criteria (in a random series must be $p > 6$)	The homoscedasticity (is estimated by the graph of the residues e_i, picture 2)	Criteria based on the coefficient of autocorrelation (the hypothesis of autocorrelation is rejected if $\|r_1\| < 0.299$)	RS - criteria (the hypothesis of a normal distribution is accepted if $3.18 < RS < 4.49$)
9	Yes	0.082	4.17

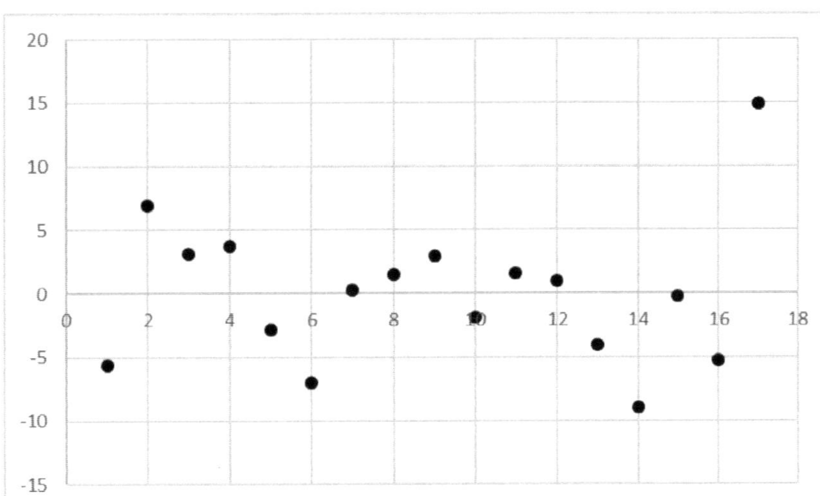

Figure 2. The graph of the residues

The adequacy check showed that the multiple linear regression model (1) is adequate and can be used to forecast total passenger traffic. In this regard, we will not check non-linear regression models.

To forecast factors it is proposed to approximate time series by one of the types of functions:

- polynomial function: $\psi(t) = a_0 + a_1 t + a_2 t^2 + \dots + a_{n-1} t^{n-1}$; (2)
- exponent: $\psi(t) = ab^t$; (3)
- logarithmic parabola $\psi(t) = ab^t c^{t^2}$. (4)

Using a variety of functions to approximate time series allows us to improve the quality of forecasting.

IOP Conf. Series: Materials Science and Engineering **760** (2020) 012032 doi:10.1088/1757-899X/760/1/012032

The least squares method is used for estimating the parameters of models, when the values of the parameter vector minimize the function [4-7]:

$$S(\alpha)=\sum_{t=1}^{m}(y_t-(\alpha,\psi(t)))^2 . \quad (5)$$

The trend model of the time series is considered to be adequate if it correctly reflects the systematic components of the time series. This requirement is equivalent to requiring that the residual component $e(t) = y_t - \psi(t), t = 1, 2, \dots, m$ satisfies the properties of the random component of the time series:

1. randomness of the residual level sequence;
2. the correspondence of the distribution of the random component to the normal distribution;
3. the equality of the mathematical expectation of the random component to zero;
4. independence of the values of the levels of the random component.

The results of experiments of modeling to forecast factors are presented in Table 7 (the values in parentheses are true values). The estimates of model parameters are given in table 8.

A retrospective forecast for 2016 is used according to data from 2000 to 2015 to estimate forecast properties of models.

The best model was chosen as a result (forecast factors for 2016 for this model are shown in bold in table 7).

Table 7. Forecast of factor values for 2016

№	The name of the model (the method of estimating the parameters)	I_3 **(291.70)**	I_6 **22846.40)**	I_7 **(368297.00)**
1	parabola of the second power	305.25	25492.39	390676.91
2	parabola of the third power	225.81	24280.86	**373783.25**
3	exponent	431.67	36440.11	393557.15
4	logarithmic parabola	**281.01**	**22575.12**	396325.41

Table 8. Model parameter estimates

№	The name of the model	Parameter	I_3	I_6	I_7
1	parabola of the second power	b_0	8.61	218.70	158680.41
		b_1	34.00	772.05	7794.76
		b_2	-0.97	42.04	344.24
2	parabola of the third power	b_0	88.05	1430.22	175574.07
		b_1	-14.83	27.30	-2590.17
		b_2	5.99	148.31	1826.14
		b_3	-0.27	-4.17	-58.11
3	exponent	a	78.88	1768.66	157911.61
		b	1.11	1.19	1.06
4	logarithmic parabola	a	51.35	1095.71	159022.35
		b	1.28	1.40	1.05
		c	0.99	0.99	1.00

An adequacy check was carried out for the selected factor models. The selected models have been tested for adequacy for all four criteria and will be used to forecast factors.

The forecast value of passenger transportation I_1 for 2016 is determined by substituting forecast values of factors in the equation of multiple linear regression (1). True value of $I_1 = 92.90$ (table 1).

The calculated values of the model (7) are:
I_1=89.30, absolute forecast error 3.6, relative forecast error 3.9 %.

3. Conclusions

The developed model will be used to forecast the volume of passenger traffic in the Transbaikal region in the aspect of updating the "Strategy of social and economic development of the Transbaikal region until 2030" in conditions of ensuring the competitiveness of the region transport system.

4. References

[1] Shiryaev D V, Litvinenko I L, Rubtsova N V, Kolpak E P, Blaginin V A and Zakharova E N 2016 Economic clusters as a form of self-organization of the economic system *International J. of Economics and Financial Issues* **6(S1)** 284–8

[2] Bulatova N N, Korodyuk I S and Sandakova N Y 2015 Model of innovative development of regional transport infrastructure *Int. J. of Applied Engineer. Res.* **10(20)** 40834–40

[3] Tarsitano A and Lombardo R A 2013 Coefficient of correlation based on ratios of ranks and anti-ranks *J. of Economics and Statistics* **233 (2)** 206–24

[4] Uvarov N D, Kuznetsov M P, Malkova A S, Rudakov K V and Strijov V V 2018 Selecting the Superpositioning of Models for Railway Freight Forecasting *Moscow University Computational Mathematics and Cybernetics* **42(4)** 186–93

[5] Gorbatenko E O A methodological approach to the evaluation and prediction of bankruptcy in the region 2018 *European Social Sci. J.* **4** 167–76

[6] Fedotov D Y 2017 Forecasting of regional economic development and budget based on the example of Irkutsk oblast *Studies on Russian Economic Development* **28(4)** 416–22

[7] Krakovsky Y M, Luzgin A N and Mikhailova E A 2018 Interval forecasting of cyberattack intensity on informatization objects of industry using probability cluster model *J. of Physics: Conf. Ser.* **1015** 032088

IOP Conf. Series: Materials Science and Engineering **760** (2020) 012033 doi:10.1088/1757-899X/760/1/012033

Reactive power sources control in railroad power supply systems

A V Kryukov [1, 2], **A P Kutsyi** [1], **A V Cherepanov** [1]
[1] Irkutsk State Transport University, 15 Chernyshevsky Str., Irkutsk, 664074, Russia
[2] Irkutsk National Research Technical University, 83 Lermontov Str., Irkutsk, 664074, Russia

E-mail: smart_grid@mail.ru

Abstract. This article presents computer modeling results for AC railroads power supply system modes. It considers various methods of connection of controlled reactive power sources (RPS). It is shown that the various problems of stabilizing the voltage levels in the traction network and reducing electrical energy losses can be solved. The computer-aided modeling indicates that to install RPS into all phases of traction substations is the best way to stabilize voltage levels in traction networks and reduce electrical energy losses. When the reactive power sources are present in the free phase, traction network losses increase in comparison with the option when the RPS are not present. The installation of sources in the retarding or leading phase allows stabilizing voltage in the load utilizing equipment, reducing losses and non-symmetry in supply mains. The best option in terms of electrical energy losses is connecting the RPS to the leading phase. In terms of non-symmetry in supply mains, the best performance is produced when the RPS are installed in the retarding phase.

1. Introduction.

Some problems related to low power quality and low energy efficiency [1 – 6] occur in AC railroads traction power supply systems (TPSS). These problems can be resolved using smart grids technologies with comprehensive facilities to control modes [7, 8].

These facilities include reactive power sources (RPS), the application of which in TPSS is discussed in works [9 – 16]. Results obtained in these works serve as a methodological basis to carry out further research aimed at the implementation of technologies of the effective use of controlled RPS in traction power supply systems.

Smart grid technology introduction is impossible without the development of the tools of modeling of TPSS modes equipped with active devices for controlling modes [2, 8]. The Fazonord software system designed in Irkutsk State Transport University can be used effectively as such a tool [3]. This article provides the results of modeling of TPSS modes based on this software with respect to the rational placement of adjustable RPS in traction networks.

2. Problem formulation.

Reactive power sources can be placed at different points of the traction network [16, 17]. Below are the modeling results and the comparative analysis of the TPSS efficiency, which use the following options of RPS installation (Fig. 1):

1. without the RPS installation;

IOP Conf. Series: Materials Science and Engineering **760** (2020) 012033 doi:10.1088/1757-899X/760/1/012033

2. RPS at all terminals of the 27.5 kV traction substation (TS);
3. RPS in the retarding phase (BC);
4. RPS in the leading phase (AC);
5. RPS in the free phase (AB).

Reactive power generation and consumption limits were set from –5 to 5 Mvar for each RPS installation option.

Controlled sources of reactive power can be used to solve the following problems:

1. the stabilization of traction network (TN) voltage levels.
2. the reduction in electrical energy losses in TPSS and the external power supply system;
3. the reduction of non-symmetry at high voltage buses of traction substations;

The first problem is the most important for transport. Its formalized record can be represented in the following way:

$$\sum_{j=1}^{n}\sum_{k=1}^{m}\left(U_j(t_k)-U_j^Z\right)^2 \to \min ; \qquad (1)$$

while having limitations

$$\mathbf{F}(\mathbf{X}(t_k),\mathbf{Y}(t_k))=\mathbf{0}; \qquad (2)$$

$$Q_i^{(\min)} \le Q_i \le Q_i^{(\max)}, \qquad (3)$$

where $U_j(t_k)$ is the voltage on load utilizing equipment of the jth train in time moment t_k, kV; U_j^Z is the voltage level set at the load utilizing equipment of the jth train, kV; $\mathbf{F}$ is the non-linear vector-function, corresponding to equations describing the kth mode of the TPSS instant scheme; $\mathbf{X}(t_k)=[x_1(t_k) \;\; x_2(t_k) \;\; \ldots \;\; x_l(t_k)]^T$ is the uncontrolled parameters' vector for the kth mode of the TPSS instant scheme; $\mathbf{Y}(t_k)=[y_1(t_k) \;\; y_2(t_k) \;\; \ldots \;\; y_l(t_k)]^T$ are uncontrolled parameters vector for the kth mode of the TPSS instant scheme; $Q_i^{(\min)}, Q_i^{(\max)}$ are RPS reactive power change limits.

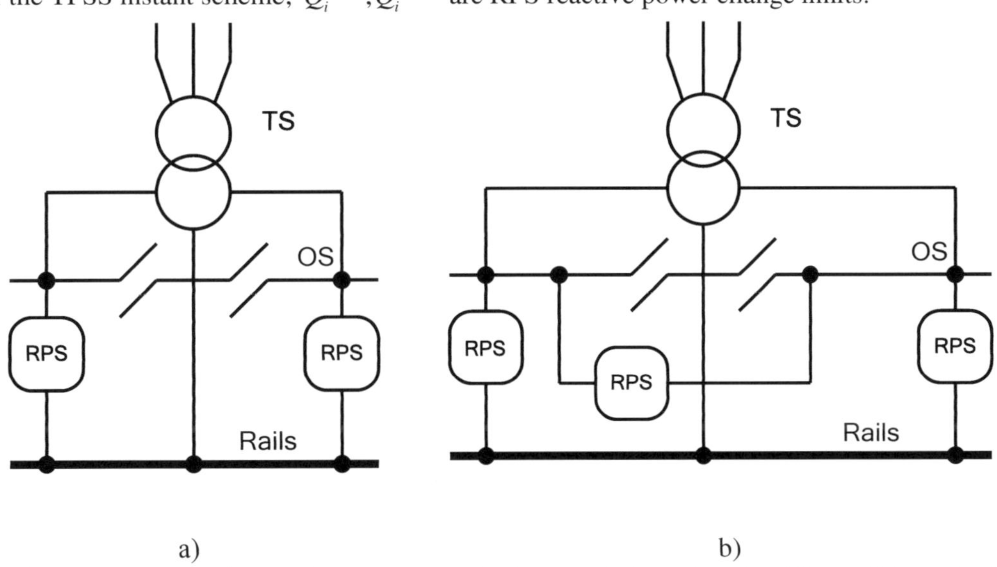

Figure 1. RPS arrangement schemes by phases: a – RPS at supply branches; b – RPS at TS terminals

Modeling of TPSS modes where TPSS are equipped with RPS with objective function (1) can be performed based on the Fazonord software application [3]. For this purpose, swing-buses are set at RPS terminal points in terms of reactive power which correspond to the following equations [2, 3]:

$$\left(U_k^{'}\right)^2+\left(U_k^{''}\right)^2-\left(U_k^Z\right)^2=0; \; Q_k^{(\min)} \le Q_k \le Q_k^{(\max)}.$$

3. Modeling results.

The design model scheme corresponding to Fig.1 is shown in Fig. 2. Maintaining voltage level of 27.5 kV at its terminal point is accepted as the aim of reactive power sources' control. During the modeling, the operation of five trains weighing 6,000 tonnes each was considered in even direction, and the same number of trains each weighing 3,200 tons were considered as moving in odd direction. The train schedule is shown in Fig. 3. The current profiles of the trains are shown in Figs. 4 and 5.

RPS installation options were compared by the following parameters:

- voltage levels under the train load utilizing equipment (Table 1, Fig. 6);
- non-symmetry coefficients for negative sequence at the high-voltage buses of the traction substations (Fig. 2, Fig. 7);
- losses in active energy in TPSS and the external power supply system (Table 3);
- powers generated (consumed) by RPS (Table 4).

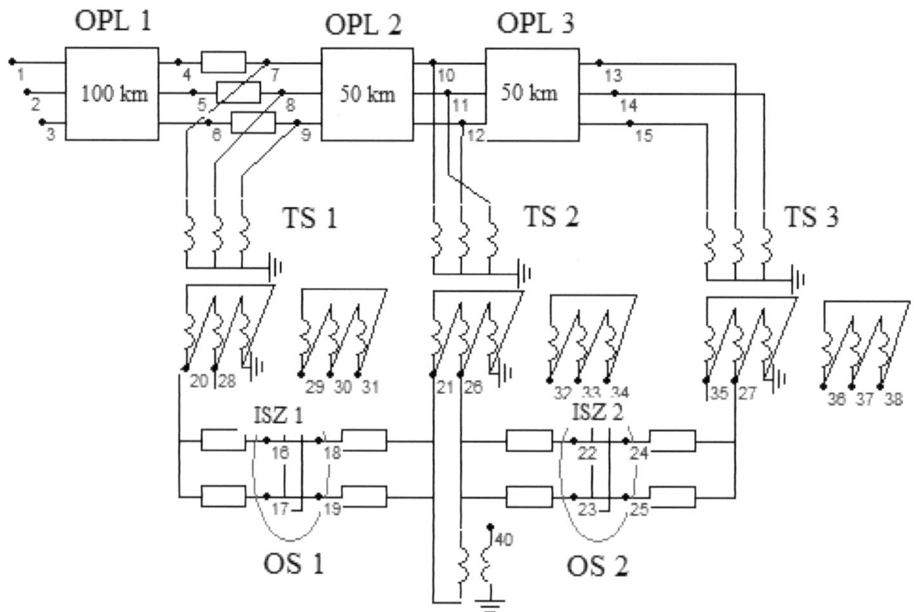

Figure 2. RPS at TS medium voltage buses: OPL is the overhead power line; ISZ is the inter-substation zone; OS are the overhead system sections

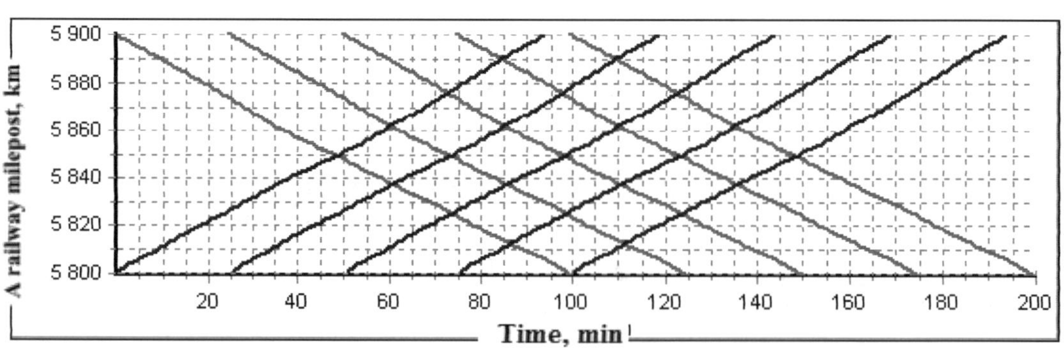

Figure 3. Train schedule

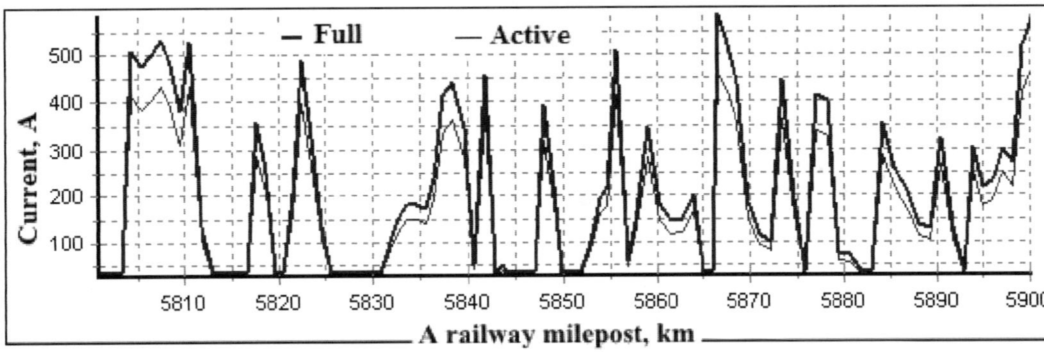

Figure 4. Even train current profile

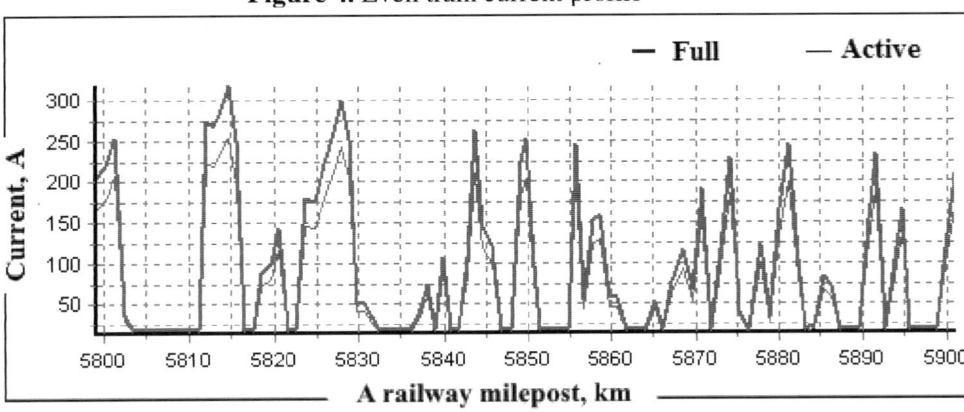

Figure 5. Odd train current profile

Table 1. Voltage levels under the train load utilizing equipment, kV

	Train number										For all trains
	1	**2**	**3**	**4**	**5**	**6**	**7**	**8**	**9**	**10**	
without RPS installation											
Min	22.87	22.87	22.84	22.84	23.02	21.87	20.50	20.50	20.91	22.06	20.5
Mid	26.55	26.11	25.91	25.98	26.35	25.76	25.32	25.29	25.45	26.03	25.875
Max	27.96	27.80	27.88	27.95	28.01	27.85	27.81	27.76	27.75	27.94	28.01
RPS installation in all TS phases											
Min	23.73	23.73	24.32	24.15	24.15	23.16	22.24	22.24	22.56	23.32	22.24
Mid	26.89	26.66	26.52	26.48	26.68	26.24	25.98	25.97	26.10	26.43	26.395
Max	27.71	27.64	27.69	27.74	27.68	27.63	27.60	27.60	27.53	27.60	27.74
RPS installation in the retarding phase (BC, node 26)											
Min	23.10	23.10	23.58	23.94	23.94	22.83	21.82	21.82	22.18	22.70	21.82
Mid	26.75	26.50	26.39	26.46	26.76	26.15	25.81	25.79	25.90	26.30	26.281
Max	27.92	27.82	28.05	28.16	28.25	27.89	27.96	28.04	28.04	28.23	28.25
RPS installation in the leading phase (AC, node 21)											
Min	23.64	23.64	22.88	22.88	23.03	21.93	20.62	20.62	20.99	21.97	20.62
Mid	26.79	26.38	26.13	26.07	26.29	25.88	25.57	25.56	25.71	26.17	26.055
Max	27.92	27.78	27.66	27.61	27.66	27.63	27.61	27.61	27.53	27.64	27.92
RPS installation in the free phase (AB, node 40)											
Min	22.71	22.71	22.51	22.51	22.74	21.32	20.11	20.11	20.52	21.68	20.11
Mid	26.33	25.87	25.66	25.73	26.10	25.54	25.10	25.07	25.23	25.78	25.641
Max	27.79	27.64	27.79	27.80	27.84	27.75	27.71	27.65	27.56	27.79	27.84

Analysis of Table 1 data allows making the following inferences:

– the highest voltage fluctuations difference under the electric locomotive load utilizing equipment is observed in the absence of RPS.

– the least voltage stabilization effect is observed when RPS are connected to AB linear voltage (in the free phase);

– the best stabilization effect is observed when three reactive power sources are installed (option 2 which is the most expensive).

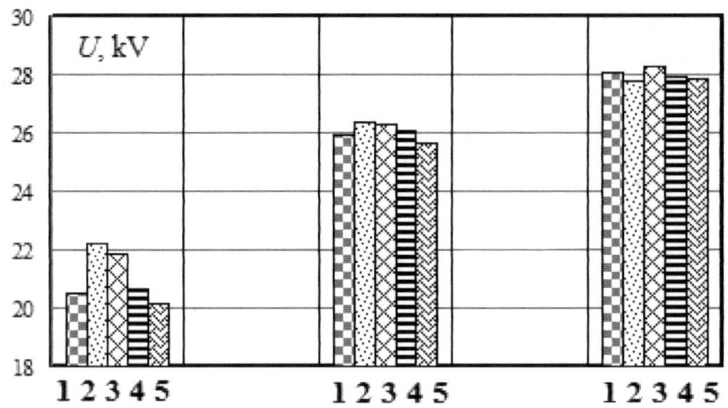

Figure 6. The minimal, mid, and maximal voltage level under the load utilizing equipment subject to RPS installation option.

Table 2. Negative sequence non-symmetry coefficients k_{2U} at TS high voltage buses

Parameter	Traction substation		
	TS 1	TS 2	TS 3
without the RPS installation			
Min	0.04	0.04	0.04
Mid	1.04	1.38	1.52
Max	2.52	3.78	4.41
RPS installation in all phases			
Min	0.05	0.06	0.08
Mid	0.58	0.72	0.85
Max	1.95	2.66	3.1
RPS installation in the retarding phase (BC, node 26)			
Min	0.26	0.29	0.32
Mid	0.81	1.08	1.24
Max	2.12	3.17	3.74
RPS installation in the leading phase (AC, node 21)			
Min	0.11	0.13	0.11
Mid	0.99	1.38	1.58
Max	2.51	3.79	4.45
RPS installation in the free phase (AB, node 40)			
Min	0.07	0.10	0.11
Mid	0.98	1.43	1.67
Max	2.59	3.92	4.57

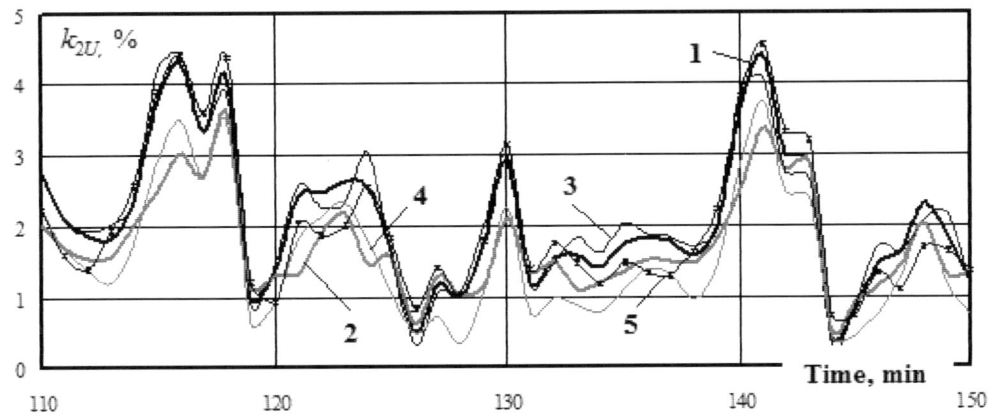

Figure 7. Negative sequence non-symmetry coefficients at TS3 high-voltage buses: numbers designate RPS installation options

Results provided in Table 2 and illustrated in Fig.7 indicate that it is the most expensive option No.2 that ensures the best effect in non-symmetry reduction at high voltage buses at the traction substations (maximum k_{2U} is reduced to 3.1 %). When RPS are installed in the retarding phase, k_{2U} coefficient maximum is reduced to 3.7 %. The growth in non-symmetry coefficient maximal value corresponds to option 4. A similar result is observed in option 5.

Table 3. Active energy losses in OPL, CT, OS, kW

The mains element	The RPS installation option				
	1	**2**	**3**	**4**	**5**
OPL-1	500.3	408.8	489.0	469.4	430.7
OPL-2	122.4	90.2	125.2	101.6	143.7
OPL-3	28.0	22.7	28.2	23.3	32.4
Amount	**650.7**	**521.7**	**642.4**	**594.3**	**606.8**
Transformer 1	263.6	254.8	258.1	260.3	260.5
Transformer 2	281.6	258.8	276.2	267.1	309.8
Transformer 3	249.4	243.8	250.1	244.8	251.5
Amount	**794.6**	**757.4**	**784.4**	**772.2**	**821.8**
OS-1	658.5	635.9	643.3	643.3	659.7
OS-2	561.1	519.3	558.5	517.7	584.6
Amount	**1219.6**	**1155.2**	**1201.8**	**1161**	**1244.3**
Total:	**2664.9**	**2434.3**	**2628.6**	**2527.5**	**2672.9**

The second option is the best in terms of electrical energy losses; in this case the total losses are reduced approximately by 9%. The 4th option is characterized by a good performance yielding this index reduction by 5%. When RPS is installed in free phase, a little increase in losses occurs.

Reactive power generation (consumption) values are provided in Table 4.

Table 4. Summary data for RPS reactive power generation (consumption), Mvar

Factor	Option					
		2		**3**	**4**	**5**
Node number	*21*	*26*	*21*	*21*	*26*	*40*
Min	–5	–2.71	–3.44	–3.44	–2.71	–5
Max	5	5	5	5	5	–2.48

Analysis of Table 4 data indicates that the broadest range for adjustment from –5 to +5 Mvar is required for option 2, whereas the least (from –2.7 to +5) is required for option 4.

Contemporary multiagent control systems [1, 8, 18] make it possible to set more complicated RPS control tasks. To these one can refer the problem of RPS optimal setpoints choice, where RPS ensures electrical energy losses minimizing in traction network along with voltage (1) levels stabilization:

$$dW = \sum_{k=1}^{s}\left(\int_{0}^{t_c} dP_k(t)dt\right) \rightarrow \min , \qquad (4)$$

where s is the number of TN elements considered; $dP_k(t)$ are power losses in the kth element of TN.

The formulated control task is reduced to the search of the RPS setpoint vector:

$$\mathbf{U}^{Zrps} = \left[U_1^{Zrps} \quad U_2^{Zrps} \quad \ldots \quad U_n^{Zrps}\right]^T ,$$

which ensures minimization of functionals (1) and (4) for each discrete time moment in case of limitations (2) and (3).

4. Conclusion

The modeling method of traction power supply systems modes proposed in this article, where traction power supply systems are equipped with adjustable reactive power sources, can be used to solve the following urgent practical issues:

– rational RPS arrangement in traction network;

– development of actions aimed at voltage levels stabilization under the electric locomotive load utilizing equipment;

– enhancing power quality in 110-220 kV mains supplying traction substations;

– raising energy efficiency of traction power supply systems.

References

[1] Tretyakov E A 2013 *Power quality management in railroads distribution networks* (Omsk: Omsk State Transport University).

[2] Zakaryukin V P, Kryukov A V, Ushakov V A, Alekseenko V A 2012 *Ongoing management in railroads power supply systems* (Irkutsk: Irkutsk state transport university).

[3] Zakaryukin V P, Kryukov A V 2005 *Complex asymmetrical modes of electrical systems* (Irkutsk: Irkutsk University Publishing House).

[4] Kryukov A V, Alekseenko V A 2011 *Improving the efficiency of operational management in traction power supply systems* (Modern technologies System analysis Modeling No 4). pp. 158–164.

[5] Zakaryukin V P, Kryukov A V, Avdienko I M 2016 *Elimination of asymmetry in electrical networks supplying traction substations of railways* (Modern technologies System analysis. Modeling No 1). pp. 189-195.

[6] Zakaryukin V P, Kryukov A V 2013 *Multifunctional approach to modeling of electric power systems* (Modern technologies System analysis Modeling No 4). pp. 100-107.

[7] Dorofeev V V, Makarov A A 2009 *Active-adaptive network – a new quality of Russia's Single Power System* (Energoexpert No. 4) pp. 29-34.

[8] Zakaryukin V P, Kryukov A V, Cherepanov A V 2015 *Intelligent technologies of power quality*

management. (Irkutsk: Irkutsk national research technical university).
[9] German L A, German V L 2014 *Automation of AC traction network's power supply* (Moscow: Russian Transport University).
[10] German L A, Serebryakov A S 2013 *Adjustable units of capacitive compensation in railroads traction power supply systems* (Moscow: Transport book).
[11] Molin N I, Zinoviev N D 1979 *Static adjustable devices designed for reactive power compenasation* (Omsk: Enhancing electrical energy quality in traction substations) pp. 44 – 51.
[12] Zinoviev N D, Molin N I, Pavlov I V 1984 *Multifunctional devices for reactive power compensation in electrified AC railroads* (Khabarovsk: Issues of VL80R electric locomotives efficiency and raising their reliability in traction and recuperative modes) pp. 57-60.
[13] Borodulin B M, German L A, Nikolaev G A 1983 Condenser *units of electrified railroad* (Moscow: Transport).
[14] Mamoshin R R, German L A 1990 *The choice of compensating device in the traction power supply system* (Vestnik of the Railway Research Institute. No. 5) pp. 26-29.
[15] Bardushko V D 2013 *Parametric synthesis of parallel capacitive compensating devices systems in traction network in contemporary conditions* (Transport electrification. No. 6) pp. 8-13.
[16] Kondratyev Yu V, Tarasenko A V 2015*The choice of power and placement for reactive power shunt compensation devices* (Journal of Transib Railway Studies. No. 2(22)) pp. 79-87.
[17] Zakaryukin V P, Kryukov A V, Cherepanov A V 2018 *Intelligent Traction Power Supply System* (Advances in Intelligent Systems and Computing vol. 692) pp. 91-99.
[18] Russel S, Norvig P 2007 *Artificial intelligence contemporary approach* (Moscow: 'Vilyams' Publishing House).

IOP Conf. Series: Materials Science and Engineering **760** (2020) 012034 doi:10.1088/1757-899X/760/1/012034

Simulation of non-sinusoidal modes in railway power supply systems during movement of high-speed trains

A V Kryukov[1,2], A V Cherepanov[1], A R Shafikov[1] and E S Bezridnyj[1]

[1]Transport Electric Power Engineering department, Irkutsk State Transport University, 15, Chernyshevsky St., Irkutsk, 664074
[2]Power Supply and Electrical Equipment department, Irkutsk National Research Technical University, 83, Lermontov St., Irkutsk, 664074

E-mail: and_kryukov@mail.ru, smart_grid@mail.ru, alshaficov1@mail.ru, bezrik4471@mail.ru

Abstract. Rectifier locomotives have non-linear current-voltage characteristics and generate higher harmonics (HH) in the power supply networks, which is accompanied by the following negative effects: reduced equipment service life, distortion of electricity metering, occurrence of resonant processes, etc. Therefore, the problem of reducing the levels of harmonic distortion in the networks adjacent to the traction substations of the AC railways is of increased relevance. The level of HH generation depends on the design features of electric locomotives. Electric stock with zone-and-phase regulation creates high harmonic distortions, which results in increased losses, increased electromagnetic effects on adjacent power and telecommunication lines and reduced reliability of power supply. Electric locomotives with asynchronous traction motors (ATM) connected via 4q – S converters virtually do not distort the sinusoidality of the current curve. To measure the degree of reduction of harmonic distortion during the movement of locomotives with ATM, simulation modeling of non-sinusoidal modes of a typical 25 kV traction power supply system was carried out for two variants of the locomotives used: 1) VL-80R with DC motors; 2) UTY –1 with asynchronous traction motors. The methodology for determining non-sinusoidal modes is based on the technology of modeling electric power systems (EPS) in phase coordinates; at the same time, the models of EPS elements were formed as lattice schemes characterized by a fully connected topology. The parameters of these circuits can be recalculated to the frequencies of HH. The proposed approach is universal and can be used to study non-sinusoidal modes in existing and future traction power supply systems. The simulation results showed that the problem of increased harmonic distortion in the networks feeding the traction substations can be completely solved by replacing the acting locomotives with new-generation electric locomotives with asynchronous electric drives and four-quadrant converters.

1. Introduction

Rectifier locomotives have non-linear current-voltage characteristics and generate higher harmonics (HH), which can penetrate into external networks and areas of power supply to non-traction consumers [1]. The flow of HH currents is accompanied by the following negative effects [2–16]:

- additional heat insulation of transformer windings, leading to their service life reduction;
- accelerated thermal aging of the insulation of stator windings of asynchronous electric motors;

• increase in measurement errors of power consumption parameters, which can result in an unacceptable degradation in the accuracy of electricity metering;

• a breakage of the neutral wire due to contact burning-off with a significant level of harmonics with numbers that are multiples of three; this can result in the failure of some power consuming units due to the emergence of increased voltages at their terminals;

• resonant effects at higher harmonic frequencies, creating severe problems for electrical equipment;

• increased levels of induced voltages in power and communication lines passing near the railway.

Therefore, the problem of reducing the levels of harmonic distortion in the networks adjacent to the traction substations (TS) of the AC railways is of increased relevance.

The level of HH generation depends on the design features of electric locomotives. To illustrate this position, Fig. 1 shows the harmonic spectra of VL-80R electric locomotives with zone-and-phase regulation and most advanced UTY-1 with asynchronous engines and four-quadrant converters [17].

Electric stock (ES) with DC engines and zone-and-phase regulation creates significant harmonic distortion; this increases the power loss in the electric traction network (ETN), which results in additional heating of the equipment and the deterioration of energy performance.

Electric locomotives with ATM, equipped with $4q - S$ converters, virtually do not distort the sinusoidal current, have lower reactive power consumption and reduce the levels of harmonic voltage components at the nodal points of the supply networks, which contributes to their energy efficiency.

To measure the degree of reduction of harmonic distortion during the movement of locomotives with ATM, simulation modeling of non-sinusoidal modes of a typical 25 kV traction power supply system was carried out for two variants of the locomotives used: VL-80R with DC motors and zone-and-phase regulation; UTY–1 with asynchronous traction motors connected to four-quadrant ($4q - S$) converters.

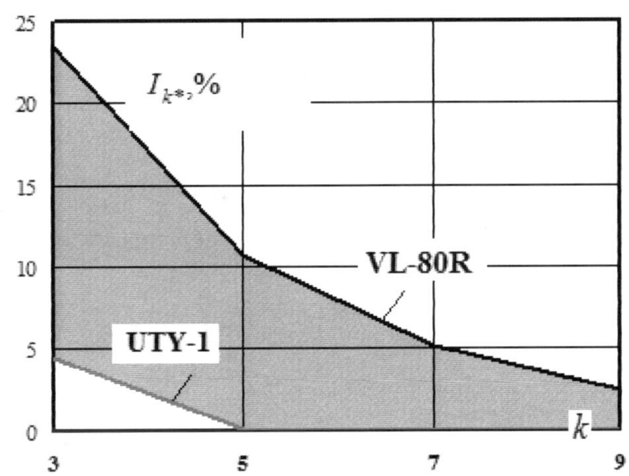

Figure 1. VL-80R and UTY-1 spectra comparison

2. The simulation technique

The method of determining non-sinusoidal modes is based on the technology of modeling electric power systems (EPS) in phase coordinates [18–23]; at the same time, the models of EPS elements were formed as lattice schemes were characterized by a fully connected topology. The parameters of these circuits can be recalculated to the frequencies of HH.

During simulation modes at HH frequencies, there are systems of equations of fairly high dimension to be solved [20]:

$$\mathbf{F}[\mathbf{X}(f_1)] = \mathbf{0}; \ \underline{\mathbf{Y}}(f_2)\dot{\mathbf{U}}(f_2) = \dot{\mathbf{I}}(f_2); \ \underline{\mathbf{Y}}(f_3)\dot{\mathbf{U}}(f_3) = \dot{\mathbf{I}}(f_3); \ \ldots \ \underline{\mathbf{Y}}(f_{40})\dot{\mathbf{U}}(f_{40}) = \dot{\mathbf{I}}(f_{40}),$$

where f_i is the higher harmonic frequencies, $i = 2, 3, .., 40$; $f_i = if$; $f = 50$ Hz; $\underline{\mathbf{Y}}(f_i)$ is the matrix of the conductivities of the design model of the power supply system, whose elements are determined by the frequencies f_i ; $\dot{\mathbf{U}}(f_i)$ is the desired voltage at the nodal points of the network; $\dot{\mathbf{I}}(f_i)$ is the vector of current HH sources.

Determination of short circuit (SC) and idle run (IR) parameters of transformers at higher harmonic frequencies has some unique features. The scattering reactances that are defined for $f = 50$ Hz must be recalculated for the HH frequencies. Short-circuit losses are also corrected in proportion to the frequencies f_i. Idle run losses are inexpedient to increase, as induction decreases in the magnetic circuit with frequency growth, which leads to loss reduction.

One of the schemes for the loads proposed in [8] is used when calculating HH. It is parallel with the resistances, which are determined by the formulas:

$$R_p = \frac{U^2}{P}; \; X_p = \frac{\nu U^2}{Q},$$

where U is the voltage at the load terminals at $f = 50$ Hz, ν is the harmonic number.

To calculate the currents of stationary sources of harmonics, one can use the following expression [8]:

$$I_\nu = \frac{2\sqrt{3} I_d}{\pi k_T \nu}$$

where $\nu = kp \pm 1$; $p = 6$, 12 or 24 is the number of transducer pulsations, $k = 1, 2, 3$ ....

For a single-phase rectifier converter $(p = 2)$ harmonic currents are calculated based on the ratio:

$$I_\nu = \frac{4 I_d}{\pi k_T \nu},$$

where I_d is the rectified current, which is determined based on the set value of the main harmonic current; phase angles of current harmonics are calculated as follows:

$$\psi_\nu = \nu \psi ,$$

where ψ is the phase of the main frequency current.

It is possible to set the currents of stationary sources of harmonics on the basis of reference data or measurement results.

In accordance with the requirements of GOST 32144-2013, 40 harmonics are taken into account when calculating non-sinusoidal modes. The voltage and current waveforms are calculated using the expression:

$$u(t) = \sum_{k=1}^{40} \sqrt{2}\, U_k \sin(k\omega t + \psi_k).$$

3. Simulation results

The simulation was carried out using the Fazonord software package for a typical 25 kV traction power supply system (Fig. 2). A double-track section with a length of 120 km was considered. The traction power supply system included three intersubstation zones 40 km long.

The simulation was carried out with the movement of four up and down freight trains weighing 6384 tons in two options:

1) locomotives VL-80R with DC engines and zone-and-phase regulation are used for the movement of trains;

2) UTY-1 locomotives with ATM and four-quadrant converters are used [17].

Schematic diagrams of electric stock are shown in Fig. 3. The train schedule is shown in Figure 4, and the dependences characterizing the current consumption of trains are shown in Fig. 5. Non-sinusoidal indicators were determined on 110 kV buses of the traction substation number 4.

The results based on the modeling are shown in table 1 and in fig. 6, 7.

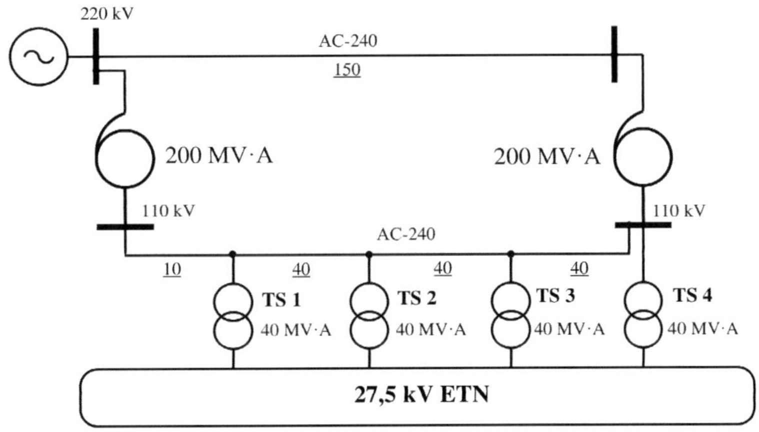

Figure 2. Railway power supply system scheme

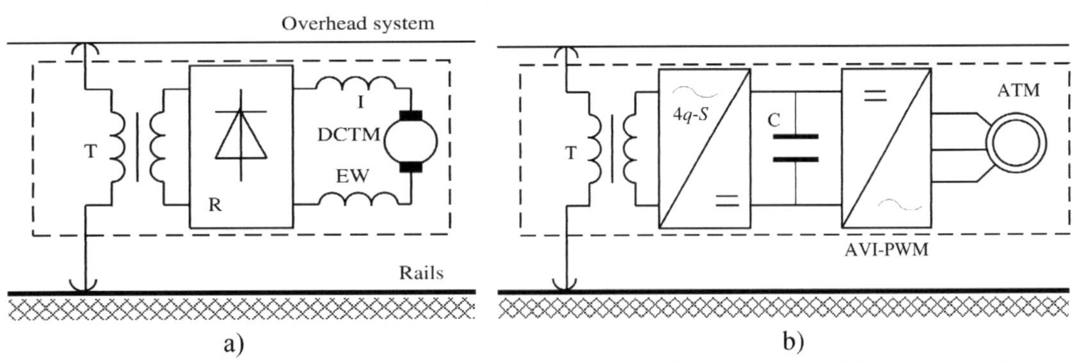

Figure 3. ES schemes [17]: a – VL-80R; b – UTY-1; T – transformer; B – rectifying installation; I – smoothing inductor; DCTM is DC traction motor; ATM – asynchronous traction motor; AVI-PWM – autonomous voltage inverter with pulse-width modulation; 4q-S – four-quadrant converter; C is capacitive filter

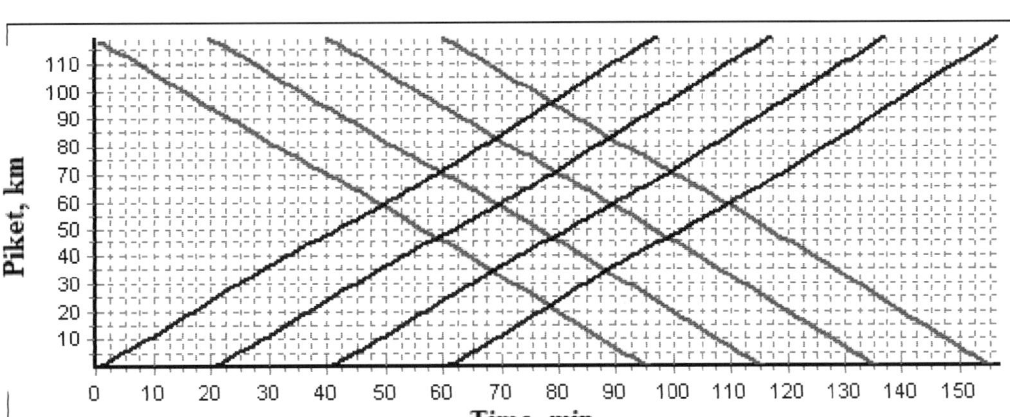

Figure 4. Train movement schedule

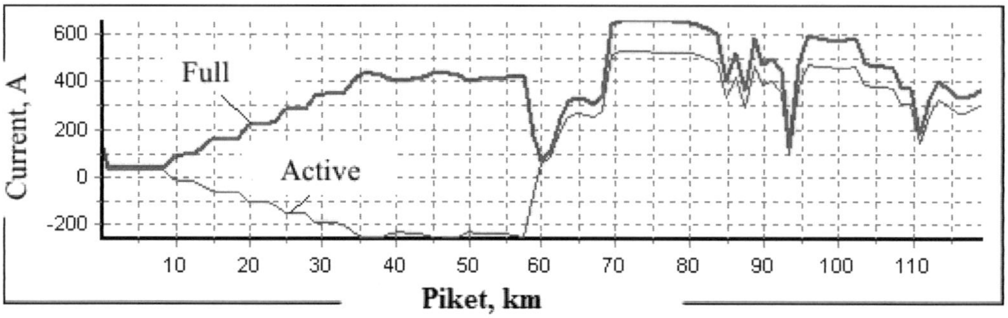

a)

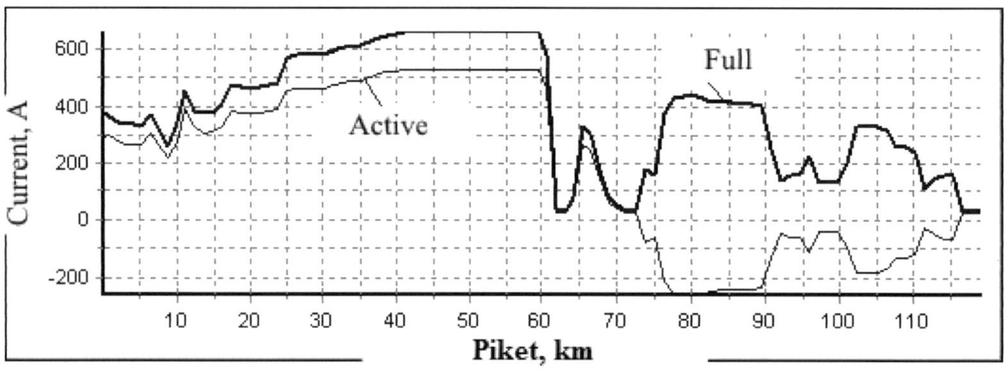

b)

Figure 5. Current consumption of trains weighing 6384 t: a – down trains; b – up trains

Fig. 6 shows the time dependences of the total harmonic coefficients on 110 kV buses of the TP-4 traction substation.

Table 1. Total harmonic coefficients on 220 kV buses TP -4

Phase	VL-80R		UTY-1		$\gamma k_U = \frac{k_U^{(\text{ВЛ80Р})}}{k_U^{(\text{UTY1})}}$	
	Average value	Maximum	Average value	Maximum	Average value	Maximum
A	6.61	9.25	0.43	0.81	15.39	11.48
B	13.21	25.61	0.70	1.68	18.74	15.24
C	12.24	21.44	0.53	1.14	23.11	18.82

The analysis of results presented in table 1 and fig. 7 allows for the following conclusions:

1. During train traffic with locomotives VL-80R, the levels of harmonic components on 110-kV TP-4 buses far exceed the maximum permissible values of 3 %. To reduce the non-sinusoidality, the use of expensive means is required, such as passive and active filters [18].

2. During train traffic with UTY-1 electric locomotives, the values k_U do not exceed the values of 2 % normally acceptable by GOST 32114-2013. Harmonic distortion levels, in comparison with the movement of electric locomotives VL-80R, are reduced 11...25 times.

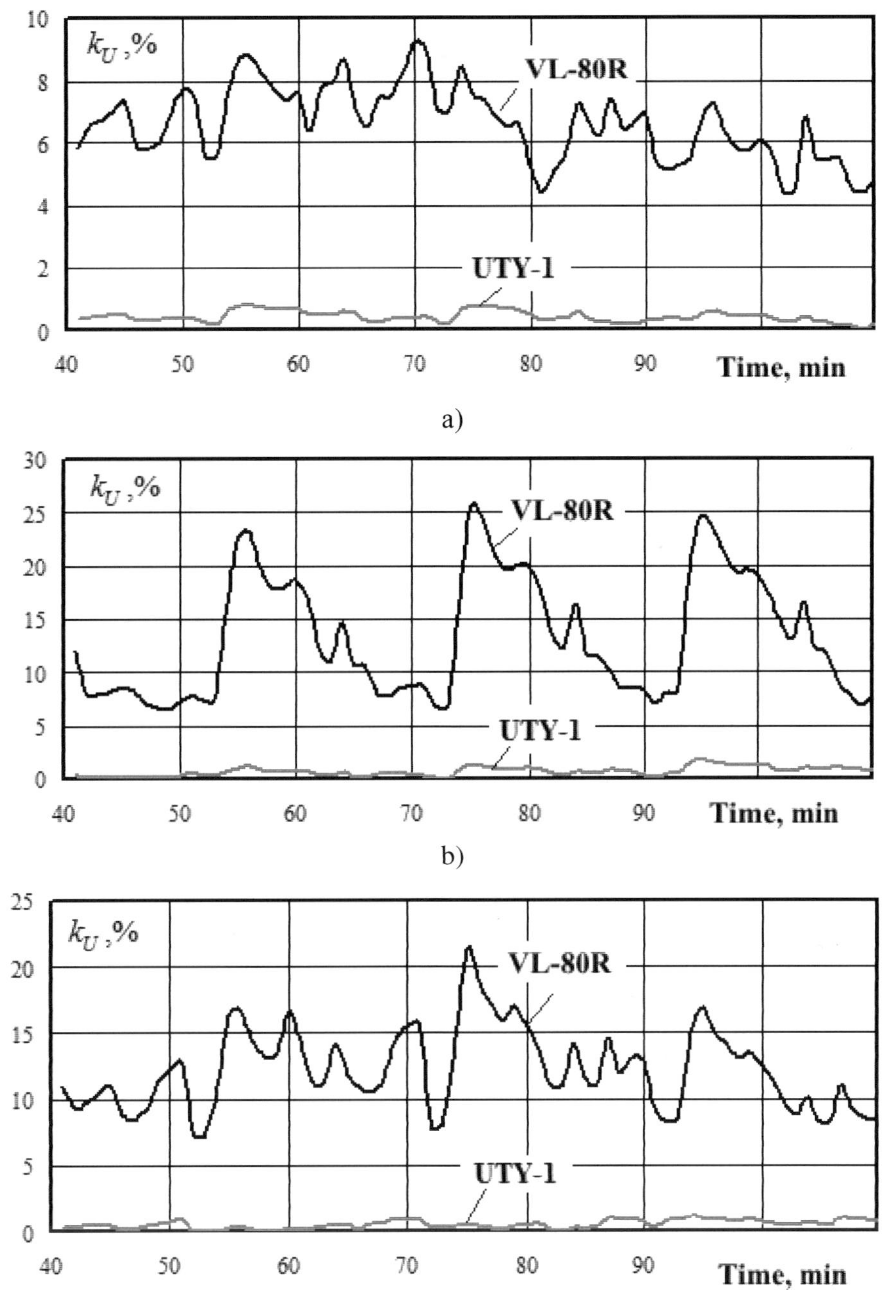

Figure 6. Coefficients k_U on 110kV TP-4 buses in time: a – phase A; b – phase B; c – phase C

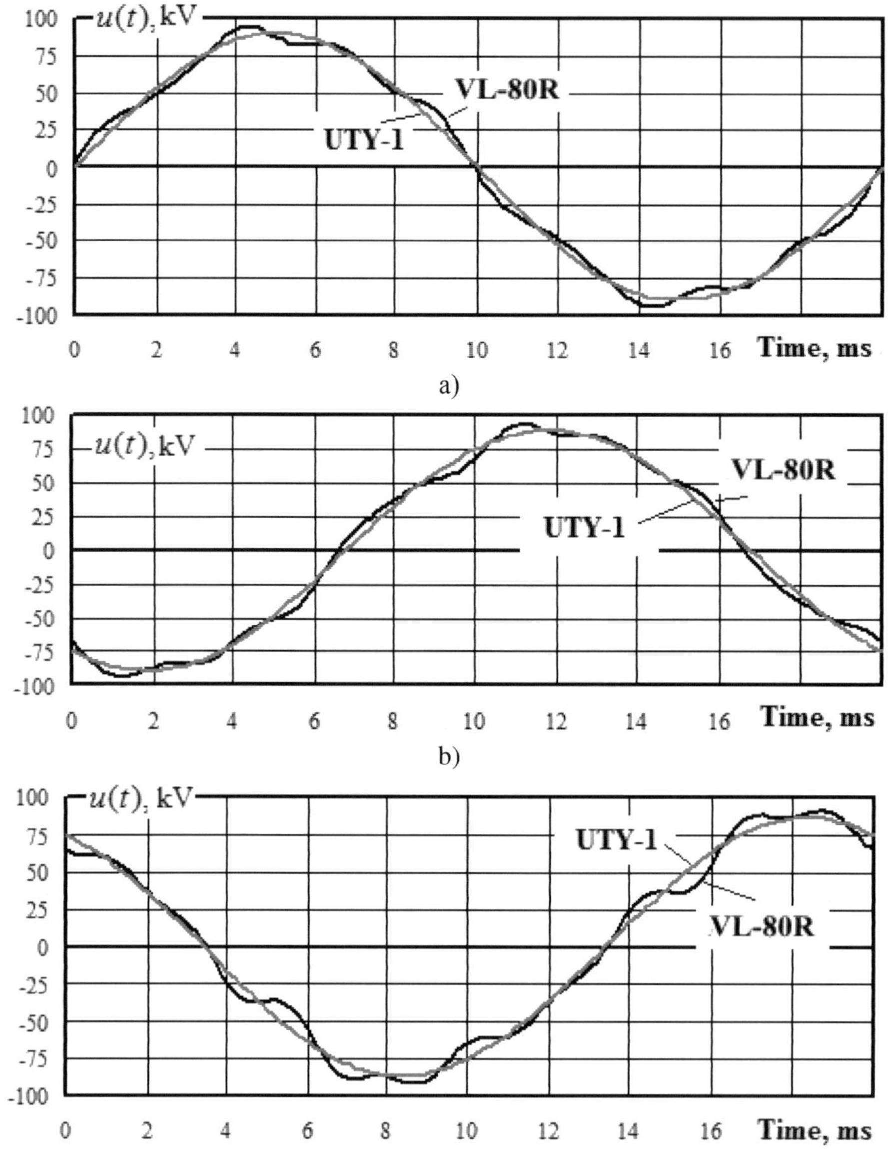

Figure 7. Voltage waveforms on 110 kV TP-4 buses: a – phase A; b – phase B; c – phase C

Thus, the problem of increased harmonic distortion level in the networks adjacent to the traction substations of main railways can be completely solved by replacing the existing electric stock to most advanced electric locomotives with asynchronous electric drives and four-quadrant converters.

4. Conclusion

Based on computer simulation of non-sinusoidal modes of a typical 25 kV power supply system in the Fazonord-APC software package, it was shown that, using most advanced electric stock with asynchronous motors and four-quadrant converters, the harmonic distortion levels in the networks, adjacent to the traction substations, do not exceed standard values.

References

[1] Marquardt K G 1982 *Electric power supply of electrified railways* (Moscow: Transport Publ.)
[2] Timofeev D V 1965 *Modes in electrical systems with traction loads* (Moscow, Leningrad: Energiya Publ.)
[3] Steimel A 2008 *Electric traction motive power and energy supply. Basics and practical experience.* (Munchen: Oldenbourg Industrieverlag)
[4] Biesenack H, Braun E, George G et al 2006 *Energieversorgung elektrischer bannen* (Wiesbaden: B.G. Teubner Verlag)
[5] Arrilaga D, Bradley D and Bodger P 1990 *Harmonics in electrical systems* (Moscow: Energoatomizdat Publ.)
[6] Dolinger S Yu et al 2013 Assessment of additional power losses from the reduction of the quality of electric energy in the elements of power supply systems *Omsk Scientific Herald* **2(120)** 178–83
[7] Zhezhelenko I V, Saenko Yu L and Gorpinich A V 2008 The impact of power quality on electrical equipment service life reduction and the reliability deterioration *Electrical engineering* **3** 14–20
[8] Zhezhelenko I V 2000 *Higher harmonics in power supply systems of industrial enterprises* (Moscow: Energoatomizdat Publ.)
[9] Kargin S V, Krasnova A N and Bekbulatov R R 2012 *Energy quality management in general purpose distribution networks* (Moscow: Energoprogress Publ.)
[10] Kartashov I I and Zuev E N 2000 Power quality in power supply systems. Ways to control and ensure it. (Moscow: MEI Publ.)
[11] Kartashov I I et al 2006 Electricity Quality Management (Moscow: MEI Publ.)
[12] Tretyakov E A 2013 *Quality management of electrical energy in the railway distribution networks* (Omsk: OmGUPS Publ.) 195 p
[13] Shidlovsky A K and Kuznetsov V G 1985 *Improving the quality of energy in electrical networks* (Kiev: Naukova Dumka Publ.)
[14] Shidlovsky A K, Novsky V A and Kaplychnyy N N *Stabilizing electrical energy parameters in distribution networks* (Kiev: Naukova Dumka Publ.)
[15] Dolgov A P, Kandakov S A and Zakaryukin V P 2011 *Improving the electric power quality in the systems of external power supply of the Eastern Siberia railways* (St. Petersburg: Electrification and development of the infrastructure for train traction power supply in rail transport) pp 37–8
[16] Sudnova V V 2000 Electric power quality (Moscow: Energoservis Publ.)
[17] Burkov A T and Mirsaitov M M 2016 Modes of electric traction network in operation of electric locomotives VL-80R and UTY-1 *Modern technologies for transport* **2** pp 146–60
[18] Zakaryukin V P, Kryukov A V and Cherepanov A V 2018 Intelligent Traction Power Supply System *Advances in Intelligent Systems and Computing* **692** 91–9
[19] Cherepanov A V Zakaryukin V P and Kryukov A V 2018 Modeling of traction power supply systems for movement of high-speed trains *MATEC Web of Conf.: Polytransport Systems-2018* **216(02006)** 1–7
[20] Zakaryukin V P and Kryukov A V 2005 *Complex non-symmetrical modes in electrical systems* (Irkutsk: Irkutsk university Publ.)
[21] Zakaryukin V P, Kryukov A V, Avdienko I M and Bezridniy E S 2017 Modeling modes of traction power supply systems when driving high-speed trains *Modern technologies System analysis Modeling* **3** 126–35
[22] Zakaryukin V P, Kryukov A V and Avdienko I M 2016 Elimination of asymmetry in electrical networks supplying traction substations of railways *Modern technologies System analysis Modeling* No **1** 189–95
[23] Zakaryukin V P and Kryukov A V 2013 Multifunctional approach to modeling of electric power systems *Modern technologies System analysis Modeling* **4**100–7

IOP Conf. Series: Materials Science and Engineering **760** (2020) 012035 doi:10.1088/1757-899X/760/1/012035

Diagnostics of high-voltage insulation of the railway transport overhead system by the method of spaced antennas

S M Kutsenko[1], N N Klimov[1]

[1] Irkutsk State Transport University, 15 Chernyshevsky Street, Irkutsk, 664074, Russia
E-mail: kutsenko_s@irgups.ru

Abstract. The growth in freight traffic transported throughout the entire network of Russian Railways, and in the Eastern Polygon in particular, is constantly growing, so an increase in train traffic reduces the time for technological work at railway infrastructure facilities. Consequently, it is necessary to develop methods for assessing the state of infrastructure facilities virtually without using "track possession". In this regard, the article proposes to improve the existing remote method for diagnosing high-voltage insulation of the overhead system (registration of ultraviolet and thermal radiation with a special device), enhancing it with additional measurement. The proposed solution is to assess the state of high-voltage insulation by measuring electromagnetic radiation, which will increase the accuracy of diagnostics. As for a technological solution – to record a signal (partial discharges and surface partial discharges) with three antennas (induction electrometric converters), it is aimed at improving the accuracy of localization of a faulty string of insulators.

1. Introduction

On electrified railways, much attention is given to the diagnostics of high-voltage insulation, since when it breaks down during operation, a dangerous failure can occur, which results in a failure of the train schedule, which ultimately leads to losses. Due to the constantly growing number of trains that pass through the railway section per day, it is extremely difficult to isolate the track possession for diagnostics. Therefore, in recent years, remote methods have been developed that allow conducting studies of the state of high-voltage insulation in motion without using a track possession. Currently, a method based on the registration of ultraviolet and thermal radiation from a faulty insulation is widely used [1, 2]. At the same time, when diagnosing high-voltage insulation of the overhead system, the measuring equipment is installed directly into the laboratory car [3]. Observations of ultraviolet radiation can only reveal discharges that are formed on the surface of the insulation, so-called surface partial discharges (SPDs). Internal discharges or partial discharges (PDs) cannot be identified by this method, and it is these discharges that gradually "destroy" the insulation from the inside. This was previously mentioned by many researchers [4–9]. The thermal imager is good at detecting a fault in the event of a current flowing through the insulation that heats it. At the initial stage, defects of high-voltage insulation inside are insignificant, the current is small in the presence of PDs and does not cause an increase in temperature recorded by the thermal imager. At the same time, there is already pulsed electromagnetic radiation of the PDs. In this regard, we propose to apply a comprehensive diagnosis, i.e. to supplement the results of observations of ultraviolet and thermal radiation by recording electromagnetic radiation of the PDs.

Earlier, the authors have already proposed using remote acousto-electromagnetic diagnostics of the state of linear isolation of an AC overhead system of railway transport [10]. In this work it is proposed to use the recording of electromagnetic radiation by several spaced sensors to determine the location of the source of the PD.

2. Experiment

To demonstrate the possibility of determining the time delay of an electro-magnetic pulse on spaced antennas, an experiment was conducted, where the radiation source is an electric discharge, and its recording was performed by a digital two-channel oscillograph using two induction electrometric converters (IECs) of the same size [12]. The antennas are connected to the first and second channels of the oscillograph, respectively. The experiment diagram is represented in Fig.1.

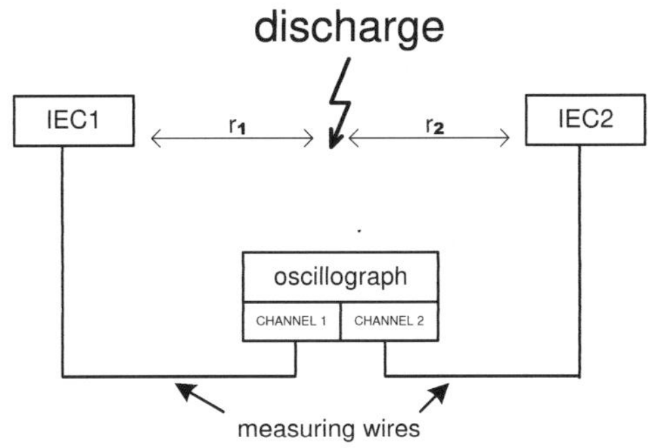

Figure 1. The experiment diagram, r_1 and r_2 are the distance between the discharge and the IEC.

In standby mode, the oscillograph records a signal, for example, through the first channel (Fig. 2). It is first necessary to determine the response level of the measuring device to eliminate electromagnetic interference. The experiment showed that with the same distance r_1 and r_2 (the length of the measuring coaxial cable is also the same), the signals are identical to each other, i.e. the amplitudes and pulse recording time coincide (Fig. 2).

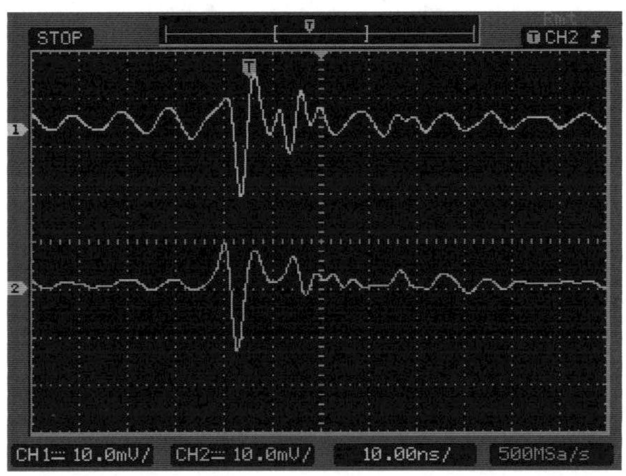

Figure 2. Oscillograms of signals recorded from IEC1 (the first channel, the upper oscillogram), from IEC2 (the second channel, the lower oscillogram). Scale: on the X axis – 10 ns / div, on the Y axis – channel 1 and 2 – 10 mV / div.

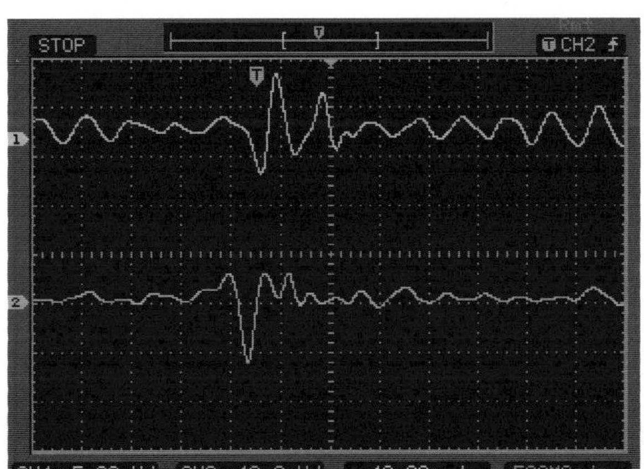

Figure 3. Waveforms of signals recorded from IEC1 (the first channel, the upper oscillogram), from IEC2 (the second channel, the lower oscillogram). Scale: on the X axis – 10 ns / div, on the Y axis – channel 1 – 5 mV / div and channel 2 – 10 mV / div.

If the distances r_1 and r_2 are not equal to each other, then there are changes in the amplitude and, most importantly, the delay of the recorded pulses relative to each other. This effect can be clearly seen in Fig. 3.

The formed delay time must be correlated with the distance, i.e. in the future, the authors of the article propose to conduct a series of experiments to bind the delay of the pulses relative to each other and the distance to the radiation source.

It should be noted that the registration of such pulses is possible with an oscillograph that captures signals in the frequency range of more than 300 MHz. When using an oscillograph with a smaller measurable frequency range, there is a high probability of recording distorted parameters of PD pulses (the temporal shape of the signal, as well as the time of its registration), which makes it impossible to determine the location of a faulty high-voltage isolation. Also, the reliability of the experimental results is affected by such a characteristic of the digital oscillograph as the frequency (rate) of the sampling of the measured signal. In our case, acceptable results were achieved only when using an Agilent Technologies oscillograph with the sampling frequency (rate) of each channel equal to 2 GSa/s.

It is also possible to increase the information content of the experiment using a second two-channel oscillograph (in this case, it is necessary to make one oscillograph the main one and start the second oscillograph from the external synchronization channel), or a four-channel oscillograph. In this case it becomes possible to connect three or four antennas, which will improve the accuracy of determining the location of the radiation source.

The shape of the signals and its polarity depend on the location of the IEC, therefore, when registering pulses of a similar shape, but of different polarity, it is necessary to adjust the position of the IEC in space. The IEC consists of two metal copper plates that are placed in parallel. The dimensions of the plates are 12x3.7 cm, the plates are placed at a distance of h = 8.5 cm. The IEC is shown schematically in Fig. 4, its equivalent circuit is shown in Fig. 5.

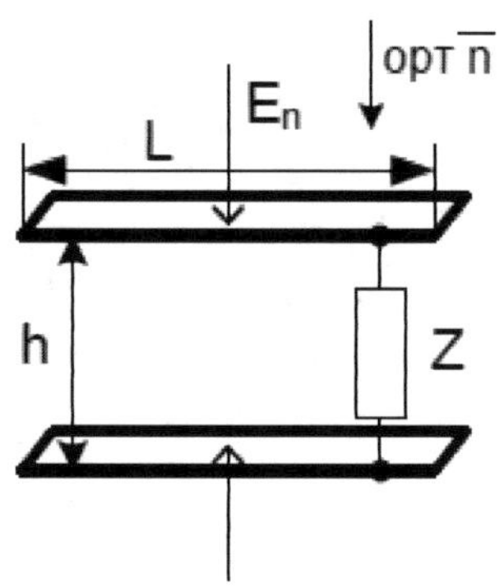

Figure 4. An induction-electrometric converter

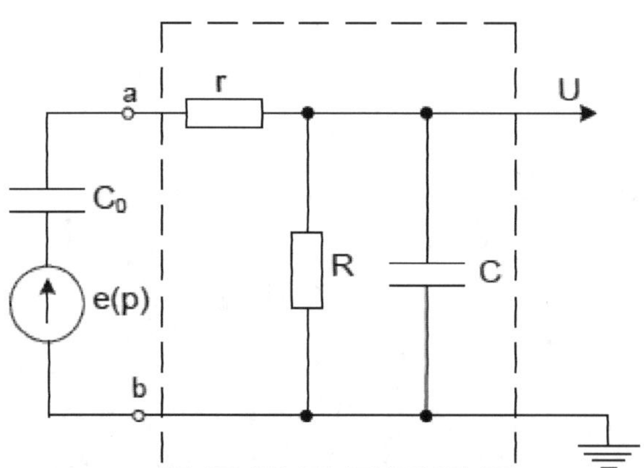

Figure 5. An IEC equivalent circuit

The IEC is made according to the recommendations [13]. The logarithmic amplitude-frequency characteristic of the transfer function of the IEC, manufactured according to the above dimensions, was published in [12]. In this case the operating frequency range of the IEC ranges from 100 kHz to 100 MHz, which allows for reliable measurements of the PD and SPD, most of which are concentrated in this range (rather than anywhere else).

This was previously reported by researchers in their works [5, 9]. It should be noted that a number of researchers have observed the PD and SPD at frequencies above 100 MHz.

During the experiment, each IEC must be located equally in space. This will eliminate the change in polarity of the signals observed by means of the IEC, as well as distortion of the signal waveform.

3. Conclusion

In practice, it is proposed to conduct an experiment by installing two three-channel recorders on the rolling stock, which will capture the same pulse and determine the location of the radiation source, i.e. in this case, the insulator of the overhead system design. The complexity of the method lies in the exact binding of the train location and the measurement time. It is also necessary to take into account that a number of powerful electromagnetic interferences are present on the railway, which interfere with the measurement [16–18]. To increase the signal-to-noise ratio, it is necessary to increase the size of the IEC, which, unfortunately, will somewhat limit the operating frequency range of the measuring complex. To do this, it is necessary to make additional calculations of the IEC [13]. It can be assumed that combining the methods of ultrasonic, thermal and electromagnetic control will significantly increase the information content of diagnostics, which will ultimately reveal a dangerous failure and prevent the failure of the train schedule.

References

[1] Vikhrov M A 2015 *Khimicheskaia tekhnika* 15
[2] Plotnikov Y I, Skorokhodov D A, Gerasimov V P, Fedorishin Y M, Grachev V F 2005 *Energetik* **4** 46-48
[3] Kutsenko S M, Fedorov M E 2014 *Sovremennye tekhnologii. Sistemnyi analiz. Modelirovanie* **44** 107-110
[4] Vdoviko V P, Ovcyannikov A G, Pospelov A I 1995 *Energetik* 16-18
[5] Ovsyannikov A G 1999 *Nauchnyi vestneyk NGTU* **5** 123-136
[6] Rezinkina V V, Rezinkin O L 2001 *IEEE International Conference on Conduction and Breakdown in Solid Dielectrics* 404-407
[7] Rezinkina V V, Rezinkin O L, Nosenko M I 2001 *Technical Physics* **46 (3)** 339-341
[8] Kupershtokh A L, Stamatelatos C P, Agoris D P 2006 *Technical Physics Letters* **32 (8)** 680-683
[9] Kuchinsky G S 1979 *Energiya. Leningr. otd-nie* 224
[10] Kutsenko S M, e.a. A method of remote acoustoelectromagnetic diagnostics of the state of linear insulation of the AC overhead system of railway transport. Patent RF, no. 2365928, 2013
[11] Bazanov V P Spirin M V, Turayev V A 2000 *Energetik* **4** 16-17
[12] Kutsenko S M, Bashkuev Yu 2014 *Datchiki i sistemy* **3 (178)** 7-9
[13] Panin V V 1987 *Energoatomizdat* 120
[14] Kutsenko S M, Klimov N N, Muratov V I 2006 *Izvestiya TPU* **2** 82-87
[15] Kutsenko S M, Muratov V I, Klimov N N 2011 *Sovremennye tekhnologii. Sistemnyi analiz. Modelirovanie* **30** 146-149
[16] Klimov N N, Cheremisin S V, Shurygin S A, Kutsenko S M 2018 *International Conference on Aviamechanical Engineering and Transport (AVENT)* 186-189
[17] Kutsenko S M, Bashkuev Yu B, Muratov V I 2013 *Sovremennye tekhnologii. Sistemnyi analiz. Modelirovanie* **37** 167-170
[18] Matyayev Yu S 1997 *Avtomatika, telemekhanika i sviaz* **9** 23-24.

IOP Conf. Series: Materials Science and Engineering **760** (2020) 012036 doi:10.1088/1757-899X/760/1/012036

Determination of the area of robust stability of the system with a PID controller

N N Kutsyi [2], **T V Malanova** [2], **A P Kutsyi** [1, 2]

[1]Irkutsk state transport university, 15 Chernyshevsky St., Irkutsk, 664074, Russia
[2]Irkutsk national research technical university, 83 Lermontov St., Irkutsk, 664074, Russia

E-mail: tosha1993irk@yandex.ru

Abstract. An approach has been proposed to determine the area of change in the parameters of the controlled plant by the example of a system with PID controller. The search for ranges of parameter changes is based on the solution of the nonlinear programming problem, which is obtained as a conclusion of V.L. Kharitonov's theorem on the robust stability of linear systems.

1. Introduction.

The parameters of automatic control systems (ACS) may change over time for reasons such as aging, wear, heating, deformation, etc. In addition, the parameters of a plant may not be determined accurately enough when developing controllers. In the general case, traditional methods of analysis and synthesis of automatic control systems suggest that, when considering a specific ACS with set nominal values of the parameters, small changes in these parameters do not significantly change the nature of the movement. However, the parameters may not always vary within sufficiently small limits. Then it becomes necessary to study the ACS to maintain its predetermined properties for all parameter values defined within any limits. In this case, they speak of the analysis of the robustness of the considered property of automatic control systems [1,2,3]. The most important among the properties of an ACS is its stability [4, 5, 6]. Thus, this paper discusses the robust stability of an automatic control system [7, 8]. In the conditions of changing the parameters of the controlled plant, it is possible to assign the following problems:

First problem. Along with the known model of an automatic control system with its nominal parameters, the maximum deviations of the real parameters of the controlled plant are known. We need to determine how the change in the parameters will affect the stability of the ACS.

Second problem. There is a known model of an automatic control system. It is necessary to determine in which ranges it is possible to change the parameters of the controlled plant so that the ACS remains stable. It is important to determine on the basis of what criteria stability is considered. So, for example, considering the Nyquist criterion, it is necessary to study the behavior of an amplitude-phase-frequency characteristic. The use of the root criterion involves the study of the relative position of the roots and poles of the transfer function, etc. Also, from a practical perspective, it should be borne in mind that the search for ranges of changes in the controlled plant parameters is not limited only to stability criteria. The specification of parameter tolerances makes its own adjustments to the solution of this problem. On the one hand, the value of these tolerances should not be small, since this requires precision equipment, and on the other hand, a large value of the tolerances will not be able to provide interchangeability of parts. It should not also be forgotten about the quality indicators of the transient process (overshoot, settling time, etc.).

There is no single approach to the analysis of robust stability of automatic control systems; depending on the nature of the system under study and the type of disturbances, the problem is solved in different ways. In the general case, the following approaches can be distinguished:

- a root approach based on the mutual arrangement of zeros and poles [9];
- a frequency approach associated with the construction of frequency characteristics [10];
- an algebraic approach, for example, a method based on V.L. Kharitonov's polynomials [2];
- using methods of the theory of sensitivity [11];
- using the concept of system structural stability in phase space [10];
- using the H-theory [12, 13].

In this paper, we solve the problem of determining the range of values of the parameters of the controlled plant within which the ACS should be robustly stable. Among the many approaches listed above, we chose the simplest approach based on the theorem of V.L. Kharitonov, who determined the formation of the theory of robustness as a whole.

Let us consider this approach by a specific example – a system with a PID controller.

2. Materials and methods

We consider an automatic control system with the following block diagram, Figure 1 [14, 15, 16] .

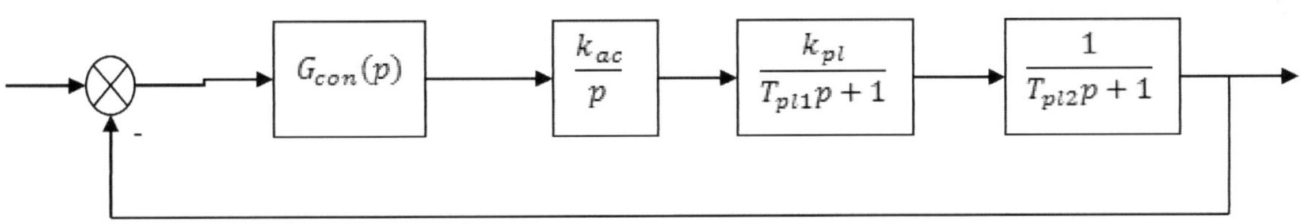

Figure 1. The block diagram of the ACS with a PID controller

Here k_{ac}, k_{ob} are the gain factors of the links of the controlled plant; T_{pl1}, T_{pl2} are time constants;

$G_{con}(p) = k_p + k_i\frac{1}{p} + k_d p$,

where p is the differentiation operator; k_p, k_i, k_d are PID controller parameters.

The transfer function of a closed-loop system with a PID controller is written as:

$$W(p) = \frac{(k_p p + k_i + k_d p^2)k_{ac}k_{pl}}{p^2(T_{pl1}p+1)(T_{pl2}p+1) + (k_p p + k_i + k_d p^2)k_{ac}k_{pl}}. \quad (1)$$

The characteristic polynomial of the considered ACS:

$$Q(p) = T_{pl1}T_{pl2}p^4 + (T_{pl1} + T_{pl2})p^3 + (k_d k_{ac}k_{pl} + 1)p^2 + k_p k_{ac}k_{pl}p + k_i k_{ac}k_{pl}.$$

Let $p = j\omega$, then:

$$Q(j\omega) = T_{pl1}T_{pl2}\omega^4 - j(T_{pl1} + T_{pl2})\omega^3 - (k_d k_{ac}k_{pl} + 1)\omega^2 + jk_p k_{ac}k_{pl}\omega + k_i k_{ac}k_{pl}.$$

Hence, the imaginary and real parts of $Q(j\omega)$ are respectively equal:

IOP Conf. Series: Materials Science and Engineering **760** (2020) 012036 doi:10.1088/1757-899X/760/1/012036

$$u = T_{pl1}T_{pl2}\omega^4 - (k_d k_{ac} k_{pl} + 1)\omega^2 + k_i k_{ac} k_{pl};$$

$$v = -(T_{pl1} + T_{pl2})\omega^3 + k_p k_{ac} k_{pl}\omega.$$

We find the ranges in which the parameters of the controlled plant can vary:

$$\begin{aligned} &\underline{T}_{pl1} < T_{pl1} < \overline{T}_{pl1}; \\ &\underline{T}_{pl2} < T_{pl2} < \overline{T}_{pl2}; \\ &\underline{k}_{pl} < k_{pl} < \overline{k}_{pl}; \\ &\underline{k}_{ac} < k_{ac} < \overline{k}_{ac}. \end{aligned} \tag{2}$$

Then the coefficients of the polynomial $Q(j\omega)$, given there is positivity of all the parameters of the controlled plant (a necessary condition for stability), change within the following ranges:

$$\underline{T}_{pl1}\underline{T}_{pl2} < T_{pl1}T_{pl2} < \overline{T}_{pl1}\overline{T}_{pl2};$$

$$\underline{T}_{pl1} + \underline{T}_{pl2} < T_{pl1} + T_{pl2} < \overline{T}_{pl1} + \overline{T}_{pl2};$$

$$(k_d \underline{k}_{ac} \underline{k}_{pl} + 1) < (k_d k_{ac} k_{pl} + 1) < (k_d \overline{k}_{ac} \overline{k}_{pl} + 1);$$

$$k_p \underline{k}_{ac} \underline{k}_{pl} < k_p k_{ac} k_{pl} < k_p \overline{k}_{ac} \overline{k}_{pl};$$

$$k_i \underline{k}_{ac} \underline{k}_{pl} < k_i k_{ac} k_{pl} < k_i \overline{k}_{ac} \overline{k}_{pl}.$$

Hence, the lower and upper boundaries of the change in the imaginary and real parts of the characteristic polynomial are expressed by the formulas:

$$\begin{aligned} &\underline{u} = \underline{T}_{pl1}\underline{T}_{pl2}\omega^4 - (\overline{k}_d \overline{k}_{ac} \overline{k}_{pl} + 1)\omega^2 + k_i \underline{k}_{ac} \underline{k}_{pl}, \\ &\underline{v} = -(\overline{T}_{pl1} + \overline{T}_{pl2})\omega^3 + k_p \underline{k}_{ac} \underline{k}_{pl}\omega, \\ &\overline{u} = \overline{T}_{pl1}\overline{T}_{pl2}\omega^4 - (k_d \underline{k}_{ac} \underline{k}_{pl} + 1)\omega^2 + k_i \overline{k}_{ac} \overline{k}_{pl}, \\ &\overline{v} = -(\underline{T}_{pl1} + \underline{T}_{pl2})\omega^3 + k_p \overline{k}_{ac} \overline{k}_{pl}\omega. \end{aligned} \tag{3}$$

We compose the Kharitonov polynomials [2]:

$$\begin{aligned} &Q_1 = \overline{u} + j\underline{v}, \\ &Q_2 = \overline{u} + \overline{v}, \\ &Q_3 = \underline{u} + \overline{v}, \\ &Q_3 = \underline{u} + \underline{v}. \end{aligned} \tag{4}$$

Given there are (3) and (4), we obtain:

$$Q_1(\lambda) = k_i \overline{k}_{ac} \overline{k}_{pl} + k_p \underline{k}_{ac} \underline{k}_{pl}\lambda + (k_d \underline{k}_{ac} \underline{k}_{pl} + 1)\lambda^2 + (\overline{T}_{pl1} + \overline{T}_{pl2})\lambda^3 + \overline{T}_{pl1}\overline{T}_{pl2}\lambda^4;$$

$$Q_2(\lambda) = k_i \overline{k}_{ac} \overline{k}_{pl} + k_p \overline{k}_{ac} \overline{k}_{pl}\lambda + (k_d \underline{k}_{ac} \underline{k}_{pl} + 1)\lambda^2 + (\underline{T}_{pl1} + \underline{T}_{pl2})\lambda^3 + \overline{T}_{pl1}\overline{T}_{pl2}\lambda^4;$$

$$Q_3(\lambda) = k_i\underline{k}_{ac}\underline{k}_{pl} + k_p\overline{k}_{ac}\overline{k}_{pl}\lambda + \left(\overline{k}_d\overline{k}_{ac}\overline{k}_{pl} + 1\right)\lambda^2 + \left(\underline{T}_{pl1} + \underline{T}_{pl2}\right)\lambda^3 + \underline{T}_{pl1}\underline{T}_{pl2}\lambda^4;$$

$$Q_4(\lambda) = k_i\underline{k}_{ac}\underline{k}_{pl} + k_p\underline{k}_{ac}\underline{k}_{pl}\lambda + \left(\overline{k}_d\overline{k}_{ac}\overline{k}_{pl} + 1\right)\lambda^2 + \left(\overline{T}_{pl1} + \overline{T}_{pl2}\right)\lambda^3 + \underline{T}_{pl1}\underline{T}_{pl2}\lambda^4.$$

For robust stability of the system, it is enough that the Kharitonov polynomials are stable. By definition, a polynomial is called a stable polynomial if all its zeros are in the left semi-plane on the complex plane [2].

We use the Hurwitz matrices and write them for each polynomial (4):

$$\Delta_{Q_1} = \begin{pmatrix} \left(\overline{T}_{pl1} + \overline{T}_{pl2}\right) & k_p\underline{k}_{ac}\underline{k}_{pl} & 0 & 0 \\ \overline{T}_{pl1}\overline{T}_{pl2} & \left(k_d\underline{k}_{ac}\underline{k}_{pl} + 1\right) & k_i\overline{k}_{ac}\overline{k}_{pl} & 0 \\ 0 & \left(\overline{T}_{pl1} + \overline{T}_{pl2}\right) & k_p\underline{k}_{ac}\underline{k}_{pl} & 0 \\ 0 & \overline{T}_{pl1}\overline{T}_{pl2} & \left(k_d\underline{k}_{ac}\underline{k}_{pl} + 1\right) & k_i\overline{k}_{ac}\overline{k}_{pl} \end{pmatrix}$$

$$\Delta_{Q_2} = \begin{pmatrix} \left(\underline{T}_{pl1} + \underline{T}_{pl2}\right) & k_p\overline{k}_{ac}\overline{k}_{pl} & 0 & 0 \\ \overline{T}_{pl1}\overline{T}_{pl2} & \left(k_d\underline{k}_{ac}\underline{k}_{pl} + 1\right) & k_i\overline{k}_{ac}\overline{k}_{pl} & 0 \\ 0 & \left(\underline{T}_{pl1} + \underline{T}_{pl2}\right) & k_p\overline{k}_{ac}\overline{k}_{pl} & 0 \\ 0 & \overline{T}_{pl1}\overline{T}_{pl2} & \left(k_d\underline{k}_{ac}\underline{k}_{pl} + 1\right) & k_i\overline{k}_{ac}\overline{k}_{pl} \end{pmatrix}$$

$$\Delta_{Q_3} = \begin{pmatrix} \left(\underline{T}_{pl1} + \underline{T}_{pl2}\right) & k_p\overline{k}_{ac}\overline{k}_{pl} & 0 & 0 \\ \underline{T}_{pl1}\underline{T}_{pl2} & (\overline{k}_d\overline{k}_{ac}\overline{k}_{pl} + 1) & k_i\underline{k}_{ac}\underline{k}_{pl} & 0 \\ 0 & \left(\underline{T}_{pl1} + \underline{T}_{pl2}\right) & k_p\overline{k}_{ac}\overline{k}_{pl} & 0 \\ 0 & \underline{T}_{pl1}\underline{T}_{pl2} & (\overline{k}_d\overline{k}_{ac}\overline{k}_{pl} + 1)\lambda^2 & k_i\underline{k}_{ac}\underline{k}_{pl} \end{pmatrix}$$

$$\Delta_{Q_4} = \begin{pmatrix} \left(\overline{T}_{pl1} + \overline{T}_{pl2}\right) & k_p\underline{k}_{ac}\underline{k}_{pl} & 0 & 0 \\ \underline{T}_{pl1}\underline{T}_{pl2} & (\overline{k}_d\overline{k}_{ac}\overline{k}_{pl} + 1) & k_i\underline{k}_{ac}\underline{k}_{pl} & 0 \\ 0 & \left(\overline{T}_{pl1} + \overline{T}_{pl2}\right) & k_p\underline{k}_{ac}\underline{k}_{pl} & 0 \\ 0 & \underline{T}_{pl1}\underline{T}_{pl2} & (\overline{k}_d\overline{k}_{ac}\overline{k}_{pl} + 1) & k_i\underline{k}_{ac}\underline{k}_{pl} \end{pmatrix}.$$

Since the coefficients of the polynomials are positive, for the stability of the Kharitonov polynomials it is sufficient to apply the Lienar-Shipar criterion, according to which the following inequalities are necessary and sufficient:

$$\left(\left(\overline{T}_{pl1} + \overline{T}_{pl2}\right)\left(k_d\underline{k}_{ac}\underline{k}_{pl} + 1\right) - \overline{T}_{pl1}\overline{T}_{pl2}k_p\underline{k}_{ac}\underline{k}_{pl}\right)k_p\underline{k}_{ac}\underline{k}_{pl} -$$

$$-\left(\overline{T}_{pl1} + \overline{T}_{pl2}\right)k_i\overline{k}_{ac}\overline{k}_{pl}\left(\overline{T}_{pl1} + \overline{T}_{pl2}\right) > 0$$

$$\left(\left(\underline{T}_{pl1} + \underline{T}_{pl2}\right)\left(k_d\underline{k}_{ac}\underline{k}_{pl} + 1\right) - \overline{T}_{pl1}\overline{T}_{pl2}k_p\overline{k}_{ac}\overline{k}_{pl}\right)k_p\overline{k}_{ac}\overline{k}_{pl} -$$

$$-\left(\underline{T}_{pl1} + \underline{T}_{pl2}\right)k_i\overline{k}_{ac}\overline{k}_{pl}\left(\underline{T}_{pl1} + \underline{T}_{pl2}\right) > 0$$

$$\left(\left(\underline{T}_{pl1} + \underline{T}_{pl2}\right)\left(k_d\overline{k}_{ac}\overline{k}_{pl} + 1\right) - \underline{T}_{pl1}\underline{T}_{pl2}k_p\overline{k}_{ac}\overline{k}_{pl}\right)k_p\overline{k}_{ac}\overline{k}_{pl} -$$

$$-(\underline{T}_{pl1}+\underline{T}_{pl2})k_i\underline{k}_{ac}\underline{k}_{pl}(\underline{T}_{pl1}+\underline{T}_{pl2})>0$$

$$\left((\overline{T}_{pl1}+\overline{T}_{pl2})(k_d\overline{k}_{ac}\overline{k}_{pl}+1)-\underline{T}_{pl1}\underline{T}_{pl2}k_p\underline{k}_{ac}\underline{k}_{pl}\right)k_p\underline{k}_{ac}\underline{k}_{pl}-$$

$$-(\overline{T}_{pl1}+\overline{T}_{pl2})k_i\underline{k}_{ac}\underline{k}_{pl}(\overline{T}_{pl1}+\overline{T}_{pl2})>0.$$

We complete the replacement. Let:

$$\underline{T}_{pl1}=T_{pl1}-h_{T11};\quad \overline{T}_{pl1}=T_{pl1}+h_{T12};$$

$$\underline{T}_{pl2}=T_{pl2}-h_{T21};\quad \overline{T}_{pl2}=T_{pl2}+h_{T22};$$

$$\underline{k}_{pl}=k_{pl}-h_{kpl1};\quad \overline{k}_{pl}=k_{pl}+h_{kpl2};$$

$$\underline{k}_{ac}=k_{ac}-h_{kim1};\quad \overline{k}_{ac}=k_{ac}+h_{kim2},$$

where

$$h_{T11},h_{T12},h_{T21},h_{T22},h_{kpl1},h_{kpl2},h_{kim1},h_{kim2}\geq 0. \tag{5}$$

As a result, we obtain the system of inequalities:

$$\begin{aligned}
&\Big((T_{pl1}+h_{T12}+T_{pl2}+h_{T22})\big(k_d(k_{ac}-h_{kim1})(k_{pl}-h_{kpl1})+1\big)\\
&\quad-(T_{pl1}+h_{T12})(T_{pl2}+h_{T22})k_p(k_{ac}-h_{kim1})(k_{pl}-h_{kpl1})\Big)k_p(k_{ac}\\
&\quad-h_{kim1})(k_{pl}-h_{kpl1})\\
&\quad-(T_{pl1}+h_{T12}+T_{pl2}+h_{T22})^2k_i(k_{ac}+h_{kim2})(k_{pl}+h_{kpl2})>0\\
&\Big((T_{pl1}-h_{T11}+T_{pl2}-h_{T21})\big(k_d(k_{ac}-h_{kim1})(k_{pl}-h_{kpl1})+1\big)\\
&\quad-(T_{pl1}+h_{T12})(T_{pl2}+h_{T22})k_p(k_{ac}+h_{kim2})(k_{pl}+h_{kpl2})\Big)k_p(k_{ac}\\
&\quad+h_{kim2})(k_{pl}+h_{kpl2})\\
&\quad-(T_{pl1}-h_{T11}+T_{pl2}-h_{T21})^2k_p(k_{ac}+h_{kim2})(k_{pl}+h_{kpl2})>0\\
&\Big((T_{pl1}-h_{T11}+T_{pl2}-h_{T21})\big(k_d(k_{ac}+h_{kim2})(k_{pl}+h_{kpl2})+1\big)\\
&\quad-(T_{pl1}-h_{T11})(T_{pl2}-h_{T21})k_p(k_{ac}+h_{kim2})(k_{pl}+h_{kpl2})\Big)k_p(k_{ac}\\
&\quad+h_{kim2})(k_{pl}+h_{kpl2})\\
&\quad-(T_{pl1}-h_{T11}+T_{pl2}-h_{T21})^2k_i(k_{ac}-h_{kim1})(k_{pl}-h_{kpl1})>0\\
&\Big((T_{pl1}+h_{T12}+T_{pl2}+h_{T22})\big(k_d(k_{ac}+h_{kim2})(k_{pl}+h_{kpl2})+1\big)\\
&\quad-(T_{pl1}-h_{T11})(T_{pl2}-h_{T21})k_p(k_{ac}-h_{kim1})(k_{pl}-h_{kpl1})\Big)k_p(k_{ac}\\
&\quad-h_{kim1})(k_{pl}-h_{kpl1})\\
&\quad-(T_{pl1}+h_{T12}+T_{pl2}+h_{T22})^2k_i(k_{ac}-h_{kim1})(k_{pl}-h_{kpl1})>0
\end{aligned} \tag{6}$$

Next, we introduce the objective function:

$$I = \left(h_{T11} + h_{T12} + h_{T21} + h_{T22} + h_{kpl1} + h_{kpl2} + h_{kim1} + h_{kim2}\right)^{-1} \quad (7)$$

Since the problem of searching for ranges (2) is reduced to finding the maximum values of $h_{T11}, h_{T12}, h_{T21}, h_{T22}, h_{kpl1}, h_{kpl2}, h_{kim1}, h_{kim2}$ taking into account restrictions (5) and (6), we obtain the problem of minimizing the function I under the indicated restrictions:

$$\min_{h_{T11}, h_{T12}, h_{T21}, h_{T22}, h_{kpl1}, h_{kpl2}, h_{kim1}, h_{kim2}} I. \quad (8)$$

Thus, in order to determine the ranges of variation of the parameters of the controlled plant of an automatic system with a PID controller, it is sufficient to solve the nonlinear programming problem (8), (7), (6), (5).

3. Results.

Let the initial automatic system with a PID controller be set by the parameters defined in Table 1.

Table 1. System parameter

Parameter	k_{ac}	k_{pl}	T_{pl1}	T_{pl2}	k_p	k_i	k_d
Value	0.1	1	3	2	0.5	0.04	0.01

The transient process for this system is shown in Figure 2.

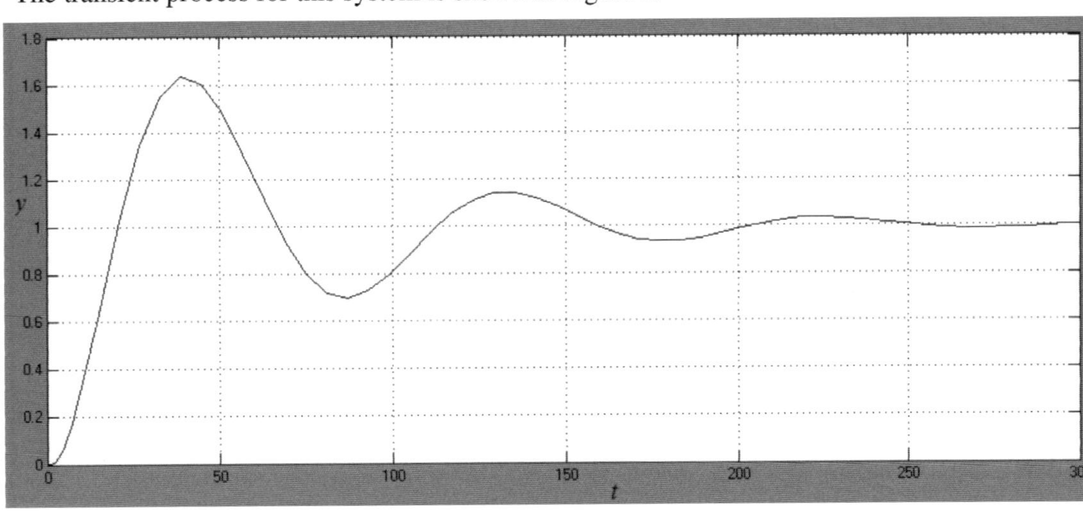

Figure 2. The transient process of the automatic system with a PID controller

Having solved problems (8), (7), (6), (5), we obtained the boundaries of the change in the parameters of the controlled plant, which are presented in Table 2.

Table 2. Limits of change of parameters of object of regulation

	Lower boundary	Upper boundary
k_{ac}	0	0.1000
k_{pl}	0	1.0000
T_{pl1}	1.1779	9.4802
T_{pl2}	0	2.0000

From Table 2 we can observe that the lower boundaries of the gain factors are zero. This is because the smaller the gain factor is, the less the oscillation is. With the values of the parameters corresponding to the upper boundaries, the system is at the stability boundary.

4. Conclusion

Thus, an approach has been developed to determine the permissible values of the parameters of the controlled plant at which the automatic system will be robustly stable, based on the reduction of the original problem to the nonlinear programming problem. The system of restrictions is obtained on the basis of the theorem of V.L. Kharitonov on robust stability.

The condition on which the approach is based is sufficient but not necessary for the system under consideration, since the coefficients of the polynomial are interdependent.

In general, any automatic control system must satisfy not only the requirement for stability, but also meet a number of other quality criteria. In this case, one can add some quality criterion to problem (8), (7), (6), (5), and, having solved the multicriteria problem, obtain solutions taking into account other characteristics of the transient process.

References

[1] Dorf R, Bishop R 2012 *Modern control systems* (Moscow: Laboratory of Basic Knowledge Publ.) 832.

[2] Kim D P 2003 *The theory of automatic control. Volume 1. Linear systems* (Moscow: FIZMATLIT Publ.) 288.

[3] Oshchepkov A Yu 2013 *Automatic control systems: theory, application, modeling in MATLAB* (St. Petersburg: Lan' Publ.) 208.

[4] Lukyanov AV, Krakovsky Yu M, Arshinsky L V and Kutsy N N 2018 The development of software for controlling a safety system of the machines using vibration analysis *Far East Journal of Mathematics 1 Sciences (EJMS).* **103(2)** 441-450.

[5] Kucyi N N, Lukyanov A V, Kargapol'cev S K and Tikhii I I 2018 Training of neural network based PWM controllers *Advances and Applications in Discrete Mathematics* **19(4)** 359-371.

[6] Galyaev A A, Lysenko P V 2019 Energy-Optimal Control of Harmonic Oscillator *Automation and Remote Control,* **80(1)** 16–29.

[7] Kogan M M 2016 Design of optimal and robust control with H ∞/γ 0 performance criterion *Automation and Remote Control* **77(8)** 1317–1333.

[8] Zhirabok A N, Suvorov A Yu 2014 A method for constructing robust diagnostic observers *Automation and Remote Control* **75(2)** 208–218.

[9] Efimov S V, Zamyatin S V, Sukhodoev M S and Gaivoronsky S A 2008 Determination of the desired location region of the dominant poles of a closed system taking into account its zeros *The News of Tomsk Polytechnic University* **312(5)** 57-61.

[10] Andronov A A, Witt A A and Khaikin S E 1959 *Theory of oscillations* (Moscow: Fiz.-mat. literature Publ.) 916.

[11] Rosenwasser E N, Yusupov R M 1971 *Methods of the theory of sensitivity in automatic control* (Leningrad: Energiya Publ.) 345.

[12] Egupov N D 2000 *Methods of the modern theory of automatic control* (Moscow: Bauman MGTU) 748.

[13] Egupov N D 2011 *Methods of robust, neuro-fuzzy and adaptive control* (Moscow: Bauman MGTU) 744.

[14] Kutsyi N N, Livshits A V 2018 Searchless algorithm for parametric optimization of a PI-controller with semi-permant integration *Advanee des in Differential Equation and Control Processes* **19(2)** 69-82.

[15] Morozov M V 2019 On Small Perturbations of a Periodic Homogeneous Differential Inclusion with an Asymptotically Stable Set *Automation and Remote Control* **80(5)** 834–839.

IOP Conf. Series: Materials Science and Engineering **760** (2020) 012036 doi:10.1088/1757-899X/760/1/012036

[16] Aleksandrov A. G. 2018 Design of Controllers by Indices of Precision and Speed. III. Control-Stable Multidimensional Plants *Automation and Remote Control* **79(2)** 241–257.

IOP Conf. Series: Materials Science and Engineering **760** (2020) 012037 doi:10.1088/1757-899X/760/1/012037

Diagnostic operation of fabrications of rolling stock by high-frequency method

A G Larchenko, N G Filippenko, Livshits A.V., Kargapoltcev S.K.

Irkutsk State Transport University, Chernyshevsky, 15, 664074, Russia

E-mail: Larchenkoa@inbox.ru

Abstract. In this scientific work, the authors consider the issue of non-destructive testing. In the article, the authors describe the process of improving the performance properties of polymeric products of complex configuration. In the article, the authors choose the method of high-frequency control as the most suitable for assessing the quality of products in the manufacture and during repair work. The scientific work describes the process of diagnosing products made from polymeric materials on the developed diagnostic device. This article describes the development of methods for designing a technological additional device for diagnosing products of complex configuration. In this paper, the authors solved the problem of manufacturing electrodes for experimental diagnostics. The authors drew conclusions and set goals for further research.

1. Introduction

Today, the engineering industry is the largest industry in the consumption of polymeric materials. The introduction of composite materials has significantly saved on the use of expensive metals; reduce the cost; increase the warranty period. But, at manufacturing plants and repair profile enterprises, the control of complex configuration products is carried out only by a visual method. The visual method cannot reveal hidden structural defects. Hidden defects lead to additional consumption of resources, reduced quality of products, and emergency situations. The purpose of a scientific article is the organization of the process of diagnosing products of complex spatial form.

2. High Frequency Method. Control device. Electrodes

Theoretical and experimental studies have shown that the most appropriate method of diagnosis in terms of safety and process efficiency is the method of high-frequency control. This method is based on the detection of defects in products made of polymeric materials, as a result of exposure to an electric field. This method can identify metallic and foreign inclusions, pores, voids [1-7]. The authors have developed a device for diagnosing polymeric materials. Figure 1 shows a diagnostic device. The device detects defects without changing the structure of the material. But without the use of electrodes it is impossible to diagnose products of complex spatial form. In the article, the authors developed a technique for the manufacture of electrodes for diagnosis. To design the electrodes, it is necessary to divide the product into sections, which are a sequence of elementary capacitors. The calculated gap between the electrode and the product was also introduced. For uniform energy impact on the elements of the product. The authors calculated the thickness of this gap. The calculations are described by the example of wear-resistant polymer linings of freight cars [8-11].

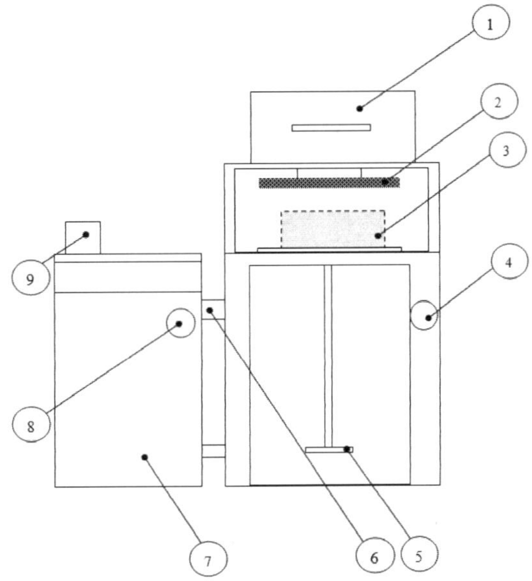

Figure 1. Biagnostic device. 1 is the protective screen; 2 is the top high-potential plate; 3 is the product; 4 is the timer; 5 is the drive; 6 is the cable, 7 is the high-frequency radiation generator, 8 is the variable capacitor controller, 9 is the automation unit.

The product is placed between the electrodes, thereby creating working capacitors, where the polymer acts as a dielectric. For uniform heating of the product throughout the volume, it is necessary to ensure equal values of the capacitance in each created condenser. Figure 2 shows a diagram of the dissection of a part into elementary geometric surfaces. The authors calculated the capacitor at each site.

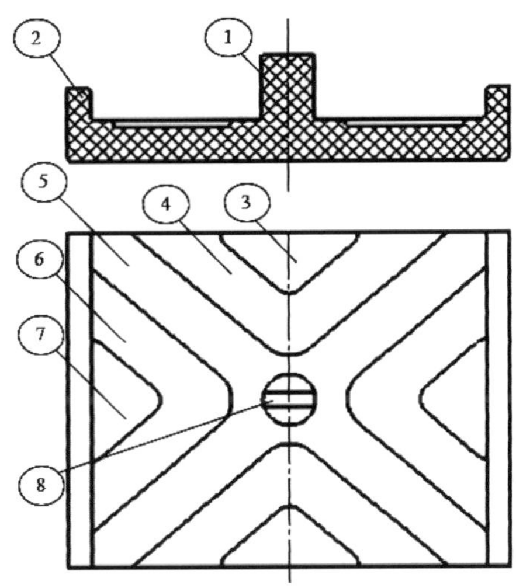

Figure 2. Dissection of a part into elementary geometric surfaces. 1 - cylindrical part; 2 - side surface; 3,5,7 - ledges; 4,6- grooves; 8- technological groove

The value of capacitors on each elementary surface is:

$$C = \frac{\varepsilon \varepsilon_0 S}{d} \tag{1}$$

IOP Conf. Series: Materials Science and Engineering **760** (2020) 012037 doi:10.1088/1757-899X/760/1/012037

where C is the electric capacity, F; ε is the dielectric constant; *S* is the plate area, m^2; d is the distance between the plates, m; ε_0 is the electric constant.

Next, the authors calculated the capacitance of the capacitors, consisting of the capacity of the product and the air gap. Figure 3 shows the serial connection of the capacitor capacitance, according to the expression:

$$1/C_1=(1/C_2)+(1/C_{air}) \quad (2)$$

Air gap capacity:

$$C_{air}=(C_1C_2)/(C_2-C_1) \quad (3)$$

Air gap value:

$$d_{air}= \varepsilon\varepsilon_0 S/C_{air} \quad (4)$$

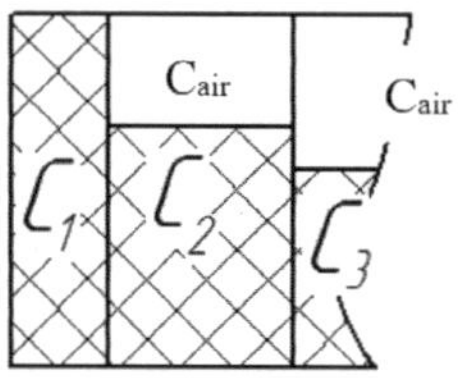

Figure 3. Capacitors circuit diagram.

An important step in creating the process of diagnosing and improving the quality of products is the development of a methodology for calculating the parameters of the electrode design gap. This gap ensures uniform energy impact on the elements of the product. Since the surface of the polymer product in question (Figure 1) has a cylindrical shape and a technological groove, it is necessary to take into account the uneven air gap when designing the electrodes. To diagnose this area, it is necessary to determine the direction of the electric field strength on different planes of the product (Figure 4) when exposed to high-frequency radiation. Studies have shown that the tension in the process of diagnosis arises in different directions. Therefore, the heating of the product is uneven.

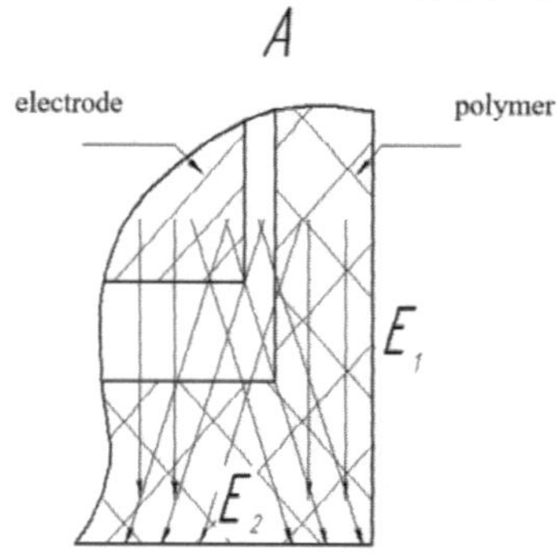

Figure 4. Electric field strength

The system electrode – air gap – the polymer can be represented as a capacitor circuit (Figure 5).

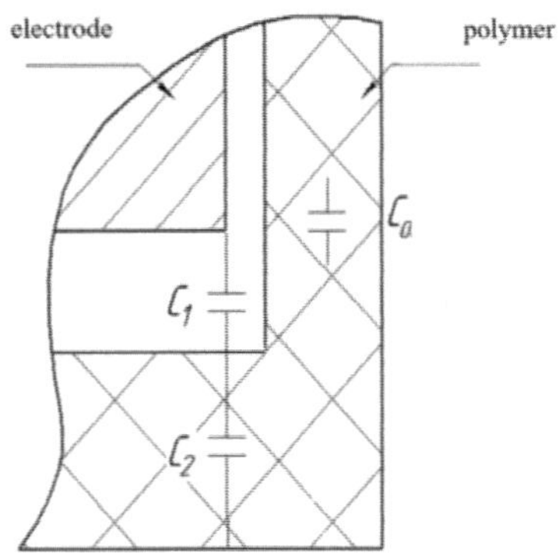

Figure 5. The system of replacing the electrode-air gap – polymer.

Based on figure 5:

$$C_o = (\varepsilon\varepsilon_0 S/\varepsilon_2) \,/\, (d_2\,\varepsilon_1 + d_1\,\varepsilon_2) \quad (5)$$

expression 5 determines the total capacitance of any system of series-connected capacitors during diagnosis. But, since the product has a complex geometric shape, it is necessary to take into account the change in the air gap and the polarity of the capacitors (Figure 6).

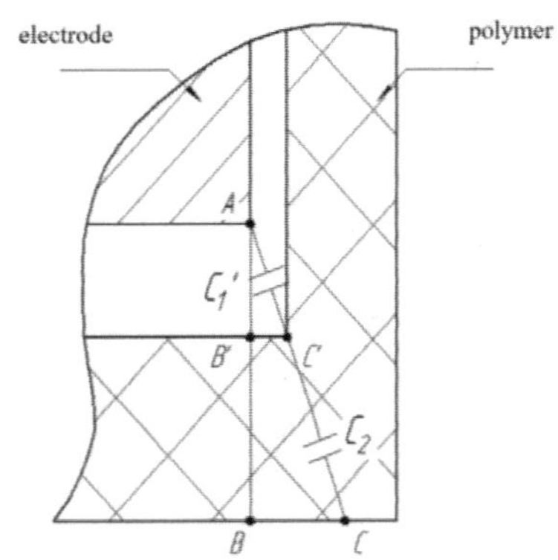

Figure 6. Air gap distance and capacitors polarity

To determine the distance between the plates, depending on the width of the air gap, an auxiliary triangle ABC was considered. Figure 7 shows the triangle.

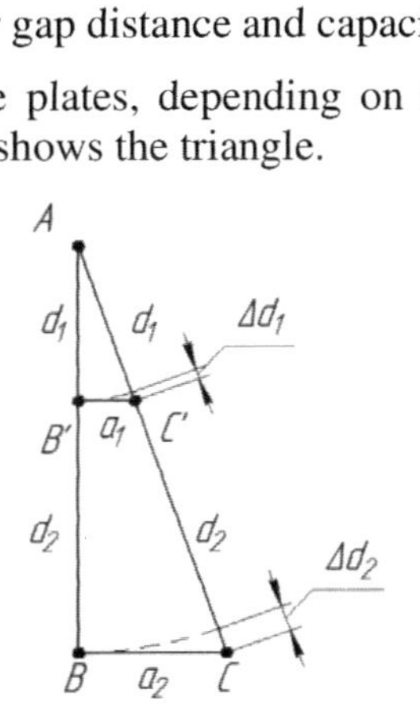

Figure 7. ABC triangle.

Air gap size Δd_1:

$$\Delta d_1 = \sqrt{d_1^2 + a_1^2} - d_1 \tag{6}$$

Δd_2 in the polymer product:

$$\Delta d_2 = \sqrt{d_2^2 + a_2^2} - d_2 . \tag{7}$$

Capacitor values for air gap:

$$C_1' = \frac{\varepsilon\varepsilon_0 S}{d_1 + \Delta d_1} \tag{8}$$

$$C_2' = \frac{\varepsilon\varepsilon_0 S}{d_2 + \Delta d_2} . \tag{9}$$

The total value of the capacitance of the serial connection of elementary capacitors:

$$C_s = (\varepsilon\varepsilon_0 S/\varepsilon_2) / ((d_2+\Delta d_2)\,\varepsilon_1) + ((d_1+\Delta d_1)\,\varepsilon_2). \tag{10}$$

The calculations made it possible to design electrodes for diagnosing a polymer product by uniformly heating with a high-frequency field. For all surfaces of the polymer product, together with the electrodes, the factors are calculated capacitance. This is necessary to ensure a constant and equal distribution of high-frequency radiation exposure over the entire surface of the part. Electrodes allow high-frequency diagnostics of all elements of the product connected to the electrodes simultaneously in one cycle. The developed technique is applicable to polymers of various grades. The authors conducted a series of diagnostic experiments. Figure 8 shows the electrodes made.

Figure 8. Electrodes

3. Conclusion

Experimental studies have allowed us to confirm the decisions made and theoretical studies. The results of the diagnostics showed that the high-frequency field affects the object of control evenly, the use of the developed equipment allows for the diagnosis of products in an industrial environment without damage.

References

[1] Livshits A V, 2015 Automated system of scientific researches high-frequency electrothermics *Problems of mechanical engineering and automation* **3** 120-126.

[2] Larchenko A G, 2016 Automated Device for Diagnosing Polymeric Products of Complex Configuration by the Method of High-Frequency Radiation *Control. Diagnostics* **2** 61–65.

[3] Livshits A V, 2015 Control of technological processes of high-frequency electrothermy of polymers *Problems of mechanical engineering and automation* **3** 120-126.

[4] Butorin D V, Filippenko N G, Filatova S N, Livshits A V, Kargapol'tsev S K, 2015. Automation of control of structural transformations in polymeric materials during electrothermal treatment *Sovremennyye tekhnologii. Sistemnyy analiz. Modelirovaniye* **4 (48)** 80-86.

[5] Popov S I, Filippenko N G, Livshic A V, 2013 Determination of physical and mechanical parameters of polymeric materials at high-frequency dielectric heating in electrothermal installations *Sovremennye tekhnologii. Sistemnyj analiz. Modelirovanie* **2 (38)** 50 – 55.

[6] Livshits A V, 2014 Investigation of the process of high-frequency electrothermy of thermoplastics on the basis of its mathematical modeling *Sistemy. Metody. Tekhnologii.* **2 (22)** 90 - 98.

[7] Livshic A V, Filippenko N G, Popov S I, Larchenko A G, 2013 Study of the effect of dielectric elements of the working capacitor in a high-frequency electrothermal installa-tion on the processing of polymeric materials *Sovremennye tekhnologii. Sistemnyj analiz. Modelirovanie* **3 (39)** 270-275.

[8] Filipenko N G, 2012 Matematicheskaya model' protsessa vysokochastotnoy obrabotki polimernykh materialov *Sovremennye tehnologii. Sistemnyj analiz. Modelirovanie* **1** 76-79.

[9] Larchenko A G, Filippenko N G, Popov S I, Livshits A V, 2013 Investigation of the influence of dielectric elements of the working capacitor of a high-frequency electrothermal installation on the processing of polymer materials *Modern technologies. System analysis. Modelling* **3 (39)** 270-275.

[10] Livshits A V, 2017 Mathematical modeling of the processes of the high-frequency heating of thermoplasts and quality improve-ment of welded polymeric items *JP Journal of Heat and Mass Transfer Pushpa Publishing House, Allahabad, India.*

[11] Livshic A V, 2015 Automated system of scientific research of high-frequency electrothermia *Problems of mechanical engineering and automation* **4** 54-60.

IOP Conf. Series: Materials Science and Engineering **760** (2020) 012038 doi:10.1088/1757-899X/760/1/012038

Automated assessment of the low-rigid composite parts influence on the product assemblability in the GePARD system

D S Leonovich[1], D A Zhuravlev[1], Yu I Karlina[1], A S Govorkov[1], A I Karlina[1]

[1]Irkutsk National Research Technical University, 83 Lermontov Street, Irkutsk, 664074, Russia

E-mail: karlinat@mail.ru

Abstract. Composite components tend to have higher deviations from the nominal dimension in conjunction with metal components. Composites production processes simulation methods have already been developed before long time ago. This article presents the possibility of integrating the technique of non-rigid dimensional chains, taking into account the elastic properties of composite materials to assess the products assemblage with machinery parts from composite materials, on the basis of three-dimensional GePARD system. The fundamental points of the dimensional chain calculating method with non-rigid parts, as well as methods of assessing the stiffness for some types of composite materials, which take into account some peculiarities of its structure, are also examined. The assemblability of an assembly unit is represented by the parameter of the functional assembly requirement depending on the critical accuracy characteristics.

1. Introduction

Geometry and design of any industrial product must match characteristics that were laid by a design-engineer in a part construction, i.e. must perform all the dimensional parameters and provide expected functions. In the ideal model, a competently designed and manufactured product or part of it does not need to be modified at the assembly level and operation stage. In modern production, these aspects are solved through the use of automation of the computer-aided design and life cycle processes modeling, which require more detailed information about nominal geometry with permissible deviations in the electronic layout of the product. It should be noted that these deviations are a kind of important macro-information that forms the electronic model of the product.

Advanced high-tech products of our time have a different number of nodes containing low-rigid parts, which will be discussed in this article. Mechanical engineering technology in the field of interchangeability methods in machining can't be used in an assembly process of this products type, otherwise there is no guarantee of required quality with acceptable costs parts reception.

The theoretical basis of the products assembly processes began to form a scientific concept of engineering technology emergence and in practice there are various options for assessing and calculation of the final product assembly, which have their effectiveness and their advantages/disadvantages [1-4]. These provisions are based on the following fundamental points introduced by B. S. Balakshin: method of achieving the accuracy of a closing link, the dimensional chains theory and the basing theory [4]. One of the most significant methods is the method of dimensional analysis, set out in the Russian Federation in the form of state standard GOST 16319-80 and in the form of guidelines RD 50-635-87. Dimensional analysis of engineering products provides an opportunity to calculate optimal geometric parameters-tolerances accuracy. This factor should be considered as factor of the final product efficiency and quality determination. The dimensional chains analysis can be both design and technological, therefore, these process of design and technological activity cycle form tasks that demands the use of automated facilities.

2. Chapter 1: Dimensional analysis of assemblies with non-rigid components

The assembly stage is a labor-intensive final technological process, which particularly affects the final characteristics, cost and competitiveness of any machine-building product, including the problems of the manufactured products quality improvement, reducing production and reducing the preparation time for serial production costs.

The assembly with low-rigid composite materials parts can be considered as a dimensional chain designing process which represents the mathematical model of the real final product, where inaccuracies of the parts are summed according to certain laws. According to the classical dimensional chain theory, dimensional error does not take into account parts rigidity, which even with correctly performed dimensional calculations in some cases does not guarantee quality of the assembly. In this regard, the accuracy, which is inherent in the design documentation and is achieved in the machine assembling process, is largely conditional: a number of the classical dimensional chain theory assumptions in practice give an error, which requires individual selection methods during assembly, including adjustments and fits for composite parts [5].

Non-rigid dimensional chains - dimensional chains where under the influence of internal and external factors in the dimensions of constituent links there are reversible/irreversible changes, which affects the value of the closing link formation. As a result, the acquired products size depends on different factors such as the elastic/plastic characteristics of the materials, the actual dimensional chain links values, the surface layer quality, the deviations of the contacting surfaces, the values of the forces applied to the parts during the conjunction process, the value of the other parts deviations. They form the dimensional chain not directly, but affect the part rigidity and deformation at its final assembly level and at the exploitation stage. Thus, since the existing theory of dimensional chains does not allow to fully evaluate and to study the dimensional connections during the assembly process or in the operated machine, the output parameters are properly controlled only on a fully assembled product.

Ensuring the assembliability of engineering products with composite components without additional fitting work is an urgent task. In addition to high labor costs, the composite materials fitting work is a source of mass microfine dust that require special conditions to protect worker's health. Such fitting operations may demand the partial disassembly of the product or its units to clean it from the resulting dirt.

Production performance and assembly quality increase in modern conditions can be achieved by creating a quality-controlled assembly process based on the reliable analytical methods of real non-rigid dimensional chains calculation and by developing new methods of the required assembly quality insurances. The emerging dimensional chains accuracy insurance problem of the specified non-rigid assembly can be solved by calculating such dimensional chains, taking into account all the significantly influencing factors.

To solve the problem of non-rigid dimensional chains calculation, it is necessary to identify and to analyze dimensional and physical connections which arrise during the assembly process. The non-rigid dimensional chain equation of an assembly in general[6] is:

$$Y = f_1(x_i) + f_2(x_i, z_i, \mu, E_d, p_a) + f_3(x_i, z_i, \mu, E_d, p_c, N, R_{\max}, R_p, \rho, b, v, H_B, R_B, \sigma_T) \quad (1.1)$$

where Y – the controlled size which also is the closing link of the dimensional chain; x_i – the actual dimensions of parts included in the dimensional chain; z_i – the actual dimensions of parts which are not included in the dimensional chain, but that affect total volumes and contact stiffness; μ – the part material Poisson's ratio; E_d – the part material elasticity modulus; p_a and p_c – the nominal and contour pressure on the contact area; $N=H_\mu surf/H_\mu nom$ – metal surface layer hardening degree; $H_\mu surf$ and $H_\mu nom$ – treated surface and workpiece raw material microhardness; R_{max} – the maximum height of the workpiece surface roughness profile; R_p – the distance from the ledge line to the middle profile line (asperities smoothing height); ρ – the asperities peaks rounding radius; b and ν – the exponential approximation parameters of the initial area curve contact surface asperities; R_B – the surface waves radius; H_B – the surface waves height; σ_T – the component material yield strength.

In the formula (1.1): (*f*1) is the real part size in the considered dimensional chain link (in the classical dimensional chain theory, the dependence $Y= f(x_i)$ is known); (*f*2) takes into account all insufficient rigidity deformations of dimensional chain parts (the dependence of these deformations value on the above parameters is considered in the science of materials resistance); the (*f*3) considers the dimensional chain deformations due to the insufficient stiffness of the mating parts joints. The contact connection displacements types that are relevant to the above parameters in the set of an assembly parts joints are considered in the works of N.B. Demkin, I.V. Kragelsky, A.G. Suslov, E.V. Ryzhov and many other authors [7-10].

As a result of a variety of reasons in non-rigid dimensional chains there are constantly significant changes in the constituent units dimensions and relative positions, it is necessary to consider these dimensions as variables:

$$x_i = f(E,\sigma,v,T,\tau,F) \tag{1.2}$$

where E is the magnetic field; σ is the magnitude of the residual stresses in the component surface layer; *v* – the details voltage, T is the ambient temperature; τ – the part work time; F – the external force acting on the part [6].

3. Chapter 2: Elastic properties of fibrous composite materials

Composite materials should be considered as an anisotropic element, which establishes a special kind of connection between stresses and strains. Thus, for thin-walled multilayer structures, the plane stress state and bending are typical, i.e. the transition from general relations for a linearly elastic anisotropic body to particular forms of their recording for these stress states is important. Then Hooke's law for such elements according to [11]:

$$[\varepsilon]=[F][\sigma] \tag{2.1}$$

where the matrix [F] is defined by the dependency:

$$[F]=\frac{1}{E}\begin{bmatrix} 1 & -\mu & -\mu & 0 & 0 & 0 \\ -\mu & 1 & -\mu & 0 & 0 & 0 \\ -\mu & -\mu & 1 & 0 & 0 & 0 \\ 0 & 0 & 0 & 2(1+\mu) & 0 & 0 \\ 0 & 0 & 0 & 0 & 2(1+\mu) & 0 \\ 0 & 0 & 0 & 0 & 0 & 2(1+\mu) \end{bmatrix}. \tag{2.2}$$

For an isotropic three-dimensional body, the matrix [F] is invariant to the choice of the coordinate system and is formed using two constants E and μ, which entirely determine the elastic properties of the isotropic body. In the complex stress state of the isotropic body, the elongations ε_{ij} are independent of the tangent stresses σ_{ij}, when the shear angles Υ_{ij} are dependent on the corresponding tangent stresses σ_{ij}. In this regard, for the elastic isotropic body, the main stress state axes coincide with the deformed state main axes.

By solving the system of equations (2.1) with respect to stresses, Hooke's law can be rewritten as:

$$[\sigma]=[k][\varepsilon] \tag{2.3}$$

where $[k]=[F]^{-1}$ is the material stiffness matrix:

$$[k]=\frac{E}{(1+\mu)(1-\mu)}\begin{bmatrix} 1-\mu & \mu & \mu & 0 & 0 & 0 \\ \mu & 1-\mu & \mu & 0 & 0 & 0 \\ \mu & \mu & 1-\mu & 0 & 0 & 0 \\ 0 & 0 & 0 & (1-2\mu)/2 & 0 & 0 \\ 0 & 0 & 0 & 0 & (1-2\mu)/2 & 0 \\ 0 & 0 & 0 & 0 & 0 & (1-2\mu)/2 \end{bmatrix}. \quad (2.4)$$

An elastic body is considered as anisotropic if its elastic characteristics differ in different directions, then any components of the stress tensor σ_{ij} can contribute the appearance of all components of the strain tensor. The malleability coefficients of an anisotropic material depend on the coordinate system position. In general, they form a completely filled matrix [*f*] of size 6x6. According to the body compliance on reciprocity theorem for the matrix components, when the condition $F_{mn}=F_{nm}$ is satisfied only when six diagonal and half non-diagonal components are independent – a total of 21 coefficients. Then Hooke's law can be rewritten in expanded form:

$$\begin{bmatrix} \varepsilon_{11} \\ \varepsilon_{22} \\ \varepsilon_{33} \\ \gamma_{12} \\ \gamma_{23} \\ \gamma_{31} \end{bmatrix} = \begin{bmatrix} F_{11} & F_{12} & F_{13} & F_{14} & F_{15} & F_{16} \\ & F_{22} & F_{23} & F_{24} & F_{25} & F_{26} \\ & & F_{33} & F_{34} & F_{35} & F_{36} \\ & & & F_{44} & F_{45} & F_{46} \\ & \text{sym} & & & F_{55} & F_{56} \\ & & & & & F_{66} \end{bmatrix} \begin{bmatrix} \sigma_{11} \\ \sigma_{22} \\ \sigma_{33} \\ \sigma_{12} \\ \sigma_{23} \\ \sigma_{31} \end{bmatrix}. \quad (2.5)$$

The general form of anisotropy in real materials is rare. Most often, the structure of the material has identical elastic properties in some directions. Then the number of independent coefficients in the matrix [*f*] and [*k*] is reduced, as well as with a rational arrangement of the coordinate system, the recording of Hooke's law is simplified.

So in the source [11] are considered special cases for an orthotropic material, for a material with the presence elastic properties symmetry plane, for a transversal-isotropic body with an isotropy plane.

At the present moment, a lot micromodels related to the composite materials reinforced with straight fibers are developed, but it should be noted that all these models mostly agree with each other and with experiment only in the longitudinal modulus of elasticity values, and in relation to other obtained modules may differ from each other. That happens because the structure of the real material in fact may be different from the idealized, which is considered in the models. So when designing and calculating structures made of composites, it is rational to apply an experimental method for determining the physical characteristics of the material. In this case, obtained experimental characteristics guarantee the possibility of taking into account all the manufacture features of particular composite material.

4. Chapter 3: The integration features of the presented methodology with GePARD system spatial analysis

Single time assembly process refers to the design principles and assembly/adjustment implementation processes. That implementation guarantees required output characteristics of the product without additional work in the serial production process. An important criterion for the implementation of one is an adequate mathematical model of the assembly process and its design tools, for example, used in the GePARD system.

In the GePARD system developed at the Irkutsk National Research Technical University is applied own development concept of "the functional requirement" to an assembly on the tolerances model basis [12-15]. The functional requirements for the area is expressed in such nominal value deviations that provide functional dimensions of the product, and is expressed in the function [17]:

$$[F_r] = f(G,T,L) \tag{3.1}$$

where $G = \{g_1, g_2, \dots g_k / k \in N\}$ – the assembly geometry representing set; $T = \{t_1, t_2, \dots t_n / n \in N \cup \{0\}\}$ – the specified tolerance set; $L = \{l_1, l_2, \dots l_m / m \in N \cup \{0\}\}$ – the assembly constraints set. The function *f* for each functional requirement is determined by a mathematical model corresponding to its type. The mutual deviation vectors calculation algorithm for these mathematical models is stated in [18], and forms the configuration process of the direct modeling [16] – where the critical accuracy characteristics definition is applied.

A systematic approach to the assembly processes is determined by the final product functional quality and its components internal state linking. Taking particular features of the actual details basing (as a part of assembly units), components basing surfaces reception inaccuracies lead to change of their stress-strain state. In addition to the theoretical operating and the real operating loads, product is reorganized as an independent assembly system, which characterizes the difficulty of predicting high-tech products behavior. The system approach methodology explains the emergence of the above-mentioned assembly techniques as opportunities to compensate negative consequences of the idealization and the unaccounted natural interaction relationships set of the parts during the assembly process.

Summarizing the above, this development is dedicated to the adequate tolerances assignment/verification problem solution. It is also dedicated to identification of "critical places" in the characteristics of the assembly units, i.e. leads to the electronic models quality improvement and reduction of the engineering products testing time.

5. Conclusion

Returning to the question of the reliable methods and programs for the low-rigid dimensional chains calculation lack is expressed in the lengthening of designed product stages and constructive-technological preparation refinement of the serial production costs increase.

Accepting the global computer-aided analysis development experience to identify the methods and tools for solving different industrial problems such as optimal flexibility fitment insurances in the assembly [19,20], tolerances analysis [21] and synthesis [22], flexible components standardization [23], the assembly process optimal sequence choice [24] or pre-calculation of the adjustment processes [25], the partial calculation methods application of the dimensional chains with low-rigid components on the basis of the composite parts stiffness calculation provisions and reference [11], complementing the three-dimensional automated analysis of the GePARD system capabilities, we can control the process of composite parts assembly by applying information about the actual assembly parts set dimensions.

This scientific research result can be implemented in the GePARD system in the future: i.e., using information about the assembly units "critical parameters", we can deduce dependences of the controlled non-rigid composite part dimensions, and, ultimately, increase the level of the direct part assembliability control.

References

[1] Chase K W, Magleby S P, Glancy C G 1997 *Proc. 5th CIRP Int. Seminar on Computer-Aided Tolerancing (Toronto, Canada)* URL: http://adcats.et.byu.edu/Publication/97-4/cirp_2_7_97a.PDF

[2] Polini W 2011 *Geometric Tolerances. Impact on Product Design,Quality Inspection and Statistical Process Monitoring* (London: Springer) pp 39-68

[3] Pasupathy T M K, Morse E P, Wilhelm R G 2003 *Journal of Computing and Information Science and Engineering* vol. 3 pp 64-75

[4] Balakshin B S 1969 *Fundamentals of engineering technology* (Moscow: Mechanical Engineering) p 559

[5] Graham Warwick 1997 *European Project Tackles Composites Assembly Hurdle To Production*

Ramp-up (Toronto, Canada) URL: https://aviationweek.com/technology/european-project-tackles-composites-assembly-hurdle-production-ramp
[6] Suslov A G (Et al.) 2012 *High technology in mechanical engineering* (Moscow: Engineering) p 528
[7] Demkin N B 2004 *Mechanics and physics of frictional contact and boundary layers* (Tver: TSTU) p 136
[8] Ryzhov E V 1962 *Fundamentals of calculating the butt surfaces of machine parts for contact stiffness* (Moscow: Mashgiz) p 143
[9] Kragelsky I V 1968 *Friction and wear* (Moscow: Mechanical Engineering) p 480
[10] Suslov A G 1977 *Technological support for contact stiffness of joints* (Moscow: Nauka) p 102
[11] Skvortsov Yu V 2013 *Mechanics of composite materials* (Samara: Samara State Aerospace University) p 94
[12] Gaer M A 2004 *ISTU Bulletin* **4** p 177
[13] Gaer M A, Zhuravlev D A 2005 *ISTU Bulletin* **1** 116-125
[14] Gaer M A, Zhuravlev D A, Shabalin A V, Yatsenko O V 2009 *The collection of materials of the scientific and technical seminar "Advanced technologies and equipment for mechanical assembly production"* pp 103-107
[15] Gaer M A, Zhuravlev D A, Yatsenko O V 2011 *ISTU Bulletin* **10** pp 32-36
[16] Gaer M A, Zhuravlev D A 2012 *ISTU Bulletin* **11** pp 44-48
[17] Shabalin A V, Zhuravlev D A, Gaer M A 2013 *ISTU Bulletin* **12** pp 69-73
[18] Breteau P, Thiebaut F, Lartigue C, Fricero B, Falgarone H, Moufle GE 2007 Assembly simulation of flexible parts through the fitting of linkage devices. 10th CIRP International seminar on computer Aided Tolerancing: specification and Verification for Assembly, (Erlangen, Germany)
[19] Fricero B, Falgarone H, Chevassus N, Breteau P, Thiebaut F, Lartigue C 2011 Method for optimising adjusments of an assembly of parts under stress. Patent. Application number EP20090796727
[20] Fricero B, Thiebaut F, Stricher A, Champaney L 2014 Method for optimizing the tolerancing of a set of flexible parts subjected to forces. Patent. Application number US201214350067
[21] Andolfatto L, Lartigue C, Thiebaut F, Douilly M 2014 *Journal of Manufacturing Systems* **33(1)** pp 103-115
[22] Lartigue C, Thiebaut F, Bourdet P, Anwer N 2006 *In Advanced mathematical & computational tools in metrology VII. World scientific publishing* pp. 196-203
[23] Mounaud M, Thiebaut F, Bourdet P, Falgarone H, Chevassus N 2010 *International Journal of Production Research* 1366-588X pp 1-23
[24] Lacroix C, Mathieu L, Thiebaut F, Douilly M, Falgarone H 2014 Numerical process based on measuring data for gap prediction of an assembly. *13th CIRP Conference on Computer Aided Tolerancing* (Hangzhou, China) **27** pp 97-102

IOP Conf. Series: Materials Science and Engineering **760** (2020) 012039 doi:10.1088/1757-899X/760/1/012039

Applying information technologies to a bulldozer design

I V Leskovets[1]

[1] Belarusian-Russian University, 43, Mira, Mogilev, 212000, Belarus

E-mail: le@bru.by

Abstract. An analysis of the application of information technology in transport and technological machines is presented. The features of the use of information technology in the design and operation of transport equipment are given. The objectives of applying information technology at the stage of machine design are determined. The information model of the bulldozer is presented as the system comprising external environment – working equipment – caterpillar track - frame - transmission - engine - operator. A dynamic model of the bulldozer has been developed. Mathematical models of the elements of the dynamic model are presented. Also, characteristics are given which reflect the processes occurring during technological operations when using the bulldozer for carrying out excavation works. It has been established that the information model of the bulldozer makes it possible to obtain data to calculate loading of the bulldozer elements and to optimize selection of the best machine parameters at the stage of functional design.

1. Introduction

Currently, information technology is becoming increasingly important in all areas of human activity. Transport and technological machines are no exception. Computing systems are widely introduced at all stages of the life cycle of products or sets of products [1]. Models of different levels are used, depending on the complexity of the structures or processes under investigation. The complexity of the model is virtually independent of the macro- or micro-level of the process or product, but depends on the degree of its detailed consideration.

As a rule, information technology uses mathematical models of different degree of complexity. In the middle of the 20th century, mathematical models of transport machines were based on static dependencies, or systems of differential equations with a small number of degrees of freedom, but today's computer applications use systems of differential equations with a number of degrees of freedom exceeding 10^6 and more [2].

In structural design, systems based on the finite element or the boundary element methods have found widest application [3]. The application of such methods enables engineers to find the best solutions, to optimize structures, etc. [4].

In the operation of transport machines, as well as in the traffic flow management, information technologies based on the use of simulation models are widely used. The application of such technologies makes it possible to optimize transport processes taking into consideration random factors, and to solve problems that do not have an analytical solution [5].

To diagnose transport systems, information technology is used to detect symptoms of faults and to search for causes and techniques for their elimination.

In the design and operation of transport systems, it is important to correctly determine external loads acting on the load-bearing systems and working equipment of transport and technological machines. Loads are functions not only of the parameters of the external environment, but also of the parameters of the machine to be designed, since the operation of all mechanisms is connected with implementing the working operations and overcoming the forces of inertia.

2. Methods and techniques

The desire to improve the competitiveness of machines makes the manufacturers improve the quality of products. To solve problems arising from this it is necessary to resolve conflict situations; like, when, on the one hand, it is required to increase strength of structures, their reliability and durability, but on the other hand, an improvement in the quality of materials and an increase in the mass of products lead to a rise in cost and loss of competitiveness. One way is to improve the quality of calculations to order to clarify loads acting on a machine, system or unit. The solution of such problems is achievable at the stage of functional design, when it is possible to select machine parameters that most fully satisfy the target functions established [6].

The basis for the functional design of transport and technological machines is a set of dynamic, mathematical and simulational models compiled with a sufficient degree of detail. Such models should be developed on the basis of the systems approach [7]. The level of detail is set by the requirements for solving problems.

It is proposed to use information technologies for the functional design of a crawler bulldozer equipped with a front blade. The dynamic model of the bulldozer is presented in Figure 3. In accordance with the system approach, the bulldozer is considered as EXTERNAL ENVIRONMENT – WORKING EQUIPMENT - CATERPILLAR TRACK - FRAME - TRANSMISSION - ENGINE - OPERATOR. Figures 1 and 2 show dynamic models of tracks located under the road wheel and between the road wheels. The purpose of the simulation is to determine the loads arising in the systems and mechanisms of the bulldozer in the process of performing technological operations.

The dynamic model of the bulldozer presents weights of an engine, a transmission, driving and driven disks of a clutch (pump and turbine wheels of a torque converter), as well as a drive sprocket, an idler wheel, road wheels, support rollers, tracks, working equipment, and elements of conditionally mobile and conditionally stationary parts of the dragging prism.

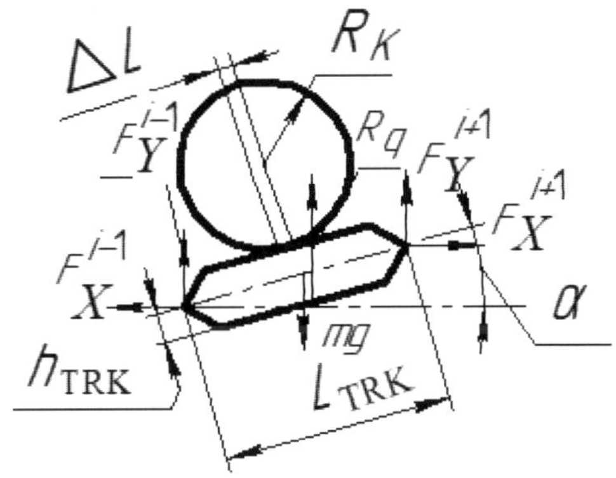

Figure 1 Dynamic model of a track located under the road wheel

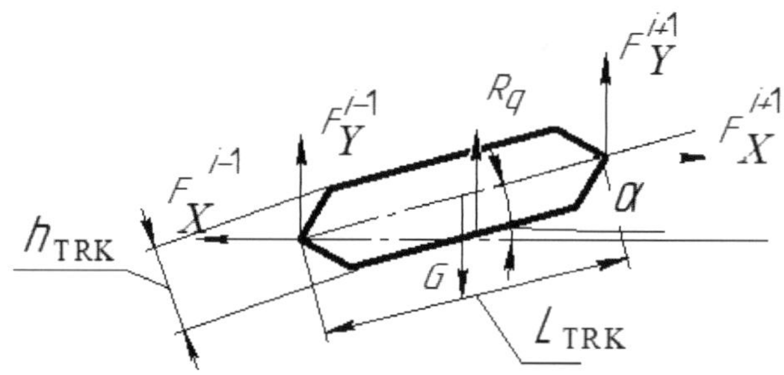

Figure 2 Dynamic model of a track located between the road wheels

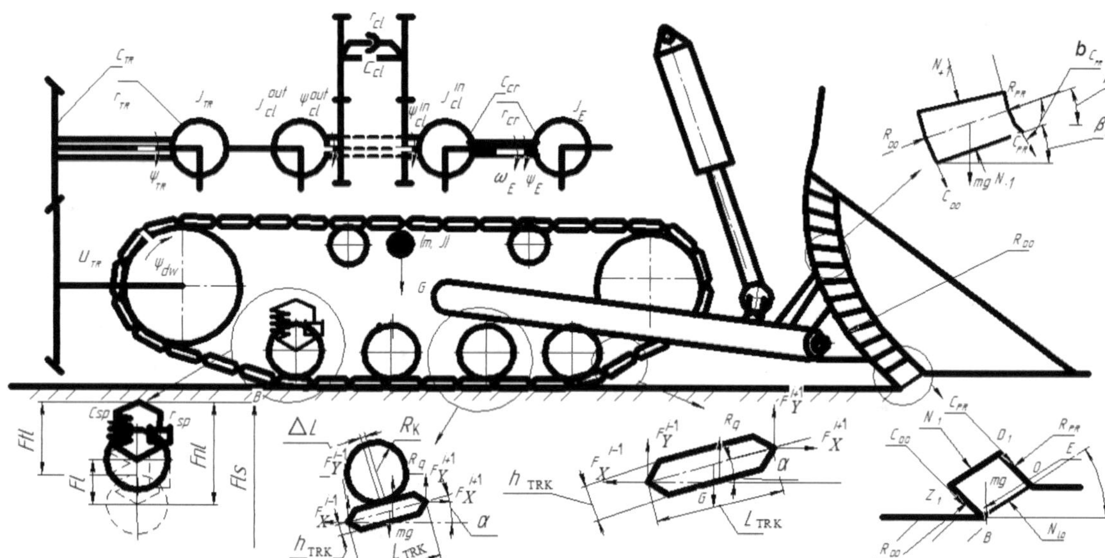

Figure 3 – Dynamic model of the bulldozer.

The model shown in Figure 3 is a complex structured system consisting of simple and complex elements. Each of the elements is presented as a dynamic model. All elements interact with the adjacent ones. This interaction is described by mathematical models.

For example, if a track is located under the road wheel, the mathematical model takes into account the interaction with adjacent tracks and the road wheel. The track has three degrees of freedom.

In case the track is located between the road wheels, the interaction occurs between the adjacent tracks and the soil.

The equations for the interaction with the adjacent tracks, the soil and the road wheel are drawn up based on the theory of elasticity.

$$\frac{dv_{TRK}^{x}}{dt} = (F_x^{i+1} - F_x^{i-1}) / m_{TRK}, \tag{1}$$

$$\frac{d\ v_{TRK}^{y}}{dt} = (F_y^{i+1} + F_Y^{i-1} - m_{TRK}g + R_q - R_k\cos(\alpha)) / m_{TRK}, \tag{2}$$

$$\frac{d\omega_{TRK}}{dt} = (F_y^{i+1} \cdot sin(\alpha) \cdot \frac{L_{TRK}}{2} - F_y^{i-1} \cdot sin(\alpha) \cdot \frac{L_{TRK}}{2} - \\ -F_x^{i+1} \cdot cos(\alpha) \cdot \frac{h_{TRK}}{2} + F_x^{i-1} \cdot sin(\alpha) \cdot \frac{h_{TRK}}{2} - R_x \cdot \Delta L) / J_{TRK} , \quad (3)$$

where v_{TRK}^x, v_{TRK}^y are the velocities of the track along the abscissa and ordinate axes, ω_{TRK} is the angular velocity of the track, F_x^{i+1}, F_x^{i-1} are the forces resulting from the interaction with subsequent and previous tracks, m_{TRK} is the track weight, g is the gravitational acceleration, R_q is the microprofile reaction, R_k is the reaction between the track and the wheel, L_{TRK} is the track length, α is the inclination angle of the track relative to the horizontal axis, h_{TRK} is the track height, ΔL is the distance between the track center and the point of application of the wheel reaction, J_{TRK} is the moment of inertia for the track.

When the track is located between the road wheels, the equations describing it by the adjacent tracks have the forms:

$$\frac{dv_{TRK}^x}{dt} = (F_x^{i+1} - F_x^{i-1}) / m_{TRK} , \quad (4)$$

$$\frac{dv_{TRK}^y}{dt} = (F_y^{i+1} + F_y^{i-1} - m_{TRK} g) / m_{TRK} , \quad (5)$$

$$\frac{d\omega_{TRK}}{dt} = (F_y^{i+1} \cdot sin(\alpha) \frac{L_{TRK}}{2} - F_y^{i-1} \cdot sin(\alpha) \frac{L_{TRK}}{2} - \\ -F_x^{i+1} \cdot cos(\alpha) \frac{h_{TRK}}{2} + F_x^{i-1} \cdot sin(\alpha) \frac{h_{TRK}}{2}) / J_{TRK} \quad (6)$$

The equations of the mathematical model for other locations of the tracks are composed in a similar way. The tracks can be located on the drive sprocket, the idler wheel and the support rollers, between the support rollers, on the guide and traction parts of the caterpillar assembly [8].

Different models can be used to simulate the process of soil digging [9]. The mathematical model of the process of soil digging used in this work is based on the theory of layer-by-layer soil excavation with the blade of E. I. Berestov [10]. This model makes it possible to determine the resistive forces acting against digging experienced by the bulldozer while performing technological operations, to simulate the penetration and raising of the blade during the bulldozer motion, as well as to calculate the magnitude and direction of the resistive forces acting against digging in the process of forming the dragging prism. The magnitudes of shear resistance of soil during excavation work are found by the equations that determine maximum E_{max} and minimum E_{min} values of the magnitudes of shear forces.

$$E_{max} = \frac{G\,sin(\alpha+\omega) + R_{DO}^0\,sin(\omega+\rho) - C_{OB}\,cos(\alpha+\omega+\varphi) + C_{PR}^0 + N^0\,cos(\omega)}{sin(\alpha+\omega+\varphi)} , \quad (7)$$

$$E_{min} = \frac{G\,sin\left(\frac{\pi}{2} - \alpha - \omega\right) + R_{DO}^0\,sin(\omega+\rho) + C_{PR}^0 + N^0\,cos(\omega)}{sin(\alpha+\omega+\rho)} \quad (8)$$

where α is the cutting angle, ω is the soil-blade friction angle, R_{DO}^0 is the friction force of the zero segment of the mobile part of soil volume in the dragging prism , ρ is the angle of internal friction of soil, C_{OB} is the cohesion of soil in the shear zone, φ is the inclination angle of the zone of big shear, G is the weight of the segment of the mobile part in the dragging prism , N^0 *is* the reaction between the soil and the zero segment of the mobile part in the dragging prism.

The friction force of the zero segment of the mobile part of soil volume in the dragging prism is determined by the formula:

$$R_{DO}^{0}=\frac{R_{PR}\cdot sin(\beta_{C_{PR}})+(C_{PR}+N_{i+1})sin(\beta_{C_{PR}})}{sin(\beta_0+\omega)}, \qquad (9)$$

where R_{np} is determined by the formula:

$$R_{PR}=\frac{G\cdot sin(\rho)+C_{PR}^{X}\cdot sin(\rho)+C_{PR}^{Y}\cdot cos(\rho)}{cos(\beta_{PR}-\rho)} \qquad (10)$$

where C_{PR}^{x} and C_{PR}^{y} are the horizontal and vertical components of cohesion forces between mobile and stationary parts of the dragging prism.

The magnitude of the cohesion force is determined by the formula:

(11)

where h_{PR} is the height of the selected segment in the mobile part of the dragging prism , C_r is the specific residual soil cohesion, B is the blade width.

The magnitude of the reaction between the soil and the zero segment of the dragging prism is determined by the formula:

$$N_{-1}=\frac{G\cdot sin(\beta_{C_{PR}}+\omega)+R_{PR}\cdot sin(\omega+\rho)+(C_{PR}+N_{+1}^{i})\cdot cos(\omega)}{cos(\beta_0-\beta_0^{i-1}+\omega)}, \qquad (12)$$

where G is the weight of the soil segment, N_{+1}^{i} is the normal force of action from the subsequent segment, β_0^{i-1} is the inclination angle of the previous segment.

The calculated characteristic of the resistive forces acting against digging has a specific "sawtooth" form, obtained during the experimental studies. It has periodic rises and falls of forces, which are caused by periodic displacements of the soil relative to the bulldoze blade. The results obtained by using this model have been verified experimentally and have good convergence. The model takes into account the properties of the soil, the blade and the dragging prism. The calculation process takes into account the loading, which is formed on the excavated soil, as the dragging prism increases [11].

3. Results

Based on the mathematical models, the software has been developed which makes it possible to conduct a series of computational experiments at the stage of functional design, as well as to obtain characteristics of the processes taking place in the main systems and mechanisms of the bulldozer while performing technological operations [12]. The obtained characteristics allow optimization activities to be carried out in order to select the best combination of parameters of the systems under investigation and to achieve the specified performance indicators. In addition, the information obtained can be used for the indicators of loading, reliability and durability of individual mechanisms.

Figures 4, 5 and 6 show some of the results obtained while carrying out the simulational computational experiments to model the process of soil digging with a caterpillar bulldozer based on a 12-ton tractor.

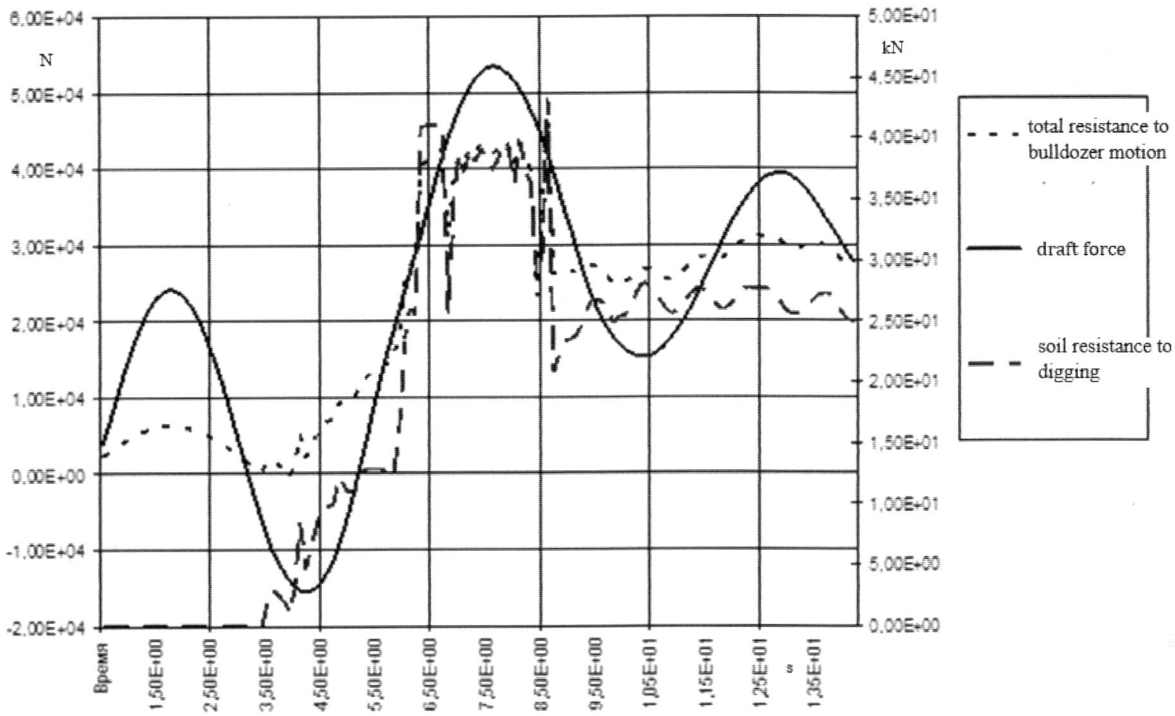

Figure 4 Draft force and total resistance to bulldozer motion are given on the left axis, soil resistance to digging is on the right axis

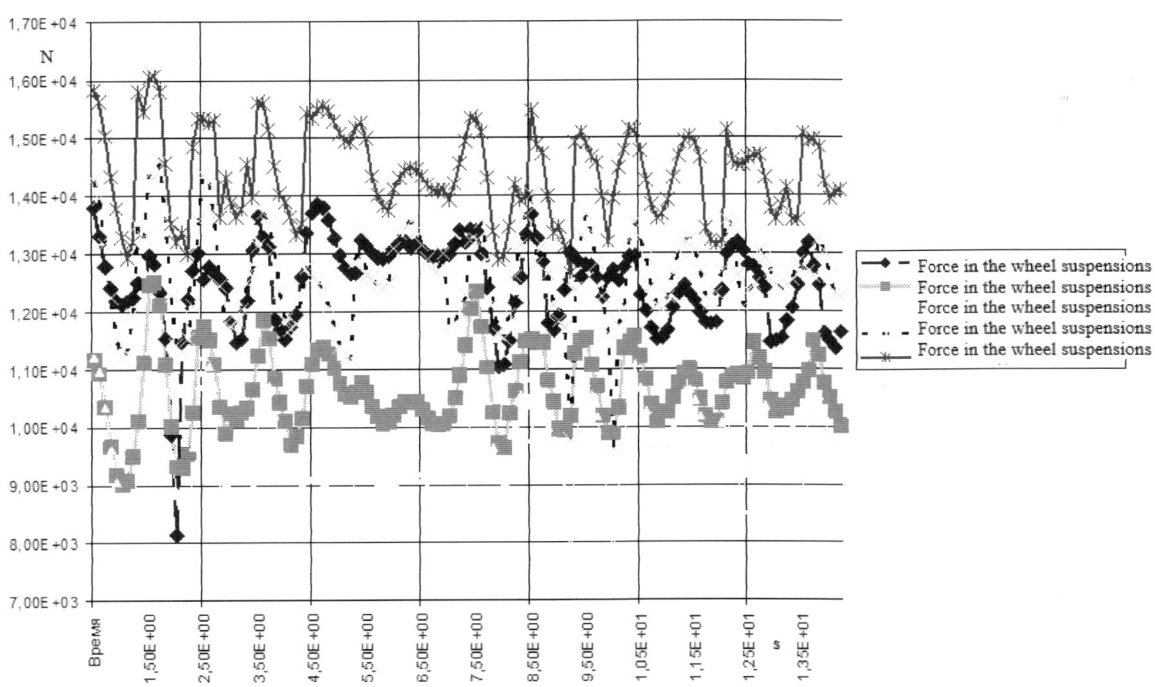

Figure 5 Forces in the wheel suspensions according to the legend.

4. Discussion

The study of the characteristics obtained during the simulation experiments with use of the information model of the bulldozer makes it possible to establish patterns that arise during technological operations and to identify the cause-and-effect relationship between quantitative and qualitative indicators of the processes. There is a possibility of conducting virtual studies of the processes occurring in the working equipment, the suspension system, the transmission and the engine. The simulation model determines the magnitudes of dynamic factors, the magnitudes and the direction of forces acting in the working equipment and the caterpillar track. The information obtained provides a wealth of material for carrying out strength calculations, analyzing the reliability and durability of the structure, as well as for determining the ways of upgrading the machine and conducting the analysis with the aim of increasing the efficiency of bulldozers.

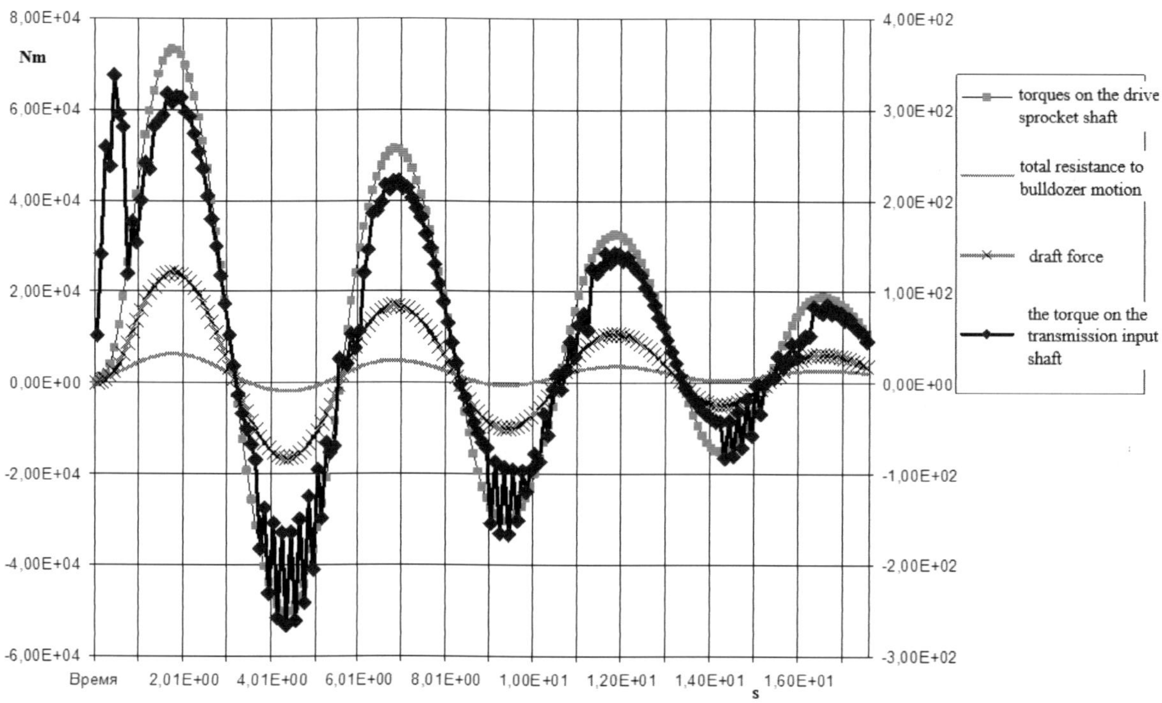

Figure 6 Torques on the drive sprocket shaft, the transmission input shaft, total resistance to bulldozer motion, draft force. The torque on the transmission input shaft is on the auxiliary axis.

5. Conclusion

Information technology is becoming increasingly important in the design of machines. It provides reduction of costs required to run experimental research, and makes it possible to reduce costs of the development workThe application of information technology in functional design allows determining the effect of parameters of the systems under study on their characteristics. The application of information technology in the form of software simulation systems provides information about the effect of parameters and relative position of parts and mechanisms on the characteristics of the machine as a whole. When modeling a crawler bulldozer, it becomes possible to theoretically determine the magnitudes of forces and their direction in the interacting systems, which is impossible, or requires considerable material costs while conducting full-scale experiments.

When designing tracked vehicles for earthworks, it is important to consider the machine as a single model, which presents the systems and mechanisms having a significant impact on the characteristics and indicators under study.

IOP Conf. Series: Materials Science and Engineering **760** (2020) 012039 doi:10.1088/1757-899X/760/1/012039

References

[1] A P Galvão, S T Fernandes, P M L Moura de Oliveirac, A BanerjeeaJosé, A B Montevechic 2018 *Computers & Industrial Engineering* **124** 474-492.

[2] Zhang L, Zeigler B P, Yuanjun Li 2019 - *Introduction to Model Engineering for Simulation Model* Engineering for Simulation.

[3] A Armin, R Fotouhi, W Szyszkowski 2014 *Finite Elements in Analysis and Design* **92** 1-11.

[4] *Finite Element Method.* A Practical Course, available at: https://www.sciencedirect.com/book/9780750658669/ finite-element-method.

[5] Micro-simulation modelling for traffic safety: A review and potential application to heterogeneous traffic environment *IATSS Research,* available at: https://www.sciencedirect.com/science/article/pii/S0386111217302133.

[6] Y-M Deng, G A Britton S B Tor 2000 *Computer-Aided Design* **32** 889-899.

[7] Tarasik V P 2004 *Matematicheskoe modelirovanie tekhnicheskikh sistem* (Minsk : Novoe znanie).

[8] Leskovets I V, Leonenko O V, Buzhinskij A D 2014 *Mexanizaciya stroitel`stva* **№ 9** 39-44.

[9] Y Qinsen, S Shuren 1994 *Journal of Terramechanics* **31** 55-65.

[10] Berestov E I 2017 *Nauchny`e osnovy` vzaimodejstviya rabochego oborudovaniya zemlerojno-transportny`x mashin s gruntom* (Mogilev: Belorus.-Ros. un-t).

[11] Berestov E I, Leskovecz I V 2010 *Vestnik Poloczkogo gosudarstvennogo universiteta. Seriya V: Promy`shlennost`. Prikladny`e nauki* **3** 45-52.

[12] Leskovets I V 2011 *Stroitel`naya nauka i texnika* **5** 48-51.

Generation of a rational training sample when predicting power consumption for train traction

V.E. Osipova[1], D.A. Yakovlev[1]

[1] Trans-Baikal Institute of Railway Transport (Irkutsk State University of Railway Transport), 11 Magistralnaya Str., Chita, 672040, Russia

E-mail: Valeri3307@mail.ru

Abstract. The paper considers peculiarities of forecasting in the field of railroad facilities based on approximation of time series and neural networks. The objects of traction power supply of the Transbaikal Railway are considered as the object of the study. Forecasting on the basis of approximation for consumers with maximum power consumption increases the accuracy of forecasting the electric energy consumption by 4...7%. The application of neural networks in forecasting of power consumption allows reducing the value of error to 2%, thus leading to significant reduction of costs for fuel and energy balance of a structural division or an enterprise as a whole.

1. Introduction

At present, the existing system of JSC Russian Railways is increasingly aimed at the formation of a unified information space for freight transport and logistics in order to ensure sustainable competitiveness of Russian Railways by increasing the attractiveness of transport and logistics services provided to clients through digital technologies [1, 2].

Digitalization of railways is primarily aimed at reducing the impact of the human factor on the condition of railway facilities and is quite attractive by the speed of its services provided to the consumer.

One of the first stages of digitalization is the digitalization of economy, namely the application of digital intelligent technologies in management in order to cut expenses, increase security and reduce costs.

One of the most effective ways to reduce costs in the railway sector is proper formation of the fuel and energy balance of a structural unit, which mainly depends on the accuracy of load forecasting.

Load forecasting is an important component of the energy management system within the power system. Accurate load forecasting helps power supply enterprises to make decisions on mandatory installation, reduces power margin during rotation, and correctly designs the maintenance plan of a device [3].

2. Analysis of power consumption data for hauling operations

Forecasting in the field of railway facilities represents a multilevel of the structure and uncertainty of factors describing the system status. Let us consider the peculiarities of power forecasting using the example of objects of the traction power supply of the Transbaikal Railway [4].

Figure 1 shows the flow chart of fuel and energy balance formation of the Transbaikal Railway.

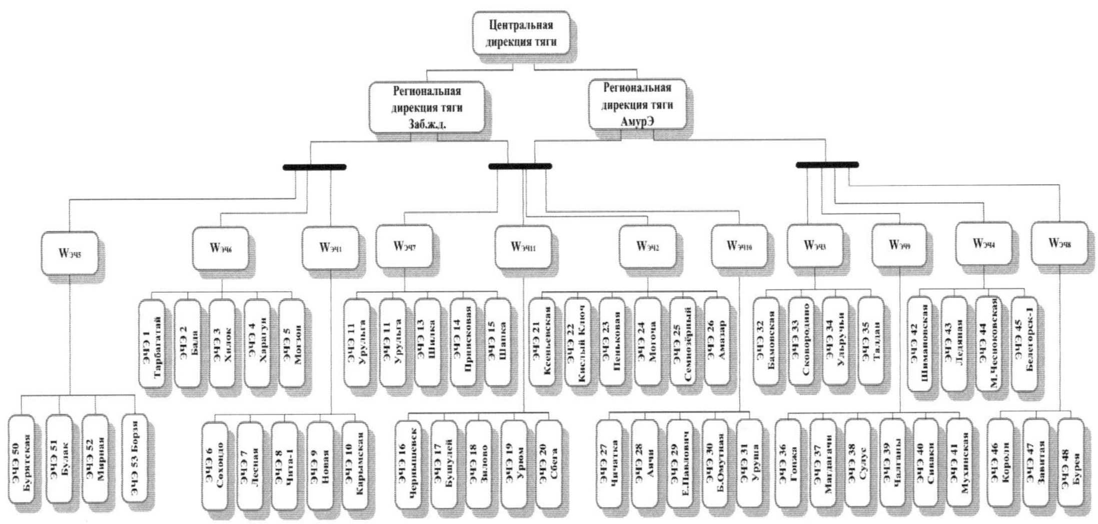

Figure 1. Flow chart of fuel and energy balance formation of the Transbaikal Railway

According to the analysis of statistical data, the average annual productivity of the Transbaikal Railway makes 1641118.1 thousand t/km, annual linear range – 14187344 km, rail traffic – 1019.30 million t/km, and energy consumption – 62195720 kWh. Figure 2 shows the dynamics of power consumption of the Transbaikal Railway section.

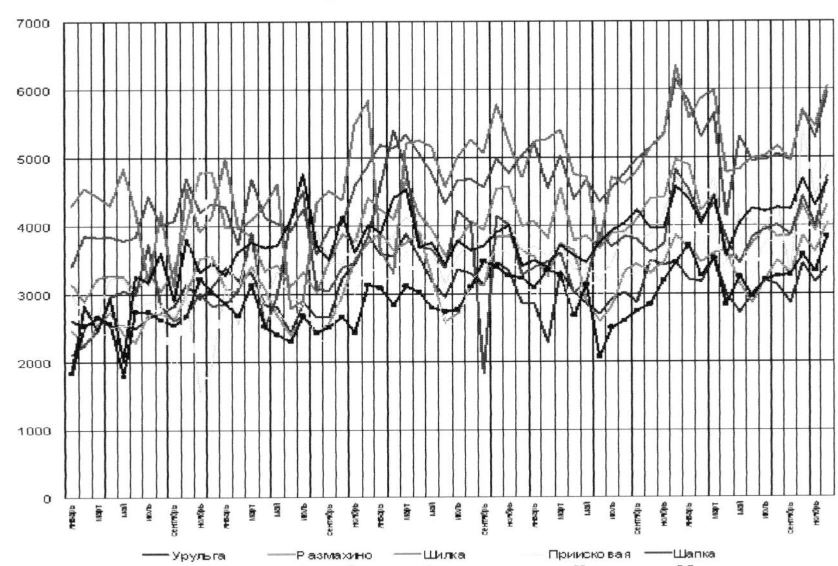

Figure 2. Dynamics of power consumption of the Transbaikal Railway section.

The analysis of statistical data on power consumption dynamics allows building the interdependence between the main parameters involved in the formation of fuel and energy balance and determining the main factors affecting the consumption of fuel and energy resources of a structural division of the Transbaikal Railway.

All factors affecting power consumption on train traction can be described by the following function

$$W_{\mathrm{I}}(t) = \mathbf{S} \cup \mathbf{T},$$

where $W_{\mathrm{I}}(t)$ – vector function characterizing a set of influencing factors;

S – vector characterizing the value of freight transport;

T – vector characterizing the state of the traction power supply system of the section.

In turn, factors affecting power consumption on train traction can be divided into constant and time-variable factors. Constant factors can be described by the following function

$$T_n(t) = [TT, L(t), F(t)], \; T_n(t) \subseteq T$$

where $T_n(t)$ – vector function characterizing the set of influencing factors that do not change over time.

The factors that vary over time can be described by the following function

$$S_n(t) = [m(t), t(t), Q(t), L_г(t)], \; S_n(t) \subseteq S$$

where $S_n(t)$ – vector function characterizing the set of influencing time-variable factors.

If T → min and S → min, then $W_l(t)$ → min

$$W_l(t) = \left\{ L_г(t) \in \begin{bmatrix} L_{г.min} \rightarrow L_{г.max} \\ L_{г.min} \leftarrow L_{г.max} \end{bmatrix}, m(t) \in \begin{bmatrix} m_{max} \rightarrow m_{min} \\ m_{max} \leftarrow m_{min} \end{bmatrix}, Q(t) \in \begin{bmatrix} Q_{min} \rightarrow Q_{max} \\ Q_{min} \leftarrow Q_{max} \end{bmatrix} \right\}$$

where $m(t)$ – vector function characterizing the weight of a train set

$t\ (t)$ – vector function characterizing the size of a train-to-train interval;

$Q\ (t)$ – vector function characterizing cargo turnover size;

$L г\ (t)$ – vector function characterizing the size of a linear run – type of traction power supply;

$L\ (t)$ – vector function characterizing section way profile;

$F\ (t)$ – vector function characterizing distance between traction substations.

3. Time series approximation

The analysis of information describing the state of the studied system was carried out on the basis of the time series approximation $W = W_1(t)$ [5].

To approximate the time series $W = W_1(t)$ the following function was used

$$W_1(t) = A + t \cdot [B + C \cdot \sin(\omega t + \varphi)].$$

Parameters A, B, C, ω, φ were determined by minimizing the error function using the modification of a gradient method described below.

To shorten the entry, let us denote the required vector of parameters through $\mathbf{P} = [p_1 \;\; p_2 \;\; p_3 \;\; p_4 \;\; p_5]^T$. Let some initial value of this vector be given $\mathbf{P}^0 = [p_1^0 \;\; p_2^0 \;\; p_3^0 \;\; p_4^0 \;\; p_5^0]^T$. The selection of this initial value determines which local minimum will be found as a result of calculations using the proposed algorithm. Let us also give the initial value of the vector step length (the meaning of which will be disclosed below) $\mathbf{L}^0 = [\ell_1^0 \;\; \ell_2^0 \;\; \ell_3^0 \;\; \ell_4^0 \;\; \ell_5^0]^T$ [6].

A sufficiently small increment d>0 is required to calculate the components of the gradient vector. The value d is selected as small as possible, in compliance with the requirement, so that the difference in values of the function to be minimized $s(p_1^n, \ldots, p_j^n + d, \ldots, p_5^n) - s(p_1^n, \ldots, p_j^n, \ldots, p_5^n)$ can be calculated with the accuracy up to 2-3 significant figures for any j.

The components of the gradient vector at point $\mathbf{P}^n$

$$\mathbf{G}(\mathbf{P}^n) = [g_1(\mathbf{P}^n) \;\; g_2(\mathbf{P}^n) \;\; \ldots \;\; g_5(\mathbf{P}^n)]^T$$

represent values of partial derivatives

$$g_j(\mathbf{P}^n) = \frac{\partial s(\mathbf{P}^n)}{\partial p_j}.$$

and with sufficient accuracy can be calculated numerically using the following expression

$$g_j(\mathbf{P}^n) = \frac{s(p_1^n,\ldots,p_j^n + d,\ldots,p_5^n) - s(p_1^n,\ldots,p_j^n,\ldots,p_5^n)}{d}.$$

Next, the normalized gradient vector is calculated

$$\mathbf{T}(\mathbf{P}^n) = [t_1(\mathbf{P}^n) \quad t_2(\mathbf{P}^n) \quad \ldots \quad t_5(\mathbf{P}^n)]^T,$$

representing a unit vector in the direction of the fastest growth $s(\mathbf{P})$, wherein the components of vector **T** are calculated through the components of vector **G** using the following expression:

$$t_j(\mathbf{P}^n) = \frac{g_j(\mathbf{P}^n)}{\sqrt{\sum_{k=1}^{5}[g_j(\mathbf{P}^n)]^2}}.$$

The predicted value **P**, for which the function s is expected to be smaller than the previous value, is calculated based on equality

$$\mathbf{P}^{n+1} = \mathbf{P}^n - \mathbf{L}^n\mathbf{T}^n,$$

where $\mathbf{L}^n\mathbf{T}^n$ – vector "composed" of component products of vectors $\mathbf{L}^n$ and $\mathbf{L}^n$, i.e.

$$\mathbf{L}^n\mathbf{T}^n = [\ell_1^n t_1^n \quad \ell_2^n t_2^n \qquad \ell_5^n t_5^n]^T.$$

After obtaining a new value of vector **P**, the function s(**P**) is calculated. In case all components of the vector are multiplied by q>1, i.e. т.е. $\mathbf{L}^{n+1} = q\mathbf{L}^n$, the vector length of a pitch **L** is increased in order to speed up the process of finding the minimum s(**P**).

Otherwise, i.e. if $s(\mathbf{P}^{n+1}) \geq s(\mathbf{P}^n)$, **L** is corrected according to the following algorithm:

1. $\mathbf{T}^{n+1}$ is calculated.
2. $\mathbf{T}^{n+1}$ is compared by each component of vector $\mathbf{T}^n$ and components of vector **T**, which signs have changed, are determined.
3. The components of vector **L** corresponding to the numbers of **T** components that changed the sign are multiplied by b, 0 < b < 1, i.e. if $\text{sign}(t_j^{n+1}) \neq \text{sign}(t_j^n)$, then $\ell_1^{n+1} = b\ell_1^n$, otherwise $\ell_1^{n+1} = \ell_1^n$.
4. The actions according to the described algorithm are repeated with a reduced vector step length $\mathbf{L}^{n+1}$, and $\mathbf{P}^n$ is taken again as a starting point, i.e. $\mathbf{P}^{n+1} = \mathbf{P}^n$.

Computational experiments showed that the best values of q and b are q=1.1 and b=0.1. The initial vector step length does not significantly affect the calculation process, as usually 5...10 steps are sufficient for dynamic adjustment of **L**.

The most convenient criterion for completing a computational process is inequality $|\mathbf{L}^{n+1}| \leq d$, which corresponds to the reduction in a vector step length by finding the minimum s(**P**) to values, at which the numerical definition of the gradient G(**P**) loses its meaning.

Figure 3 shows the forecasting results.

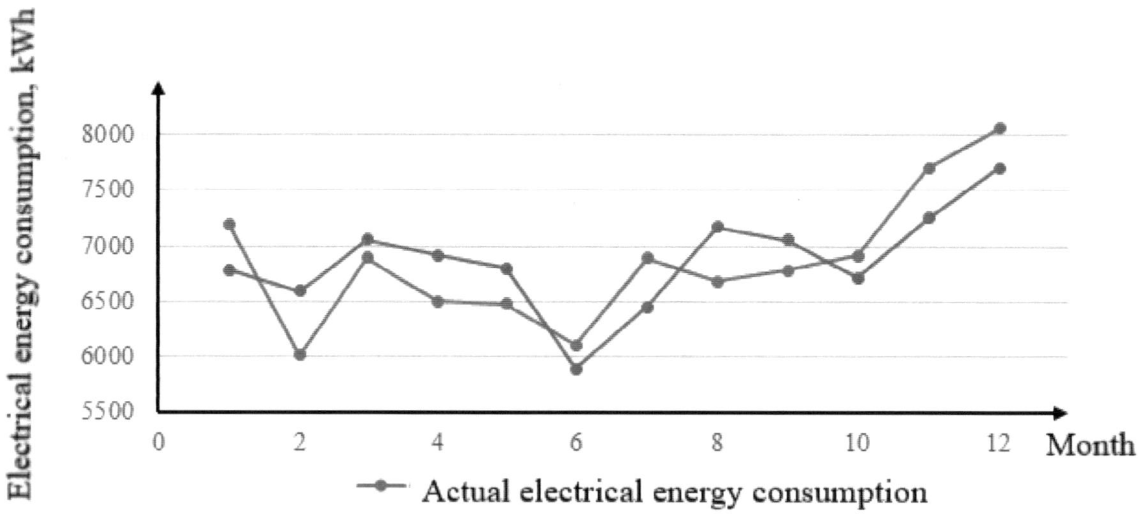

Figure 3. Maximum power consumer forecast

Calculations showed that this variation of the algorithm increases the accuracy of electric energy consumption forecast by 4...7%.

4. Neural networks in forecasting power consumption

Today, hybrid systems based on complex mathematical machines of fuzzy logic [7, 8] and neural networks [9, 10] are becoming ever more popular along with traditional methods of formation of fuel and energy balance of an enterprise and forecasting that provide for more accurate description of processes capable of self-learning in the process of operation and producing more accurate results.

The advantages of such systems are caused by a number of factors:

1) complexity of mathematical models of real control systems and processes related to the desire to improve their adequacy and to take into account a growing number of different factors influencing the decision-making process;

2) focus on models that take into account incomplete and inaccurate source data;

3) possibility of application in the description of technical systems, which have uncertainty and accuracy of initial data, makes it difficult or even impossible to use accurate quantitative methods and approaches.

Next, let us consider possible options for generating a learning sample in order to build a learning curve that is most adequately solved. The essence of neural networks learning process is to perform the following multi-step procedure [7].

Step 1. The learning set (set of problems) is specified

$\{(\mathbf{X}_l,\mathbf{D}_l);(\mathbf{X}_2,\mathbf{D}_2);...;(\mathbf{X}_L, \mathbf{D}_L)\}$,

The elements of which represent learning pairs ($\mathbf{X}_l$, $\mathbf{D}_l$), (l=1, 2, ..., L,). In this case $\mathbf{X}_l = \left[x_1^{(l)}x_2^{(l)}...x_n^{(l)}\right]^T$ is the 1st input vector (or the 1st input image) proposed by neural networks; $\mathbf{D}_l = \left[d_1^{(l)}d_2^{(l)}..d_m^{(l)}\right]^T$ – vector of reference (required) neural networks reactions in response to the 1st input vector $\mathbf{X}_l$; L – number of different learning pairs of the sampling.

Step 2. The initial state of neural networks is set by assigning some random small values to all its weights w_{ij}^k. Here w_{ij}^k – link weight connecting the output of i neuron of k layer with the input of j neuron of k+1 layer.

Step 3. The input vector $\mathbf{X}_l$ is sent to network input and reactions $y_i^{(l)}$ (i=1,2,..., m) of the output layer neurons are determined.

Step 4. The difference $\boldsymbol{\varepsilon}^{(l)}$ between desirable reaction of network $\mathbf{D}_l$ and its actual exit $\mathbf{Y}_l = \left[y_1^{(l)} y_2^{(l)} \ldots y_m^{(l)}\right]^T$, i.e. $\boldsymbol{\varepsilon}^{(l)} = \mathbf{D}_l - \mathbf{Y}_l$ and a total squared error $E^{(l)} = 0.5\sum_{j=1}^{m}\left(d_j^{(l)} - y_j^{(l)}\right)^2$ is calculated.

Step 5. The weights w_{ij}^{k} of neural networks are corrected to reduce error $E^{(l)}$.

Step 6. Steps 3...5 are repeated for each pair of the learning set ($\mathbf{X}_l$,$\mathbf{D}_l$) until the error in the whole set does not reach small predetermined value E^*.

The learning result is such adjustment of synaptic link weights in neural networks when each input vector the network generates the required (or close to it) output. The obtained forecast values are shown in Table 2.

Table 2. Forecasting of power consumption within the neural network model

Month	Actual electrical energy consumption, kWh	Forecast value of the consumer with maximum power consumption		Forecast power consumption of the neural network model	
		kWh	Error, %	kWh	Error, %
January	7199.5	6784.5	4.15	7098.591	1.00909
February	6007.5	6591.2	-5.837	6157.888	-1.50388
March	6885.3	7055.5	-1.702	7005.913	-1.20613
April	6496.1	6918.2	-4.221	6601.133	-1.05033
May	6471.8	6796.2	-3.244	6601.492	-1.29692
June	6098.6	5887.4	2.112	6192.307	-0.93707
July	6890.5	6450.1	4.404	6997.888	-1.07388
August	6685.3	7171.2	-4.859	6770.667	-0.85367
September	6781.1	7054.4	-2.733	6884.785	-1.03685
October	6917.1	6717.9	1.992	7057.237	-1.40137
November	7708.8	7250.1	4.587	7802.307	-0.93507
December	8066.6	7712.1	3.545	8196.959	-1.30359

Analyzing the early obtained data (Table 1), let us carry out a comparative analysis on the error level for two forecast options. The result is presented as a comparative diagram in Figure 4.

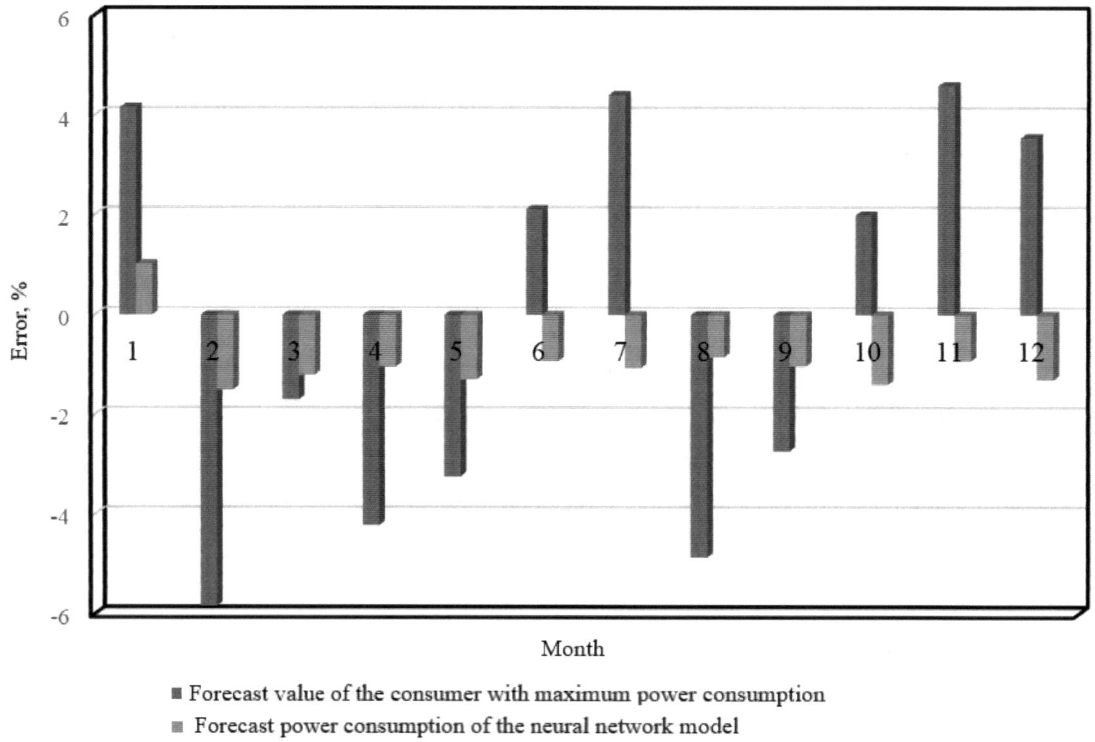

Figure 4. Comparative diagram of error value in power consumption forecast

5. Conclusion

Based on the above diagram it can be concluded that the error of forecast based on approximation for the 1st level electric consumer made from 4 to 6%. The application of neural networks in forecasting of power consumption allows reducing the value of error to 2%, thus leading to significant reduction of costs for fuel and energy balance of a structural division or an enterprise as a whole.

References

[1] Rawlinson S, Pike C, Starmans M 2017 *Digital railway* Digital Rail UK.
[2] Timur Bek 2018 Digital Road Economy Gudok Issue 84.
[3] Zaria Mohammed Shamseldin 2015 *Short Term Electrical Load Forecasting Using Fuzzy Logic*, A Dissertation Submitted In Partial Fulfillment for the Requirements of the Degree of M.Sc. In Electrical Engineering (Microprocessor and Control).
[4] Q. Ren, L. Baron, and M. Balazinski 2006 *Type-2 takagi-sugeno-kang fuzzy logic modeling using subtractive clustering in Fuzzy Information Processing Society*Annual meeting of the North American 120-125.
[5] Zakaryukin V.P., Krukov A.V., Raevsky N.V., Yakovlev D.A. 2017 Modeling and forecasting of power consumption processes of railway transport.
[6] Hianikova T 2008 *Application of Fuzzy lodic in financial time series analysis.*
[7] K. Bataineh, M. Naji and M. Saqer 2011 *A comparison study between various fuzzy clustering algorithms* Editorial board 335.
[8] Osipova V.E, Yakovlev D.A. 2019 *The Methodology of the Department of Fuel and Energy Resources of the Enterprise Based on Fuzzy Logic* Journal of Siberian Federal University. Engineering & Technologies.
[9] Saha S 2013 *Comparison of Performance Analysis using Different Neural Network and Fuzzy*

Logic Models for Prediction of Stock Price.
[10] Osipova V.E., Komogortsev M.G. 2016 Automated calculation of train-to-train intervals using the apparatus of artificial neural networks Transsib bulletin 4 (28).

IOP Conf. Series: Materials Science and Engineering **760** (2020) 012041 doi:10.1088/1757-899X/760/1/012041

Planning of qualitative indexes of railroad operational work in polygon technologies

E V Malovetskaya[1], R S Bolshakov[1], A V Dimov[1], A A Byshlyago[1]

[1] Irkutsk State Transport University, 15, Chernyshevsky Str., Irkutsk 664074, Russia

E-mail: katerina8119@mail.ru

Abstract. Increase of transit traffic volume demands application of new technological solutions structured in accordance with the tight schedule of the movement of freight trains. The forms of organizing traffic volumes based on the tight schedule are the intersectional and group trains and technical routes of regular service requiring specially organized reception of groups of wagons at their railroad yard.
Now there are quite a few optimizing models aimed at describing and researching of the uniform object – a freight transportation control system. Many of them give the individual formal description of the system, while being closed to further specification and development.
One of the main goals of transport processes control consists in optimum performance of cargo transportation of the predefined volume determined at a certain time interval, using the existing technical, information and steering complexes.

1. Introduction

Currently, railroads across the entire network of the Russian Federation are switching to polygon technologies. Their use has come a long way from management of traction resources at limited sections between single stations to coordination of interactions of several roads. Furthermore, these technologies have proven their feasibility and the ability to quickly respond to changes in both external influences and internal factors.

The carrying capacity of all the polygon in general directly depends on the number of the limiting sites in independence regardless of a type of these restrictions (technical, technological, organizational). Thus, the occurrence of any problematic situation in the Far East, Transbaikal or the East Siberian Railway will quickly affect other railroads of the Russian Railways network [1].
Let us dwell in more detail on the Eastern polygon, the most advanced in the whole network.

The operational length of the railroads of the Eastern polygon is 17207 km, including the length of the electrified lines - 9196 km (53.4%). The share of two-acceptable routes is 7525 km (43.7%). 12144 kms (70.6%) of the main directions are equipped with automatic blocking. 29.4% of lines are equipped with semi-automatic blocking. 9669 kms (56.2%) are equipped with the centralized traffic control. There are 810 railway stations at the polygon, including 8 classifying sections, 6 passenger stations, 58 sectional stations, 109 cargo stations, 629 intermediate stations.

There are 15 operational car repair sheds at the polygon. Sections of warranty passage of freight wagons reach to 1300 - 1400 km. There are 4 infrastructure maintenance control centers (ICC) at a polygon.

IOP Conf. Series: Materials Science and Engineering **760** (2020) 012041 doi:10.1088/1757-899X/760/1/012041

Transportation work at the Eastern polygon is characterized by permanent increase in freight turnover, with the main freight flows subsequently directed to sea trade ports, such as Vostok, Nakhodka, Vanino, Vladivostok, Posyot [2].

Further growth of loading volumes is provided in ports of the Far East with increase in a cargo flow at the Trans-Siberian Railway and BAM for 30% in the 2020 perspective. Providing subdivisions with resources needs to be supported by implementation of the corresponding technologically sufficient organizational structures within the Eastern polygon borders.

The handling of heavy trains with the increased length in large numbers will be implemented after lifting restrictions of infrastructure, first of all, on traction power supply and overhead system devices [3].

It is also necessary to mention the fact of decrease in sectional speed and performance of locomotives during the current year at the Eastern polygon, which can be objectively justified. Fig. 1 shows the dynamics of accomplishment of local speed in 2018.

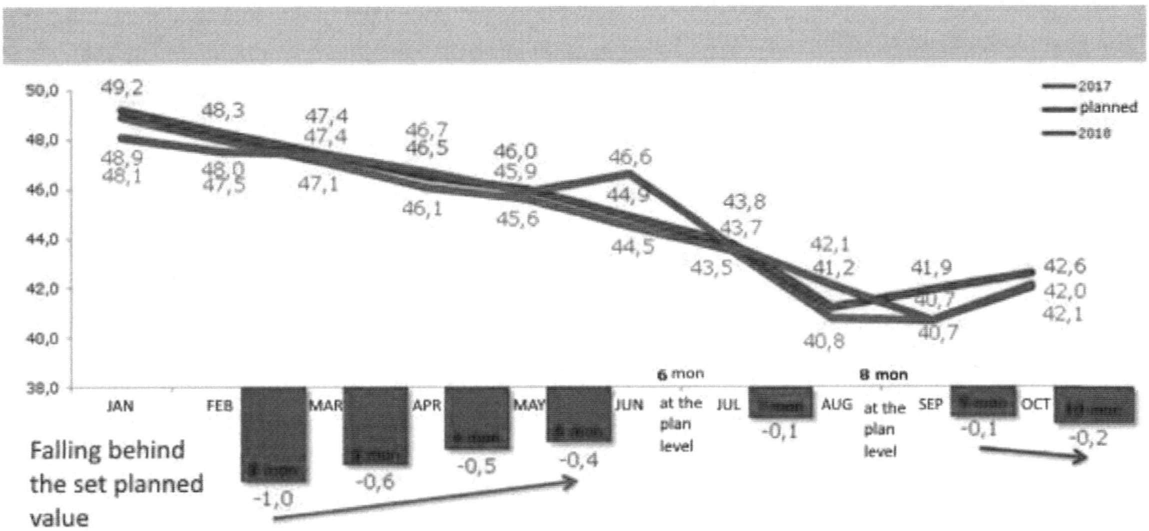

Figure 1. The dynamics of accomplishment of local speed in 2018 at the Eastern polygon.

Since 2010 Russian Railways have been administering a policy which is directed to increase the train mass and use of heavy trains in cargo transportations. In 2015 the company presented the program for development of heavy train service till 2020. Its goal is to increase the number of heavy trains on the lines and introduce the trains with weight of 9 thousand tons to various directions.

If we speak objectively about the infrastructure of the Russian Railway, it should be recognized that possibilities of increase in the train-handling capacity are exhausted and we need new technology solutions supported with the information component. Transition to the organization of the movement of freight trains according to the schedule is one of options. While organizing the movement of freight trains according to the schedule, local and group trains (including three or four selected groups) and regular technical routes can become the main forms [4], [9], [10].

However, the development of a system of making up of two- and three-group trains will promote reduction of classification work, personnel decrease, acceleration of freight delivery and increase in speed of trains. Group trains allow saving costs for accumulation of wagons and rationally distributing interstation classification work.

Since 2011 the increase of average mass of freight cars made up about 4%, and the number of heavy trains was increased on various highways. For example, for 6 months in 2018, 49 thousand trains were handled on the Southern Ural railway, which is 20% more as opposed to 2017. With all aforesaid, there wasn't a single day on the railway without equipment failures. Just in 10 months of 2018, 37,024 warnings were issued on the railway to limit the speed of freight trains by 29,620.7 km, which is 3168 restrictions and 5,805.6 km higher than the same period in 2017.

The increasing complexity of management functions in railway transport led to the restructuring of the sector's purposes and tasks. Within the Comprehensive program of optimization of operational work, a number of the major tasks have been formulated, among which are the creation and introduction of efficient methods of organizing and controlling freight transportation.

Today the train handling reserves are almost exhausted on the Russian Railways network. In turn, it demands application of the new technological solutions based on the tight schedule of movement of freight trains. The forms of the organization of train service which are based on the tight schedule are: group trains (including mutually coordinated exchange of groups at stations) and the technical routes of regular service requiring strictly organized reception of groups of wagons at their railroad yard [5]. A coordinated (controlled) supply of wagons with a scheduling of movement of specialized transfers, scheduling of the loading by destination as well as controlled supply of empty wagons under the planned loading should be provided for this purpose.

At the same time, upon transition to the organization of movement of freight trains according to the schedule there is an opportunity not only to keep, but also to increase average weight of the freight train by 5–10% taking the following measures:

• the organization of movement of trains of the increased weight and length using specially developed "train paths", including the organization of make-up and further handling of train sets on the short tracks;

• replacing individual trains with group trains;

• organizing the delivery of local cargo with the possibility of the attachment of wagons to transit trains, provided that all technological times for processing and the norms of weight and length on the section are maintained;

• the automatic calculation of train make-up process using the seat booking system in train sets with the efficient application of empty "train paths".

The models and algorithms developed during the planned management of transportations do not allow making practical decisions to achieve the goal in modern conditions.

Therefore in recent years much attention was paid to a research of technological process of freight transportation, creation of new criteria of effectiveness, optimizing models of scheduling and control, the analysis of organizational and structural, economic and other process flow diagrams.

As a result, at present there are a fairly large number of the optimizing models aimed at describing and a researching of a uniform object - a freight transportation control system. Many of them give the individual formal description of system, at the same time being closed for further development.

The main goal of managing transport processes is the optimal implementation of freight transportation of the given volume determined at a specific time frame, with application of the existing technical, informational and operating systems.

To achieve this goal, various models are used in railway transport to optimize the main technological process of operational work – the making-up of trains.

The making-up of trains is one of the major processes in the operation of the station. Its parameters are of great importance, both in the organization of work of the station, and for all directions. It necessitates a sufficiently detailed description of regularities of train accumulation.

2. Problem definition

In railway transport there are also standard control models that can optimize the transportation process in terms of operational work.

Issues of organizing car flows in trains can be considered and implemented by various methods, but two are most widely used:

• a method of modeling the sequence of intervals between the moments of the end of accumulation of train sets on the basis of the considered cumulative distribution function;

• a method of modeling the decomposition of the broken-up train sets according to the plan of formation with the subsequent summation of groups of wagons to the whole trains.

In the first method, there is no process of the train make-up. This makes it impossible to use the model to study the operation of stations in the conditions of high loadings of its elements.

The considered model implements the second method to describe the process of train make-up. One track (in case of applying strict specialization), or group of tracks (assigned to each appointment in the wagon yard when using of the fluctuating specialization).

It is known that the processing of wagons at the station is random, due to a number of the external and internal reasons, such as non-uniform arrival of trains at the station, non-uniform accumulation of wagons and processing of trains [6], [11], [12].

So emergence of wagons of various purposes in train sets is considered random events.

With that, the current value of magnitudes of separate groups of wagons is determined by the following formula:

$$m'_{gr}=m_{avi}\left| \ell nF(m'_{gr}) \right| \quad (1)$$

$$F(m'_{gr})=Z_i \quad (2)$$

where Z_i is a randomly distributed number from the set of random numbers which are evenly distributed in the range from 0 to +1;

m_{avi} is the average magnitude of a group of wagons of i destination equal to:

$$m_{av}=Ni/n_{ar}=(19-0.1k_{dest})\sqrt{N_i}/0.4n_{reh} \quad (3)$$

where N_i is a wagon flow under consideration of i destination;

k_{dest} is the total number of destinations according to the plan of the train make-up;

n is the total number of trains arriving for the breaking-up.

All the daily fluctuations of the arrival of train flows and intraday non-uniformity, as a number of researches showed, are described by the normal distribution law. Model operation of number of the broken-up trains happens every time upon transition to imitation of the process for the next days. Its current value turns out with use of a random number of Zi and integral function of distribution:

$$N=N_{av}+kvN_{av}(\Sigma Z_i-3), \quad (4)$$

N_{av} is an average daily number of the trains broken-up at the station;

kv is a variation factor.

On the basis of the above provisions, the algorithm of the making-up of group trains is constructed. In the course of accumulation and on its completion, the following magnitudes by each destination are counted: the parameter of accumulation of structures – c, the number of the accumulated train sets – ΣN_{acci}, the time of accumulation of structures – ΣT_{acci}, accumulation freight wagon-hours – ΣB_{acci}.

Using this algorithm, the values of the accumulation parameter at the make-up stations and the expectation parameter at the wagon group re-attachment stations for the following categories of group trains have been obtained by a method of modeling:

- group trains without constant weight of groups and not attached to permanent schedules;
- group trains with constant weight of groups and not attached to permanent schedules;
- group trains of the constant schedule with a variable of groups;
- group trains with constant weight of groups and attached to particular schedules.

In the course of measurements, further data processing as well as formalizations of results in the form of a mathematical model, errors occur and a part of information contained in the basic data is lost. Application of planning methods of an experiment makes it possible to define an error of a mathematical model and to judge its adequacy.

When scheduling an experiment, the complete factorial experiment was used in this work. The complete factorial experiment allows estimating effects of interaction quantitatively.

3. Conclusions

Following the results of processing of experimental data, the regression equations for the coded values of separate factors were obtained. The influence of each factor on the optimization parameter was determined by the available coefficients of interaction. In addition to the regression equations for the coded values of factors, the regression equations for natural values were retrieved.

According to the results of the experiments, the dependences of accumulation parameters and expectation parameter became available.

The obtained adequate linear model takes the form of a polynomial of the first degree. The coefficients of the polynomial are partial derivatives of the response function on the corresponding variables. Based on the obtained coefficients of interaction of bij, a measure of the influence of each of the factors on the optimization parameter was established. In addition to the regression equations for the coded values of factors, the creation of the equations of regression is of interest for natural values.

On the basis of the obtained dependences, the specified formulas were proposed to calculate the costs of freight wagon-hours and locomotive-hours for the group and single make-up of trains helping to make adjustments of the train make-up plan.

It should be noted that today the issues of development of the step-by-step instruction for the decision-making on the efficient updating of the make-up plan by the dispatching office of the stations and regional directorates with the list of an operations procedure on the development of offers and their coordination are especially topical.

It should be noted that in new working conditions it is necessary to analyze and identify the list of indicators, the value of which can be adjusted depending on a situation during this time for each railroad at the polygon [7], [13], [14].

For polygons, it is necessary to consider the following operation indicators:

- train traffic schedule performance regulations;
- locomotive run;
- the extent of equipment failures;
- overall wagon fleet.

The following actions should also be taken:

- to organize work on the redistribution of locomotive crews between structural divisions;
- to establish a task and to provide control of the running of locomotive crews at the extended service sections;
- to organize replenishment of trains at the local stations;
- to provide monitoring of expectation of locomotives.

The complex of the listed indicators well characterizes the transportation process organized in new working conditions on polygons and reflects market requirements in work of the main carrier of Russian Railways, however, their application demands further research, justifications and additions [8], [15].

The full range of actions for modernization of the general system of indicators of operational work should include the development of process models of the production block of the Russian Railway holding and also the formation of a new system of the indicators of the operational activity of polygons. In the nearest future it is necessary to define a set of indicators which are subject to monitoring and control in new conditions and allow providing the decision-making process with adequate and sufficient information.

References

[1] Sotnikov E A, Shenfeld K P 2011 Freight transportation in modern conditions and their influence on promotion of trains *The Messenger of VNIIZhT* **5** 3-9 ISSN - 2223-9731

[2] Sharov V A 2016 New risks during the realization of the uniform integrated planning on a public rail transport *Science and technology* **2** 87-93.

[3] Zubkov V V, Sirina N F 2019 The analysis of impact of participants of the transport providing functions on quality of providing complex transport service and a technique of determination of extent of their influence on criteria of quality. – *Modern technologies. Systems analysis. Modeling* **1 (61)** 47 - 55.

[4] Makhutov N A, Kulagin V A 2019 Modern technologies of fuel and energy complex and possibility of development of a rail transport of Siberia. *Modern technologies. Systems analysis. Modeling* **1 (61)** 64-73.
[5] Sharov V A 2011 The integrated technology of a traffic control of freight trains for the schedule *Railway – transport* **8** 11-21.
[6] Malovetskaya E V 2017 Modeling of process of forming of group trains upon transition to tough strings of the schedule. – *Modern technologies. Systems analysis. Modeling* **1 (53)** 184-188.
[7] Sharov V A 2016 New scratches at realization of the uniform integrated scheduling on public railway *transport transport Science and technology* **2** 87-93.
[8] Shenfeld K P 2013 Development of a system of qualitative indexes of transportation process in the conditions of structural reform of a rail transport *The innovation factors of development of the Trans-Siberian Railway at the present stage: collection of materials* (Novosibirsk: SGUPS) **2** 151-153. - ISBN 978-5-93461-592-6.
[9] Zubkov V V, Smol'yaninov A V 2012 The Concept of Interaction between Regional Directorate Infrastructures on the Borders of Railways. *Transport of the Urals* **2** 18 – 21.
[10] Zubkov V V. 2012 Formation of a model of region management for improving the maintenance of the railway **3** 40 – 43.
[11] Sirina N F, Zubkov V V 2012 Management of the transportation process based on the theory of forward-thinking systems *Intelligent control systems in railway transport: collection of scientific articles of the 1st sci.-pract. conf.* (Moscow: NIIAS Publ) 138 – 140.
[12] Lapidus B M 2014 On the contribution of Russian Railways in the formation of the country's GDP and the economic tasks of the company in the context of tariff restrictions Vestnik VNIIZhT **1** 3-7.
[13] Krakovskii Yu M, Tamir D, Yakhina A S 2016 Multifactorial estimation of indicators of the transportation process on the basis of second-order models *Modern technologies. System analysis. Modeling*. **2** 82-85.
[14] Krakovskii Yu M, Dombrovskii I A 2013 Prediction of road traffic on the basis of stat. and exp. information *DAAD Fellows Bulletin*. **1 1-1(10)** 18-25.
[15] Krakovskii Yu M, Dombrovskii I A 2013 Prediction of freight traffic indicators of the Ulan-Bator railway *Modern technologies. System analysis. Modeling* **4 (13)** 225-228.

The correction system for the direction of a locomotive spotlight's luminous flux

A S Manikovskij[1], M G Komogorcev[1], V V Stepanov[1]

[1]Transbaikal Institute of Railway Transport, 11, Magistral'naya ave., Chita, 672040, Russia

E-mail: andrey-18@mail.ru

Abstract. In this article the authors propose their concept of a correction system for the direction of a locomotive spotlight's luminous flux in the horizontal plane during train movement. This solution will keep the railway tracks well-lit while the locomotive moves on curved track during night time and in limited visibility conditions and drastically increase the railway traffic safety.

1. Introduction

Traffic safety is one of the key performance indicators of railway transport. Today a plethora of automated systems provide it [1, 2]. Nevertheless, making the most important decisions still falls to humans. The direct responsibility for accident-free train movement lies on the train driver's shoulders, and freight and passenger transport safety depends on his correct actions [3, 4, 5]. During a large portion of working time, taking the geographic location of Russian railways and climate conditions into account, drivers operate their train during nighttime and in limited visibility condition (rain, snow, fog, etc.). To operate in such conditions all locomotives are outfitted with lighting devices [4]. One of the drawbacks of the optical system currently used with locomotives is the fact that during the movement on curved track the spotlight doesn't light the way ahead of the locomotive and the driver keeps moving practically «blind» (fig. 1 a). Needless to say, this circumstance directly influences the train traffic safety and often serves as the cause of numerous emergency situations.

2. Methods

The authors propose a correction system for the locomotive spotlight's luminous flux that will allow to keep the railroad tracks well-lit by the spotlight when the train moves on curved track (fig. 1 b).

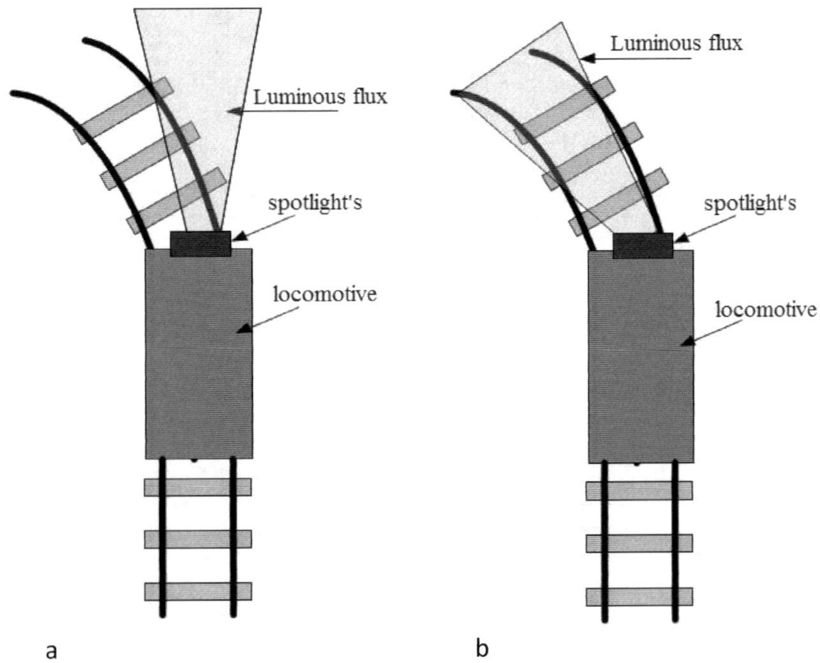

Figure 1. The direction of a locomotive spotlight's luminous flux a) no correction, b) using the automatic correction system.

The system consists of a locomotive spotlight, a gear mechanism, a stepper motor, a stepper motor driver, a central computer, and a dashboard camera (fig. 2). This equipment is located in the locomotive cabin and is protected from any atmospheric influences. Power is supplied by the on-board electrical network (DC voltage 12 V) with a step-down converter.

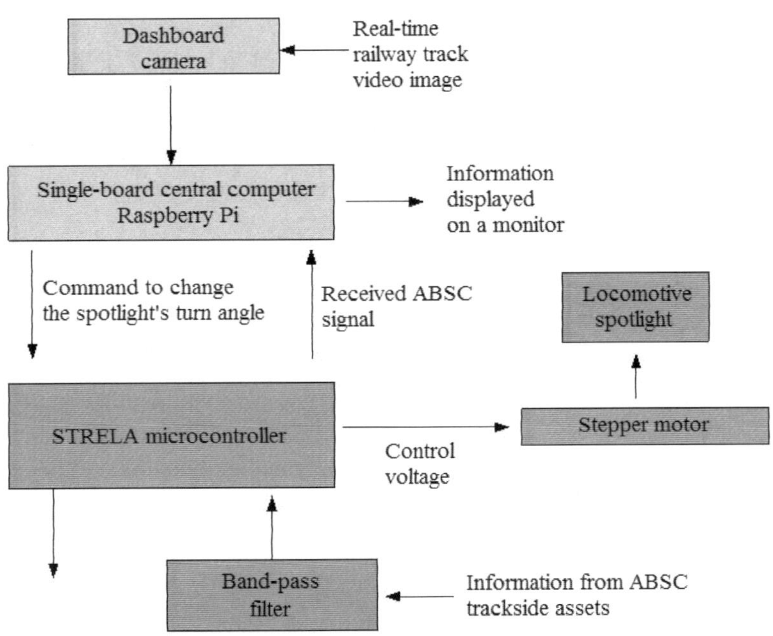

Figure 2. Structure and the operating principle of the correction system's hardware.

The correction system for the direction of a locomotive spotlight's luminous flux functions as follows . The spotlight is connected by the gear mechanism to the stepper motor's crankshaft that is controlled by the CPU through the motor's driver. The dashcam (set on the locomotive dashboard) records the rail tracks ahead of the locomotive. While the train moves in a straight direction, the rail tracks remain in the same position on the video, but once it reaches a curve the rails start moving to the right or to the left in the camera's field of view (fig. 3).

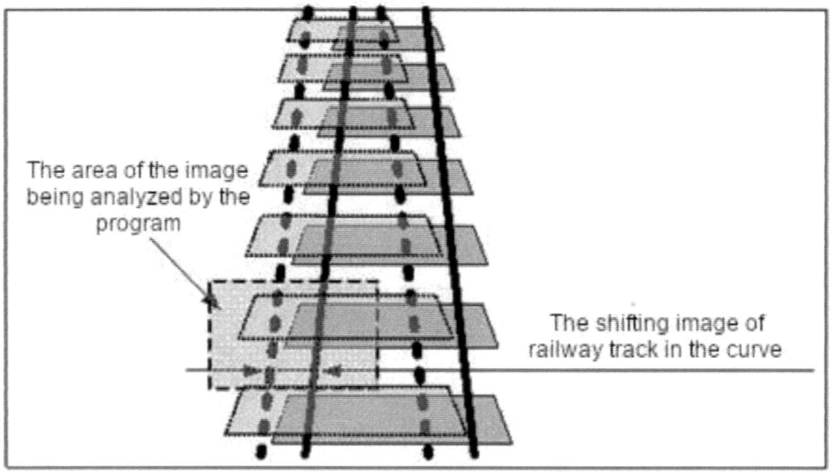

Figure 3. The image of railway tracks ahead of the locomotive, recorded by the dashcam

The system's CPU constantly processes the video images. If the tracks on the image shift from their original position, a command is sent to the stepper motor's driver to turn the motor's crankshaft on a required number of steps (to an according angle on a horizontal plane) (fig. 3). Similar solutions were discussed in concepts of the correction system for the direction of automobile headlights' luminous flux [6, 7, 8].

While approaching a railroad station the correction system receives signals from automatic brake control system's (ABCS) trackside assets and locks the spotlight, fixating it in the center position in the direction of the locomotive's axis. After the train leaves the station the correction system still receives signals from ABCS trackside assets and unlocks the spotlight's gear mechanism.

In case of software malfunctions the spotlight pivots to the central position automatically. The emergency manual control of the gear mechanism is also envisioned in this concept.

3. Results and Discussion

Currently there exists a working model of the correction system for the direction of a locomotive spotlight's luminous flux (fig. 4). Test drives are planned to take place in Transbaikal Railway testing area in 2019/2020. The video image processing and the stepper motor control are performed by means of original software developed for Raspberry Pi (a microcomputer used to analyze the video images) and Arduino (a microcontroller used to control the stepper motor) platforms. The Raspberry and Arduino platforms are used due to their small size, high reliability and low cost [9, 10].

The force transfer from the stepper motor to the locomotive spotlight is performed by a special gear mechanism developed by the authors that is compatible with any type of locomotive spotlights.

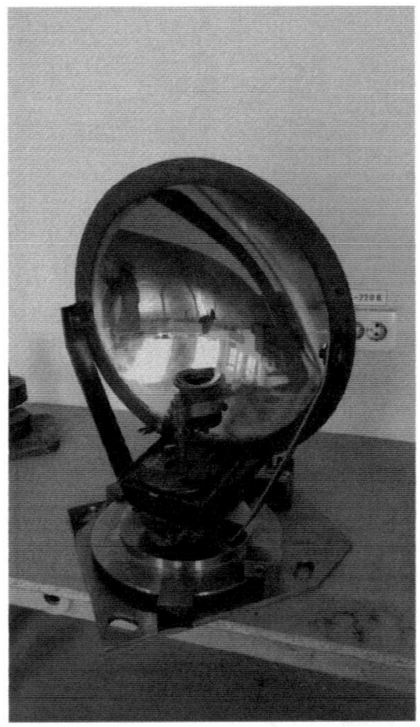

Figure 4. A locomotive spotlight outfitted with the gear mechanism

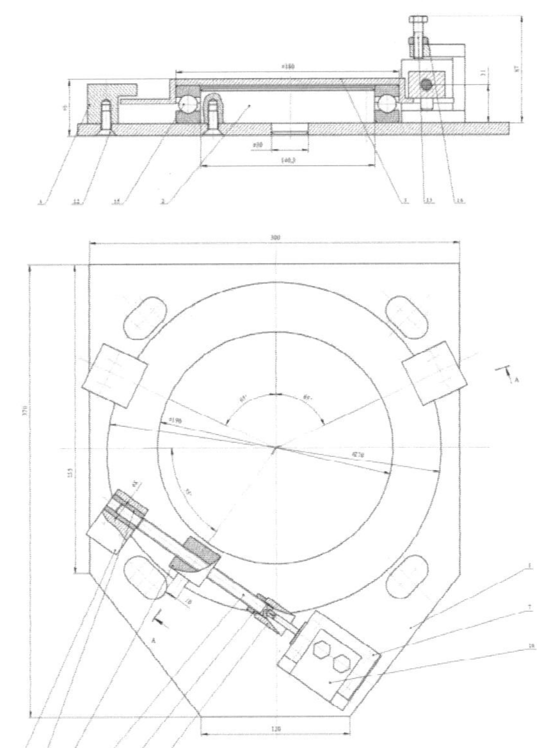

Figure 5. The gear mechanism's blueprint.

This mechanism is a round rotating base placed on a journal bearing. The spotlight is placed on the rotating base. Once the luminous flux correction system is active the stepper motor receives commands from the microcomputer. The motor's crankshaft is connected to the base with a screw-and-nut type drive and pivots the spotlight to a required angle by rotating.

4. Conclusion

The insufficient illumination of railway tracks ahead of the locomotive makes the driver strain their eyes and concentrate their attention, which leads to increased fatigability [11]. The correct rail track lighting in curves will improve the train drivers' working conditions and decrease their fatigue. In certain cases, curve tracks illumination will help prevent collisions with other objects or drop the train's speed, minimizing the damage sustained by the locomotive.

The use of the correction system for the direction of a locomotive spotlight's luminous flux will raise the railway traffic safety during nighttime and in limited visibility conditions.

References

[1] Zhang Y, Zhang H, Wang H 2017 *Security analysis on railway network time protocol based on colored petri nets* **39(10)** 82-88

[2] Wang Y H, Sun Q, Jia L M, Cai G Q, Li J 2011 *Research on safety evaluation of rail freight vehicle operating status & its application* **33(4)** 38-45

[3] Pavlov V I, Aksenova T V, Aksenov V V 2015 *Efficiency of information support of operator when controlling power engineering facilities* **3** 70-75

[4] Thompson N., Goffman M 1988 *Electric experience with online generator monitoring and diagnostics* **50** 468-471

[5] Zinser K., Welsang C 1990 *Knowledge-based systems in distributed control* **1** 63-67

[6] Method of and device for dynamic correction of position of motorcycle headlight beam at cornering RU. Patient No. 2304250 C2 (2007.08.10)

[7] Brightness LED Headlights Automatic Control System for Railway Transport KR. Patient No. 20160108877 (2016.10.04).

[8] Apparatus for controlling swivel angles of on-vehicle headlights U.S. Patient No. 2008103661 A1 (05.01.2008).

[9] Kazanskiy N L, Kotlyar V V, Soifer V.A 1994 *Computer-aided design of diffractive optical elements* **33(10)** 3156-3166

[10] Song K, Liu H, Cheng Z, Li Y, Ren J, Wang Z 2017 *Design and experimental study of three-dimensional magnetic memory testing system based on arduino* **30(6)** 962-967

[11] Shaisoultanov K Sh 2007 *Health state of locomotive crew members in railway transport* **7** 33-35

IOP Conf. Series: Materials Science and Engineering **760** (2020) 012043 doi:10.1088/1757-899X/760/1/012043

Increasing efficiency of braking control algorithm for a two-wheeled motorcycle

A S Melnikov[1], V A Kim[1], and A A Melnikov[1]

[1]Belarusian-Russian University, Mogilev, 212000, Republic of Belarus

E-mail: as_mlun@mail.ru

Abstract. The paper presents a new method for improving the efficiency of the braking control algorithm for a two-wheeled motorcycle. The method is based on the use of high quality information sources - direct measurement and analysis of forces in the wheel-road contact. The criterion for generating control signals by actuators is based on the negative sign of derivatives of lateral forces acting on wheels with respect to time. Experimental studies have shown that the negative sign of lateral force derivatives with respect to time determines side slip of wheels relative to the road surface. The consequence of the sideslip is loss of vehicle stability. The algorithm for generating vehicle control signals, which is based on the negative sign of the lateral force derivative with respect to time, is, to some extent, analogous to modern ESP systems (electronic stability control systems), which are currently being actively installed in two-wheeled vehicles (motorcycles).

1. Introduction

The risk of death for motorcyclists in road traffic accidents (RTA) is 29 times higher than that for motorists. This conclusion has been drawn by Ian Savage, Professor of Northwestern University, USA. Having extensively studied the statistics on RTA from 2001 to 2009, he found that the average motorcycle trip distance is 24 km per day throughout the year, and the probability of death for motorcyclists is 1/860.

Mistakes made by motorcyclists result in falling, skidding and running off the road, most often when cornering at high speed. The main cause of these accidents is loss of stability due to sideslip of motorcycle wheel contacts relative to the road surface. It is the side slip of wheel contacts that leads to a decrease of the wheel-road friction coefficient in the lateral direction, and, as a consequence, to loss of motorcycle stability [1-8].

The statistics on road traffic accidents involving motorcycles in Moscow for the year 2013 is presented in Fig. 1. The data in the figure confirm that the issue of development of active safety systems for two-wheeled vehicles is highly relevant.

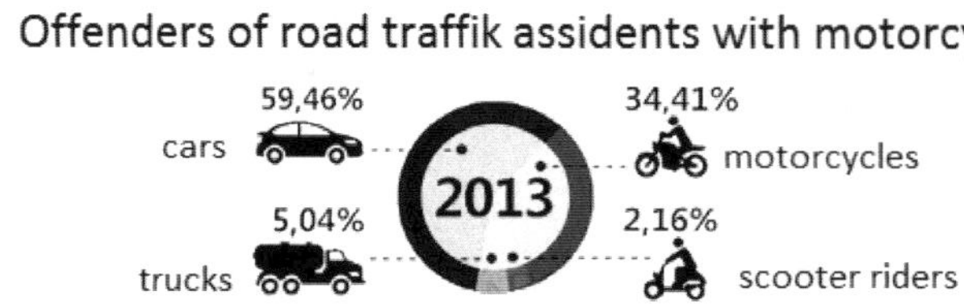

Fig 1. Statistics on road traffic offenders [1]

It is known that modern active safety systems (ASS) are based on measuring and analyzing kinematic parameters (angular velocities and accelerations of wheels, angular velocities and accelerations of vehicle frames), which are necessary for indirect calculation of forces in the wheel-road contact.

Based on indirect calculations, identification of the road surface is performed. Control algorithms become more complicated, and the quality of control is significantly reduced.

In this regard, it is obvious that the solution to the problem of direct measurement and analysis of forces in the wheel-road contact will significantly improve the efficiency of control algorithms for wheeled vehicles under all modes of their movement.

Primarily, ASS solves the problem of maintaining stability and controllability of wheeled vehicles. Modern ESP systems include an anti-lock braking system (ABS) and a traction control system (TCS).

The ABS algorithm is based on tracking a predetermined threshold value of the relative slip ratio of braking wheels, and TCS is based on tracking the difference between threshold values of angular velocities of drive wheels. The specified threshold values are indirect estimates of the grip of wheels on the road surface — these are forces or coefficients of friction between wheels and the road surface.

It should be noted that all ASS systems, without exception, assess the wheel grip in the longitudinal direction of the wheel rolling. In this regard, longitudinal slip of wheel contacts is used for indirect estimation of the friction coefficient in the lateral direction, since stability and controllability of a wheeled vehicle are mostly determined by the friction coefficient in the lateral direction [4 - 8].

For example, if all wheels lock up during braking, but there are no lateral responses, then there is no reason for ABS to engage, as there is no loss of stability. The same can be attributed to the traction mode of motion (TCS).

2. Theory

The aim of the research is to improve the efficiency of the braking control algorithm for a two-wheeled motorcycle by using a high quality informative source — direct measurement and analysis of lateral responses of the road surface.

We have created devices for measuring lateral responses of the road surface for a two-wheeled motorcycle. The results of field tests of the device for measuring lateral forces on various road surfaces (asphalt, wet asphalt, compacted snow cover, ice) at different speeds at the start of motorcycle braking are presented in Fig. 2. The analysis of the results of the motorcycle emergency braking showed that, in the absence of side slide, the derivative of the lateral response $R_{б}$ with respect to time t has the positive sign, i.e., $\frac{dR_{б}}{dt} > 0$.

The lateral response reaches its maximum value at the point A, and the time derivative of the lateral response with respect to time is zero, i.e., $\frac{d\mathbf{R}_b}{dt} = 0$.

A noticeable decrease in the lateral response is observed on the interval AE, $\frac{dRb}{dt} < 0$.

Subsequently, an oscillatory character of the change in the lateral response with a change in the signs of the derivatives is observed on the interval EB.

On the BD interval, the lateral response reaches a constant value.

In addition, it can be seen from the graph of the change in the lateral response (Fig. 2) that the sign of the lateral response derivative changes before the braking wheel locks up.

This means that the change of the sign of the lateral response derivative is a priority criterion for generating control signals, as it makes it possible for braking to occur before wheels lock up, i. e., without sideslip, which improves stability of a two-wheeled motorcycle while maintaining its trajectory.

Thus, a pattern of changes in the lateral response during emergency braking of the wheel has been established.

Experimental studies have shown that the wheel contact slip in the lateral direction can be identified by the negative sign of the derivative of the lateral response with respect to time. Therefore, the negative sign of the lateral force derivative with respect to time can be used as a criterion for generating control signals for brake actuators.

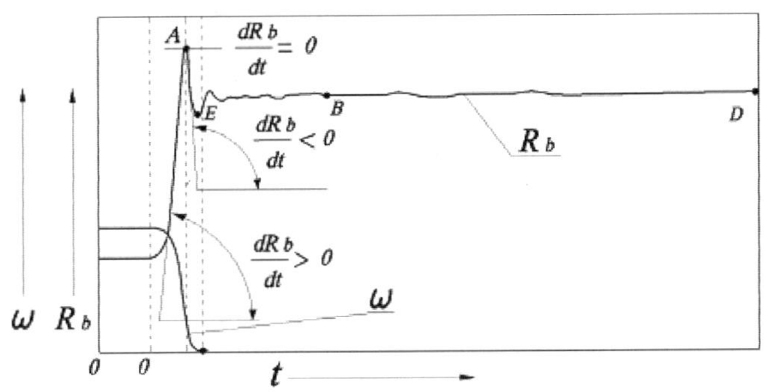

Fig 2. The change in the lateral force acting on the wheel and the change in the force derivative with respect to time during emergency braking of a two-wheeled motorcycle: ω is the angular velocity of the front wheel with the disc brake; R_b is the lateral response (road surface - wet asphalt, speed at the start of braking V = 22 m/s)

The braking control algorithm for a motorcycle, which is based on identification of the negative sign of the time derivative of the lateral force acting on wheels, is presented in Fig. 3. The control algorithm ensures the maximum use of friction coefficients of wheels taking into account conditions of their grip, which ensures stability and controllability of a motorcycle.

The main task of the algorithm is to provide stability control of a motorcycle during braking. Therefore, negative signs of derivatives of lateral responses of motorcycle wheels are proposed as priority criteria for generating control signals in the active safety system algorithm.

The algorithm makes maximum use of the wheel-road friction coefficient by controlling the effort that presses brake pads against the brake disc to bring actual wheel torques to the required values.

The algorithm must ensure the operation of the active safety system taking into account actual magnitudes of the forces arising in the contact patch of both front and rear wheels of a motorcycle and redistribution of road surface responses and braking torques during braking.

IOP Conf. Series: Materials Science and Engineering **760** (2020) 012043 doi:10.1088/1757-899X/760/1/012043

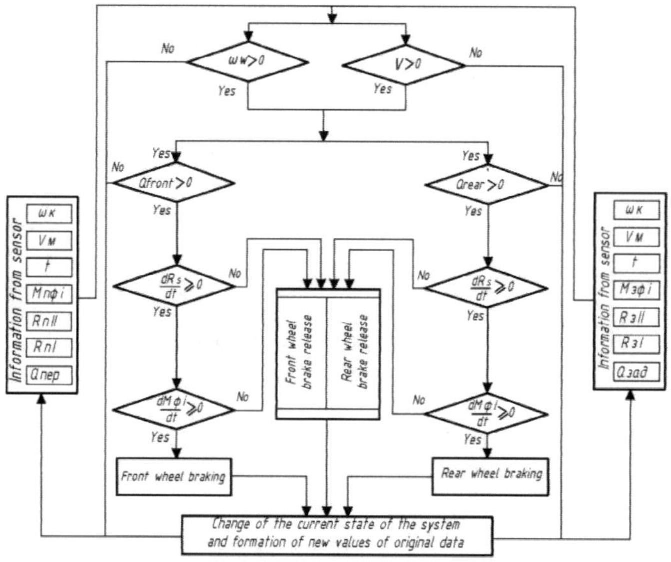

Fig 3. Algorithm of the active safety system of a two-wheeled motorcycle based on the force analysis

In this regard, the problem of complete dissipation of kinetic energy, which takes place in the brake mechanism itself and not in the wheel-road contact, is solved. After each calculation of the brake release force, the signs of the derivatives of the lateral response *dRб/dt* and the braking torque *dMφi/dt* are checked.

The principle of generating motorcycle braking control signals can be used in traction control systems. The joint operation of the braking control algorithm and the traction control algorithm will ensure the operation of the automatic dynamics control system of motorcycles.

A diagram of a two-wheeled motorcycle control system during braking, which uses the principle of control based on tracking the negative sign of the derivative of the lateral response arising in the wheel-road contact, is shown in Fig. 3.

The developed system can be adapted for use in mechanical disk brakes.

The system involves coordinated control of braking torques on the front wheel (1) and the rear wheel (12) of a motorcycle depending on the value of the lateral responses detected by the sensors (2) and (16), as well as the magnitude of the actual braking torque on the front wheel detected by the sensor (18) and on the rear wheel detected by the sensor (17), redistribution of road responses between the front wheel (1) and the rear wheel (12) detected by the sensors (2) located in left and right tubes of the front fork, as well as in the bottom supporting elements of the rear wheel shock absorbers (16). When the brake lever (5) and the rear brake pedal (7) are operated, the brake mechanism (9) of the front wheel (1) and the brake mechanism (13) of the rear wheel (12) generate brake torques.

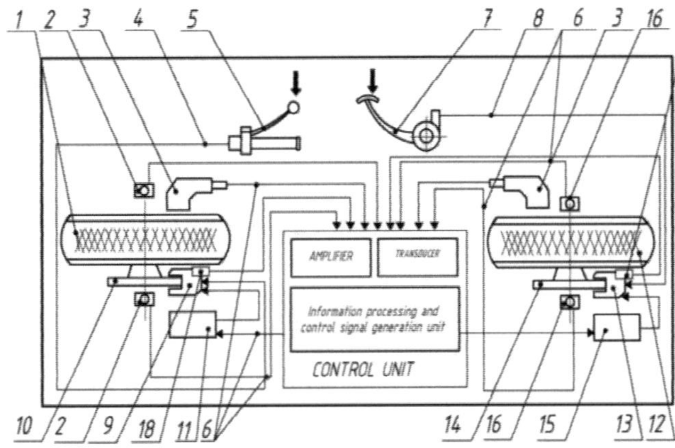

Fig 4. Schematic of the braking control system for a two-wheeled motorcycle: 1 - front wheel; 2 – supporting element for the front fork with a lateral force-measuring sensor; 3 – front wheel RPM sensor; 4 – cable drive of the front wheel brake; 5 – front wheel brake lever; 6 – transmission channels; 7 – rear wheel brake pedal; 8 – cable drive of the rear wheel brake; 9 – front wheel brake; 10 – front wheel brake disc; 11 – actuator – tractive electromagnet of the front brake; 12 – rear wheel; 13 – rear wheel brake; 14 – rear wheel brake disc; 15 – actuator – tractive electromagnet of the rear brake; 16 – bottom supporting element for rear wheel shock absorbers with a lateral force-measuring sensor; 17 – sensor of the actual braking torque of the rear brake; 18 – sensor of the actual braking torque of the front brake.

The control unit handles the information about values of the braking torques on the front wheel (1) and the rear wheel (12) and the magnitudes of the lateral responses on the front and rear wheels, performs redistribution of the road responses between the front and rear wheels and their RPM. In addition, it is responsible for simulation of control action directed at the actuator (in this system it is a tractive electromagnet), at the brake release of the front or rear wheels in the case of the negative sign of the derivative of the lateral forces and braking torques of the front and rear wheels of a motorcycle. Redistribution of road responses for control signal generation is taken into account as well.

The developed braking control system for a two-wheeled motorcycle, which is based on the principle of analysis of lateral forces arising in the wheel-road contact, makes it possible to ensure stability of a two-wheeled motorcycle and improve its braking efficiency.

3. Conclusion

1 Wheel slip relative to the road surface can be monitored by identifying the negative sign of the derivative of forces in the wheel-road contact with respect to time.

2 The experimental studies have shown that the negative sign of the force derivative with respect to time is associated with a decrease in the wheel friction coefficients when the wheel contact is slipping in the longitudinal or lateral direction.

3 The algorithm for generating control signals for actuators of wheeled vehicles, which is based on identification of the negative sign of the force derivative in the wheel-road contact, can be used in the active safety system of any wheeled vehicle.

4 The results of tests of braking torque measurement devices show that they can be used to improve brake mechanisms and as sources of information for onboard diagnostic systems for monitoring braking performance of wheeled vehicles.

References

[1] Radac M B, Precup R E 2018 Data-driven model-free slip control of anti-lock braking systems using reinforcement Q-learning *Neurocomputing*. **275** 317-329

[2] Andrikova D, Andrikov D, Mecapeu C P 2017 Design of flat wheel braking control system with three modes of motion: rolling, sliding, locking *Procedia Computer Science* **103** 466-469

[3] Rishabh B, Sangram P, Ramesh K S 2003 Surface prediction and control algorithms for anti-lock brake system *Transportation Research Part C: Emerging Technologies* **21(1)** 181-195

[4] Song J, Che W S 2009 Comparison between braking and steering yaw moment controllers considering ABS control aspects *Mechatronics* **19(7)** 1126-1133

[5] Rajendran S, Spurgeon S, Tsampardoukas G, Hampson R 2017 Time-Varying Sliding Mode Control for ABS Control of an Electric Car *IFAC-Papers On Line* **50(1)** 8490-8495

[6] Sazonov I S, Amelchenko P A, Kim V A, Melnikov A. S, Dubovik D A 2011 *Wheeled vehicle braking systems* (Mogilev: Belarusian Russian University)

[7] Melnikov A S, Sazonov I S, Kim V A, Bilyk O. V, Vasilevskiy V I 2012 Algorithm of active security system of a two-wheeled mobile car *Bulletin of the Belarusian Russian University* **4** 40-49

[8] Melnikov A S, Sazonov I S, Kim V A, Vasilevskiy V I 2015 Motorcycle brake control *Current issues of mechanical engineering, Collection of research papers* **4** (Joint Institute of Mechanical Engineering of the National Academy of Sciences of Belarus Publ.) 88-90

The development of a method to reduce wear in the wheel-rail system

E A Milovanova[1] and M A Egorova[2]

[1] Railway Rolling Stock. Locomotives, Irkutsk state transport university, 15 Chernyshevskogo Street,
Irkutsk, 664074, Russia
[2] Department of Foreign Languages, Baikal State University, 11 Lenin Street, Irkutsk, 664003, Russia

Abstract. The article deals with the issue of friction reduction. Application of lubrication to decrease the intensity of wear in the wheel-rail system is widely-spread in rail transport. However, the lubricants used in this technology, even those which are made according to contemporary significant requirements relating to the environment, cause irreparable harm to the nature. The article suggests the application of lubrication based on reducing the temperature in the contact area by applying a refrigerant.

1. Introduction

Numerous experimental studies and operational tests carried out in many developed countries with a wide branched network of railways have shown that greasing the ridges of wheels and rails can increase their service life and reduce energy consumption for train traction. The task of reducing the intensity of wear of the side of wheels and rails surfaces is one of the most acute technical problems in railway transport [1, 2]. According to American experts, every US dollar invested in antifriction of rails saves 13 dollars [3]. The urgency of this problem has increased especially due to switching to electric traction and increasing the weight norm of trains. In our country, it is also a burning problem which is caused by the transition to new market relations [4], which affect the cost of rails and their acute shortage, the rise in the cost of turning wheel pairs, as well as due to other problems [5] associated with the operation of tracks and a rolling stock [6].

Antifriction substances used for this purpose in railway transport must meet a number of requirements [7]:

1. High antifriction efficiency, that is, a sharp decrease in the friction coefficient of the wheel ridges sliding on the rails when applying the lubrication, taking into account its minimum consumption.
2. The used friction modifiers should be applied easily to a contact point and kept on the side surface of a rail. This requirement is related to the design of a modifier, and it is complicated by the fact that it should not stick to the elements of the rail lubricator, but adhere to the surfaces of the wheel and rail, between which there are high contact pressures (700-3500 MPa) [2,3]. Under such pressures, many friction modifiers are simply squeezed out of the contact area, causing slippage.
3. Lubricant components should be accessible and inexpensive materials.
4. Friction modifiers must maintain quality during storage, transportation, and under specified operating conditions.

IOP Conf. Series: Materials Science and Engineering **760** (2020) 012044 doi:10.1088/1757-899X/760/1/012044

5. Friction modifiers themselves and their components must be non-toxic to humans [8] and have no impact on the environment [9].
6. Friction modifiers should not have side effects such as corrosion of metal parts, destruction or intensification of rotting of sleepers, rail edging in the curve, etc.
7. Preparation of an antifriction composition, application of it must meet the requirements of fire safety. The modifier itself should have the highest temperature flash point, as it hits cross-sleepers when abrading the rails, you increase the probability of ignition of cross-sleepers from sparks arising from this process.
8. Friction modifiers must meet certain operating conditions (climate change (winter - summer), precipitation, etc.).

A large number of modifiers are currently in operation at the world's railways and under development. Due to the toughening of the above-mentioned requirements intensive scientific research is carried out in this direction in most countries of the world.

2. Discussion

The used friction modifiers are divided into liquid (oil-solid) and plastic (salve like grease or thick) greases according to their aggregate state.

Oil lubricants are most easily applied to the contact area, but they are retained poorly on the side surface of the rail head. In addition, when they are applied, wind loads and air movements caused by the movement of a train have a great influence. A great impact is also a sharp change in ambient temperature (especially in spring and autumn). When developing these types of lubricants, the specifics of the friction unit under consideration is not taken into account, where the antifriction surface of the lubrication zones (the wheel edge - the side surface of the rail) passes into the friction pair (the wheel bandage - the surface of the rail head rolling), where the lubricant should not fall. In the latter case, there is a significant decrease in the coefficient of adhesion of the wheels to the rails and to the appearance of the probability of wheel slipping, which is especially dangerous on long climbs and large slopes. As a rule, using liquid lubricants there is an increased consumption, which also contributes to the surface of the rolling surface of the rail head. Thus, oil lubrication does not allow realizing fully the effect of rail lubrication [3]. However, this type of lubricant is sometimes used due to its low cost, ease of application, non-toxicity, low fire hazard and rapid biodegradation in the environment. For example, one of such lubricants is waste diesel oil, which is used at the East-Siberian railway since July 1997 both in pure form and with the addition of graphite or its substitutes [10].

The above disadvantages of oil lubricants can be leveled by introducing various additives, that is, the use of lubricant compositions. This type of liquid lubricants includes rail lubricant recommended for Russian Railways, PC - 6B. The lubricant consists of a polymer film-forming agent and finely crushed graphite, which are diluted with gasoline A-76. However, the use of gasoline leads to a sharp increase in the fire and explosion hazard while preparing the lubricant composition for its further usage, and to the necessity of balanced system of ventilation. In addition, the resulting composition does not have the necessary stability (when it is applied, the substance should be mixed constantly), and its components are quite expensive materials. The modification of the lubricant RS-6V, by introducing additional components (for example, molybdenum disulfide and anti-corrosion additives), leads to an improvement in the operational properties of the lubricant, but makes it even more expensive.

Oil-solid grease includes crystalline substances having a layered structure and, consequently, having lubricating properties: graphite, molybdenum and tungsten disulfide, boron nitride, tin bromide and cadmium, silver sulfate, bismuth iodide, nickel and cadmium, phthalocyanines, selenides and tellurides of tungsten, titanium, etc. [7]. Some of them are also used in railway transport. Graphite and, to a lesser extent, molybdenum disulfide are mostly widespread [11].

Graphite has self-lubricating properties, high adhesion to various surfaces and ease of sliding on planes. The usual thickness of the graphite layer at the point of contact is 0.1 - 0.32 mm. Graphite is applied to the contact area usually with the use of a rod that is gradually worn against the side surface of the rail or wheel edge. The main drawback of this lubrication, the possibility of breakage of the graphite rods, which requires a special approach to the design of wheel flange lubricator leads to a decrease in the operating life of the rods [11, 12]. This often leads also to destruction of the attachment of wheel flange lubricator. In addition, the used lubricants are expensive and difficult to access products.

Solid metal modifiers are also referred to as solid lubricants [6]. The most preferred method of application is friction metallization of wheels or rails. This technology has been introduced, in particular, at the West Siberian road. Metallization of a wheel is carried out using additional anti-friction pads made of the applied material and having a profile of the bandage. The metallization of the rails can be carried out with the help of "collecting shoes". Non-ferrous metals and antifriction graphitized cast irons are used as friction modifiers.

However, the use of solid metal compounds do not sufficiently apply at the railways, apparently because of the complexity of the technological process of metallization and because of the lack of effective lubrication. In addition, the worn-out metal of the lubricant gets into the environment, which contributes to its pollution, including heavy metals.

Plastic or grease modifiers are of the greatest use at the railways. As a rule, they are complex compositions containing oil liquids, thickeners and finely dispersed solid lubricant (graphite molybdenum disulfide, etc.). According to the physical structure greases are dispersed systems, in which liquid oils (usually oil) serve for dispersion, and the dispersed phase - solid lubricants. Thickeners create a structural framework in the lubricant volume, inside which the liquid is retained by the interaction between molecules and particles of the dispersed phase and the thickener.

The thickener molecules usually have a strand - or ribbon-like shape. Long-chain salts (C_{12}-C_{25}) of fatty acids (lithium aluminum, sodium, calcium, etc.), called "soaps", are used as thickeners [13]. Like all soaps, thickeners of this type are surface-active substances (surfactants). Along with the obvious advantages of using surfactants in lubricating compositions (high stability of the lubricant, easy consistency adjustment, improved adhesion properties of the lubricant, etc.), they have a number of disadvantages. First of all, they are quite expensive. In addition, they are easily washed off the rail by precipitation and pollute the environment, especially natural reservoirs near the railways. When the thickeners are removed, the lubricant composition loses its performance (for example, it is easily exfoliated from the rail surface).

The use of surfactants in lubricating compositions is complicated by the fact that getting into microcracks and cracks, these substances have a wedging effect [14] and can contribute to destruction, especially in places that experience mechanical stress, but are not intended for lubrication.

Polymer materials are also used as thickeners [15-18]. The disadvantage of their use is the poor solubility of polymers in oil, forming dispersion. This can make it difficult to use the lubricant when applying it.

Oil bitumen is used in some lubricant compositions [19]. They can be used simultaneously with a polymeric thickening agent [20]. Bitumen has an adverse impact on the environment and, as a rule, it has high acidity and, thus, can provoke corrosion damage (outside the friction zone).

Some inorganic substances are also used as thickeners, aqueous solutions of which have a high viscosity, such as liquid glass [21]. However, thickeners of this type have an alkaline reaction and, consequently, they can cause corrosion cracking of steels.

Sometimes auxiliary components are added into a lubricating composition in order to enhance the anti-corrosion, anti-wear and other performance properties of lubricants [22]. For example, sulfur (1%), caustic soda (1%) and surfactants in the form of sodium-calcium soaps are added to the lubricant of LR to strengthen the metal surface [23]. However, firstly, such a grease has a disadvantage, typical of greases containing a surfactant. In addition, it contains free alkali, which is washed away by precipitation (without alkali, sulfur does not give a strengthening effect), and causes corrosion.

Along with the complexity of the issue about the composition and performance properties of lubricants, the problem of lubrication is also associated with the problem of designing devices that supply lubricant to the contact zone. In some cases, the high dignity of a friction modifier can be negated by the poor design of a rail lubricating car (and vice versa). Currently, railways use stationary lubrication systems (lubricating waypoints), mobile lubricators, as well as wheel flange lubricator installed on a rolling stock.

IOP Conf. Series: Materials Science and Engineering **760** (2020) 012044 doi:10.1088/1757-899X/760/1/012044

Locomotive wheel flange lubricators using liquid lubrication require a systematic adjustment and monitoring work in the operation, and it is often difficult to fulfill in conditions of intensive use of locomotives and their maintenance by loose parties [24]. The main drawback of such wheel flange lubricators is high probability of getting grease on the running surfaces of wheel pairs, which is especially unacceptable on heavy sections of the railway.

Track lubricators lubricate rather satisfactorily the section of the way near their location, which often coincides with the place of the most active use of sand to increase traction and prevent slippage. The combination of lubrication with sand on the rail surface increases wheel and rail wear. In addition, in winter such lubricators give frequent failures. Stationary lubricators are usually placed on stretches of railroad track where the intensity of side wear exceeds 3 mm per 100 million tons of gross transported cargo [25]. They are installed on the roadside at the beginning of a curve. They also require quite complex maintenance (periodic filling of the oil-air tank, adjustment, cleaning, etc.). This is especially difficult to do in winter. These lubricators often give increased consumption of lubricant. That is why a part of it can get onto the friction surface of a rail. Simultaneously, a slug feed can give ungreased sections of the friction pair “a wheel flange lubricator - rail”.

Side rail lubricators installed on a motor vehicle are more optimal [25]. They can ensure the application of a strictly calculated thickness, an even layer of grease on the side surface of the rail head. However, these designs are often far from perfection. When using them, high requirements are imposed on the physical and chemical properties of the lubricant, especially on the viscosity, which depends significantly on the temperature. In order to prevent the solid component from settling (and sometimes to ensure its heating), it is necessary to provide its circulation [24]. And these facts complicate the design greatly.

The most complex element in the design of a wheel flange lubricator is the one for the immediate supply of lubricant in the contact zone, which requires constant monitoring of the operating conditions [26].

The complexity of the problem of lubrication is quite obvious due to the wide variety of lubricants and rail lubricators market both in our country and abroad [27]. And in this respect, we can note the fact that the company "Abba" from the United States has developed a portable lubricator that can be carried by a track worker alone. The structure of the lubricator includes a lubricant tank, the contents of which (O, kg) is sufficient to lubricate 1.6 km of the way in the curves.

Thus, one of the most important problems in railway transport, the problem of rail lubrication, cannot be considered to be solved in no way. Most lubricants and lubricant compositions, as well as devices for their supply to the contact zone of the wheel rail, do not meet the complex requirements for them.

3. Results

An innovative approach may be a possible way to cope with this problem.

The Irkutsk institute of railway engineers patented a method of reducing wear in the wheel-rail friction pair by reducing the temperature directly in the contact zone [28]. The idea of the method is based on seasonal changes in the wear of wheels and rails, from which it follows that at lower temperatures, wear occurs to a lesser extent. It follows that if a cooling agent is given into the contact zone, this will lead to a sharp decrease in temperature in this zone, and therefore will help reduce wear.

The choice of device for cooling the contact area is directly related to the choice of refrigerant. Chilled salt solutions, freons, liquid ammonia, solid carbon dioxide and liquid nitrogen are used as refrigerants in the chemical and food industry [29]. Salt solutions (mainly $CaC1_2$) and ammonia are used only for closed cooling systems, because they cause corrosion of metals, have an adverse effect on the environment, and ammonia is also a toxic gas and has an unpleasant smell. The use of freons is continuously reduced even in closed cooling systems, because according to many scientists, they destroy the ozone layer of the Earth which is inappropriate [30].

Solid carbon dioxide is a very fragile material and can hardly be used as rods (like graphite). The relatively low coefficient of heat transfer from solid carbon dioxide to solid surfaces should also be taken into account as a gas layer of sublimated acid, which prevents the transfer of heat, is formed constantly between them. When a moving locomotive needs cooling it is difficult to achieve it.

4. Conclusion

The most acceptable and promising cooling agent seems to be liquid nitrogen. It is an absolutely harmless, incombustible substance, the main component of the Earth's atmosphere. It has a boiling point of 75 K (-198°C), it can quickly evaporate in the air cooling the surface on which evaporation occurs. Its getting on the surface of the rolling wheel pairs will not affect the movement of a train. At present, rather compact units have been developed to produce liquid nitrogen from the air, which can be installed onto a locomotive. Due to the low viscosity its supply to the contact area can be provided by a simple pipeline system.

Thus, the wear reduction system using refrigerants can successfully solve the problem of rail lubrication.

References

[1] Ermolenko I Ju, Zheleznyak V N 2016 Mechanics of formation of defects on the surface of the wheel and rail rolling during the passage of the mountain surface relief. *Modern technology. System analysis. Modeling* **3** (**51**) pp 193-199

[2] Gozbenko V E, Tuvshintur B 2017 Reduction of wear of ridges of wheel pairs of freight cars *Modern technology. System analysis. Modeling* **3** (**55**) pp 159-168

[3] Shapovalov V V, Bogdanov V M and Fendrikov A I 1992 Rail lubricators *Railway transport* **11** pp 42-44

[4] Astrahantseva A S 2018 Approaches to the use of innovative technologies on railway transport as a tool for efficient management of logistics costs (Moscow: non-profit Foundation of assistance in development of economic science and education "Economy") 4 pp 147-151

[5] Goroduk I S 2017 Transport complex of Irkutsk region as an integral part of international transport corridors *Proc. Int. Conf. on Eurasian cooperation, Proceedings of the international scientific and practical conference* (Irkutsk: Baikal state university) p 160

[6] Andreev A I, Komarov K L and Karpushhenko N I 1997 Wear of rails and wheels of rolling stock, *Railway transport* **7** pp 31-36

[7] Obelnicky A M, Egorushkin E A and Chernyavsky Ju N 1997 Fuels, lubricants and coolants (Moscow: IPO Poligran) 270 p

[8] Rybinskaya E T 2017 Environmental and legal problems of energy resources use current trends in socio-economic and humanitarian Sciences: theory and practice of scientific works (Irkutsk) p 238-243

[9] Jakimova E M, Sirina-Lebua N V and Chuksina V V 2017 State regulation of the sphere of environmental audit as a tool to improve environmental safety *Proc. Int. Conf. on Problems of ensuring national security in the context of the changing geopolitical situation* (Irkutsk: Baikal state university) pp 260-267

[10] Moroz B A, Korotaev E N and Marutin K A 1999 A complex system of resource saving of wheels and rails on the East-Siberian railway **5** *New technologies on the East-Siberian railway* (Novosibirsk) pp 96-99

[11] Buinosov A P and Nagovitsin V S 1996 New wheel flange lubricating machine based on graphite-containing material, *Railway transport* **2** pp 39-41

[12] Buinosov A P and Dibrov S A 1995 An important factor in reducing wheel and rail wear *Railway transport* **6** p 39

[13] Lubricating work at the contact point of wheels and rails: laboratory and field tests (SShA) 1992 *Express-information, Railway transport abroad,* **5** *Track and track facilities***1** pp 5-17

[14] Voiucky S S 1975 *The course of colloid chemistry* (Moscow: Chemistry) 511p

[15] Shibryaev S B and others 1995 Pat. RU №2048506, Lubrication, Bul. invention **32** p 194

[16] Shapovalov V V and others 1995 Patent application №93007533/04 A lubricator for oiling of railway rails Bul. invention **23** p 53

[17] Skripec V I, Samoilenko S P and others 1993 A.S. SSSR №1810384 The lubricant composition Bul. invention **15** p 56

IOP Conf. Series: Materials Science and Engineering **760** (2020) 012044 doi:10.1088/1757-899X/760/1/012044

[18] Kononenko Ju K and others 1993 A.S.SSSR №1807075 A lubricant composition for heavily loaded friction units Bul. invention **1** DshhSD87
[19] Shapovalov V V 1996 Pat. RU №2065484 A lubricator for oiling of railway rails Bul. invention **23** p 171
[20] Kapustin A M and others 1995 Pat. RU №2036223 Bituminous grease for heavily loaded friction units Bul. invention **15** p 159
[21] Evdokimov Ju A 1996 and others Pat. RU №2067110, A lubricator for oiling of rails Bul. invention **27** p 169
[22] Shastin V I, Kargapoltsev S K, Permyakov A G 2018 Modification of surfaces of structural materials: equipment and methods of the research of tribological parameters *Modern technology. System analysis. Modeling* **4** (**60**) pp 31-38
[23] *Surface phenomena and surfactants. Handbook* 1984 Red. A.A. Abramzon and others (Leningrad: Chemistry) p 364
[24] Neglinsky V V, Atamanuk A M and Shherbakov K F. 1993 The new technology of rails greasing *Railway transport* **11** pp 52-55
[25] Shapovalov V V, Evdokimov Ju A, Bogdanova V M and Majba I A 1993 Improvement of the lubrication effectiveness *Railway transport* **7** pp 40-41
[26] Buinosov A P and Dibrov S A 1994 Ensuring effective lubrication of locomotive wheel bands *Railway transport* **11** pp 60-61
[27] Rail lubricators and materials for rail lubrication on railways SShA 1993 *Express-information. Railway transport abroad* Issue IV *Track and track facilities* **4** pp 28-35
[28] Korotaev B V, Milovanov A I and Ostromensky P I 1999 Pat. RU № 2142890 A method for reducing the wear of the side surface of rails and ridges of wheels of a railway vehicle Bul. **5**
[29] Kasatkin A G 1973 *Main technological processes and devices of chemical technology* (Moscow: Chemistry) pp 646-678
[30] Shobodoeva A V 2018 Ecological safety of the Russian Federation: theoretical and methodological problems *Bulletin of Irkutsk state university Issue: Political studies. Religion study* (Irkutsk: Irkutsk state university) **25** pp 35-42

IOP Conf. Series: Materials Science and Engineering **760** (2020) 012045 doi:10.1088/1757-899X/760/1/012045

Thermodynamic estimination of adsorption ability of modified zeolites

M V Obuzdina[1], **E A Rush**[1]

[1] Federal Agency of Railway Transport, Irkutsk State Transport University (ISTU), 15, Chernishevskogo str., Irkutsk, 664074, Russia

E-mail: obuzdina_mv@mail.ru, lrush@mail.ru

Abstract. The article presents the results of original investigations of the thermodynamic laws of sorption extraction of Ni^{2+}, Zn^{2+}, $Cu^{2+}$from wastewater. As an adsorbent, a natural zeolite from the Holinsky deposit of the klinoptilolite type, modified with a sulphur polymer, obtained from waste in the production of epichlorohydrin was proposed. The limiting parameters of the adsorption process are described by the Langmuir equation. The graphical method the limiting value of adsorption, the adsorption equilibrium constant, the differential heat of adsorption, the adsorption rate constant, the activation energy, the Gibbs energy was founded. It is proved that the investigated processes obey the laws of physical adsorption and are spontaneous.

1. Introduction

The energy of adsorption processes is one of the most important components in their investigations. The investigations of structure of natural and modified zeolites, the identification of adsorption process mechanism, the direct investigation of energy heterogeneity and other problems require the use of modern precision physicochemical methods. Heats of adsorption can be determined either directly by measurement using a calorimeter or by calculation based on various experimental data [1].

In microporous systems, during adsorbent – adsorbate interactions, bulk filling of the adsorption space takes place. Therefore, the volume of micropores becomes the main geometric parameter, and the specific surface loses its physical meaning for microporous adsorbents [2]. Representatives of microporous adsorbents with an extremely narrow pore distribution are dehydrated zeolites. A remarkable feature of zeolites is the presence of exchangeable cations in their adsorption cavities, which create high gradients of electrostatic fields [3]. The adsorption of molecules on the adsorbent limits, inhibits their mobility; the interaction between the adsorbate and the adsorbent reduces the mobility of the molecules [4]. A decrease in the degree of mobility of molecules as a result of adsorption leads to a decrease in entropy [5].

2. Results and discussion

In previous investigations, conducted by us, we investigated the thermodynamic laws of sorption extraction of oil products and heavy metal ions from industrial wastewaters by zeolites from the Holinsky deposit, modified by high organosilicon compounds HMDS – hexamethyldisilazane $[(CH_3)_3Si\text{-}]_2NH$ and TEOS - tetraethoxysilane $(C_2H_5O)_4Si$ [6,7,8].

The actual direction of modern development of physical chemistry is to increase the reactivity of solids by chemical modification to obtain new sorbents from them [9, 10]. Therefore, the goal of our further research was defined as a directional change in the physicochemical properties of the natural zeolitive rocks of klinoptilolite type of Holinsky deposit [11] by means of modification with sulfur polyme, obtained from epichlorohydrin production wastes. The main component of the waste is 1,2,3 - trichloropropane [12].

The obtained kinetic characteristics of the extraction of Ni^{2+}, Zn^{2+}, Cu^{2+} from industrial wastewater of railway transport enterprises are presented in detail in our previous works [13]. Let us consider the adsorption capacity of klinoptilolites, modified by sulfur polymer from the point of view of thermodynamic estimation.

The limiting parameters of the adsorption process can be qualitatively obtained by describing the isotherm by the Langmuir equation:

$$A = A_{\infty}\frac{K_p \cdot C}{1 + K_p \cdot C}, \quad (1)$$

where A – current value of adsorption, mg/g; A_{∞} – maximum value of adsorption, mg/g; K_p – adsorption equilibrium constant; C - concentration of solution, mg/l.

The isotherm in the coordinates of the Langmuir equation is linearized. Dividing the unit on the left and right sides of the equation, we obtain the equation of a straight line in coordinates 1/A=f(1/C):

$$\frac{1}{A} = \frac{1}{A_{\infty}} + \frac{1}{A_{\infty} \cdot K_p} \cdot \frac{1}{C}. \quad (2)$$

Equation (2) allows graphically finding the constants A_{∞} and K_p in the Langmuir equation. Figure 1 shows the adsorption isotherm of heavy metal ions by a zeolite, modified by a sulphur polymer, for example Ni^{2+} and Cu^{2+}. The line cut off by the straight line on the y-axis is $1/A\infty$. According to the tangent of the angle of a straight line, the adsorption equilibrium constant *K* can be found by the formula:

$$K_p = \frac{1}{A_{\infty} \cdot tg\alpha} \quad (3)$$

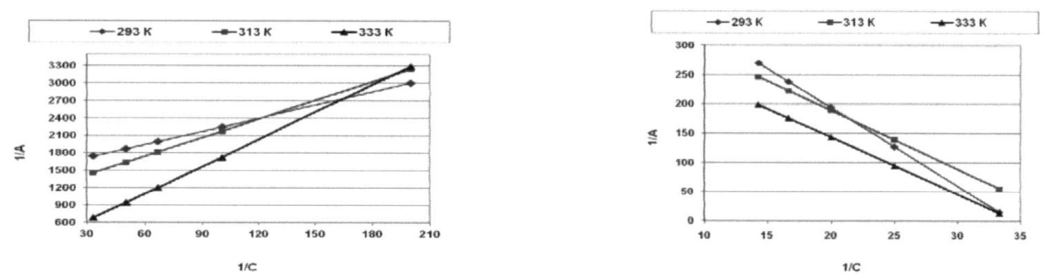

(a) (b)

Figure 1 Adsorption isotherms of Ni^{2+} (a) and Cu^{2+} (b) with zeolite, modified by sulfur polymer in Langmuir coordinates at various temperatures

Table 1 presents the constants A_{∞} and K of the Langmuir equation for the sorption of Ni^{2+}, Zn^{2+}, Cu^{2+} ions on a zeolite, modified with a sulfur polymer at different temperatures. The data in Table 1 indicate that with increasing temperature, the value of A_{∞} increases, and the equilibrium constant K_p decreases.

Table 1. Constants of the Langmuir equation for the adsorption of Ni^{2+}, Zn^{2+}, Cu^{2+} of zeolite, modified by sulphur polymer

Temperature, K	Ni^{2+}		Zn^{2+}		Cu^{2+}	
	K_p	$A_\infty*10^{-3}$, mg/g	K_p	$A_\infty*10^{-3}$, mg/g	K_p	$A_\infty*10^{-3}$, mg/g
1	2	3	4	5	6	7
293	975	0,67	4,92	0,12	972	2,17
313	515	0,91	3,98	0,26	601	2,57
333	245	6,15	2,77	0,55	345	2,96

Adsorption is a spontaneous process. The thermal effect in the range of changes in the degree of filling of the adsorption space is characterized by the differential heat of adsorption [14]. To calculate the isosteric heat of adsorption, the Klaiperon – Clausius equation is applied at A = const:

$$\frac{\Delta \ln C}{\Delta(1/T)} = -\frac{Q}{R} \tag{4}$$

$$Q = -R \cdot \frac{\Delta \ln C}{\Delta(1/T)} \tag{5}$$

where A – the sorbent capacity, mg/g; T - temperature, K; R – the gas constant; R = 8.31 J/mol·K; C – the equilibrium concentration of heavy metal ions in solution, mg/l.

Isostere of adsorption is an important function that carries direct information about the heat of adsorption. When A = const:

$$\ln C = f(\frac{1}{T}). \tag{6}$$

The differential heat of the sorption of metal ions (see Figure 2) can be calculated from the angles of inclination of the isoster, converting equation (5) into the following form:

$$Q = -R \cdot tg\alpha` \text{ when A = const} \tag{7}$$

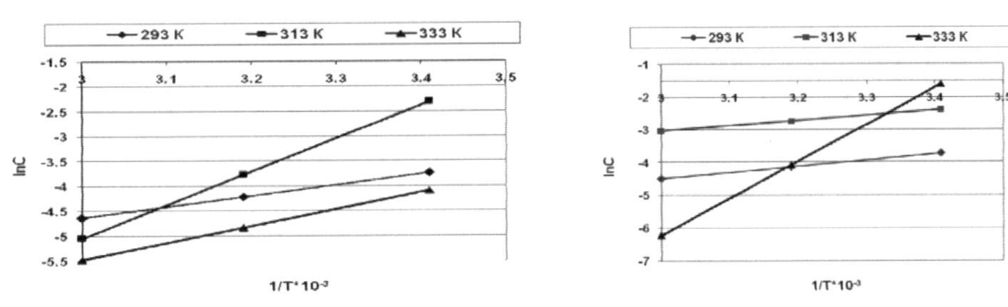

(a) (b)

Figure 2 Isostera adsorption Ni^{2+} (a) and Cu^{2+} (b) of zeolite, modified by sulphur polymer, at different temperatures

The results of the calculation are presented in table 2. The adsorption rate constant *K* was calculated using the first-order adsorption rate equation. The equation of the rate of adsorption of the first order in differential form is:

$$\frac{dc}{dt} = K \cdot C. \tag{8}$$

The equation for the rate of adsorption in integral form:

$$K = \frac{2{,}303}{t} \cdot \lg \frac{C_0}{C} \tag{9}$$

where C_0 – the initial concentration of metal ions in solution, mg/l, C – the equilibrium concentration of metal ions in solution, mg/l, t - time. From equation (9) it is easy to get:

$$tg\alpha = 0{,}434 \cdot K \tag{10}$$

where $\acute{\alpha}$ – the angle of inclination of the tangent to the kinetic curve to the positive direction of the x-axis. The results of the calculation of the adsorption rate constant are presented in Table 2.

Table 2. The results of calculation of the rate constants, Gibbs Energy, the heat of adsorption of Ni^{2+}, Zn^{2+}, Cu^{2+} of zeolite, modified by sulfur polymer

Temperature, T, K	Ni^{2+}			Zn^{2+}			Cu^{2+}		
	K, sek^{-1}	ΔG, kJ/mol	Q, kJ/mol	K, sek^{-1}	ΔG, kJ/mol	Q, kJ/mol	K, sek^{-1}	ΔG, kJ/mol	Q, kJ/mol
1	2	3	4	5	6	7	8	9	10
293	0,69	-16,76	-5,39	0,44	-3,88	-4,64	0,57	-16,75	-3,11
313	1,29	-16,24	-7,37	0,74	-3,59	-6,01	1,21	-16,64	-4,1
333	1,82	-15,22	-8,42	1,18	-2,82	-6,65	1,63	-16,17	-5,49

To find the adsorption rate constant K, it was necessary to construct the kinetic sorption curves of Ni^{2+}, Zn^{2+}, Cu^{2+}. The results of the calculation are presented in Table 2. An analysis of the data in Table 3 shows that an increase in the adsorption rate K is observed as the temperature increases. This can be explained by the fact that with increasing temperature the movement of molecules becomes more intense, and they more quickly interact with the surface of the adsorbent.

The appearance of the kinetic curves of dependence of the adsorption of metal ions by a zeolite, modified by sulfur polymer, indicates an increase in the rate of adsorption with increasing temperature. The curves obtained at higher temperatures lie above the curves obtained at low temperatures. Thus, the process is activated [15].

Only active molecules with activation energy, i.e. excess energy, in accordance with the Arrhenius equation can penetrate into the pores and adsorb [16]. The reaction rate constant is associated with the activation energy by the following relationship:

$$K = A \cdot e^{-E/RT} \tag{11}$$

where K – reaction rate constant, s^{-1}; A – constant; E_a – activation energy, kJ/mol; R – the universal gas constant; T – temperature, K.

From formula (11), we obtain the following expression for calculating the activation energy:

$$E_a = -2{,}303 \cdot R \cdot \frac{\Delta \ln K}{\Delta(1/T)} = -2{,}303 \cdot R \cdot tg\alpha \tag{12}$$

where tgα – the slope angle of graphical dependence lgK= f (1/T).

If the adsorption obeys the Langmuir equation, i.e. the equilibrium constant K_p does not depend on the degree of filling of the sorbent surface, then the Gibbs energy (see table 2) of adsorption can be calculated using the Vant-Hoff isotherm equation:

$$\Delta G = -R \cdot T \cdot \ln K \qquad (13)$$

where K_p – the constant of adsorption equilibrium in the Langmuir equation; R – the universal gas constant; T - temperature, K; ΔG – Gibbs energy, kJ/mol.

The activation energy of sorption of Ni^{2+}, Zn^{2+}, Cu^{2+} of zeolite, modified by sulfur polymer, is 10.24 kJ/mol; 5.75 kJ/mol and 14.03 kJ/mol, respectively. The temperature dependence and the values of the sorption activation energy indicate that investigating processes obey the laws of physical adsorption. An analysis of the Gibbs energy calculation results indicates its negative value. This confirms the fact that adsorption is a spontaneous process.

3. Conclusion

In the course of the experimental investigations, a thermodynamic estimate was made of the adsorption capacity of the modified klinoptilolites of the Holinsky deposit in Eastern Transbaikalia with respect to heavy metal ions. A sulfur polymer obtained from waste in the production of epichlorohydrin was used as a modifier. The degree of sorption purification with respect to heavy metal ions is close to 100%.

Adsorption isotherms are presented in the coordinates of Langmuir, which allowed us to find the maximum value of adsorption and the constant of adsorption equilibrium. For the solution of further thermodynamic problems, the isosteric method was chosen. The differential heat of adsorption of $Ni2^{+}$, Zn^{2+}, Cu^{2+} was determined depending on the angle of inclination of the isoster.

As a result of the analysis of the constructed kinetic curves of adsorption of heavy metal ions, the adsorption rate constants were calculated. The results of the calculation of the activation energy and Gibbs energy, determined by the graphical method, are presented. It is proved that the investigated processes obey the laws of physical adsorption and are spontaneous.

References

[1] Devdariani O G 1980 Aut. dis. ... cand. technical sciences (Tbilisi: Tbilisi Order of the Red Banner of Labor State University) p 24.
[2] Dubinin M M 1987*Adsorption and porosity* (Moscow: VACh) p 172
[3] Chelishchev N F, Berenshtein B G, Volodin V F 1987 *Zeolites is a new type of mineral raw materials* (Moscow: Nedra) p 176
[4] Brek D 1976 *Zeolitic molecular sieves* (Moscow: World) p 781
[5] Vernov A V and Lopatkin A A 1979 *Zh. physical chemistry* **53 12** 3161-3166
[6] Obuzdina M V 2011 *Modern technologies. System analysis. Modeling* **31 3** 209 213
[7] Rush. E A and Obuzdina M V 2013 *Izvestia Transsib* (Omsk: Omsk State Transport University) **13 1** 27-34
[8] Patent of the Russian Federation Patent No. 2524111 (07.27.2014)
[9] Dabija O N 2012 *Sorption and chromatographic processes* **6 12** 860-866
[10] Solovyova Y V, Krasnova T A, Yustratov V P 2010 *Ecology and Industry of Russia.* **4** 54-58.
[11] Kotov P A 1995 *Fields of Transbaikalia* (Moscow: Geoinformmark) 280
[12] Patent of the Russian Federation Patent No. 2624319 (07.03.2017)
[13] Obuzdina, M V, Rush E A, Korchevin N A, Shaluch L V 2017 *Bulletin of the Technological University.* **20 7** 149-155
[14] Timofeev D P 1962 *Adsorption kinetics* (Moscow: Publishing House of the Academy of Sciences of the USSR) p 252
[15] Greg S and Singh K 1984 *Adsorption, specific surface area, porosity* (Moscow: Mir) p 306
[16] Prigochin I and Condepudi D 2002 *Modern thermodynamics* (Moscow: Mir) p 464

Investigation of sorption characteristics of modified klinoptilolites by the method of infra-red energy and differential X-ray spectroscopy

M V Obuzdina[1], **E A Rush**[1]

[1] Federal Agency of Railway Transport, Irkutsk State Transport University (ISTU), 15, Chernishevskogo str., Irkutsk, 664074, Russia

E-mail: obuzdina_mv@mail.ru, lrush@mail.ru

Abstract. The article presents investigations of adsorption of heavy metal ions Ni^{2+}, Zn^{2+}, Cu^{2+} by modified zeolites. As a modifier of zeolites of the Holinsky deposit, a sulfur polymer is proposed that is obtained from waste during the production of epichlorohydrin, the main components of which is 1,2,3 - trichloropropane. The mechanisms of adsorption and the results of investigations by the methods of infra-red pectroscopy and energy differential X-ray spectroscopy of the initial samples of materials are described in detail. On the IR- spectres new absorption bands are considered. These bands reflect the nature of fixation of modifier on the surface of sorbent.

1. Introduction

Heavy metal ions and oil products are the main components of wastewater, which are formed at the enterprises of railway transport [1]. Adsorption as a method of wastewater treatment is the most effective, allowing to reach 98–99.5% of the degree of purification and, accordingly, the residual concentrations of pollutants in the treated wastewater to standard values [2, 3]. The sorption method allows removing pollutants of different origin, while avoiding the formation of secondary pollutants. The degree of purification reaches 97-99%, with the possible adsorption of pollutants of multicomponent mixtures [4]. In presented investigations, zeolites of the Holinsky deposit of Zabaykalsky region were considered [5].

Chelishchev N.F. and others [6], in the middle of the 20th century, revealed that zeolites have a large number of voids in their structure, caused by a framework in the form of $[Al, Si]O_4$ tetrahedra, which joined together. These voids are grouped into open channels in which cations are located [7]. At the same time, water molecules are relatively weakly bound to both cations and the skeleton. Thus, they can be removed and replaced with pollutant molecules. In this case there will be no destruction of the bonds of the crystal lattice [8]. Coordination-related exchange cations water can be removed by heating the zeolite.

IR-spectroscopy helps to assess the sorption activity and establish the mechanism of sorption. Analyzing the IR-spectra, it is possible to obtain the necessary information about the structure and other properties of zeolites, to identify what changes they undergo during a chemical reaction, taking into account various treatments. Analysis of the IR-spectra helps to describe which functional groups are present on the zeolite surface as a result of a particular modification.

IOP Conf. Series: Materials Science and Engineering **760** (2020) 012046 doi:10.1088/1757-899X/760/1/012046

2. Results and discussion

The method of IR-spectroscopy for investigations of zeolites of various crystal structures was first used in the works of P.N. Kiselev [9]. E.M. Flatilen with employees [10] conducted a more detailed systematic investigation of the frame structure of a number of synthetic zeolites by the method of IR-spectroscopy. Each zeolite has its own IR-spectrum. In the range from 200 to 1300 cm^{-1}, the frequencies of fundamental vibrations of the TO_4 tetrahedra (T = Si,Al) of the frameworks of all zeolites are displayed [8]. To obtain the IR-spectrum of the natural zeolite of the Holinsky deposit, the Fourier FTIR spectrometer was used at the Baikal Analytical Center for Collective Use of the SB RAS. Figure 1 shows the dependence of the transmission T (%) on the wave number V.

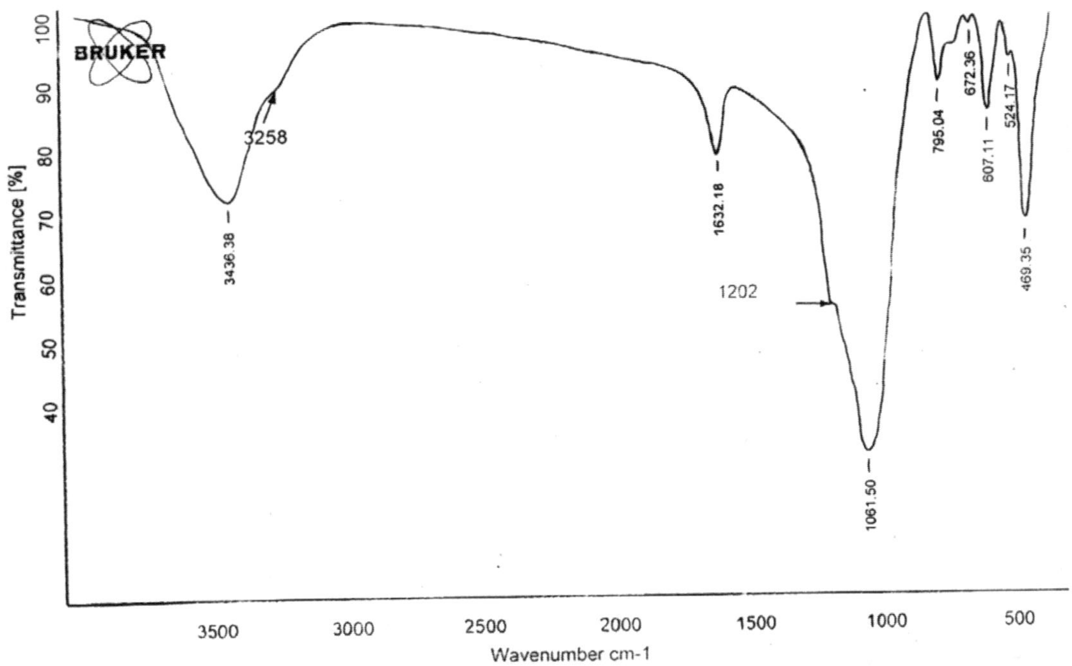

Figure 1. IR- spectrum of zeolitive tuff of Holinsky deposit of natural origin

Analyzing the IR-spectrum (see Figure 1), it can be noted that the absorption bands at 610 cm^{-1} and 1220 cm^{-1} are related to the vibrations of the external tetrahedral Si–O–Al bonds. At the same time, in the middle part of the IR-spectra, absorption bands are presented that characterize the main Al, $Si–O_4$ ($T–O_4$) tetrahedra, which are the basis of the framework of natural zeolites of klinoptilolite type. The deformation vibrations of OH-bonds in the spectrum of natural zeolite appear at 1632,18 cm^{-1} (Fig. 1). Valence vibrations of OH-bonds are at 3436.38 cm^{-1}. When zeolite is dehydrated, polyvalent cations become localized. Due to the presence of an electrostatic field associated with polyvalent cations, deformation may occur and even, perhaps, the dissociation of water molecules coordinated to these ions. In this case, cations of $MeOH^+$ and H^+ surface compounds are formed [11].

The absorption bands, related to vibrations along the external bonds of the Al,$Si–O_4$($T–O_4$) tetrahedra, are characterized by absorption bands in the region of 750–820 cm^{-1} and 1050–1150 cm^{-1}. The ratio of SiO_2/Al_2O_3 in a zeolite affects the magnitude of the frequencies of the absorption bands in the region of skeletal vibrations, but not the general form and their number. In this case, the antisymmetric vibrations of Si-O-Si (Al), which are traced in the middle region of the IR-spectra with the absorption band at 1000-1100 cm^{-1}, are most dependent on this ratio. The frequency of the maximum of this band increases with increasing SiO_2/Al_2O_3 modulus. However, cases of deviation from this pattern are possible with a high sensitivity of vibrations of the Si-O-Si (Al), bonds, and taking into account the topology of the zeolite crystal structure [9].

For klinoptilolite, the frequency of the antisymmetric stretching vibrations of the Si-O-Si (Al) chains is traced as a function of SiO_2/Al_2O_3 modulus. When the ratio of SiO_2/Al_2O_3 = 8,3, the atomic fraction of silicon is 0,81, which is characterized by an absorption band of 1061.50 cm^{-1}. In the IR-spectra obtained, the stretching vibrations of the Si–O bonds are displayed as absorption band at 1061.50 cm^{-1}. At the same time there is a small shoulder at 1202 cm^{-1}. Analyzing Figure 1, we can note the fact that the deformation vibrations of the Si–O bonds are traced at 469.35 cm^{-1} and 524.17 cm^{-1}. The silicon-oxygen tetrahedron of the zeolite lattice has Si-O-Si bonds, which are displayed in the IR-spectra as absorption bands at 795.04 $cm^{-1}$and shoulders at 702 cm^{-1}. Si-O-Al bonds with deformation vibrations can be traced along absorption bands at 672.36 $cm^{-1}$and 607.11 cm^{-1}.

We were [2] conducted investigations on the heat treatment of zeolite tuffs from the Holinsky deposit. IR-spectra were obtained reflecting changes in the structure in the case of processing at a temperature of 350-600 °C. It was found that the optimal calcination temperature of the zeolite is 350 °C, since this removes water, coordinated with exchange cations and oxygen scaffold. The sharp decrease in the sorption capacity of the initial sample in the case of calcination at 600 °C is explained by the destruction of the functional active centers in the structure of klinoptilolite. Taking into account the fact that the frequency of 1637.88 cm^{-1} is characteristic of water molecules, coordinated with calcium cations [9], it can be concluded that the residual H_2O molecules in the zeolite, which are most strongly associated with it, are retained by calcium cations.

As a result of joint research [1,12,13,14] hydrophobization of the zeolite surface occurs in the case, when high silicon organics, hexamethyldisilazane $[(CH_3)_3Si\text{-}]_2NH$ (HMDS) and tetraethoxysilane $(C_2H_5O)_4Si$ (TEOS), is using as a surface modifiers of natural zeolite of Holinsky deposit. It is recommended that the preliminary heat treatment of the sorbent at 350 °C. This contributes to a significant increase in the adsorption capacity of the investigating samples with respect to Ni^{2+}, Zn^{2+} and oil products. To prove the nature of the fixation of modifiers on the surface of zeolites and to explain the sorption process, the IR-spectra of samples of modified zeolites were measured and investigated in detail before and after sorption.

Next, we [13, 14] investigated the energy-dispersive X-ray spectra of the investigated samples of zeolites, modified by HMDS and TEOS after sorption of heavy metal ions. The data obtained confirm the formation of Ni^{2+} and Zn^{2+} complexes on the zeolite surface. Analysis of the spectra confirms that the modified zeolites contain K, C, Ca, O, Al, Si. The fact of metal cations fixation of on the surface in the cavities of zeolites is confirmed by the presence of Ni^{2+} and Zn^{2+} peaks, respectively, 7,71% and 5,14%.

The next stage of our research was the investigation of sorption of heavy metal ions such as Ni^{2+}, Zn^{2+} and Cu^{2+} on the natural zeolite of the Holinsky dsposit of klinoptilolite type, modified by sulfur polymer, obtained from epichlorohydrin production waste [14]. The fixation of the modifier on the zeolite surface was confirmed by energy dispersive X-ray spectroscopy (see figure 2, table 1).

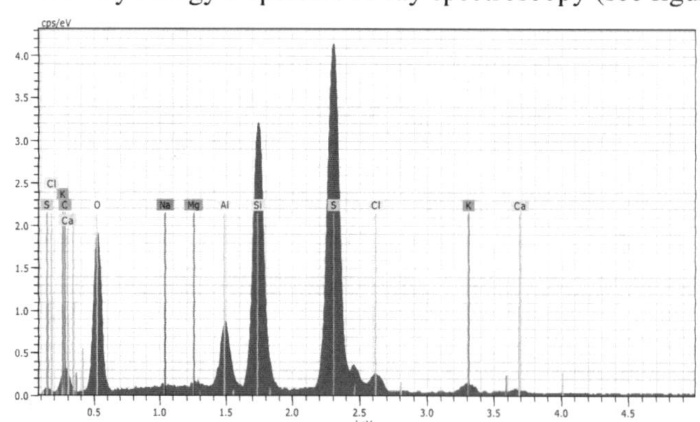

Figure 2. Energy dispersive X-ray spectrum of natural zeolite, modified by sulfur polymer

As can be seen from the presented spectrum, the modified zeolites contain O, S, C, Si, Al, Cl, K, Ca, Na, Mg. The fact of sorption, for example, of copper ions on the surface of modified zeolites, is characterized by the presence in the spectrum of the peak of this metal, the content of which is 1,3% by weight. After desorption, there are no peaks in the corresponding spectra characterizing metal ions, which confirms the destruction of the Me-O bonds. The nature of the IR-spectrum also indicates the change in the structure of the zeolite as a result of modification with a sulfur polymer (see Figure 3). The fixation of modifying substances on the surface of natural zeolite is characterized by the appearance of new absorption bands at 2954.89 cm^{-1} and 2903.08 cm^{-1} (C–H bonds from sulfur polymer) on the IR-spectra.

Table 1. Data of energy differential X-ray spectrum of a natural zeolite, modified by sulfur polymer

Element	AN	Series	norm. C [wt.%]	Atom, C [at.%]	Error [%]
Oxygen	8	K-series	38.32	45.00	4.6
Sulfur	16	K-series	23.88	14.00	0.8
Carbon	6	K-series	18.17	28.43	2.9
Silicon	14	K-series	12.82	8.58	0.5
Aluminium	13	K-series	2.74	1.91	0.1
Chlorine	17	K-series	1.82	0.96	0.1
Potassium	19	K-series	1.48	0.71	0.1
Calcium	20	K-series	0.59	0.28	0.0
Sodium	11	K-series	0.09	0.07	0.0
Magnesium	12	K-series	0.08	0.06	0.0
		Total:	100.0	100.0	

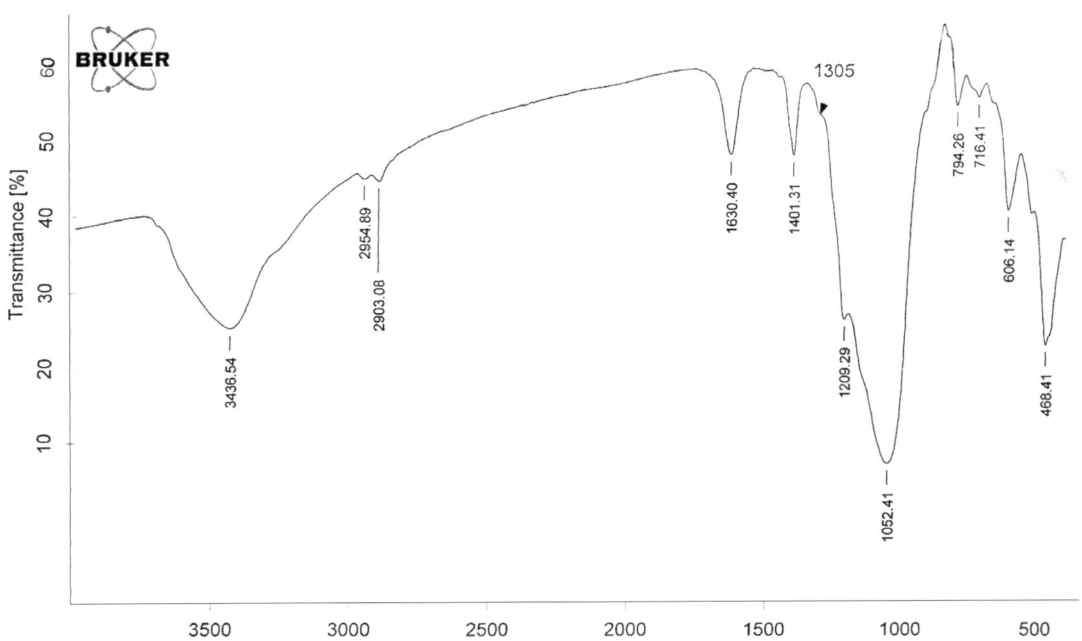

Figure 3 IR-spectrum of natural zeolite, modified by sulfur polymer

The adsorption of metal ions Ni^{2+}, Zn^{2+}, Cu^{2+} forms stable complexes with sulfur-containing ligands on the surface of the sorbent [15]. To prove these interactions, we analyzed the IR-spectra of sorbent samples with an absorbed metal [16]. The absorption band at 1052.41 cm^{-1} of a natural zeolite, modified by sulfur polymer, is shifted to the high-frequency region and appears at 1074.15 cm^{-1}, 1077.16 cm^{-1}, 1071.85 cm^{-1} in the spectra of the zeolite after sorption of metal ions copper, nickel and zinc, respectively. Thus, a donor – acceptor sorption mechanism can be assumed. During sorption of metal ions from wastewater, sorption of the water molecules themselves also takes place, which spectrally manifest as very weak oscillations of OH-groups at 3624 cm^{-1}; 3624 cm^{-1}; 3622.36 cm^{-1} after sorption of Ni^{2+}, Zn^{2+}, Cu^{2+}, respectively. As a result of the sorption process, the absorption band of 716.41 - 723.29 cm^{-1} for copper ions, 716.41 - 720.33 cm^{-1} - for nickel ions, 716.41 - 720.33 cm^{-1} - for zinc ions confirms the formation of sulfides.

3. Conclusion

As a result of experimental investigations on the extraction of Ni^{2+}, Zn^{2+}, Cu^{2+} from industrial wastewaters of railway transport enterprises, it has been proved in principle that zeolites from the Holinsky deposit of the Eastern Transbaikalia can be used. In order to improve the adsorption characteristics of the raw materials, we have proposed a modification of the natural zeolites of the sulfur polymer, obtained from waste in the production of epichlorohydrin. In this case, it is recommended to thermally pretreat the initial zeolite at a temperature of 350°C. Using IR-spectroscopy and energy dispersive X-ray spectroscopy, the high adsorption capacity of the presented modified zeolites was proved. The obtained IR-spectra show the appearance of new absorption bands, which characterize the fixation of modifiers on the surface of the zeolite. In this case, after sorption of heavy metal ions, a shift in the absorption bands appears as evidence of the adsorption process.

References

[1] Obuzdina, M V and Rush E A 2011 *Modern technologies, system analysis and modeling* **29** **1** 142-145
[2] Dubinin M M 1987 *Adsorption and porosity* (Moscow: VACh) p 172
[3] Tarasevich Yu I and Ovcharenko F D 1975 *Adsorption on clay minerals* (Kiev: Naukova Dumka) p 351
[4] Smirnov A D 1982 *Sorption water purification (*Moscow: Chemistry) p 168
[5] Kotov P A 1995 *Fields of Transbaikalia* (Moscow: Geoinformmark) p 280
[6] Chelishchev N F, Berenshtein B G, Volodin V F 1987 *Zeolites is a new type of mineral raw materials* (Moscow: Nedra) p 176
[7] Greg S and Singh K 1984 *Adsorption, specific surface area, porosity* (Moscow: Mir) p 306
Prigochin I and Condepudi D 2002 *Modern thermodynamics* (Moscow: Mir) p 464
[8] Brek D 1976 *Zeolitic molecular sieves* (Moscow: World) p 781
[9] Kiselev A V and Lygin V I 1966 *Infrared spectrums of adsorbed molecules* (Moscow: Mir) p 459
[10] Flanigen E M 1980 *Chemistry of zeolites and catalysis on zeolites* (Moscow: Mir) pp. 104-147.
[11] Smith A 1982 *Applied IR- spectroscopy* (Moscow: Mir) p 327
[12] Patent of the Russian Federation Patent No. 2524111 07.27.2014.
[13] Makarov A V, Rush E A, Ignatova O N 2012 *Modern technologies, systems analysis and modeling* **33** **1** 153-159
[14] Patent of the Russian Federation Patent No. 2624319 (07.03.2017)
[15] Redinova A B 2013 *ISTU Bulletin.* **1** 113-116
[16] Obuzdina, M V, Rush, E A, Shaluch L V 2017 *Ecology and Industry of Russia.* **21** **8** 20-25

Voltage regulation characteristics of a two-pack synchronous generator with combined excitation

D M Toporkov[1], I L Popov[1], A A Blinov[1], D P Kravchenko[1], M A Kovtun[1]

[1] Novosibirsk State Technical University, 20, Karla Marksa ave., Novosibirsk, 630073, Russia

E-mail: Select12345@yandex.ru

Abstract. The design, principle of operation, method for determining the inductance and the algorithm for plotting the voltage regulation characteristics of a two-packet synchronous generator with a combined excitation are described in this article. Characterization and inductance calculation were carried out for a prototype having a power of 11 kW, number of phases - 3, frequency of induced EMF in the AC winding - 550 Hz, number of pole pairs - 11.

1. Introduction

In recent years, due to the development of unmanned vehicles, electric batteries and electronics, the demand for autonomous sources of constant voltage has increased.

DC voltage generators are capable of generating constant voltage. These machines have a commutator. Due to the complexity of the design of the commutator and frequent switching plates, the commutator significantly reduces the reliability of the machine, increases its coast, and complicates maintenance.

Also synchronous machines operating with a rectifier can be used as direct current sources. Classical synchronous generators are equipped with an excitation winding located on the rotor of the machine. It's possible to regulate the main magnetic flux by having an excitation winding. However, in that case, the machine has a sliding contact, which reduces its reliability and complicates maintenance.

The generator considered in this article has a combined excitation, which allows it to work with the excitation winding turned off, and with it turned on. The excitation winding is located on the stator between the packages and has no sliding contacts (figure 1). Thus, this machine has the following advantages:

1. Increased reliability relative to DC machines and synchronous machines in the classic version.
2. Lower power of the excitation winding due to the presence of permanent magnets.
3. Increased utilization of magnetic induction compared with an axial inductor machines [2].
4. The ability to control the main magnetic flux.

2. Problem statement

Voltage regulation characteristics are necessary for the development of an automatic excitation system. It is assumed that this generator will work with the rectifier and its load will be only active. But because of its own inductance, the load of the generator will be active-inductive. Due to the unusual design and the presence of the constant component of magnetic induction in the magnetic conductor, there are no expressions for valuation the inductance along two axes. This article presents one of the methods for valuation of the inductance for this generator.

IOP Conf. Series: Materials Science and Engineering **760** (2020) 012047 doi:10.1088/1757-899X/760/1/012047

3. Theory

3.1. Description of the design

Figure 1 shows the constructions of the generator. The stator consists of laminated packages fixed in a frame and an excitation winding installed between the packages. The rotor also consists of two laminated packages in the slots of which magnets with radial magnetization are installed. Magnets have opposite magnetization vectors in different packages. Packages mounted on a molded liner and are shifted relative to each other on the pole pitch. So that both rotor packages have the same polarity. Magnetic flux does not pass through the end shield and parasitic air gaps, as is the case in a single-package construction [2].

This generator has the following characteristics: 11 pole pairs, power 11 kW and rotor speed of 3000 rpm. Making of this generator with classical type winding will lead to diminishment teeth-slots area, which will cause a decrease in specific values (this is due to the need to a lower fill factor) and complicates the production technology [5]. Therefore, the machine is made with a fractional slot concentrate winding.

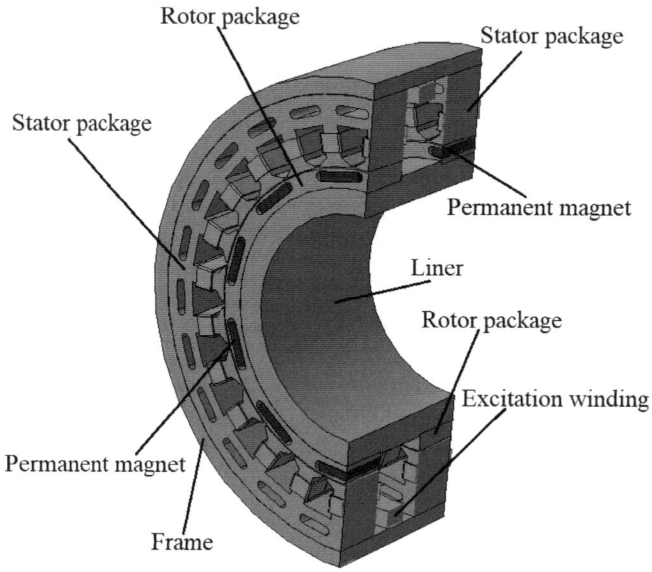

Figure 1. Construction of the generator

3.2. Principles of operation

When the excitation winding is turned off, most of the flow generated by the magnets passes along the contour: the magnet of the first rotor package - air gap – teeth of the first stator package – frame – teeth of the second package of the stator – air gap – iron pole of the second package of the rotor – liner. A smaller part of the flow is closed radially, as in classical synchronous machines with permanent magnets. When the excitation winding is turned on, the magnetic flux created by DC winding is directed oppositely to the flow from the magnets and passes along the contour: the frame – teeth of the first package of the stator – iron pole of the first package of the rotor – liner – iron pole of the second package of rotor – teeth of the second package of the stator. The equivalent circuit of the magnetic circuit is shown in Figure 2. Thus, the flux of permanent magnets decreases slightly, but the flux passing through the pole increases and the constant component of the magnetic induction in the air gap decreases. By changing the MMF of the excitation winding, it is possible to control the magnetic flux passing through the pole.

3.3. Valuation of inductance

Generator inductance is calculated by numerical simulation of the magnetic field. Inductance – the coefficient of proportionality between the current flowing in a circuit and the flux linkage created by this current through the surface, the edge of which is the circuit [1]:

$$L=\frac{\psi}{I} \quad (1).$$

The d axis coincides with the axis of the pole, in other words the magnetic resistance in this case is minimal, respectively the magnetic resistance of the q axis is maximum. Figure 3 and 4 show the location of the rotor relative to the stator, corresponding to the minimum and maximum magnetic resistance. By knowing the value of the phase flux linkage at given rotor positions and taking into account the flux created by permanent magnets, it is possible to determine the inductance of the generator along the d and q axes:

$$L_A=\frac{\psi'_A-\psi''_A}{2I_A} \quad (2),$$

In equation (2) ψ'_A - phase flux linkage created by permanent magnet and phase current, ψ''_A -phase flux linkage created by only permanent magnet, I_A - amplitude value of the phase current.

By calculating the three-dimensional magnetic field by the finite element method in the ANSYS program, it is possible to determine these flows and calculate the inductance.

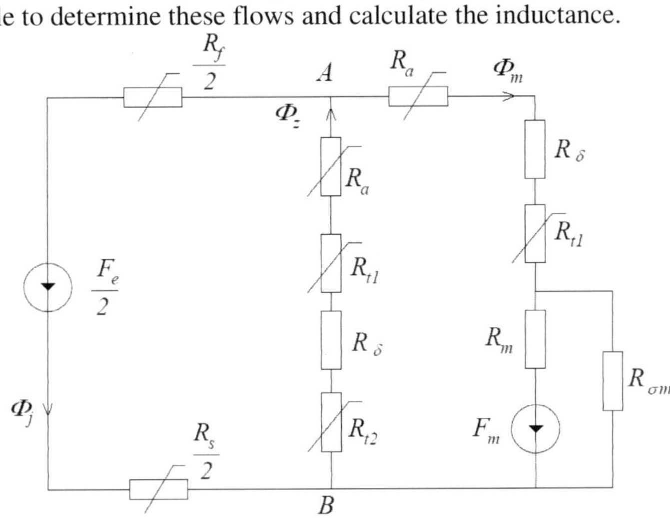

Figure 2. Equivalent circuit of the magnetic circuit

In accordance with Figure 2 R_f and R_s - reluctance of the frame and the sleeve, R_{t1} and R_{t2} - reluctance of the teeth of the stator and rotor, R_δ - reluctance of the air gap, R_m and $R_{\sigma m}$ - reluctance of the permanent magnet and its stray flux, R_a - reluctance of the stator back, F_m and F_e - MMF of the permanent magnet and excitation winding.

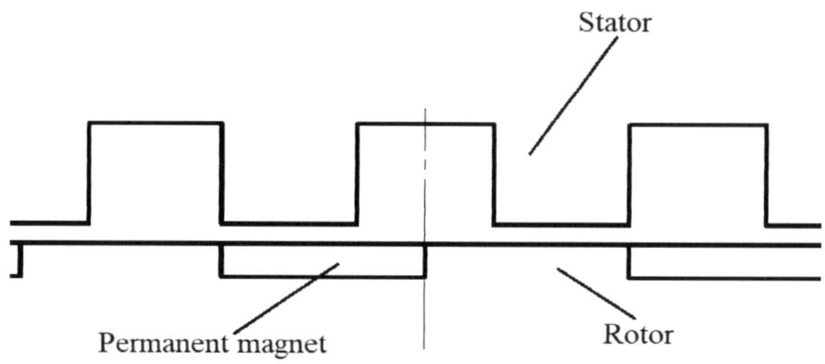

Figure 3. The position of the rotor, providing minimal reluctance to the flux

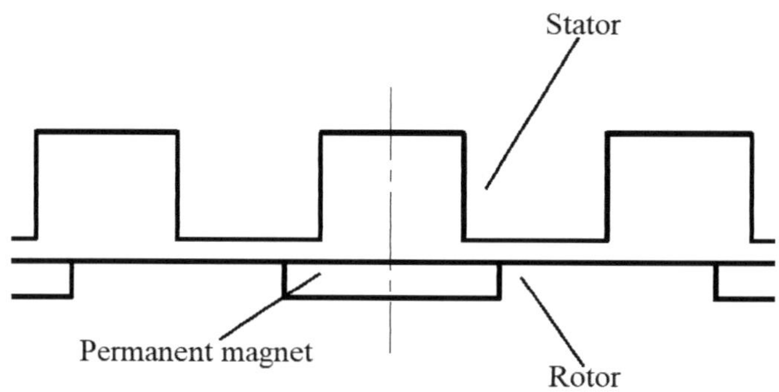

Figure 4. The position of the rotor, providing maximum reluctance to the flux

3.4. Mathematical description of the generator

Creating a mathematical description of any electric machine begins with the compilation of a system of differential equations that determine the behavior of the processes inside the machine. The calculation and formulation of equations of the machine were carried out in a rotating coordinate system. In this case, for a synchronous machine with a combined excitation description has the following form:

$$\begin{cases} \frac{d}{dt}i_d = \frac{1}{L_d}U_d + \frac{R}{L_d}i_d + \frac{L_q}{L_d}p\cdot\omega\cdot i_q, \\ \frac{d}{dt}i_q = \frac{1}{L_q}U_q + \frac{R}{L_q}i_q + \frac{L_d}{L_q}p\cdot\omega\cdot i_q - \frac{\psi\cdot p\cdot\omega}{L_q}, \\ M_e = 1.5\cdot p\cdot\left(\psi\cdot i_d + \left(L_d - L_q\right)\cdot i_d\cdot i_q\right). \end{cases} \quad (3).$$

In accordance with the system of equations (3): $L_{d,}$ L_q - generator inductance along the d and q axes; R - active phase resistance of the stator winding; ψ - flux linkage of the phase; p - number of pole pairs; ω - rotary speed of the rotor.

With the help of this system of equations, a structural scheme of the machine was compiled in the Matlab/Simulink software package. A feature of the scheme is that it is implemented using both standard blocks of the Simulink library and Simscape. In scheming of the machine, it was assumed that the speed of the engine driving the generator rotor is constant over time.

IOP Conf. Series: Materials Science and Engineering **760** (2020) 012047 doi:10.1088/1757-899X/760/1/012047

4. Experimental results

Simulation of the three-dimensional magnetic field was carried out in the ANSYS program. According to the simulation results, the values of the flux linkages necessary for the calculation of inductance were determined. The results are shown in Table 1.

The distribution curves of the magnetic induction in the air gap and their harmonic compositions were also obtained (Figures 5-6).

Table 1. Parameters for inductive calculation

Parameter	Value
Phase flux linkage at maximum magnetic reluctance created by phase current and magnets	0.0094 Wb
Phase flux linkage at maximum magnetic reluctance created by phase current and magnets	0.00011 Wb
Phase flux linkage at maximum magnetic reluctance created by phase current and magnets	0.0017 Wb
Phase flux linkage at maximum magnetic reluctance created by phase current and magnets	-0.0069 Wb
Amplitude value of armature winding current	309.1 A

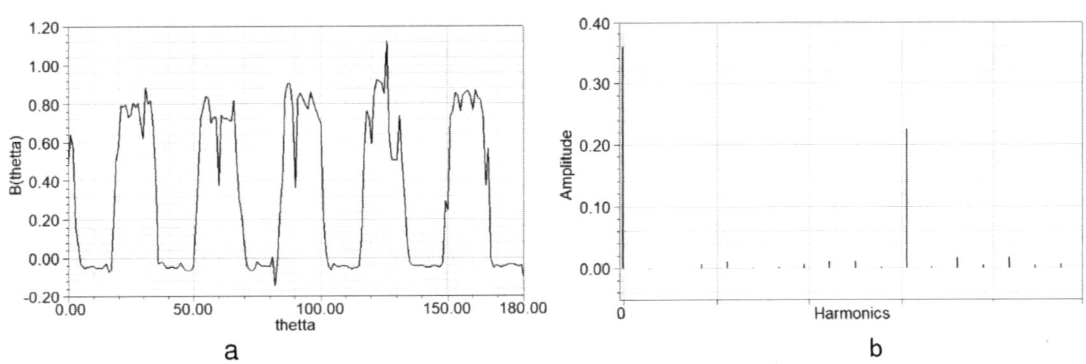

Figure 5. a – the distribution curve of the magnetic induction in the air gap with excitation winding is off; b – the harmonic composition of this curve.

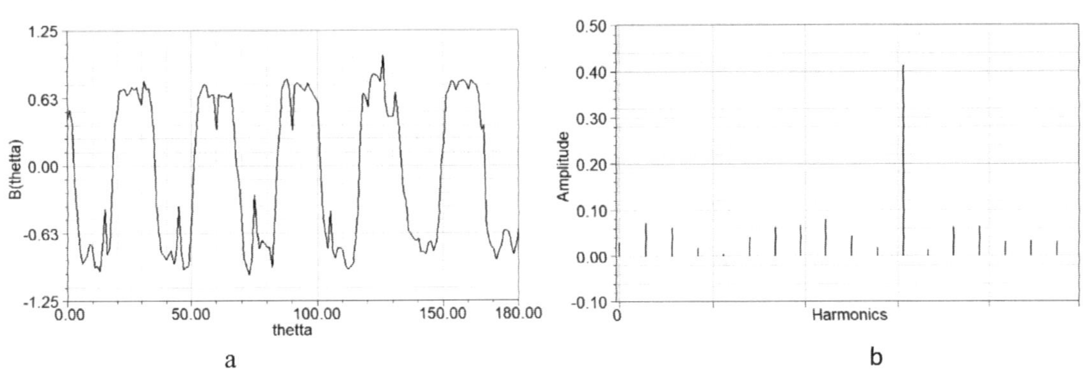

Figure 6. a – the distribution curve of the magnetic induction in the air gap with excitation winding is on; b – the harmonic composition of this curve.

5. Discussion of results

As seen in figures 5-6, with the excitation winding off, the constant component of the magnetic induction in the air gap is larger than the amplitude of the main harmonic. When the excitation winding is turned on, the constant component is almost completely compensated, and the amplitude of the main harmonic increases. According to the values of flux linkage (Table 1), inductances were determined along the q and d axes ($L_d = 13.79 \cdot 10^{-6}$ H, $L_q = 15.02 \cdot 10^{-6}$ H) and a set of voltage regulation characteristics was constructed (Figure 7).

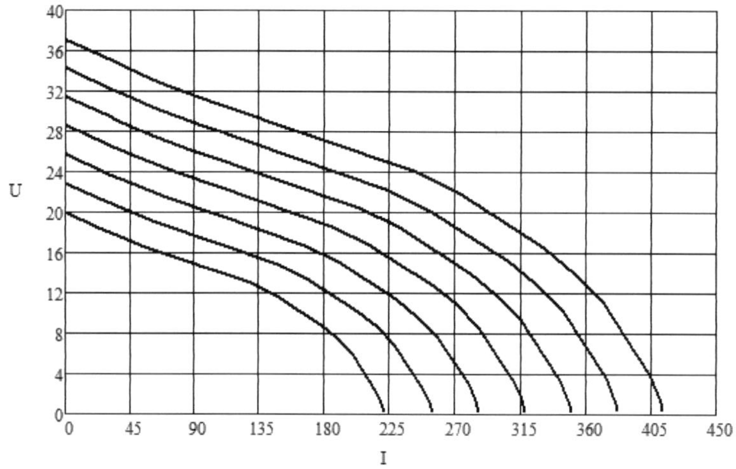

Figure 7. Set of the voltage regulation characteristics

6. Conclusion

The article describes the design and principle of operation of a two-pack synchronous generator with combined excitation. The method for calculating inductances along the q and d axes is presented. The calculation of the three-dimensional magnetic field in the ANSYS program was also carried out, the results of the calculation are presented in the graphs of the distribution of magnetic induction in the air gap. For the obtained inductance values, voltage regulation characteristics were constructed.

To construct regulating characteristics, it is necessary to construct the characteristic curve of the main harmonic of the flux linkage on the excitation current. When constructing this plot, one may encounter a number of difficulties caused by the design features of the generator and the presence of the constant component of magnetic induction. The development of an algorithm for constructing this graph may be the goal of further research.

References

[1] Voldek A I 1978 Electrical machines (Leningrad: Energiya) 832
[2] Dombur L E 1984 Axial inductor machines (Riga: Zinatne) 247
[3] Shevchenko A F, Pristup A G 2016 Electric machines with permanent magnets (Novosibirsk: Izdatelstvo NGTU) 64
[4] Kroneberg U N, Gomzyakov V B, Zhibinov A S 1971 "Two-Pack Generator with Combined Excitation *News of Tomsk Polytechnic University* **212**
[5] Bukhgolc U G, Komarov A V, Shevchenko A F, Shevchenko L G 1996 Multipolar synchronous machines with fractional slot windings part 1, (NGTU: Novosibirsk)
[6] Bukhgolc U G, Komarov A V, Shevchenko A F, Shevchenko L G 1996 Multipolar synchronous machines with fractional slot windings part 2, (NGTU: Novosibirsk)
[7] Bespalov V Ya, Kotelenec N F 2006 Electrical machines (Moskow: Akademiya) 320
[8] Kacman M M 2003 Electrical machines (Moskow: Vysh. shk.) 469

[9] Tapia J A, Leonardi F, Lipo T A, 2001 Consequent Pole Permanent Magnet Machine With Field Weakening Capability *IEEE International Electric Machines and Drivers Conference, IEMDC* 126-131.

[10] Tapia J A, Leonardi F, Lipo T A 2003 Consequent-Pole Permanent Magnet Machine With Field-Weakening Capability *IEEE Transactions on Industry Applications* **39** **6** 1704-1709.

IOP Conf. Series: Materials Science and Engineering **760** (2020) 012048 doi:10.1088/1757-899X/760/1/012048

Accuracy estimate of the paved surface of an automobile road after its reconstruction

L A Pronina[1], E N Kupreyeva[1]

[1]Department of geodesy and remote sensing, Omsk state UNIVERSITY, 4 Sibakovskaya St., Omsk, 644008, Russia

E-mail: la.pronina@omgau.org, en.kupreeva@omgau.org

Abstract: The article outlines the importance of engineering and geodetic survey during the reconstruction of automobile roads. The main phases of engineering and geodetic survey, such as the creation of a horizontal and vertical control, topographic plotting, technologies for the detailed staking of circular curves, and softwares for the execution of data processing, are taken into consideration. Possible tolerance limits in building and geodetic operations are reported. Research on the evenness of the upper surface of the asphaltic coating of an automobile road after its reconstruction is carried out.

1. Introduction

Automobile roads are part of the so-called vital support systems; therefore they are of great importance for the social and economic development of a country. Roads with paved surface are of crucial importance, as without their development the recovery of a country's economy as a whole and of individual areas in particular, as well as the successful implementation of economic reforms, is impossible. At the present time, engineering and geodetic surveys (EGS) and up-to-date applied technologies for the purposes of automobile road reconstruction are particularly relevant, which allows for high performance and efficiency. Automobile roads leading to rural settlements in the Omsk Region are in critical condition and, subsequently, require reconstruction, for this purpose a whole set of engineering and geodetic surveys is being carried out. Engineering surveys for the design planning and reconstruction of automobile roads are understood as an integrated production process on the working site which provides the design planning with source data for the adoption of further decisions. The main task of engineering surveys is to obtain the necessary materials for the development of economically viable and technically grounded solutions for the design planning, construction and reconstruction of working sites.

The initial stage of EGS (engineering and geodetic surveys) is the creation of a horizontal and vertical control. The creation of a reference geodetic network (RGN) on the working site is performed using GNSS technologies. Surveying with the support of GNSS technologies has a whole range of advantages: globality, operativity, all-weather capacity, optimum accuracy, and efficiency. GNSS technologies ensure the accuracy of differential determinations at a level of +(3-5 mm + 1 mm/km), which fully satisfies the accuracy conditions for the creation of a surveying control. Important advantages of such method of determining the coordinates are: fast result reception (i.e. in real time), possibility to determine the coordinates of points at any time of the day, possibility of operation under difficult meteorological conditions, and calculation capacity at significant distances between the source and determined points [1]. Dual frequency and dual system equipment sets GPS/GLONASS Trimble R7 and Trimble R8 were used on the working site. The horizontal and vertical control of points for the determination of fixed solutions was obtained by means of accumulation over 60 minutes with a recording interval of 15 seconds. The results were processed using the Trimble Geomatics Office software. The accuracy estimate of the determination of the coordinates and heights of the reference points is shown in Table 1.

The measurement of angles and distances in traverses was performed with an electronic total station SOKKIA Set 230RK. Angles were measured applying the ordinary method. Distances were measured applying the repetition method, measurement results were recorded in an electronic storage device.

The vertical study was obtained by means of technical leveling traversing of the points of the geodetic reference network along the points of horizontal justification. In the production of the trigonometric leveling, an electronic total station was used with the following conditions: measurements were taken in forward and reverse directions with two guidances to the reflector, the maximum distance between the total station and the reflector did not exceed 300 m. The equalization of the surveying network was performed using the CREDO_DAT software. The technical characteristics of the traverses are shown in Table 1.

Table 1. Technical characteristics of the traverses.

Traverse	Point traverse	Distances , m	N	Fb fact.	Fb add.	Fx, m	Fy, m	Fs, m	[S]/Fs
1	Rp.4, T15, ..., Rp.5	1005,636	5	-0°00'32,65"	0°01'29,44"	-0,009	-0,046	0,047	1:21500
2	Rp.4, T12, ..., Rp.8	1443,847	6	-0°00'03,47"	0°01'37,98"	-0,009	0,040	0,041	1:35300

The factual discrepancies in the leveling traverses were as follows: traverse No. 1 «Rp.4, T15, ..., Rp.T5» -0,038 m, with f_h add = 0,051 m; traverse No. 2 « Rp.4, T12, ..., Rp.8» -0,061 m, при with f_h add = 0.066 m. The resulting discrepancies in theodolite and leveling traverses are within tolerance.

The next stage of engineering surveying consisted in the execution of the topographic plotting. The topographic plotting of the road was carried out on a 1:2000 scale with a cross section of the relief at 0.5 m using an electronic total station SOKKIA Set 230RK. Part of the plotting was performed by the specified geodetic satellite equipment in kinematic mode. Engineering and topographic plans were drawn up electronically using the CREDO software package with further export in AUTOCAD format.

During construction, design planning and thorough overhaul of automobile roads, the Construction Norms & Regulations 3.06.03-85. Automobile roads [3] are applied. The term “tracing” should be understood as a complex of engineering and geodetic surveys on the choice of the route according to its technical and economic conditions, while distinguishing cameral and field tracing. The accuracy limits of works on the placement of the route and the main axes of the structure when transferring the construction projects to the site must comply with the design ones. The accuracy of the placement of the individual parts and axes of the structure between themselves and with respect to the main axes and the geodetic support network must comply with the applicable construction tolerances. The processed data were transmitted to form a DSM in the CREDO Linear surveys system.

The method of geometric leveling transfers the main points of the grade line or plane located at all angles of rotation of the grade plane in plan, as well as at all breaks of the grade line and fractures of the plane in profile. The staking and fixing of the route axis involves the determination of the position of the route, the turning points, and the characteristic points of the situation and relief on site. The determination of the position of the route axis is obtained through perpendiculars and linear crossings from solid contours on site. Staked and fixed elements of the route are marked with temporary signs on site. The staking of cross sections is made on straight lines along the perpendiculars to the road axis and on curves along the normals to the curves. The staking of the boundaries of the road embankment slopes (embankments beds and excavation edges) is made separately at each project site along the cross sections or normals to the curves, continued at all major turning points of the terrain. The staking of the base and coating of the roadway begins with the placement of markers (survey pegs), which serve as reference points. They indicate the thickness of each structural layer of the base and coating. The staking

IOP Conf. Series: Materials Science and Engineering **760** (2020) 012048 doi:10.1088/1757-899X/760/1/012048

of the markers is performed within a uniform area on each cross section. The limiting accuracy of the staking of the shape and size of individual elements of the structure must be 2-3 times higher than the deviations that are set for them upon acceptance of the structure into operation. The limiting relative errors of the line deposition in case of detailed staking of trough-shaped road formation and road embankment and the limiting errors in excess in road embankment staking should not exceed the values shown in [4].

2. Materials and methods

The values of limiting errors should serve as a guide also while taking into account the accumulation of errors in vertical staking that occur when transferring design elevations from one cross section to another. With a detailed staking of the ground embankment, bases and coatings, deviations of factually determined values from design data are allowed within the limits specified in [4].

Trigonometric leveling trials using up-to-date geodetic equipment are presented in [6-11]; in these publications objects which constitute non-linear structures are taken into consideration. In the present article, the evenness of the upper surface of the coating is investigated using an electronic total station by means of trigonometric leveling method.

After the completion of the reconstruction process of an automobile road leading to a rural settlement, leveling of the upper surface of the road covering more than one kilometer in length was performed. The leveling was performed along the road axis in order to determine the marks of the surface of the coating, applying a leveling step of 20 meters. According to the marks, using formula 1, the amplitude differences were calculated in line to [5]; the amplitude series is presented in Table 2.

$$\delta H_{i+1} = \frac{H_i + H_{i+2}}{2} \quad H_{i+1} \tag{1}$$

where H_i, H_{i+1}, H_{i+2} – the marks of nearly points of the coating.

Table 2. Amplitude series of the upper surface of the automobile road coating.

No.	values	No.	values	No.	values	No.	values	No.	values
1	-21	11	4	21	-14	31	-24	41	-15
2	4	12	2	22	2	32	19	42	-17
3	13	13	10	23	-9	33	15	43	-4
4	-32	14	-24	24	13	34	-11	44	-28
5	7	15	18	24	-15	35	12	45	-14
6	14	16	-11	26	16	36	18	46	-14
7	24	17	4	27	-7	37	3	47	15
8	-32	18	-2	28	-6	38	24	48	29
9	13	19	-16	29	-20	39	16	49	-5
10	-28	20	28	30	-3	40	-4	50	17
-	-	-	-	-	-	-	-	51	22

According to the Set of Rules 78.13330.2012 [6], the values of the amplitude differences were mathematically processed and the coating evenness after the reconstruction of the automobile road was investigated. At the same time, as far as the research component is concerned, the interval difference h of the series was calculated using formula (2), according to the procedure given in [12].

$$h = \frac{R}{1 + 3{,}32\, lg\, N}, \tag{2}$$

IOP Conf. Series: Materials Science and Engineering **760** (2020) 012048 doi:10.1088/1757-899X/760/1/012048

where R is the peak-to-peak amplitude dispersion, N is the amplitude number, x_{min} and x_{max} the characteristics of amplitude variation.
The quantity of intervals in the amplitude series is defined using the formula:

$$k = R/h \,. \tag{3}$$

The beginning of the first interval is defined through the expression: $x_n = (x_{min} - k/2)$.

The quantitative characteristics that denote the number of occurrences of the x_i values in the interval is determined by frequency and labeled as n, whereas its relation to the total number of the amplitude series values is called relative frequency and determined, following [12], by the relation:

$$W_i = n_i / N \,. \tag{4}$$

The interval series and it relative frequencies constitute the statistical distribution of the amplitude series. The frequency of the interval is defined as the sum of the frequencies of the amplitudes included in the interval, and their sum is equal to their total number in the series.

The average value is obtained from the thus acquired amplitudes using the formula:

$$\bar{x} = \frac{\sum_{i=1}^{N} n_i x_i}{N} \tag{5}$$

where N is the number of amplitudes in the estimated series; n_i is the amplitude frequency in the estimated series; and x_i is the average value in each interval.

Using the according formula, the root-mean-square error (RMSE) for the obtained amplitude series is obtained through formula 6:

$$m = \sqrt{\frac{\sum_{i=1}^{k} n_i (x_i - \bar{x})^2}{N - 1}} \,, \tag{6}$$

Following the magnitude of the error obtained using formula 6, the error of the average value of the amplitude series is determined using formula (7); the error value calculated through formula 6 can be determined with an error the limits of which are determined by formula 8:

$$M = \frac{m}{\sqrt{N}} \,; \tag{7}$$

$$m_m = \frac{m}{\sqrt{2(N - 1)}} \,. \tag{8}$$

The accuracy estimate of the series obtained using the presented formulas is called a point estimate. The interval estimate of the amplitudes, which determines not just a single parameter, but allows us to determine two values between which the amplitudes obtained can vary, is taken into consideration. "Confidence intervals" for the mean (mathematical expectation a) from the obtained amplitudes and for the standard deviation value are obtained using formulas 9 and 10. The confidence probability is equal to $P=0.95$, and the estimate precision to $q=0.05$.

$$\bar{x} - t_q \bullet m/\sqrt{N} < a < \bar{x} + t_q \bullet m/\sqrt{N}\,, \tag{9}$$

$$m(1-g) < \sigma < m(1+g)\,, \tag{10}$$

where t is the coefficient applied in the transition to RMSE, determined the following Tables [13] and [14] with a probability value of $P=0.95$ and an amplitude number N; m is the RMSE of the amplitude variations; g is the coefficient obtained from Tables [14] and [13] depending on the number of the estimated amplitudes.

3. Research of quality of the asphalted covering of the highway after its reconstruction.

Table 3 presents the statistic characteristics of the series according to formulas 2-8 indicated above; below the table the interval estimate obtained according to formulas 9, 10 is reported.

Table 3. Statistic characteristics of the amplitudes of the upper surface of the coating.

Interval, mm a	b	Frequency, n_i	Frequency, W_i	Average value., x_i mm	nx_i, mm	$x_i - \bar{x}$, mm	$n_i(x_i - \bar{x})$ mm	$n_i(x_i - \bar{x})^2$, mm	t_1	t_2	$\frac{1}{2}\Phi(t_1)$	$\frac{1}{2}\Phi(t_2)$	$P(x_i)$
-36	-27	4	0,078	-31,5	-126	-31,59	-126,35	3991,27	-2,09	-1,57	-0,4817	-0,4418	0,0399
-27	-18	4	0,078	-22,5	-90	-22,59	-90,35	2040,91	-1,57	-1,05	-0,4418	-0,3531	0,0887
-18	-9	9	0,176	-13,5	-121,5	-13,59	-122,29	1661,76	-1,05	-0,53	-0,3531	-0,2019	0,1512
-9	0	8	0,157	-4,5	-36	-4,59	-36,71	168,42	-0,53	-0,01	-0,2019	-0,0040	0,1979
0	9	7	0,137	4,5	31,5	4,41	30,88	136,25	-0,01	0,52	-0,0040	0,1985	0,2025
9	18	11	0,216	13,5	148,5	13,41	147,53	1978,63	0,52	1,04	0,1985	0,3508	0,1523
18	27	6	0,118	22,5	135	22,41	134,47	3013,72	1,04	1,56	0,3508	0,4406	0,0898
SUM		51	1,0		4,5			14964,35					0,9629

$$\bar{x} = 4{,}5/51 = 0{,}09\ mm \qquad M = 17{,}30/\sqrt{51} = 2{,}42\,mm$$

$$m = \sqrt{14964{,}35/50} = 17{,}30\,mm \qquad m_m = 17{,}30/\sqrt{2\cdot(51-1)} = 1{,}73\,mm$$

Confidence interval for "a":

$$\bar{x} - t_q \cdot M < a < \bar{x} + t_q \cdot M, \qquad \text{где}\ t_q(N=51;\ P=0{,}95) = 2{,}01$$

$$0{,}09 - 2{,}01 \cdot 2{,}42 < a < 0{,}09 + 2{,}01 \cdot 02{,}42 \qquad 4{,}77\ mm < a < 4{,}95\,mm$$

Confidence interval for "σ":

$$m \cdot (1-g) < \sigma < m \cdot (1+g), \qquad \text{где}\ g(N=51, P=0{,}95) = 0{,}21$$

$$17{,}3 \cdot (1-0{,}21) < \sigma < 17{,}3 \cdot (1+0{,}21) \qquad 13{,}66\ mm < \sigma < 20{,}57\,mm$$

The step in each interval amounted to ± 9, while their number was equal to seven. The error values turned out to be 17.3. The maximum value of the standard was approximately 21, which does not contradict the normative document on road construction [5]. According to the obtained characteristics, a theoretical and practical amplitude distribution was outlined, as shown in Fig. 1.

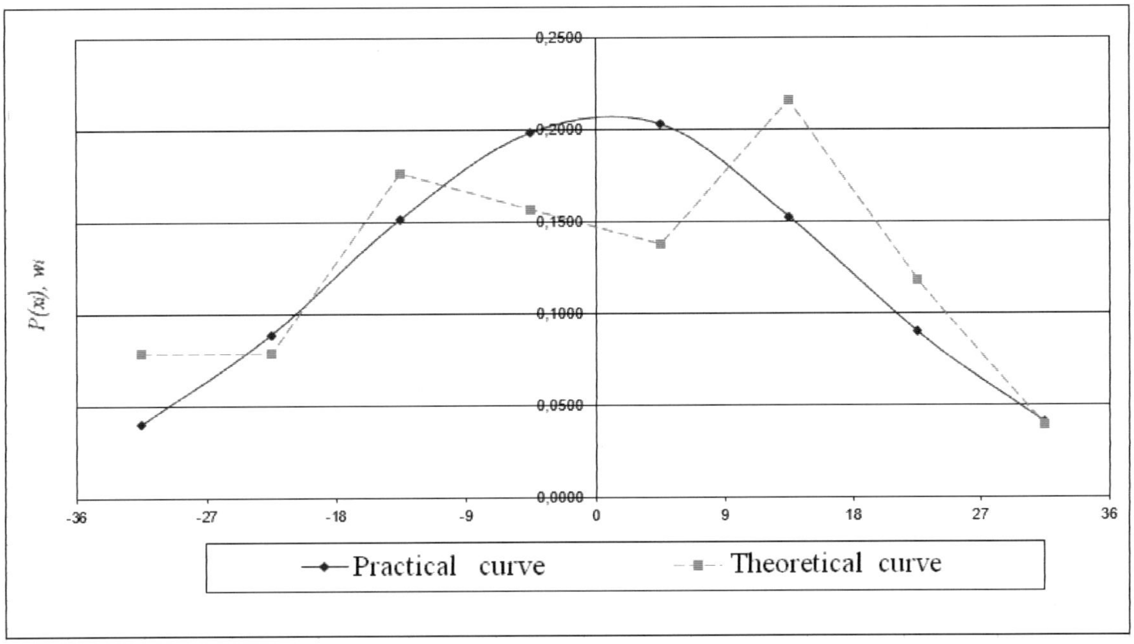

Figure 1. Practical and theoretical amplitude distribution of the upper surface of the coating

The null hypothesis on magnitude distribution, which characterizes their normal distribution, was adopted. According to Laplace's formula 11, the probability is calculated:

$$\Phi(t) = \frac{1}{\sqrt{2\pi}} \int_0^t e^{-t^2/2} \cdot dt, \tag{11}$$

where $t = (a - \bar{x})/m$ or $t = (b - \bar{x})/m$; a and b are the maximum and minimum values in the interval; $\bar{x}$ is the average mathematical value of the amplitudes; m is the RMSE in amplitude determination.

The probability Φ(t) is obtained from Tables [11] and [12]; their sum must be equal to one. The subsequent amplitudes according to the results of statistical observations comply with normal distribution. As exemplified by the amplitude series, the hypothesis is plausible. The value χ^2 (chi-square) of Pearson's test, which is given by formula 12, is considered as the measure of deviation of the hypothesis:

$$\chi^2_{nabl} = \sum_{i=1}^{\kappa} \left(n_i - NP(x_i)\right)^2 / NP(x_i), \tag{12}$$

where n are the empirical frequencies; $NP(x_i)$ are the theoretical frequencies; κ is the number of intervals; $P(x_i)$ is the theoretical probability.

In the investigated series of amplitude variation, the value of Pearson's test χ^2 amounted to 5.45; the calculations are presented in Table 4; the obtained characteristics do not exceed the critical value of 11.1.

Table 4. Definition of Pearson's test χ^2 for the estimate of the coincidence of the trial and normal distribution of the amplitudes of the upper surface of an automobile road coating.

Interval,mm *a*	*b*	Frequency n_i	Probability $P(x_i)$	Theoretical frequency $NP(x_i)$	$n_i - NP(x_i)$	$[n_i - NP(x_i)]^2$	$\frac{[n_i - NP(x_i)]^2}{NP(x_i)}$
-36	-27	4	0,0399	2,0349	1,965	3,862	1,90
-27	-18	4	0,0887	4,5237	-0,524	0,274	0,06
-18	-9	9	0,1512	7,7112	1,289	1,661	0,22
-9	0	8	0,1979	10,0929	-2,093	4,380	0,43
0	9	7	0,2025	10,3275	-3,328	11,072	1,07
9	18	11	0,1523	7,7673	3,233	10,450	1,35
18	27	6	0,0898	4,5798	1,420	2,017	0,44
SUM		51	0,9629				$\chi^2_{выч} = 5,45$

4. Conclusion

In conclusion, the set of EGS performed for the reconstruction of an automobile road with the accuracy estimate of the upper surface of its paved coating was examined. The applied technologies and the conducted research were analyzed, the normative requirements for the accuracy of geodetic works were met, and the appropriate findings were taken into account. In case of reconstruction of an automobile road, technological processes which might ensure the specified accuracy and efficiency were established.

5. Acknowledgments

The study was carried out under the theme "Development of new technologies for geodetic monitoring of engineering structures and natural objects using geodesy and remote sensing methods" state registration number AAAA-A18-118070290063-7

Reference

[1] SNiP 1985 *SniP 3.06.03-85.Avtomobil'nye dorogi. Gosstroy SSSR*(Moscow : TsITP Gosstroya SSSR)
[2] Sitema dopuskov 1981 *Sistema dopuskov v stroitel'stve* (Moscow: Stroyizdat)
[3] Stolbov Yu V and Stolbova S Yu 2015 O reglamentatsii tochnosti geometricheskikh parametrov vysotnogo polozheniya osnovaniy i pokrytiy avtomobil'nykh dorog v normativnykh dokumentakh *Vestnik SibADI* **3(37)** 59–63
[4] Instruktsiya 2001 *Instruktsiya po razbivochnym rabotam pri stroitel'stve, rekonstruktsii i kapital'nom remonte avtomobil'nykh dorog i iskusstvennykh sooruzheniy (*Moscow: GUP TsPP)
[5] SP 2012 *CP 78.13330.2012. Aktualizirovannaya redaktsiya SNiP 3.01.03-84. Geodetic works in construction* (Moscow: Minregion Rossii)
[6] Gercog V N, Dolgikh G V and Kuzin N V 2015 Calculation criteria for road pavement evenness. Part 1: substantiating the flatness standards of asphalt concrete pavement. Magazine of Civil Engineering **5(57)** 45–57
[7] Privalov F V 2016 Research of accuracy of elevation measurements made by total station Leica TS06plus. New technologies - to the oil and gas region: materials of the International. scientific and practical conf.**2(258)** 41–44
[8] ZouJ, ZhuY, XuY, Li Q, MengL and LiH 2017 Mobile precise trigonometric levelling system based on land vehicle: an alternative method for precise leveling *Survey Review***49(355)** 249–58

[9] Nestorovic Z and Delcev S 2014 Comparison of height differences obtained by trigonometric and spirit leveling method*Geonauka***2(4)** 30–37

[10] Shults R and Roshchyn-Kyiv O 2016 Preliminary determination of spatial geodetic monitoring accuracy for free station method*Geodetski List***70(4)** 355–70

[11] Takalo M andRouhiainen P On *Digital Levelling Technique Applied in Water Crossing*Retrieved from: http://www.fig.net/resources/proceedings/fig_proceedings/fig2006/papers/ts53/ts53_03_takalo_rouhiainen_0507.pdf

[12] Stolbov Yu V 1981 *Osnovy rascheta i analiza tochnosti vozvedeniya sbornykh zdaniy i sooruzheniy (*Omsk: SibADI)

[13] Venttsel' E S and Ovcharov L A 2003 *Teoriya veroyatnostey i ee inzhenernoe prilozhenie (*Moscow: Akademiya)

[14] Gmurman E V 1977 *Teoriya veroyatnostey i matematicheskaya statistika*(Moscow: Vysshaya shkola)

IOP Conf. Series: Materials Science and Engineering **760** (2020) 012049 doi:10.1088/1757-899X/760/1/012049

Calculating the stress-strain state of connecting nodes in pre-fabricated structures from composite materials

S L Aleskovsky[1], A A Pykhalov [1], Van Lam Zyong [1]

[1]Irkutsk State Transport University

E-mail: asl-rassoha@mail.ru

Annotation. The paper numerically studies adhesive-mechanical composite joints with metal and composite connecting elements (rivets and fasteners). The results of comparative experimental tests of metal and composite connectors are presented. The stress-strain state in various layers of composite plates reinforcing composite fibers and in an epoxy matrix is investigated.

1. Introduction

Composite materials (KM) are increasingly used in various fields of technology. One of the most common are polymer composite materials (PCMs), which are actively introduced into highly loaded structural parts [1, 2]. As a rule, highly loaded structural elements from PCMs form metal-composite contact pairs at attachment points, which require the solution of design, strength and technological problems. The calculation of the stress-strain state (SSS) in the metal-composite joint zone causes significant difficulties associated with the lack of a developed theory, the presence of a volumetric stress state in the contact zone and the need to experimentally determine a significant number of PCM characteristics [3, 4].

Currently, mechanical fasteners, metal bolts, rivets and small fasteners are actively used [5, 6].

Making and machining of holes for metal bolts and rivets causes the PCM fibers to be cut and requires precise and expensive tools; the technological process for installing such fasteners is associated with the occurrence of significant stresses due to tightening and increased stress concentration in the region of the hole.

One of the drawbacks of metal fasteners is that when two or several parts made of PCMs are connected at the moment of formation of the closing head, micro-cracking (destruction of the binder) of the connected parts occurs due to a shock or other mechanical stresses that occur during their formulation, due to the small relative elongation of a number of composites (for carbon plastics it is 0.5–0.15%). It should also be borne in mind that when exposed to moisture, such compounds are subject to corrosion, and this decreases the strength properties of the compound.

Analyzing the area of lap-connected plates using mechanical fasteners, it can be noted that when transferring forces from one element to another through riveting or a bolt, a complex VAT is formed in the composite around the bolt [5, 7, 8].

The part of the material located between adjacent holes is stretched; the zone located behind the fastener body is loaded with contact pressure, which causes compression and shear stresses in the PCM.

The destruction of the mechanical connection of composite elements can occur in a weakened section, as a result of shear, crushing of the composite, separation of heads and a cut of fasteners. A combination of these forms of destruction is also possible. Therefore, the strength of mechanical joints is affected by such physicomechanical characteristics of the composite material as the tensile, shear and collapse strengths determined experimentally.

Unlike metal rivets, a more complex SSS arises in a composite fastener, which is determined by the structure of the arrangement of fibers in the matrix and their joint strength.

The behavior of the PCM fastening mechanical element under the action of external loads is determined by the properties of the reinforcing fibers and the matrix, as well as the structure of the fiber-matrix boundary layer.

In composites, high-strength fibers are known to absorb the main stresses arising in the composition under the action of external loads and provide rigidity and strength in the direction of fiber orientation. The mechanical properties of the matrix should ensure the effective joint work of reinforcing fibers under various types of loads. Strength characteristics of the matrix material are determinative for shear loads, structural loading in directions different from fiber orientation, and also for cyclic loading [2, 4, 9, 10].

As a result of the analysis of the available fasteners, a composite rivet was developed [11] (Fig. 1).

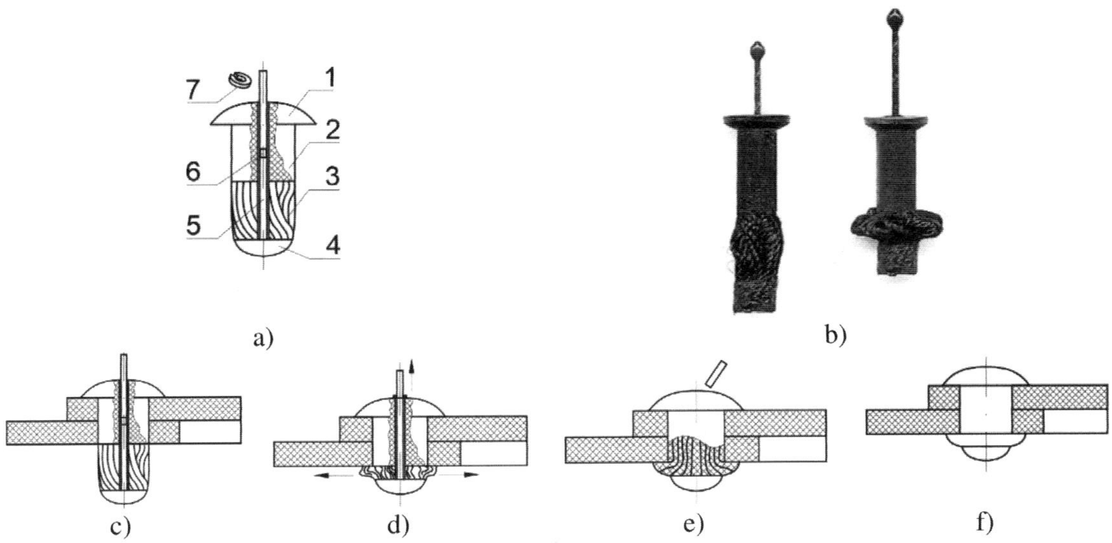

Figure 1 – PCM-based fastener: a) composite rivet; b) photo of the rivet; c) positioning; d) fixation; e) polymerization; e) finished product

The setting head (1) and the shank (2) are made of polymerized fiber composite. The length of the shank (2) is equal to the thickness of the connected package of PCMs. The length of unpolymerized fibers (3) is determined from the conditions for the formation of the height and diameter of the setting head, the ends of the fibers are polymerized into the secondary head (4) with a diameter equal to the diameter of the rod.

In the center of the shank (2) along its axis, there is a hole in which the polymer mandrel (5) freely moves, one end of which is polymerized in the center of the secondary head (4). Before the rivet is inserted into the mounting hole, unpolymerized fibers (3) are impregnated with an epoxy composition. After the rivet mandrel is inserted into the mounting hole, the mandrel (5) is pulled towards the shank, and a locking head is formed due to the bending of unpolymerized fibers (3) in the radial direction. The rod is fixed in an extended position due to a groove on its body (6) and a split washer (7). After polymerization of the locking head, the mandrel is broken off along the groove.

In [12], with the participation of one of the authors, comparative experimental studies were carried out to determine the strength properties of metal and composite rivets. Two pre-glued carbon fiber plates KMU-4l were connected with rivets, with simultaneous making of an adhesive-mechanical joint. For testing, 5 samples of joints with standard rivets were prepared: with a plunger (aluminum body and steel core); rivnuts (high-strength aluminum alloy V-65); small fasteners of two types (d = 1 mm, aluminum alloy AMg-5 and steel 1Cr18Ni9Ti), as well as with the developed composite rivets. The riveting by metal rivets was carried out at the Irkutsk Aviation Plant. Experimental studies were carried out in the Central Plant Laboratory of the plant using the R-5 rupture machine.

The following mechanical characteristics of the materials were accepted:

- for the ***KMU-4l monolayer:*** $E_1 = 135000$ MPa; $E_2 = 5000$ MPa; $G_{12} = 4500$ MPa; $\sigma_{t1} = 850$ MPa; $\sigma_{t2} = 95$ MPa; $\tau_{12} = 600$ MPa; $\mu_{12} = 0{,}2$; $\mu_{21} = 0.0155$;

- for the ***KMU-4l package*** with a thickness of 3 mm and with layer laying $[0;+\_ 45; 90]_{30}$: E_1 = 34856 MPa;

E_2 = 30219 MPa; E_3 = 43846 MPa; μ_{12} = 0.273; μ_{21} = 0.237; G_{12} = 11921 MPa; G_{xy} = 4500 MPa; σ_v = 400 MPa; τ_{12} = 80 MPa; μ_{31} = 0.343; μ_{12} = 0.343;

- for rivets with a plunger: aluminum body, E =70000 MPa;

σ_v = 400 MPa; μ = 0.32; steel core, E = 200000 MPa; σ_v = 500 MPa; μ = 0.26;

- for ***rivnuts*** (aluminum alloy B-65): E = 72000 MPa; σ_v = 400 MPa; σ_t =280 MPa;

μ = 0.32;

- for small-diameter fasteners (steel 1Cr18Ni9Ti): E = 212000 MPa; σ_v = 655 MPa; σ_t = 355 MPa; μ = 0.26;
- for ***composite rivets***: E = 34856 MPa; σ_v = 400 MPa; μ = 0.373.

Composite rivets were tested with three patterns of reinforcing fiber winding along the mandrel.

Type A (*longitudinal arrangement*). The fibers are located along the rivet mandrel; shear and shearing forces are suffered by the matrix and, to a lesser extent, by fibers; the force of head separation is made by fibers in interaction with the matrix (Fig. 2a).

Type B (*transverse arrangement*). The fibers on the shank are arranged transversely, and in the central part along the rivet mandrel. The shearing and crushing forces are experienced mainly by the matrix and part of the fibers on the outer side of the mandrel, the forces of rivet head tear off are experienced by longitudinal fibers in the central part of the rod (Fig. 2b).

Type C (*cross arrangement at an angle*). The outer layers of fibers are located at an angle φ along the rivet mandrel. Angle φ equals 45 degrees or so. The shearing and crushing forces are perceived by the matrix and, for the most part, of fibers located at an angle φ to the axis of the rod, the riveting heads are torn off by longitudinal fibers located in the central part (Fig. 2c).

In winding types **B** and **C** in the central part of the mandrel, there is a small part of the fibers (about 15...20%) arranged along the longitudinal axis.

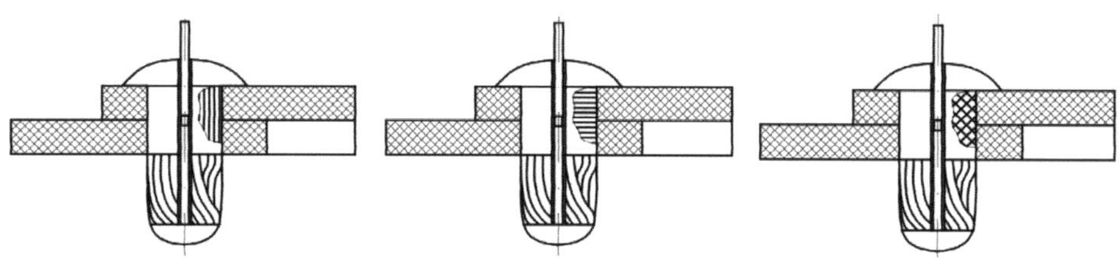

Figure 2 Layout of the fibers in the rivet

The material used for rivets was “Grapan-27”, which has the following characteristics (TU 1916–204–51385208–2001):

- • density - 1.75;
- • linear density - 724 tex,
- • ultimate compressive strength σ_c = 1100 MPa;
- • ultimate tensile strength σ_t = 1300 MPa;
- • modulus of elasticity on bending E_b = 183 GPa.

As a binder, high-strength epoxy adhesive, σ_v = 400 MPa, was used.

In technology, about 90% of all fasteners work in shear, while one of the main types of rivet fracture is a cut along the plane of connection of the plates [1, 2]. Based on this, samples with rivets were subjected to tensile strength tests on a bar shear using KMU-4l material as the basis for the joint (Fig. 3).

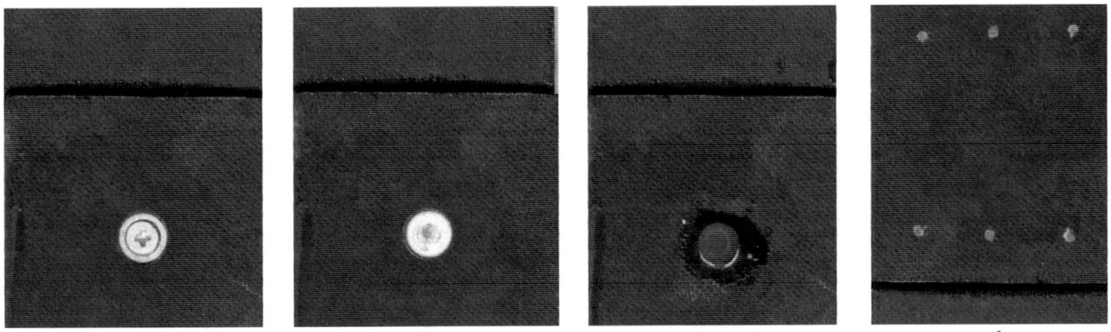

Figure 3 Samples of the studied joints: a) with a rivnut; b) with a rivet with a plunger; c) with PCM rivet; d) with small-diameter fasteners

The test results for some fasteners are summarized in Table 1. Figure 4 shows the results for rivets made from PCM. The average values for five trials are presented.

Table 1 Test results for all types of rivets

Sample type	F_t [N]	τ_{av} [MPa]
PCM rivet, longitudinal reinforcing of fibers in mandrel (type **A**)	16110	410.4
PCM rivet, lateral reinforcing of fibers in mandrel (type **B**)	18500	471
PCM rivet, cross reinforcing of fibers in mandrel (type **C**)	19060	485.4
Rivet with plunger	4000	203.8
Rivnut	4660	237.5
Small-diameter fasteners	19450	497

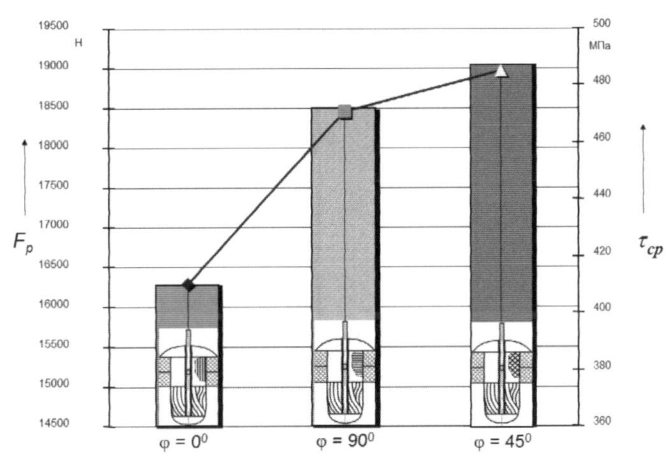

Figure 4 Test results for PCM rivets of **A**, **B** and **C** types

According to the experimental results, the best result was shown by the winding pattern of reinforcing fibers of type **C** with a cross arrangement. In addition, there was a high strength of the adhesive-mechanical connection with small diameter fasteners made of 1Cr18Ni9Ti steel. Connections with metal rivets turned out to be the least durable due to the fact that riveting creates a high level of interference. In contrast to the tightening along the axis of the rivet, the interference in the transverse direction relative to the axis of the rivet in a composite package can lead to cracking of the walls of the holes for the rivets and reduce the strength of the connection.

Next, numerical modeling was carried out using the FEM of loading composite rivets of types **A**, **B** and **C** in order to confirm the experimental results for these patterns of winding reinforcing fibers.

At the first stage, the elastic characteristics of the composite were determined by the technique presented in [13, 14]. For this, a typical scheme of a representative element of a composite material with the dimensions is selected shown in Fig. 5a. Then a quarter of the representative element is modeled (due to the symmetry of the problem) by the finite-element method (FEM) (Fig. 5b). With a representative element, numerical experiments were carried out, consisting in a sequential predetermined movement of the faces of the representative element along the coordinate axes while fixing the remaining faces. As a result of the FEM solution, average stresses on the faces of the element were obtained, and using the generalized Hooke's law in elastic form, the elastic characteristics of the composite were determined.

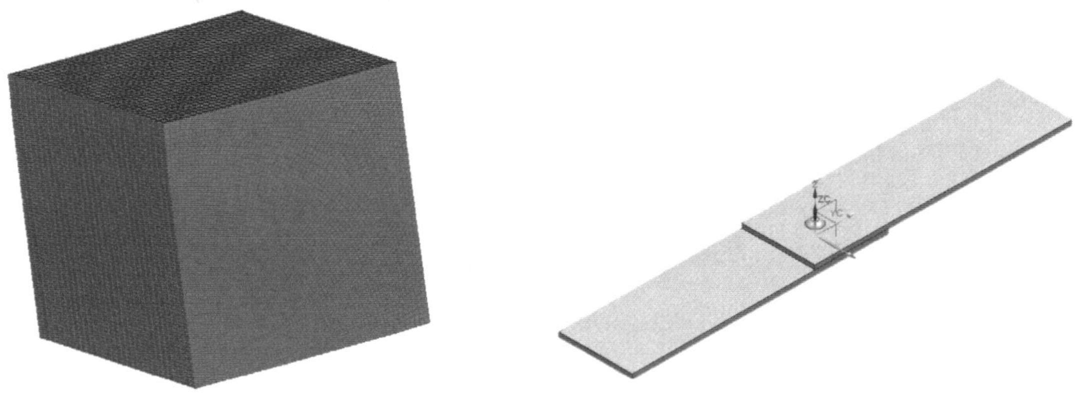

a) b)

Figure 5 a) finite element model of the representative element of the composite; b) modeling the connection of composite plates and composite fasteners

Figure 6 Models of composite fasteners of types **A**, **B** and **C** (matrices not shown)

a) b) c)

Figure 7 FEM model of fibers and matrix of composite fasteners of type **C** (a, b) and fiber model for composite fasteners of type **B** (c)

This technique has been widely used recently, because it gives more accurate results in determining the elastic characteristics of the composite than the calculation according to the mix rule or other analytical formulas.

At the second stage, finite element modeling of composite rivets (Figs. 6 and 7) and general loading of the adhesive bond with such rivets (Fig. 5c) was carried out.

The calculation results—the acting stresses in MPa and strains—are presented in Tables 2 and 3.

Table 2 shows the von Mises contact stresses between the layers of reinforcing fibers and the composite connection plates for type B composite fastener. Table 2 in each column shows the interaction of the corresponding layer (1st, 3rd, and 4th) from top to bottom reinforcing fiber with composite plate material. The first row of the table shows the deformed state of the corresponding fiber layer; in the second line, the deformation of this layer in the plane of contact of composite plates; in the third row, contact stresses in composite plates. The right column of the table shows the magnitude of the acting contact stresses according to von Mises in MPa.

Table 2 Contact von Mises stresses in the layers of composite plates for composite rivet type **C** (MPa)

1st composite layer (+45°)	3rd composite layer (90°)	4th composite layer (0°)	MPa

Section view of layers

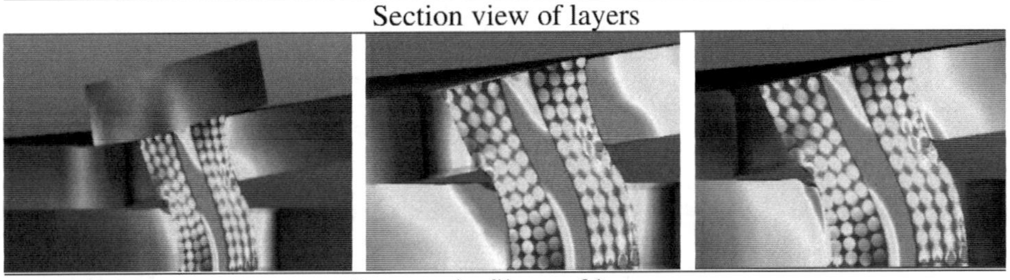

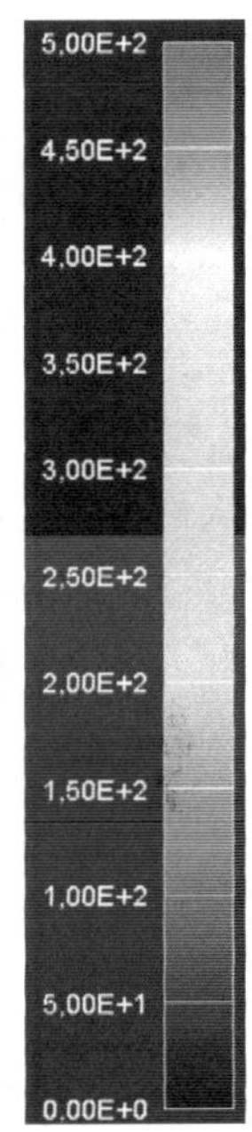

Stresses in fibers of layers

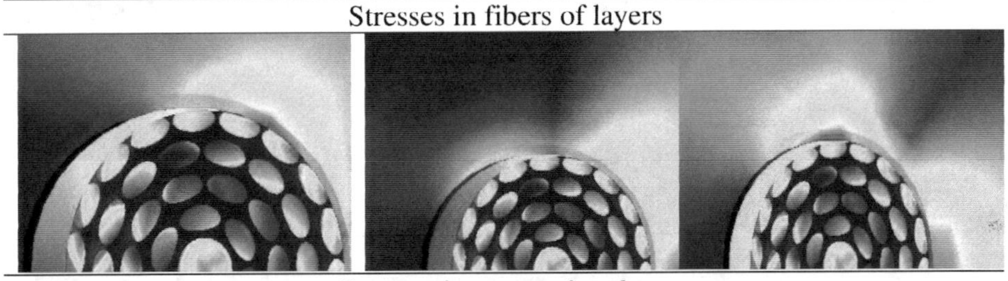

Stresses in composite plates

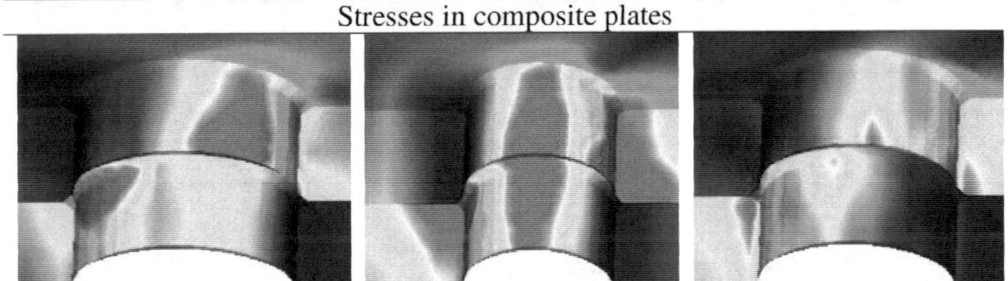

Table 3 illustrates the stresses in the layers of reinforcing fibers for a composite type C fastener (cross reinforcement). Von Mises stresses in 1...4 fiber layers are shown from top to bottom in each column. For the 1st and 2nd fiber layers, the first row of the table shows the stresses in opposite diametrical sides. The second row of the table shows the stresses at the section of the corresponding layers of fibers, while the epoxy matrix is not shown. The right column of the table shows the magnitude of the acting von Mises stresses in MPa.

Layers of fibers are numbered from 1st inner to 4th outer.

IOP Conf. Series: Materials Science and Engineering **760** (2020) 012049 doi:10.1088/1757-899X/760/1/012049

Table 3 Stresses in the fiber layers for composite type **B** fastener

1st fiber layer	2nd fiber layer	3rd fiber layer	4th fiber layer	MPa

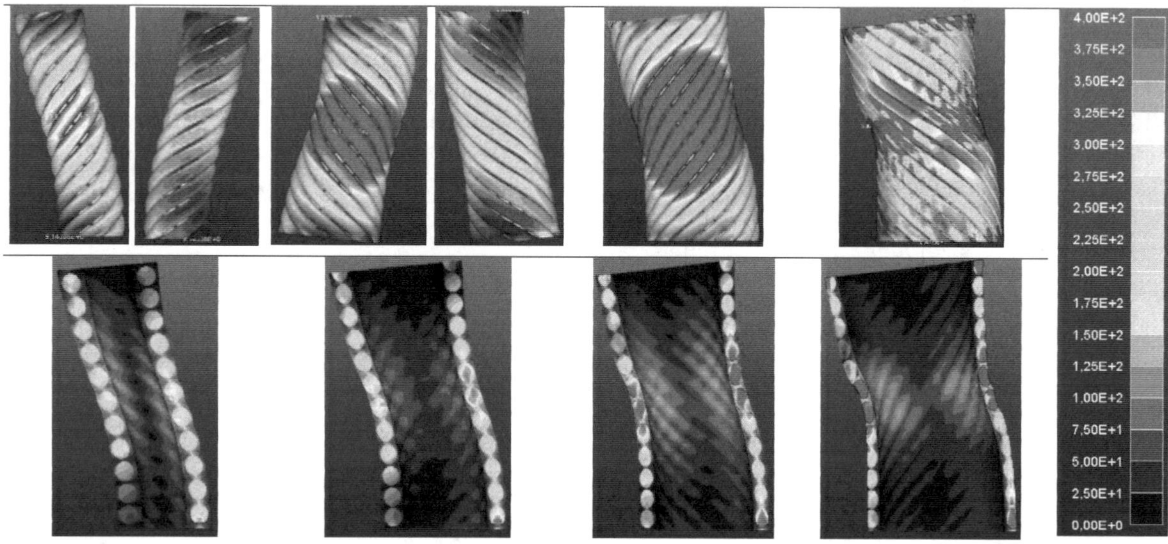

At the next stage, we compared the maximum von Mises stresses that occur in the outer layers of composite fastener fibers of types **A**, **B**, and **C** with the same external load on the simulated adhesive-mechanical joint. In addition, sections with maximum shear stresses (at the junction of composite plates) were selected in composite rivets. Stresses were averaged over finite elements located in this section along the outer rows of fibers (Fig. 8).

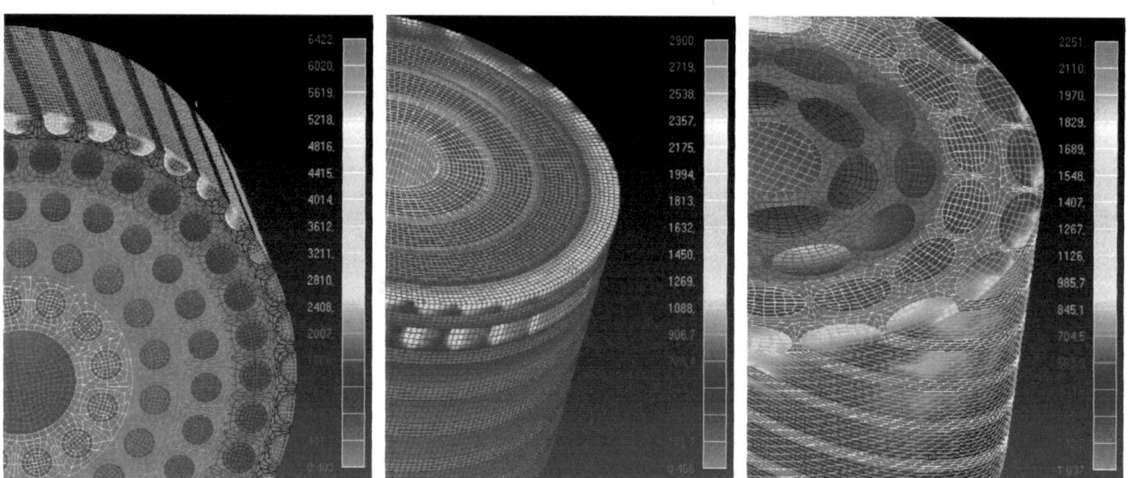

a) b) c)

Figure 8 Von Mises stresses on the external fibers of composite fasteners: a) type **A**, b) type **B**, c) type **C**

The average values of maximum von Mises stresses were as follows:
type **A** – 4017 MPa, type **B** – 2820 MPa, type **C** – 1132 MPa. The stress distribution by types of composite fasteners is shown in Fig. 9.

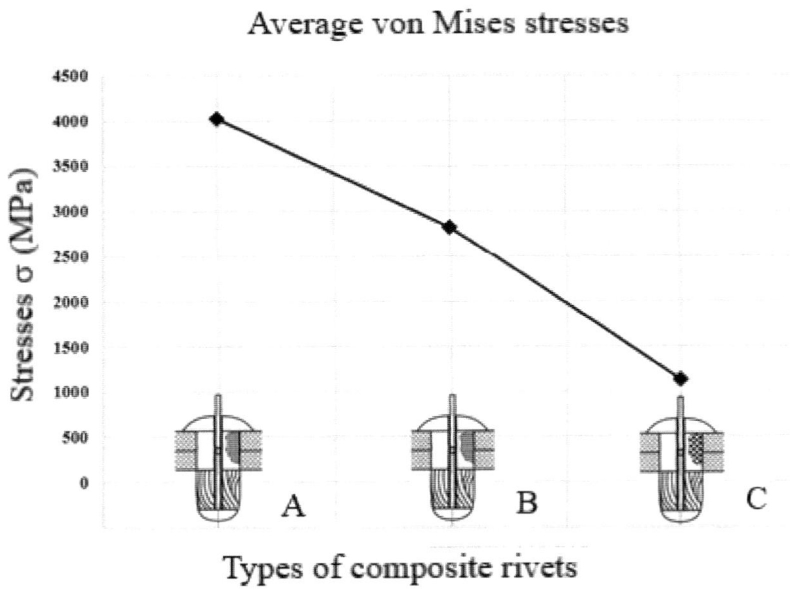

Figure 9 Von Mises stress dependence in composite rivet fibers

Evidently, type **C** composite rivets (with cross-fiber reinforcement) provide the lowest level of stress, and, therefore, are the most durable of the considered types of rivets.

Conclusions

1. The strength of composite fasteners is largely due to the angle of winding of fibers along the rivet core. The destruction of the connection with the type **C** fastener (cross-reinforcement of fibers in the rivet mandrel) occurs at a force level of 15...25% more than for metal rivets and with type **A** fasteners (longitudinal reinforcement of fibers in the rivet mandrel).

2. The margin of static strength for adhesive-mechanical joints with metal rivets is lower than for adhesive-mechanical joints with PCM fasteners with cross-fiber reinforcement, the effectiveness of which reaches 100...120% of the initial strength. Samples with metal rivets withstand a load of 70...90% of the initial strength.

3. The comparative tests of composite rivets of three types showed that rivets of type **C**, with cross-reinforcement of fibers, have the highest strength characteristics. Taking into account numerous studies [15], confirming the higher resistance of composites to the effects of adverse environmental factors compared with various types of metal rivets, type **C** composite rivets can be recommended for use as fasteners of structures.

4. Comparison of theoretical and experimental values of the research results shows that for 90% of the experiments, the error did not exceed 15...20%. This allows speaking about the adequateness of the FEM model used to determine the strength properties of the developed types of composite fasteners, as well as to calculate the SSS in the region of the rivet hole in composite plates.

References

[1] Vorobei V V, Sirotkin O S 1985 *Connection of structures from composite materials* (Leningrad: Mashinostroenie) p. 168

[2] Karpov Ya S 2006 *Joints of composite parts and assemblies* (Kharkov: Kharkiv Aviation Institute) p. 359

[3] Tsarakhov Yu S 1980 *Design of joints of composite aircraft elements: Teaching guide* (Moscow: MIPT) p. 82

[4] Karpov Ya S 2001 Principles of designing joints of highly loaded parts of aircraft from composite materials *Questions of design and production of aircraft: Coll. Sci. Papers* 122

[5] Vashukov Yu A, Peresypkin V P, Peresypkin K V, Nikonorov N N 2005 Finite element stress

state modeling when reinforcing holes in structural elements from polymer composite materials *Bulletin of the Samara Scientific Center of the Russian Academy of Sciences* **7 2** 436-441.
[6] Pykhalov A A, Pashkov V P 2010 Application of the finite element method and the contact problem of a solid deformable body in modeling of fractured bone fixation *Modern technologies. System analysis. Modeling* **3 (27)** 27-33
[7] Pykhalov A A, Pashkov V P, Zotov I N 2014 Modeling of mechanical systems with vague material properties using the finite element method and computed tomography. *Modern technologies. System analysis. Modeling* **2 (42)** 44-50
[8] Korzhenevsky A V, Veselsky S I, Makarenko V A 1985 et al. Studying crushing strength of composite materials *Problems of design and production of aircraft structures: Coll. Sci. Papers. – Kharkov: Kharkiv. Aviation Institute* 112-116.
[9] Karpov Ya S, Makarenko V A, Marchenko V G 1989 Studying the anisotropy of the crushing strength of composite materials under fasteners *Calculation and design of structures of aircraft: Coll. Sci. Papers* 82-91.
[10] Pykhalov A A, Fedotov P K 2013 Numerical modeling of boundary conditions and load in the analysis of the stress strain state of a rock during its crushing between rolls in a roller press. *Modern technologies. System analysis. Modeling* **1 (37)** 27-32
[11] Aleskovsky S L, Vlasov S V, Kuznetsov S N, Natsubidze S A, Shevchik A P RF Patent No. 2219390 (20 December 2003)
[12] Tanis C, Poullos M 1980 Composite Fasteners – A Compatible Joining Technique for Fibrous Composites in Structural Design *Fibrous Composites in Structural Design* 645–657.
[13] Kuznetsov S N 2004 *Technique for repair of airframe design units for military aircraft with composite fasteners, taking into account the influence of operating conditions.* Disertation of Cand. Sci. (Engineering) Irkutsk Higher Military Aviation School 189
[14] Andrianov I V, Danishevskyy V V, Guillet A, Pareige P 2005 Effective properties and micromechanical response of filamentary composite wires under longitudinal shear // *European Journal of Mechanics* **24 2** 196–206.
[15] S. Dariya Zade. 2013 Numerical technique for determining the effective characteristics of unidirectional composites *Bulletin of NTU "KhPI". – Series: Dynamics and Strength of Machines,* **58 (1031)** 71-77.

IOP Conf. Series: Materials Science and Engineering **760** (2020) 012050 doi:10.1088/1757-899X/760/1/012050

The use of electrohydropulse treatment of waste solutions in refining production for the recovery of platinum group metals

D Yu Parfenov [1], V S Ratushnyak [1], E S Ilyin [1], A V Yuryev [1], V O Kolmakov [1]

[1]Krasnoyarsk, Krasnoyarsk Railway Institute, branch of the Irkutsk Railway Engineering University

Abstract. The high cost of precious metals and environmental requirements lead to the need for the completeness of their extraction from primary and secondary raw materials.
During the processing of primary raw materials, waste solutions are produced that contain low-concentration precious metals, which must be removed from the process. To prevent losses and comply with environmental protection requirements, it is necessary to extract precious metals from waste solutions. The complexity of this task lies in the low concentration and chemical forms of the presence of precious metals in aqueous systems. The cementation method, currently used for the extraction of precious metals, has several disadvantages: the duration of the process, high-energy consumption, etc. To increase the efficiency of the method used, it is necessary to pre-treat the solution with the goal of labilizing the noble metal complexes. The purpose of the experiment: determination of the optimal modes of the electrohydropulse (EHP) method for intensifying the process of additional extraction of platinum group metals (PGM) and heavy non-ferrous metals from industrial solutions of refining production. As a result of the study, the necessary materials were selected to create the reactor and electrode systems, the volume of the laboratory reactor was determined. Trial launches of the installation were carried out, according to the results of which a pulse current generator (PCG) was configured. A series of tests was carried out, based on the results of which conclusions were drawn about the possibility of using this method to activate stock solutions.

1. Methods for the recovery of precious metals from waste process solutions

The following methods for the recovery of precious metals from process solutions are widely used:
1 Cementation.
2 Sorption.
3 Electrolysis.
4 Galvano-coagulation.

Cementation is one of the first and most common methods for isolating PGMs from effluent solutions. The driving force of the cementation process is the difference between the real redox potential (RRP) of the cementing metal and the ions of the metal being deposited to the solution [1, 2, 3].

This method is effective with a high concentration of precious metals in solution (more than 20 mg / l). The process of cementation lasts about 20 - 24 hours, since the destruction of the durable complex compounds of precious metals requires a long time, which implies high energy costs. The temperature of the solution is 80 - 90 °C during the whole cementation process, the solution usually is heated with direct steam.

Sorption is widely used at gold mining enterprises. This operation is most often used for the following types of solutions:

- productive solutions of heap and in-situ leaching;
- discharge of the thickeners containing cyanide complexes of gold and silver;
- stock solutions after gold electrolysis;
- filtrates after the filtration of cyan-containing sorption tailings;
- solutions of the industrial sewage of the plants after the preliminary separation of the solid phase [4].

The disadvantages of this method are the large dependence of sorption on the forms of valuable components in the solution, the relatively long duration of sorption and the small capacity of sorbents, and in relation to precious metals - the complexity, and in some cases, the impossibility of desorption.

Electrolysis is the process of the isolation of the constituent parts of solutes or other substances on the electrodes that are the result of the secondary reactions on the electrodes, which occurs when an electric current passes through the solution or the electrolyte melt. The disadvantages of electrolysis are high energy consumption, the need for frequent replacement of the electrolyte, cleaning of the electrodes and the fundamental impossibility of the complete extraction of valuable components.

Galvano-coagulation is the method based on the effect of a galvanic iron (aluminum) -cox or iron (aluminum, copper) -cox element, placed in the solution to be purified. Due to the difference in electrochemical potentials, iron is polarized anodically and goes into solution, followed by precipitation in the form of oxyhydrates, on the surface of which there are adsorbed the negatively charged anions. Coke or copper in the galvanic couple is polarized in the cathode way, as a result of which there is electrodeposition of cations on the cathode [5, 6]. Thus, the metals are extracted from solutions. The disadvantage of this method is a relatively low degree of extraction of metals (about 80-90%) from the solution and the formation of a significant amount of slime sediment.

2. Method to improve the efficiency of precious metals extraction from solutions

Stabilization of precious metal complexes is the most promising method for increasing the efficiency of extraction. Chemical and physical methods can be used for labialization: complexing agents, electrochemical activation, activation of iridium complexes by autoclave chlorination, microwave radiation, UV radiation, and other methods.

The electro-impulse effect is one of the ways to activate the solution.

EHP is a method based on breaking complex compounds and speeding up chemical reactions. This method does not replace the existing methods, but is an addition to them. The EHP method is actively used to accelerate the chemical reactions in various areas of the national economy (in metallurgy — activation of the lime mortar; in agriculture — activation of bio solves (bacterial explosion); in chemical industry — acceleration of reactions, in particular, to intensify the leaching process of platinum-palladium catalysts). The use of EHP method speeds up processes significantly (2-5 times) when replacing the traditional types of activation, that is, increasing the temperature of the solution, expensive catalysts, mechanical and ultrasonic methods.

The main properties of the EHP discharge, affecting the activation of solutions:

- significant (tens of kV / cm) electric field strength;
- values of the current in a pulse are up to tens of kA and, as a result, the large values of the magnetic field;
-pressure in the pulse is 10-100 atm;
- the temperature in the discharge zone is up to 30 000 ° C;
- the ultraviolet radiation;
-the X-rays (the level of XR depends on the voltage);
-the cavitation effect.

The possible variants of the technology are electrohydropulse activation - cementation with iron; electrohydropulse activation - galvanic-coagulation on the pair of coke - iron or on the filler of coke – iron - aluminum. In both cases, reducing agents may be added.

The studies of the possibility to use pulse technologies to intensify the leaching process of platinum-palladium catalysts [6] showed the recovery up to 98% by treating a catalyst with an energy of 15 kJ; the catalyst is most active in the first 30 minutes of treatment.

At Krasnoyarsk Institute of Railway Transport, patent studies were conducted, covering more than 130 documents (58 applications and 76 abstracts of the Russian inventions) devoted to the extraction of precious metals. Most of the inventions were connected with cementation or sorption [7,8].

The considered methods have a number of disadvantages, such as, the long duration of the process, a significant increase in the volume of effluent, high energy consumption, the high cost of facilities and sorbents. In this connection, in the research conducted with the help of the EHP method, the method was chosen as the most optimal one. It extends the existing methods and is easily integrated into the existing system; it is characterized by relatively low power consumption [9].

3. Installation description

The installation includes the pulse current generator and the reactor with an electrode system.

The main technological characteristics of the developed equipment.

1) Pulsed current generator.

IOP Conf. Series: Materials Science and Engineering **760** (2020) 012050 doi:10.1088/1757-899X/760/1/012050

The PCG should have the following characteristics: the maximum energy per pulse is 0.3 kJ; voltage on electrodes is from 5 to 10 kV; current between the electrodes is from 2 to 10 kA; the repetition rate of micro pulses is from 0.5 to 2 Hz; the maximum power consumption of the installation is 2 kWh; the power supply of the generator from the network of alternating voltage is 220 ± 22 V [10,11,12,13].

2) Reactor (Figure 1).

Figure 1. Photo of the reactor

The reactor has a housing (working chamber), a cover, an electrode system, and it is positioned on a support platform with fasteners. The system of electrodes consists of the units: negative and positive electrodes.

The volume of the reactor is 1 l. The working chamber of the reactor was made of HDPE as per GOST 22689.2-89, the cover and the electrode system are made of grade VT1-0 titanium as per GOST 19807-91 and of titanium alloy OT4 as per GOST 26492-85. The choice of these materials was due to the use of the reactor for working with chemically aggressive liquids. Other elements of the reactor that are not in contact with the medium being processed (washers, nuts, tuckers, etc.) are made of general-purpose St3 steel, GOST380-2005. The overall dimensions of the reactor are 310x110x140 mm.

3) Electrode system.

It consists of two electrodes: positive and negative. The electrode material is polyurethane, titanium, iron.

4. Manufacture of reactor and electrode systems

Some parts of the reactor (metal elements, mainly: electrodes, washers, glasses, lugs, covers, tuckers, studs, etc.) are produced according to the design documentation (DD) at the industrial enterprises of Krasnoyarsk. Figure 2 shows the assembly drawing of the reactor.

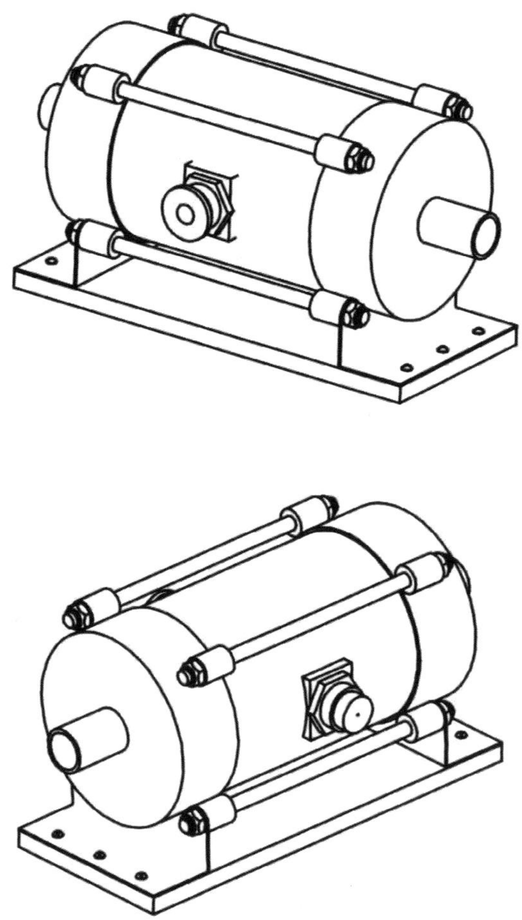

Figure 2. Schematic of the reactor

Some structural elements (washers for electrodes, an insulator of a positive electrode) were made in the Student Design Bureau of "Engineering Solutions". The technology for manufacturing parts from molded polyurethane (PMC780) was developed and mastered for this purpose. This technology has the following operations:

1) Preparing the surface of the form (cleaning, applying a release agent to prevent the form from sticking to the model).
2) Mixing A and B components in a ratio of 2A: 1B.
3) Mixing the components.
4) Degassing under vacuum.
5) Pouring the mixture into the form.

6) Cure for 48 hours.

As a result, washers and electrode insulator with the specified insulation and strength characteristics were produced.

5. Test methodology

For the experiments it is necessary to simulate the process of cementation. The existing cementation process is carried out as follows. The containers designed for the cementation process were loaded with scrap metal to 2/3 of their volume, then they pour the stock liquor and heat it for 20-24 hours and partially mix the solution with direct steam, the temperature of which is up to 160°C. The solution in the process of cementation is heated to 90°C, the temperature is controlled by steam consumption.

To simulate the process, 3 electric tiles were taken (to heat the solution), 3 cups for grouting with a volume of 1 liter (as a replacement for cementing baths), 3 kg of steel nails.

A 1.5-liter beaker was used for pretreatment of the solution (pH adjustment, addition of additive reagents, etc.).

• Before the experiment, the nails were degreased with a detergent and boiled for 10 minutes in a solution of hydrochloric acid.

• All three cementation cups, with a volume of 1 liter, were numbered in order and each one was loaded with 1 kg of nails.

• The reactor was installed in a cabinet with the subsequent connection to the PCG.

• 20 liters of solution were delivered after the first cementation to a canister.

• 1 kg of the reducing agent of Na2S2O3 was taken to add it to the solution in order to exclude the possibility of recovery of the complex compound after its destruction.

• The process was carried out at pH = 1. Acidification was performed with sulfuric acid.

Preparation of the solution for cementation, acidification of the solution, namely, mixing and addition of the reducing agent were done in a 1.5-liter beaker. Subsequently, the prepared solution in equal volumes was sent to the beaker for cementation and to the reactor, where it was treated with an electro pulse effect. The treated solution was poured into a cementation beaker. Thus, dehumidification by cementation was compared when the solution was pre-treated by an electric pulse and when it was not pre-treated.

During the experiments, the number of pulsed discharges and in some cases the discharge energy varied. During the experiment, the expediency of acidification was checked before or after the EHP treatment. The addition of the reducing agent also varied in time; the addition of the reducing agent was also done with conventional cementation, in order to assess the effect of the EHP effect on the cementation process.

6. Plan of experiments

The program of the experiments was designed for 2 types of solutions brought to the shop for extraction by cementation.

For solution number 1.

1. One liter of the initial solution was taken into a measuring cup heated to 80-90 °C; steel of a 3-5 steel grade was used as a cementator filling 2/3 of the volume of the glass. The solutions are taken for the analysis of precipitation-based chemical analytical method (CAM) in Central Factory Laboratory (CFL). Two samples are taken for analysis: initial one and that one after cementation. The samples were analyzed with CAM. The experiments were done for varying periods of time.

2. One liter of the initial solution was taken into the measuring cup; it was prepared for the EHP treatment. It was heated to 80–90 ° C; steel of the 3-5–5 grade was used as the cementator metal filling 2/3 of the glass volume. The solutions were taken for analysis with CAM - CFL. The experiments were done for varying periods of time and number of pulses.

3. The initial solution (V = 2 l), to which a reducing agent is added (allowing one to exclude the restoration of the original properties of the solution), was divided. One liter was taken for electro pulse treatment, and then for cementation; the rest was taken for the cementation process. The solutions were taken for the analysis with CAM – CFL. The experiments were done with varying periods of time and number of pulses.

For solution number 2.

1. One liter of the initial solution was poured into a measuring cup heated to 80-90 ° C. Steel of a 3-5 grade was used as a cementator filling 2/3 of the volume of the glass. The solutions were taken for the analysis of CAM - CFL. Two samples were taken for the analysis: initial one and that one after cementation. The samples were analyzed by CAM. The experiments were done for different periods of time.

2. One liter of the initial solution was poured into the measuring cup prepared for the EHP treatment. It was heated to 80-90° C; steel of the 3-5–5 grade was used as a cementator metal filling 2/3 of the glass volume.

The solutions were taken for the analysis of CAM - CFL. The experiments were done for different periods of time and number of pulses.

3. A reducing agent was added to the initial solution (V = 2 l), allowing one to exclude the restoration of the initial properties of the solution. One liter was taken for electro pulse treatment and then – for cementation; the other was taken for cementation. The solutions were taken for the analysis of CAM – CFL. The experiments were done for different periods of time and number of pulse.

7. Analysis of the results of the experiment

During the project, the following experiments were conducted:

– preliminary EHP treatment before cementation;

– preliminary EHP treatment with the addition of a reducing agent (sodium thiosulfate) before cementation;

– EHP treatment followed by sludge solution;

– EHP treatment with the subsequent sludge and additional processing before cementation. Steel shavings, nails and graphite were used as the cementing agents.

The analyses of the samples were done with WSS (working standard solution) and CAM methods. The EHP treatment has a beneficial effect on the recovery of base metals (BM) during cementation, as shown in Figure 3.

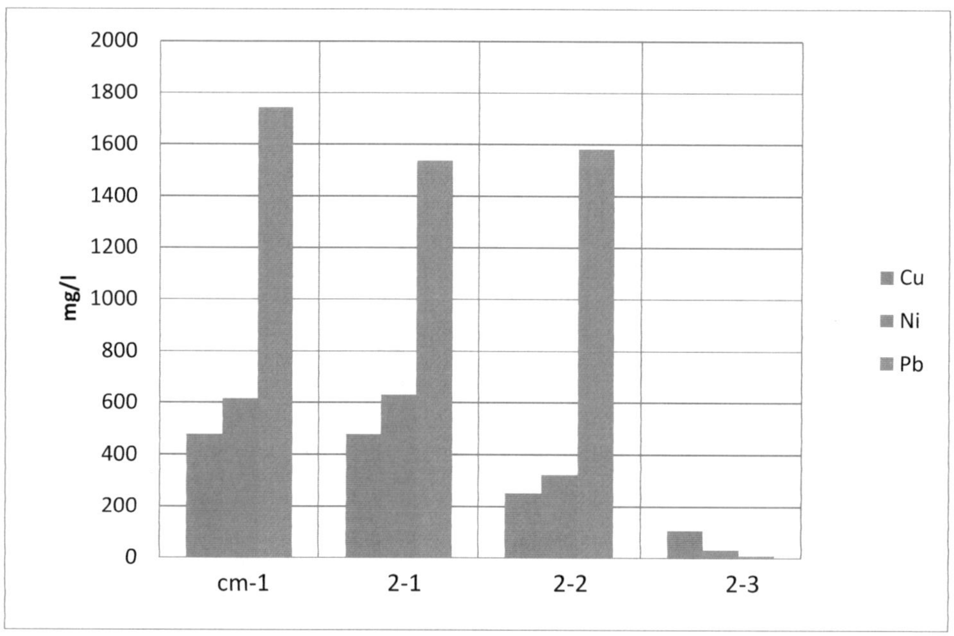

Figure 3. Chart of the EHP effect on the NB extraction.

CM-1 Initial solution, 2-1 conventional cementation (heating + stirring), 2-2 EHP treatment (8 kV, 100 imp., Hea ting + stirring), 2-3 EHP processing + reducing (8 kV, 100 imp., heating + mixing). All the samples had 2-hour cementation, which is 22 hours less than the usual cementation time.

The results of the experiments are presented in Table 1.

Table 1. Results of additional metal extraction of PGM by cementation with and without EGP treatment.

№ Samples	Pt		Pd		Rh		Ir		Time h
	content, mg / l	deviation from init. s-s.%	content, mg / l	deviation from init. s-s.%	content, mg / l	deviation from init. s-s.%	content, mg / l	deviation from init. s-s.%	
Initial solution	30	0.0	0.5	0.0	3.7	0.0	4.8	0.0	-
Conditional cementation	7.2	-76.0	<0.5	0.0	1	-73.0	0.7	-85.4	3
EHP processing (8 kV, 50 imp., heating + mixing)	3.4	-88.7	<0.5	0.0	1	-73.0	<0.5	-89.6	3
Cementation with reducing agent (heating + mixing)	9.3	-69.0	<0.5	0.0	0.7	-81.1	0.9	-81.3	3
EHP processing + reducing agent (8 kV, 50 imp., heating + stirring)	5	-83.3	<0.5	0.0	0.3	-91.9	1.3	-72.9	3
Initial solution	27.3	0.0	<0.5	0.0	3.2	0.0	5	0.0	-
Without EHP, day of sediment	25	-8.4	<0.5	0.0	2.3	-28.1	3.6	-28.0	2
EHP 200 pulses 8.2 kV, sludge day	23.6	-13.6	<0.5	0.0	0.6	-81.3	3.1	-38.0	2

Table 1 shows that the influence of the EHP method increases the extraction of PGM from the stock liquors during the cementation time within 2 and 3 hours. Also, two-hour solutions were not cemented, as a result of which the sediment rate fell in the samples, while in the sample processed with EGP, the sediment was rather marked, and the sampling showed the decrease in PGM in these samples.

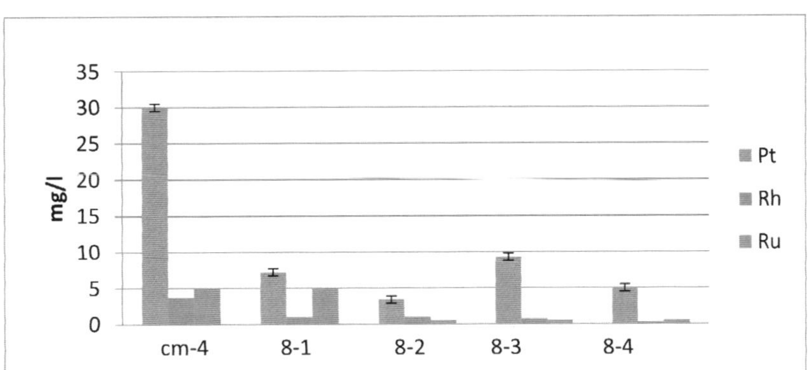

Figure 4. The effect of solution treatment on the cementation process.

Figure 4 shows the influence of the EHP treatment with and without a reducing agent on the additional extraction process. The comparison is made against the conventional cementation with and without a reducing agent. CM-4 initial solution, 8-1 ordinary cementation, 8-2 EGP treatment (8 kV, 50 imp., heating + mixing), 8-3 cementation with a reducing agent (heating + mixing), 8-4 EGP processing + reducing agent (8 kV, 50 imp., heating + mixing), all cementation processes lasted for two hours.

8. Conclusion

The experiments have shown the effect of EHP on the recovery of PGM and the possibility of its use in industrial production. To establish the optimal processing parameters, it is necessary to upgrade the power section of the power generation plant to 2 kJ, and also to change the reactor design so that it can withstand 2 kJ loads. In connection with the change in properties (characterized by turbidity) of the solution within an hour, development of a method of rapid analysis is in demand.

References

[1] Barchenkov V V, 2016 Sorption of gold from technological solutions. Gold mining. **206** 122
[2] Boyko N I, Odaryuk V A, Safonov A V, 2015 Study of the possibility of using the discharge-pulse method in the treatment of surface and industrial wastewater. *Civil Security Technologies*. **12** 39
[3] State standards R 52793-2007. Precious metals. Terms and Definitions.
[4] Dzyubinsky F A, Feofanov V A, Zhdanovich L P, 2009 Conditioning of circulating water by galvanic-coagulation method. Water treatment. *Water supply: production-technical and scientific-practical journal* **5** 32
[5] Korostovenko V V, 2008 Electric methods in combined technologies of mineral raw materials. (Krasnoyarsk: IPK Sib. Federal University).
[6] Korostovenko V V, Gron V A, Shakhrai S G, Kaplichenko N M, Galayko A V, 2013. Application of the electric pulse method for wastewater treatment in coal deposits. *Modern high technology*. **10-1** 164
[7] Orlenko A I, Tsuglenok N V, Yuryev A V, 2013.About the use of the electro hydrodynamic effect in agriculture. *Bulletin of the KrasSAU*. **1** 143.
[8] Decree of the Government of the Russian Federation of August 17, 1998 No. 972. On approval of the operating procedure of organizations engaged in the refining of precious metals, and a list of organizations authorized to carry out refining of precious metals.
[9] Ilyin E S, Orlenko A I, Ratushnyak V S, Sergienko T S, Yuryev A V, 2017 Electrohydropulse method for wastewater treatment. *Ecology of production: a scientific and practical journal* **2** 57.
[10] Ilyin E S, Orlenko A I, Ratushnyak V S, Trukhina. I S, Yuryev A V, 2017 Electrohydropulse method for the disposal of reinforced concrete products. *Ecology of production: a scientific and practical journal* **11** 60
[11] Ratushnyak V S, Yuryev A V, Ilyin E S, Trukhina I S Russia Patient No. 2660260 (15 September 2017)
[12] Ratushnyak V S, Garanin A E, Yuryev A V, Ilyin E S Russia Patient No. 2666225 (07 March 2017)
[13] Ilyin E S, Ratushnyak V S, Yuryev A V, Parfenov D Yu. 2018. lncrease of the Treatment Facilties Efficiency Liu Means of Electrohydropulse Treatment of Sewage. Proceedings of the 6th Intertional Symposium on Innovation and Sustainabllity of Modem Railway ISMR, Irkutsk, Russia. 805-808.

Methods and algorithms of development of a CAD system to design automated process control systems in the transport sector

D I Sachkov[1], A V Dimov[1], R U Upyr[1]

[1]Irkutsk State Transport University, Irkutsk, Russia

E-mail: leninb@yandex.ru

Abstract. The article considers questions of creation of an intelligent CAD (Computer-Aided Design) system to design APCS (Automated Process Control System) in the transport sector. The CAD system should represent a graphic 2D engine adapted for the solution of problems of design of APCS, surrounded with an information support cloud in the form of a set of relational databases of the engineering information necessary for effective and error-free work of the APCS designers. Programs of processing information that appears on drawings, schemes and tables of the APCS projects on the basis of the algorithms adopted in this subject domain should become an essential part of information support. A CAD system should be equipped with artificial intelligence programs for processing and sorting of external information and search of mistakes in the project.

1. Introduction.

The main direction of the innovation development of the Russian industry is to universally develop and implement computerized APCS (Automated Process Control Systems), as well as to replace old analogue monitoring and control systems (MCS) with digital APCS. At the same time, human participation in process flows is minimized (the proverbial "human factor" is eliminated), the volume of automatic regulation of the process flow expands and its quality improves, providing reliable protection of equipment from all kinds of accidents, simplifying the analysis of emergency situations, improving the personnel labor quality, etc. The APCS is required to be pre-designed.

This article discusses an intelligent CAD (Computer Aided Design) system for the designing of automated process control systems. The basis of this CAD is a cloud of information support in the form of a set of relational databases of engineering information necessary for the efficient and error-free operation of the designers of the automated process control system. An essential part of the information support should be the programs of processing information that appears on the drawings, diagrams and tables of the APCS projects. They contain a general part, applicable to all types of drawings, and an individual component, applicable only to drawings, diagrams and tables of a particular type.

As practice shows, the need to develop information assurance to support CAD systems is extremely high. The available graphic programs, such as AutoCAD, AutoCAD-electrical, Compass, CAD-Alpha, E-Plan, and others do not fully satisfy the needs of designers, as they solve only some part of the APCS design tasks and leave too much manual work as a consequence.

IOP Conf. Series: Materials Science and Engineering **760** (2020) 012051 doi:10.1088/1757-899X/760/1/012051

2. Distinctive features of the design of APCS

The design of automated process control systems is different from the design of machines, mechanisms, buildings and structures:

1. The graphic part of the APCS project consists mainly of a set of flat (2D) drawings and diagrams with its specific set of graphic elements connected by lines and arrows of various types.
2. Quite a number of alphanumeric designations - by orders of magnitude more than in the schemes of machines and mechanisms, buildings and structures - are plotted to the drawings.
3. The project of the automated process control system has a tabular part, the amount of which exceeds 50% of the total number of project volumes. The information contained in the tabular part is complexly connected with that which is put on the drawings.

This article discusses the development of intelligent CAD for the automated process control system. A CAD system like this should contain a project management program, a 2D drawing program, a relational database, a generator of output forms and a significant amount of information management and processing programs.

The methodological aspects of developing a database to support CAD for the automated process control system, as well as the specific content of the CAD information support are elaborated in [1,2,3,4]. It is shown that, conceptually, works in the field of databases for CAD systems have a direct link to the problem of artificial intelligence and have not been implemented earlier [5, 6].

3. An approach based on engineering information databases

The result of the work is a new type of CAD systems for the APCS design. The graphic engine of such a CAD system is immersed in a cloud of engineering information databases arising in a specific APCS project and accumulated from various projects. Work with the CAD is performed in a general purpose program through a special interface. Engineering information in such a CAD system is processed using "artificial intelligence" technologies, i.e. special well-considered design algorithms that implement a modular or object-oriented approach to the formation of the APCS structure.

An intelligent CAD for designing automated process control systems will make it possible to:

1. Reduce the time to design systems of APCS class by three or four times.
2. Reduce the number of engineering and technical personnel engaged in theAPCS design in design organizations.
3. Reduce the time spent on the APCS project adjustment, project checking and finding errors up to 10 times.
4. Minimize the number of errors in the APCS project arising due to the "human factor".
5. Semiautomatically sort and place in the database external engineering and economic information with minimal involvement of the designer's brain costs.
6. Automatically find and correct errors in numerous codings and markings of signals, equipment, cable cores, terminal arrangements, etc.
7. To give hints to the designer during the collection of primary information and block possible errors.
8. Automatically generate output design forms, complex and bulky cable schedules, cable records, tables of connections of cable cores to terminal arrangements, etc.
9. Automatically generate a field information database for broadcasting to a SCADA system.

The developed CAD system contains several software modules: AutoCAD module, general purpose C ++ program module, Access database module, Word module, AutoCAD interaction module (VBA, VB.NET macros). We purposefully chose the AutoCAD-14 module for implementation, since it has a powerful and well-developed graphics system. Naturally, the entire graphic service of AutoCAD-14 will be available to the user.

A CAD system for automated process control systems must be intelligent. CAD for APCS belongs to the category of "intelligent support" for computer systems. Its development includes: system-wide solutions, information support, mathematical support, programming support, organizational support.

In the "System-wide solutions" section, the general CAD system software architecture, the general-purpose program architecture, and the AutoCAD add-in architecture are developed.

The information support of CAD system includes the development of: common interfaces, specialized interfaces, information structure of the database, structure of reference information.

The mathematical support may contain (a) a description of the mathematical and logical formulas used by the program, (b) an algorithmic part that describes the logic of the operation of various software modules. In the case of CAD system, the algorithmic part is very extensive. Without the development of algorithmic support, the CAD programming support cannot be written. In the section of "algorithmic support" a set of algorithms is developed, on the basis of which the CAD system works. It includes the following algorithms:

1. construction of project document identifiers, technological installation paths, field devices for the automated process control system, electric signals for field devices, and other APCS elements;
2. operation of AutoCAD built-in windows intended for the construction of paths, the application of field devices to circuits, the construction of electrical circuits;
3. automatic formation of frames and stamps for documents; algorithms for the formation of special tables for schematic electrical diagrams and processing of these tables;
4. processing of electrical circuits necessary to switch to the external wiring diagrams; distribution algorithms for field devices in control cabinets, intermediate terminals and controllers (automation cabinets);
5. automatic generation of project documents of the second group (list of field devices, list of signals of hardware and software complex, cable schedules);
6. formation of reference catalogs;
7. other algorithms.

It is advisable to select the AutoCAD design system as the CAD graphics engine. This system has a powerful 2D graphics editor with advanced service. On the other hand, AutoCAD is essentially an advanced development environment, which makes it easy to write your own services and connect to a standard field. In the third, almost all design institutes and industrial enterprises leading innovative activity have the AutoCAD system.

The standard CAD system database contains the parameters of graphic elements, notes to them, text data for stamps (not always). Additional products (see, for example, AutoCAD-Electrical) reflect domestic designations, terminology and standards. With that, there are practically no products on the market that implement the distribution of devices and circuits for some units, cabinets, boxes and other similar devices. All this results in the absence of effective means of automatic control and taking into consideration the embedding structures and communication of units. Thus, the problem of supplementing the standard parameters of AutoCAD elements with a description of their location and switching with the infrastructure of the structure is considered relevant.

Description of additional parameters can be similarly performed and effectively implemented in relational databases. In these conditions, the standard SQL query language can be used to solve various table formation problems. AutoCAD provides ample opportunities to access tabular data [8], including those placed in databases [9].

Using SQL allows one to implement various queries to design data in a database management system (DBMS). One can use a variety of data formats to select a DBMS. MS Access DBMS is selected to implement this project. It provides the processing of the required data volumes for a large project of automated process control systems (within the limits of 1000 field devices). This DBMS is widely distributed and is part of the standard office software. The stem-forming technology provides storage of tabular data and running queries to them, and also allows the database to be included in the project as an additional, free file, avoiding the need to manage and maintain the SQL server.

As a result, in the work on the basis of the technology of additional databases the following labor-intensive non-standard tasks of automatic electrical design can be implemented within the framework of a modular (object) approach: Construction of tables for distribution of field devices over APCS cabinets; Construction of tables for distribution of field devices over the APCS subsystems;

IOP Conf. Series: Materials Science and Engineering **760** (2020) 012051 doi:10.1088/1757-899X/760/1/012051

Construction of tables for connecting cables to cabinets; Construction of cable project schedules for the system installation; Construction of a cable record; Construction of project specifications; Construction of the project register; Construction of lists of project field devices and their signals; Generation of engineering tables in the drawings; Providing reference tables.

Each of these technologies has its own advantages and disadvantages. Using Object-ARX allows one to write the most efficient and small-footprint code. However, in programs in C ++, one have to do manually a lot of things that are automated in higher level languages. As a result, the speed of development and especially debugging may fall unnecessarily. It makes sense to use this technology when resource-heavy calculations are expected, such as 3D modeling and non-standard analysis and data processing. In our case, this need arises when writing a general-purpose program. At that, one can use simpler approaches when interacting with AutoCAD.

Auto-LISP is the oldest mechanism for expanding the capabilities of AutoCAD [14], but this technology has not been actively developed lately. In particular, it is not very convenient to develop forms and other user interface mechanisms by using Auto-LISP. It is also hard enough to find qualified developers with knowledge of the LISP language, which can increase development and support costs.

The Visual Basic for Applications (VBA) technology is widely used to automate AutoCAD after Microsoft began using the same environment in its Microsoft Office suite. It compares favorably with Object-ARX by automatic memory management, user-friendly user interface creation and database access. The disadvantages of this technology are the lowest software code efficiency, as well as the recent trend of development of this environment, slowed down by both Autodesk and Microsoft [16].

The .NET technology, on the contrary, is actively developing. With virtually every version of Windows, new versions of the .NET Framework are released, while maintaining compatibility with previous versions. The convenience of creating a user interface and database access has only increased compared to VBA. The unique advantage of this environment is that the same program (the compiled code, not the source code) can be run on both 32-bit and 64-bit versions of AutoCAD [16]. The speed of the programs on .NET occupies an intermediate position between VBA and C ++. Considering the fact that it is the user interface and the work with the database that are critical in our task, the choice of .NET seems to be the most reasonable. This technology, unlike the others described, allows the developer to choose among several programming languages. The most common are VB.NET and C #, there are also less common options: F #, IronPython, IronRuby and others, but choosing them is worthwhile in rather specific cases. VB.NET and C # are almost equivalent in all characteristics, so one can leave the choice between them to the discretion of the developers.

One of the important technical issues that needed to be resolved to accomplish the task, is a method of permanent storage of information.

It is advisable to store the main body of information about the paths, devices, etc. in the database in terms of ease of access. However, it is required to associate this information with the elements of the drawing. Two options are possible:

1. Links to the drawing objects are kept in the database.
2. The AutoCAD document stores information that allows one to link drawing objects and records to the database.

Drawing objects can be uniquely defined by an identifier — a 64-bit integer. To implement the first option, it would be enough to store identifiers of related objects in the database. However, there is a possibility that the user will edit the drawing without downloading our extensions. At the same time, he may, inadvertently as well, perform actions that change the identifiers of objects — for example, cut out a fragment of a drawing and paste it onto another page.

In this regard, the second option seems to be more reliable: the drawing stores information about which element of the drawing corresponds to which record in the database. AutoCAD presents two mechanisms for storing this kind of information: XCODE and XRECORD.XCODE is an older mechanism that is not recommended for use in new projects due to limitations on the amount of stored data [17]. Working with XRECORD entries is somewhat more complicated than with XDATA, but safer due to the absence of the above limitations. Additionally, in the XRECORD record it is possible to

IOP Conf. Series: Materials Science and Engineering **760** (2020) 012051 doi:10.1088/1757-899X/760/1/012051

duplicate the main part of the information stored in the database. This will be the key to preventing unnecessary access to the database while working with the drawing.

Report generation is an important task for software developers. There is both specialized software for performing design and report generation tasks, as well as utilities, and libraries included in other software and systems (for example, in the DBMS). Examples of such generators are: Crystal Reports; Fast Report; Gen Rep; List and Labels; Active Reports; Stimul soft Reports.

The presented generators have powerful capabilities, in particular: a large amount of different data drivers; access to sources of heterogeneous data; xml support; visual report designer; experts and masters; drawing diagrams and schematics; custom templates; powerful formula language; export to various formats; customizable report viewers and more.

One of the main tasks of information systems is to promptly present information necessary for decision-making. At the same time, the structure of modern database management systems, on which information systems are based, is focused primarily on compact, secure and consistent storage of information, rather than on optimizing arbitrary sampling and data presentation. The main problem is not storing information, but providing it to the end user in the form of a report in the necessary context.

A special module - a report generator - is responsible for presenting user-friendly information in information systems. The report generator is a module (or a separate program) that allows one to present information in a readable structured form, in other words, to make information from the data (document, report, table), which can be printed or saved in various electronic formats.

Taken together, all of the above disadvantages resulted in the decision to develop our own generator, designed to create reports of the information system for the automated process control system. Reports are created in the form of MSWord documents in accordance with the assigned technical requirements: requirement on the formation of the project title page; requirement on the formation of a statement of the project; requirement on the formation of a list of paths; requirement on the formation of the list of equipment; requirement on the formation of lists of signals.

The generator is based on Microsoft Visual Studio, Visual Basic. NET. ADO Net technology is used for data connection. Connection to various databases is possible. Due to the fact that the APCS uses the Microsoft Access database, all queries are created statically “inside” the generator. The creation of two versions is envisaged: as an embedded library and as a separate software product.

4. Conclusion

This paper considers some issues of developing an intelligent CAD system to design an automated process control system in the transport sector. At the moment, there is no CAD system that satisfies the needs of designers in all respects. The main problem is the lack of advanced intelligent support to form the APCS project. Above, there are main problems that arise when creating a CAD system like this. These include: development of relational databases of project information, development of mathematical algorithms for designing automated process control systems, additional specific service to the graphic engine of CAD, IT technologies of generating report forms using databases, developing a general-purpose control program. These issues are addressed in the article.

Further promising scientific developments include:

the study of various options for the technical structures of the APCS and their formalization necessary for the best use of IT technologies, such as relational databases, a modular or object-oriented approach in the APCS design;

the analysis and further development of optimal algorithms for generating APCS design structures;

development of problems of artificial intelligence to search for and prevent errors of designers.

References

[1] Sachkov D I, Bratischenko V V, Arkhipova Z V 2013 Evaluating the effectiveness of information and telecommunications systems on the basis of the free software: a monograph. (Irkutsk: BSUEL Publ) p 150

IOP Conf. Series: Materials Science and Engineering **760** (2020) 012051 doi:10.1088/1757-899X/760/1/012051

[2] Sachkov D I, Arkhipova Z V, Bratischenko V V 2013 Modern information and telecommunication technologies in the management of socio-economic systems: a monograph. (Irkutsk: BSUEL Publ.) p. 196
[3] Database of electrical symbols for AutoCAD-Electrical, available at: http://www.iesoft.ru/cad-po-napravleniyam/elektrotehnika/baza-dannyh-elektrotehnicheskih-oboznachenii-dlya-autocad-electrical
[4] Cherkashin E A, Orlova I V Instrumen-tal tools for construction of the digital archives of the documents based on linked data *Modern Technologies. System Analysis. Modeling*. **56 4** 8–16.
[5] AutoCAD 2012 available at: http://www.autodesk.ru/adsk/servlet/item?siteID=871736&id=18221914#link7
[6] The C ++ language, advantages and comparison, a brief description. available at: http://www.cplusplus.com/info/description/
[7] Wohlkinger W, Vincze M, 2011 Ensemble of shape functions for 3d object classification, *IEEE International Conference on Robotics and Biomimetics* 2987-2992
[8] Lazebnik, Schmid C and Ponce J 2003 A sparse texture representation using affine-invariant regions", 2003 IEEE Computer Society *Conference on Computer Vision and Pattern Recognition* **2** II-319-II-324
[9] Rusu R B 2010 Semantic 3d object maps for everyday manipulation in human living environments *KI - Künstliche Intelligenz,* **24 4** 345-348
[10] The Qt library, description sections. available at: https://qt-project.org/doc/qt-5/overviews-main.html
[11] The Review of Delphi. available at: https://www.embarcadero.com/ru/products/delphi
[12] Description of the Visual Studio environment available at: http://msdn.microsoft.com/ru-ru/library/vstudio/bb514232%28v=vs.100%29.aspx
[13] A list of technologies for the empowerment of AutoCAD available at: http: //www.autodesk.ru/adsk/servlet/index? Id = 22740301 & siteID = 871736
[14] AutoLISP: available at: http://ru.wikipedia.org/wiki/AutoLISP#AutoLISP
[15] Konevtsov V A, Verteshev S M and Poletaev I A 2014 Eigenschaften von Complex SDSDC *European Science and Technology: 7th International scientific conference. Germany, Munich.* **I** 493-497
[16] Konevtsov V A, Poletaev I A 2013 Methods of Software Developing of Complex SDSDC *European Science and Technology: 4th International scientific conference. Germany, Munich* 377-380
[17] Zastrow D 2011 Automatisierung mit SPS. Theorie und Praxis.- Vieweg + Teubner, 5. Auflage, 870.
[18] Limited support for VBA in new versions of AutoCAD. available at: http://knowledge.autodesk.com/support/autocad/troubleshooting/caas/sfdcarticles/sfdcarticles/VBA-support-in-AutoCAD.html
[19] AutoCAD 2008 – 64-bit application migration. available at: http://through-the-interface.typepad.com/through_the_interface/2007/02/autocad_2008_64.html
[20] Comparison of the data storage mechanisms. available at: http://through-the-interface.typepad.com/ through_the_interface / 2007/04 / adding_xdata_to.html

IOP Conf. Series: Materials Science and Engineering **760** (2020) 012052 doi:10.1088/1757-899X/760/1/012052

Research on ecologically safe flokulants in clay fluid fluids

E V Averkina[1], **E V Shakirova**[1], **L A Butakova**[1]

[1]Irkutsk National Research Technical University, Irkutsk, Russian Federation

E-mail: viva160@mail.ru

Abstract. Currently, a large number of flocculants differing in properties and composition and working differently in various conditions is manufactured. Drilling fluids are complex systems differing in component composition, thus, it is not possible to recommend a universal flocculant. The present study carries out the analysis with the use of two methods with the objective to assess reagents' flocculating ability. Having analyzed the results obtained, the following conclusions can be drawn: the reagents BEN-EX, BENTOPLUS and DRB-9 can be used to increase the clay solution viscosity provided that the additives BENTOPLUS and DRB-9 do not exceed the concentration of 0.2 kg/m^3; in a concentration of up to 0.1 kg/m^3 flocculant SPECFLOC A 7950-20 can be used to increase a clay solution viscosity, at a concentration of 0.2 and 0.3 kg/m^3 the additive affects a drilling fluid mainly as a flocculant, thus, reducing the static shear stress of the solution and increasing filtrate flow. If dynamic shear stress is the only determining parameter in drilling with drilling fluid with SPECFLOC N6919 additive, then it can be recommended for being used. In all other cases additional research is required.

1. Introduction

Organic synthetic high molecular weight polymers called flocculants are used to accelerate and improve the separation of solid and liquid phases. When introduced into dispersed systems, these polymers are adsorbed or chemically bonded to the surface of the dispersed phase particles, combine the particles into agglomerates (flocculas) and facilitate their rapid deposition. A wide range of flocculants is presented due to the variety of drilling fluids. The flocculants have different composition and technical characteristics. The optimal type is selected depending on the properties of a suspension, which is to be cleaned, the cleaning method and the desired result. The number of flocculants entering the market is increasing every year. Various tests are carried out with the objective to select a flocculant suitable for the given conditions. In this case, it is advisable to conduct laboratory tests in order to limit the number of grades for further testing. The choice of a test method depends on the problem statement with regards to production.

Flocculants were tested on the basis of IRNITU, the research and development laboratory of drilling fluids and well fixing, for being used in clay wash fluids. The list of the applied chemicals is presented in table 1.

Table 1. Chemical reagents for comparative analysis

№	Chemical	Manufacturer	Chemical role	Recommended concentration, kg/m³
1	Bentonite clay powder	Kurgan	gelling additive	50.0
3	DRB-9	-	Selective flocculant and bentonite additive	0.5 – 1.5
4	BENTOPLUS	Lamberti	Synthetic polymer, bentonite additive	0.1 – 1.0
5	BEN-EX	M-I Drilling Fluids	Bentonite clay modifier	from 0.14
6	SPECFLOC N6919	Corporate group "Himalyans"	Polyacrylamide with a basic substance content of at least 90%, nonionic series	-
6	SPECFLOC A 7950-20		Polyacrylamide with a basic substance content of at least 90%, anionic series	-

Laboratory studies were carried out in a room with a temperature of 20 ± 2° C and a relative humidity of at least 50%.

2. Research Methodology

Analysis with the use of two methods was carried out with the objective to observe the effect of chemicals on the properties of the drilling fluid.

It was decided to add SPECFLOC N6919 and SPECFLOC A 7950-20 in the same amount as the rest reagents due to the fact that the accompanying documents for these chemicals do not contain recommendations for their introduction into the drilling fluid (see Table 1).

3. Methodology

Bentonite suspension preparation

1. Weigh 100 g of bentonite clay powder.
2. Pour two liters of distilled water into a glass and place it in a mixer.
3. Introduce the clay powder with a stirrer on. Continue mixing until the clay powder has completely dissolved.
4. The resulting suspension is divided into 4 equal parts.
5. Treat the samples of solution 2, 3, 4 with the test chemical while constant stirring on a laboratory stirrer in compliance with a concentration presented in table 2.
6. Measure the rheological parameters of the obtained samples of the solution (1 – 4), fill table 2 with the results. To ensure the results reliability, the drilling fluid was measured twice (due to the limited duration of the test) on each prepared sample of the solution. Table 2 presents the arithmetic mean values of the indicators.
7. Build a diagram of the dependence of the solution sample dynamic shear stress on the chemical's concentration.

Table 2 Research results

№	Solution composition, kg/m3	Chemical concentration, kg/m3	Rheological parameters			
			at 600 RPM, DR	at 300 RPM, DR	PV, cPs	Dynamic shear stress, daPA
1	Bentonite suspension	50.0	31,1	23,5	7,6	76,3
	DRB-9	0.1	30,7	24,1	6,6	84,0
		0.2	37,0	30,4	6,6	114,2
		0,3*	47,4	37,2	10,2	129,6
2	Bentonite suspension	50,0	30,1	22,8	7,3	74,4
	BENTOPLUS	0.1	34.1	26,5	7,6	90,7
		0.2	38,6	30,3	8,3	105,6
		0.3*	42,2	29,1	13,1	76,8
3	Bentonite suspension	50.0	31,8	24,0	7,8	77,8
	BEN-EX	0.1	35,2	27,2	8,0	92,2
		0.2	39,1	30,9	8,2	109,0
		0.3	43,9	34,1	9,8	116,6
4	Bentonite suspension	50.0	31,6	23,6	8,0	74,9
	SPECFLOC A 7950-20	0.1	42,3	35,7	6,6	139,7
		0.2*	67,7	58,9	8,8	240,5
		0.3*	26,8	19,0	7,8	53,8
5	Bentonite suspension	50.0	29,8	22,6	7,2	73,9
	SPECFLOC N6919	0.1	27,8	25,5	2,3	111,4
		0.2	32,7	28,9	3,8	120,5
		0.3	49,3	41,6	7,7	162,7

* **Note:** Solutions in selected concentrations were characterized by a separation of the suspension into two phases, flocculation.

The diagrams of the dependence of the sample solution dynamic shear stress on the concentration of the chemical were constructed with the objective to determine the optimal concentration of the additive to bentonite, at which the maximum value of the sample solution dynamic shear stress is achieved.

The results presented in Figure 1 and in Table 2 show that the value of a dynamic shear stress parameter is maximum for the SPECFLOC A 7950-20 reagent at a concentration of 0.2 kg/m^3. However, at this saturation, the solution was divided into phases, which can affect other parameters of the washing liquid. Thus, the maximum dynamic shear stress value has a solution with SPECFLOC N6919 additive at a concentration of 0.3 kg / m3.

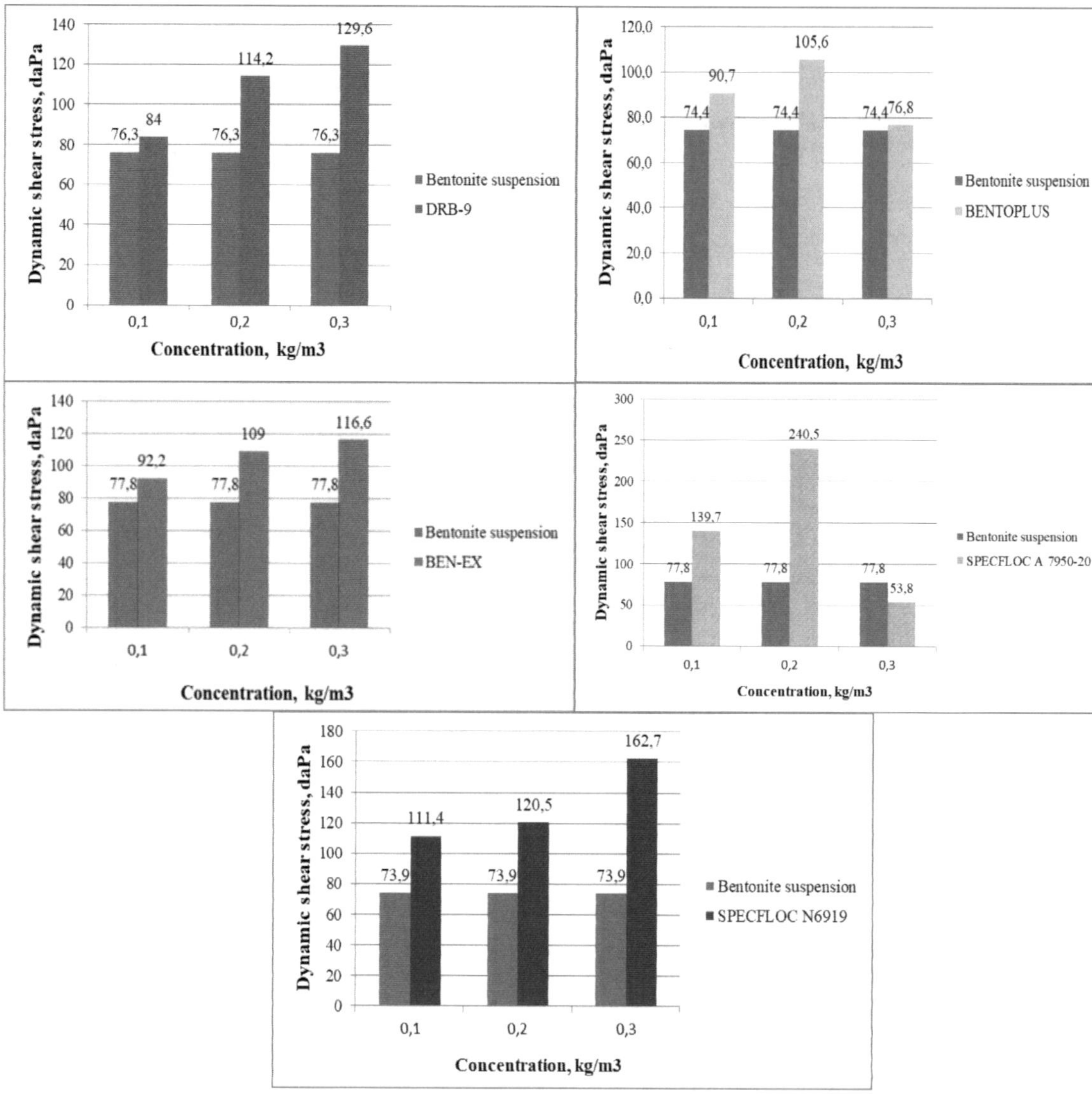

Fig. 1 Diagrams of the dependence of dynamic shear stress on the concentration of a chemical reagent in a bentonite suspension

In accordance with the Provisional regulations concerning the preparation and processing of drilling mud during the construction of production wells in Eastern Siberia, mud was mixed according to the formulation for drilling for direction and conductor (Table 3).

Table 3. Composition of initial solution

№	solution composition	Amount of chemical, kg/m^3
1	Sodium carbonate Na_2CO_3	1.5
2	Bentonite clay powder	50.0

Bentonite suspension preparation

1. Prepare all chemicals included in the drilling fluid.
2. Make a sample of each chemical in accordance with table 1.
3. Pour 800 cm^3 of distilled water into a measuring cup and place it under a mixer.
4. Turn on a stirrer at a stirring speed of 600 rpm.
5. Insert all components of the solution into a beaker while the stirrer is on. Stir it at constant speed.
6. Introduce sodium carbonate Na_2CO_3 to remove Ca^{2+} ions.
7. Introduce clay powder while vigorous stirring.
8. Continue mixing until a homogeneous suspension is obtained.
9. Divide the obtained solution into 4 equal parts.
10. Treat the samples of solution 2, 3, 4 with the test chemical while constant stirring on a laboratory stirrer in a concentration presented in table 4.
11. Maintain solution samples with stirring on a laboratory stirrer at low speeds for 2 hours.
12. Measure the parameters of the obtained samples of the solution (1–4); enter the results in table 4.
13. For the reliability of the results obtained, the drilling fluid was measured twice (due to the limited duration of the test) on each prepared sample of the solution. Table 5 presents the arithmetic mean values of indicators.

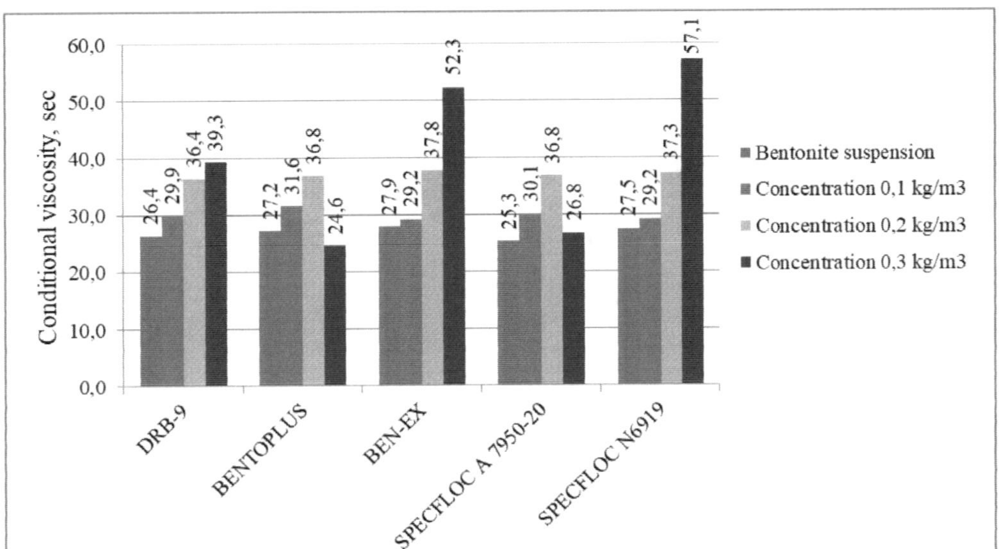

Fig. 2 Diagram of changes in conditional viscosity depending on the concentration of the chemical

IOP Conf. Series: Materials Science and Engineering **760** (2020) 012052 doi:10.1088/1757-899X/760/1/012052

Table 4 Research results

№	Solution composition	Chemical reagent concentration, kg/m³	Indicators							
			ρ, g/cm³	Assumed viscosity$_{700/500}$, sec.	PV, cPs	Dynamic shear stress, daPA	Gel static shear stress $_{10/10}$, aPA	F_{30} on API, sm³	Cake, mm	pH
1	base solution		1,026	26,4	6,8	87,0	75,0/127,5	20,0	2,0	10,66
	DRB-9	0,1	1,026	29,9	5,6	116,0	103,0/149,0	19,0	2,0	10,53
		0,2	1,026	36,4	7,9	136,0	91,0/85,0	21,5	2,0	10,64
		0,3	1,026	39,3	10,6	140,0	54,5/52,0	25,5	2,0	10,61
2	base solution		1,026	27,2	6,9	92,5	94,0/156,0	18,6	2,0	10,63
	BENTOPLUS	0,1	1,026	31,6	7,4	110,0	97,0/197,0	18,0	2,0	10,59
		0,2	1,026	36,8	8,4	121,0	75,0/78,0	19,8	2,0	10,62
		0,3	1,026	24,6	12,6	80,0	63,0/89,5	73,0	2,0	10,59
3	base solution		1,026	27,9	7,0	91,5	78,0/127,5	19,2	2,0	10,63
	BEN-EX	0,1	1,026	29,2	5,3	108,5	105,0/134,0	18,0	2,0	10,57
		0,2	1,026	37,8	7,1	129,0	101,0/104,0	17,0	2,0	10,70
		0,3	1,026	52,3	7,3	161,0	104,0/112,0	16,5	2,0	10,63
4	base solution		1,026	25,3	6,5	83,5	91,0/140,0	19,0	2,0	10,56
	SPECFLOC A 7950-20	0,1	1,026	30,1	7,5	116,5	93,0/101,0	17,5	2,0	10,66
		0,2	1,026	36,8	10,8	103,0	40,0/41,0	62,0	2,0	10,53
		0,3	1,026	26,8	6,7	24,5	18,0/36,0	46,0	2,0	10,64
5	base solution		1,026	27,5	6,9	88,0	83,0/132,0	19,0	2,0	10,66
	SPECFLOC N6919	0,1	1,026	29,2	4,9	119,0	101,0/155,0	18,0	2,0	10,64
		0,2	1,026	37,3	6,7	145,0	100,0/89,0	16,5	2,0	10,69
		0,3	1,026	57,1	7,9	184,0	79,0/87,0	18,5	2,0	10,65

Diagrams of the dependence of the parameters of the washing liquid on the concentration of the concentration of additives in the bentonite suspension were constructed on the basis of the results obtained (Fig. 2-7).

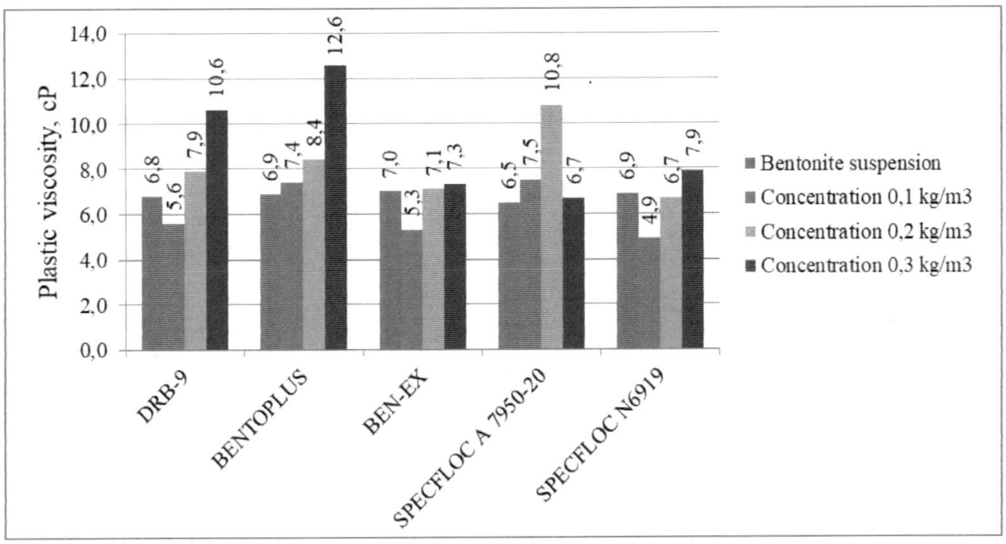

Fig. 3 Diagram of changes in plastic viscosity depending on the chemical concentration

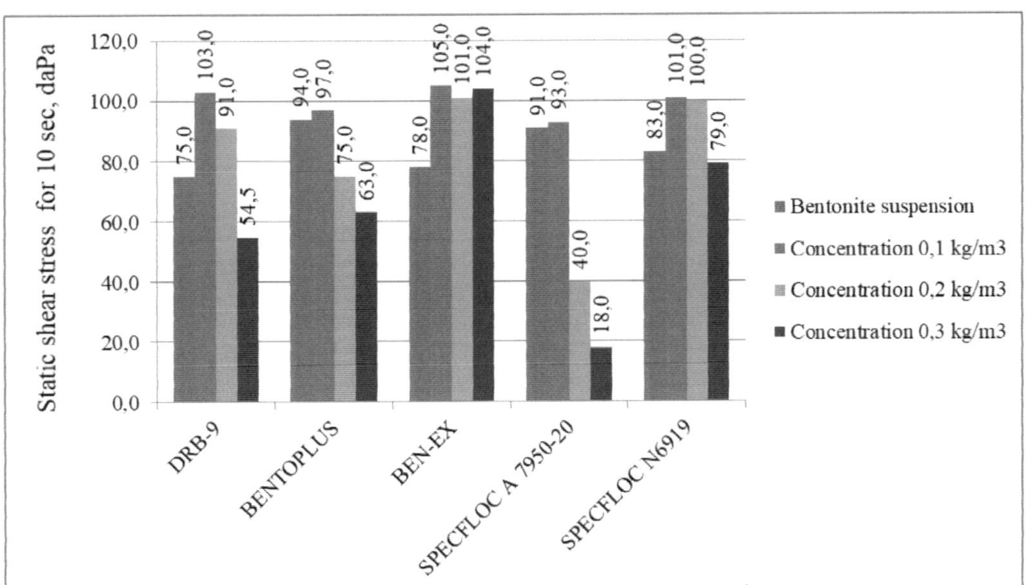

Fig. 4 Diagram of changes in static shear stress during 10 seconds depending on chemical concentration

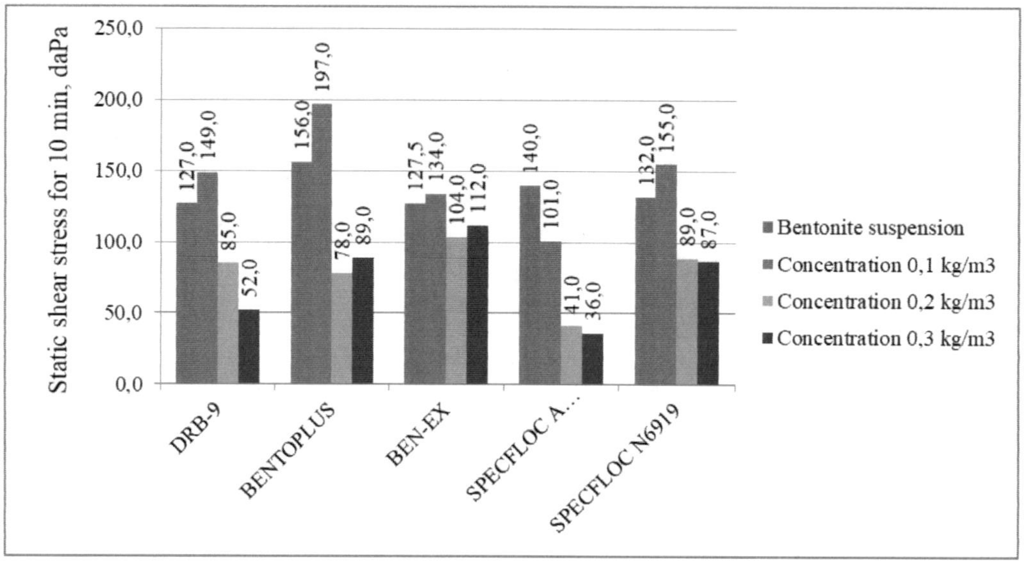

Fig. 5 Diagram of changes in static shear stress for 10 min depending on chemical concentration

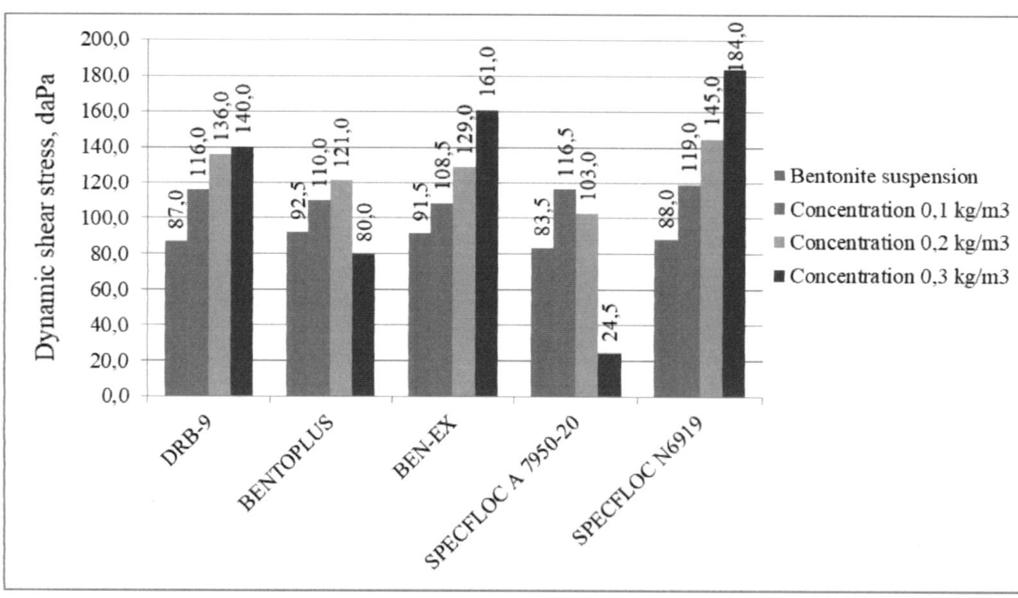

Fig. 6 Diagram of changes in dynamic shear stress depending on chemical concentration

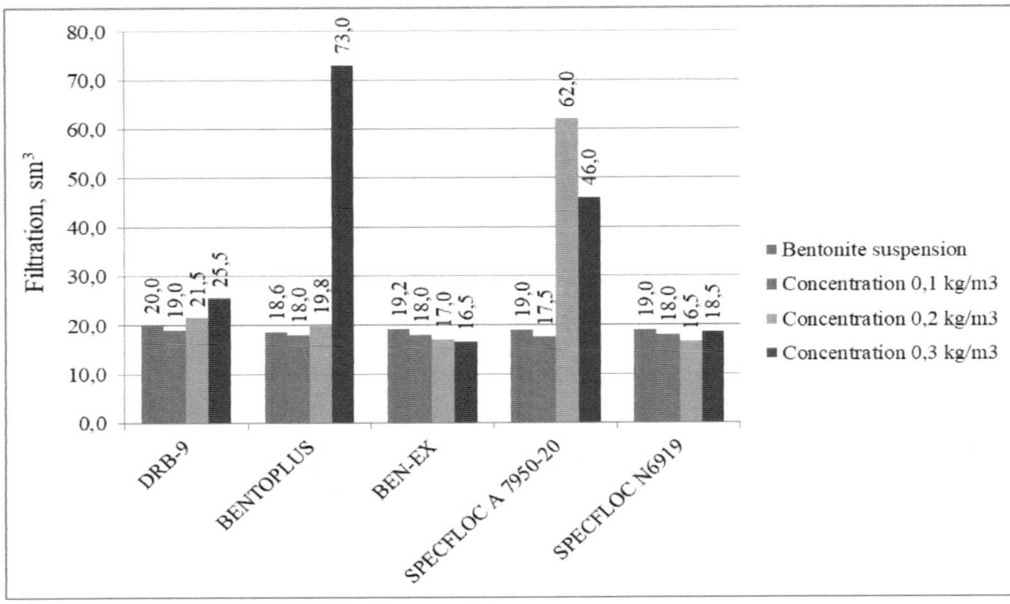

Fig. 7 Diagram of changes in filtration depending on chemical concentration

Laboratory tests enable to make the following conclusions:
1. All the chemicals provided do not affect the density and pH of the drilling fluid.
2. The main function of reagents BEN-EX and BENTOPLUS is to improve the technological properties of a bentonite suspension.
BEN-EX conforms to the properties declared in the accompanying documents. When it is added in various concentrations to a bentonite suspension, the rheological properties and conditional viscosity of the latter increase and the mud filtration slightly decreases (Fig. 2-7).
If taken at 0.1 and 0.2 kg/m^3 concentration, BENTOPLUS improves the properties of bentonite. If it was added in the amount of 0.3 kg/m^3, the solution began to separate into layers (photo 1), its rheological characteristics decreased, and the filtration increased sharply (see Fig. 7). Thus, it can be assumed that if the reagent is used to improve the technological properties of a washing liquid, it is not recommended to exceed a concentration of 0.2 kg/m^3.
3. Reagents SPECFLOC A 7950-20, SPECFLOC N6919 are flocculants with different functional groups, while DRB-9 is a selective flocculant. According to literature, a selective flocculant should flocculate cuttings, but not increase the viscosity of a bentonite solution. The test results show that when DRB-9 is added to the wash liquid, an increase in the viscosity at all concentrations is observed due to an increase in the size of the flocculas. With the addition of a larger amount, specifically, up to 0.3 kg/m^3, individual molecules form flocculas with an even larger size, which leads to an increase in the filtrate yield of the solution due to an increase in the amount of free water.
The flocculants SPECFLOC A 7950-20 and SPECFLOC N6919 at a concentration of 0.1 kg/m^3 have the same effect, that is, they increase the rheological and lower filtration properties of the washing liquid. With an increase in the concentration of the SPECFLOC A 7950-20 additive to 0.2 and 0.3 kg/m^3, the solution was stratified and a clarified liquid appeared on the surface (photo 2). At a concentration of 0.3 kg/m^3, the filtration increased sharply and the rheological properties decreased.
When SPECFLOC N6919 was added at a concentration of 0.2 and 0.3 kg/m^3, the conditional viscosity and dynamic shear stress increased. However, a decrease in the static shear stress of the washing liquid was observed. This may indicate that the reagent began to manifest itself as a flocculant. In addition, low values of static shear stress can lead to the fact that the solution will not cope with its most important functions of keeping the cuttings in suspension [9].
Important indicators of the drilling fluid are gel strength and bearing capacity, which are responsible for keeping the drilled particles in suspension and cleaning the borehole from the cuttings, respectively.

Conclusion

Thus, basing on the foregoing, it can be stated that the following are distinguishing characteristics of this type of bentonite clay powder:

1. Reagents BEN-EX, BENTOPLUS and DRB-9 can be used to increase viscosity of a clay solution, provided that the additives BENTOPLUS and DRB-9 do not exceed the concentration of 0.2 kg/m^3.
2. Flocculant SPECFLOC A 7950-20 can be used to increase the viscosity of a clay solution in a concentration of up to 0.1 kg/m^3, but only after additional studies aimed at obtaining a drilling fluid with a given set of technological parameters by varying the initial concentrations of the base reagents. At a concentration of 0.2 and 0.3 kg/m^3, the additive affects the drilling fluid mainly as a flocculant, reducing the static shear stress of the solution and increasing the filtration yield.
3. If dynamic shear stress is the only determining parameter in the process of drilling a drilling fluid with SPECFLOC N6919 additive, then it can be recommended for being used. In all other cases, additional research is needed.

References

[1]. Aksenov V I, Anikin Yu V, Galkin Yu A, Nichkova I I, Ushakova L I, 2008, *Application of flocculants in water management systems. Manual.* (Yekaterinburg, USTU - UPI).

[2]. Shakirova, E V., Averkina, E V., Sabirov, T R., 2016, Influence of additives on the characteristics of drilling mud used in drilling wells in Eastern Siberia *News of the Siberian Branch of the Earth Sciences Section of the Russian Academy of Natural Sciences. Geology, prospecting and exploration of ore deposits.* **3 (56)** 86–94.

[3]. Averkina E V, 2009, Analysis of rapperating wells in gas-condensate fields of the Irkutsk *region News of the Siberian Branch of the Earth Sciences Section of the Russian Academy of Natural Sciences. Geology, prospecting and exploration of ore deposits*, **2 (35),** 152-157.

[4]. Averkina E V, Shakirova E V, Fokin Yu V, 2017, Investigation of antifoam reagents in the composition of the drilling fluid *News of the Siberian Branch of the Earth Sciences Section of the Russian Academy of Natural Sciences. Geology, exploration and development of mineral deposits.* **40, 3 (60),** 90-98.

[5]. Shakirova E V, Averkina, E V, Sabirov, T R, Peryshkina, K O, 2018, Use of oil as a lubricant additive in drilling mud (on the example of the Yarakta oil and gas condensate field) *Oil and Gas Business,* **16. 2.** 12-19.

[6]. Lambin A I, Ivanishin V M, Siraev P U, Averkina E V, Shakirova E V, Korotkov A V, 2015, Study of the effect of the composition of emulsion drilling fluids on their performance *News of the Siberian Branch of the Earth Sciences Section of the Russian Academy of Natural Sciences. Geology, prospecting and exploration of ore deposits*, **4 (53),** 58–66.

[7]. Makhmudova M M, Koroleva A M, Shakirova E V, Efimenko E L, 2016, Investment attractiveness of the Northern region *News of higher educational institutions. Sociology. Economy. Politics.* **1 (48),** 27-32.

[8]. Khairullin V A, Zentsov V N, Shakirova E V, 2015, Evaluation of the economic effect of the introduction of new constructive solutions in construction *News of Higher Educational Institutions: Sociology. Economy. Politics*, **1 (44),** 75-79.

[9]. Kyazimov E A, Suleymanov A B, 2016, Investigation of inhibitory properties of drilling fluids with copper nanoparticles *SOCAR Proceedings* **2,** 11-15.

IOP Conf. Series: Materials Science and Engineering **760** (2020) 012053 doi:10.1088/1757-899X/760/1/012053

Analytical design of control system mathematical models for mobile robots based on the methods of inverse problems of dynamics and modal PID controllers

V N Sizykh[1], S B Antoshkin[1], R A Daneev[2], M V Bakanov[1], A V Livshits[1] and A A Aleksandrov [1]

[1]Irkutsk State Transport University, 15, Chernyshevskogo str., Irkutsk, 664074, Russia
[2]East Siberian Institute of the Ministry of Internal Affairs of Russia, 110 Lermontov Str, Irkutsk, 664074, Russia

Abstract: Mathematical models of a three-wheeled mobile robot based on the variable state apparatus and in the operator's form are presented. Based on these mathematical models, the synthesis of its adaptive control system is carried out according to the method of constructing Ziegler-Nichols PID controllers, as well as modal PID controllers. The design method of proportional controllers with double differentiation (PDD) of the autonomous robot is justified. The rules of wheel pair control are synthesized on the basis of reference models using a simple gradient scheme of the inverse problems method of dynamics in the formulation of P. D. Krutko for the problem of stabilizing (blanking) the angular velocities for a three-wheeled mobile robot. Simulink-models of robot movement are developed for cases with the use of PID-controllers which are adjusted by the Ziegler-Nichols method at synchronous deviation of wheels. Simulink-models of robot motion have been developed for cases using PID controllers set up according to the Ziegler-Nichols method for synchronous wheel deflection, modal and real PID controllers for differential wheel deflection, as well as double differentiation controllers. The analysis of modeling results is considered.

1. Introduction

There are hardly more scientific publications in the field of the theory and practice of automatic control as compared to those devoted to PID controllers [1, 2] and various correction devices. They are the most widely used industrial controllers as they can provide high control efficiency despite the different dynamic characteristics of an object. In the world about 90 ... 95% of controllers [3, 4], which are currently in operation, use PID algorithms. Nevertheless, the number of journal publications on the theory of PID controllers constantly increases [5-10]. The reason for this is the fact that the problem of synthesizing even ordinary industrial controllers still hasn't been solved at the level of modern requirements.

At the same time, it is desirable to have any regular procedure for the synthesis of an automatic control system that would allow, given the known structure of the control algorithm, obtain specific laws for changing the parameters of the PID controller or a synthesized correction device based on the results of an experimental study of the time characteristics of an object's free movement. The most formalized procedures for the synthesis of PID controllers and their comparative analysis on the example of solving a navigation problem (the method of inverse problems of dynamics, the criteria of the optimal module and symmetric optimum, algebraic method) are presented in the works [11, 12]. Modern methods of tuning the parameters of PID controllers based on microcontrollers and ADCs are carried out according to the Ziegler-Nichols and Chien-Hrones-Reswick methods [13, 14], but have a number of significant disadvantages [1, 3, 4].

Despite the long history of development and a large number of publications on the theory of PID controllers, there are still problems of objects control with hysteresis and nonlinearities, automatic tuning and adaptation under the conditions of high-frequency noise. The problems are complicated by the fact that the dynamics of the object is often unknown. The object is multi-linked itself (there are cross-links between control loops) and measurements are very noisy (filtering of signals is necessary).

In the article on the example of object with one input and one output (e.g., technological modules in the APCS [15-18] and a mobile robot with a synchronous deviation of wheels) is studied modal method of determining the PID controller parameters based on the principle of dynamic compensation [1, 19].It is required to determine the parameters of the PID controller according to the experimental characteristics of the free movement of the object so that the characteristics of the system closed by the controller correspond to the desired properties of the aperiodic link with the minimum control time. It is assumed that the control object does not contain any non-reducible zeros.

On the example of a multidimensional multiply-connected system - a mobile three-wheeled robot using the inverse problems method of dynamics (REM), a double differentiation controller (PDD controller) is investigated.

Thus, the task is posed to search for the parameters of industrial controllers that are self-adjusting according to the operating modes of a dynamic object. The main idea of this research is design the controller in a way when Modal control zeros approximately corresponded to poles of the approximated object with first and/or second-order polynomials, and at control on REM the trajectory of movement of the robot would precisely follow in the course of time on the appointed trajectory defined by the right parts of reference models (RM).

2. Dynamic model of the mobile robot

To describe the behavior of a mobile robot, it is necessary to develop its mathematical model. This will allow to analyze the work of the control system being created without hardware implementation of the mobile robot and it will allow to make necessary corrections. The mathematical model of the mobile robot consists of kinematic and dynamic models. The kinematic model is the simplest description of the mobile robot's behavior and it allows studying its properties. The dynamic model is a more detailed description of the mobile robot and it takes into account the force moment-related effects produced by the actuators.

The following mathematical description of the mobile robot movement is valid for the case of the robot movement on the horizontal plane.

In order to describe the mobile robot mathematically, it is necessary to introduce a number of conditions formulated in the form of the following assumptions [20]:

- mobile robot will be considered under the condition that its mechanism is tightly coupled;
- wheels are non-deformable and they are in point contact with the surface;
- robot moves without slippage;
- platform is considered as a solid body on which the wheel system is mounted.

Taking into account the accepted assumptions, the nonlinear mathematical model of the autonomous mobile robot is described by the system of differential equations:

$$\begin{aligned} \dot{x} &= V\cos\varphi \\ \dot{y} &= V\sin\varphi \\ \dot{\varphi} &= \omega \\ \dot{V} &= \frac{1}{\rho m}(\tau_R + \tau_L) \\ \dot{\omega} &= \frac{1}{2\rho J_C}(\tau_R - \tau_L), \end{aligned} \tag{1}$$

here x, y, φ – generalized linear and angular coordinates of the platform position; V, ω – Linear and angular velocities of the platform; m – mass; ρ – two-wheel drive radius; J_C – the torque of inertia of the platform relative to the center of mass; τ_R, τ_L – Electromagnetic moments produced by electric motors (EM).

Further, the equations of motion of the mobile robot are determined taking into account the dynamics of the servo drives.

Consider the case when the wheels of the platform through the gears are directly controlled by DC motors with independent excitation.

Suppose that the account of presence of following drives does not involve principal complexities.

It is known that the equation of voltage balance of one electric motor (EM) of a direct current made according to the II law of Kirchhoff looks like

$$L\dot{I} + RI + K_W \omega_W = U \tag{2}$$

here: где L – the inductance of the anchor circuit; I – the current flowing through the anchor winding; R – armature resistance; ω_W – angular velocity of the EM shaft; U – armature voltage.

The angular rotation speeds of the ED shaft ω_m and the wheels are related by the ratio

$$\omega_m = \omega_W i_p, \tag{3}$$

here i_p – the gear ratios of the gearboxes.

The expression for the torque τ developed by each EM is

$$\tau = K_M I. \tag{4}$$

here K_M, K_W – the given design constants of EM on the electromagnetic moment and its angular speed of rotation.

Taking into account relations (3), (4) equation (2) takes the form

$$\dot{\tau} = -\frac{R}{L}\tau - \frac{K_M K_W i_p}{L}\omega_W + \frac{K_M}{L}U. \tag{5}$$

Since two DC EMs are used separately for the right and left wheels:

$$\begin{aligned} \dot{\tau}_L &= -\frac{R}{L}\tau_L - \frac{K_M K_W i}{L}\omega_L + \frac{K_M}{L}U_L, \\ \dot{\tau}_R &= -\frac{R}{L}\tau_R - \frac{K_M K_W i}{L}\omega_R + \frac{K_M}{L}U_R, \end{aligned} \tag{6}$$

here U_L, U_R – Voltages to the electric circuits of the right and the left wheel.

Now let's express angular velocities of rotation of wheels ω_L и ω_R through phase coordinates. For this purpose, the following relationships are used

$$\begin{aligned} \omega &= \frac{1}{l}(V_R - V_L), \\ V &= \frac{1}{2}(V_R + V_L), \end{aligned} \tag{7}$$

here V_L, V_R – linear speeds of the right and left wheels; l – axle length between wheels.

Given the fact that $V_L = \omega_L \rho$ и $V_R = \omega_R \rho$, it turns out

$$\begin{aligned} \omega_R &= \frac{1}{\rho}\left(V + \frac{\omega l}{2}\right), \\ \omega_L &= \frac{1}{\rho}\left(V - \frac{\omega l}{2}\right). \end{aligned} \tag{8}$$

After substitution of equations (8) in equations (6), there determine the complete mathematical model of mobile robot dynamics taking into account the following drives

$$
\begin{aligned}
\dot{x} &= V\cos\varphi, \\
\dot{y} &= V\sin\varphi, \\
\dot{\varphi} &= \omega, \\
\dot{V} &= \frac{1}{\rho m}(\tau_R + \tau_L), \\
\dot{\omega} &= \frac{1}{2\rho J_C}(\tau_R - \tau_L), \\
\dot{\tau}_L &= -\frac{R}{L}\tau_L - \frac{K_M K_W i}{\rho L}\left(V - \frac{\omega l}{2}\right) + \frac{K_M}{L}U_L, \\
\dot{\tau}_R &= -\frac{R}{L}\tau_R - \frac{K_M K_W i}{\rho L}\left(V + \frac{\omega l}{2}\right) + \frac{K_M}{L}U_R.
\end{aligned}
\qquad (9)
$$

The model of dynamics (9) represents a system of nonlinear differential equations of the seventh order with a phase vector $(x, y, V, \varphi, \omega, \tau_L, \tau_R)$ and a control vector. (U_L, U_R). It can be used directly when modeling in any algorithmic programming language. However, to synthesize the laws of robot control other forms of representation of its mathematical model are necessary.

Note that in the system of equations (9) only the last four equations are dependent on each other, and the first three equations are related equations. Differentiating the sixth and seventh equations in system (9) and taking into account the substitution of the fourth and fifth equations in them,

$$
\begin{aligned}
\ddot{\tau}_L + a_{11}\dot{\tau}_L + \left(a_{12}\frac{1}{\rho m} - \frac{l^2}{4\rho J_C}\right)\tau_R + \left(a_{12}\frac{1}{\rho m} + \frac{l^2}{4\rho J_C}\right)\tau_L = b_{11}\dot{U}_L, \\
\ddot{\tau}_R + a_{11}\dot{\tau}_R + \left(a_{12}\frac{1}{\rho m} + \frac{l^2}{4\rho J_C}\right)\tau_R + \left(a_{12}\frac{1}{\rho m} - \frac{l^2}{4\rho J_C}\right)\tau_L = b_{11}\dot{U}_R,
\end{aligned}
\qquad (10)
$$

denote: $a_{11} = \frac{R}{L}$, $a_{12} = \frac{K_M K_W i_p}{\rho L}$, $b_{11} = \frac{K_M}{L}$, $\dot{U}_L, \dot{U}_R$ – supply voltage change rate (control parameter).

Design parameters of a tricycle autonomous mobile robot (Figure 1) are presented in Table 1.

Figure 1. Robot visual appearance

Table 1. Constructive parameters of the mobile robot

Parameter	J_C	ρ	m	l	K_W	R	L	i_p
Units of measure	$kg \cdot cm^2$	m	kg	m	-	ohm	H	-
Numerical value	$25 \cdot 10^{-5}$	0.0325	1	0.135	0.0080	0.36	$1.23 \cdot 10^{-3}$	55.74

The robot's mathematical model (10) is taken as a control object. The matrix equation system (10) looks like

$$\ddot{X} + C\dot{X} + DX = Bu, \tag{11}$$

here $X = (\tau_L, \tau_R)$ – system condition vector; $u = (\dot{U}_L, \dot{U}_R)$ – control vector; $C = \begin{bmatrix} a_{11} & 0 \\ 0 & a_{11} \end{bmatrix} = diag\{a_{11}, a_{11}\}$ – diagonal matrix; $B = \begin{bmatrix} b_{11} & 0 \\ 0 & b_{11} \end{bmatrix} = diag\{b_{11}, b_{11}\}$ – diagonal control efficiency matrix; $D = \begin{bmatrix} a_{12}\dfrac{1}{\rho m} + \dfrac{l^2}{4\rho J_C} & a_{12}\dfrac{1}{\rho m} - \dfrac{l^2}{4\rho J_C} \\ a_{12}\dfrac{1}{\rho m} - \dfrac{l^2}{4\rho J_C} & a_{12}\dfrac{1}{\rho m} + \dfrac{l^2}{4\rho J_C} \end{bmatrix}$ – symmetrical matrix $(D^{-1} = D^T)$;

Thus, the mobile robot as a control object is a multi-linked (number of inputs $m = 2$), multi-dimensional ($n = 2$) system.

3. Mobile robot movement control using PID control method

To apply the technique of PID controller synthesis and Simulink-models construction in Matlab programming environment, it is necessary to bring the robot mathematical model (10) to the operator's view.

a) Common case of a robot operator's model

Suppose that in the initial model (10) the condition of synchronization of wheel movement is satisfied: $\tau_R = \tau_L$. Then the angular velocity of the robot platform is equal to given one $\omega = \omega_{giv} = const$, and the angle of rotation of the wheels is equal to $\varphi = \omega t$. This condition is met if the equations (10) of the robot model include

$$d_{12} = a_{12}\frac{1}{\rho m} - \frac{l^2}{4\rho J_C} = 0.$$

There are two possible approaches to robot design:

1) Standard DC EMs are used. Then it is advisable to select the axis length of the robot's front drive wheels from the synchronization condition

$$l = \sqrt{\frac{4K_M^2 i_p J_C}{\rho L m}}.$$

2) The axle length of the front drive wheels l is rigid and you need to select the DC EMs characteristics. Then the winding factor and the design parameters of the EM are selected from the synchronization condition

$$K_M = K_W = \frac{l^2}{4 i_p J_C}.$$

In both cases, the equations in the model (10) can be considered independently $\tau = \tau_R = \tau_L$

$$\ddot{\tau} + a_{11}\dot{\tau} + \left(a_{12}\frac{1}{\rho m} + \frac{l^2}{4\rho J_C} \right)\tau = b_{11}\dot{U},$$

here $\dot{U}_R = \dot{U}_L = \dot{U}$.

Further, the last equation through the change of variables is reduced to the Cauchy normal form $\tau = x_1, \dot{\tau} = x_2$

$$\dot{x}_1 = x_2,$$

$$\dot{x}_2 = -\left(a_{12}\frac{1}{\rho m} + \frac{l^2}{4\rho J_C} \right)x_1 - a_{11}x_2 + b_{11}\dot{U},$$

or in matrix form

$$\dot{x} = Ax + Du, \tag{12}$$

here $x = (x_1, x_2)$, $A = \begin{bmatrix} 0 & 1 \\ -\left(a_{12}\frac{1}{\rho m} + \frac{l^2}{4\rho J_C} \right) & -a_{11} \end{bmatrix}$, $B = (0, b_{11})$, $u = \dot{U}$.

Apply to equations (12) continuous Laplace transformation

$$A^*(p)x(p) = Bu(p), \tag{13}$$

here $A^*(p) = (Ip - A) = \begin{bmatrix} p & -1 \\ a_{12}\frac{1}{\rho m} + \frac{l^2}{4\rho J_C} & p + a_{11} \end{bmatrix}$, I – unit size matrix 2×2.

The solution of the system of algebraic operator equations is determined by Kramer equation

$$x_{ij}(p) = \frac{\Delta_{ij}}{\Delta(p)}u_j(p),\ i = \overline{1,n};\ j = \overline{1,m}.$$

In our case, the system is multidimensional ($n = 2$), but single-linked ($m = 1$), and its determinants are not equal to zero:

$$\Delta(p) = \begin{vmatrix} p & -1 \\ a_{12}\frac{1}{\rho m} + \frac{l^2}{4\rho J_C} & p + a_{11} \end{vmatrix} = p^2 + a_{11}p + a_{12}\frac{1}{\rho m} + \frac{l^2}{4\rho J_C} \neq 0.$$

Therefore, the system has the only solution ((the matrix $A^*(p)$ is undegenerate).

Private determinants and transfer functions are equal:

$$\Delta_{11}(p) = \begin{vmatrix} 0 & -1 \\ b_{11} & p + a_{11} \end{vmatrix} = b_{11},\ \Delta_{21}(p) = \begin{vmatrix} p & 0 \\ a_{12}\frac{1}{\rho m} + \frac{l^2}{4\rho J_C} & b_{11} \end{vmatrix} = pb_{11},$$

$$W_{11}(p) = W_{\frac{\tau}{\dot{U}}}(p) = \frac{b_{11}}{p^2 + a_{11}p + a_{12}\frac{1}{\rho m} + \frac{l^2}{4\rho J_C}},$$

$$W_{21}(p) = W_{\frac{\dot{\tau}}{\dot{U}}}(p) = pW_{11}(p) = pW_{\frac{\tau}{\dot{U}}}(p),$$

$$\dot{V} = \frac{2}{\rho m}\dot{\tau},$$

$$W_{\frac{\tau}{U}}(p)=\frac{\frac{2}{\rho m}b_{11}}{p^2+a_{11}p+a_{12}\frac{1}{\rho m}+\frac{l^2}{4\rho J_C}}.$$

b) General case of a robot operator's model

Similarly to the special case, the matrix equations (11) are reduced to normal Cauchy form by replacing the variables $x=x_1, \dot{x}=x_2$

$$\dot{x}_1=x_2,$$
$$\dot{x}_2=-Dx_1-Cx_2+bu.$$

here $x_1=(\tau_L,\tau_R)$, $x_2=(\dot{\tau}_L,\dot{\tau}_R)$.

In block form, the latest equations take the form

$$\dot{Y}=AY+Bu, \tag{14}$$

here $Y=(x_1,x_2)$ – generalized vector of state, size 4×1; $u=(U_L,U_R)$ – control vector, size 2×1; $A=\begin{bmatrix} O_{(2\times2)} & I_{(2\times2)} \\ -D_{(2\times2)} & -C_{(2\times2)} \end{bmatrix}$ – block matrix at state vector, size 4×4; $B=\begin{bmatrix} O_{(2\times2)} \\ b_{(2\times2)} \end{bmatrix}$ – block matrix at control vector, size 4×2.

Through the Laplace transform, the system of differential equations is transformed into a system of algebraic equations of the form

$$A^{**}(p)Y(p)=Bu(p),$$

here $A^{**}(p)=(Ip-A)=\begin{bmatrix} I_{(2\times2)}p & -I_{(2\times2)} \\ D_{(2\times2)} & I_{(2\times2)}p+C \end{bmatrix}$, $I_{(2\times2)}$ – single matrix.

As $\Delta(p)=\det A^{**}(p)\neq 0$, the system is non-degenerate and has a unique solution defined by the greatest common zero divisor (GCD) [21]

$$Y_{ij}(p)=\frac{\Delta_{ij}(p)GCD_j(p)}{\Delta(p)}u_j(p),\ i=\overline{1,n};\ j=\overline{1,m}.$$

In proposed case $n=4$, $m=2$. Therefore, the robot is a multivariate and multi-linked (multi-circuit) control object.

Here, the private identifier is equal to $\Delta_{ij}(p)=\det A_{ij}^{**}(p)$, and the matrix $A_{ij}^{**}(p)$ is determined by the substitution in the matrix $A^{**}(p)$ $i-\text{th}$ column on $j-\text{th}$ free-member column. The overall solution for m controlling impacts is obtained as a superposition of private solutions

$$x_i(p)=\sum_{j=1}^{m}x_{ij}(p)=\sum_{j=1}^{m}\frac{\Delta_{ij}(p)GCD_j(p)}{\Delta(p)}u_j=\sum_{j=1}^{m}W_{ij}(p)u_j,$$

here $GCD_j(p)$ – largest common denominator by $j-\text{th}$ control input.

Thus, there are generalized controls and matrix transfer functions of the numerator and denominator.

The transfer functions from the control actions to the moments of the right and the left wheel are determined by equations:

$$W_{\frac{\dot{\tau}_R}{\dot{U}_R}}(p)=\frac{\Delta_{11}GCD_1(p)}{\Delta(p)},$$

here $\Delta_{11}=b_{11}$;

$$GCD_1(p)=GCD_L(p)=p^2+a_{11}p+d_{11}-d_{12}=p^2+a_{11}p+\frac{l^2}{2\rho J_C}.$$

$$W_{\frac{\dot{\tau}_L}{\dot{U}_L}}(p)=\frac{\Delta_{12}GCD_2(p)}{\Delta(p)},$$

here $\Delta_{12}=b_{11}$;

$$GCD_2(p)=GCD_R(p)=p^2+a_{11}p+d_{11}+d_{12}=p^2+a_{11}p+\frac{2a_{12}}{2\rho m},$$

$$\Delta(p)=GCD_1(p)GCD_2(p)=\left(p^2+a_{11}p+d_{11}-d_{12}\right)\left(p^2+a_{11}p+d_{11}+d_{12}\right)=$$

$$=\left(p^2+\frac{R}{L}p+\frac{2K_W^2 i_p}{\rho^2 mL}\right)\left(p^2+\frac{R}{L}p+\frac{l^2}{2\rho J_C}\right).$$

$$W_{11}=W_{\frac{\dot{\tau}_L}{\dot{U}_L}}(p)=\frac{b_{11}}{p^2+a_{11}p+d_{11}-d_{12}}=\frac{b_{11}}{p^2+a_{11}p+\frac{l^2}{2\rho J_C}},$$

$$W_{12}(p)=pW_{11}(p).$$

$$W_{22}(p)=W_{\frac{\dot{\tau}_R}{\dot{U}_R}}(p)=\frac{b_{11}}{p^2+a_{11}p+d_{11}+d_{12}}=\frac{b_{11}}{p^2+a_{11}p+\frac{2K_W^2 i_p}{\rho^2 mL}},$$

$$W_{21}(p)=pW_{22}(p).$$

4. Modal control method for the second-order object

On the basis of PID controller transfer functions (TF) with ideal and real differentiating links on dominant characteristics of free movement of the object is enough simply defined its parameters. The method of designing PID controllers is applicable to systems of the first, second and higher orders on the basis of their approximation by systems of the first or second order with dominant poles

$$G_{PID}(s)=\frac{K_D\left[s^2+\frac{K_P}{K_D}s+\frac{K_I}{K_D}\right]}{s}=\frac{K_D\left[s^2+2\xi\omega_n s+\omega_n^2\right]}{s}, \tag{15}$$

$$G_{PID}=K_P\left(1+\frac{1}{T_I s}+\frac{T_D s}{1+\frac{T_D s}{N}}\right), \tag{16}$$

here $\omega_n^2=\frac{K_I}{K_D}$, $2\xi\omega_n=\frac{K_P}{K_D}$, constant N determines the gain K_{HF} of PID controller in the high frequency range. The gain K_{HF} should be limited because the noise measurement signal often contains high-frequency components. As a rule, the divider is selected in the range from 2 to 20 [14].

The TF of the controller (15) is determined after a constant time for the differential and integral components in the following form

$$G_{PID}=K_P\left(1+\frac{1}{T_I s}+T_D s\right). \tag{17}$$

Here $T_I = \frac{K_P}{K_I}$ – time constant of the integral component; $T_D = \frac{K_D}{K_P}$ – time constant of the differential component. That's why we have: $K_I = \frac{K_P}{T_I}$, $K_D = K_P T_D$.

It is required to determine the parameters of the PID controller according to the free movement characteristics of the reduced object ω_n, ξ so that the zeros of its PF approximately coincide with the poles of the PF of the reduced object, that is, that the equalities:

$$\omega_n^2 = \frac{K_I}{K_D},\ 2\xi\omega_n = \frac{K_P}{K_D},\ K_D = X_{\text{giv}}.$$

General standard form of description of second-order systems with a dominant pole in the form of PF oscillator link with a relative attenuation coefficient ξ and the frequency of the object's free oscillations ω_n is determined by the equation

$$G(s) = \frac{K\omega_n^2}{s^2 + 2\xi\omega_n s + \omega_n^2} \quad <=> \quad G_{PID}(s) = \frac{K_D\left[s^2 + 2\varsigma\omega_p s + \omega_p^2\right]}{s},$$

which is equivalent to the description of the PF PID controller by the expression (15). If the parameters of the damped oscillations are taken equal to the parameters of free oscillations: $\omega_p = \omega_n$ and $\varsigma = \xi$, then $G_{PID}(s)G(s) = \frac{KK_D\omega_n^2}{s}$, and the TF of the closed loop system will look like an aperiodic link $F(s) = \frac{1}{Ts+1}$ with time-constant $T = \frac{1}{KK_D\omega_n^2}$.

Then the parameters of the PID controller K_P, K_I, K_D can be found by the equation (15) through the relative attenuation coefficient ξ and the frequency of free oscillations ω_n of the object:

$$\omega_n^2 = \frac{K_I}{K_D} \ => \ K_D = \frac{K_I}{\omega_n^2}, \tag{18}$$

$$2\xi\omega_n = \frac{K_P}{K_D} \ => \ K_D = \frac{K_P}{2\xi\omega_n}. \tag{19}$$

Assuming in proportional component of the PID regulation law the gain equal to one ($K_P = 1$), from the expressions (18), (19) the gains K_I, K_D of the integral and differential components are determined:

$$\frac{K_P}{2\xi\omega_n} = \frac{K_I}{\omega_n^2} \ => \ \frac{1}{2\xi\omega_n} = \frac{K_I}{\omega_n^2} \ => \ K_I = \frac{\omega_n^2}{2\xi\omega_n} \ => \ K_I = \frac{\omega_n}{2\xi},$$
$$K_D = \frac{K_P}{2\xi\omega_n} \ => \ K_D = \frac{1}{2\xi\omega_n}. \tag{20}$$

However, calculation of PID controller coefficients by formulas (20) leads to initial overshooting and to the transition process tightening, because for a stable system the frequency of undamped (free) oscillations ω_n determines the number of these fluctuations and has a direct impact on the quality of transition processes. Therefore, in order to speed up the attenuation of processes and reduce overregulation, smoothing coefficient $\varepsilon = \frac{\xi}{\varsigma}$ and damping $\alpha = \frac{\omega_n}{\omega_p}$, with the help of which the parameters of the integral (coefficient K_I) and less often the differential (coefficient K_D) components of the PID controller are corrected.

Thus, all three components of the PID controller parameters can be selected through the free movement properties of the object:

$$K_P = 1\,,\; K_D = \alpha\frac{1}{2\xi\omega_n}\,,\; K_I = \varepsilon\frac{\omega_n}{2\xi}\,. \tag{21}$$

Researches have shown that the introduction of smoothing coefficient ε into the integral component allows to eliminate the oscillation of the transition process in the range of its change from 0.1 to 2. Further, increasing the value ε will reduce the time of regulation, but will lead to some oscillatory transient process. With the help of the setting damping coefficient α in the differential component the minimum time of process control is provided. The summary equations for calculations and the limits of the parameters of the PID controllers are shown in Table 2.

Similar formulas for calculating parameters of PID controllers through time constants T_D и T_I are obtained from the equations (16), (17):

$$T_I = \frac{K_P}{K_I} = \frac{2\xi K_P}{\varepsilon\omega_n} = \frac{2\xi}{\varepsilon\omega_n}\,,\; T_D = \frac{K_D}{K_P} = \frac{\alpha\frac{1}{2\xi\omega_n}}{K_P} = \frac{\alpha}{2\xi\omega_n} = K_D.$$

Table 2. Equations for calculation of PID controllers coefficients

Object		Parameters of PID controller					
		K_P	K_I	K_D	T_D	T_I	N
ξ	ω_n	1	$\frac{\omega_n}{2\xi}$	$\frac{1}{2\xi\omega_n}$	$\frac{1}{2\xi\omega_n}$	$\frac{2\xi}{\omega_n}$	$2\div 20$
Limits		1	$\varepsilon\frac{\omega_n}{2\xi}$ $\varepsilon = 0.1..2$	$\alpha\frac{1}{2\xi\omega_n}$ $\alpha = 0.58..1.5$	$\frac{\alpha}{2\xi\omega_n}$	$\frac{2\xi}{\varepsilon\omega_n}$	

5. Mobile robot control algorithm using the inverse dynamics problem method

Since in the equation system (9) the first three equations are coupling equations, it is sufficient to use the last four equations in formula (9) to study the robot motion dynamics.

By differentiating the sixth and seventh equations in system (9) and after substituting the fourth and fifth equations in them, equations (10) are obtained.

In the synthesis of control laws by the method of inverse problems of dynamics, the mathematical model of the robot (10) is taken as the control object

$$\begin{aligned}\ddot{\tau}_L + a_{11}\dot{\tau}_L + d_{11}\tau_L + d_{12}\tau_R &= b_{11}\ddot{U}_L\,,\\ \ddot{\tau}_R + a_{11}\dot{\tau}_R + d_{11}\tau_R + d_{12}\tau_L &= b_{11}\ddot{U}_R\,.\end{aligned} \tag{22}$$

Therefore, the mobile robot as a control object is a multi-linked ($m = 2$) and multidimensional ($n = 2$) system.

The problem of synthesizing controls for stabilizing (damping) the robot's angular velocities in the time interval $[t_0, t_k]$ (t_k – non-fixed value) is formulated as follows.

Suppose that the current time $t = t_0 = 0$ torques-controlled $\tau_1 = \tau_L$, $\tau_2 = \tau_R$ system state (22) is determined by the values of $\tau_i(0) = \tau_{i0}$, $i = 1, 2$.

It is necessary to define the control actions $\dot{U}_i(\dot{U}_1 = \dot{U}_L, \dot{U}_2 = \dot{U}_R)$, at which the system (22) switches from the state τ_{i0} to the new state τ_{igiv}, where $\tau_{igiv} = \tau_{igiv}(t)$ – the set torques of rotation of the system (22) must be defined. It is necessary that the moments of controlled motion on each degree of freedom with required accuracy followed the moments defined by the reference model

$$\ddot{X}_{\mathrm{ref}\,i} + C_i\dot{X}_{\mathrm{ref}\,i} + D_i X_{\mathrm{ref}\,i} = D_i\tau_{igiv}\,,\; i = 1, 2, .. \tag{23}$$

IOP Conf. Series: Materials Science and Engineering **760** (2020) 012053 doi:10.1088/1757-899X/760/1/012053

with the previously calculated coefficients C_i and D_i.

The specified change in supply voltage at the terminals of the armature winding of the right and left electric motor determines the specified reference movement

$$B_i \dot{U}_{i\,\mathrm{giv}} = D_i \tau_{i\mathrm{giv}},\; i = 1,2,3... \tag{24}$$

The PWM controller is used to change the supply voltage. The reference movement is stable $X_{\mathrm{ref}\,i} \to \tau_{i\mathrm{giv}}$ при $t \to \infty$.

Let's consider a reference model (23) in scalar form

$$\begin{aligned} &\ddot{\tau}_{\mathrm{ref}\,L} + a_{11}\dot{\tau}_{\mathrm{ref}\,L} + d_{11}\tau_{\mathrm{ref}\,L} = d_{11}\tau_{L\,\mathrm{giv}}, \\ &\ddot{\tau}_{\mathrm{ref}\,R} + a_{11}\dot{\tau}_{\mathrm{ref}\,R} + d_{11}\tau_{\mathrm{ref}\,R} = d_{11}\tau_{R\mathrm{giv}}. \end{aligned} \tag{25}$$

From equation (24) the change of supply voltage on the armature clamps of the right and left wheel winding in scalar form will be written as

$$\dot{U}_{L\mathrm{giv}} = \frac{d_{11}}{b_{11}}\tau_{L\mathrm{giv}}, \quad \dot{U}_{R\mathrm{giv}} = \frac{d_{11}}{b_{11}}\tau_{R\mathrm{giv}}. \tag{26}$$

By minimizing the functional

$$G(\dot{U}) = \frac{1}{2}\sum_{i=1}^{2}\left[\ddot{X}_{\mathrm{ref}\,i} - \ddot{X}_i(t,\dot{U}_i)\right]^2 \tag{27}$$

the degree of approximation of the controlled process to the reference model is assessed $x_{\mathrm{ref}\,i}(t) \to \tau_{i\mathrm{giv}}$.

Functional characterizes the change of the moment of rotation of the mobile robot wheels in the vicinity of the change of the standard moment of rotation.

The gradient method scheme defines the laws of control $\dot{U}_i(x)$ for each degree of freedom of rotation of wheels

$$\frac{\partial \dot{U}_i(X)}{\partial t} = \ddot{U}_i(x) = -\sum_{j=1}^{2} r_{ij}\frac{\partial G(\dot{U})}{\partial \dot{U}_j}, \quad r_{ij} = const. \tag{28}$$

From equations (23), (27), the gradient components of (28) are defined

$$\frac{\partial G(\dot{U})}{\partial U_J} = -b_{11}\left(\ddot{X}_{\mathrm{ref}\,j} - \ddot{X}_j\right), \; j = 1,2... \tag{29}$$

It follows that

$$\ddot{U}_i(x) = \sum_{j=1}^{2} r_{ij} b_{11}\left(\ddot{X}_{\mathrm{ref}\,j} - \ddot{X}_j\right). \tag{30}$$

The required control law (voltage regulation in the circuit of the armature winding of motor) is determined by integrating over time of both parts of the expression (30)

$$U_i(x) = \sum_{j=1}^{2} r_{ij} b_{11}\left(X_{\mathrm{ref}\,j} - X_j\right). \tag{31}$$

Here, the required torque values $x_{\mathrm{ref}\,j}$ are calculated using the equation (23).

The laws of control (31) of the robot's wheel set in scalar form look like:

$$\begin{aligned} U_L &= r_{11}b_{11}\left(\tau_{\mathrm{ref}\,L} - \tau_L\right) + r_{12}b_{11}\left(\tau_{\mathrm{ref}\,R} - \tau_R\right), \\ U_R &= r_{21}b_{11}\left(\tau_{\mathrm{ref}\,L} - \tau_L\right) + r_{22}b_{11}\left(\tau_{\mathrm{ref}\,R} - \tau_R\right), \end{aligned} \tag{32}$$

and in matrix form they define a known procedure of analytical design of controllers in the next form

$$U = -R\,B\Delta\tau, \tag{33}$$

here $\Delta\tau = \begin{bmatrix} \tau_L - \tau_{\text{ref}\,L} \\ \tau_R - \tau_{\text{ref}\,R} \end{bmatrix}$ – vector of change of moments developed by electric motors; $B = \text{diag}(b_{11}, b_{11})$ – diagonal matrix. The structure of matrix is determined by the type of cross-linkages in a variable x_j in equations (30) or (31), $R = \begin{bmatrix} r_{11} & r_{12} \\ r_{21} & r_{22} \end{bmatrix}$, moreover $r_{12} = r_{21}$.

Sustained $R = R^T > 0$. Then $G(\dot{U}) < 0$ and closed loop system (31) (22) asymptotically stable in Lyapunov $\ddot{X}_{\text{ref}j} \to \ddot{X}_j$ at $t \to \infty$.

6. Modeling and discussing

a) PID controllers tuned according to Ziegler-Nicols methodology

On the basis of mathematical model (9) and obtained transfer functions, Simulink-models of controlled mobile robot with parameters of table 1 were developed. For a particular case of the Simulink robot operator's model, the model is shown in Figure 2.

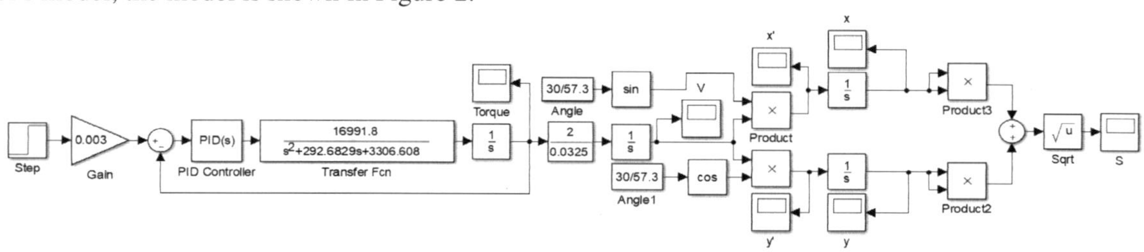

Figure 2. Simulink-model at synchronous deviation of mobile robot wheels

PID controller coefficients were selected according to the Ziegler-Nickols method: $K_P = 9.5681$, $K_I = 6.3948$, $K_D = 0.6863$. Wheel deflection angle $\varphi = 30°$.

The control torque τ transient graphs from time to time t are shown in Figure 3.

Speed coordinate change graph V of the robot and the robot's track S, depending on the time t, are shown in Figures 4 and 5.

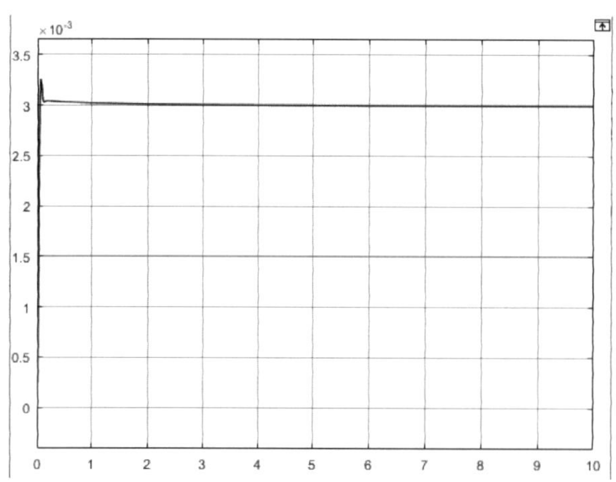

Figure 3. Control torque τ of the mobile robot's wheel set

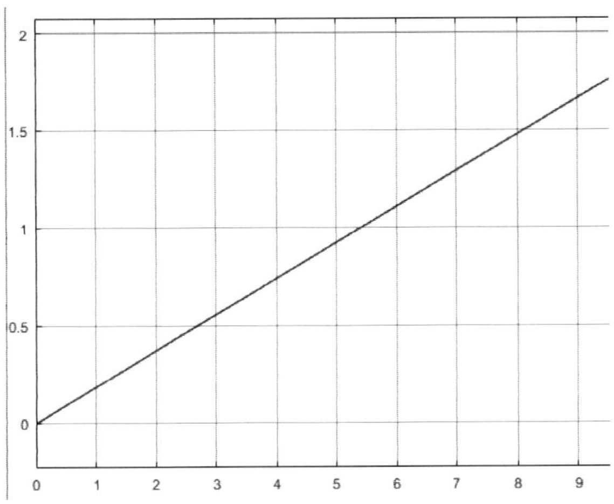

Figure 4. Changing the speed of the robot V depending on the time

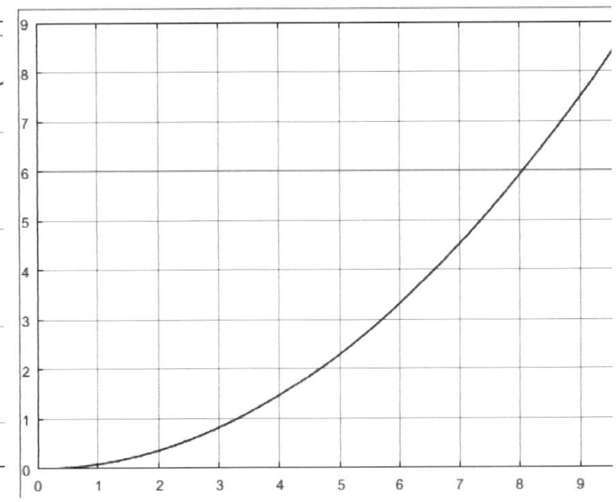

Figure 5. Changing the track S, passed by the robot depending on the time

On the basis of the above method of modal PID control Simulink-model (Figure 6) of three-wheeled robot as a multi-dimensional multi-linked object ($n = 4$, $m = 2$) for the cases of stabilization of moments with the help of ideal and real (with a real differentiating link) PID controllers was constructed. The stability of the control system was ensured by the introduction of external negative feedback on the angular velocity of the robot's platform, as its initial dynamic model contains two consecutively included integrative links (structural sign of instability).

The nominal robot platform angular velocity values specified in the technical specifications were used as the determining factors, converted into changes in the supply voltage of the electric circuit of the wheel set.

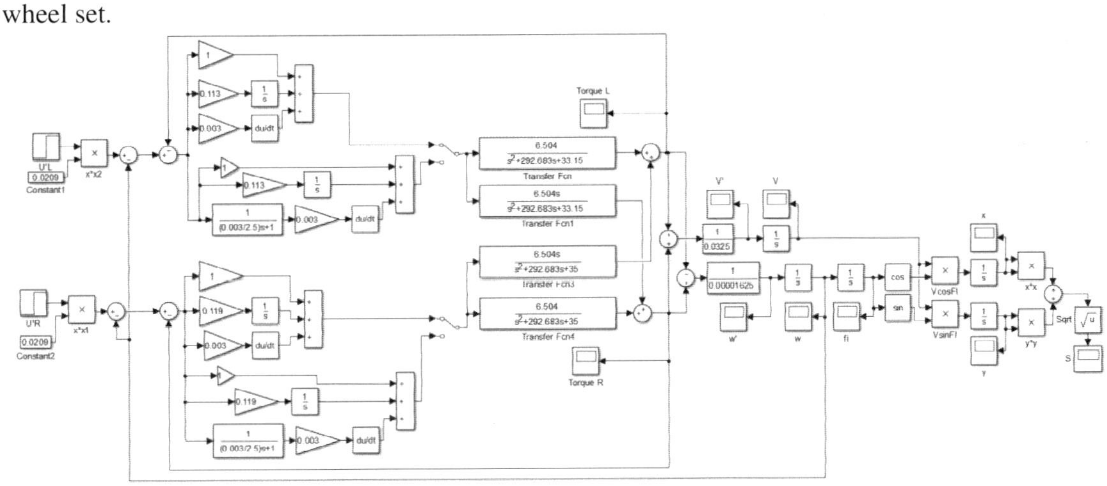

Figure 6. Simulink-model of mobile robot using modal method of PID controller adjustment

The simulation results are shown in Figures 7-16 for two cases:

1) Using ideal modal PID controllers:

PID1 coefficients for the left wheel: $K_P = 1$, $K_I = 0.113$, $K_D = 0.003$;

PID2 coefficients for the right wheel:: $K_P = 1$, $K_I = 0.119$, $K_D = 0.003$.

2) Using real modal PID controllers::

PID3 coefficients for the left wheel: $K_P = 1$, $K_I = 0.113$, $K_D = 0.003$, $N = 2.5$;

PID4 coefficients for the right wheel: $K_P = 1$, $K_I = 0.119$, $K_D = 0.003$, $N = 2.5$.

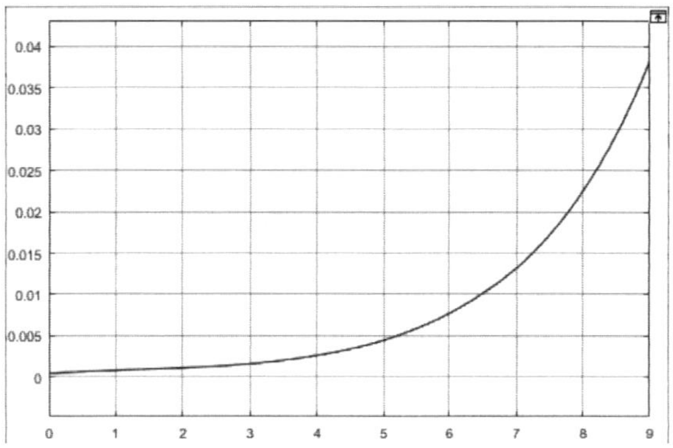

Figure 7. Left wheel torque (ideal PID1)

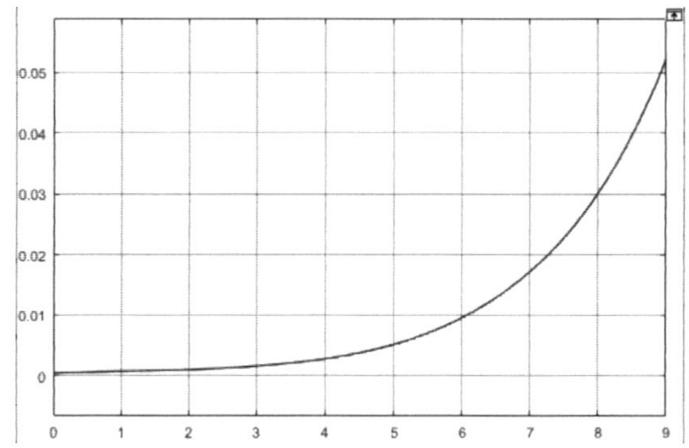

Figure 8. Left wheel torque (real PID3)

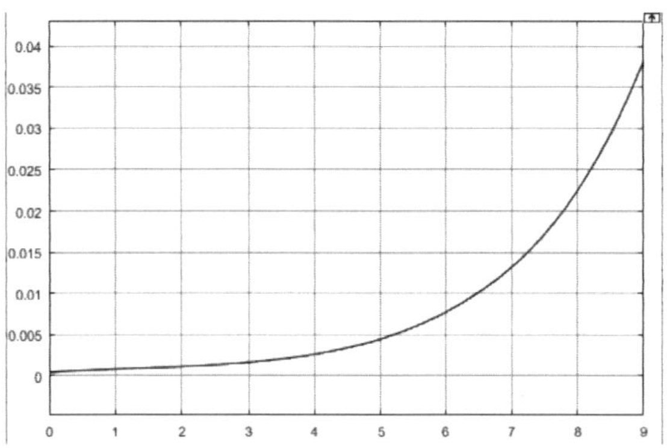

Figure 9. Right wheel torque (ideal PID2)

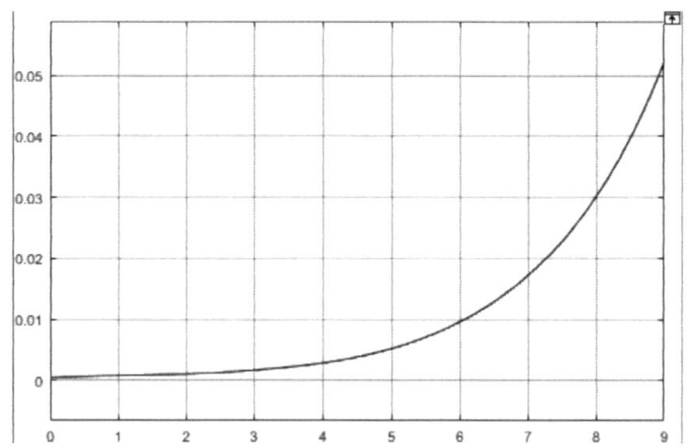

Figure 10. Right wheel torque (real PID4)

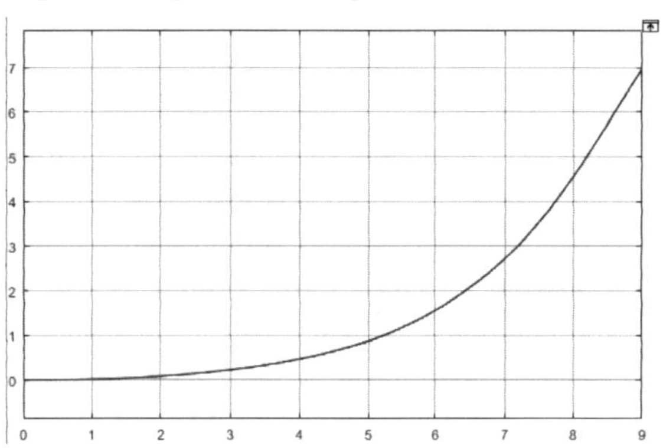

Figure 11. Change of x coordinate (ideal PID controllers)

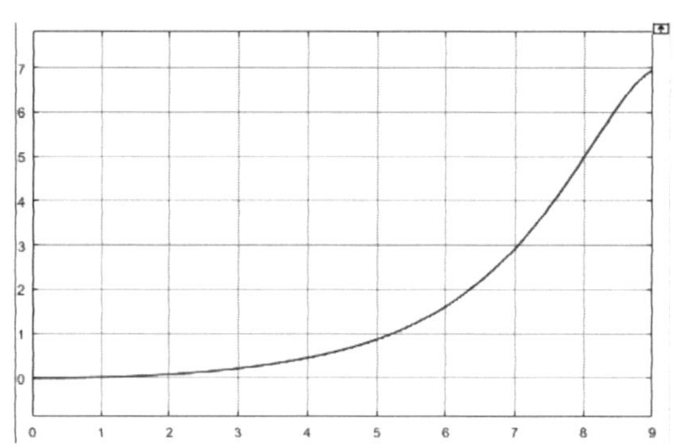

Figure 12. Change of x coordinate (real PID controllers)

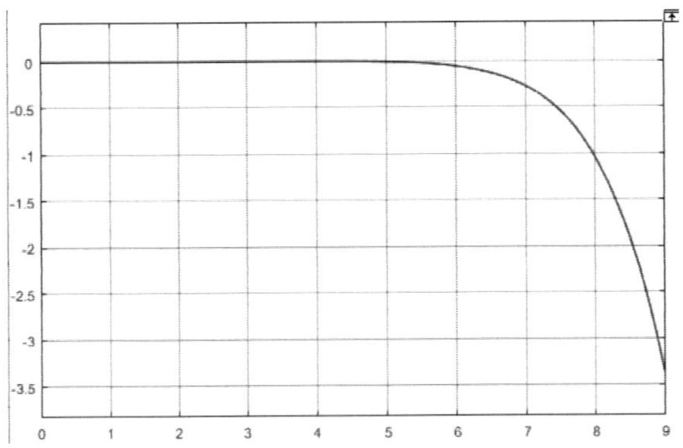

Figure 13. Change of y coordinate (ideal PID controllers)

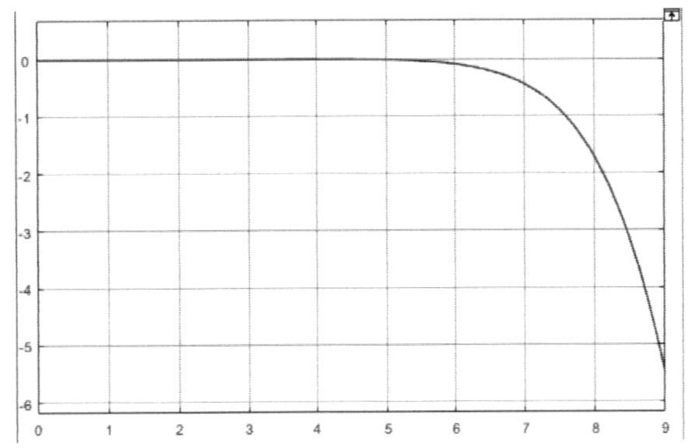

Figure 14. Change of y coordinate (real PID controllers)

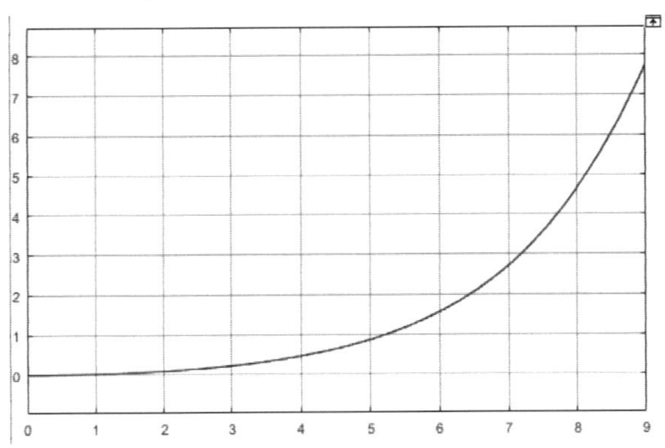

Figure 15. Changing the robot's position in space (ideal PID controllers)

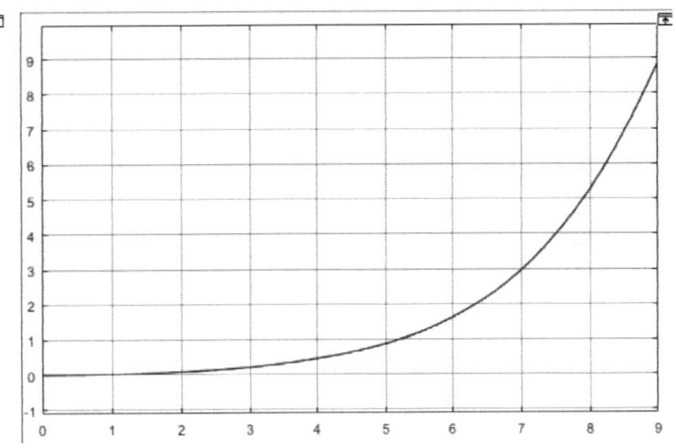

Figure 16. Changing the robot's position in space (real PID controllers)

Analysis of the results of the simulation shows that the experimental characteristics of the free movement of the robot can be used to determine the parameters of the PID controllers in such a way that the characteristics of the closed loop adaptive control system correspond to the desired properties of the aperiodic link with the minimum control time. At the same time, it is quite predictable that the direct indicators of quality (time of regulation) become worse for the control system with PID controllers containing real differentiating links due to inertia of the forming high-frequency filter.

b). Double differentiation controllers (PDD)

On the basis of the equations (22) of inverse problems of dynamics method [22, 23, 24, 25] which is described above, the mathematical model of the control system of the autonomous robot is made. The general structure of Simulink-model of robot control is presented in Figure 17. The model consists of the following main blocks:

1) a block that implements the robot's angular position and trajectory motion (Figure 17);
2) a block of multivariate mathematical model of robot angular motion (Figure 18);
3) a block implementing a double differentiation controller (PDD controller) based on reference wheel set models (Figure 19).

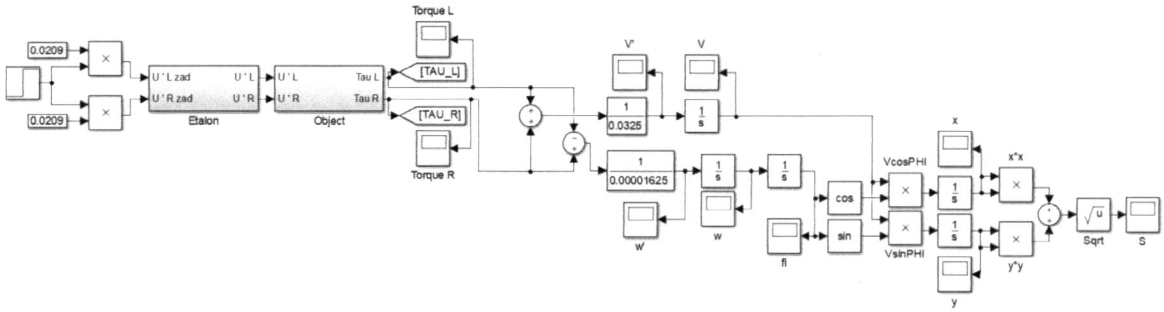

Figure 17. General structure of Simulink-model of mobile robot control

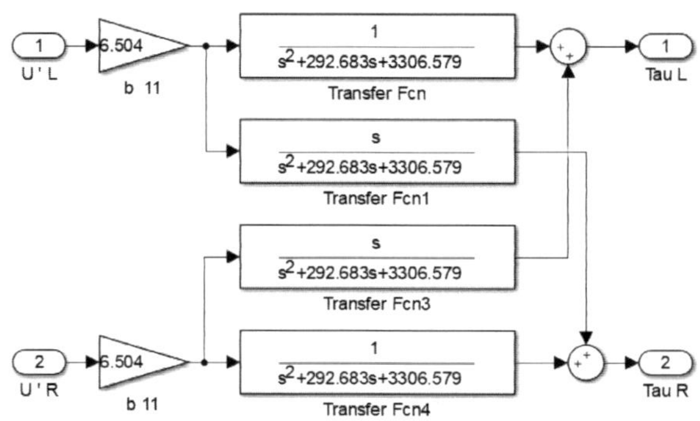

Figure 18. Multivariate mathematical model of robot angular motion

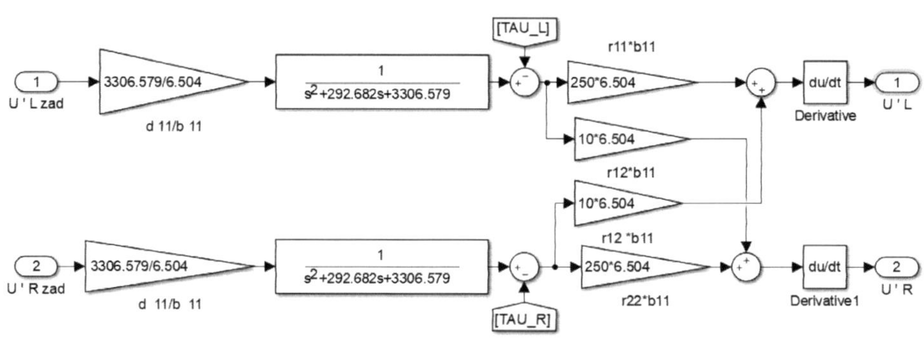

Figure 19. PDD controller based on a reference model

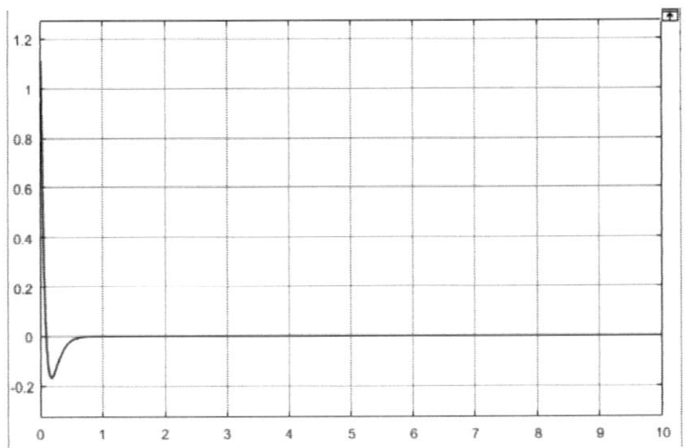

Figure 20. Electromagnetic moment of left wheel motor

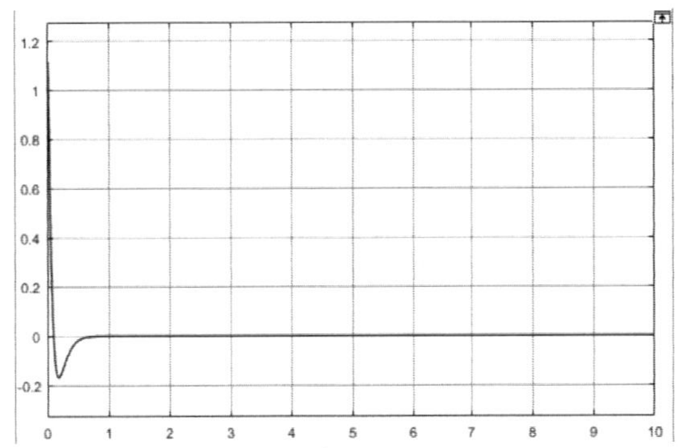

Figure 21. Electromagnetic moment of right wheel motor

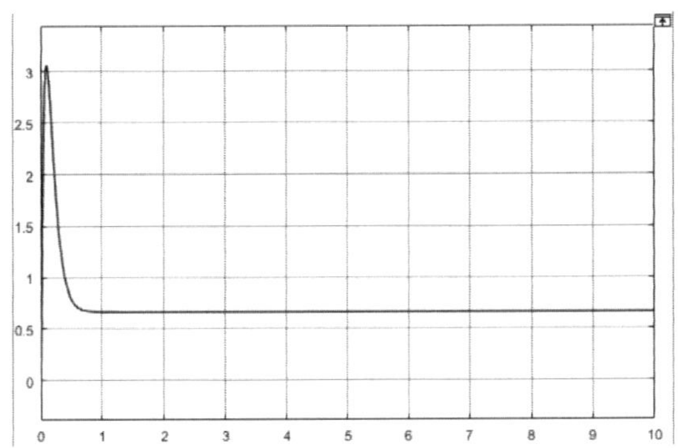

Figure 22. Linear robot speed

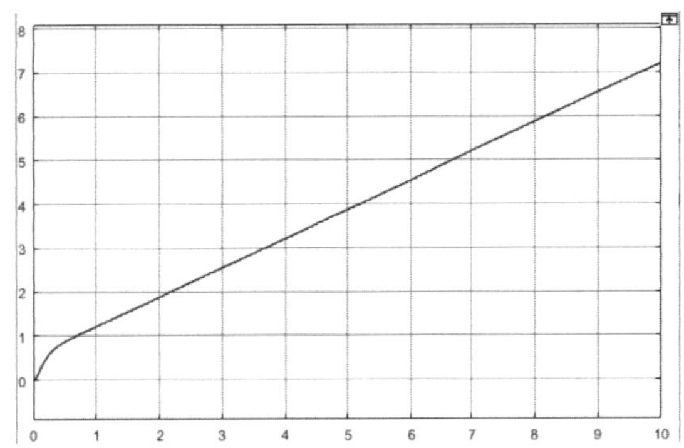

Figure 23. The track, passed by the robot depending on the time

Analysis of modeling results shows that the synthesis of the control system on the basis of PDD-controller provides high synchronism of the wheelset deflection. According to the fifth equation of the system (9), the angular velocity in the mode of stabilization of robot movement ($\omega_{\text{giv}} = 0$) also tends to zero, and the robot moves in a straight line.

7. Conclusion

Thus, there were obtained the following main results:

1) Mathematical models of the mobile robot in the state space and in the operator's form have been developed.
2) Synthesized algorithms of robot motion control on the basis of modal PID method - control at synchronous and differential deviation of the wheel set.
3) On the basis of the principle of dynamic compensation the method of construction of PDD controllers of the autonomous robot synthesized on the basis of reference models by a method of inverse problems of dynamics is proved.
4) Simulink-models of the controlled mobile robot are developed on the basis of the method of modal PID control at synchronous and differential deviation of the wheel set, and on the method of inverse problems of dynamics.

References

[1] Under edit by K.A. Pupkova and N.D. Egupova 2004 *Methods of classical and modern theory of automatic control. T.3: Synthesis of the controllers of automatic control systems* (M.: Bauman Moscow State Technical University)

[2] Amarendra Reddy B., Anil Kumar M., Anusha K., Ananda Rao G. 2018 PID Controller Design for Non-Minimum Phase Time-Delay Systems Using Optimization Methods *Helix* **8(2)** 3100-3106.
[3] Ang A. H, Chong G., Li Y. 2005 PID control system analysis, design and technology *IEEE Trans. on Control Syst. Tech* **13, 4** 559-576.
[4] Astrom K. J., Hagglund T. 2006 *Advanced PID control Instrumentation, Systems and Automation Society* 406
[5] Quevedo J., Escobet T. 2000 Digital control: past, present and future of PID control *Proc. IFAC Workshop* (Terassa, Spain, Apr. 5-7).
[6] Astrom K. J., Panagopoulos H., Hugglund T. 1998 Design of PID Controllers based on Non-Convex Optimization *Automatica* **34(5)** 585-601.
[7] Ba H., Li. F. A. 1954 General Method for Time Domain Network Synthesis *IRE Transactions – Circuit Theory* **1(3)** 21-28.
[8] Gorez R. 1997 A survey of PID auto-tuning methods *Journal Autom.* **38. 1** 3-10.
[9] Huba M., Menini L., Zaccarian L., Abdallah Ch. T., Edts. 2006 Constrained pole assignment control *Current Trends in Nonlinear Systems and Control* (Boston, Birkhauser) 163-183.
[10] Umland J. W., Safiuddin M. 1990 Magnitude and symmetric optimum criterion for the design of linear control systems: what is it and how does it compare with the others? *IEEE Transaction on Industry Application* **26(3)** 489-497.
[11] Bulgakov V.V., Kulabukhov V.S. 2013 Comparative analysis of the formalized methods of the following system controller synthesis *Devices* **1 (151)** 39-44
[12] Kim D.P. 2011 Algebraic method of synthesis of the linear continuous control systems *Mechatronics, automation, control* **1** 9-15
[13] Ziegler J.G., Nichols N.B. 1942 Optimum setting for automatic controllers *Trans. ASME* **64** 759-768.
[14] Farhan A.S. 2013 New efficient model-based PID design method *European Scientific Journal Edition* **9. 15** 181-190.
[15] Sizykh V.N., Muhopad A.Yu. 2014 Association automaton of the adaptive control of the technological processes on the basis of the neural networks *Scientific Bulletin of the Novosibirsk State Technical University* **1 (54)** 34-45
[16] Ageev A.M., Sizikh V.N. 2014 Synthesis of the optimal aircraft control system controllers through the solution of the inverse task of the AKOR *Scientific Bulletin of the Novosibirsk State Technical University* **3 (56)** 7-22
[17] Aleksandrov A.A., Livshits A.V. 2016 Influence of the billet stretching on the level of the thermal residual stresses *Modern technologies. System analysis. Modeling* **4 (52)** 66-69
[18] Aleksandrov A.A. 2014 Prediction of the cooling dynamics of the aluminium alloy blanks at the heat treatment *Modern technologies. System analysis. Modeling* **1 (41)** 140-145
[19] Leva A., Cox C., Ruano A. 2001 Hands-on PID autotuning: a guide to better utilization *IFAC Professional Brief, http://www.ifac-control.org* 84
[20] Krutko P.D. 2004 *Inverse problems of dynamics in the theory of the automatic control* (Moscow, Mashinostroenie) 576
[21] Bukov V.N. 2006 *System investment. Analytical Approach to the Analysis and Synthesis of the Matrix Systems* (Kaluga, Research Literature of N.F. Bochkareva) 720
[22] Kruglov S.P. 2003 Interrelationship between two approaches to analytical design of the optimal controllers *Automation and Remote Control* **64. 4** 564-575
[23] Bronnikov A.M., Kruglov S.P. 1998 Simplufied adaptability conditions for the control systems with identifier and reference model *Automation and Remote Control* **59 7 2** 993-1001.
[24] Livshits A.V., Filippenko N.G., Popov S.I., Larchenko A.G. 2013 Research of influence of dielectric elements of the working capacitor of high-frequency electrothermal installation on process of processing of polymeric materials *Modern technologies. System analysis. Modeling* **3 (39)** 270-275

IOP Conf. Series: Materials Science and Engineering **760** (2020) 012053 doi:10.1088/1757-899X/760/1/012053

[25] Filippenko N.G. 2012 Mathematical model of high-frequency processing of polymer materials *Modern technologies. System analysis. Modeling* **(33)** 76-79

IOP Conf. Series: Materials Science and Engineering **760** (2020) 012054 doi:10.1088/1757-899X/760/1/012054

The braking service of the linear induction motors with a compound-equalized magnetic flux for magnetic-levitation transport

A V Solomin[1]

[1]Rostov State Transport University (RSTU), 2, Rostovskogo Strelkovogo Polka Narodnogo Opolcheniya sq., Rostov-on-Don, Russia, 344038,

E-mail: ya.solomin2016ya.ru@yandex.ru

Abstract. In recent years, more and more attention has been paid to the development of new promising high-speed modes of the transport, especially, magnetic and vacuum pipelines. The linear induction motors will be used as traction machines in the early stages of the development of the magnetic-levitation and vacuum transport. The linear induction motors with compound-equalized magnetic flux for lines of the force, which, in addition to traction, can create lateral stabilization forces of high-speed carriages on a magnetic suspension, will be used in a magnetic vehicle in the nearest future. The effective and safe braking issues are of particular importance for high-speed magnetic-levitation transport. To reduce the speed and stop high-speed transport carriages on a magnetic suspension, the linear traction motors can operate as eddy current principle of the brakes. The article discusses the braking modes of a linear induction motor with a compound-equalized magnetic flux for lines based on the equivalent circuit of the machine. The analytical relationships are obtained for the calculation of the braking forces.

1. Introduction

The development of the world economy requires the improvement of the transport communications. The increase of the passenger traffic speed and cargo movement involves usage of the new types of the rolling stock, in particular, the use of the high-speed magnetic-levitation transport (MLT). The scientists and engineers from different countries are working to create the high-speed MLT systems for both passenger and freight traffic. In the Russian Federation, the scientists from transport universities and research organizations are working on the problems developing MLT systems. The contribution of the scholar staff of the Petersburg State Transport University to the development of new transport systems is noticeable [1-7]. Various types of the linear induction motors, which convert electricity directly into the forward movement of a transport carriage, are considered as traction machines of a magnetic-levitation transport. The linear motors can be DC machines, synchronous or induction-based. At the present time of the MLT systems` development, the traction linear induction motors (LIM) seem to be the most promising. The linear induction motors because of their design can be made with compound, equalizing and compound-equalized magnetic flux for lines of a force. The main range of the movement speeds of the mass transport vehicles is in the range from 350 to 500 km/h and the study of braking conditions is of particular importance and relevance. The traction LIM can be able to work in the brake mode. The article discusses the braking modes of the linear induction motors with compound-equalized magnetic flux for lines, the LIM with compound-equalized magnetic flux for lines are multifunctional and are able to create, in addition to traction and side stabilization efforts, which increase the safety of the MLT movement. The linear induction motors of this type will develop the efforts of the lateral stabilization when it stops the high-speed carriages suspended in a magnetic field.

2. The purpose of the research

We investigated the brake modes of a linear induction motor with compound-equalized magnetic flux for lines of the force.

3. The diagram of the replacement of the LIM phase with compound-equalized magnetic flux

The use of the equivalent circuits for characterizing and analyzing modes of the induction machines is widespread in electrical engineering [8, 9]. The use of the detailed equivalent circuits allows increasing the accuracy of the calculation of linear induction engines [10-12]. To analyze the processes in the LIM, to determine the parameters and characteristics of the engine, equivalent circuits and a number of other researchers are used [13-16]. In the systems of the magnetic-levitation transport, when studying processes in traction linear induction motors, the influence of the magnetic fields of the carriage's suspension on the characteristics of the LIM should be taken into account [17–20].

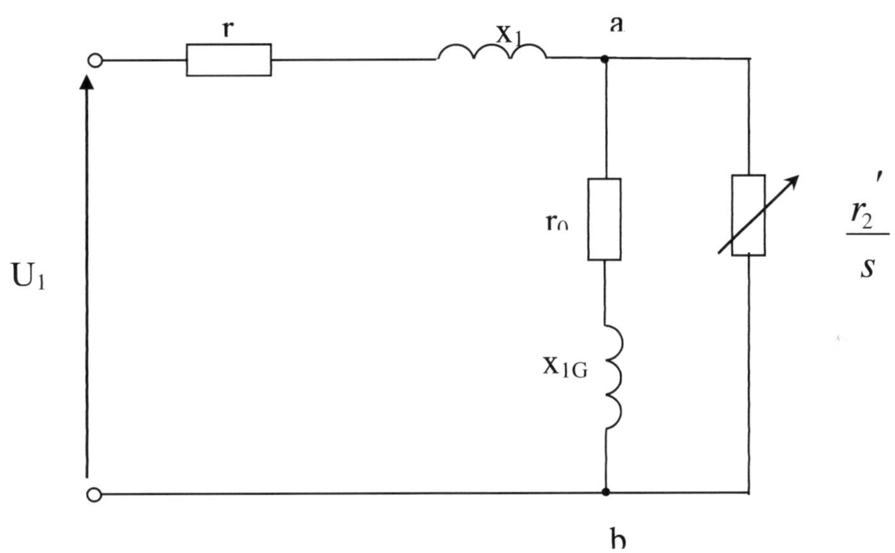

Figure 1. The LIM replacement circuit with compound-equalized magnetic flux for lines of the force

The equivalent circuit of a linear induction motor with a compound-equalized magnetic flux for lines of the force contains a branch with active r_1 and inductive x_1 inductor resistances, a branch with active resistance of the secondary element (SE) r_2'/s (the electrically conductive part of the secondary element is a bus-bar, therefore we ignore the inductive resistance) and a magnetizing branch with resistances r_0 (it takes into account the power loss in steel) and x_{1G}. On the basis of the LIM replacement scheme with compound-equalized magnetic flux for lines of the force (Figure 1), we consider the braking modes of the operation.

4. The braking modes of the LIM operation with compound-equalized magnetic flux for lines of the force

In the braking mode of the opposition LIM with compound-equalized magnetic, flux for lines of the force passes without the use of the special equipment. The LIM control circuit must contain a device for reversing the phase order. The braking force of the LIM with a compound-equalized magnetic flux for lines of the force in the opposition mode depends on the amount of the slip that exceeds one. At large slips, the frequency of the electromotive force (EMF) and current in the secondary element, induced by a running magnetic flux, increases to such values that it becomes necessary to take into account the influence of the surface effect (current displacement) when determining the active resistance of the SE. Therefore, when it determines the mechanical braking force, the effect of the current displacement in the secondary element should be taken into account by calculating the depth of penetration of an electromagnetic wave of a running magnetic field into a flat conducting bus-bar based on relation (1):

$$Z_0 = \sqrt{\frac{2}{\omega\mu\mu_0\gamma}}, \tag{1}$$

where Z_0 is the depth of the penetration of the electromagnetic wave (current) into the secondary element;

ω is the angular frequency of the EMF and current in the electrically conductive element;

μ - the relative magnetic permeability;

μ_0 - the absolute magnetic permeability of the vacuum;

γ - a conductivity of the material of the secondary element.

The need to take into account the effect of the current displacement is determined by the condition $Z_0 < \Delta_2$, where Δ_2 is the thickness of the secondary element, m.

To determine the braking effort, the reduced active resistance value of the secondary element is introduced into the calculation, taking into account current displacement:

$$r_2'' = r_2' \cdot k_r, \tag{2}$$

where $k_r = \Delta_2 / Z_0$ is the coefficient of the increase in the active resistance of the secondary element of the LIM due to the displacement of the current in the braking mode.

Since the angular frequency of the EMF and current in the secondary element of a linear induction motor with a compound-equalized magnetic flux for lines of the force depends on the slip, the k_r value is determined for the entire slip range actually possible in the deceleration mode by inhibition. The determination of these factors precedes the calculation of the brake characteristics of the LIM.

The braking with opposite connection can be considered as a logical continuation of the motor mode in the area of the slips exceeding one. In this regard, to calculate the braking characteristics of a linear electric motor, a replacement circuit with the same parameters is used as for the motor operation mode (Figure 1).

The braking forces of the LIM with a compound-equalized magnetic flux for lines of the force in the opposition mode are determined in the following order:

1) by the slip in a given range of the speeds;

2) by the expression (1) for the depth of the current penetration into the secondary element of the linear motor for each value of the angular frequency within a given speed range;

3) the calculated parameters of the equivalent circuit LIM are determined by the resistance values:

$$r_{emo} = \frac{x_{1G}^2 \cdot \frac{r_2''}{s}}{\left(\frac{r_2''}{s}\right)^2 + x_{1G}^2}; \quad x_{emo} = \frac{x_{1G} \cdot \left(\frac{r_2''}{s}\right)^2}{\left(\frac{r_2''}{s}\right)^2 + x_{1G}^2}, \tag{3}$$

with connection of which is equivalent to the impedance of the parallel-connected elements of the equivalent circuit x_{1G} and r''_2/s;

4) by the total equivalent resistance of the equivalent circuit of the phase of a linear induction motor with a compound-equalized magnetic flux for lines of the force:

$$Z = \sqrt{\left(r_1 + r_{emo}\right)^2 + \left(x_1 + x_{emo}\right)^2}; \tag{4}$$

5) the phase current inductor LIM:

$$I_{1ph} = \frac{U_{1ph}}{Z}; \tag{5}$$

6) the current in secondary element connected with inductor:

$$I_2' = I_{1ph} \frac{x_{1G}}{\sqrt{\left(\frac{r_2''}{s}\right)^2 + x_{1G}^2}}; \tag{6}$$

7) the electromagnetic power transmitted by the magnetic field from the inductor of the linear motor through the air gap to the secondary element:

$$P_{EM} = m_1 \left(I_2'\right)^2 \cdot \frac{r_2''}{s}; \tag{7}$$

8) the braking force in opposition mode:

$$F = \frac{P_{EM}}{V_1}, \tag{8}$$

where V_1 is the synchronous speed of the traveling magnetic field of a linear induction motor with a compound-equalized magnetic flux for lines of the force, m / s.

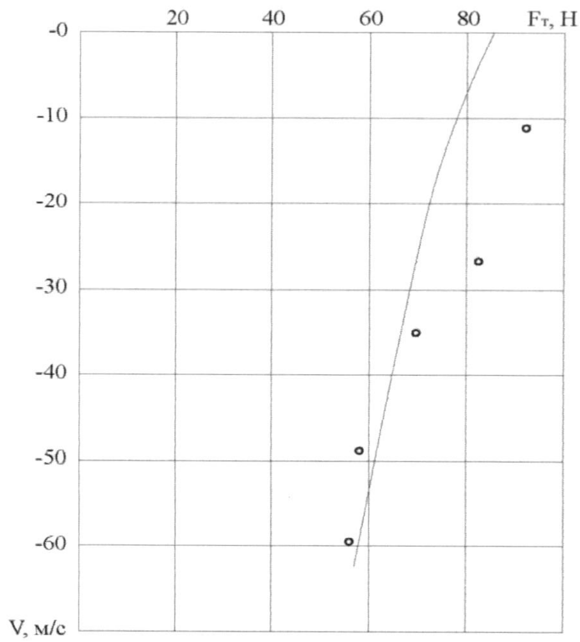

Figure 2. The mechanical characteristics of LIM in mode braking with opposite connection: - calculation; ° - experiment.

In order to test the methodology for calculating the LIM braking forces in the opposition mode and to evaluate the braking properties of the linear machines of this type, the mechanical characteristics of an experimental laboratory engine model were calculated. The mechanical characteristic of a linear induction motor in the mode of the braking with opposite connection, obtained by calculation, is shown in Figure 2. A pilot study of the braking of the LIM with the opposite connection produced a number of the test points, shown in Figure 2 in the form of the circles. The convergence of the calculated and experimental values of the braking forces can be considered quite satisfactory. The maximum discrepancy of the results was 15%.

The regenerative braking is used at high speeds of the carriages for uniformity of motion (for improving dynamic qualities) or for smoothing the transitional regime with significant changes in the speed of movement from higher to lower. In the first case, the wiring diagram of the inductor winding remains unchanged, while in the second, a change in the pole division value of the inductor is required.

With the regenerative braking, the linear induction machine goes into the generator mode of the operation. The LIM operates in an asynchronous generator mode connecting in parallel to the network and giving it electrical energy. At the same time, the linear movement speed does not decrease to "0", but at the same time, the linear machine prevents external accelerating factors acting on the carriages and gives the movement a more uniform character smoothing the transition process. It is noted that while it operates in regenerative braking mode, a linear induction machine with a compound-equalized magnetic flux for lines of the force consumes the reactive energy from the network. The equivalent circuit of a linear induction motor with a compound-equalized magnetic flux for lines of the force operating in regenerative braking mode is shown in Figure 3. The main feature of this equivalent circuit is the EMF representation of the secondary element in regenerative braking in the form given to the inductor as follows:

$$E'_{2B} = -I'_2 \cdot r'_2 \frac{1-s}{s}. \tag{9}$$

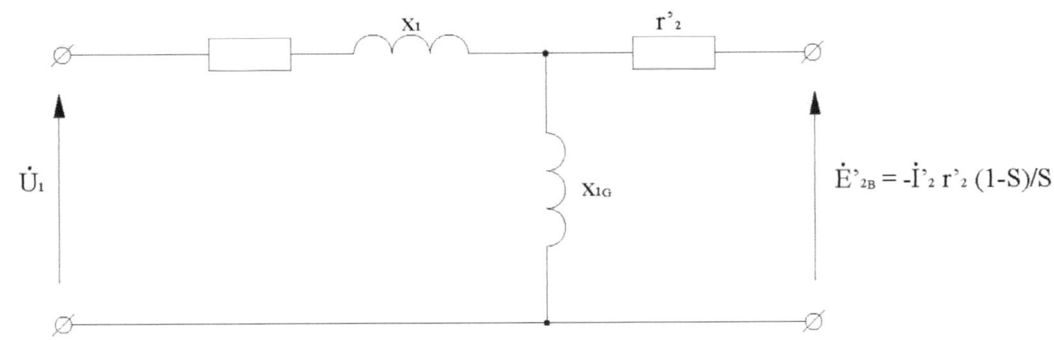

Figure 3. The equivalent circuit of a linear induction motor mode for regenerative braking mode

Given there are directions of the currents of the phase of the inductor LIM and to the inductor with the current of the secondary element that applies the method of the loop currents, we obtain a system of the initial equations for calculating the braking characteristics:

$$\begin{cases} \dot{U}_1 = -\dot{I}_1(r_1 + jx_1) + (\dot{I}_2'' - \dot{I}_1)\,jx_{1G}; \\ \dot{E}_{2B} = \dot{I}_2' r_2' + (\dot{I}_2' - \dot{I}_1)\,jx_{1G} = -\dot{I}_2' r_2' \dfrac{1-s}{s}. \end{cases}, \tag{10}$$

Having performed a series of the transformations of the second equation system (10), we obtain:

$$\dot{I}_1 = \dot{I}_2' \cdot \frac{\frac{r_2'}{s} + jx_{1G}}{jx_{1G}}. \tag{11}$$

Substituting the result in the first equation of the system (10), we obtain the following expression:

$$U_1 = -\dot{I}_2' \left[\frac{\frac{r_2'}{s} + jx_{1G}}{jx_{1G}} (r_1 + jx_1) + \frac{r_2'}{s} \right]. \tag{12}$$

After a series of simple conversions, we present the reduced current of the secondary element of the LIM operating in regenerative braking mode in the following form:

$$\dot{I}_2' = \frac{\dot{U}_1 jx_{1G} \left[\left(x_{1G}x_1 - \frac{r_2' r_1}{s} \right) + j \left(\frac{r_2' x_1}{s} + \frac{r_2' x_{1G}}{s} + r_1 x_{1G} \right) \right]}{\left(x_{1G}x_1 - \frac{r_2' r_1}{s} \right)^2 + j \left(\frac{r_2' x_1}{s} + \frac{r_2' x_{1G}}{s} + r_1 x_{1G} \right)^2}. \tag{13}$$

Performing a series of transformations, we obtain the modulus of the current value of the current in the secondary element:

$$I_2' = \frac{U_1 x_{1G}}{\sqrt{R^2 + X^2}}, \tag{14}$$

where $R = x_{1G}x_1 - \frac{r_2' r_1}{s}$; $X = \frac{r_2'}{s}(x_1 + x_{1G}) + r_1 x_{1G}$.

Substituting $\dot{I}'_2$ into the expression that determines the current in the phase of the LIM inductor for the generator operating mode:

$$\dot{I}'_2 = \frac{\dot{U}_1\left(\frac{r'_2}{s} + jx_{1G}\right)\left[\left(x_{1G}x_1 - \frac{r'_2 r_1}{s}\right) + j\left(\frac{r'_2 x_1}{s} + \frac{r'_2 x_{1G}}{s} + r_1 x_{1G}\right)\right]}{\left(x_{1G}x_1 - \frac{r'_2 r_1}{s}\right)^2 + j\left(\frac{r'_2 x_1}{s} + \frac{r'_2 x_{1G}}{s} + r_1 x_{1G}\right)^2}. \quad (15)$$

Having made several transformations, we obtain the module of the current value of the inductor phase current in the form:

$$I'_2 = \frac{U_1\sqrt{\left(\frac{r_2}{s}\right)^2 + x_{1G}^2}}{\sqrt{R^2 + X^2}}; \quad (16)$$

The obtained results made it possible to create a methodology for calculating the LIM braking forces with a compound-equalized magnetic flux for lines of the force in the mode of the energy recovery to the network. The mechanical characteristic of a linear machine in regenerative braking mode is calculated in the following sequence:

1) the range of slides is set, in which the linear machine will be transferred to regenerative braking mode (usually the slip changes from 0 to -1);

2) equivalent resistance values are determined:

$$R = x_{1G}x_1 - \frac{r'_2 r_1}{s}; \quad X = \frac{r'_2}{s}\left(x_1 + x_{1G}\right) + r_1 x_{1G}; \quad Z = \sqrt{R^2 + X^2}; \quad (17)$$

3) reduced current of the secondary element (during regenerative braking, it is actually the primary current):

$$I'_2 = \frac{U_1 x_{1G}}{Z}; \quad (18)$$

4) reduced the phase EMF of the secondary element of the LIM operating in the generator mode:

$$E'_{2B} = -I'_2 \cdot r'_2 \frac{1-s}{s}. \quad (19)$$

5) total power of the linear machine in the brake mode:

$$P_{B1} = m_1 \cdot E'_{2B} \cdot I'_2; \quad (20)$$

6) electrical losses in the secondary element:

$$P_{EL2} = m_1 \left(I'_2\right)^2 r'_2; \quad (21)$$

7) electromagnetic braking power transmitted by the electromagnetic field from the secondary element to the inductor:

$$P_{EM} = P_{B1} - P_{EL2}; \quad (22)$$

8) braking force generated by the linear machine:

$$F_B = \frac{P_{EM}}{V_1}; \quad (23)$$

where V_1 is the synchronous speed of the motor mode of the LIM;

9) inductor phase current:

$$I_1 = I_2' \frac{\sqrt{\left(\frac{r_2'}{s}\right)^2 + x_{1G}^2}}{x_{1G}}; \tag{24}$$

10) power delivered to the power supply:

$$P_{B2} = P_{EM} - m_1 I_1' r_1. \tag{25}$$

As for countering braking, the calculation is made for the number of the slips. The results are combined into a regenerative braking system. It is highlighted that this technique does not take into account the magnetic losses in the steel of the inductor core. But as the effective theoretical research showed, if these losses are not taken into account, the error does not exceed 3-5% and in engineering practice the losses in the inductor steel can be neglected.

5. Conclusion

Analytical relations have been obtained for calculating the braking characteristics of a high-speed magnet levitation transport based on the equivalent circuit of a linear induction motor with a compound-equalized magnetic flux for lines of the force. Methods have been developed for calculating the braking modes with opposite connection and recuperative for traction the LIMs with compound-equalized magnetic flux for lines of the force. The results of the effective theoretical studies are confirmed by experimental data. The comparison showed that the discrepancy between the theoretical and experimental data does not exceed 15%.

6. References

[1] Antonov Yu. F. Zaitsev A. A. 2014 *Magnetic-levitation transport technology: monograph* (Moscow: FIZMATLIT)

[2] Zaitsev A. A. Talashkin G. N. Sokolov Ya. V. 2010 *Transport on a magnetic suspension: monograph* (St. Petersburg: PSTU)

[3] Antonov Yu. F. 2015 *Magnetic-levitation transport: scientific problems and technical solutions* (Moscow: FIZMATLIT)

[4] Zaitsev A. A. Morozov E. I. Talashkin G. N. and Sokolov Ya. V. 2015, *Magnetic-levitation transport in a unified transport system of the country: monograph* (St. Petersburg: Publishing house LLC)

[5] Zaitsev A. A. 2016 About the modern stage of the development of magnetic-levitation transport *Railway transport* **12** 62–65

[6] Zaitsev A. A. 2014 Magnetic-levitation systems and technologies *Railway transport* **5** 69–73

[7] Antonov Yu. F. Zaitsev A. A.and Morozov E. I. 2014 Investigation of magnetodynamic levitation and electrodynamic braking of a cargo transport platform *News PSTU* **4(41)** 5–15

[8] Woldek A. I. 1974 *Electric cars* (L .: Energy)

[9] Woldek A. I. 1970 I*nduction magnetohydrodynamic machines with a liquid-metal working medium* (L .: Energy)

[10] Konyaev A. Yu.and Bagin D. N. 2018 Simulation of an electrodynamic separator based on a linear inductor *Electrical Engineering* **3** 34–40

[11] Sarapulo F. N. Friezen V. E. and Shvydky E. L. 2018 Mathematical modeling of linear induction motor based on detailed equivalent circuits *Electrical Engineering* **1** 58–63

[12] Bakhvalov Yu. A. Gorbatenko N. I. Grechikhin V. V. and Yufanova A. L. 2017 Designing the optimal electromagnets of magnetic levitation systems and lateral stabilization of land transport based on the solution of inverse problems *Electrical Engineering* **1** 44–48

[13] Kalnin T. K. 1980 *Linear induction machines with transverse magnetic flux: a monograph* (Riga: Zinatne)

[14] Solomin A. V. 2008 *Mathematical modeling of linear induction motors for traction and brake devices of high-speed transport: monograph* (Rostov on/D)

[15] Solomin A. V 2004 Linear induction motor for high-speed Maglev systems *News of the RSTU* **4** 41–44

[16] Solomin A. V. 2008 Mathematical modeling of the current density of the inductor of a linear induction motor for high-speed transport *News of the RSTU* **1** 127–135

[17] Jaewon Lim, Jae-Hoon Jeong, Chang-Hyun Kim, Chang-Wan Ha, Doh-Young Park 2017 Analysis and experimental evaluation of normal force of linear induction motor for Maglev vehicle *IEEE Transactions on Magnetics* **53 11**

[18] Woo-Young Ji, Geochul Jeong, Chan-Bae Park, Ik-Hyun Jo, Hyung-Woo Lee 2018 A study of non-symmetric double-sided linear induction motor for Hyperloop All-In-One System (propulsion, levitation, and guidance) *IEEE Transactions on Magnetics* **54 11**

[19] Gang Lv, Zhiming Liu, Shouguang Sun 2016 Analysis of torques in single-side linear induction motor with transverse asymmetry for linear metro *IEEE Transactions on Energy Conversion* **31 1** 165-173

[20] Takenori Yonezu, Ken Watanabe, Erimitsu Suzuki, Takashi Sasakawa 2017 Study on electromagnetic force characteristics acting on levitation/guidance coils of a superconducting Maglev vehicle system *IEEE Transactions on Magnetics* **53 11**

IOP Conf. Series: Materials Science and Engineering **760** (2020) 012055 doi:10.1088/1757-899X/760/1/012055

The effect of asymmetry of the loading cycle on heat dissipation in metal structures

L Yu Solovyov[1] and A L Solovyov[1]

[1]Siberian Transport University, 191, D. Kovalchuk st., Novosibirsk, 630049 Russia

E-mail address: lys111@yandex.ru

Abstract. The paper is devoted to the experimental study of heat generation processes in various metallic materials under high-cycle loading with different values of asymmetry and amplitudes of cycles. Tests were performed on flat, smooth specimens subjected to cyclic stretching with increasing amplitude. Change in thermal radiation detected by an infrared detector, by the value of the thermal sensitivity of the device NETD was taken for the beginning of a metal sample self-heating. The stresses corresponding to the beginning of self-heating, are called critical. It has been established that Smith and Haigh diagrams of critical stresses of the onset of self-heating are similar to the same diagrams of ultimate fatigue stresses.

1. Introduction

It is known that during the destruction of a metal structure under mechanical loads, the accumulated energy is released and dissipates largely as heat [1, 2] – i.e. the material is self-heating. This phenomenon accompanies any type of fracture – transition to a plastic state, rupture upon reaching the strength limit, fatigue damage under low- and high-cycle loads. It is obvious that energy dissipation can be used to characterize the state of metal structures, to identify and assess the degree of damage.

To register the amount of dissipated energyб infrared detectors (thermal imagers) having a temperature sensitivity of not more than 0.05 ° K are used. The main feature of IR detectors is the measurement of infrared radiation, characterizing the intensity of energy dissipation, not in a single point (as in pyrometers or thermocouples), but over a certain field of points. Such a field includes both a damaged area (apex of a visible or hidden crack) and intact areas. The thermal radiation gradient between the points of the damaged and intact areas gives information about the intensity of the development of destruction [3], and, therefore, indirectly allows us to make a prediction about the durability of the structure.

Assessment of structures durability is an important task in the operation of facilities. To identify damage that reduces durability, various methods of non-destructive testing are used during periodic inspections (acoustic emission, ultrasound, etc.) or strain gauge and vibration diagnostics with constant monitoring [4].

IR thermography is one of the relatively new methods of non-destructive testing. The advantages of IR thermography are the possibility of remote detection of damage, the minimum amount of preparation for measurements (sometimes measurements can be carried out without any preparation), obtaining results directly at the time of measurement, and ease of use.

IOP Conf. Series: Materials Science and Engineering **760** (2020) 012055 doi:10.1088/1757-899X/760/1/012055

Known successful application of the method of IR thermography to assess the fatigue life of bridges [3, 5, 6]. Used in the more general thermoelastic stress analysis (TSA) [7], this approach is referred to as self-reference lock-in IR thermography.

Many parameters have a significant effect on fatigue life – a complex (multiaxial) stress state, random stresses (for example, residual assembly from welding or caused by surface defects) [7, 8], asymmetry of the loading cycle. To use these parameters in the TSA method, it is necessary to conduct a special experimental study of this effect using IR thermography. In particular, residual stresses were studied by IR thermography in [10].

This article shows some results of experimental studies of the effect of the asymmetry of the loading cycle on the intensity of heat release in various materials under high-cycle loading.

2. Experimental

The development of fatigue damage in metals is known to be accompanied by the dissipation of the mechanical energy of deformation in the form of heat. The phenomenon of metal self-heating under cyclic loads, accompanied by an increase in the intensity of infrared radiation, is observed where the total stresses in the material exceed the critical ones, i.e. $\sigma > \sigma_{cr}$. By saying critical stress, we denote the minimum value of the maximum stress in a cycle at intense heat dissipation, leading to an increase in material temperature by a value of ΔT, begins. The value of ΔT is the sensitivity of the device NETD $\Delta T_{NETD} = 0.05$ ° K.

At low cyclic stresses, the material of metal structures elements, even in dangerous places, is deformed almost elastically and there is no significant increase in heat dissipation during an oscillation cycle. The amount of heat released is small and is spent on heat exchange with neighboring slightly loaded sections, on convective heat exchange of the oscillating metal construction element with the environment and on local material self-heating. At higher stress levels, as a result of structural heterogeneity of the material, local microplastic deformations occur in locally stressed zones [5], which leads to a significant increase in heat release.

The resistance of many materials to fatigue failure depends on the asymmetry of the loading cycle. Therefore, in order to reliably estimate the fatigue strength of materials operating under asymmetric loading cycles, stress limit diagrams or lines of equal durability are constructed in coordinates $\sigma_m \sim \sigma_{max} / \sigma_{min}$ (Smith diagram) or in coordinates $\sigma_m \sim \sigma_a$ (Haigh diagram). Here, σ_m, σ_{max}, σ_{min}, and σ_a are the mean, maximum, minimum, and amplitude values of the stresses of a specimen loading cycle, respectively. We shall note the great complexity of constructing these diagrams. For example, in order to construct these diagrams for a single material for different durabilities, it is necessary to test 7-10 batches of samples, approximately 20 samples in a batch until destruction at different asymmetry coefficients, then construct Weller's curves and build Smith or Haigh diagrams afterwards.

Each point on the limiting envelope of these diagrams corresponds to the fatigue limit for the selected durability in the case of loading a sample with a loading cycle, the parameters of which correspond to the coordinates of the point characterizing this cycle: $\sigma_m \sim \sigma_{max} / \sigma_{min}$ or σ_m, σ_a. Since in this paper the fatigue process is associated with the critical stress of the beginning of self-heating under cyclic loading, the impact of the loading cycle asymmetry on the magnitude of such a critical stress was investigated.

Heat release, or more precisely, the intensity of the infrared radiation flux off the surfaces of test specimens from various alloys, was measured using thermograms obtained with a thermal imager, which has a temperature sensitivity of at least 0.05 ° C.

Samples made of steels 08X18H10T (chemical composition C ≤ 0.08 %, Si ≤ 0.8%, Mn ≤ 2 %, Cr ≤ 17-19 %, Ni ≤ 9-11 %, Ti ≤ 0.7 %, S ≤ 0.02 %, P ≤ 0.035 %, Cu ≤ 0.3 %, the basic element is – Fe) and 30HGSA (C ≤ 0.28 – 0.34 %, Si ≤ 0.9 – 1.2 %, Mn ≤ 0.8 - 1,1 %, Ni ≤ 0.3 %, S ≤ 0.025 %, P ≤ 0.025 %, Cr ≤ 0.8 – 1.1 %, Cu ≤ 0.3 %, Fe ~ 96 %), as well as alloy BT-20 (Fe ≤ 0.3, C ≤ 0.1, Si ≤ 0.15 %, Mo ≤ 0.5 – 2 %, V ≤ 0.8 – 2.5 %, N ≤ 0.05%, Ti ≤ 84.94 – 91.7%, Al ≤ 5.5 – 7 %, Zr ≤ 1.5 – 2.5 %, O ≤ 0.15 %, H ≤ 0.012 %) were tested. A number of samples were made of 30HGSA steel with heat treatment corresponding to $\sigma_{ult} = 635$ MPa, $\sigma_{ult} = 837$ MPa, and $\sigma_{ult} = 1470$ MPa, where σ_{ult} is the ultimate strength of the material. The samples were flat with thickness $\delta = 5$ mm (Fig. 1).

The test method was as follows. To improve emissivity, the working part of each sample was covered with matte paint. For fatigue testing, the specimen was fixed in a machine clamp. After that, the specified average stress σ_m was created in the sample and the sample was subjected to cyclic loading, at which the amplitude component of the cycle – stress σ_a – was stepwise increased. At each stage of loading, the sample was held for 10–20 seconds, during which the temperature at the stage stabilized.

The amplitude of the cycle was raised until at some stage of loading the self-heating in the working zone of the sample occurred. At the time of the start of self-heating, the stress was noted, which was taken as the critical σ_{cr}. Then another medium stress was created in the sample and the test cycle was repeated.

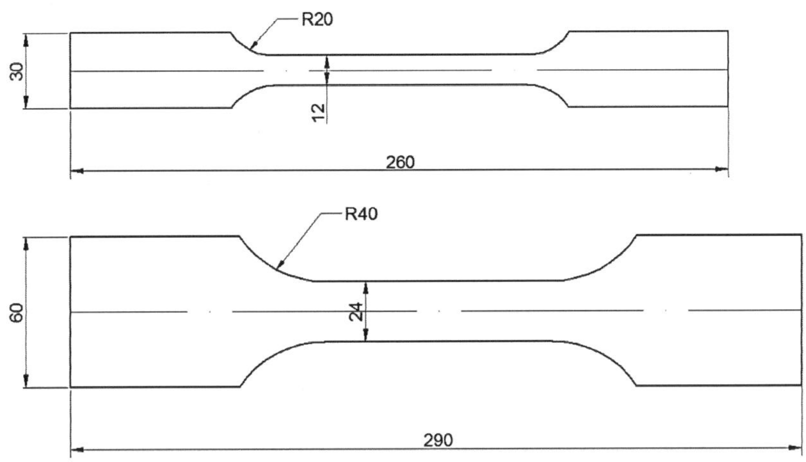

Fig. 1. Samples for testing

3. Results and discussion

The test results are shown in Fig. 2-4 in the coordinates $\sigma_m \sim \sigma_{max}$ (σ_{min}), taken in the construction of the Smith's diagram of the ultimate stress, and in the coordinates $\sigma_m \sim \sigma_a$, taken in the construction of the Haigh's diagram. Here, the experimental points correspond to the parameters of the loading cycle, at which self-heating of some point of the working part of the sample begins, i.e. these lines of self-heating start correspond to lines of equal temperature rise on ΔT_{NETD}.

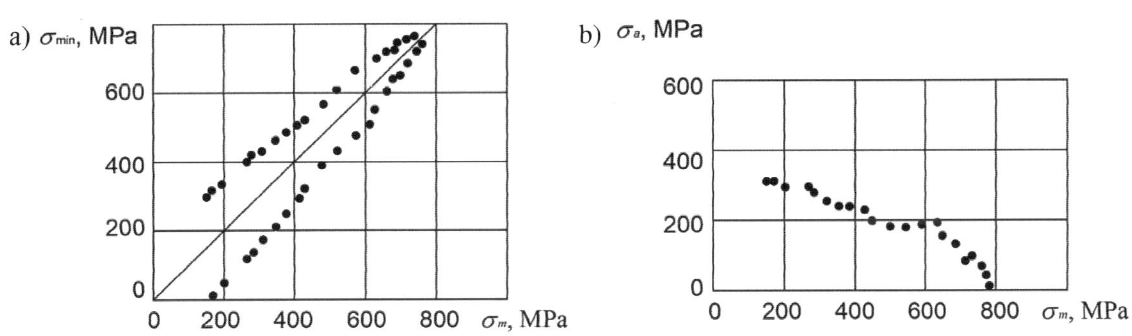

Fig. 2. Smith (a) and Haigh (b) diagrams of the critical stresses σ_{cr} of the beginning of self-heating for steel 08X18H10T

IOP Conf. Series: Materials Science and Engineering **760** (2020) 012055 doi:10.1088/1757-899X/760/1/012055

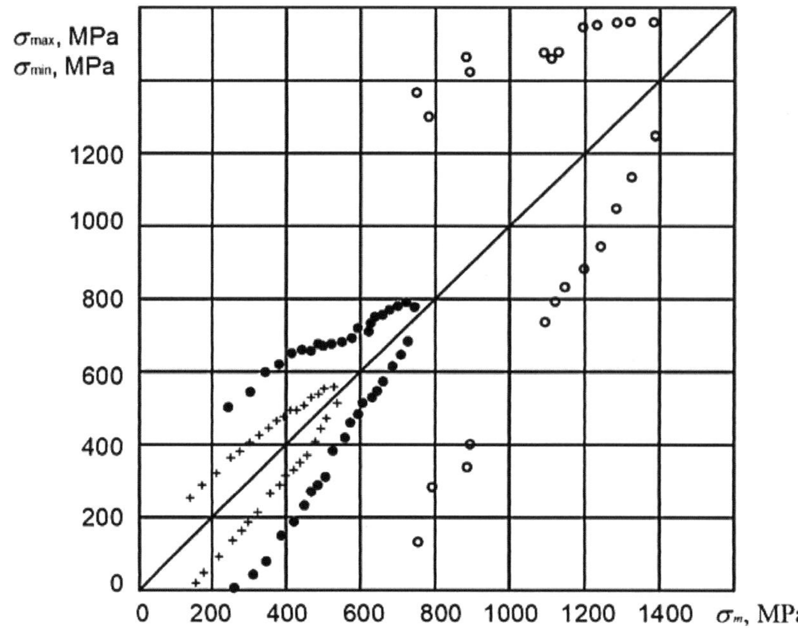

Fig. 3. Smith's diagram of critical stresses σ_{cr} of the beginning of self-heating for steel 30HGSA with "+" - σ_{ult} = 635 MPa, "•" - σ_{ult} = 837 MPa, "o" - σ_{ult} = 1470 MPa.

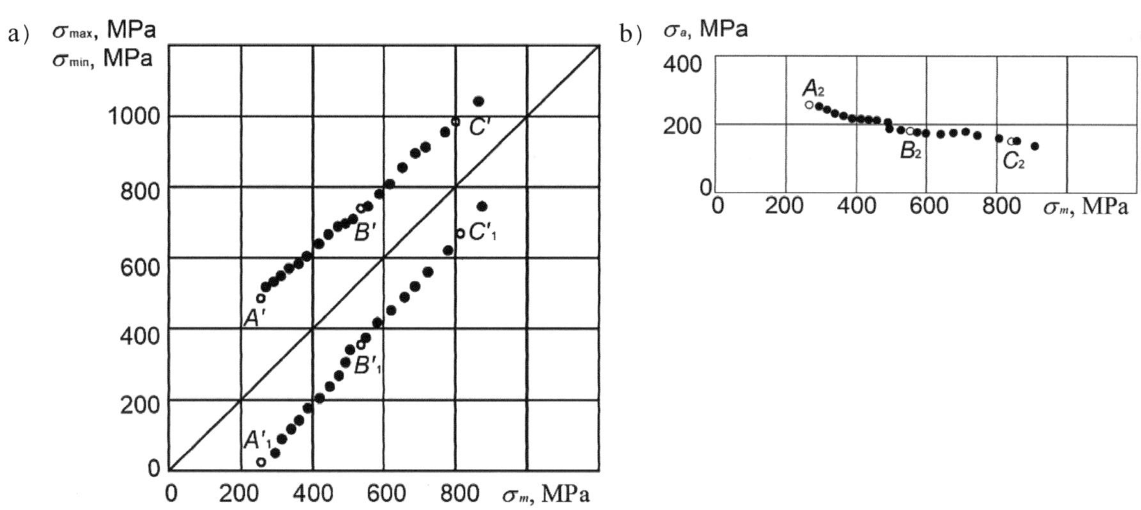

Fig. 4. Smith (a) and Haigh (b) diagrams of the critical stresses σ_{cr} of the beginning of self-heating for alloy BT-20

Points "o" correspond to the experimental parameters of the loading cycle A, B, C shown on the Weller curves (Fig. 5)

Fig. 5 shows Weller's fatigue curves for samples of the BT-20 alloy, obtained experimentally at different cycle asymmetry coefficients using the Weller method. Samples tested for fatigue were fully consistent with samples tested for self-heating. Using these fatigue curves, you can determine the number of cycles to failure, corresponding to the critical stress diagrams σ_{cr} shown in Fig. 4. For example, the parameters of the loading cycle of the envelope of the critical stresses for points A_1, A'_1; B_1, B'_1; C_1, C'_1 (Fig. 4, a) and similar points A_2, B_2, C_2 (Fig. 4, b) correspond to points A, B, C on the fatigue curves shown in Fig. 5. It can be seen that points A, B, C on the fatigue curves correspond practically to the same durability $N_k \approx 2.65 \cdot 10^6$ cycles.

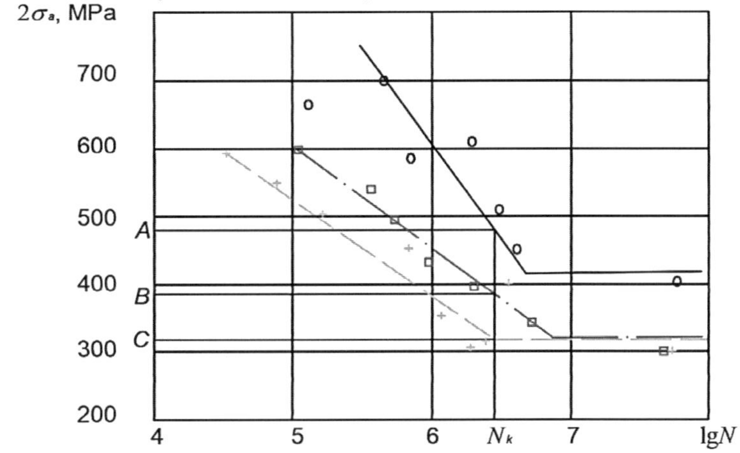

Fig. 5. Weller curves for samples made of alloy BT-20; points on curves A, B, C correspond to points A_1, A'_1; B_1, B'_1; C_1, C'_1 (fig. 4, a) and points A_2, B_2, C_2 (fig. 4, b)

Thus, the critical stress diagrams σ_{cr} shown in Fig. 4 are lines of equal rise in self-heating temperature or lines of equal durability N_k, which for material BT-20 corresponds to $N_k = 2.65 \cdot 10^6$ cycles.

4. Conclusion

1. Using structural steels samples under a uniaxial stress state and asymmetric loading cycles, the heat release in the sample material with a stepwise increase in the operating time was investigated.
2. It has been stated that each material for any given loading cycle and selected type of deformation has its own critical stress. In the places of the sample, where the stresses are greater than or equal to the critical, heat dissipation is observed, leading to the self-heating of these sites.
3. For each sample, tested without damaging it, the diagrams of critical stresses of the beginning of self-heating are constructed in coordinates $\sigma_m \sim \sigma_a$ (Haigh diagram), and $\sigma_m \sim \sigma_{max} / \sigma_{min}$ (Smith diagram).
4. The fatigue tests carried out for the control samples of the BT-20 alloy before failure showed that the points of these diagrams in the coordinates of Haigh and Smith correspond to stresses equal to the values of the critical stresses σ_{cr} of the beginning of self-heating. Thus these points fit well on a line that has the same self-heating temperature or a constant critical stress σ_{cr} onset of self-heating, or the same durability of these samples, under fatigue loading with their cycle, the parameters of which correspond to this accepted critical stress σ_{cr}. It is logical to assume that the same pattern will take place in other materials.

Acknowledgments

The publication was carried out within the framework of the grant of JSC Russian Railways for the development of scientific and pedagogical schools in the field of railway transport.

5. References

[1] Naderi M, Amiri M and Khonsari M M 2010 On the thermodynamic entropy of fatigue fracture *Proc. R. Soc. A* **466** 423–38

[2] Fedorova A Yu, Bannikov M V and Plekhov O A 2012 *Frattura ed Integrità Strutturale* **21** 46-53

[3] Sakagami T, Izumi Y and Kubo S 2010 Application of infrared thermography to structural integrity evaluation of steel bridges *Journal of Modern Optics* **57** 1738–46

[4] Yashnov A and Kuzmenkov P 2018 Innovative aspects in developing bridge monitoring systems *MATEC Web of Conferences* **216** (*Polytransport Systems – 2018*) 01009
[5] Sakagami T, Mizokami Y, Shiozawa D and Izumi Y 2017 TSA based evaluation of fatigue crack propagation in steel bridge member *Procedia Structural Integrity* **5** 1370–76
[6] Bremond P 2007 New developments in thermoelastic stress analysis by infrared thermography available online at: http://www.ndt.net/article/panndt2007/papers/138.pdf
[7] Marsavina L and Tomlinson R A 2014 *Frattura ed Integrità Strutturale* **27** 13–20
[8] Malgin M G, Kirian V I and Dvorecky V I 2012 Calculation of local stresses in welded joint zones of large-sized space structures *The Paton Welding Journal* **4** 2–5
[9] Wolchuk R 2007 Discussion of "Consistent approach to calculating stresses for fatigue design of welded rib-to-web connections in steel orthotropic bridge decks" by Robert J. Connor and John W. Fisher *Journal of bridge engineering ASCE* **6** 811–15
[10] Amjad K, Asquith D, Patterson E A, Sebastian C M and Wang W-C 2017 The interaction of fatigue cracks with a residual stress field using thermoelastic stress analysis and synchrotron X-ray diffraction experiments. *R. Soc. open sci.* **4**: 171100

IOP Conf. Series: Materials Science and Engineering **760** (2020) 012056 doi:10.1088/1757-899X/760/1/012056

Advanced technologies of international cargo correspondence in railway transport

V V Zubkov[1], N F Sirina[2]

[1] Deputy Head of the Production Infrastructure Department of Federal Freight Company, Ekaterinburg
[2] Ural State University of the Railway Transport, Ekaterinburg

E-mail: nsirina@usurt.ru

Abstract. The technologies of formation of transportation documents and preliminary informing in the international traffic are the primary directions of improvement of information technologies of interstate transfer points both for the information systems of state control and regulation and for the information systems of the subjects of transportation services. In the railway transport system of the Russian Federation the model of formation of transportation documents and preliminary information has a number of distinctive features from the existing model of the world practice. The article analyzes and determines the possibility of development and interaction of information systems of state authorities, owners and consumers of transport services in a single information space, methods of their consolidation and integration into international transport corridors. Automation of technological operations on processing of transport documents and preliminary informing by creation and introduction of innovative developments in the field of management of process of technological interaction of participants of transport - providing services is considered. The automated system of processing of transportation documents for imported cargo with barcoding and the automated system of informing cargo owners about the approaching expiration date of the phytosanitary certificate are presented and proposed for use at the interstate junction points. The proposed information systems improve and optimize these technologies, accelerate trade turnover and ensure the implementation of projects for the development of a common information space in the segment of international traffic and the development of transport international relations in general. The systems have undergone states registration.

1. Introduction

The active development of international trade and industrial relations requires the search for new, more qualitative developments in the digital economy of the Russian Federation [1]. One of the ways to improve it is to increase the efficiency of the existing economy through the automation of technologies and processes of processing information data. The development of this segment is a priority for the railway transport, since the majority of cargo correspondence on the domestic and international markets of transport services falls on its share [2]. The universal development and implementation of automation of the technology of formation of transport documents and preliminary information in international traffic are the primary directions of improvement of information technologies of interstate docking points, both for information systems (IT - systems) of state control and regulation, and for IT - systems of subjects of transport services [3]. In the railway transport system of the Russian Federation, the model of formation of transport documents and preliminary information has a number of distinctive features from the existing model of world practice [4]. Studies show that in the modern information space of the country's transport system there is no interaction and data exchange between the information systems of the transport control bodies providing the complex and information systems of the subjects of transport services. Implementation of the project on creation of information support of cargo correspondence in international transport corridors is at the initial stage of development. Analysis of the work of interstate junctions shows that with the increase in the volume of export cargo transportation, the number of delayed trains and wagons with export cargo increases proportionally for such reasons as waiting for processing of transportation documents and waiting for the formation of transportation documents. In addition, there is a negative growth rate of unhitched wagons at border transfer stations (BTS) due to the expired validity of the phytosanitary certificate (PSC) for export cargo, discrepancies in the information about the goods in the shipping documents and the lack of shipping documents [3,4]. The solution to these problems can be found through the implementation of the concept and projects of the digital economy, methods of consolidation of IT systems and their integration into international transport corridors (Fig. 1). The aim of the work is to improve document management and preliminary information (PI) technologies in the organization of transportation of foreign trade cargoes by creating and implementing innovative developments in the field of managing the process of technological interaction between participants of transport services.

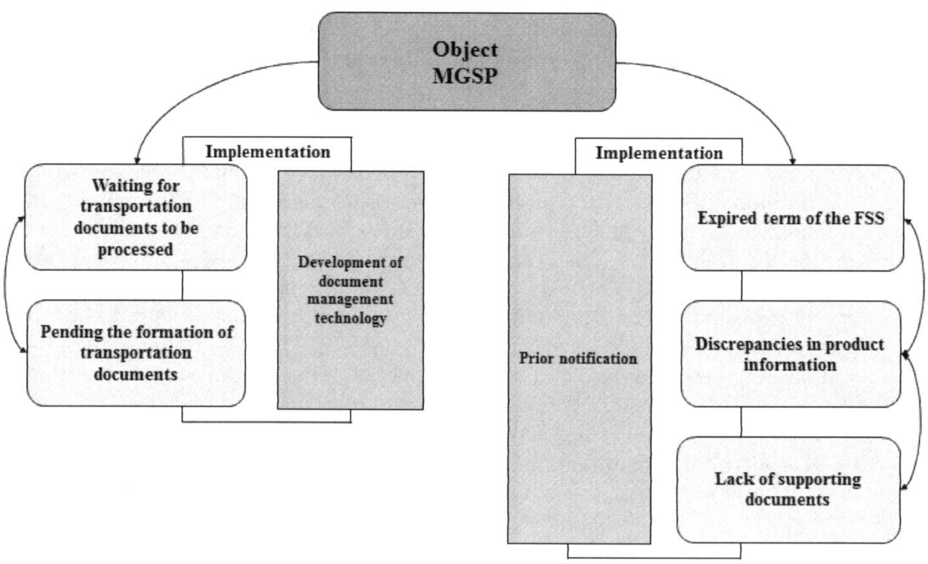

Fig. 1. Structure of reasons for delays of export cargoes on the IJP and ways to eliminate these reasons

In this paper, the authors propose a new effective way of development and implementation of the information resource of the railway transport system in the field of international communications. The IT system for processing of transport documents for imported goods with the use of barcoding (AC barcode) and IT system for informing cargo owners about the approaching expiration date of the phytosanitary certificate (AC KSDFS) are presented and proposed for use at the MSSP IT [5,6].

The main purpose of creation of the presented IT-systems is realization of the project of development of digital economy in space of creation of a uniform information field of a transport complex of the country:

– systems of accounting, processing and analysis of statistical information of the single transport complex;

– systems of analysis and planning of correspondence of export cargoes;

– systems of interaction of subjects of management of transport branch and subjects of transport services;

– systems of electronic document flow and preliminary informing of market users in the segment of transport services.

Achieving the goal is possible by solving the following tasks:

Ensuring transparency of statistical data.

Ensuring information consolidation and integration of producers of goods, transport companies and consumers of transport services on the basis of and using information resources of railway transport.

3) Ensuring automation of the processing of transport documents and preliminary information on imported and exported goods and their preliminary translation from the language of the transmitting party into the language of the receiving party.

4. Introduction of automated bar-coding technology for identification of import and export documents.

5. Introduction of automated IP technology.

6. Formation of a database for storing documents in foreign and Russian languages (translations).

7. Creation of an automated system to search for documents for imported and exported goods with the possibility of feedback from the shipper and correction of errors in design.

8. Introduction of changes in the existing technology of work, personnel training and support of the system operation.

Creation of conditions allowing to realize the listed problems promote formation of system of optimization of expenses at transportation of cargoes in the international communication, the account, processing of the information data of a transport complex and modeling of variants of interaction of controls and subjects of transport services in a uniform information field [7].

2. Methods of construction and structure of IT systems

The reliability of IT systems depends on the principles of governance in the areas of coordination, monitoring and exchange, and data management [8]. This area is within the scope of the activities of the transport complex management bodies. The quality and direction of information flows of IT systems depend on the users - subjects of transport services. In general, both the former and the latter are IT systems integrators, which provide system information consolidation and integration in the organization of transportation of foreign trade goods, the method of formation of information and technological and using the mechanism of automation of functional IT systems with self-determination and maintenance. At construction of IT - systems of AC barcode and AC KSDFS, methods of identification, transformation, coding, accounting, analysis and protection of information flows were applied [8]. The mechanisms of modeling of variants of perfection and interaction of controls and subjects of a transport complex (fig. 2) were used.

Using methodological principles of formation of adaptive systems with self-maintenance the volume of information flow at the input (ΣP_i) in IT systems is determined:

$$\sum P_i = \left\{ P_{1i}, P_{2i}, P_{3i} \right\}, \tag{1}$$

the volume of information flow (ΣP_e) at the exit from IT systems looks like:

$$\sum P_e = \left\{ P_{1e}, P_{2e}, P_{3e} \right\}, \tag{2}$$

the preferred information flow conversion function (ΣP_{ij}) is as follows:

$$\Sigma P_{ij} = \left\{ P_{1ij}, P_{2ij} \ldots P_{5ij} \right\}, \tag{3}$$

When creating the IT - barcode AC systems and AC KSDFS systems, it is assumed that the purpose of the information space unity function is to create a predictive model of the mechanism that represents the maximum value of the target function, or preferred for system integrators of information data for a period of time (T_{ij}). As a result, the solution at the initial given forecast (X), on the set of mechanisms (Q_x), is defined as follows:

$$Q_x = \Sigma x = \left\{ X, P_{in}, P_{en}, P_{ijn}, T_{ijn} \right\}. \tag{4}$$

Thus, the target function of the synthesis of self-contained systems in a single information space is to ensure the development of all system integrators, without exception, included in the model of the market of transport services in international traffic, and to maintain a transportation process that will ensure maximum performance of the target function with a variety of preferred options.

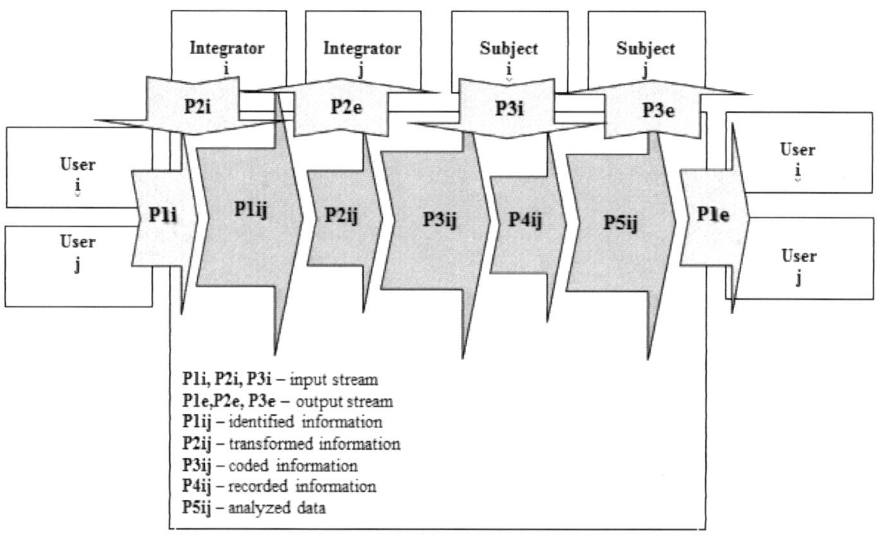

Fig. 2 shows the scheme of transmission, transformation, coding of accounting and analysis of information flows in AC barcode and AC KSDFS systems

The structure of IT systems. The structure of AC barcode - code and AC KSDFS includes the Server (WEB-server, application server), database server, personal computers of IT-systems users. The server of systems is intended for installation of the software realising functionality of AC a bar code, AC KSDFS. Database server is designed to install software that implements the functionality of the system of traffic safety management, placement and storage of IT-systems data [9]. Personal computers of users are designed to provide a user interface for access to IT systems. Server components of the System (WEB-server, application server, database server) provide the possibility of scaling in the event of increased load when connecting additional users and / or expanding the functionality of IT - systems. Components are implemented on the basis of industrial software. The software is documented and provides the necessary and sufficient for the implementation of basic functions set of subsystems dedicated to the functional and technological principle. The software provides a user interface in Russian and other languages (by decision) and has the property of extensibility of functionality [10,11].

IT systems function in a single information space on the basis of a single database and reference information systems. Data exchange is carried out in accordance with the approved data exchange regulations [10,11].

According to the functions and the level of tasks solved by users, for each IT workplace - barcode AC systems, AC KSDFS provide the following access rights:

– the consignor's workstation for inputting primary documents for import and export cargoes;
– an interpreter's workstation to ensure control over electronic document translations;
– specialist workstation for efficient processing of freight trains, with access to transportation documents and the possibility of correcting electronic documents;
– the system administrator's workstation, which allows to add and improve IT systems, change and supplement regulatory and reference information to form a database.

AC barcode - a barcode and AC KSDFS represent an information, software and hardware complex, which is based on standard technical means and software and hardware platform used in JSCo "Russian Railways".

Reliability of IT systems. IT systems remain operational and provide restoration of their functions in case of the following emergency situations [11]:
– power failure;
– malfunction of IT systems software;
– violation of the technical means complex operability.

Reliability of IT-systems operation is provided at the expense:
– data integrity control;
– provision of backup copying of information data;
– organization of hot backup of information data on logical and individual physical media (software/hardware RAID) levels [12];
– localization of program codes components of their configuration and results of work in separate catalogues on corresponding servers;
– use of own means of protection of standard software for data protection and access control;
– use of certified computer facilities;
– connection of technical means to uninterruptible power supplies to protect equipment and data from power failures [11,12].

3.Results

Results of AC barcode development - code and AC KSDFS are issued in the form of computer programs, a set of technical documentation, user's instructions and methodical materials in the volume necessary for implementation and use. IT systems are designed to work in the information space of JSCo "Russian Railways", railways of participants of international communications and interested government agencies of the Russian Federation. Access is provided both by the subjects of transport services and by the employees of the transport complex management bodies (border and customs services and other federal supervisory services) [13,14]. The work of IT systems is based on the principle of displaying the required accompanying documents for export and import cargoes through the WEB interface at the workplace of users. Data exchange is carried out in accordance with the approved regulations. The software is placed on the server and does not require installation of additional software on the user's workstations. The software and data storage are located on the server. The software is implemented on the personal computers, united by standard means of communication via the Internet [15]. IT systems are built on the principle of displaying screen forms at the workplace with the possibility of adding files in electronic form, further work with transportation documents, adding language translations, and displaying the required information [15].

The input data for the AC barcode is the AC barcode:

IOP Conf. Series: Materials Science and Engineering **760** (2020) 012056 doi:10.1088/1757-899X/760/1/012056

– Selection of operation mode (subject receiving cargo, subject transmitting cargo, translator);
– Selection of transportation documents;
– accompanying documentation (list of electronic documents);
– accompanying documentation (work with translation).

The output data of the AC barcode are:
– output of forms of electronic documents;
– Barcode - the code of the consignment (it allows one to read the data and search for the accompanying information, including translation into the user's language), examples of AC barcode visualization - the code is shown in Fig. 3, 4.

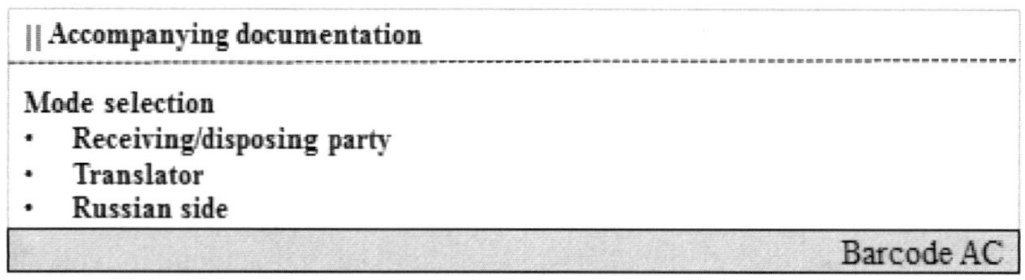

Fig. 3. Example of the screen form for inputting the operating mode in the IT - AC system barcode

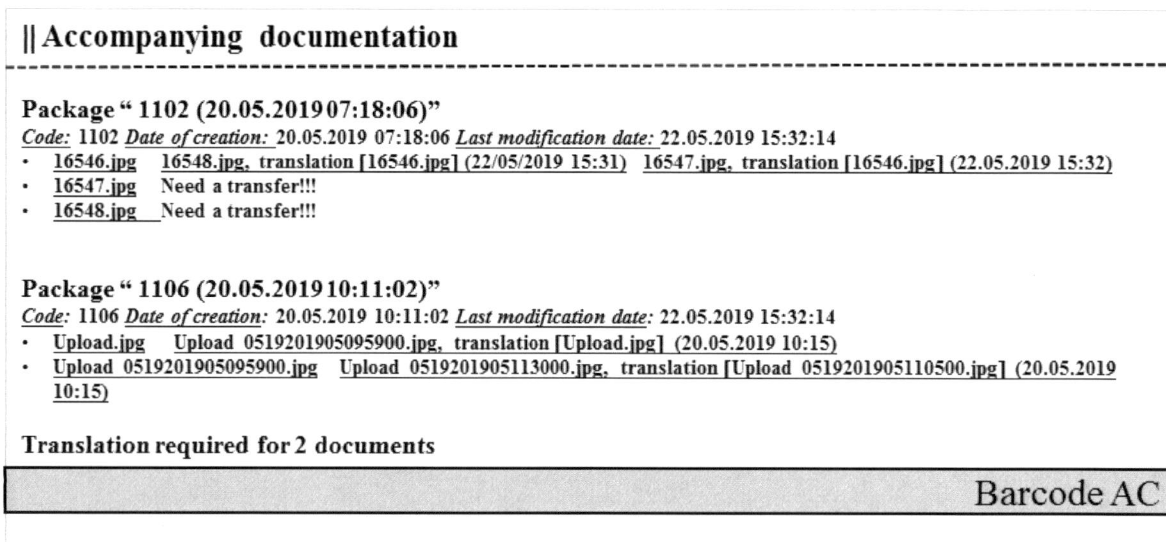

Fig. 4. Example of a screen form for displaying electronic shipping documents on an export cargo consignment in IT - AC barcode system

IT system AS KSDFS provides preliminary informing of cargo owners about the end of the FSS validity period and improves the process of processing of initial incoming information on the export cargo. The algorithm of this process is that with the help of the information resource of the global network "Internet", the consignor, as a system integrator, receives a notification in the PI mode about the period of validity of the FCC with the indication of the reason and the guilty in this subject of the transport complex. A link is sent with the notification containing the form of an electronic application to form the FSS again. The integrator sends the newly formed electronic application form to the relevant supervisory authorities for reissuance. As a result, having preliminary information, the accompanying documents for export cargo are reissued in advance, except for the restarting of the process at the MSME. The IT system functionality is being developed in stages:

– collection of statistical data;

– accounting and processing of statistical data;

– computer analysis of information data;

– visual display of the analyzed data;

– Formation of managerial and corrective decisions;

– user notification.

Examples of visualization of KSDFS AC are shown in Fig. 5, 6.

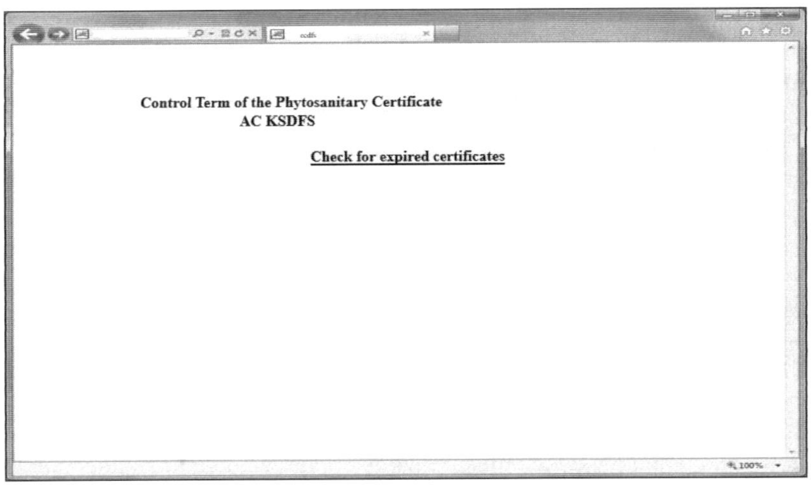

Fig. 5. Example of the screen form of request for the analyzed information about overdue FSS in IT - KSDFS AC system

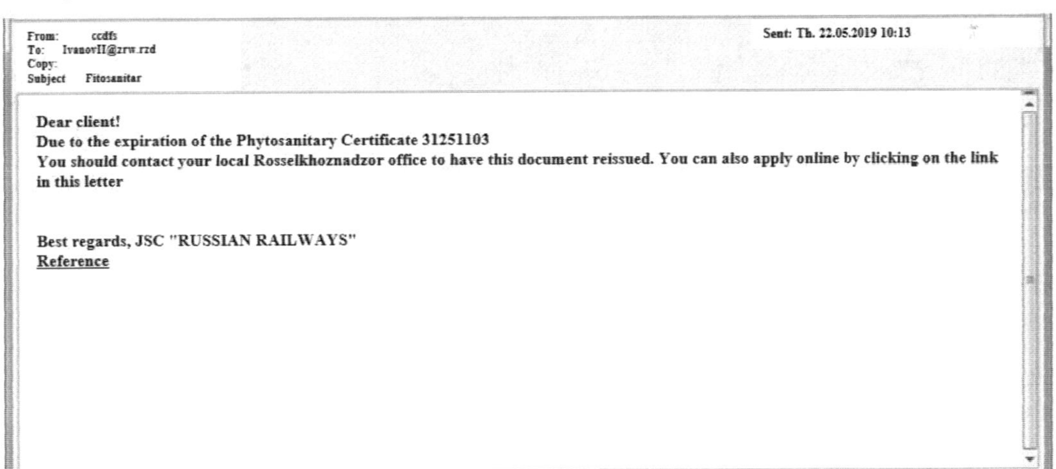

Fig. 6. Example of a screen form for notifying the consignor of the FSS expiration date in the IT system of the KSDFS AC

Thus, the developed IT systems provide the unification of information statistical data from other systems working in a single information space that provides the possibility of filling them with the necessary flow of information, which creates the conditions for adaptation and self-maintenance of the transport production process. IT - barcode AC systems and KSDFS AC were registered with the Federal Service for Intellectual Property of the Russian Federation.

4. Discussion

The IT systems presented by the authors allow for implementation:
– preliminary translation of electronic documents into the language of users, as well as creation of conditions for even loading of translators;
– preliminary informing transport service providers about the status of accompanying transport documents;
– Possibility of using electronic documents on consignments of export and import cargoes by the state management bodies of the transport complex and subdivisions of JSCo "Russian Railways" to speed up the technological process of their processing, reception and transfer;
– Assigning a unique bar code to copies of electronic transportation documents, i.e., a code that ensures identification of the cargo throughout its delivery;
– automation of the process of preparation of accompanying documents for imported cargo, prompt search for any document and its translation;
– Preservation of electronic documents in the database, protection from distortion;
– visualization of the task for the employees of IHR;
– forecasting and planning of the processing of transportation documents for the forthcoming period;
– generation of barcodes and search for transportation documents in information systems by barcode;
– adaptation of the technological process to changing internal and external conditions;
– maintenance of the main technological process for processing electronic transportation documents in international traffic.

Calculation parameters show that when using the IT - barcode AC systems and KSDFS AC, the prerequisites for excluding unproductive technological periods from the single technological process, such as waiting for processing of transportation documents for export and import cargoes, waiting for the formation of transportation documents, are created at the MSSP [16]. Preliminary information on the state of accompanying documents excludes the idle time of cargoes and rolling stock at border crossing points and approaches to them [16].

5. Conclusion

Widespread introduction and further development of the proposed IT systems will make it possible to accelerate the implementation of projects to create information support for freight correspondence in international transport corridors and to improve the unified information space of the Russian Federation. The application of these software products will ensure the formation of foreign trade relations with the consolidating interaction in the space of international communications.

The obtained results due to automation of technologies and processes of information data processing increase the efficiency of the existing economy, customer focus and improve the digital economy of the country.

The issue of further development of the IT potential of the AS barcode and AS KSDFS systems reinforced by the information flow, accounting and analysis of statistical data sent to the personalized complex of information transport services, is the subject of future research.

References

[1]. Lapidus B, Macheret D, Miroshnichenko O 2011 On increasing the productivity of resource use and efficiency of the railways. *Railway economy* **6** 12–22
[2]. Lapidus B, Macheret D 2012 *The use of rail transport for transcontinental transport is a tool to improve the efficiency of the world economy. Problems of rail transport. Tasks and ways of their solution: collection of works by scientists of JSC «VNIIZhT»* (Moscow: Intext) 260

[3]. Zubkov V, Sirina N 2014 Developing a transportation management model. *Transport of the Urals* **1(40)** 12–17

[4]. Sirina N, Zubkov V 2012 Formation of a regional management model to improve railway service. *Transport: science, technology, management. RAS VINITI* **3** 40–43

[5]. Zubkov V, Amelchenko O 2017 Automated system of processing of transportation documents for imported goods with the use of barcodes - coding. Certificate of state registration of the computer program 20176618679

[6]. Sirina N, Zubkov V, Amelchenko O 2018 Automated system of informing cargo owners about the approaching expiration date of the phytosanitary certificate. Certificate of state registration of the computer program 2018619816

[7]. Mitrofanov V, Grishina T, Feofanov A 2015 Management of automated technological systems and modeling of efficiency when making decisions. *Tekhnologiya Mashinostroeniya* **8** 43–45

[8]. Zubkov V, Raevskaya P, Amelchenko O 2018 *The Automated system of planning, management, accounting United (ASOPUS). Modern technologies: current issues, achievements and innovations: collection of papers of XVII International scientific. - prakt. Conf.* (Penza: ICSN «Science and Enlightenment») 288

[9]. Zubkov V, Amelchenko O *2018 Automated control system of industrial enterprise of complex transport service. Proceedings of the III all-Russian scientific-practical conference with international participation* (Omsk) 258

[10]. Zubkov V, Sirina N, Amelchenko O 2018 *Information Technologies in the Area of Intersectoral Transportation. International Scientific Conference Energy Management of Municipal Facilities and Sustainable Energy Technologies* (EMMFT 2018 Advances in Intelligent Systems and Computing, **982** Springer, Cham) 922

[11]. Zhou C, Yao K, Jiang Z, Bai W 2017 *Research on the Application of NoSQL Data base in intelligent Manufacturing. Wearable Sensors and Robots. Lecture Notes in Electrical Engineering* **399** 423–434

[12]. Agarwal R, Dhar V 2014 Editorial – big data, data science, and analytics: the opportunity and challenge for is research. *Information Systems Research* **25(3)** 443–448

[13]. Zubkov V, Sirina N 2019 Analysis of the impact of participants of transport-supporting functions on the quality of delivering integrated transport services and the technique of determining the degree of their influence on the quality criteria. *Modern Technologies. System Analysis. Modeling* **61(1)** 47–55

[14]. Zubkov V 2019 Specification of impact structure of participants of the complex transport service on the service quality improvement. *Modern Technologies. System Analysis. Modeling* **61(1)** 131–139

[15]. Sirina N.F., Zubkov V.V. 2018 Application of information technologies in international cargo carriage. *Modern Information Technology and IT-education* **14 (3)** 743–748

[16]. Zubkov V, Sirina N 2018 Stages of formation of the target model of the complex transport service in the sphere of freight transportation. *Vestnik VNIIZhT* **77 (6)** 368–374

IOP Conf. Series: Materials Science and Engineering **760** (2020) 012057 doi:10.1088/1757-899X/760/1/012057

Rationalization of parameters of hydraulic dampers of a carriage

A A Tarmaev[1], G I Petrov[2] and V N Filippov[2]

[1] Irkutsk State Transport University, 15 Chernishevsky st., Irkutsk, 664074, Russia
[2] Russian University of Transport (MIIT), 9 Obraztsova st., build. 9, Moscow, 127994, Russia

E-mail: t38_69@mai.ru

Abstract. The possibility of the car dynamic performance improvement is studied through structural changes (modernization) of the passenger bogie. Based on the calculations of the natural oscillations of the car, the critical values were determined and rational values of the damper resistance coefficient were chosen for vertical and lateral vibrations. It was concluded that with the forced spatial fluctuations of a carriages on KCCW-CRI bogies the required damping value of the vertical oscillations should be twice as large as for the lateral ones; important factors in damping of the car oscillations are the central suspension dampers. The rational angle of installation of the hydraulic damper in the central suspension and the use of separate action in the central suspension of hydraulic dampers are justified. The proposed version of the modernization of the bogie can improve the smoothness indicators of the car by 25-30%.

1. Introduction

Bogies are the main nodes of cars in traffic safety. Therefore work is continuously carried out for creating bogies, which are more reliable and convenient for maintenance and repair [1-11]. Passenger bogies use a two-stage spring suspension which allow increasing its static deflection and flexibility, which is very important to ensure the necessary speed and smoothness. Two steps of suspension allow various elements of spring suspension to be placed in a rational combination: springs, rubber springs, hydraulic and friction dampers.

Most of the cars in the fleet are equipped with KCCW-CRI bogie. Studies conducted by JSC Railwaycar-Building Works Tver, Tver Institute of Car Building, Joint Stock Company Railway Research Institute and others, showed that the KCCW-CRI bogie with its dynamic driving characteristics meets the traffic safety criteria in straight and curved sections of the track, it has an indicator of smoothness course at the level of "satisfactory" in the range of speeds of movement up to 160 km/h. Analysis of the experimental data showed that friction dampers in the axle suspension stage worsen the smoothness performance. In addition, hydraulic dampers of the central suspension do not provide optimal damping characteristics of the vertical and lateral vibrations and the angle of location in the bogie.

Therefore, as an option to upgrade the bogie, it was proposed to determine the rational resistance parameters of the hydraulic damper [12] in the central suspension and determine its installation angle in the bogie.

When determining the rational values of damper resistance coefficients separate damping of oscillations in the central suspension in the vertical and lateral directions was considered.

The tasks related to the specification of the mathematical description were solved:

- the angular displacements of the crowning suspension under the action of transverse and vertical forces and taking into account the friction in the hinges;
- dynamic deformations of elastic pads, located under the springs, which are the inertia-free elements but with elastic and damping properties;
- operation of a stopped hydro-damper damping which performs oscillations in two planes and having a pressure relief valve of fluid pressure;

- limitations of lateral movements of the bogie bar of the bogie with rubber shock absorbers located on the frame.

When making calculations related to the modernization of a KCCW-CRI bogie, the following structural elements should be taken into account in the mathematical model of the car:
- thrust cradle suspension committing pendulum oscillations under the action of lateral and vertical forces;
- hydraulic dampers, which are installed obliquely and have safety valves for fluid pressure;
- elastic pads located under the spring springs;
- limiters of transverse movements of the bolster relative to the bogie frame.

2. Results and Discussion

2.1. Determination of critical values of damper resistance coefficients for vertical and lateral vibrations

As it is known, the critical value of the damper resistance coefficient is its minimum value at which the process of its own damped oscillations becomes aperiodic (it decays exponentially). It is extremely important to know the critical value of the damper resistance coefficient for vertical and lateral vibrations, since it is recommended to take rational values [13] within 25% of the critical values.

The critical values of damper drag coefficients were determined on the basis of calculations of the natural vertical and lateral oscillations of the car [14-21].

Differential equations describing the oscillations of all bodies of the design scheme of the car can be more compactly represented in a vector-matrix form

$$[M]\cdot\ddot{\vec{U}} + [B]\cdot\dot{\vec{U}} + [C]\cdot\vec{U} + \vec{F} + \vec{Q} = 0, \quad (1)$$

where $\vec{U}$ is the coordinate vector of the computational scheme, defined by the components of expression (1);

$[M]$ is an inertial matrix;
$[B]$ is a damping matrix;
$[C]$ is a stiffness matrix;
$\vec{F}$ is a nonlinear vector defined by accepted physical and geometric hypotheses;
$\vec{Q}$ is a cargo vector determined by external perturbations.

In determining the critical value of the damper resistance coefficient for the body's vertical oscillations, the general system of differential equations (1) was integrated with initial conditions corresponding to the initial deviation of the body from the position of static equilibrium in the absence of external disturbances. As a result of the series of calculations, it was found that for vertical body oscillations, the critical value of the drag coefficient of the damper is $\beta_{z\ cr}$ = 20 t·s/m. Figure 1 shows the graphs of damped processes for three cases when $\beta_z < \beta_{z\ cr}$, $\beta_z = \beta_{z\ cr}$ and $\beta_z > \beta_{z\ cr}$.

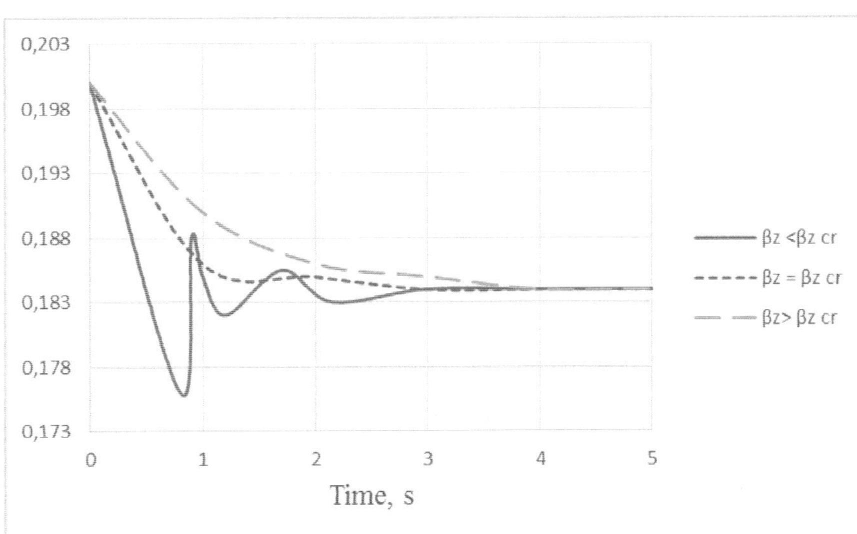

Figure 1. Vertical body travel (m).

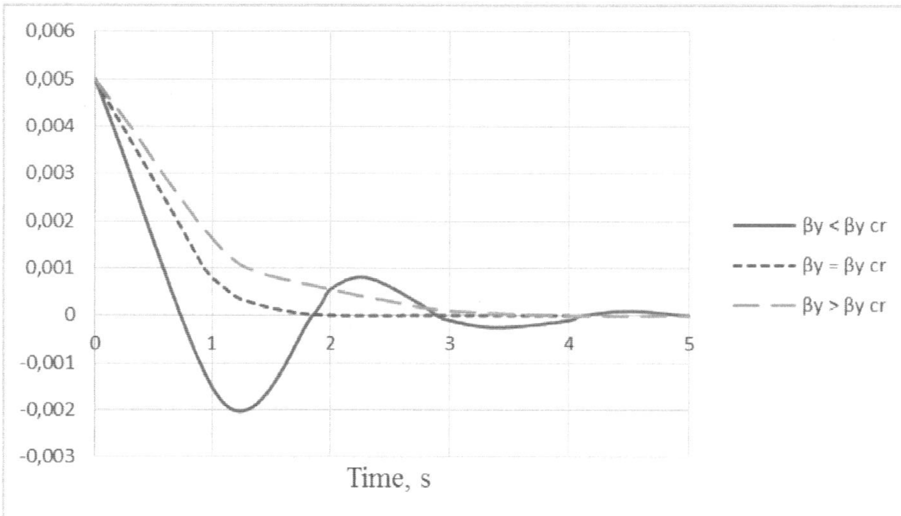

Figure 2. Transverse movements of the body (m).

In the same way the critical value of the damper resistance coefficient for lateral oscillations was determined. When integrating the system of differential equations (1), the initial lateral deviation of the body from the equilibrium position was set and the value β_y was selected by calculation so that the process of its own damped oscillations became aperiodic. Fragments of these calculations are shown in figure 2 graphs. As a result of the calculations performed, it was found that for lateral body vibrations, the critical value of the damper resistance coefficient is $\beta_{y\ cr} = 10$ t·s/m.

From the above calculations it follows that with the forced spatial oscillations of a carriage on KCCW-CRI-type bogies the necessary amount of vertical oscillation damping should be twice as large as for the lateral ones.

This circumstance must be taken into account when determining the rational angle of installation of the damper in the trolley for simultaneous damping of the vertical and lateral vibrations, as well as when choosing the resistance coefficients of the vertical and lateral dampers with separate damping of the oscillations.

2.2. Selection of damping drag coefficients with separate vibration damping in the central suspension and the absence of friction dampers in the axle box gear

It has been suggested that side-oscillation damping can be carried out by friction in the hinges of the cradle suspension without setting up an additional damper.

According to this variant, the series of calculations were carried out which showed that the absence of side oscillation damper in the central suspension leads to an unacceptable level of lateral and frame forces, as well as lateral accelerations on the frame and body.

The friction in the hinges of the cradle suspension is insufficient for damping lateral vibrations.

In order to establish the necessary amount of damping of lateral oscillations calculations were carried out at $\beta_z = 5$ t·s/m and varying the value of β_y.

From the results of calculations it follows that such indicators as lateral forces in the central suspension, frame forces, lateral accelerations of the body and lateral accelerations of the bogie frame have satisfactory values for βy not less than 2.5 t·s/m. In other words when extinguishing separately the damping resistances of the dampers in the vertical and lateral directions should be within 0.25 of their critical values. Moreover, the overestimation of the resistance coefficient of the vertical absorber over 0.25 $\beta_{z\ cr}$ does not lead to a noticeable change in the dynamic performance of the car in a vertical plane. At the same time, an increase in the drag coefficient of a horizontal absorber above 0.25 $\beta_{y\ cr}$ significantly impairs the side dynamics: lateral and frame forces and lateral accelerations.

Along with this, a series of calculations was carried out to determine the significance of the central and axle suspension dampers for the ride quality of the car. It turned out that the decisive factor in damping of the car oscillations are the central suspension dampers. When they are cut off, axle suspension dampers cannot provide a safe course even when placing additional hydraulic dampers in the axle box.

With the right choice of damper parameters in the central suspension, the placement of dampers in the axle box does not lead to a significant improvement in the dynamic performance of the car.

2.3. Determination of rational installation angles of hydraulic dampers for joint damping of vertical and lateral vibrations

Since preliminary calculations of the dynamic indicators of the car on bogies with separate damping of oscillations showed that the rational values of damping resistance coefficients in the vertical and lateral directions are different, it is very important to choose the angle of the damper in the bogie in which the inclined damper damps oscillations in two planes. In this regard, in the next series of calculations, the dynamic indicators of the car were analyzed at damper installation angles of 45°, 60°, 75° to the horizontal plane and various coefficients of fluid resistance. The maximum values were analyzed as dynamic indicators of the car:

- total vertical forces in the central suspension acting on the bolster;
- vertical forces acting on the spring set of the axle box springs;
- lateral forces acting on the spring sets of the central suspension;
- lateral forces acting between the wheel and the rail;
- frame forces acting on wheelsets;
- vertical acceleration of the body and frames of carriages;
- lateral accelerations of the body and frames of carriages.

The graphs of these dynamic indicators at the angle of installation of absorbers 45° are shown in figures 3 and 4.

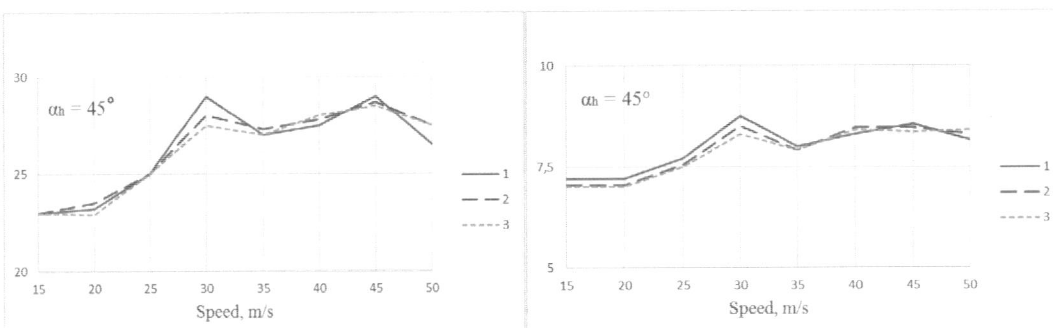

Figure 3. Vertical reaction of the central suspension of the bogie (a) and vertical reaction of the spring set of the axle box suspension (b)

1 – β_h = 5.0; 2 – β_h = 6,0; 3 – β_h = 7.0.

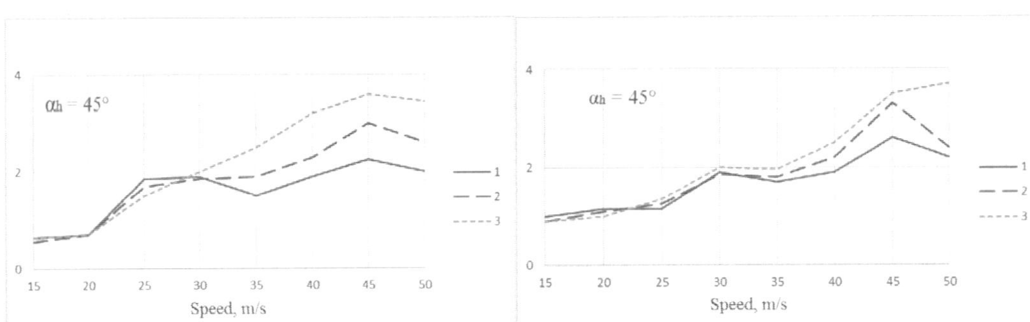

Figure 4. Side reaction in the center suspension of the trolley (a) and the frame force on the wheel pair (t): 1 – β_h = 5.0; 2 – β_h = 6,0; 3 – β_h = 7.0.

Similar dependences were obtained at installation angles of 60° and 75°. Comparison of the calculation results showed that the 45° installation angle of the damper is the most unfavorable, since at the same time, approximately the same damping forces are received both in the vertical and in the lateral directions. Preliminary calculations of the values of critical damping with vertical and lateral oscillations showed that they are two times different and therefore there is not enough damping force for vertical oscillations at a 45° angle, and for lateral oscillations, it turns out to be excessive.

Analysis of the calculated data showed that the most rational is the angle of installation of the damper 60-75°.

An increase in the drag coefficient of absorbers above 0.25 β_{cr} does not significantly affect the vertical forces however it noticeably worsens the forces and accelerations in the lateral direction.

Therefore, with an increase in the damping resistance coefficient, it is necessary to increase the angle of the damper installation to the horizontal plane to about 60° -75°.

The calculations have also shown that the best dynamic indicators occur when the damping resistance coefficient is 6-8 t·s/m and the installation angle is 60-75°. So accelerations on the body are 0.06-0.08g, and on the frame do not exceed 0.5-0.6g in the range of speeds of movement from 15 to 50 m/s (54-180 km/h). With such damper resistance coefficients and its installation angles the required ratio of damping forces in the vertical and transverse directions is ensured, and the values of the drag coefficients are close to 0.25 of βcr.

The standard damper of the KCCW-CRI bogie has a drag coefficient determined from a bench chart of the order of β = 12 t·s/m. The cut-off diagram of the safety valve is about 1.5 tons. For the indicated damper parameters calculations were made for three installation angles of 60°, 75° and 80°. From the results of the calculations, it follows that for a standard hydraulic damper the rational value of the installation is 65° -75° since at the same time all the considered dynamic indicators of the car take on minimal values.

2.4. Indicators of the smoothness of the car for typical and upgraded versions of trucks

Ultimately, the effectiveness of measures to modernize the central stage of spring suspension is determined by the indicators of smoothness in the vertical and lateral directions.

In this paper the indicators of smoothness were determined by calculation when integrating the system which describes the dynamic state of the car. The indicator of smoothness of motion according to [21] was determined by the formula

$$W = 2{,}7 \cdot K \cdot \sqrt[10]{z_0^3 \cdot \omega^5} \qquad (2)$$

where z_0 is the amplitude of oscillation;

ω is the oscillation frequency;

K is an empirical coefficient depending on the frequency and the plane in which oscillations occur.

In the calculations the values of the coefficient K were determined according to the schedule [21] showing the change in its values from the frequency and direction of oscillations.

The results of calculations of smoothness in the vertical W_z and horizontal W_y directions are presented in table 1 and figure 5.

Table 1. Typical option.

V, м/с	ω, с$^{-1}$	z_0, м	y_0, м	W_z	W_y	k_z	k_y
10	2.512	0.0117	0.0161	1.106	1.328	1.10	1.20
15	3.678	0.0136	0.0142	1.550	1.830	1.20	1.40
20	5.024	0.0171	0.0095	2.155	2.090	1.21	1.40
25	6.280	0.0227	0.0081	2.590	2.260	1.22	1.45
30	7.536	0.0590	0.0137	3.698	2.662	1.21	1.35
35	8.792	0.0257	0.0102	3.055	2.500	1.20	1.30
40	10.048	0.0175	0.0099	3.020	2.560	1.19	1.20
45	11.300	0.0081	0.0069	2.430	2.337	1.15	1.18
50	12.560	0.0062	0.0071	2.340	2.500	1.10	1.15

Variants were calculated for the following path condition: the lengths of the vertical and horizontal irregularities were 25 m; amplitudes of vertical and horizontal irregularities of 0.02 m; track width 1520 mm.

A typical bogie variant (Table 1) had a hydraulic damper angle of 45° and friction dampers with a friction force of 0.07 tons. A modernized version of the trolley with separate damping of vertical and lateral oscillations in the central suspension (Tables 2, 3) was investigated.

The data of the Table 1 shows that the maximum values of the smoothness indicators in the vertical direction are 3.698 and in the horizontal 2.662 at a speed of 30 m/s (108 km/h) which does not fit the norms of satisfactory performance equal to 3.25.

The upgraded version (Tab. 2, 3) with separate damping of oscillations in the central stage has the best smoothness. This option has indicators of smoothness in the vertical and lateral directions respectively

2.847 and 2.088 at speeds of 30 and 45 m/s. In other words when using the upgraded version of the bogie, it is possible to reduce the maximum values of the smoothness indicators by 25-30% compared to the standard version.

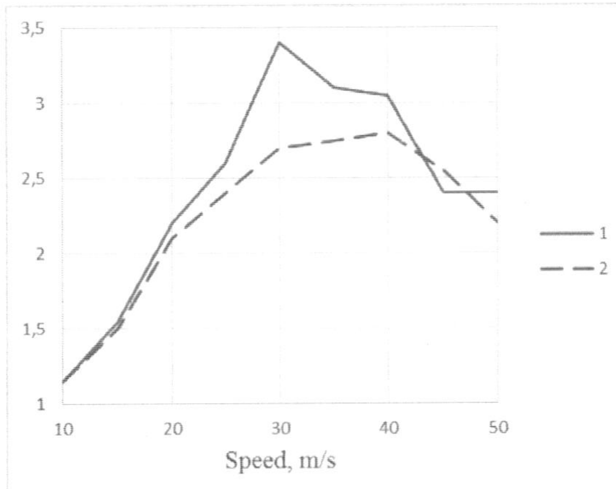

Figure 5. Vertical ride index:
1 - standard option; 2 - separate cancellation.

Table 2. Separate cancellation (I option, βz = 5 t·s/m and βy = 2.5 t·s/m).

V, м/с	ω, c^{-1}	z_0, м	y_0, м	W_z	W_y	k_z	k_y
10	2.512	0.0117	0.0333	1.105	1.651	1.10	1.20
15	3.678	0.0139	0.0111	1.556	1.696	1.20	1.40
20	5.024	0.0179	0.0055	2.185	1.774	1.21	1.40
25	6.280	0.0298	0.0057	2.812	2.035	1.22	1.45
30	7.536	0.0318	0.0061	3.072	2.088	1.21	1.35
35	8.792	0.0219	0.0055	2.912	2.084	1.20	1.30
40	10.048	0.0139	0.0051	2.817	2.102	1.19	1.20
45	11.300	0.0094	0.0049	2.539	2.142	1.15	1.18
50	12.560	0.0063	0.0053	2.250	2.230	1.10	1.15

Table 3. Separate blanking (II variant, βz = 7 t · s/m and βy = 2.5 t·s/m).

V, м/с	ω, c^{-1}	z_0, м	y_0, м	W_z	W_y	k_z	k_y
10	2.512	0.0116	0.0302	1.103	1.603	1.10	1.20
15	3.678	0.0136	0.0118	1.545	1.728	1.20	1.40
20	5.024	0.0169	0.0062	2.247	1.839	1.21	1.40
25	6.280	0.0220	0.0054	2.567	2.000	1.22	1.45
30	7.536	0.0247	0.0055	2.847	2.024	1.21	1.35
35	8.792	0.0201	0.0049	2.838	2.013	1.20	1.30
40	10.048	0.0146	0.0049	2.859	2.078	1.19	1.20
45	11.300	0.0095	0.0045	2.547	2.088	1.15	1.18
50	12.560	0.0054	0.0049	2.148	2.181	1.10	1.15

3. Conclusions

1. The calculations have established that the critical value of the drag coefficient of the hydraulic damper in the central suspension of the trolley is 20 t·s/m with vertical oscillations and 10 t·s/m with lateral ones.

2. The rational resistance coefficients for separate oscillations are 25% of their critical values.

3. When inclined installation of hydraulic dampers in the bogie, the angle of inclination of the damper to the horizontal plane should be 60-70 degrees for joint damping of vertical and lateral oscillations.

4. In case of separate damping of vertical and lateral oscillations in the central stage rational values of the resistance coefficients of hydraulic dampers should be in the vertical direction 5-7 t·s/m, in the lateral direction 2.5 t·s/m.

5. The proposed version of the modernization of the bogie made it possible to obtain smoothness indicators by 25-30% less than the allowable equal to 3.25 and determining the level of «satisfactory» progress.

References

[1] Pradhan S, Menezes A, Samantaray A K and Bhattacharyya R 2014 Performance evaluation of steering bogies on various tracks *Proc. Int. Conf. on Theoretical, Applied, Computational and Experimental Mechanics* (IIT Kharagpur, India) ICTACEM-2014/221
[2] Voron O A, Bulavin Ju P and Volkov I V 2018 K voprosu vybora hodovyh chastej dlja perspektivnogo izotermicheskogo podvizhnogo sostava [On the choice of running gears for prospective isothermal rolling stock] *Vestnik RGUPS* iss 4 pp 63–70
[3] Zhang T, True H and Dai H 2018 The influence of the perturbation of the wheel rotation speed on the stability of a railway bogie on steady curve sections of a track *Vehicle Syst Dyn* doi: 10.1080/00423114.2018.1469778
[4] Levin A V, Pavljukov A Je and Smol'janinov A V 2014 Ocenka dempfirujushhih svojstv buksovogo podveshivanija mnogoosnyh telezhek gruzovyh vagonov [Evaluation of the damping properties of the axle suspension of multi-axle freight car bogies] *Transport Urala* [*Transport of the Urals*] iss 2(41) pp 27-32
[5] Ushkalov V F, Mokrij T F and Mashhenko I A 1996 Vlijanie parametrov hodovyh chastej passazhirskih vagonov na processy vzaimodejstvija i iznos koles i rel'sov [Influence of parameters of chassis parts of carriages on interaction processes and wear of wheels and rails] *Theses of the reports of the IX Int. Conf. "Problems of railway mechanics"* (Dnepropetrovsk) p 134.
[6] Hohlov A A 1981 Parametry perspektivnyh dvuhosnyh telezhek vagonov [Parameters of promising two-axle wagon bogies] *Trudy VNIIZhT [Proceedings of the Railway Research Institute*] iss 639 pp 51-60
[7] Romen Ju S 2015 Faktory, obuslovlivajushhie processy vzai-modejstvija v sisteme koleso-rel's pri dvizhenii poezda v krivyh [Factors contributing to the interaction processes in the wheel-rail system when the train moves in curves] *Vestnik VNIIZhT* [*Vestnik of the Railway Research Institute*] no 1 pp 17-26
[8] Pevzner V O, Romen Ju S, Orlova A M and Zavertaljuk A V 2010 Jeksperimental'nye issledovanija po ocenke vlijanija shiriny kolei i sostojanija hodovyh chastej na uroven' bokovyh sil [Experimental studies to assess the influence of gauge width and condition of the running gears on the side curveslevel] *Vestnik VNIIZhT* [*Vestnik of the Railway Research Institute*] no 2 pp 39-41
[9] Vuong Q T, Ermoshenko Yu V 2018 Opredelenie dinamicheskih reakcij v soedineniyah elementov podvesok: novye podhody [Determination of dynamic responses in connections of suspension elements: new approaches] *Sovremennye tekhnologii. Sistemnyj analiz. Modelirovanie* [*Systems. Methods. Technologies*] no 2(58) pp 118-125 doi: 10.26731/1813-9108.2018.2(58).118-125
[10] Kargapoltsev S K, Kuptsov Y A, Novoseltsev P V, Gozbenko V E 2018 Sposob opredeleniya koefficienta scepleniya kolesnoj pary s rel'sami pri tormozhenii [A method for determining the coefficient of wheel pair adhesion with rails at braking] *Sovremennye tekhnologii. Sistemnyj analiz. Modelirovanie* [*Systems. Methods. Technologies*] no 2(58) pp 112-117 doi: 10.26731/1813-9108.2018.2(58).112-117
[11] Akhmadeeva A A, Gozbenko V E and Kargapolcev S K 2014 Vertikal'naja dinamika vagona s uchetom nerovnostej kolei [Vertical dynamics of the carriage), *Sovremennye tehnologii. Sistemnyj analiz. Modelirovanie* [*Systems. Methods. Technologies*] no 3(23) pp 57-62.
[12] Burchak G P., Savos'kin A N., Fradkin G N and Kossov V S 1997 Modelirovanie vozmushhenija v vide gorizontal'noj nerovnosti osi puti dlja issledovanija izvilistogo dvizhenija rel'sovogo jekipazha [Simulation of disturbances in the form of horizontal irregularities of the axis of the path for the study of the winding movement of the rail carriage] (Moscow: Proc. MSTU) iss 912 pp 23-29
[13] State standard 34093-2017 Carriage of locomotive thrust. Requirements for strength and dynamic qualities
[14] Petrov G I and Tarmaev A A 2018 Modeling of railway vehicles movement having deviations in the content of running parts *Proc. Int. Conf. "Aviamechanical Engineering and Transport" (AVENT 2018)* (Atlantis Press, Series: Advances in Engineering Research) vol 158 pp 410-415 doi: 10.2991/avent-18.2018.79

[15] Vershinskij S.V., Danilov V.N., Husidov V D 1991 *Dinamika vagona* [*Dynamics of the car*] ed S.V. Vershinskij (Moscow: Transport)

[16] Ushkalov V F , Reznikov L M, Ikkol V S et al 1989 *Matematicheskoe modelirovanie kolebanij rel'sovyh transportnyh sredstv* [*Mathematical modeling of oscillations of rail vehicles*] ed V F Ushkalov (Kiev: Naukova dumka)

[17] Hohlov A A 1981 Optimal'nye zakony upravlenija dinamicheskimi processami vagonov [Optimal laws for controlling dynamic processes of cars] *Trudy MIIT* [*Proceedings* MIET] iss 679 pp 42-60

[18] Hohlov A A 1982 Postroenie edinoj matematicheskoj modeli kolebanij mnogoosnyh jekipazhej [Building a unified mathematical model of oscillations of multi-axle crews] *Vestnik VNIIZhT* [*Vestnik of the Railway Research Institute*] no 3 pp 23-25

[19] Husidov V D, Zaslavskij L V, Chan F T, Husidov V V 1995 Tsifrovoe modelirovanie kolebaniy passazhirskogo vagona pri dvizhenii po pryamym i krivolineynym uchastkam puti [Digital simulation of carriage oscillations while driving along straight and curved sections of the track] *Vestnik VNIIZhT* [*Vestnik of the Railway Research Institute*] no 3 pp 18-25

[20] Husidov V V 1995 Komp'juternoe modelirovanie dvizhenija passazhirskih vagonov po prjamym i krivolinejnym uchastkam [Computer simulation of the movement of carriages in straight and curved sections] *Theses of the Reports of the Republican Sci. and Methodical Conf. "The use of computers in the educational process and scientific research"* (Gomel) pp 81-82.

[21] Chelnokov I I 1975 *Gidravlicheskie gasiteli kolebanij passazhirskih vagonov* [*Hydraulic dampers for carriages*] (Moscow: Transport)

The application of relational interactive logic in control operation problems by the example of monitoring the server equipment of transport systems

D V Zakharov[1], O V Kuzmin[1]

[1] Irkutsk State University, 1, Karla Marksa st., Irkutsk, 664003, Russia

E-mail: zakharov-danyl@mail.ru

Abstract. The article demonstrates the use of relational interactive logic in solving the problem of monitoring the server equipment of the transport control system. An example of such a problem is the problem of analyzing log records of monitoring server hardware. The server is considered as a special case of the physical control system. The conditions for the correctness of the current state of the equipment are defined in the form of systems of logical equations and inequalities. Solving a system of logical equations by executing a SQL query on a set of tables describing the elementary control systems in the server allows us to judge the deviation of the hardware parameters of the equipment. Recording the correctness conditions in the form of systems of logical equations provides the ability to take into account both the peculiarities of the functioning of the server hardware and some external context within which the server operates. The considered solution demonstrates the applicability of relational interactive logic in solving the problem of controlling individual classes of control systems within the scope of the system analysis discipline.

1. Introduction

With the ubiquity of automation tools, the problem of ensuring the reliability and control of various kinds of automated systems is becoming more and more urgent every day. There are a significant number of various techniques [1, 2] to ensure reliability and increase the resiliency of such systems as, for example, server equipment. The use of servers — highly targeted computers — can involve either the usual storage of files or their use for building high-load real-time systems.

It is worth noting that in a number of industries heavy demands are placed on the work of that kind of equipment with respect to stability and reliability [3]. For example, in transport systems, the life of passengers may depend on the performance of the server equipment. In addition, even a short-term failure in these systems can result in significant economic losses.

As part of this work, an approach to solving the problem of monitoring server hardware by analyzing log entries using relational interactive logic is demonstrated. An important feature is that the server is considered here as a special case of a physical control system, and the monitoring task is considered as the control task of this control system.

In paragraph 2, the concept of a control system is introduced, which is studied in the framework of cybernetics, and the main features are indicated.

The third paragraph describes the subject and method of relational interactive logic (RIL), introduces the necessary concepts and terminology.

IOP Conf. Series: Materials Science and Engineering **760** (2020) 012058 doi:10.1088/1757-899X/760/1/012058

Next, we consider the formalized model of the server in terms of the concept of a physical control system.

The fifth paragraph describes an example of analyzing the correctness of the state of the server through RIL. Special attention is paid to the representation of the model of the control system in terms of RIL and the relational data model (RDM).

2. The concept of control system

Server automation tools have been firmly anchored in railway transport systems. Both ticket services and train traffic monitoring function on the basis of a variety of server hardware platforms [4]. These systems must meet the highest standards of reliability. Various techniques are used for this in practice. [5]. This section discusses the general concept of a control system.

The control system (CS) is one of the central concepts of cybernetics [6]. This term has a very broad interpretation by its very nature. Cybernetics considers the control system as the unity of the scheme, function and coordinates. The scheme CS defines the structure and interconnection of the system elements. An image of the arrangement of pieces on a chessboard or a directed graph indicating the structure of computer network nodes can be used as examples of the scheme. The function, in turn, reflects possible transformations of the system, taking into account its informational nature. In the case of chess pieces, the function defines a set of possible moves of the pieces from the current position, for a computer network, the function implements possible packet routes in the network, according to the specified addressing scheme. Coordinates allow you to set and record the current state of the system.

The implementation of the scheme and functions for the same control system are ambiguous and depend on which side of the object in question is studied within the scope of a specific task.

Control systems considered in cybernetics have a number of peculiar features. These are relativity and discreteness. Relativity in this context means that the constructed model CS always reflects a real-life object with some approximation. In solving the problem, the individual, key characteristics of the object under consideration are taken into account. For example, in the context of a separate problem, the key characteristic may be the topology of a computer network — the mutual arrangement of individual nodes, but not the type of protocol used for packet transport. Discreteness is determined by the fact that there is no continuous transition between the schemes and functions. In some cases, some parameters of the system can take on values from a certain segment of real numbers. However, as a rule, for the functioning of the system, it is only important that the value of the parameter remain within some specified range, therefore we are not interested in all the information about the parameter, but only in the fact of its deviation from the indicated boundaries. The discrete nature of control systems largely determines the tools used in their study. Set-theoretic, combinatorial [7], probabilistic [8] and especially logical methods are prevalent here [9].

When solving practical problems, it is convenient to consider CS as some set of elementary control systems. For example, when considering a computer server as a control system, it can be divided into such elementary control systems as a processor, random access memory and read-only memory.

When studying CS, we can distinguish a number of tasks: the task of analyzing the CS, the task of synthesizing the CS, and also the task of controlling the CS. The task of control is to establish and localize changes that the system undergoes under the influence of external factors. In this paper, we consider the problem of detecting incorrect states when monitoring server hardware, which can be regarded as a special case of a control task for a CS.

The next paragraph describes the subject and method of relational interactive logic, and also introduces the terminology required when describing the methodology for analyzing server logs using RIL.

3. Relational interactive logic

This section discusses the basic concepts and terms of relational interactive logic [10]. The concepts of a logical equation, a logical inequality and their systems are introduced.

IOP Conf. Series: Materials Science and Engineering **760** (2020) 012058 doi:10.1088/1757-899X/760/1/012058

RIL is one of the branches of formal logic. A special feature of RIL is that the representation of formulas is implemented using relational tables and SQL queries, and the processing of tables through the execution of queries is an act of inference.

We consider the basic terms and concepts used in RIL.

3.1 Relational table
Each relational table consists of columns (they are also called fields or attributes) and rows (they are called records or tuples). Relational tables must have a number of properties [11]. Relational tables are an integral part of the tools of relational logic.

3.2 Field term
The field term is used in the construction of logical formulas and exists solely in the context of RIL. On the one hand, a term denotes a variable as a part of a logical formula, and, on the other hand, it is used in the construction of a relational query. With the help of a relational term, reference is made to relational tables and their fields. A term is a construction of the form [T].[F], where T is the table identifier, and F is the field identifier.

3.3 Constant
The constant specifies some fixed value. This value can be in the format of a text string, numbers, dates, etc. Constants are used when creating formulas for comparison operations, when restrictions are imposed on the field values of relational tables. The constant is a special case of a relational term and is a table of one field and one row.

3.4 Formula
We consider the concept of a logical formula. A formula can be atomic — indivisible or composed of atomic formulas. An example of the simplest atomic formula is the logical constants true and false. Atomic formulas are constructed using logical functions and comparison operators, from relational terms and constants. Based on atomic formulas, more complex formulas are constructed using logic connectives (NOT, AND, OR, IMP, XOR, EQV).

3.5 Logical equation
By a logical equation we mean a formula of the form: [F1 EQV F2], where F1 and F2 are formulas, and EQV means ≡ is an identity. Note that formulas F1 and F2 can, in turn, consist of other formulas, constants, and field terms. The logical equation sets the equivalence relation.

3.6 Logical inequality
By logical inequality we mean a formula of the form: [F1 IMP F2], where F1 and F2 are formulas, and IMP means => implication. A logical relation specifies the relations of consequence (derivability). In this case, the relation will be understood as a multiplace predicate or predicate property.

3.7 Systems of logical equations and inequalities
An ordinary system of logical equations and inequalities (SLEI) will be understood as a finite set of logical equations and inequalities united by the AND operator, which is equivalent to the logical conjunction operator $\wedge$. The conjunction operation provides simultaneous fulfillment of conditions. Conditions are established by equations and inequalities in the system.

4. The formalized server model as a control system
Let us consider a formalized model of the server as CS. Note that the server can be viewed as a combination of hardware and software platforms. In order to create a consistent representation of the server as a control system, it is necessary to understand which parts of it we are interested in. As part of

our task, we will talk exclusively about the control of hardware parameters, such as, for example, processor temperature, voltage on the power supply unit, or the percentage of RAM load.

The server will be considered as a discrete set $S=\{s_1, s_2 ... s_n\}$ of its hardware parts where s_i is some device included in the server ($i=\overline{1...n}$, n is a number of devices). It is worth noting that each device, in turn, may have a number of parameters that must be monitored. The set of parameters for all devices ultimately determines the current state of the server. In this case, it is convenient for us to move from considering the server as a whole to its elementary control systems.

For example, a processor can be considered as an elementary control system, which has a number of its parameters. These can be current temperature, power consumption and load percentage. Since we are not interested in the design and technical features of the device, we can represent it as a tuple $s_i=\langle s_i^j \rangle, (i=\overline{1...n}, j=\overline{1...m})$, where s_i^j is the specific device parameter. The tuple of these parameters can be considered as the scheme of the elementary CS (Fig. 1).

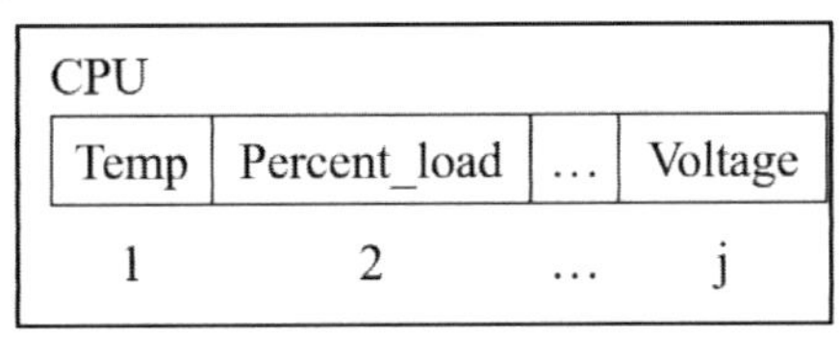

Figure 1. Representation of the processor as a tuple of parameters.

For the CS function we take the following function:

$$F_i(s_i)=\begin{cases}1, \forall s_j^i \in v_j^i, \\ 0, \exists s_j^i \notin v_j^i,\end{cases} \quad (1)$$

which takes the value 1 – true if all parameters are in the specified range and 0 –false if at least one of the components of the tuple deviates from the specified value (Fig. 2). Valid ranges are specified by a tuple v_i, each component v_i^j contains the corresponding allowable interval. The coordinates of this elementary control system will be specified by a pair $\langle s_i, t_k \rangle$, where t_k is the point in time at which the recording of device parameters occurs.

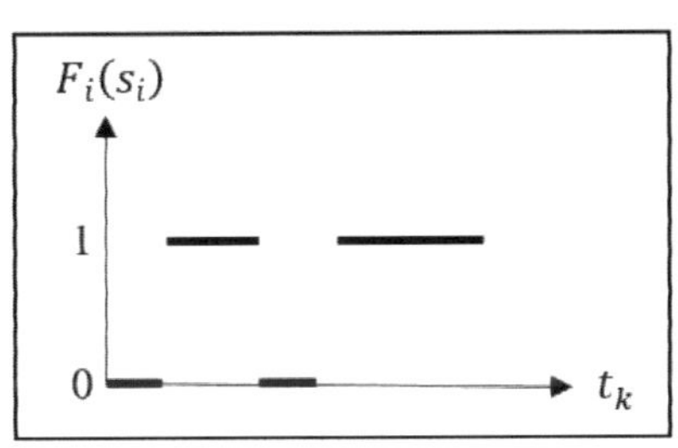

Figure 2. An example of the graph of the elementary CS state function.

In turn, the server as a set of control systems will be a set of tuples $S=s_i$, each of which contains the parameters of the elementary systems being a part of the server (Fig 3.). For the server as a management system, we define the function $P(S)=F_1 \wedge F_2 \wedge ... \wedge F_i \wedge F_{i+1} \wedge F_n$.

If the value of this function becomes false, it means that some of the server parameters deviated from the specified conditions.

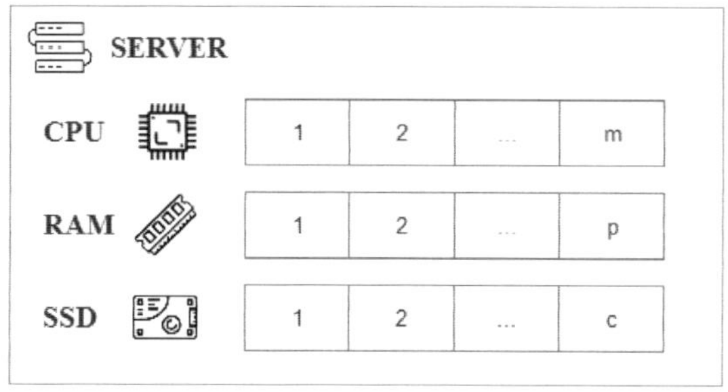

Figure 3. Representation of the server as a set of tuples of parameters of elementary control systems.

Thus, we can say that the server state is characterized by a set of relations, where each relationship is an operation of comparing the components of a tuple with a certain numerical value. For example, let s_1^1 be the temperature of the processor, then $v_1^1 = (35;70)$. That is, CPU temperature cannot exceed 70 degrees and fall below 35 degrees.

5. The examples of control ratios

Having described the model of our control system, we can describe the control procedure with the help of RIL. To work with RIL, we need to translate the model of our system within the scope of RIL terminology and the relational data model.

As mentioned above, the main element of the formulas in the RIL is the field term. A term is a collection of a table and its field, for example, [Table_one]. [Field_one]. Note that in this case we can consider the table as a representation of a particular elementary system, and the field as one of the components of the tuple. Then, s_i will be represented by the corresponding table [Table_si], and s_i^j will be represented by its field [Field_sj]. The rows of the table will represent the state of the elementary system at a particular point of time (Table 1).

Table 1 Processor parameters in the context of the points of time

t_k	Temp	Percent_load	...
1	40	12	-
2	60	45	-
3	90	89	-
4	76	60	-
...	...	...	...

Hence for each of the tuples s_i we obtain a relational table. The structure of the table columns will coincide completely with the corresponding scheme of the elementary system. So, for example, for the tuple s_1 we will create a CPU table that will contain the Temp, Percent_Load fields, and so on. When writing a specific relationship for specific parameters, we will use the corresponding relational term. We can store values of the tuple v_1 containing sets of admissible values in separate tables; within the framework of RIL, they will act as constants in the construction of formulas.

Consider an example of a ratio that imposes a limit on the CPU temperature: [CPU].[Temp] > 35 AND [CPU].[Temp] < 70 EQV FALSE.

Using this relationship, we set a condition under which we consider incorrect all processor states under which the temperature is outside the range (35;70). To describe the complete set of incorrect states

IOP Conf. Series: Materials Science and Engineering **760** (2020) 012058 doi:10.1088/1757-899X/760/1/012058

for the processor, we need to supplement the first condition with the rest. For example, we write the condition on the percentage of RAM usage: [CPU]. [Percent_Load] <75 EQV FALSE.

Next, combine the two ratios into one: ([CPU].[Temp] > 35 AND [CPU].[Temp] < 70) AND ([CPU].[Percent_Load] < 75) EQV FALSE.

We have obtained a logical equation, in the left-hand side of which there are several relations connected by a logical AND. We used the logical AND, since we consider the server state to be correct, in which all its parameters are within the specified limits. To search the database for incorrect server states, we need to solve this equation, i.e. find the values of terms for which the left-hand side of the equation will become false. From the point of view of the RDM, we need to find in the database the lines for which the condition, written using the equation, would be fulfilled. To do this, based on the written equation, you need to create a SQL query to the database:

SELECT Temp, Percent_Load
FROM CPU
WHERE ([CPU].[Temp] > 35 AND [CPU].[Temp] < 70) AND ([CPU].[Percent_Load] < 75) EQV FALSE

The expression after the SELECT statement indicates which fields we want to see in the selection, the FROM clause indicates which table we want to select, and the WHERE clause specifies the selection condition — in our case, this is a logical equation. Note that in the WHERE clause we can apply various operations and perform a selection from several tables at the same time.

If, after performing a query on a set of tables, a solution is found, i.e. some strings are returned, this will mean that there are records in the logs with incorrect parameter values, and our server has deviated from the correct state.

To write the general correctness condition for the entire server, we need to write the correctness conditions for each elementary control system (for brevity, we denote them as Cond_si) and combine them as a system of logical equations: ((Cond_s1 EQV TRUE) AND (Cond_s2 EQV TRUE) AND ... (Cond_si EQV TRUE)) EQV FALSE

It is important to note that this is a relational representation of the function discussed above.

If necessary, we can flexibly change the conditions of correctness, add new or delete existing elementary systems or their parameters, we only need to rewrite the SLEI.

Separately, it is worth noting that logical equations allow us to write fairly complex correctness conditions taking into account the context of external factors.

Imagine that we need to monitor the percentage of memory usage. Suppose that in the normal state of the server, this figure should not exceed 65%. It is also known that a daily backup procedure is performed automatically between 00:00 and 03:00, which simultaneously significantly increases the load on the server, in particular, on the RAM, increasing its load percentage [12]. In this case, exceeding the specified limit is permissible and is not a deviation from the normal state.

Using RIL, we can write a control ratio, which, in addition to the direct restriction on the indicator, will take into account some external context, under which the condition must be fulfilled or not fulfilled. In this case, that external condition is the current time.

We write this expression:

([RAM].[PERCENT_LOAD] > 65) AND (Time() < 00:00 AND Time() > 03:00) EQV TRUE

Note that in this expression, in addition to relational terms and constants, we also used the utility statement, which returns the current time. When an expression is translated into a SQL query, this statement can be converted to a SQL CURTIME() statement, which returns the current time.

This example shows that we can not only directly control the parameter values, but also take into account various extra conditions. This is especially important when solving these problems in real conditions, since very often the surrounding context plays a significant role in assessing the correctness of the system state.

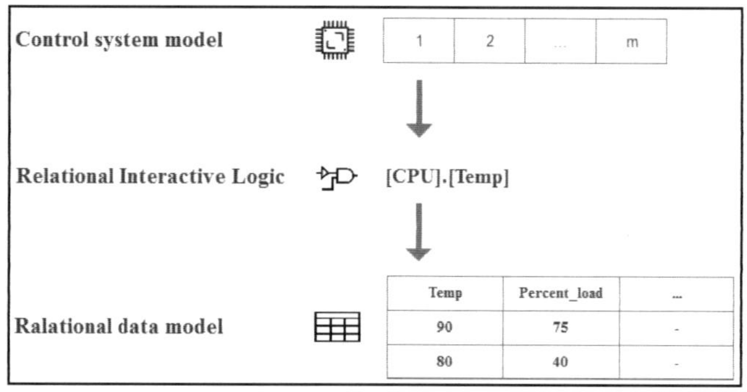

Figure 4. Representation of a CS as a formalized model in the framework of RIL and in the relational data model.

6. Conclusion

By the example of the task of analyzing the hardware logs of server equipment, this article has demonstrated the use of relational interactive logic in solving the problem of monitoring the control system.

The use of systems of logical equations and inequalities in describing the conditions of the correct state of the system makes it possible to quite flexibly consider various sets of conditions and relationships, both between parts of the system itself and between the system and external factors.

Automating the SLEI solution by means of a SQL query provides an efficient process of analyzing the CS.

The process of using RIL for tasks of this kind can be divided into four steps:

1. Preliminary analysis of the system. Identification of key parameters, characteristics and elementary subsystems.

2. Building a formalized model of CS. Depending on the task, a decision should be made on the form of representation of the scheme and function both for elementary subsystems and for the entire control system as a whole.

3. Record the formalized model of CS in terms of RIL. Formation of a set of relational terms and constants necessary for building formulas. Formation of the list of relational tables for elementary subsystems.

4. Record and decision SLEI. Analysis of the results.

It should be noted that hypothetical limitation of being applied for some classes of systems is one of the disadvantages of this method. For example, it is problematic to use this method for complex hierarchical systems with a large number of cyclic links. While implementing step 3, the question arises about the validity of the CS model transformations into the terms of the RIL.

At the moment, the authors set out to continue testing this method in other practical tasks, and also to consider the possibility of formalizing the process of converting a CS model to an RIL. In particular, it seems very promising to use this technique in analyzing the correctness of data in various automated accounting systems.

In addition, the project of the "Object Logical Design Tool" software application has been launched. It is assumed that this application will automatically generate the necessary list of relational terms, tables and constants in the interactive mode according to a predetermined scheme of the control system.

References

[1] Ali M, Bagchi S 2019 Probabilistic normed load monitoring in large scale distributed systems using mobile agents *Future Generation Computer Systems* **96** 148-167

[2] Wang T, Xu J, Zhang W, Gu Z, Zhong H 2018 Self-adaptive cloud monitoring with online anomaly detection *Future Generation Computer Systems* **80** 89-101

[3] Hamamoto A H, Carvalho L F, Sampaio L D, Abrão T, Proença M L 2018 Network Anomaly

Detection System using Genetic Algorithm and Fuzzy Logic *Expert Systems with Applications* **92** 390-402
[4] Nadezhnin A 2013 Datacenter Russian Railways, [Online] Avaliable: https://habr.com/ru/post/204122/ (May, 2019)
[5] Sapoznikov V V, Efanov D V 2015
Detection of dangerous errors in the working outputs of combinational logic circuits Automation on transport **2** 195-211
[6] Yablonsky S V 1959 Basic concepts of cybernetics *Problems of Cybernetics* 7-39
[7] Kuzmin O V, Seregina M V 2010 Plane sections of the generalised Pascal pyramid and their interpretations *Discrete Mathematics and Applications* **20** 377-389
[8] Kuzmin O V, Leonova O V 2000 Touchard polynomials and their applications *Discrete Mathematics and Applications* **10** 391-402.
[9] Kim M, Stehr M-O, Talcott C 2013 A distributed logic for Networked Cyber-Physical Systems *Science of Computer Programming* **78** 2453-2467
[10] Kurgansky V I 2014 Relational interactive logic. From logical equations and inequalities to answers to difficult questions (Lambert Academic Publishing) p 124
[11] Codd E F 1970 Relational completeness of data base sublanguages *Communications of the ACM* **13** 377-387
[12] Anderson K 2017 *Creating a Data-Driven Organization* (O' Reilly) p 336

IOP Conf. Series: Materials Science and Engineering **760** (2020) 012059 doi:10.1088/1757-899X/760/1/012059

Reconfiguration of intelligent electrical distribution networks of railways

E A Tretyakov[1], **V T Cheremisin**[1], **G E Golovnev**[1]

[1] Omsk Transport University, 35, Karla Marksa ave., Omsk, 644046, Russia

E-mail: Eugentr@mail.ru

Abstract. The Digitalization of the power supply system and the introduction of sources of distributed generation based on renewable energy create new conditions for improving the efficiency of transport and distribution of electric energy. In normal operation, the 35/10 (6) kV distribution electric networks of railways have an open-ended structure and are partitioned in such a way as to be able to connect a group of substations to another power source in emergency or repair modes. Existing technologies for managing modes in these electrical networks do not imply the possibility of changing their topology to ensure the tasks of reducing overload and electricity losses in the pace of processes of change in demand for electricity in normal modes. As a rule, at present, the task of reconfiguring the electrical network arises when normal conditions are restored after accidents, when lines are overloaded, and if necessary, electrical equipment needs to be repaired according to the n-1 criterion, while optimizing the modes. The most important task of increasing the transmission capacity of distribution grids can be solved by reconfiguring them at the pace of changing the demand for electricity in normal conditions, without resorting to expensive reconstruction and increasing the installed capacity of electrical installations and lines. An algorithm for reconfiguring the electrical network has been developed to increase its capacity based on solving problems of optimizing normal modes (static reconfiguration) and minimizing power consumption without calculating steady-state modes in the pace of electricity demand changes (dynamic reconfiguration).To ensure the admissibility of the modes of distribution electric networks, approaches to managing the supply of active (based on renewable sources) and reactive capacities and the demand of active consumers have been proposed. There are the results of the implementation of dynamic reconfiguration for the test circuitry without taking into account the power supply and demand management of active consumers, indicating the validity of the proposed approaches to improving the capacity of the electrical network

1. Introduction

At present a digitalization of railway electric power supply system opens new opportunities for transport efficiency increase and electric energy distribution.

In normal work mode distributive railways electric networks 35/10(6) kV have an open structure and are segmented so, that they have a connection opportunity of one group of substations to another power supply in emergency or repair modes.

Existing technologies of modes control in the specified electric networks do not assume change opportunities of their topology for tasks providing for overload and electric energy losses reduction at a pace of demand change processes for electricity in normal modes. As a rule, a task of an optimal electric

IOP Conf. Series: Materials Science and Engineering **760** (2020) 012059 doi:10.1088/1757-899X/760/1/012059

network configuration appears during restoration of normal mode after accidents, by lines overload, by necessity of electrics output in repair by criterion *n*-1, by modes optimizations.

In this research we will consider an approaches realization to electric schemes reconfiguration in a railways electric power supply system of non-traction and outside consumers for tasks providing for voltages stabilization, overload and electric energy losses reduction at a pace of demand change processes for electricity in normal modes. An optimization task of electric network structure by change of active and reactive capacities in last years is widely studied in foreign and domestic scientific circles.

As a whole the given task can be considered as an optimization nonlinear of mixed integer programming [1].

A part of researchers follow heuristic task solution methods of an optimal electric network reconfiguration [2-5], which allow substantially reducing computing and и temporary resources, and finding approximate solutions. An optimization task of combinatorial type can be solved by some authors with the help of salute (fireworks) algorithms, neural networks, fuzzy sets [6-9] mainly in the conditions of limited initial information.

Heuristic solution methods of combinatorial optimization tasks showed its high efficiency in comparison with c classic enumeration methods by significant number of objective function arguments, when accuracy isn't defining, and speed of calculations is defining.

Optimization methods, which are bond with search on graphs, are also used by search task solutions of optimal electric network topology first because of unambiguous solvability [10].

Although a problem of minimum tree in graphs theory is well studied [11], solution of given task isn't simple due to the fact, that for creation of an optimal topology it is necessary not only to minimize resistance of distributive network lines, but also electric energy losses, to determine capacity flows for all variants of network structure, and also to consider restrictions of mode parameters and exploitation, electric power supply reliability and other.

In this study it is proposed to solve the task of an optimal radial electric schemes reconfiguration for overload restriction and electric energy losses reduction in a perspective system of railways electric power supply. Herewith as an optimization criterion of electric network topology there is minimum electric energy consumption by operational restrictions by electric power supply schemes, permissible mode parameters, lines overloads etc.

2. Materials and methods.

Distributive railways electric networks 6-10 kV are performed open (fig. 1), as a rule, by loop or combined schemes, therefore the variants number of their reconfiguration is enough limited, as well as by operational modes.

The task is to define, which of switches must be open to provide the largest electricity consumption reduction in the electric network with specified loads.

If for all combinational switches variants we will perform a full current distribution calculation and/or solve an optimization nonlinear task of integer programming, then it will be unacceptable for operational control of electric network reconfiguration by overload and/or for electric energy losses reduction.

Therefore the purpose of this work is creation of more effective calculations ways of electric network spanning graph creation in terms of speed and calculations correctness by considered reconfiguration.

For this purpose we proposed a configuration task of an open distributive electric network to divide into two subtasks, as a dynamic and static reconfiguration, and to solve them by different methods.

An electric network reconfiguration for fast removal of elements overload is performed on base of an agent approach by method of branches and borders using the recurrent relations for electric energy losses determination for all areas without a full calculation of a fixed mode (in contrast to [12]).

In normal mode for choose of an electric network topology with minimum electric energy consumption (with minimum electric energy losses) a nonlinear task of fixed modes optimization is solved, which imitate a reconfiguration, subject to prognostic loads values.

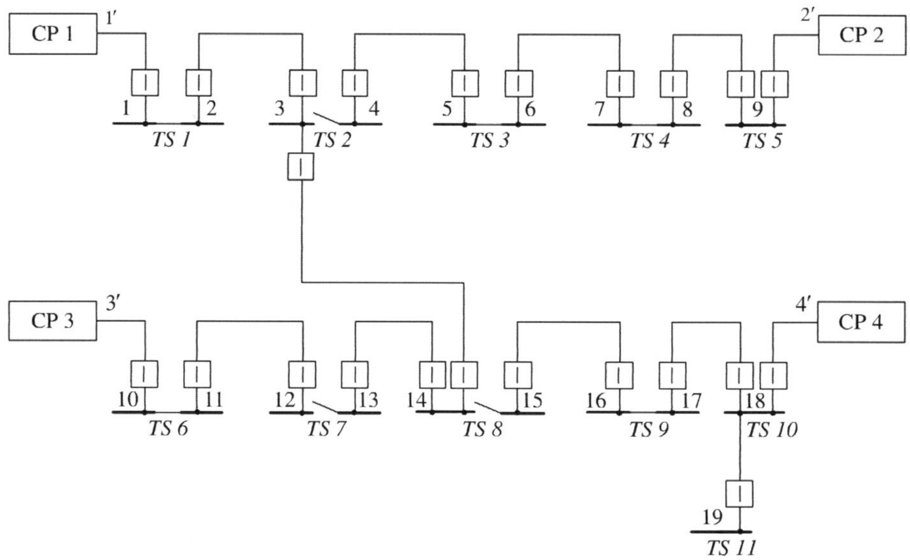

Figure 1. Distributive electric network scheme 10 kV of a railway junction

The given in fig. 1 distributive electric network we denote as a graph *G (N; E)*, where *N*– bus bars: $N = \{1, 2, ..., m\}$, and $(i, j), (i, j) \in E$ – lines (arcs, running from bus bar *j* to bas bur *j*). This graph can be divided into four sub graphs (by the number of feeding centers of an open electric network): G_1 $(N_1; E_1)$, G_2 $(N_2; E_2)$, G_3 $(N_3; E_3)$, G_4 $(N_4; E_4)$. Herewith $N_1(i) = \{j | (i, j) \in E\}$, $N_2(i) = \{j | (j, i) \in E\}$ etc.

A model of capacity flows for the radial electric network can be presented as:

$$P = \sum_{i \in N} \sum_{j \in N(i)} \delta_{ij}(P_{Lj} - P_{Gj}) + \sum_{i \in N} \sum_{j \in N(i)} \delta_{ij}(\Delta P_{ij});$$
$$Q = \sum_{i \in N} \sum_{j \in N(i)} \delta_{ij}(Q_{Lj} - Q_{Gj}) + \sum_{i \in N} \sum_{j \in N(i)} \delta_{ij}(\Delta Q_{ij}), \quad (1)$$

where P_{Lj}, P_{Gj} и Q_{Lj}, Q_{Gj} – active and reactive load capacities and injections on bus bur *j*; $\Delta P_{ij} = \frac{P_{ij}^2 + Q_{ij}^2}{U_j^2} r_{ij}; \Delta Q_{ij} = \frac{P_{ij}^2 + Q_{ij}^2}{U_j^2} x_{ij}$ – active and reactive capacity losses in lines *ij*; P_{ij}, Q_{ij} – active and reactive capacity overflows in lines *ij*; r_{ij}, x_{ij} – active and inductive lines resistances *ij*; δ_{ij} – binary variable ($\delta_{ij} = 0,1$ $(i, j) \in E$).

For considered open electric network (fig. 1) with four feeding centers an optimization task of dynamic reconfiguration based on (1) can be written as:

$$\sum_{k=1}^{4} P_k(\delta_{ij}) \to \min_{\delta_{ij}=0,1\ (i,j) \in E}, \quad (2)$$

by restrictions:

$$\sum_{\substack{j\in N_1 \\ j\in N_3}} P_{ij} - \sum_{\substack{j\in N_2 \\ j\in N_4}} P_{ji} \begin{cases} \le P_i^{ist}, \, i \in N_{1'}, N_{2'}, N_{3'}, N_{4'} \\ \le -(P_{Lj} - P_{Gj}), \, j \in N_{\text{н}}, \end{cases}$$

$$\sum P_j^{ist} > \sum (P_{Lj} - P_{Gj}), \; \Delta P_{\substack{ij \\ i=j}} = 0;$$

$$0 \le P_{ij} < P_{ij}^{\max}; 0 \le Q_{ij} < Q_{ij}^{\max}; \tag{3}$$

$$U_j^{\min} \le U_j < U_j^{\max};$$

$$\sum_{i \ne j} \delta_{ij} = a, \; a = 1, \dots, n.$$

The solution of given optimization task (2) using the method of branches and borders will allow getting values of a binary variable, which describe an electric network graph branches condition: 0 – line is open, 1 – line is closed.

Determination of capacity flows is performed without calculation of fixed mode by given injections in nodes and topology of electric network. Permissible mode parameters of electric network by voltage are estimated by module in the line end on bus bur j from date at the line start on bus bur i as [13]:

$$U_j = \sqrt{\left(U_i - \frac{r_{ij} P_i + x_{ij} Q_i}{U_i}\right)^2 + \left(\frac{x_{ij} P_i - r_{ij} Q_i}{U_i}\right)^2}. \tag{4}$$

Electric network reconfiguration for lines overload exception must provide a permissible new mode, if it is possible, otherwise to resort to demand control, up to loads shutdown.

Reconfiguration criteria:

– capacity balance provision, including reserve;
– exception of electric equipment overloads;
– exception of voltage and frequency unacceptable levels;
– provision of required electric power supply reliability level;
– provision of required reconfiguration frequency (threshold).

Proposed algorithm of considered electric network reconfiguration is given in fig. 2.

Reconfiguration control system provides a continuous control and network condition assessment, and in case of overload absence of electric network elements performs an optimization of electric network new topology and mode parameters subject to demand prognostic values and capacity proposals (statistic reconfiguration), for example, as it's shown in works [14, 15].

As a rule, an electric network reconfiguration for mode parameters optimization and electric energy losses reduction has a practical meaning with periodicity up to twice a day and isn't always justified from an economic point of view, what cannot be said about reconfiguration for electric equipment overload to except normal mode recovery in case of emergency disturbances.

Determination of electric network new topology by dynamic reconfiguration is performed on base of task solution (2) by method of branches and borders by keeping the conditions of mode admissibility (3). Method of branches and borders is based on following procedures: setting an original plenty of enumeration variants, choice of the most perspective lots by an original plenty partitioning, perspective lots branching into enumeration subsets [16]. For each border there is determined a low border of a target function, and for a final top there is an exact value of a minimized target function. Branching principle consists of an obligatory participation or non-participation in a tree of any branch that means that the original plenty of permissible solutions is divided in two no overlapping subsets: vectors with a fixed value 1 and 0. Herewith for each electric network tree branching capacity losses are calculated recursively by given in (1) formulas without a nonlinear equations calculation of fixed modes.

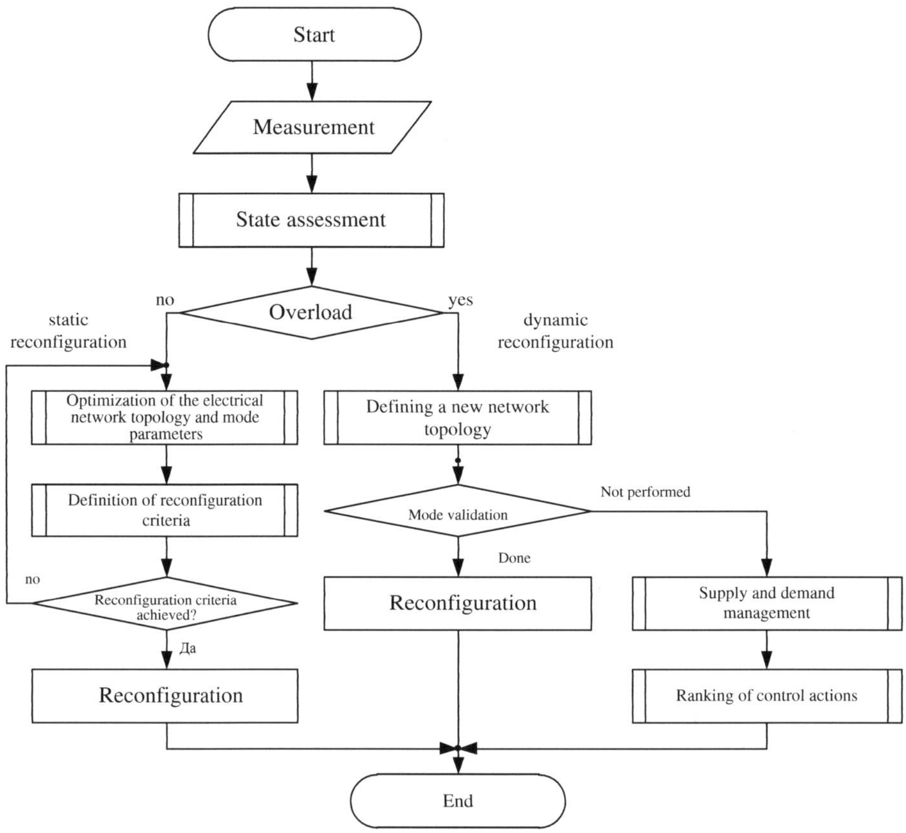

Figure 2. Electric network reconfiguration algorithm

Some authors by search of electric network flow partition point use an approach based on sign determination of capacity flow in schemas branches [16, 17] with all their enumeration. Herewith a need arises of additional electric network schema transformation for exception of blind branches branching.

According to the given in figure 2 algorithm, if an overload exception because of reconfiguration is impossible on the conditions of mode admissibility (3), then there is performed a demand control and a capacity proposal in the electric network up to the load shutdown.

As a capacity proposal there are considered sources of reactive capacity and distributive generation (in case of availability).

In this case the lines overload means an excess of not limit current load for specific wires types, but limit current values subject to provision of static stability and minimum necessary capacity reserve [13].

Along with a restriction of consumers' capacity emergency modes by appropriate automatic equipment (shutdown) there is proposed an approach with more thorough consumers' ranking in terms of a possible damage (cost) of their shutdown in real timeв and their flexibility in the demand control.

Let's consider more detailed approaches to the elements overload reduction of a distributive electric network by active capacity, as a reactive capacity balance can be provided with fixed compensation equipment of reactive capacity (a separate task outside of this research).

The main ways of lines overload reduction:

– regulation of an active capacity generator (in case of availability);
– regulation of a phase angle (by technical capability);
– setting of a capacity transit graph (import/export);
– change of an electric network topology;

– consumers' load reset;

– control of an active consumers' demand.

The load reset is the last variant, when the lines overload cannot be reduced using another methods.

Within a distributed agent control realization of capacity flows in the electric network [19, 20] by overload there are proposed the following steps of this problem solution:

– warning (availability of potential conditions for an overload appearance subject to prognostic values, structures of capacity balances and other factors);

– notification (excess of established dynamic threshold values);

– emergency mode;

– overload exception.

3. Practical part

Let us consider a realization of given approaches to the electric network reconfiguration 10 kV (fig. 1).

The main line parameters and calculated loads are given in table 1.1, 1.2.

Table 1. Lines resistances

Line	1`-1	2-3	3-14	4-5	6-7	8-9	9-2`
r, Ohm	0.329	0.266	1.119	0.419	0.464	0.354	0.291
x, Ohm	0.099	0.080	0.144	0.102	0.113	0.064	0.087
Line	3`-10	11-12	13-14	15-16	17-18	18-19	18-4`
r, Ohm	0.253	0.160	0.160	0.384	0.278	0.222	0.253
x, Ohm	0.076	0.039	0.039	0.094	0.084	0.040	0.076

Table 2. Calculated loads

Load	1	2	3	4	5	6	7	8	9	10
P, kW	1200	930	1270	1100	820	910	830	560	780	860
Q, kVar	1100	870	1140	930	620	820	730	420	650	760
Load	11	12	13	14	15	16	17	18	19	–
P, kW	760	370	1200	1270	560	715	810	1100	390	–
Q, kVar	650	290	1150	1100	430	550	725	970	300	–
Generation	3	4	5	6	7	8	12	13	14	15
P, kW	–	–	–	–	–	–	–	–	–	–
Q, kVar	400	400	–	–	–	–	–	–	–	200

According to the given in fig. 2 algorithm for electric scheme with four feeding centers in the solution result of an optimization reconfiguration task (2) there were received optimal places of lines disconnection and quantitative assessments of mode parameters (table 3).

Reconfiguration purpose is a lines 1`-1 and 9-2`overload exception.

Table 3. Capacity consumption in the electric network before and after reconfiguration

Electric network mode	Capacity consumption					
	Type	FC 1	FC 2	FC 3	FC 4	In total
Initial	P, mW	6.225	5.391	2.343	4.134	18.093
$\delta_{34}, \delta_{1213}, \delta_{1415} = 0$	Q, mVar	6.12	4.808	2.011	3.527	16.466
After reconfiguration	P, mW	5.017	4.064	4.688	3.809	17.578
$\delta_{45}, \delta_{314}, \delta_{1415} = 0$	Q, mVar	4.447	3.59	4.463	3.344	15.844

Electric network reconfiguration for the fast removal of elements overloads is performed on the basis of an agent approach (more detailed in [18]) by method of branches and borders using the recurrent relations for determination of electric energy losses for all areas without a calculation of a steady mode.

From the given in table 3 date it follows that after reconfiguration the consumed capacity in the electric network decreased by 2,9 %.

In fig. 3 there are the simulation modeling results of capacities on electric network bus bars, in fig. 4 and 5 there are lines load and capacity losses in them.

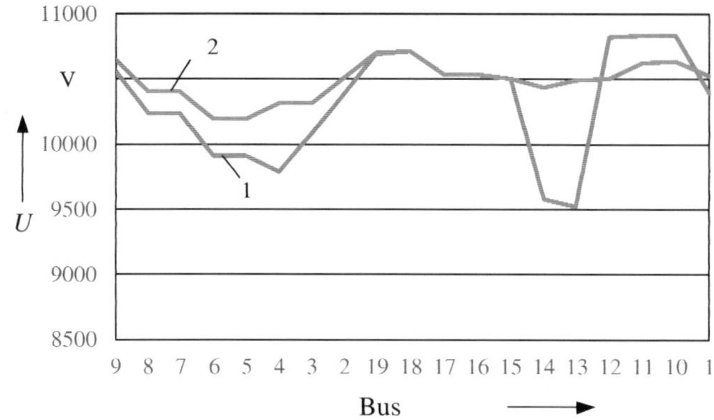

Figure 3. Capacity on electric network bus bars before (line 1) and after (line 2) reconfiguration

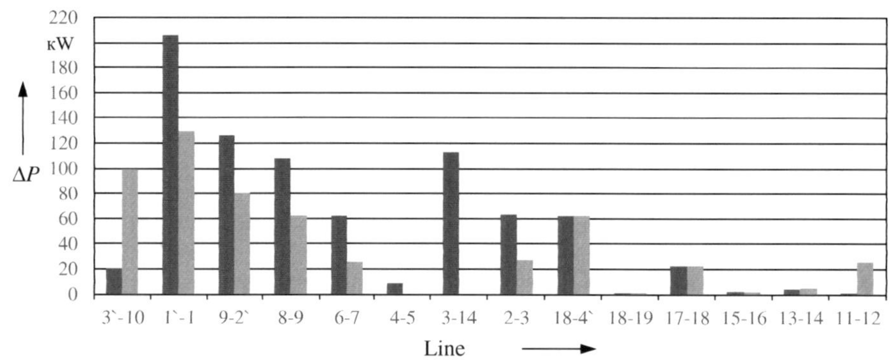

Figure 4. Active capacity losses in electric network lines before (left column) and after (right column) reconfiguration

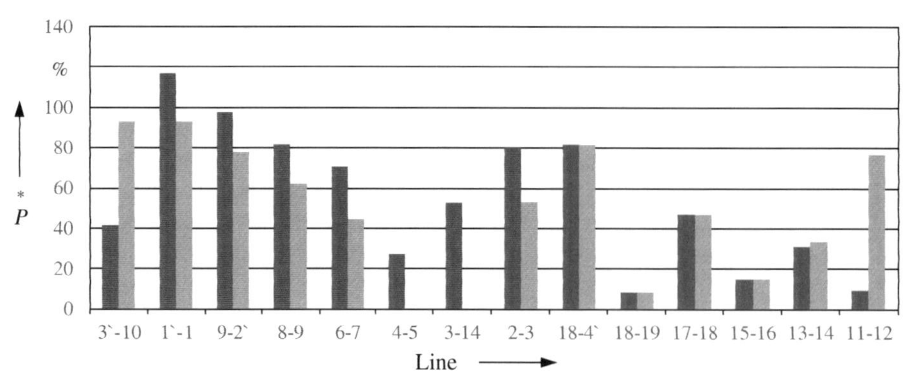

Figure 5. Load of electric network lines before (left column) and after (right column) reconfiguration

IOP Conf. Series: Materials Science and Engineering **760** (2020) 012059 doi:10.1088/1757-899X/760/1/012059

Total load losses of active capacity in electric network line are: 798,08 kW before reconfiguration and 541,9 kW after.

As it was mentioned earlier lines load is determined subject to reserve for provision of a static статической stability (it's accepted 10 %) and a minimally necessary capacity reserve (if necessary, also on condition *n*-1).

Thus, the optimal reconfiguration for the examined case provided as mode admissibility by voltage (fig.5), so as lines overload exception. Therefore in this case there is no need to resort to loads capacity restriction and demand control of active consumers that is subject of authors' research in other works.

4. Conclusions.

Research results showed validity of the proposed approach to the electric network dynamic reconfiguration for overload restriction and electric energy loss reduction without a calculation of steady modes nonlinear equations for every variant of a topology network change. Herewith as an optimization criterion of electric energy topology there is minimum electric energy consumption by operational restrictions by electric power supply schemes, acceptable modes parameters, line overloads, etc.

References

[1] Williams H P 2009 *Logic and integer programming (Heidelberg: Springer)* p464

[2] Aman M M, Jasmon G B, Naidu K 2013 Discrete evolutionary programming to solve network reconfiguration problem. *Proceedings of the TENCON Spring Conference (Sydney)* 23–29

[3] Manju M, Leena G, Saxena N S 2016 Distribution Network Reconfiguration for Power Loss Minimization Using Bacterial Foraging Optimization Algorithm. *Engineering and Manufacturing* № 2 234–239

[4] Dian N T, Hazlie M, Mohamad Sofian T, Kanendra N 2017 Parallel Power System Restoration Planning Using Heuristic Initialization and Discrete Evolutionary Programming. *Modern Power Systems and Clean Energy* . **5 6** 991–1003

[5] Huang S, Wu Q, Cheng L, Liu Z 2015 Optimal Reconfiguration Based Dynamic Tarifffor Congestion Management and Line Loss Reductionin Distribution Networks. *IEEE Transactions on Smart Grid* 10.1109/TSG.2015.2419080

[6] Santos A C, Delbem C B, London B A, Bretas N G 2010 Node Depth Encoding Andmultiobjective Evolutionary Algorithm Applied to Large Scale Distribution System Reconfiguration. *IEEE Transactions on Power Systems* **25(3)** 1254–1265

[7] Niknam T 2010 An Efficient Hybrid Evolutionary Algorithm Based on PSO and ACO for Distribution Feeder Reconfiguration. *European Transactions on Electrical Power* **20** 575–590

[8] Ola B, Saad M, Hazlie M 2017 Wardiah Dahalanc Optimal Reconfiguration of Distribution System Connected with Distributed Generations: A Review of Different Methodologies. *Renewable and Sustainable Energy Reviews* **73** 854–867

[9] Mostafa S, Reza B 2016 Optimal Multi-objective Reconfiguration and Capacitor Placement of Distribution Systems with the Hybrid Big Bang–Big Crunch Algorithm in the Fuzzy Framework. *Shams Engineering Journal* **7** 113–129.

[10] Grebanyuk G G, Krygin A A 2015 *Predel'nye grafy v strukturnoj optimizacii rezhimov raspredelitel'nyh setej* [Limit graphs in structural optimization of distribution network modes]. *Automation and Remote Control* **1** 147–162

[11] Diestel R 2005 *Graph Theory*. Springer-Verlag p 422

[12] Fishov A G, Mukatov B B 2015 Reconfiguration of electrical networks with distributed generation and multi-agent control. *Bulletin of the Tomsk Polytechnic University. Geo Assets Engineering* **326 9** 143–152

[13] Venikov V A 1998 *Elektricheskie sistemy. Elektricheskie seti* [Electrical systems. Electricity of the Net] (Moscow: Higher School Publ.) p 511

[14] Lukyanov D V, Vasilev D A, Ivashchenko V A, Lukyanov D V, Shabelnikova A YU 2010 Vybor struktury elektricheskih setej promyshlennyh predpriyatij v usloviyah avtomatizirovannogo

upravleniya elektropotrebleniem [The choice of the structure of electrical networks of industrial enterprises in the conditions of automated power consumption management]. *News of universities. Volga region. Technical science* no 2 52–61

[15] Reza J-S, Seyed-Masoud M-T, Seyed-Sattar M 2016 Microgrid Operation and Management using Probabilistic Reconfiguration and unit Commitment. *Electrical Power and Energy Systems* no 75, 328–336

[16] Peng Q, Low S H 2013 Optimal Branch Exchange for Feeder Reconfiguration in Distribution Networks. *Decision and Control (CDC): IEEE 52nd Annual Conference on. IEEE* (Italy) 2960–2965

[17] Jizhong Z 2009 Optimization of Power System Operation. *IEEE Press Series on Power Engineering* **49** 624

[18] Tretyakov E A, Golovnev G E, Galkin A G, Sidorova E A 2018 Perspektivnye metody upravleniya transportom i raspredeleniem elektroenergii v elektricheskih setyah zheleznyh dorog [Perspective methods of transport management and distribution of electricity in electric networks of railways]. *Izvestiya Transsiba* **2 (34)** 113–124

[19] Kucherov YU N, Ivanov A V, Korev D A, Utkin N A, Zhuk A Z 2018 Razvitie tekhnologij aktivnogo potrebitelya i ih integraciya v elektricheskuyu set' obshchego pol'zovaniya [The development of active consumer technologies and their integration into the public electrical network]. *Energy Policy* **5** 73–86

[20] Arefi A, Abeygunawardana A, Ledwich G 2018 A new Risk-Managed Planning of Electric Distribution Network Incorporating Customer Engagement and Temporary Solutions. *IEEE Transactions on Sustainable Energy* **7 (4)** 1646-1661

Reduction of electrical energy losses of power transformers of 25 kV traction substations

N P Grigoriev[1], M S Klykov[1], V A Tikhomirov[2], P N Trofimovich[1]

[1] Far Eastern State Transport University, 47 Seryshev Str., Khabarovsk, 690021, Russia
2 Irkutsk State Transport University, 15 Chernyshevsky Str., Irkutsk, 664074, Russia

E-mail: ptrofimovich@bk.ru

Abstract. The aim of the study is to reduce the losses of electrical energy in power transformers of AC traction substations. To reduce the losses of electrical energy, interval counterregulation of performance indicators in the traction power supply system is performed. The planning of the electrical energy losses is determined by the predictive time intervals. The electrical energy losses in the time interval are determined by the power losses of instantaneous circuits. Instantaneous circuits determine the voltage of the windings of transformers, the currents of the power supply arms of the electric traction network, the transformation ratios, the temperature of the windings and the power of regional loads. The predictive values of the instantaneous circuits are determined by the results of the monitoring of the completed train schedules, statistical analysis and simulation modeling of the operation of the traction substation. The initial data for the calculation are the predictive discrete values of the performance indicators of the traction power supply system. The calculation of power losses in steel in power transformers is performed taking into account voltage unbalance. Copper losses are determined in phases, taking into account the distribution of arm currents in the windings and the actual transformation ratios. Actual transformation ratios are determined based on the voltage control stage. The effect of winding temperature on power losses is taken into account by the temperature coefficient of resistance. The number of transformers in operation at the predictive time intervals is selected by the minimum loss of electrical energy. The accuracy of the choice of the number of transformers in operation is checked by analyzing the implementation of the forecast train schedule. The error of the initial data and simulation modeling is determined by comparing the predictive values and the monitoring results. It is proposed to carry out performance management of the traction power supply system on the basis of digital technologies.

1. Introduction

The strategy of the scientific and technical development of the Russian Railways holding for the period up to 2020 envisages an increase in the energy efficiency of the transportation process. One of the target areas is the reduction of electricity consumption for train traction. In order to improve the efficiency of the organization of transportation activities, it is necessary to move from reactive control to modeling and forecasting the development of situations [1].

It has been proposed to reduce the electrical energy loss in the traction power supply system (TPSS) by selecting the normal power supply circuits for traction loads [2]. The algorithm of choosing the number of power transformers of traction substations (traction transformers) in operation for the minimum losses is considered in [3]. The accuracy of the selected power supply circuits for traction loads is determined by matching the calculated and actual losses. This being the case, electric energy losses are determined for predictive and completed train schedules [4].

2. Problem definition

The currents of the windings of the traction transformers are determined by the currents of the arms and the power of the regional consumers. The currents of the arms are related to an interval train schedule. With that, the planned train schedule can be timely adjusted by the train dispatcher. The static display of the dynamic process of the TPSS operation at a single point in time, characterized by discrete values of the performance indicators, is called an instantaneous circuit. The monitoring system allows one to record discrete values of indicators of the TPSS performance in the form of instantaneous circuits in the database. The results obtained on the basis of DBMS and simulation modeling [5, 6] make it possible to determine the predictive integral indicators of TPSS. Power supply schemes of traction loads are determined at intervals of time [7]. It has been proposed to reduce the loss of electrical energy in the TPSS, taking into account losses in traction transformers. The number of transformers in operation is determined by [8] with interval counterregulation of performance indicators.

The extendability of the service life due to thermal wear of the insulation of the windings of traction transformers was proved in [9]. The uneven wear of the insulation occurs due to asymmetrical currents of the transformer windings. New circuits for connecting the transformer bushings to electric transmission lines and electric traction network reduce the intensity of winding wear with minimal residual life. Therefore, the calculation of power losses and electrical energy is relevant when determining the residual insulation life of the windings of the traction transformers [10, 11, 12].

As a rule, two traction transformers are used at AC traction substations. To supply traction loads, traction transformers are installed, for example, of a type TDTNZH 40000 / 220-76 U1 with a rated power of 40 MVA and a winding connection circuit Y_n/ Δ / Δ-11-11. Three-phase transformers of 220 and 110 kV voltage classes are equipped with a on-load tap changer (LTC) at the zero point of the upper voltage windings. One stage of regulation by an on-load tap changer for a traction transformer with a voltage class of 220 kV is 1% at ± 12 steps, for a voltage class of 110 kV - 1.78% at ± 9 steps, respectively. Nominal no-load losses (ΔP_{xx}) are 51 kW. Short circuit losses (ΔP_k) of 230 kV and 27.5 kV windings are 192 kW, of 230 kV and 11 kV windings are 208 kW, of 25 kV and 11 kV windings are 165 kW.

Power losses at n traction transformers of the same type (ΔP_n) are usually determined by the formula [13]:

$$\Delta P_n = n\Delta P_{xx} + \frac{\Delta P_k}{nS_{wind}^2} \times S_H^2 , \tag{1}$$

where ΔP_{xx} are the no-load losses of the power transformer, kW; ΔP_k are short circuit losses, kW; S_{wind} is the rated power of the traction transformer, kVA; S_H is the transformer load power, kVA.

Formula (1) does not take into account the asymmetry of currents and voltages, the temperature of the windings and the actual transformation ratios.

Electricity losses in the steel of a power transformer with a symmetrical voltage of the windings have been proposed [14] to be determined by the formula:

$$\Delta W_x = \Delta P_{xx} \sum_{j=1}^{k} T_{pj} \left(\frac{U_j}{U_{wind}}\right)^2 , \tag{2}$$

where T_{pj} is the operating time of the transformer at the jth actual voltage, h; k is the number of values of the actual voltage; U_j is the winding voltage at the j-th voltage value, kV; U_{wind} is the nominal voltage of the transformer winding, kV.

The special features of TPSS include a single-phase traction load. As a result, the voltages and currents of the windings of the upper, traction and regional voltage of the traction transformers of the substations are asymmetrical. Formula (2) does not take into account voltage asymmetry [15]. Determining the loss of electrical energy in steel by formulas (1) and (2) does not yield an accurate result.

3. Materials and methods

We consider the loss of electrical energy of the traction transformers at actual voltages and winding currents. Asymmetric system of voltage vectors of windings AX, BY and CZ (Figure 1.) will be obtained as the sum of symmetric systems of voltages of direct (U_{1i}) and reverse (U_{2i}) sequence [16]. Figure 1 shows the currents of the upper voltage windings ($\dot{I}_{AX}$, $\dot{I}_{BY}$, $\dot{I}_{CZ}$), of the traction voltage ($\dot{I}_{ax}$, $\dot{I}_{by}$, $\dot{I}_{cz}$) and currents of regional windings ($\dot{I}_{Pax}$, $\dot{I}_{Pby}$, $\dot{I}_{Pcz}$). The currents of the left and right power supply arms ($\dot{I}_L$, $\dot{I}_R$) are oriented with respect to the voltage of the arms ($\mathring{U}_L$, $\mathring{U}_R$). The operators of the voltage rotation of the arms ($\underline{D}_L$, $\underline{D}_R$) take into account the voltage vector in phase coordinates.

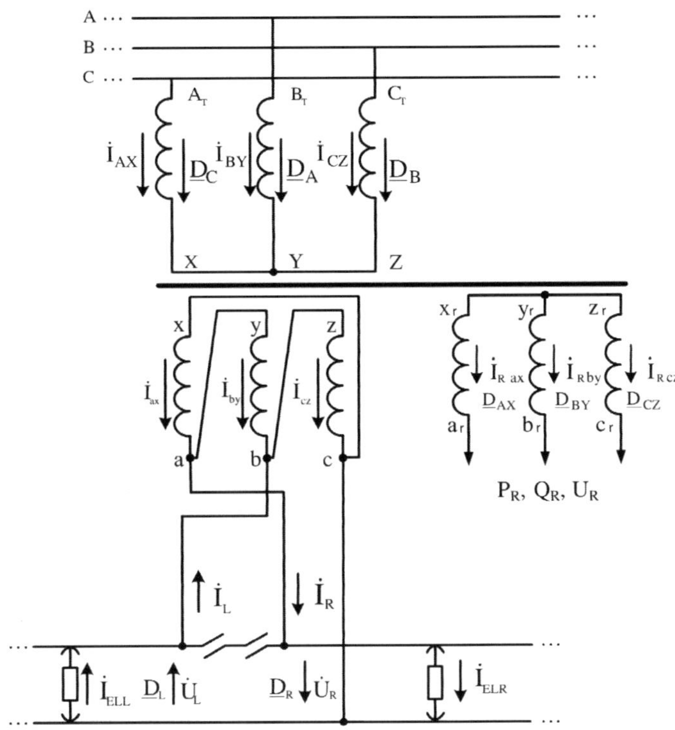

Figure 1. The power supply scheme of traction and regional loads

The actual power losses in steel from the voltage of direct (ΔP_{st1i}) and the reverse (ΔP_{st2i}) sequence are determined by the formula:

$$\Delta P_{sti} = \Delta P_{st1i} + \Delta P_{st2i}. \tag{3}$$

where i is the sequence number of the instantaneous circuit.

Power losses in steel from the voltage of direct sequence are determined by the formula:

$$\Delta P_{\text{CT}1i} = \Delta P_{xx}\,(\frac{U_{1i}}{U_{nom}})^2, \tag{4}$$

where U_{nom} is the nominal voltage of the transformer winding, kV; ΔP_{xx} are no-load losses of the traction transformer, kW.

Power losses in steel from the voltage of reverse sequence are determined by the formula:

$$\Delta P_{st2i} = \Delta P_{xx}\,(\frac{U_{2i}}{U_{nom}})^2. \tag{5}$$

Let us determine power loss in copper in traction transformers by the amount of losses on the active resistances of the windings of the upper, traction and regional voltage for each phase by the currents of the power supply arms of traction loads and power of regional loads:

$$\Delta P_{ci} = \Delta P_{Ui} + \Delta P_{Ti} + \Delta P_{Ri}. \tag{6}$$

The load power losses in the windings of the upper voltage (ΔP_{Bi}) are determined by the formula:

$$\Delta P_{Ui} = \left(I_{AXi}^2 + I_{BYi}^2 + I_{CZi}^2\right) \cdot R_{Ui} \cdot 10^{-3}. \tag{7}$$

where I_{AX}, I_{BY}, I_{CZ} is the module of currents of windings of the upper voltage AX, BY and CZ, respectively, A; R_{Ui} is the active resistance of the windings of the upper voltage, taking into account the temperature of the windings and the on-load tap-changing step, Ohm.

IOP Conf. Series: Materials Science and Engineering **760** (2020) 012060 doi:10.1088/1757-899X/760/1/012060

The effective active resistance of the winding of the upper voltage traction transformer, taking into account the temperature and the on-load tap-changing step [17], is:

$$R_{Ui} = (R_{U20} + \Delta R_{U20i})[1 + \alpha(t_{windi} - 20)]\,, \quad (8)$$

where R_{U20} is the nominal active resistance of the upper voltage windings, Ohm; ΔR_{U20i} is the increment of the resistance of the upper voltage winding, taking into account the on-load tap-changing step; α is the temperature coefficient of resistance; t_{windi} is the temperature of the windings, °C.

The modules of currents of the windings AX, BY and CZ of the transformer's upper voltage are determined by the formulas:

$$\begin{aligned} I_{AXi} &= \left|\frac{1}{n_{Ti}}\dot{I}_{axi} + \frac{1}{n_{Ri}}\underline{D}_{AX}\dot{I}_{Ri}\right| \\ I_{BYi} &= \left|\frac{1}{n_{Ti}}\dot{I}_{byi} + \frac{1}{n_{Ri}}\underline{D}_{BY}\dot{I}_{Ri}\right| \\ I_{CZi} &= \left|\frac{1}{n_{Ti}}\dot{I}_{czi} + \frac{1}{n_{Ri}}\underline{D}_{CZ}\dot{I}_{Ri}\right| \end{aligned} \quad (9)$$

where $\dot{I}_{axi}$, $\dot{I}_{byi}$, $\dot{I}_{czi}$ are the currents of the traction windings; $\underline{D}_{AX}$, $\underline{D}_{BY}$, $\underline{D}_{CZ}$ are the operators of the voltage rotation of the windings $a_r x_r$, $b_r y_r$ and $c_r z_r$ of regional loads' currents, respectively; $\dot{I}_{Ri}$ is the current of the regional load; n_{Ti}, n_{Ri} are the actual transformation ratios of the windings of the upper and traction voltage, of the upper and regional voltage, respectively.

Actual transformation ratios can be determined by formulas (10) and (11), taking into account the nominal winding ratio of the upper and traction voltage (n_{Tnom}), the windings of the upper and regional voltage (n_{Pnom}) and the on-load tap-changing step ($\Delta n_{Ti}, \Delta n_{Ri}$):

$$n_{Ti} = n_{Tnom} + \Delta n_{Ti}, \quad (10)$$

$$n_{Ri} = n_{Rnom} + \Delta n_{Ri},. \quad (11)$$

The currents of the traction windings ax, by and cz we determine through the currents of the arms and the operators of the voltage rotation of the power supply arms:

$$\begin{aligned} \dot{I}_{axi} &= \alpha_{Rax}\underline{D}_R\dot{I}_{Ri} - \alpha_{Lax}\underline{D}_L\,\dot{I}_{Li} \\ \dot{I}_{byi} &= \tfrac{1}{3}\underline{D}_R\,\dot{I}_{Ri} - \tfrac{1}{3}\underline{D}_L\,\dot{I}_{Li} \\ \dot{I}_{czi} &= \alpha_{Rcz}\underline{D}_R\,\dot{I}_{Ri} - \alpha_{Lcz}\underline{D}_L\,\dot{I}_{Li} \end{aligned} \quad (12)$$

where $\dot{I}_{Li}$, $\dot{I}_{Li}$ are the currents of the left and right power supply arms, respectively, A; $\underline{D}_L$, $\underline{D}_R$ are the operators of voltage rotation of the left and right power supply arms in phase coordinates; α_{Rax}, , α_{Lax}, α_{Rcz}, , α_{Lcz} are the distribution coefficients of currents of arms in the transformer windings.

The current of the regional load will be determined by the formula:

$$\dot{I}_{Ri} = \frac{\sqrt{P_{Ri}^2 + Q_{Ri}^2}}{k_{cx}U_{Ri}}e^{j(arctg\frac{Q_{Ri}}{P_{Ri}})}, \quad (13)$$

where P_{Ri}, Q_{Ri} is the active and reactive power of the regional load, kW and kVAr; U_{Ri} is the regional load voltage, kV; $k_{cx} = 3$ with the circuit of connecting the regional voltage windings into a "triangle"; $k_{cx} = \sqrt{3}$ with the circuit of connecting the regional voltage windings into a "star".

Power losses in the windings of the traction voltage (ΔP_{Ti}) we define by the formula:

$$\Delta P_{Ti} = (I_{axi}^2 + I_{byi}^2 + I_{czi}^2) \cdot R_{Ti} \cdot 10^{-3}, \quad (14)$$

where I_{ax}, I_{by}, I_{cz}, are the modules of currents of the transformer traction windings, A; R_{Ti} is the active resistance of traction voltage windings taking into account the temperature of the windings, Ω.

The effective resistance of the traction winding, taking into account the temperature, is calculated by the formula:

$$R_{Ti} = R_{T20}[1 + \alpha(t_{windi} - 20)], \quad (15)$$

where R_{T20} is the nominal active resistance of the traction voltage windings, Ohm.

Power losses in the windings of the regional [18] voltage (ΔP_{Pi}) we define by the formula:

$$P_{Ri} = 3I_{Ri}^2 R_{Ri} \cdot 10^{-3}, \quad (16)$$

where R_{Pi} is the effective resistance of the regional winding, taking into account the temperature and the step of the on-load tap-changer and off-load tap changer, Ohm, which we calculate by the formula:

$$R_{Ri} = (R_{R20} + \Delta R_{R20i})[1 + \alpha(t_{windi} - 20)], , \quad (17)$$

where R_{R20} is the nominal active resistance of the windings of the regional load, Ohm; ΔR_{R20i} is the increment of the regional winding resistance, taking into account the voltage tapping step .

4. Results

We determine the total power losses with one traction transformer in operation by the formula:

$$P_{1i} = \Delta P_{STi} + \Delta P_{Ui} + \Delta P_{Ti} + \Delta P_{Ri} \quad (18)$$

Power losses with two traction transformers in operation are respectively equal to:

$$\Delta P_{2i} = 2\Delta P_{STi} + \frac{1}{2}(\Delta P_{Ui} + \Delta P_{Ti} + \Delta P_{Ri}) \quad (19)$$

Thus, the losses of electrical energy for the billing period, when n traction transformers in operation, are determined by the formula:

$$\Delta W_n = (\sum_{i=1}^{m} \Delta P_{ni})\Delta t, \quad (20)$$

where Δt is the instantaneous circuit time; m is the number of instantaneous circuits in the time interval.

Reducing the electric power losses of the traction transformers is performed by selecting their number in operation for the predictive time intervals. The electrical energy losses for the billing period are determined by the time intervals, taking into account the efficient use of the resource of switching devices.

5. Conclusion

1. It is proposed to increase the efficiency of TPSS operation by choosing the number of traction transformers in operation based on digital technologies.
2. The article proposes formulas for calculating power and electrical energy losses in traction transformers with asymmetric voltages and winding currents, taking into account the winding temperature and actual transformation ratios.

References

[1] The White Book of Russian Railways 2015 The strategy of the scientific and technical development of the Russian Railways holding company for the period up to 2020 and the perspective up to 2025

[2] Vlasenko S A, Grigoriev N P, Parfianovich A P 2017 Improving the energy efficiency of the operation of electrified 25 kV AC railways by choosing the normal power supply for traction loads *Electrification, development of electric power infrastructure and electric rolling stock of rapid and high-speed rail transport: the materials of VII Intern. symposium Eltrans 2015* (St. Petersburg: FSBEI VO PGUPS Publ.) pp. 88–92

[3] Grigoriev N P 1985 The choice of transformation ratios using the computer *All-Union correspondence inst. of rail. transp. Engineers* (Scientific works vol 121) (Moscow: VZIIT Publ.) pp. 45-49

[4] Grigoriev N P, Parfianovich A P, Vukotich A A 2017 Management of electric energy losses in power transformers of traction substations *Mathematical and experimental modeling of physical processes: Coll. of mater. of international correspondence scient. and practical. conf.* ed M V Candeli (Birobidzhan: Sholom-Aleichem Priamur. state un-ty Publ) pp 132–139

[5] Zakaryukin V P, Kryukov A V 2007 *Modeling of traction power supply systems: a study guide* (Irkutsk: IrGUPS Publ)

[6] Zakaryukin V P, Kryukov A V, Litvinov E Yu Calculations of modes and load capacity of traction power supply systems in phase coordinates. Cetificate of official registr. of a computer

IOP Conf. Series: Materials Science and Engineering **760** (2020) 012060 doi:10.1088/1757-899X/760/1/012060

program No. 2005611179 (RF) "Fazonord" Registered on May 19 2005 (Federal Service for Intellectual Property, Patents and Trademarks)

[7] Grigoriev N P, Parfianovich A P, Voprikov A V 2017 The information model of the 25 kV traction power supply system *Scientific, technical and socio-economic cooperation of the countries of the Asia-Pacific region in the twenty-first century: the works of the All-Russia scientific and practical conf. of creative youth with the intern. partic* (vol 1) ed S A Kudryavtsev (Khabarovsk, DVGUPS Publ.) pp 102–107

[8] Grigor'ev N P, Klykov M S, Parfianovich A P, Shestukhina V I 2016 Increasing the power efficiency of a 25 kV alternating current traction power supply system *Russian Electrical Engineering* **87(2)** 113–115

[9] Grigor'ev N P, Voprikov A V, Parfianovich A P 2016 Increasing the operating life of power transformer winding insulation when upgrading 25 kV alternating current traction substations *Russian Electrical Engineering* **87(2)** 100–103

[10] Grigoriev N P, Voprikov A V 2014 Resource-saving traction-feeding transformer substations circuits connect to the traction power supply of AC 25 kv *Modern Technologies, System Analysis. Modeling* **3(43)** 190-194

[11] Zakaryukin V P, Kryukov A V 2008 Modeling of multi-winding transformers in phase coordinates *Elektrotekhnika* **5** 56-61

[12] Zakaryukin V P, Kryukov A V, Cherepanov A V 2017 Intelligent Traction Power Supply System *International Scientific Conference Energy Management of Municipal Transportation Facilities and Transport* (EMMFT vol 692) pp 91–99

[13] Idel'chik V I 1989 *Electric systems and networks* (Moscow: Energoatomizdat)

[14] Order of Energy Ministry of the Russian Federation of December 30, 2008, no. 326 *On setting and approving standards for technological losses of electric power during its transmission through electric networks*

[15] Cherepanov A V, Tikhomirov V A, Kutsiy A P 2017 Influence of the tensio n asymmetry on energy efficiency *Modern Technologies, System Analysis. Modeling* **4(56)** 207-217

[16] Markvardt K G 1982 *Power supply of electrified railways* (Moscow: Transport)

[17] Markvardt K G 1981 *The reference power supply of railways. Part 2.* (Moscow: Transport)

[18] Cherepanov A V, Tikhomirov V A, Kutsiy A P 2017 Decrease in asymmetry and harmonious distortions in areas of power supply of non-tractive consumers *Modern Technologies, System Analysis. Modeling* **3(55)** 145-151

Development of methods and models of computer-aided design of security system against information threats for aviation-instrument-making

V E Trushnikov[1], N N Baranova[2], M V Grishin[3]

[1]Doctor of Technical Sciences, Professor, Professor of the Department of System Analysis and Management, St. Petersburg Mining University, St. Petersburg.
[2]Designing engineer of Ulyanovsk Instrument Design Bureau JSC, Ulyanovsk
[3]Grishin Maxim, Ph.D., Designing engineer of Ulyanovsk Instrument Design Bureau JSC

E-mail: Trushnikov_VT@pers.spmi.ru

Abstract. The article discusses the problems of existing methods of computer-aided design for protecting confidential information of design enterprises of the tool manufacturing industry. The method is described on the basis of a modular, unified approach to security systems. A simulation model of information security threats was developed for the instrument design engineering bureau.

1. Introduction

In the context of information threats to manufacturing enterprises, there is no single method for the automated design of information protection systems.

The heterogeneity of information systems used at instrument-manufacturing enterprises does not allow for the end-to-end interaction of co-executing enterprises and customer organizations, which makes it difficult to implement infrastructure solutions [1].There is a lag in the level of development of Russian standards on information protection, including national standards [2] for describing the digital architecture of enterprises. There are no process management methods, legal support for the use of electronic design, technological and operational documentation as part of the life cycle of defense products.

Almost all organizations use foreign computer equipment and system-wide software. This entails the technological dependence of most industries on imported hardware and software and increases risks of cyber attacks [3].

For example, significant resources are spent on the Microsoftsoft licenses required for the operation of applied information systems, as well as on licenses for foreign application software. A common practice is the use of software by various developers which increases information processing costs [4].

Due to the transition to electronic interaction between organizations and remote access to the consolidated resources of integrated structures, information leakage or distortion are possible. Most foreign software tools interact with the software manufacturer through global data networks, including when activating and verifying the validity of licenses, updating software. The access to files can be blocked after checking licenses. In addition, licensing agreements of Western system suppliers often impose a direct ban on the use of their software products for military purposes; in crisis periods, the technical support may be discontinued.

The ubiquitous use of imported telecommunication equipment and related software makes it difficult to ensure information security of control systems and protection against cyber threats when designing the latest products using digital models containing complete information about the products.

Consulting companies often defend interests of software suppliers, mainly foreign ones.

It is an urgent scientific task to develop modular protection of confidential information for instrument manufacturing enterprises whose main idea is to design a protection system from protection modules, inside

IOP Conf. Series: Materials Science and Engineering **760** (2020) 012061 doi:10.1088/1757-899X/760/1/012061

of which information is formed and transformed, and outside of the protection modules there are information threats. The shell of the modules should resist information threats.

It is assumed that information threats cannot arise inside the modules. Otherwise, the module affected by the information intrusion should be divided into two or more protected and secure modules with the distribution and modification of the information left from the broken module or the information processing should be stopped.

The protection modules interact; and the safe transfer of information from one module to another one is carried out. At the same time, information from a more secure module cannot be transferred to a less secure module if the level of confidentiality of the transmitted information prevents this.

Information can be transferred from a less secure module to a more secure module if the level of confidentiality corresponds to the degree of security of the transmitted module.

Depending on the level of confidentiality and the level of external threats, the protection module has an appropriate shell that allows you to perform tasks. The modular system of information protection allows various related enterprises to interact at the information level, provided that they have appropriate protection modules [5].

The merger, division, reorganization of enterprises is possible if there are indestructible interacting protection modules. In case of destruction or violation of the protective shell of the modules, the generation and transmission of information must be suspended, and these functions are transferred to other workable protection modules [6].

2. Information Threat Model for Instrument Manufacturing Enterprises

Specific products of the Instrument Design Bureau are scientific and technical information [7]. The development of methods for the creation, processing, transmission and storage of information by means of computer technologies generated threats caused by losses, modification and disclosure of data belonging to various subjects: developers and consumers of finished products and systems [8, 9].

2.1 The nomenclature, classification and specifics of information threats for instrument manufacturing enterprises.

The need to resist information threats is growing in the face of a weakly monopolized domestic and international market, when the level of competitive advantage of organizations and defense security of states depend on the degree of protection of confidential information [10].

Information threat is a possibility of an adverse event (impact, process or phenomenon) that can cause direct or indirect damage to the multilateral state of security of the instrument design engineering bureau during:

- development of new products and ensuring the rights to intellectual property;
- financing of scientific and technological developments and computer-aided design;
- domestic and interstate competition;
- counteraction to international intelligence structures;
- counteraction to national and international crimes and terrorism;
- defending their own reputation;
- exchange of scientific and technical information, etc.

There are a lot of information threats and their number is increasing. [11, 12, 13]

Figure 1 shows the general classification of information threats for a scientific-production system (design enterprises of the instrument-manufacturing industry) that uses computing tools and interacts with other similar scientific-production systems during the generation, modification, and exchange of confidential information. [14]

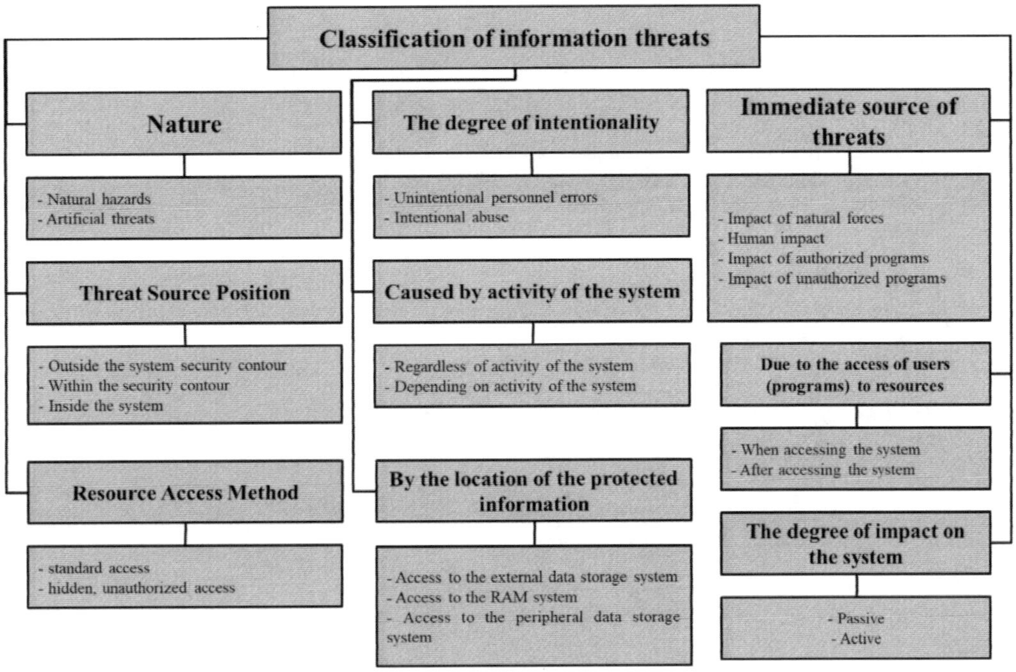

Fig. 1 General classification of information threats to the scientific and production system

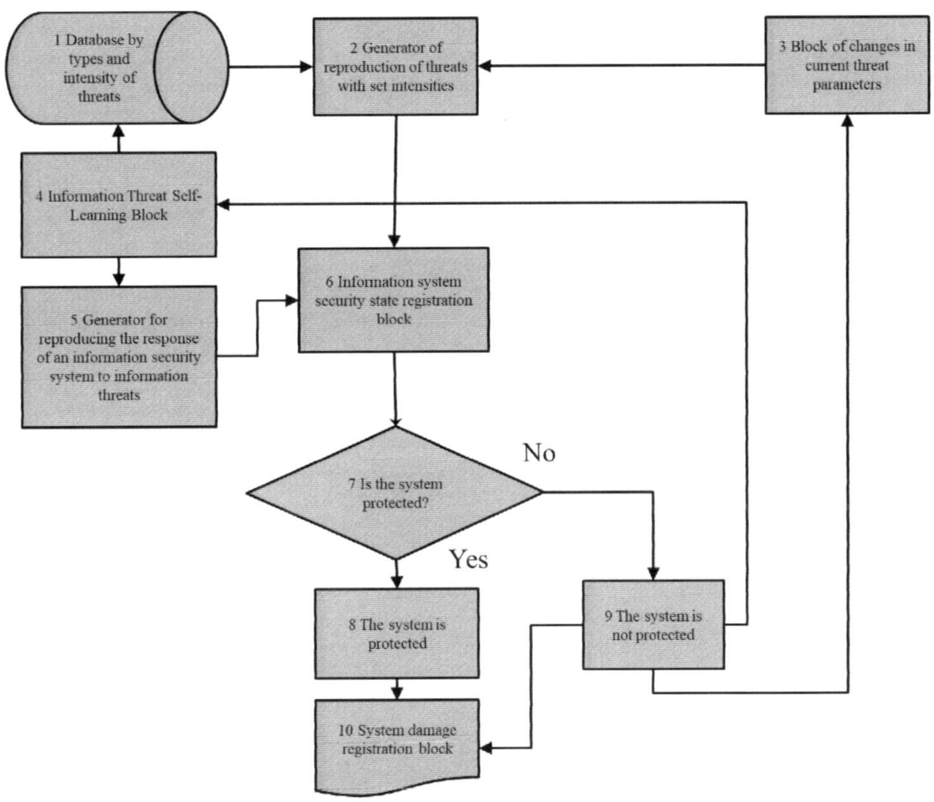

Fig. 2. The simulation model of information threats for the design bureau

2.2 Purpose, structure, input and output data of the information threat model for design enterprises of the instrument-making industry

The model of information threats for design enterprises of the instrument-making industry is intended to formalize the type, intensity of manifestation and probability of the impact of a particular threat to the information system of the enterprise, as well as to determine the damage from the implementation of a particular threat to a given enterprise.

The model's structure (Figure 2):
- The database of threats by their types and intensity;
- The generator of reproduction of threats with specified intensities;
- Changes of current threat parameters;
- Self-learning of information threats;
- The generator reproducing responses of the information security system to information threats;
- The registration of the status of security of the information system;
- The recording of damage to the system.

Input and output data:

Input data:
1. Threats by type, intensity of manifestation, potential damage to the system;
2. Probability of recognition of the types of threats.

Output data:
1. The amount of damage to the system from the types of threats;
2. Prevented damage to the system from the types of threats in response to the information security subsystem.

1.3 The algorithm of the simulation model of information threats

Step 1. The database on the types, intensity of manifestation of threats transmits information to the threat reproduction generator with specified intensity parameters.

Step 2. The generator of types of information threats is launched with the modified parameters of the types of threats.

Step 3. The generated threat parameters from the generator are fed to Block 6.

Step 4. The generator reproducing the response of the information security subsystem to information threats transmits information about probabilities of recognizing threats to Block 6.

Step 5. The security level of the system is checked. At a level higher than permissible (for example, 0.9), the conclusion is drawn: the system is protected (Block 8), otherwise it is not protected (Block 9).

Step 4. In the block registering the damage to the system, the damage to the system from the types of threats and the prevented damage to the system are calculated.

Step 5. In Block 4, new parameters of information threats are transmitted to Block 1 and Block 5.

3 Determination of security probabilities

To determine the probability of information security threats (P_i), the design enterprise needs to develop an effective information monitoring system that can monitor the intensity of various threats at various time intervals. The longer the observation intervals (t_n) and the higher the threat recognition efficiency, the more accurate the forecast for the next threat and the probability (P_i) of the threat.

The intensity of manifestation of a particular security threat to the system (λi) is equal to the ratio of the number of manifestations of such a threat (N_i) during the last observation period to the duration of the observation period of the threat (t_{ni}):

$$\lambda_i = \frac{N_i}{t_{ni}} \tag{1}$$

The intensity of manifestation of a particular threat to security of the system can be expressed as a function of the number of manifestations of threats to the time the threat was being observed.

The number of manifestations of a particular threat to security of the system ($N_{t1,t2}$) is equal to the integral of intensity (λi) of manifestations of the threat from the observation time (t_n) in the current interval [t_1; t_2] of the duration of manifestation of the threat:

$$N_{t1,t2} = \int_{t_1}^{t_2} \lambda_i dt_n \tag{2}$$

The probability of manifestation of the number N * t₁, t₂ of certain security threats to the system P (N * t₁, t₂) in the current interval [t₁; t₂] is calculated as follows:

$$P(N^*_{t1,t2}) = \begin{cases} 1, \text{если } N^*_{t1,t2} \leq N_{t1,t2} \\ \dfrac{N_{t1,t2}}{N^*_{t1,t2}}, \text{если } N^*_{t1,t2} > N_{t1,t2} \end{cases} \quad (3)$$

If the probability of one threat in the current interval [t₁; t₂] is calculated as follows:

$$P(1) = 1, \text{если } N_{t1,t2} \geq 1 \quad (4)$$

If the enterprise does not have an effective information monitoring system, it cannot immediately calculate probabilities of information security threats. However, due to the accumulation of statistical data on threats, the accuracy of such calculations can increase.

To improve the effectiveness of the information monitoring system, an enterprise lacking experience in calculating the probability of threats should use the statistics of a similar enterprise.

It is important to assess the potential damage to the system caused by a specific threat.

The information processed by the system has a value that can be expressed in monetary units. This cost includes expenses for the development and subsequent modification of information (primary costs) (M_m). For the production of serial samples of products developed on the basis of scientific and technical information, financial resources are used for production (production costs, M_p).

As a result of operation of new serial samples of instrument-making products, the level of security of objects with price (M_o) should increase. Thus, the cost of primary and production costs is added to the cost of objects, which can be devalued if the enemy side reveals scientific and technical information during the implementation of a certain information threat in the system.

The damage prevented by the information security subsystem (M_i) will be calculated as the product of recognition probabilities (P_r) and counteraction (P_c) of the threat from the information security subsystem and the sum of costs for developing and manufacturing new products and objects using new products:

$$M_u = P_r \cdot P_c.(M_m + M_n + M_o) \quad (5)$$

If different i-types of threats are directed to various j-objects, in formula (2.5), the components are summed up:

$$M_u = \sum_{i=1}^{i=F} \sum_{j=1}^{j=H} P_{ri} \cdot P_{ci}.(M_{mj} + M_{nj} + M_{oj}) \quad (6)$$

In special cases, if the developed objects are similar, or threats are similar, the summation by objects or threats is excluded from formula (2.6).

An additional tool for calculating the probability of information threats is a system simulating information threats, a training system for information monitoring of an enterprise whose elements are presented in Figure 2.2.

The system for simulating information threats can collect information on all potential threats that have ever occurred in the industry and generate threats for the enterprise being audited to evaluate the effectiveness of its information security system.

4. Conclusion

This article presents the nomenclature, classification and specifics of information threats to design enterprises of the instrument-making industry. The list of information threats and their expected consequences were formulated. The structure, input and output data of the simulation model of information threats were presented. Definitions are given and analytical expressions are suggested for calculating intensity of security threats; the amount of manifestation of a particular security threat in the time interval; the probability of occurrence of certain security risks was determined. The assessment of the potential damage to the system was presented given. The calculation of the prevented damage from activities of the information security subsystem was provided. The need of the subsystem for simulating information threats as part of the information security subsystem for design enterprises of the instrument-making industry was identified. The obtained data will be used for further work when designing the modular system for protecting the confidential information of a design enterprise of the instrument-making industry.

IOP Conf. Series: Materials Science and Engineering **760** (2020) 012061 doi:10.1088/1757-899X/760/1/012061

References

[1] Maksarov V V, Olt J, Krasnyy V A 2019 Investigation of wear resistance of bearing units engines of mining dump trucks operating in conditions of fretting corrosion *Journal of mining institute*, **235**, 77 – 83

[2] Gerasimenko V A , Malyuk A A, 1997 *Fundamentals of information* security (Moscow: Moscow Engineering Physics Institute), 537

[3] Gromov Yu Yu, Drachev V O, Ivanova O G, Shakhov N G 2010 *Information security and information protection: a training manual* (Stary Oskol: TNT), 384

[4] Olt Y; Maksarov V V, Krasnyi V A 2018 Provision of adhesion strength of gasthermal coatings on piston rings of quarry transport engines *Journal of mining institute* **229** 77-83

[5] Maksarov V V, Olt J 2017 Dynamic stabilization of machining process based on local metastability in controlled robotic systems of cnc machines *Journal of mining institute* **226** 446-451

[6] Maksarov V V, Olt J. 2015 Improving the precision of manufacturing power hydraulic cylinders of powered roof supports based on a vibration-damping tooling system *Journal of mining institute* **214** 71-84

[7] Yablochnikov E I, Molochnik V I, Fomina Yu N 2008 *Reengineering of business processes of design and production / Textbook* (St. Petersburg: St. Petersburg National Research University of Information Technologies, Mechanics and Optics) 152

[8] Alekseenko V N, Dreve Yu G 1996 *Fundamentals of building protection systems for industrial enterprises and banks: a manual* (Moscow: Moscow Engineering Physics Institute).

[9] Prokhorov S A, Fedoseev A A, Ivashchenko A V 2008 *Automation of integrated enterprise security management* (Samara: Saratov Scientific Center of the Russian Academy of Sciences) 55

[10] Krasnyy, V A, Maksarov, V V, Olt, J 2015 The use of polymer composite materials in the friction nodes downhole oil pumps *Journal of mining institute* **211** 71-79

[11] Malyuk A A 2004 *Information security: conceptual and methodological foundations of information security. Textbook manual for universities*, (Moscow: Hotline - Telecom) 280

[12] Ignatiev V A 2005 *Information security of a modern commercial enterprise: Monograph* (Stary Oskol: TNT LLC) 448

[13] Saderdinov A A, Traynev V A, Fedulov A A, *2004 Information security of an enterprise: Textbook. - 2nd ed* (Moscow, Publishing and Trading Corporation "Dashkov and Ko") 336

[14] Shangin V F 2013 *Information security of computer systems and networks: a training manual* (Moscow: Publishing House “FORUM”: INFRA-M) 416

IOP Conf. Series: Materials Science and Engineering **760** (2020) 012062 doi:10.1088/1757-899X/760/1/012062

Coordination of the microwave module used in the Doppler meter of the velocity vector components

V E Trushnikov[1], R R Valitov[1], M V Grishin[1]

[1]Doctor of Technical Sciences, Professor, Professor of the Department of System Analysis and Management, St. Petersburg Mining University, St. Petersburg
[1]Chief of the department, Designing engineer of Ulyanovsk Instrument Design Bureau JSC, Ulyanovsk.
[1]Ph.D., Designing engineer of Ulyanovsk Instrument Design Bureau JSC, Ulyanovsk.

E-mail: Trushnikov_VT@pers.spmi.ru

Abstract: The article describes a method for coordinating the microwave path and verifying the microwave module of the DMVC. The method for coordinating and adjusting the microwave module helps measure and evaluate characteristics of the device to confirm compliance with the DMVC requirements.

1. Introduction

One of the significant radio navigation devices of the onboard equipment is the Doppler meter of the velocity vector components (DMVC) [1-3]. As a rule, the DMVC includes a transceiver module of superhigh frequencies (microwave), which determines most of the characteristics and measured parameters of the device.

Microwave modules operating in the range from 3 GHz to 30 GHz are implemented as devices connected by transmission lines. The part of the system located between the initial and terminal devices is the microwave path or the microwave circuit. Such a path transmits electromagnetic energy from the transmitter to the antenna or from the antenna to the receiver, provides the required operating mode of the output and input circuits of the transmitter, performs the frequency and polarization separation of transmitted signals, etc. The most common elements of microwave circuits are transmission lines, transition and butt nodes between lines, matching and tuning elements, adders, power dividers and couplers [4].

Coordination of transmission lines is required to suppress waves reflected from the load under the equal load and wave resistance of the transmission line[5].

In case of full coordination of all elements of the transmission line and the supply generator in the transmission line, the reflected wave is absent (the traveling wave mode), the reflection coefficient is 0, and the standing wave coefficient is 1.

If the load is not consistent with the transmission line, a number of undesirable effects occur [6-9]

1) frequency and power of the generator change due to the pulling effect;
2) power supplied to the load decreases;
3) the limit value of the transmitted power decreases due to the electrical breakdown in the path;
4) the broadband transmission path decreases;
5) active losses in the transmission line increase.

It is clear that coordination in the transmission line is one of the most common and important tasks of the microwave technology.

The authors developed a method for coordinating the microwave path and verifying the microwave module of the DMVC. The method for coordinating and adjusting the microwave module helps measure and evaluate characteristics of the device to confirm compliance with the DMVC requirements.

2 Consideration of basic characteristics when coordinating microwave devices

The main source of losses is standing waves [10] They appear in circuits when the wave encounters changes in wave resistance [11]. In order to prevent standing waves and achieve the maximum value of the power flow in the circuit, it is necessary to coordinate the load with the transmission line 124.

Let us consider the main characteristics that must be taken into account when coordinating microwave devices with different wave resistances.

The wave resistance Z (Ω) is the resistance that an electromagnetic wave encounters when propagating inside a homogeneous line without reflection, i.e., provided that the transmission process is not affected by inconsistencies at the ends of the line. The wave resistance is a parameter that depends only on physical parameters of the line, such as sections, conductor shapes, thickness, material, etc.).

The reflection coefficient G shows how much of the radiation is reflected from the inhomogeneity in the propagation medium, for example, the junction of two lines with different wave impedances. It can be calculated by formula:

$$G = \frac{Z_{r}-Z_{l}}{Z_{r}+Z_{l}}, \tag{1}$$

where $Z_{г}$ and $Z_{н}$ – wave resistance and load of the generator, respectively.

The standing wave coefficient (SWC) characterizes the degree of coordination of the device. In real devices, reflected waves are always present in the transmission line, which indicates that part of the energy is reflected and does not reach the load [12]. The reflected energy is not only "lost" in the transmission line, but can cause the overheating of the entire microwave device. This can significantly change its characteristics and lead to greater reflections. The SWC is calculated by formula:

$$\mathrm{SWC} = \frac{U_{i}+U_{r}}{U_{i}-U_{r}}, \tag{2}$$

where U_i and U_r - the amplitude of incident and reflected electromagnetic waves. The SWC should be equal to 1, which corresponds to the absence of reflected waves, matching in the traveling wave mode. With a purely standing wave, the SWC tends to infinity. The admissible SWC is up to 1.5.

3. The method for coordinating the microwave path

The microwave module consists of a low-frequency part and a high-frequency part, which are structurally separated by the module housing (Figure 1).

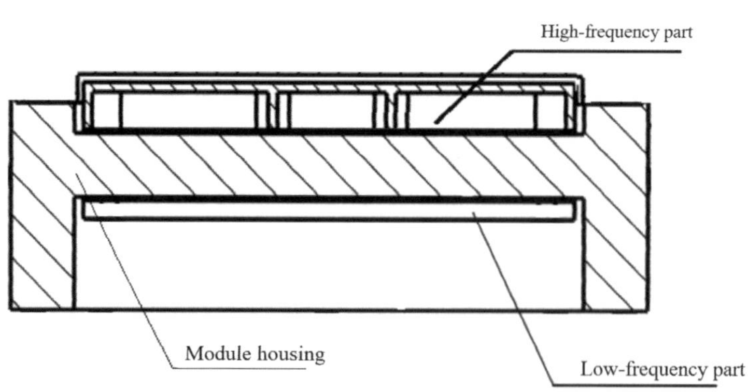

Fig. 1. The design of the microwave module

IOP Conf. Series: Materials Science and Engineering **760** (2020) 012062 doi:10.1088/1757-899X/760/1/012062

In the high-frequency part, there is a circulator (Figure 2).

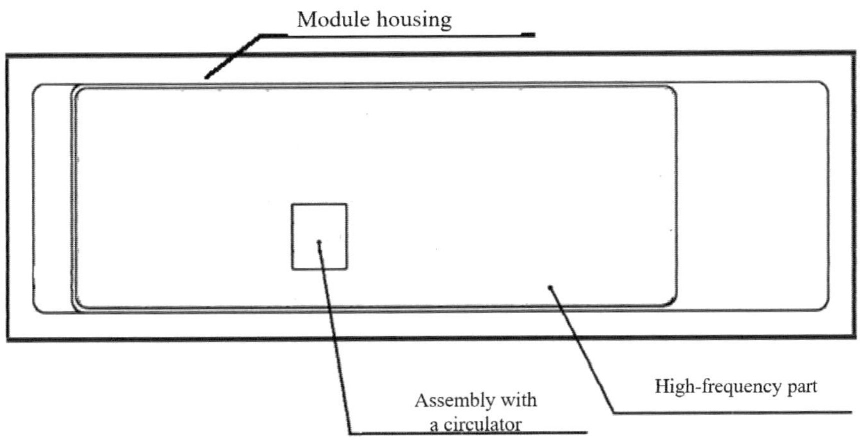

Fig. 2. Assembly with a circulator.

In the high-frequency part, there is a frequency generator, RF elements and four outputs. In the low-frequency part, there are shapers of the internal power supply ± 6 V and +4 V, a connector for input and output signals, a controller for monitoring the carrier frequency and output power, and a cascade of filters. This microwave module is responsible for formation, amplification and reception of the microwave frequency-modulated (FM) signal and allocation of the Doppler frequency. The module generates, modulates, and transmits a microwave frequency signal to the antenna input with the following parameters:

a) the central operating frequency (Fp) - (13250-13400) MHz;

b) the type of modulation is continuous frequency with a modulation index of (3.7 ± 0.5) dB.

c) the output power of the module is not less than 8 and not more than 20 dBm.

d) the module must have a level of single-band phase noises of 1000 Hz and no more than 50 dBc / Hz.

d) the module must ensure allocation of the Doppler frequency and its spectrum when receiving and transmitting the signal through the channel with a power of at least 100 dBm with suppression of spurious components with a frequency of 20 dB.

f) the transceiver of the microwave module must have an isolation between the channels of at least 23 dB.

g) the level of quadratures I, Q of the analog output signal at the output of the module should be 0-0.7 V.

The microwave module must have built-in controls for the following parameters:

- excess current consumption in the power circuit (logical "0" corresponds to normal operation, logical "1" - failure);
- the presence of the output power of the amplifier on the antenna key (logical “1” corresponds to normal operation, logical “0” - failure.).

The control results should be issued in the form of digital signals of the LVTTL level. Logical “1” corresponds to a voltage range (2.4 ... 3.6) V, logical “0” corresponds to a voltage range (0 ... 0.4).

The scheme of the workplace for the coordination and debugging of the microwave module is shown in Figure 3

IOP Conf. Series: Materials Science and Engineering **760** (2020) 012062 doi:10.1088/1757-899X/760/1/012062

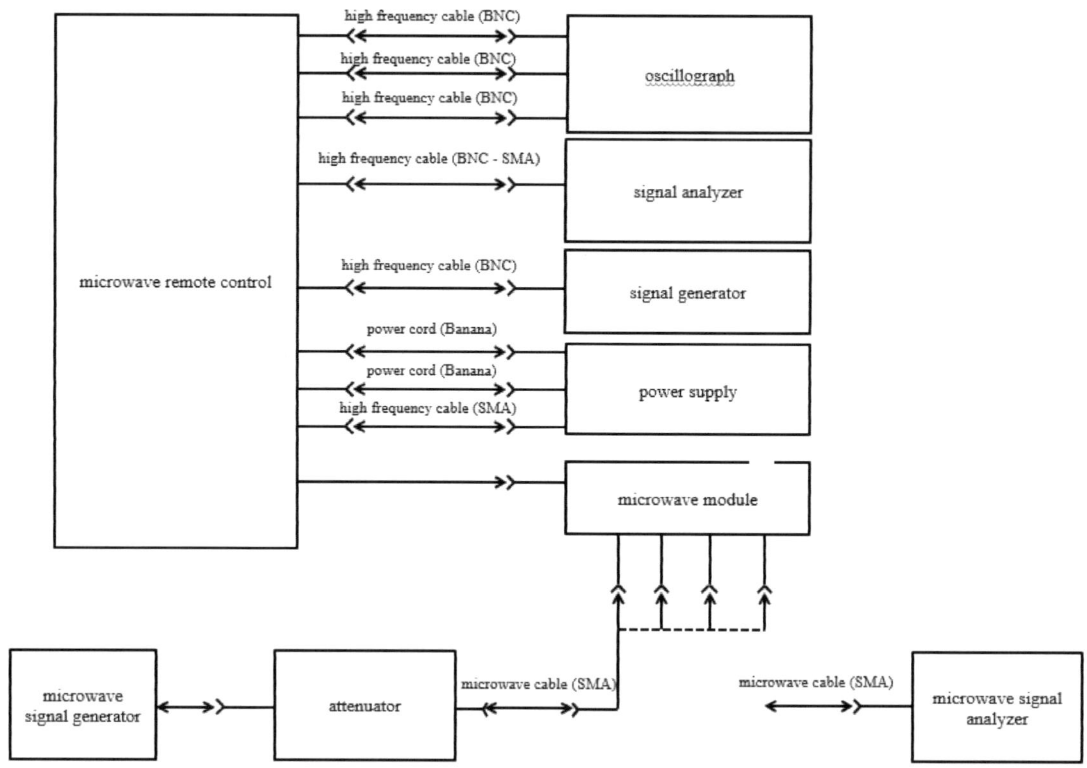

Fig. 3. Scheme of the workplace for the coordination and debugging of the microwave

Before connecting the power source (PS), the value of 6V is set on the channels, the current limit is 0.6 A.

It is necessary to control the value of the consumed current from the PS by the built-in indicator, the current should be no more than 5 mA for each of the three channels. It is necessary to measure the secondary supply voltage (3.30 ± 0.10) V, (5 ± 0.1) V, (minus 5 ± 0.1) V. If the parameters deviate, turn off the power sources, determine and eliminate the cause of the discrepancy by health checks of PS voltage stabilizers and associated circuits.

The procedure for programming the microwave module firmware (software) is carried out. The STM32 controller from STMicroelectronics located in the low-frequency part is programmed. The programmer is connected to the microwave module only when the supply voltage is off. Therefore, it is necessary to remove the supply voltage from the microwave module; to connect the ST-LINK / V2 programmer to the programming socket located in the low-frequency part; to supply voltage to the microwave module; to open the STM32 ST-LINKUtility application; to create a connection to the controller by selecting TARGET - CONNECT in the program menu; in the program window (Figure 3), to select the following menu items "File - Openfile ...".

IOP Conf. Series: Materials Science and Engineering **760** (2020) 012062 doi:10.1088/1757-899X/760/1/012062

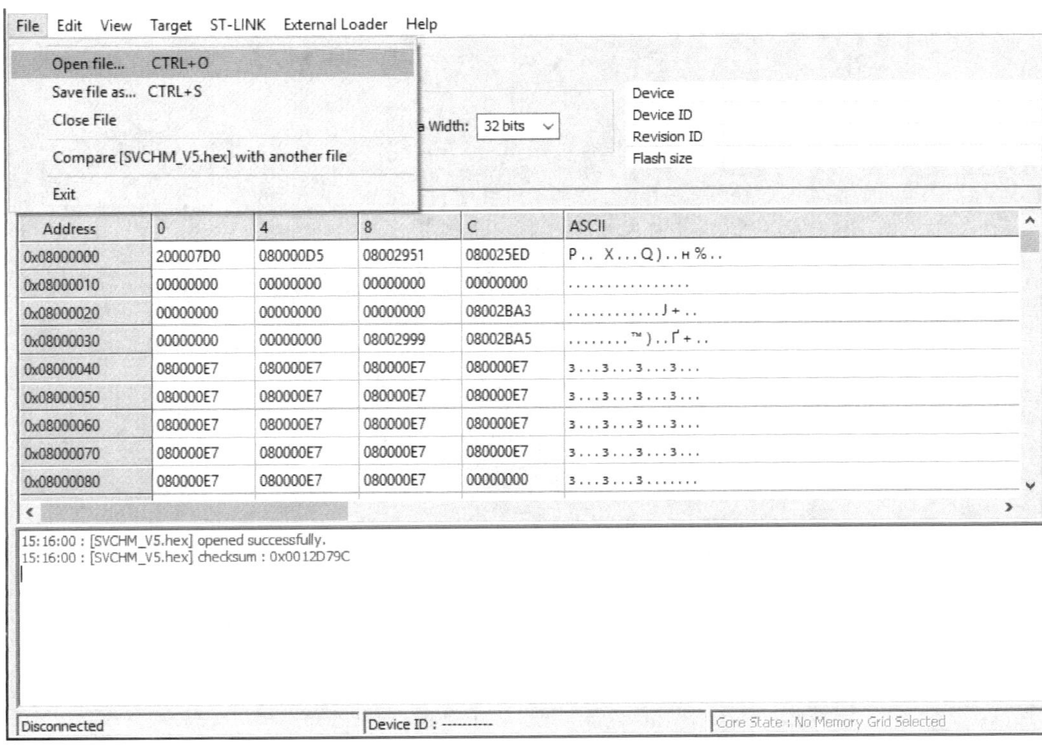

Fig. 4. Sample built-in software window

In the dialog window, select the file "SVCHM_Vx.hex" and start the programming process by selecting the following menu items "Target - Program & Verify ...". Wait for the end of the programming process, remove the supply voltage from the transmitter, disconnect the ST-LINK / V2 programmer from the programming connector and reapply the supply voltage.

After the programming stage of the microwave module, install the assembly with the circulator in the high-frequency part. To do this, remove the power from the microwave module. Connect the assembly with the WS1 circulator to the microwave path with IN-00 indium foil 0.1 mm thick with an approximate size of 1 mm * 2 mm (selected locally). The figure shows an assembly with the circulator after connecting to the microwave path. Fix the joints with AK-113 and dry them in accordance with GOST 23832-79 [2] (Fig. 4).

At the next step, it is necessary to measure the carrier frequency at four outputs of the microwave module. To do this, using a torque wrench with a size of 8 mm and a gain of 0.9 N, connect the RF cable with SMA connector to the first output of the microwave module.

Set up the microwave signal analyzer as follows: using the control devices, set FREQ - 13.325 GHz (by pressing FREQ, CENTER), SPAN - 500 kHz, BW - 2 kHz, AMPT - 20 dBm. The spectrum tab should be open (if necessary, press the mode button, select the spectrum mode).

On the microwave control panel, the first channel "1" is selected with the help of a dial switch and the "BP" key is pressed.

On the microwave signal analyzer, the carrier frequency value is read. Check that the carrier frequency is (13250-13400) MHz and meets the specified requirements. If the parameter deviates, check the operation of the 40 MHz reference oscillator and its associated circuits, as well as the correct installation of the frequency synthesizer IC and its associated circuits. These measurements must be carried out for each output channel of the microwave module.

IOP Conf. Series: Materials Science and Engineering **760** (2020) 012062 doi:10.1088/1757-899X/760/1/012062

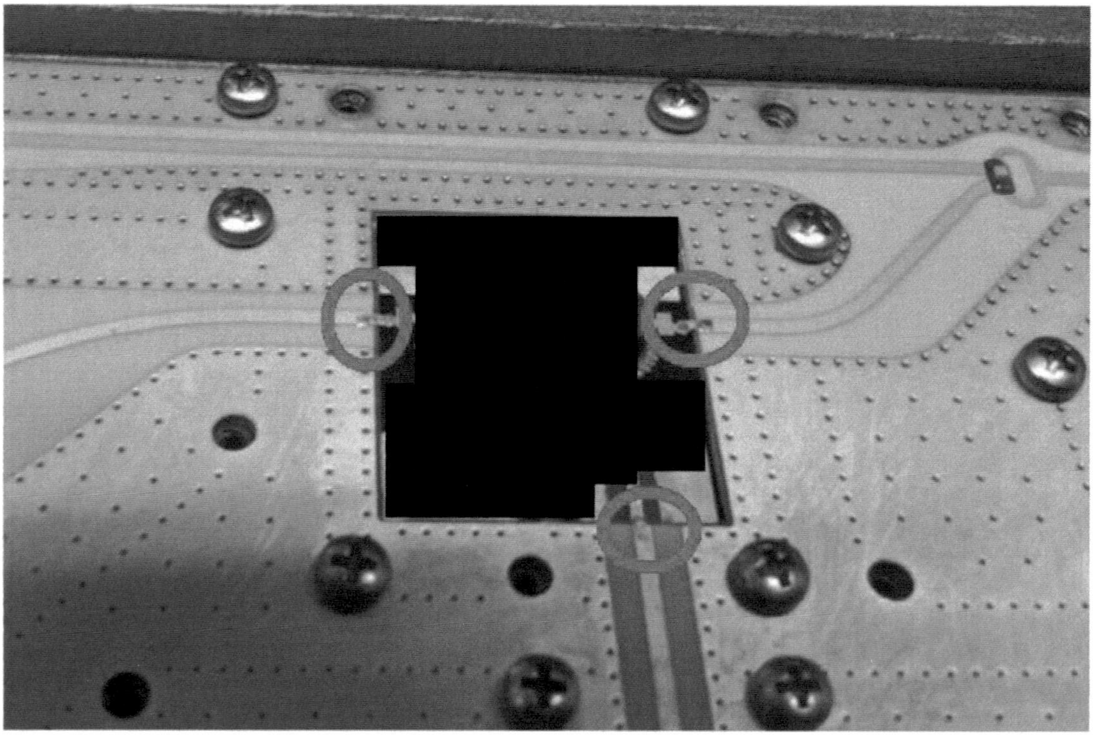

Fig. 5. Sample installation of an assembly with a circulator

The output power level of each channel of the microwave module is adjusted. The output power level is read using a microwave signal analyzer with the above settings. Before reading the power value, update the reading. Make sure that the output power level is in the range from 8 to 20 dBm, in accordance with the assigned requirements, and the "KVM" indicator is on on the microwave remote control. If the parameter is rejected or the "KVM" indication is not displayed on the microwave remote control, carry out adjustments.

Carry out adjustments at the input of the filter or at the input of the microassembly with the circulator, if the previous adjustment was not efficient. Recommended locations for trimming elements are shown in Figure 6.

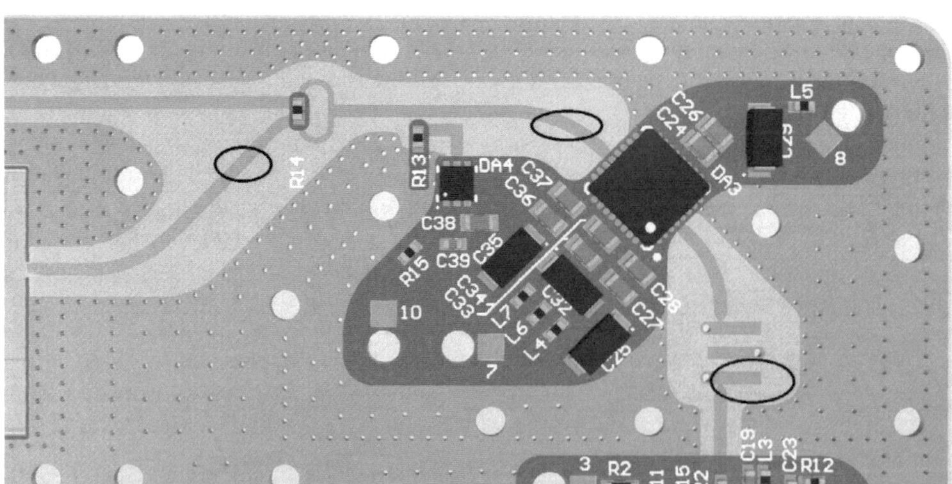

Fig. 6. Recommended Locations of tuning elements

IOP Conf. Series: Materials Science and Engineering **760** (2020) 012062 doi:10.1088/1757-899X/760/1/012062

The tuning element is indium foil 0.1 mm in thickness; the size and shape of the tuning element is determined during the tuning process. It is recommended to start the adjustment with a tuning element 1x1mm in size. Move and fix the tuning elements with any convenient tool with dielectric properties. The tuning element should fit snugly against the board. After tuning the microwave path, the tuning elements must be fixed on the printed circuit board with AK-113 varnish and dried in accordance with GOST 23832-79 [2]. Recheck the output power level in the steps above. Repeat the check for all microwave channels.

To make further adjustments, install the frames of the receiver, transmitter and assembly with the circulator to regular places and fasten with screws, as shown in Figure 7/

Check for the absence of foreign objects, unused particles of adjusting elements on the high-frequency part of the microwave module and on the faces of the frames before installation.

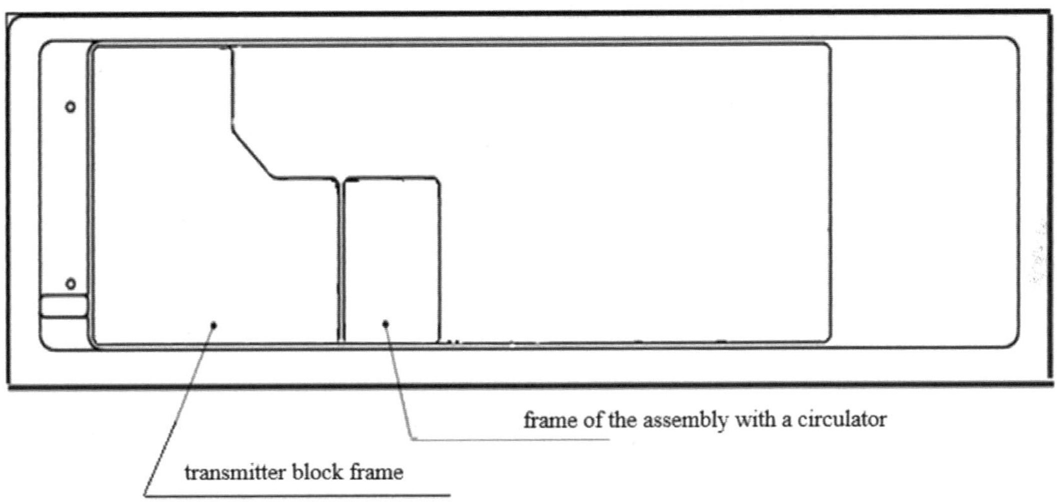

Fig. 7. Regular places of the transmitter and microassemblies with a circulator

The next parameter is the modulation index. The analyzer adjusts the center frequency so that the carrier frequency is in the center of the screen. Adjust the REF level so that the carrier peak is located at a distance of half a division from the upper border of the screen. On the generator, according to the operating instructions for the device, set the output waveform to be sinusoidal, frequency of v25 kHz, amplitude of 1.6 V. Install channel 1 on the microwave control panel.

Using the analyzer, the difference between the level of the carrier and the level of the first lateral harmonic is read out as shown in Figure 8.

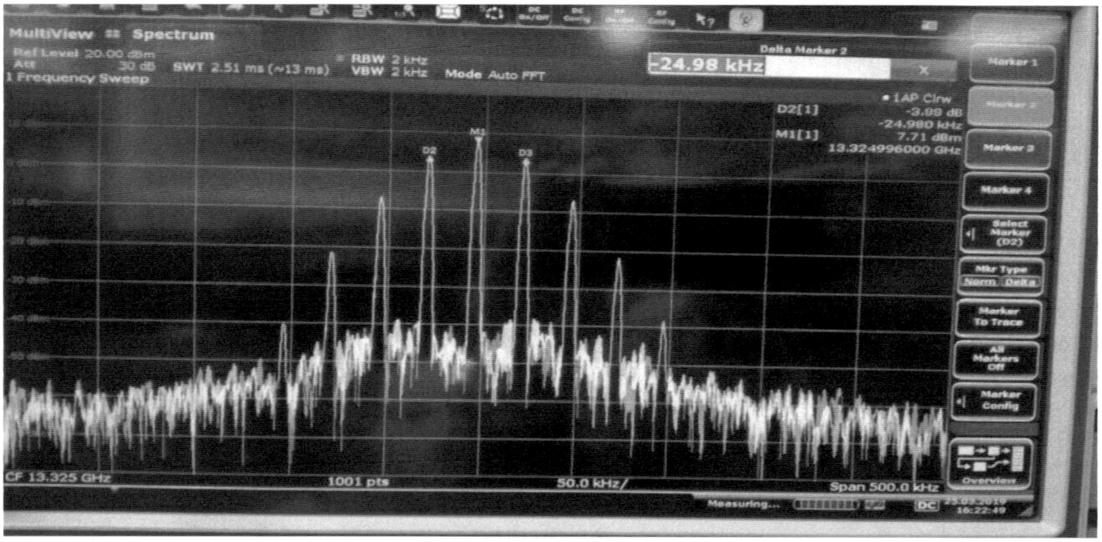

Fig. 8. Reading the difference between the level of the carrier and the level of the first lateral harmonic

Make sure that the difference is (3.7 ± 0.5) dB. If the parameter deviates, select the R84 resistor (selection range from 50 to 1000 Ohms).

Phase noise is measured at each microwave module output using an analyzer. On the analyzer, select the phase noise measurement mode by pressing the MODE button and selecting PHASENOISE from the menu, as shown in Figure 9.

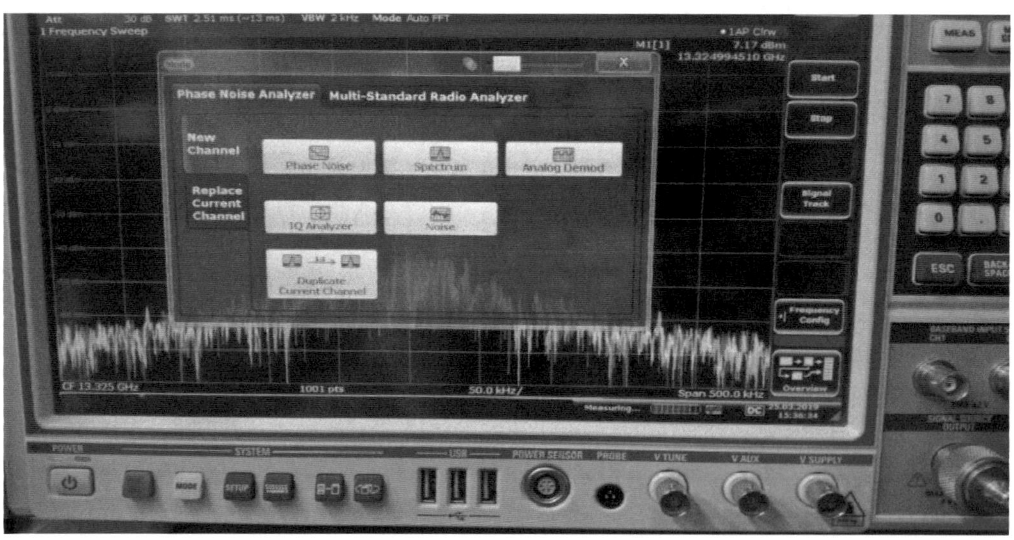

Fig. 9. Phase noise measurement mode

Make sure that the level of single-band phase noise is not higher than minus 50 dBc / Hz. The oscillogram is shown in Figure 10.

IOP Conf. Series: Materials Science and Engineering **760** (2020) 012062 doi:10.1088/1757-899X/760/1/012062

Fig. 10. Oscillogram of single-band phase noises

If the parameter deviates, check the operation of the 40 MHz reference oscillator and its associated circuits, as well as the installation of the frequency synthesizer IC and its associated circuits. Coordinate the microwave path between the assembly with the circulator and the switch. Adjustthe signal analyzer.

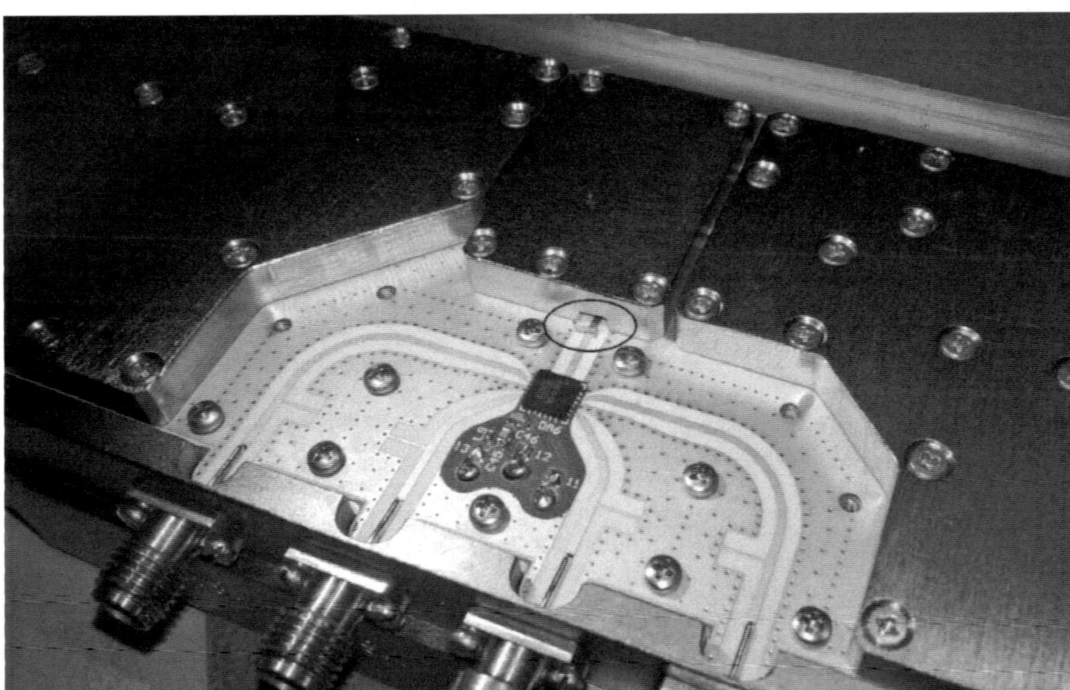

Fi g. 11. Approximate position of the tuning element

IOP Conf. Series: Materials Science and Engineering **760** (2020) 012062 doi:10.1088/1757-899X/760/1/012062

Set the following parameters on the signal analyzer: CENTRAL FREQUENCY - 25 kHz, SPAN - 50 kHz, BW - 200 Hz, AVERAGE - 200 (press TRACE, select TRACE1, AVERAGE in the menu). Connect the generator to the second channel of the microwave module through the attenuator. Set channel 2 on the microwave remote control. On the generator, set the frequency of 23 kHz higher than the previously measured carrier frequency; set the output signal level to minus 65 dBm and apply the microwave power. Apply the power to the microwave module. On the analyzer screen, the difference of the useful signal and noise is at least 20 dB. Adjust the frequency of the generator to achieve the peak (the maximum signal) of 25 kHz. Set the output level to 10 dBm. On the oscilloscope screen, observe quadrature signals. I is the in-phase component, Q is the quadrature component from the output of the microwave control panel. By introducing tuning elements into the microwave line between WS1 and the switch, coordinate the path. The approximate position of the tuning element is shown in Figure 10.

Repeat the check of the output power level as described earlier, turning off the generator.

Coordinate the microwave path between the switch and the microwave outputs of the module.

The circuits connecting the switch and the contact pads of the output connectors of the microwave module are subject to coordination, depending on the selected channel. The approximate location of the tuning elements is shown in Figure 12.

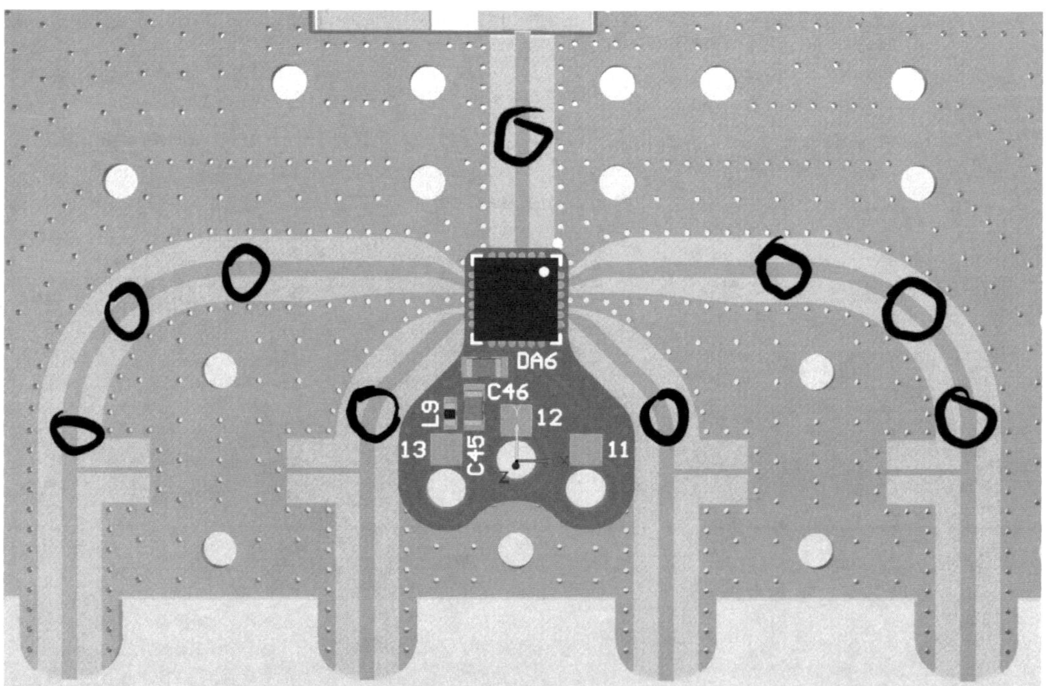

Fig. 12. Approximate location of tuning elements

Set the frequency of 25.0 kHz. Using an oscilloscope, control the level and phases of signals I and Q. Achieve the difference in amplitudes and the phase shift of 90 ° of no more than 20% (the time shift between transitions through zero 10ms ± 2ms). The type of signals I, Q before and after adjustment is shown in Figure 12.

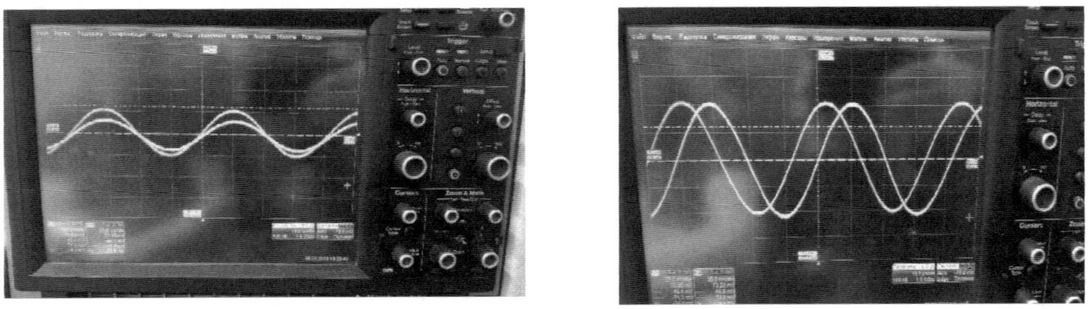

Fig. 13. Type of signals I, Q before and after adjustment

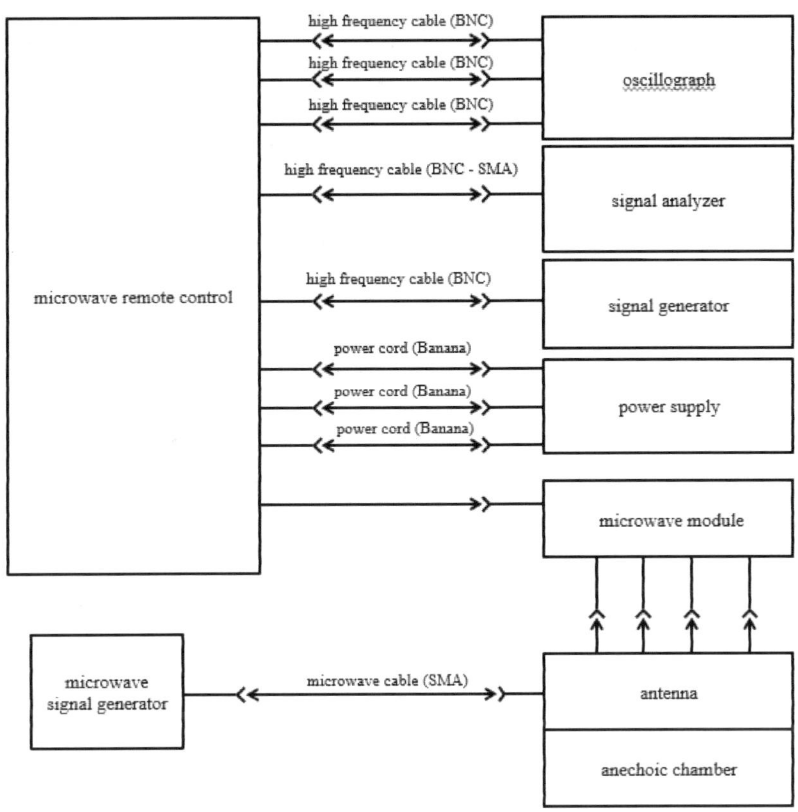

Fig. 14. Microwave module workstation assembled with an antenna

Monitor the current consumption on the 4V circuit. With the correct adjustment, a decrease in the current consumption should be observed. There should be no indication of "KTM" on the microwave control panel. Repeat the check of the output power level as described earlier, turning off the generator.

Repeat the above steps for output 2, 3 and 4 of the microwave module (alternately connecting the analyzer to the connectors and selecting the "2", "3", "4" positions with the microwave remote control switch, respectively).

IOP Conf. Series: Materials Science and Engineering **760** (2020) 012062 doi:10.1088/1757-899X/760/1/012062

Further adjustment of the microwave module is carried out as an assembly with the antenna. Four outputs of the microwave module must be connected to the antenna and collect the workplace according to Figure 13 for the following adjustment operations.

Install the antenna on the anechoic chamber by positioning the camera mark on the side of the microwave module. Connect the generator to the anechoic chamber. Connect the microwave signal analyzer to input 1 of the antenna through the adapter p / nADSC-A8A8 (female-female).

Set the power level of 0 dBm on the generator. Read the power levels from each input of the antenna. These antenna losses are subject to accounting for further adjustment and measurements.

Disable the microwave signal analyzer. Connect the antenna to input 1 of the microwave remote control. Connect it to the output I_P (Q_P) of the microwave remote control. On the generator, set the power so that the input of the microwave module is minus 100 dBm. On the analyzer, monitor the markers - noise at 25 kHz and useful signal I_P (Q_P). Install additional tuning elements to achieve the maximum difference in the amplitudes of the markers (the difference between the useful signal and noise). The waveform is shown in Figure 15.

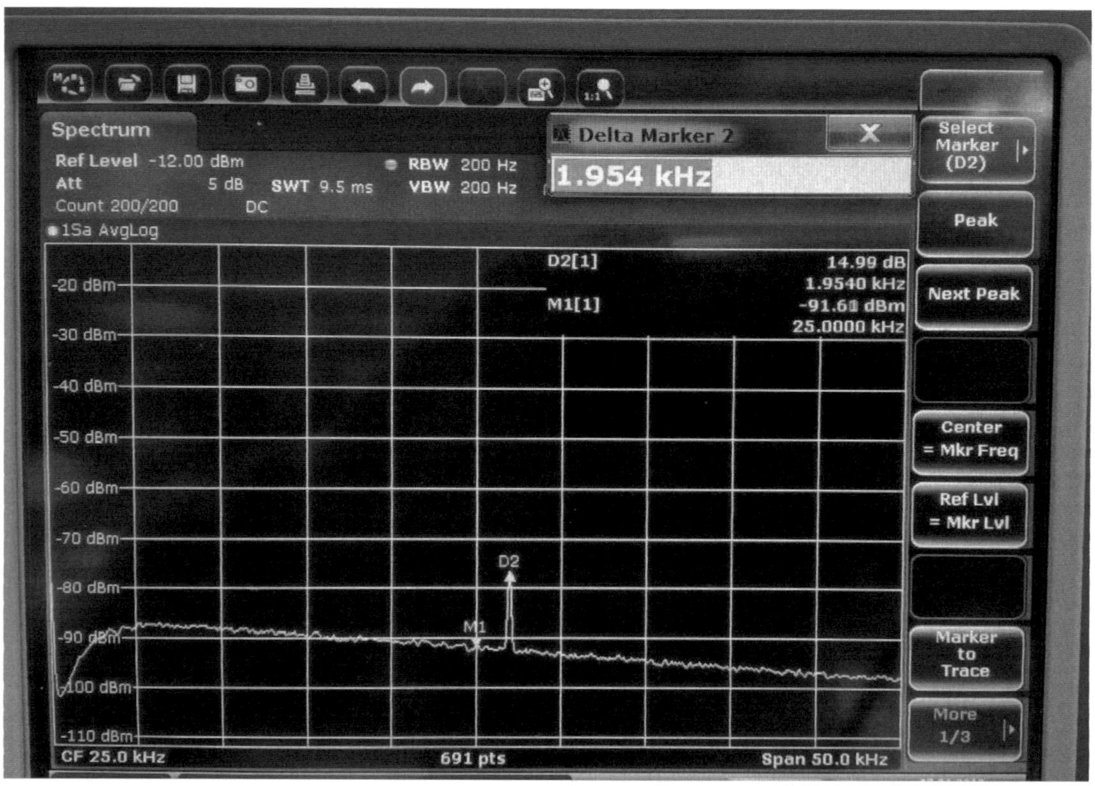

Fig. 15. Oscillogram of the maximum difference in the amplitudes of the markers

Additionally, observe a decrease in the consumption of current along the 4 V circuit. An approximate arrangement of the tuning elements is shown in Figure 16.

IOP Conf. Series: Materials Science and Engineering **760** (2020) 012062 doi:10.1088/1757-899X/760/1/012062

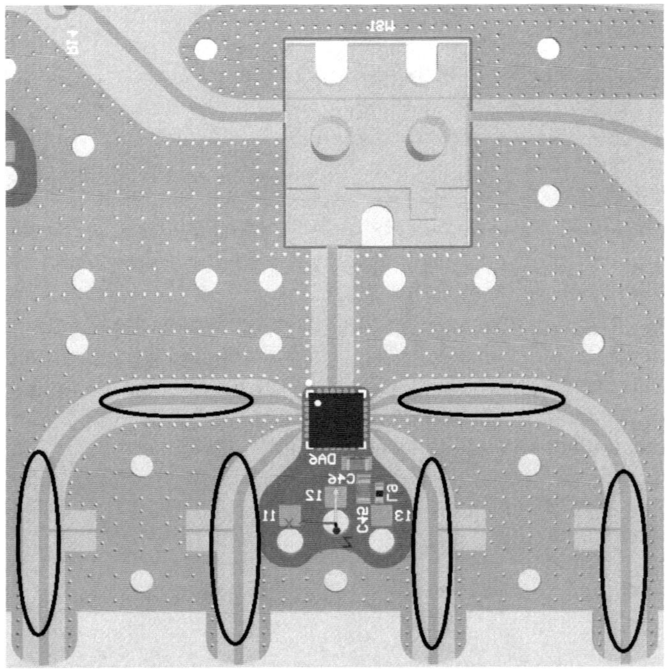

Fig. 16. The approximate arrangement of tuning elements

The signal levels of the I_P and Q_P channels (the difference between the useful signal and noise for each quadrature signal) for each channel of the microwave module must differ by no more than 2–3 dBm and be at least 13 dBm.

Repeat the procedure for the remaining channels of the microwave module. If the parameters go beyond the specified limits, repeat the setting.

4.Conclusion

The article presents a method that allows you to rebuild and adjust the microwave module in order to confirm the given requirements.

Indium foil was chosen as the main construction material, which has good adhesion to form the topology and use the "casts" method.

References

[1] Maltsev V A, Myakinkov V Yu, Rudy Yu B, Lebedev V N 2009 Microwave Transmitter Module for the Doppler Speed and Range Meter of the Phobos-Ground Program. *Electronic Engineering, Microwave Technology Series* **3 (502)** 27 - 35

[2] GOST RF 23221–78. *Microwave modules, microwave blocks. Terms, definitions and letter designations* 2005 (Moscow: Standartinform)

[3] Dzhurinsky K, Lisitsyn A 2008 Design and technological features of microwave modules. Moscow, *Modern Electronics* **1** 6

[4] Maksarov V V, Efimov A E 2018 Simulation modeling of dynamic characteristics of machining in NI LabView software environment to improve processing technique of a rod component, *IOP Conference Series: Earth and Environmental Science*, **194**, 2

[5] Maksarov V V, Keksin A I 2018 Forming conditions of complex-geometry profiles in corrosion-resistant materials *Published under licence by IOP Publishing Ltd,* **194**, 6

[6] Naimushin M P, Solovyanova I P 2005 Elements of radio-frequency transmission lines: Methodological development for the courses “Microwave devices and antennas”, “Technical electrodynamics” *Ekaterinburg: Ural State Technical University*

[7] Lyalin K S, Prikhodko D V. 2009 *Microwave electrodynamics: study guide: [in 2 parts]; Ministry of Education and Science of the Russian Federation, Federal Agency for Education, Moscow State Institute of Electronic Engineering (Technical University)* (Moscow: MIET)

[8] Maximov V M 2002 *Microwave transmission lines: textbook for universities* (Moscow: SAYNS-PRESS).

[9] Alekseenko A M, Myakinkov V Yu, Fedorov I S, Rusakova A K et al. Method for sealing and radio screening of multifunctional microwave devices, *Electronic Technology, Series 1, Microwave Electronics*, **10 (394),** 1986

[10] Grigoriev L.D. 1990 *Electrodynamics and microwave technology: Textbook for universities but specialty "Electronic devices"* (Moscow: High School)

[11] Maksarov V V, Makhov V E 2018 Studying parameters and quality of thread by optical light field recorder, *IOP Conference Series: Earth and Environmental Science*, **194**, 6

[12] Maksarov V, Khalimonenko A 2018 Quality assurance during milling of precision elements of machines components with ceramic cutting tools *International Review of Mechanical Engineering (IREME)* DOI: https://doi.org/10.15866/ireme.v12i5.14570

IOP Conf. Series: Materials Science and Engineering **760** (2020) 012063 doi:10.1088/1757-899X/760/1/012063

Development of multivariate equivalent cyclic tests of HALT-based aviation radio equipment for assessing reliability

V E Trushnikov[1], A V Komissarov[2], M V Grishin[3]

[1]Doctor of Technical Sciences, Professor, Professor of the Department of System Analysis and Management, St. Petersburg Mining University, St. Petersburg
[2]Chief designer of Ulyanovsk Instrument Design Bureau JSC, Ulyanovsk
[3]Ph.D., Designing engineer of Ulyanovsk Instrument Design Bureau JSC, Ulyanovsk

E-mail: Trushnikov_VT@pers.spmi.ru

Abstract: The article studies forced test models for determining reliability of on-board electronic equipment due to the combined effects of broadband random vibration and cyclical changes in temperature. The model of multivariate equivalent cyclic tests based on HALT (High accelerated life / limited test) procedures is described.

1. Introduction

In [1], the authors described the main methods for assessing reliability of civilian avionics. The urgent task is to develop a mathematical model of multivariate equivalent cyclic tests based on the characteristics of test equipment and procedures for conducting highly accelerated life (limit) rejection High accelerated life / limited tests [2]. Simulation of the effects of external factors using CAE-system software packages (ANSYS, ABAQUS, NASTRAN, MARC, Asonika, etc.) is based on reference and experimental (exploration) data which cause an inadequate assessment of reliability of the on-board electronic equipment.

Several sources provide options for using HALT procedures to assess the level of reliability (application of the electronic component base, design flaws, imperfections in technological operations and errors during their implementation in production, etc.) of on-board radio-electronic equipment and determine the resource indicator (reliability) of the product. The basic concept is based on the theory of fatigue accumulation.

Table 1. The model for describing forced factors.

Acceleration factor	Basic model	Test model
Cyclic changes in temperature	Arrhenius equation	$0{,}0833 \cdot \left(\frac{\Delta T_{HALT}}{\Delta T_{oscillations_i}}\right)^{-k}$
The cyclic rate of temperature changes	Coffin-Manson Equation	$\left(\frac{V_{\Delta T_{HALT}}}{V_{\Delta T_{oscillations}}}\right)^{1/3}$
Multi-axis broadband random vibration	Basquin equation	$\sqrt{\left(G_{xx}+G_{yy}+G_{zz}\right)^{b}}$
Electrical loads	The ratio of the number of on and off modes for a period of time	$\frac{N_{HALT}}{N_{Э}}$

The paper [3] describes the test equivalence model in which the limiting state during the cyclic exposure to forcing factors is the excess of a certain level of performance parameters as a result of accumulation of residual (fatigue) effects, while the mechanism of development of such effects does not make a decisive difference (phenomenological model).

The moment of achieving the limit state depends on the exposure parameters and properties of the product. The criterion combining these factors is the impact energy E required to reach the limit state during T_L. The following expressions are used for non-stationary (1) and stationary (2) random effects:

$$E = \int_0^{T_L} D(t)dt \tag{1}$$

$$E = DT_L \tag{2}$$

The impact energy applied to achieve the limit state of the product serves as the aggregate characteristic of the limit values of its destruction parameters. The convenience of the energy approach is due to the fact that the nature of the impact is not of fundamental importance; however, this takes into account both properties of the impact and properties of the product structure, thereby describing the stress state. This feature allows for the justification of equivalent tests [4].

2. The model of impact of the broadband random vibration

The concept of design models of the resource involves two main approaches to vibration exposure [5]:
1. The model of multi-cycle fatigue described by the Basquin equation:

$$\frac{\Delta\sigma}{2} = \sigma_f'(2N_f)^b \tag{4}$$

where $\Delta\sigma/2$ - voltage amplitude per cycle; N_f - the number of cycles before destruction; σ_f – the fatigue coefficient and b – material dependent fatigue strength (Basquin indicator).
In practice, the empirical relation developed by S. Crandle and W. Mark has gained wider application.:

$$N_f = N_b\left(\frac{\sigma_0}{\sigma_a}\right)^m \tag{5}$$

where N_f – the number of cycles before destruction; N_b – the basic number of cycles;
σ_0- fatigue limit;
σ_a- voltage amplitude per cycle; m – the material dependent parameter.
The low-cycle fatigue model described by the Manson-Coffin-Basquin equation:

$$\frac{\Delta\sigma}{2} = \frac{\sigma_f'}{E}(2N_f)^b + \varepsilon_f'(2N_f)^c \tag{6}$$

where $\Delta\sigma/2$- the amplitude of total deformation in the cycle; N_f- the number of cycles before the destruction; σ_f- the fatigue coefficient, b- Basquin's exponent, ε_f- the fatigue ductility coefficient; c the indicator of fatigue ductility depending on the material; E - the modulus of elasticity.

The fatigue model is the accumulation of damage characterized by the monotonous irreversible achievement of the ultimate state (failure) due to the summation of relative damage described by the Miner hypothesis of linear accumulation (it is assumed that complete destruction is achieved at D = 1):

$$D = \sum_{i=1}^{n} \frac{n_i(S_i)}{N_i(S_i)} \tag{7}$$

where D_i - accumulated fatigue damage during the stress S_i; n_i - the number of stress cycles on load with amplitude S_i; N_i – the number of stress cycles with amplitude S_i before the destruction.

[6] describes the effect of multiaxial vibration used in tests using the HALT / HASS techniques (Figure 1).

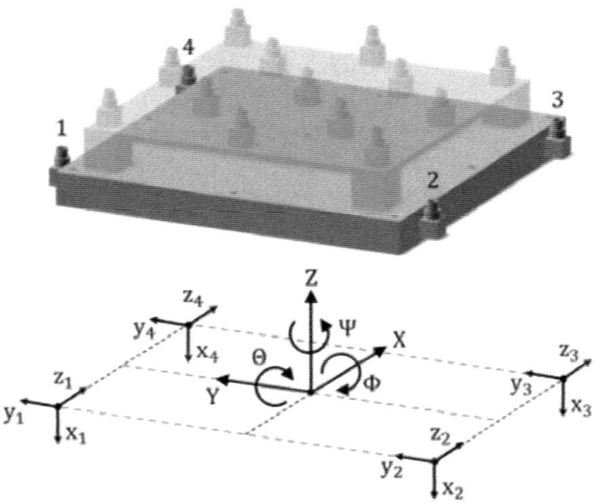

Fig. 1 The HALT multi-axis broadband random vibration

It was proved that the method of conducting equivalent cyclic tests under the sequential action of broadband random vibration along each axis does not interpret the exploratory mode. A comparative analysis conducted by ESPEC Qualmark (USA) showed that HALT multiaxial vibration tests provide a greater energy impact than traditional tests, for example, NAVMAT-9492 tests (Figure 2).

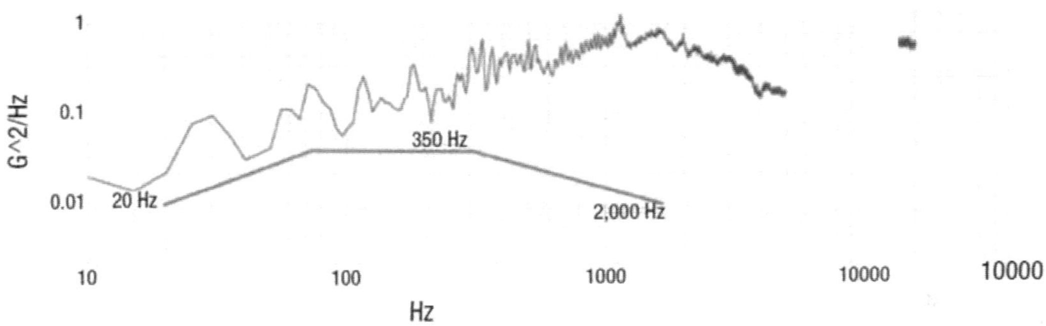

Fig. 2 Comparative graphs of spectral power densities of vibration test for NAVMAT-9492 (red) and HALT (blue)

Uisng the Huber-Mises energy theory for simulating the test process and taking into account the zero excitation of the vibration table, the voltage power spectrum equivalent to the multiaxial vibration mode can be presented as follows [7]:

$$G_M = G_{xx}+G_{yy} + G_{zz} + 3(G_\Theta + G_\Psi + G_\Phi)$$

where G_{xx}, G_{yy}, G_{zz}- voltage power spectra along each axis X, Y and Z , G_Θ, G_Ψ, G_Φ- rotation voltage power spectra. The value of the parameter of accumulation of fatigue damage during the multiaxial vibration DM in expression (9) is many times higher than the fatigue accumulated during the sequential test mode along each axis D_s in expression (8):

$$D_s = \sqrt{G_{xx}^b} + \sqrt{G_{yy}^b} + \sqrt{G_{zz}^b} \quad (8)$$

$$D_M = \sqrt{G_M^b} = \sqrt{\left(G_{xx}+G_{yy} + G_{zz} + 3(G_\Theta + G_\Psi + G_\Phi)\right)^b} \quad (9)$$

Studies of the effect of power spectra of rotational stresses on fatigue damage to products require more detailed experimental data to build theoretical models of fatigue accumulation [8]. Hence, neglecting the energy of rotation during multiaxial vibration, we obtain the fatigue accumulation parameter during testing in the HALT chamber:

$$D_{HALT} = \sqrt{\left(G_{xx} + G_{yy} + G_{zz}\right)^b} \quad (10)$$

The maximum exploratory level of spectral density, taking $G_{expl\,max} = G_{xx_э} = G_{yy_э} = G_{zz_э}$ and:

$$D_{Expl} = 3\sqrt{G_{expl\,max}{}^b} \quad (11)$$

The equivalence of tests is the equality of the energy of the impact of exploratory loads during the assigned resource to the energy of the impact of tests. Assuming that the exposure energies during the tests in the HALT chamber and the exposure energies during the exploration are equal, taking into account expressions (7) and (10) for the spectral power density and the equality of the accumulated fatigue parameter, we have the following expression

$$\frac{T_{exp}}{T_t} = \frac{3 \cdot \sqrt{G_{expl\,max}{}^b}}{\sqrt{\left(G_{xx_И} + G_{yy_И} + G_{zz_И}\right)^b}} \quad (12)$$

The acceleration factor is

$$F_{Vibration} = \frac{3 \cdot \sqrt{G^b_{expl\,max_i}}}{\sqrt{\left(G_{xx} + G_{yy} + G_{zz}\right)^b}} \quad (13)$$

The parameter b for electronic devices is usually chosen in the range from 3 to 8 [4]. Recent studies of reliability of electronic components in BGA and CMS packages give a parameter value of 4 to 6 [9].

Moreover, according to Table G.1, the mode of the basic operation cycle in terms of vibration exposure is divided into different periods depending on the duration of exposure and location on board [4] (Figure 3).

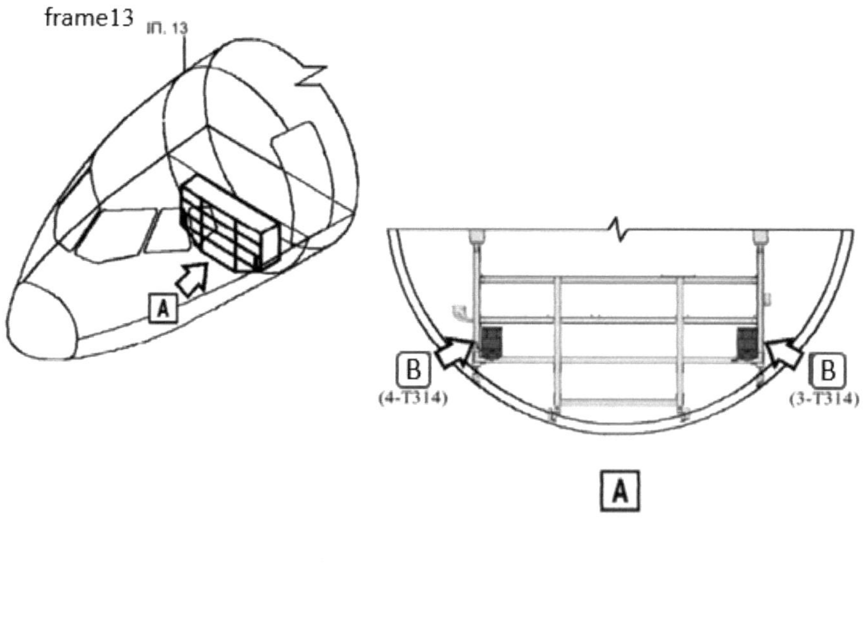

Fig. 3. Typical layout of avionics

The test time for each exploring mode N is calculated by expression:

$$T_{HALT} = \sum_{i=1}^{N} T_{exp_i} \frac{3\times\sqrt{\left(G_{expl\,max_i}\right)^b}}{\sqrt{\left(G_{xx}+G_{yy}+G_{zz}\right)^b}} \tag{14}$$

3. The effect of cyclic changes in temperature

For equivalent tests of on-board electronic equipment in a functional mode, the following formula of the base year of opeartionis used to calculate the number of test cycles for cyclic temperature changes [4]:

$$N_{ERE} = \sum_{i=1}^{12} N_{t_i} = N_{tl} \sum_{i=1}^{12} 0{,}0833 \cdot \left(\frac{\Delta T_{HALT}}{\Delta T_{oscillations_i}}\right)^{-k} \tag{15}$$

where N_{t_i}- the number of test cycles per month; N_{tl}- the number of take-offs / landings per year; ΔT_{HALT}- the absolute value of the range of temperature changes during testing; $\Delta T_{oscillations_i}$- the range of fluctuations in temperature in the aircraft compartment per month; the longevity line slope coefficient k is equal to 6 in the test mode equivalent to the flight mode.

However, in expression 15, the rate of changes in temperature oscillations during testing $V_{\Delta T_{HALT}}$ and operation $V_{\Delta T_{oscillations}}$ is not taken into account [4, 5, 12]

$$N_{\Delta T} = \left(\frac{V_{\Delta T_{HALT}}}{V_{\Delta T_{oscillations}}}\right)^{1/3} \tag{15}$$

For the base year of operation, expression (15) is as follows:

$$N_{ERE} = N_{tl} \sum_{i=1}^{12} 0{,}0833 \cdot \left(\frac{\Delta T_{HALT}}{\Delta T_{oscillations_i}}\right)^{-k} \times \left(\frac{V_{\Delta T_{HALT}}}{V_{\Delta T_{oscillations}}}\right)^{1/3} \tag{16}$$

For equivalent tests of on-board electronic equipment in aircraft storage modes during the year^:

$$N_{ESM} = \sum_{i=1}^{12} N_{x_i} \cdot \left(\frac{\Delta T_{HALT_x}}{\Delta T_{xc_i}}\right)^{-k} + \left(\frac{\Delta T_{HALT_x}}{\Delta T_{xr_i}}\right)^{-k} \tag{17}$$

where N_{x_i} – number of test cycles per one month; ΔT_{HALT_x} – the absolute value of the range of temperature changes during testing; ΔT_{xc_i} - the absolute value of the range of changes in storage temperature per day, the difference between day and night temperatures ($\Delta T_{xc_i} = T_{xд_i} - T_{xн_i}$); ΔT_{xr_i}- the absolute value of the range of changes in storage temperatures per year, the difference between the maximum value and the minimum temperature ($\Delta T_{xr_i} = T_{xr_{max}} - T_{xr_{min}}$).

Given the rate of temperature changes during testing $V_{\Delta T_{HALT}}$, the rate of changes during the day $V_{\Delta T_{oscillations_c}}$ and the year $V_{\Delta T_{oscillations_r}}$, let the number of test cycles per 1 storage year be

$$N_{ESM} = \sum_{i=1}^{12} N_{x_i} \cdot \left[\left(\frac{\Delta T_{HALT_x}}{\Delta T_{xc_i}}\right)^{-k} \left(\frac{V_{\Delta T_{HALT}}}{V_{\Delta T_{oscillations_c}}}\right)^{\frac{1}{3}} + \left(\frac{\Delta T_{HALT_x}}{\Delta T_{ys_i}}\right)^{-k} \times \left(\frac{V_{\Delta T_{HALT}}}{V_{\Delta T_{oscillations_r}}}\right)^{1/3}\right] \tag{18}$$

The acceleration factor under transportation mode tests:

$$F_{Ttransp} = \frac{\left(\frac{\Delta T_{HALT}}{\Delta T_{transp}}\right)^{-k} \times \left(\frac{V_{\Delta T_{HALT}}}{V_{\Delta T_{oscillations}}}\right)^{1/3}}{2} \tag{19}$$

4. The electric load model

The model of electrical loads is a uniform distribution of the on (during heating) and off (during cooling) modes during each testing cycle for cyclic changes in temperature and heat shock. The initial data for developing the calculated model of electrical loads are as follows: the average flight duration during one day T_P, the average annual flight T_{afr}; the average annual number of flights (takeoffs and landings) $N_{t/l}$ which determines the number of on and off modes, N_E - the total number of on / off modes for the entire time of operation, including the routine maintenance work according to the technical documents.

The calculated value of the acceleration factor is determined by the following ratio:

IOP Conf. Series: Materials Science and Engineering **760** (2020) 012063 doi:10.1088/1757-899X/760/1/012063

$$F_E = \frac{N_t}{N_E}$$

$$N_E = N_P + N_R + 1$$

$$N_P = 2 \cdot \mathrm{N}_{V-P} \text{ or } N_{\Pi} = \frac{\mathrm{T_{SH}}}{365} \cdot \frac{2}{T_P}$$

For one year

$$N_R = 2 \cdot \left(\frac{\mathrm{T}_{YH}}{\mathrm{T_{DY}}} + \frac{\mathrm{T}_{YH}}{24 \cdot \mathrm{T_{WY}}} + 3{,}65 \cdot \frac{\mathrm{T}_{YH}}{\mathrm{T}_A}\right)$$

5. The model of the combined impact of factors

[9] studies the simultaneous influence of broadband random vibration and cyclical changes in temperature; the accelerating effects of vibration and temperature can be used for the calculations independently. However, [4] developed a model in which with equal flows of failure intensities during operation and testing, the mutual influence of factors leading to failures is taken into account. The model of the combined effect (acceleration of test modes causing failures similar to the operation ones) is the total acceleration factor (acceleration coefficient) [4]:

$$F_{comb} = \frac{1}{N_F} \sum_{i=1}^{N_F} F_i \cdot \prod_k F_k \quad (19)$$

$\prod_k F_k$ – is a product of acceleration factors affecting the failure mode of the product; F_i- the acceleration coefficient for each of the forced factor.

The effective method for describing models of mutually influencing factors in models of forced equivalent cyclic tests is the physico-statistical method, according to which the test object is a thermodynamically unstable system that changes due to the global principle of entropy growth. To develop models of forced equivalent cyclic tests, a generalized mathematical model of resource consumption is used. It is developed by combining resource consumption models for individual elements of on-board electronic equipment depending on mutually affecting factors (for example, broadband random vibration and cyclic temperature change), reflecting only reliable physicochemical properties and mechanical processes.

For each stage of testing, the acceleration coefficient is calculated according to the expressions for BRV (8), for cyclic temperature changes during transportation:

1) 1) For the broadband random vibration:

$$F_{Vibration} = \frac{\sqrt{G^b_{expl\ max_i}}}{\sqrt{\left(G_{xx} + G_{yy} + G_{zz}\right)^b}}$$

2) For cyclic temperature changes during transportation:

$$F_{\mathrm{T}transp} = \frac{\left(\frac{\Delta T_{HALT}}{\Delta T_{oscillations_{transp}}}\right)^{-k} \times \left(\frac{V_{\Delta T_{HALT}}}{V_{\Delta T_{oscillations}}}\right)^{1/3}}{2}$$

3) For cyclic temperature changes during operation:

$$F_{\mathrm{T}exp} = \frac{\left(\frac{\Delta T_{HALT}}{\Delta T_{oscillations_{\text{эксп}}}}\right)^{-k} \times \left(\frac{V_{\Delta T_{HALT}}}{V_{\Delta T_{oscillations}}}\right)^{1/3}}{2}$$

4) For cyclical changes in temperature during storage:

$$F_{\mathrm{T}st} = \frac{\left(\frac{\Delta T_{HALT_{\text{x}}}}{\Delta T_{\text{x}c_i}}\right)^{-k} \times \left(\frac{V_{\Delta T_{HALT}}}{V_{\Delta T_{oscillations_{\text{c}}}}}\right)^{\frac{1}{3}}}{2} + \frac{\left(\frac{\Delta T_{HALT_{\text{x}}}}{\Delta T_{ys_i}}\right)^{-k} \times \left(\frac{V_{\Delta T_{HALT}}}{V_{\Delta T_{oscillations_{\text{r}}}}}\right)^{1/3}}{2}$$

The model of the combined broadband random vibration and cyclic temperature changes for the equivalent cyclic test program is as follows:

$$\Phi_{comb} = \left[\frac{\Phi_{Vibration} \cdot \Phi_{Ttransp}}{2}\right]_{trasnp} + \left[\frac{\Phi_{Vibration} \cdot \Phi_{Tst}}{2}\right]_{storage} + \left[\frac{\Phi_{Vibration} \cdot \Phi_{Texpl}}{2}\right]_{Expl}$$

The combined ECT model at three stages of the product life cycle is as follows:

$$\Phi_{comb} = \left[0{,}0833 \cdot \left(\frac{\Delta T_{HALT}}{\Delta T_{oscillations_i}}\right)^{-k} \times \left(\frac{V_{\Delta T_{HALT}}}{V_{\Delta T_{oscillations}}}\right)^{\frac{1}{3}} + \sum_{i=1}^{N} T_{exp_i} \frac{\sqrt{G^{b}_{expl\,max_i}}}{\sqrt{(G_{xx}+G_{yy}+G_{zz})^{b}}}\right] +$$

$$\left[\left(\frac{\Delta T_{HALT_x}}{\Delta T_{xc_i}}\right)^{-k} \left(\frac{V_{\Delta T_{HALT}}}{V_{\Delta T_{oscillations_c}}}\right)^{\frac{1}{3}} + \left(\frac{\Delta T_{HALT_x}}{\Delta T_{xr_i}}\right)^{-k} \times \left(\frac{V_{\Delta T_{HALT}}}{V_{\Delta T_{oscillations_r}}}\right)^{1/3}\right]$$

Parameter F_{comb} is the the acceleration coefficient of the MECTM..

6. Conclusion

The developed MECTM model has the following advantages:

1) The multi-axis SHS has a synergistic effect creating test conditions equivalent to the real exploring modes;

2) The vibrational energy is focused in the high-frequency spectrum;

3) The acceleration factor (acceleration coefficient) is higher due to the refined calculation of the product of the WWF, in comparison with the tests conducted according to the sequential scheme adopted in the current regulatory and technical documents.

However, the negative effects of the MECTM are inherent in the HALT: failure mechanisms associated with wear may be absent due to too high acceleration factors. HALT procedures do not provide an adequate assessment of reliability for electromechanical products [9].

References

[1] Komissarov A V., Shishkin V V. 2018 The main methods for assessing reliability of on-board electronic equipment of modern civilian planes before the mass production. *Bulletin of Samara Scientific Center of the Russian Academy of Sciences.* **20, 4 (3)** 319 - 326.

[2] Komissarov A V, Vinogradov A B 2016 Quality management methodology for on-board aircraft and ground equipment. *Bulletin of Samara Scientific Center of the Russian Academy of Sciences.* **18, 4 (3)** 571 - 577.

[3] Olt J, Maksarov V V, Efimov A E 2018 Impacts of gradient structure on the dynamic characteristics of machining process system *29th Daaam international symposium on intelligent manufacturing and automation* 190-196 DOI: 10.2507/29th.daaam.proceedings.027

[4] Guru P P, Pecht M A 2018 critique of reliability prediction techniques for avionics applications. *Chinese Journal of Aeronautics* **31(1)** 10–20

[5] Maksarov V V, Keksin A I 2018 Technology of magnetic-abrasive finishing of geometrically-complex products, *IOP Conference Series: Materials Science and Engineering*, **327**

[6] Harry McLean, Mike Silverman, 2010, From HALT Results to an Accurate Field MTBF Estimate. *Reliability Society 2010 Annual Technical Report*

[7] Maksarov V V, Krasnyy V A, Viushin R V 2018 Simulation of dynamic processes when machining transition surfaces of stepped shafts, IOP Conference Series: Materials Science and Engineering **327**

[8] Efimov A E, Maksarov V V, Timofeev D Y 2017 Modeling dynamic processes at stage of formation of parts previously subjected to high-energy laser effects, *IOP Conference Series: Materials Science and Engineering* **327**

[9] Alekseeva L B, Maksarov V V 2018 Evaluation of effect of oil film of rotor bearing, IOP Conference Series: Materials Science and Engineering **327**

[10] Harry McLean, Mike Silverman 2010 From HALT Results to an Accurate Field MTBF Estimate. *Reliability Society 2010 Annual Technical Report* DOI 10.1109/RAMS.2010.5448013

IOP Conf. Series: Materials Science and Engineering **760** (2020) 012064 doi:10.1088/1757-899X/760/1/012064

Analysis of stresses and deformations of elements of axle boxes and improvement of their shape

L B Tsvik[1], A A Tarmaev[1], I S Bocharov[1] and V K Eremeev[1]

[1]Irkutsk State Transport University, 15 Chernishevsky st., Irkutsk, 664074, Russia

E-mail: tsvik_l@mai.ru

Abstract. The development of constructive recommendations aimed at increasing the service life of roller bearings of axle boxes of railway cars is described. The elastic volumetric stress-strain state of the bearing elements of the bearings under consideration is analyzed. Numerical analysis of the corresponding mathematical models was carried out using the finite element method in the NX NASTRAN software environment. It is shown that near the edge of the roller with chamfer-rounding, a stress concentration arises, caused by insufficient smoothness of the contour line of the axial section of the roller. It is noted that this concentration is caused by the fact that the line of the contour under consideration has a finite specified curvature in the section corresponding to the chamfer and almost zero curvature in the adjacent straight section corresponding to the cylindrical surface of the roller. It is shown that the use of smoother contour lines for the marginal zone of a roller more than doubles the maximum value of the stress intensity in the boundary zone under consideration and accordingly increases the cyclic life of the bearing. The proposed smoothing of the line of the considered circuit allows one not only to unload the edge of the roller but also to reduce the contact pressure in its middle part. At the same time the stress-strain state of the roller becomes more uniform by increasing the smoothness of the contour line in the area of the roller's edge and simultaneously reducing the degree of bombing (barrel-shaped) its cylindrical surface.

1. Introduction

Currently, there are significant changes in the operating conditions of wheel sets of railway cars. The greatest contribution to these changes is made by an increase in the speeds of movement of the compositions and an increase in their weight [1–3]. At the same time, the decisive contribution to ensuring the reliability of wheelset operation is made by increasing the operational reliability of the wheels and elements of axle boxes – rollers and bearing rings [4–9]. One of the effective ways to improve them is to search for rational design options based on the use of mathematical models of their stress-strain state (SSS) and experimentally verified criteria for assessing their durability [10–12]. Under the conditions of cyclic operation of high-loaded bearing elements of structures, such a criterion, as a rule, is the intensity of stresses [13–22].

2. Results and Discussion

Let us consider for definiteness a constructive scheme for coupling a bearing roller and its rings by the example of one of the typical axlebox nodes (figure 1) [23].

The durability of the elements of the axle assembly in many cases is determined by the moment the contact fatigue shells [24] appear on the rolling contact surfaces of the bearings and their inner and outer rings (figure 2). The determining factor in the occurrence of fatigue damage is an increased level of stress intensity in a possible focus of damage. To calculate this factor, a mathematical model of elastic deformation of the elements of the axle box can be used. In the case of a homogeneous isotropic material of the body in question, this model is determined by the differential equilibrium equations of the theory of elasticity:

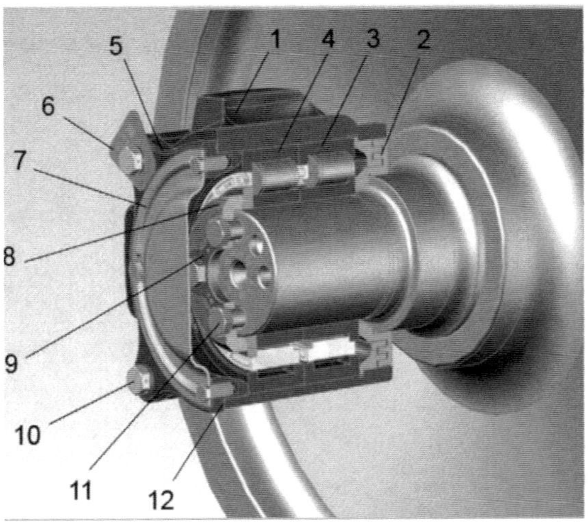

(a)

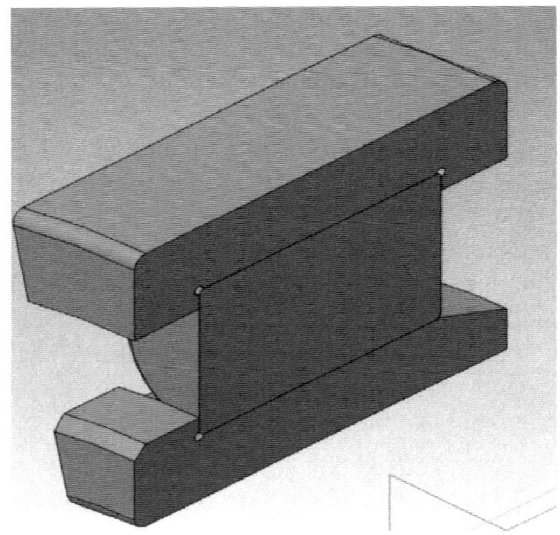

(b)

Figure 1. Constructive scheme of the coupling roller bearing and its rings: (a) - constructive diagram of a typical axle box of a freight car: 1 - box housing; 2 - labyrinth ring; 3 - rear bearing; 4 - front bearing; 5 - cover fastening; 6 - tag; 7 - viewing cover; 8 - disc washer; 9 - lock washer; 10 - M20 bolt of the disc washer; 11 - M12 cover bolt; 12 - sealing ring; (b) - axial section of a fragment of the rear bearing of an axle box corresponding to one roller

$$Lu = (\lambda + \mu) grad\ div\ \boldsymbol{u} + \Delta \boldsymbol{u} = 0, \tag{1}$$

where L is the differential operator of the theory of elasticity [25];

$\boldsymbol{u}$ is the desired displacement vector at points in the domain D occupied by the deformable elements of the structure under consideration;

λ, μ – coefficients that determine the mechanical properties of a homogeneous isotropic material of structural elements (Lame parameters).

Figure 2. Contact fatigue shells on the surface of the inner ring of the rear bearing in the zone of contact of the edge of the roller and ring

Equations (1) are considered in the domain D bounded by the surface $S = S\_u + S\_\sigma + S\_u\sigma$ under boundary conditions

$$\boldsymbol{u}\,|_{S_u} = \boldsymbol{\varphi}\,(M)\ with\ M \in S_u;\ (\boldsymbol{\nu}\cdot \boldsymbol{T})\,|_{S_\sigma} = \boldsymbol{F}(M)\ with\ M \in S_\sigma \tag{2}$$

$$(\boldsymbol{u}\cdot\boldsymbol{\nu})\,\Big|_{S_{u\sigma}} = \varphi_\nu(M), (\boldsymbol{\nu}\cdot \boldsymbol{T}\cdot\boldsymbol{\tau})\,\Big|_{S_{u\sigma}} = 0 \text{ with } M \in S_{u\sigma}, \tag{3}$$

where $S_u, S_\sigma, S_{u\sigma}$ are the surfaces of the considered parts, on which surface displacements, surface forces and mixed (contact) boundary conditions are given, respectively;

T is the stress tensor;

$\boldsymbol{\varphi}$ (M), $\varphi\nu$ (M) are given vector and scalar functions of point M, respectively;

$\boldsymbol{F}$ (M) is the vector of surface forces specified on the surface $S\sigma$;

$\boldsymbol{\nu}$, $\boldsymbol{\tau}$ are unit vectors defining the normal and tangential directions to the surface S at the considered point M.

Numerical solution of problem (1) - (3) was carried out using the finite element method (FEM), based on the preliminary discretization of the studied deformable system [26, 27]. In this case, the FEM ratios were considered from the standpoint of the variational approach, according to which the sought nodal displacements of the discrete analogue of the deformable parts of the axle box were found from the condition of minimizing its potential energy:

$$\Pi = \frac{1}{2}\int_V \boldsymbol{\varepsilon}^T \boldsymbol{\sigma} dV - \boldsymbol{U}^T \boldsymbol{R}, \tag{4}$$

where $\boldsymbol{\varepsilon}$ is the strain vector at the points of the deformed system under consideration (T is the transposition symbol) formed from the components of the strain tensor;

$\boldsymbol{\sigma}$ is a similar stress vector;

$\boldsymbol{U}$ is the vector of displacements of the discretization nodes of the deformable structure of the axle box, V is its volume;

$\boldsymbol{R}$ is the vector of external nodal effects on the considered discretized deformable system of elements.

Within a finite element, the displacement vector u can be represented using the matrix of interpolating functions $\boldsymbol{N}$:

$$u = Nd, \tag{5}$$

where $\boldsymbol{d}$ is the vector of nodal displacements in the considered finite element (FE).

The displacement vector $\boldsymbol{u} = \{u(x,y,z), v(x,y,z), w(x,y,z)\}^T$ is considered in the Cartesian coordinate system, which is global for the deformed system under consideration. When constructing a discrete deformation model, linear CEs were used, the displacements within which were interpolated by functions depending on the x, y, z coordinates. The deformation vector is determined within the CE through the vector $\boldsymbol{d}$ by the relation:

$$\varepsilon = Bd, \tag{6}$$

in which $\boldsymbol{B}$ is the corresponding matrix of differential operators (strain matrix), defined by the equality:

$$B = AN^T, \tag{7}$$

and the matrix of operators $\boldsymbol{A}$ determines the transition from displacements to deformations:

$$\varepsilon = A^T u. \tag{8}$$

Using the given relations (5) – (8), we represent the potential energy under consideration through the vector $\boldsymbol{U}$ – the global vector of the nodal displacements of the discretized deformed system:

$$\Pi = \frac{1}{2}\int_V U^T \boldsymbol{B}^T \boldsymbol{DBU} dV - \boldsymbol{U}^T \boldsymbol{R}, \tag{9}$$

where $\boldsymbol{D}$ is the matrix of elastic constants of the material: $\boldsymbol{\sigma} = D\boldsymbol{\varepsilon}$.

The conditions for minimizing the magnitude Π:

$$\frac{\partial \Pi}{\partial \boldsymbol{U}} = \left\{\frac{\partial \Pi}{\partial U_1}, \frac{\partial \Pi}{\partial U_2}, \dots, \frac{\partial \Pi}{\partial U_{N_u}}\right\}^T = 0 \tag{10}$$

and lead to a system of linear algebraic equations:

$$\boldsymbol{KU} = \boldsymbol{R}, \tag{11}$$

in which $\boldsymbol{K}$ is the global stiffness matrix of the considered discretized construction of the axle box:

$$\boldsymbol{K} = \int_V \boldsymbol{B}^T\boldsymbol{DB}dV. \tag{12}$$

The matrix $\boldsymbol{K}$ is built in the process of implementing the FEM by assembling the stiffness matrices of individual finite elements determined by the equality:

$$\frac{1}{2}\boldsymbol{d}_e^T\boldsymbol{F}_e = \frac{1}{2}\int_{V_e} \boldsymbol{\varepsilon}_e^T\boldsymbol{\sigma}_e dV_e \tag{13}$$

where V_e is the volume of the CE number e in the general array of the CE numbers of the discretized construction of the considered constructive node as a whole (the global EC number, then the index e in the given relations relating to the individual CE is omitted).

The finite element stiffness matrix $\boldsymbol{k}$ is determined by the ratio:

$$F = kd, \tag{14}$$

in which $\boldsymbol{d}$ is the vector of nodal displacements of the considered FE;

$\boldsymbol{F}$ is the corresponding vector of nodal forces.

Using relations (13), (14) for the stiffness matrix of an individual CE, we obtain:

$$k = \int_{V_e} B^T DBdV_e. \tag{15}$$

The given relations (4) – (15) completely determine the construction of system (11) – a system of linear algebraic equations (SLAE), which determines an approximate solution of problem (1) – (3). Taking into account the specified boundary displacements specified in the nodes of the axle box design under consideration allows replacing the general system (11) with the reduced SLAE:

$$K_{red}U = R_{red}, \tag{16}$$

the determinant of which is non-zero, and present the desired approximate solution of the problem in the form:

$$U = (K_{red})^{-1}R_{red}. \tag{17}$$

The FEM implementation as applied to the solution of the problem of determining the VAT of the elements of the axle box of the wheel set was carried out in the NX NASTRAN software environment. In the same system, the necessary geometric models were built, the axial section of which is shown in Fig. 3-5.

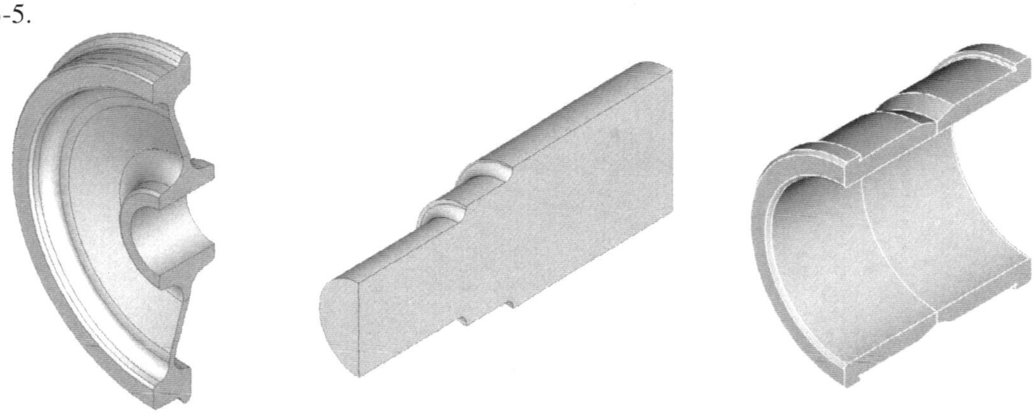

(a) (b) (c)

Figure 3. Geometrical models of a solid-rolled wheel (a), axles of a set of wheels (b) and inner rings of a bearing (c).

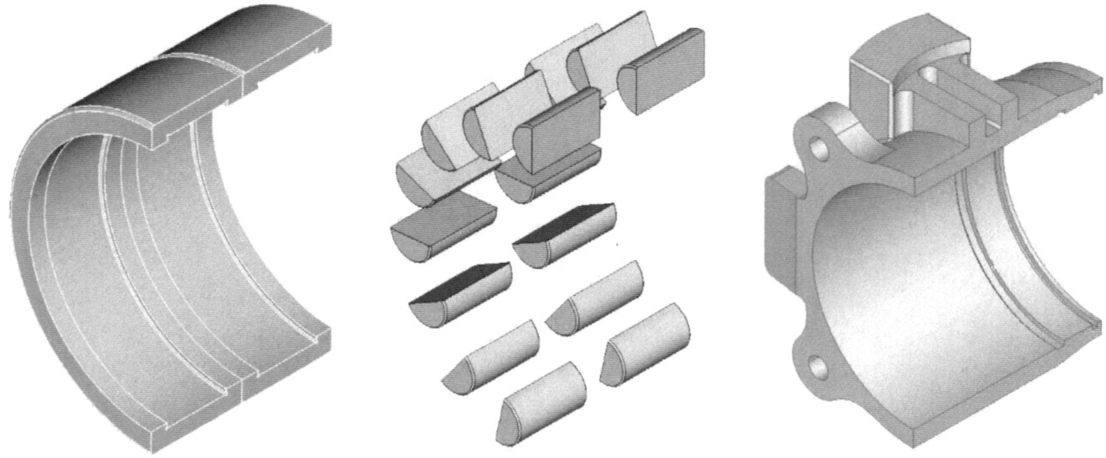

(a) (b) (c)

Figure 4. Geometric models of outer rings (a), roller bearings (b) and axle box housing (c).

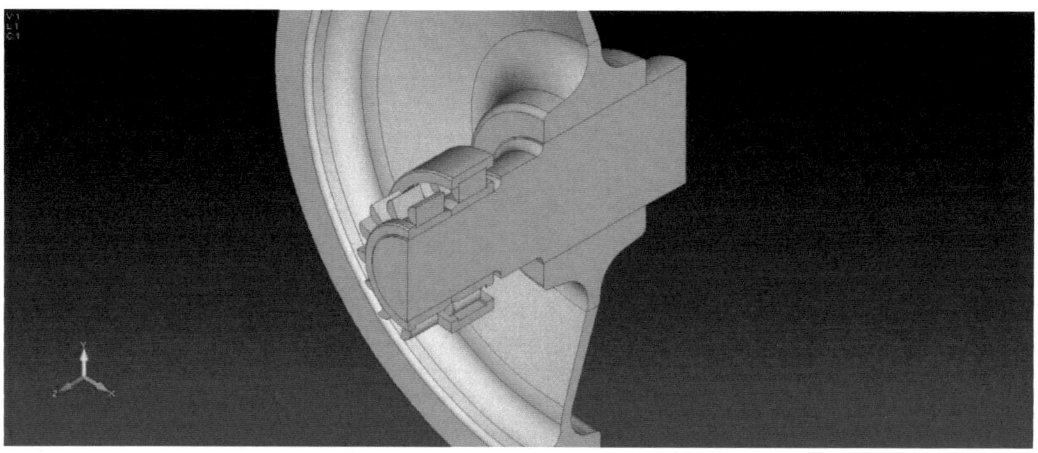

Figure 5. The location of the rollers in the front bearing box.

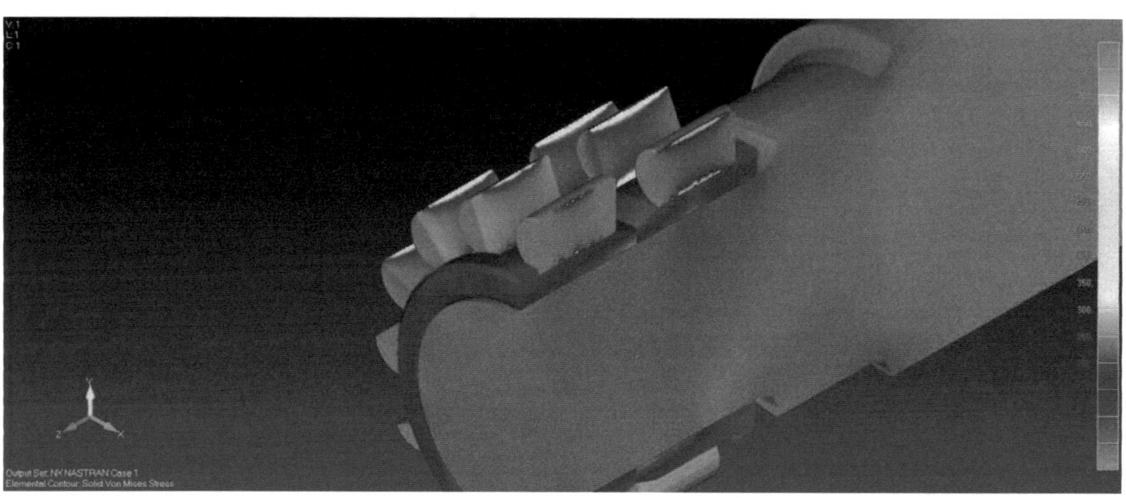

Figure 6. The pattern of stress intensity distribution that occurred when the rollers interact with the bearing rings when a load is applied.

In the process of modeling, it was assumed that a vertical uniformly distributed calculated load of 10^7H acts on the thrust surfaces of the housing of the axle box. The general picture of the stress intensity distribution arising in the roller of the axle box under this load is shown in figure 6. From the figure it can be seen that in the rollers of the rear bearing the stress level is higher than in the rollers of the front one. At the same time, the third roller (with respect to the vertical axial section of the axle box), the rear bearing roller, experiences the greatest load.

When considering the features of contact and rings, a preliminary breakdown of the rollers and bearing rings into fragments was used to achieve the required modeling accuracy (Figure *7a*). In this case, each of the fragments was split on the FE independently with increasing concentration as it approaches the contact surface and uniformly within the fragment (Figure *7b*).

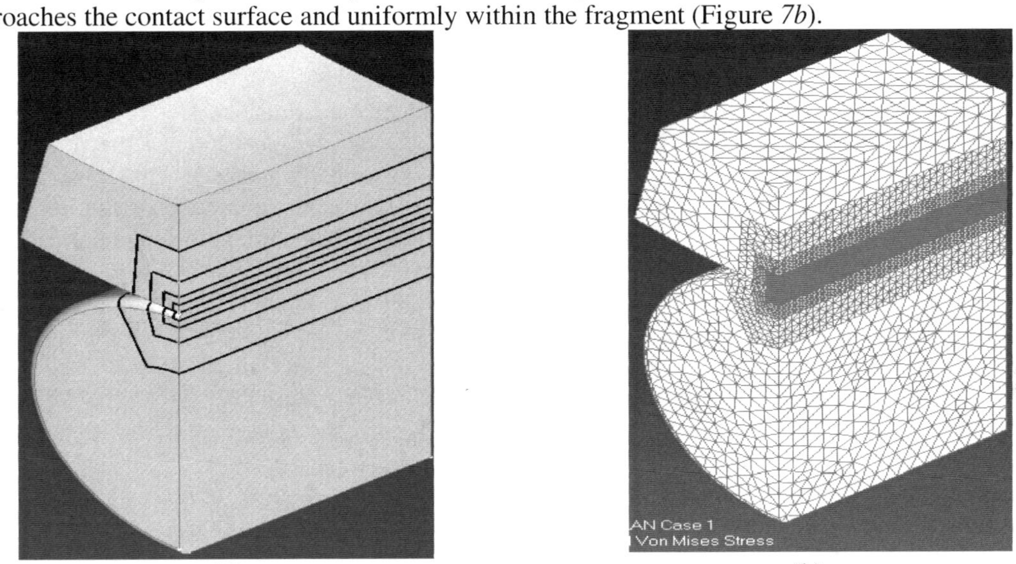

(a) (b)

Figure 7. Fragmentation of a discrete model of contact interaction between a roller and bearing rings: (a) – fragmentation scheme; (b) is a finite element breakdown with step condensation.

VAT within the structural elements of the axle box is substantially uneven. So, the size of the stress intensity concentration zone in a roller is determined by the depth of the location of the maximum stress intensity point σ_i^{max} below the roller contact surface. This size determines the depth of the contact fatigue pits (Figure 2) that appear on the contact surfaces during the operation of the bearing, and is about 0.25 mm (Figure 8).

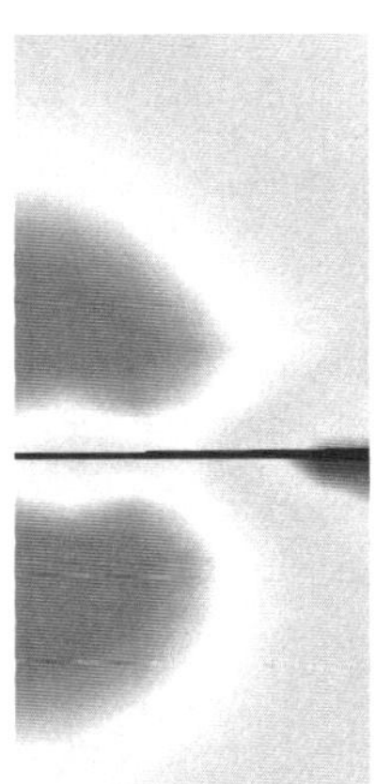

Figure 8. The stress intensity distribution below the contact surface of the roller and bearing ring

The fragments were considered as separate deformable bodies, the solutions in which were mated as follows. On the inner (non-contact surfaces of the rollers and rings) surfaces of the fragments, the solutions were mated corresponding to the full adhesion (gluing) of the fragments. On the contact surfaces of the rollers and rings, mating conditions were specified, corresponding to the ideal contact of the mating surfaces (contact without friction). The described technique allowed the use of fairly small finite elements in fragments adjacent to the contact surfaces of the rollers and rings, the size of which does not exceed 0.04 mm, and achieve the required modeling accuracy on the used computing equipment (Intel Xeon E5 central processor, 64 GB RAM).

Of considerable interest is the distribution of stress intensity near the edge of the roller. The results of the corresponding modeling for a roller with a rectilinear generatrix (without bombing) and facet rounding are shown in figure 9*a*. It can be seen from the figure that there is a significant stress concentration near the edge of the examined roller. The reason for the stress concentration near the edge of the roller in this case is the lack of smoothness of the contour of the profile of the roller (the contour of the profile of the axial section of the roller). So in the interface zone of the chamfer-rounding contour and the rectilinear generatrix of the cylindrical surface of the roller, the curvature of the contour line changes abruptly: within the chamfer, this line has a predetermined non-zero curvature, and the curvature of the rectilinear generatrix is zero. An even more significant stress concentration near the edge of the roller is caused by a conical chamfer, which creates a corner point at the edge of the contact surface of the roller and bearing ring. Let us note that the curvature of the curve is proportional to the second derivative of the function describing the contour line. From a mathematical point of view, the profile contour line with chamfer-rounding has a discontinuity of the second derivative of this function, which reduces the smoothness of the solutions of problem (1) – (3), found in the area bounded by the surface with the contour line under consideration [28]. From this circumstance, it follows that the use of smoother lines of the profile of the roller profile will lead to a significant decrease in the level of stress intensity near the edge of the roller.

To check the formulated position, the VAT simulation was carried out near the edge of the roller, the chamfer of which had in the axial section instead of rounding with a constant radius of curvature (Figure 10*a*) an ellipse shape with a ratio of 1:10 semiaxes (Figure 10*b*).

The results of stress intensity modeling for a roller with an elliptical shape of a chamfer profile are shown in Figure 9*b*.

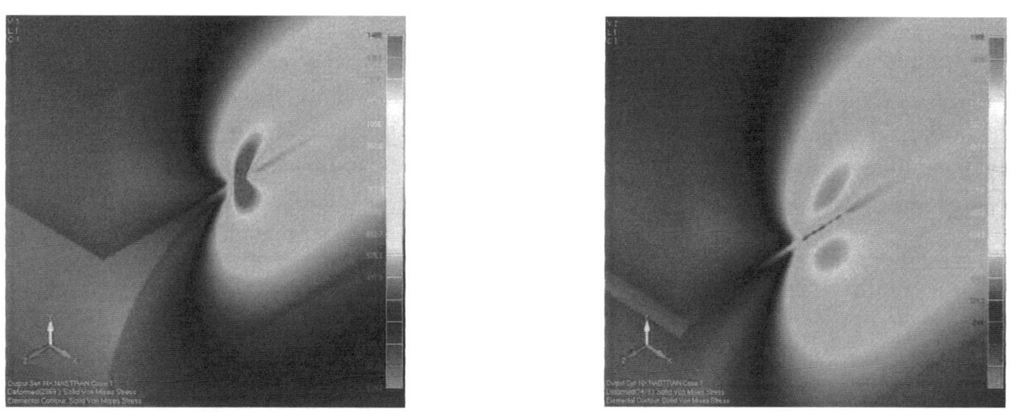

(a) (b)

Figure 9. The stress intensity distribution σ_i in the axial section of the bearing near the edge of the most loaded roller: (a) – roller with a typical shape of a chamfer profile; (b) – a roller with an elliptical shape of a chamfer profile

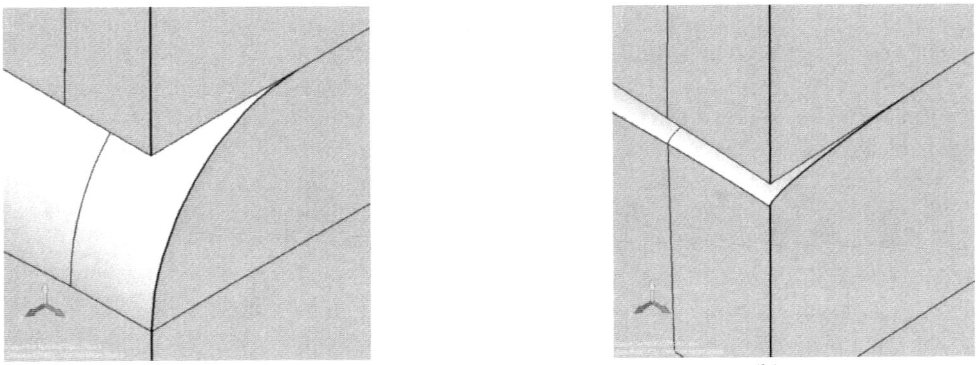

(a) (b)

Figure 10. Profiles of the axial section of the rollers: (a) – a chamfer-round roller; (b) – a roller with an elliptical chamfer profile.

3. Conclusions

From the obtained results it follows that the use of rolling elements with a smoothly varying curvature of the line of the profile of their axial section allows to significantly reduce the maximum value of stress intensity near the edge of the roller and, accordingly, increase the fatigue life of its work. The results obtained can be used to improve the profile shape of the bombed rollers. It is significant that the proposed smoothing of the contour line profile of the axial section of the roller allows not only relieving the roller edge, but also reducing the contact pressure in its middle part, making the stress-strain state of the roller more uniform by increasing the smoothness of the contour line in the area of the roller edge and simultaneously reducing the degree of bombing (barrel-shaped) of its cylindrical surface.

4. References

[1] Development strategy of the transport engineering of the Russian Federation for the period up to 2030 Approved by the order of the Government of the Russian Federation of August 17, 2017 № 1756-r

[2] Bayasgalan D 2019 Justification of the necessity of constructing the third railway line between Mongolia and Russia *Modern Technologies. System Analysis. Modeling* **61**:1, 97–101 DOI: 10.26731/1813-9108.2019.1(61).97–101

[3] Bayasgalan D 2018 The necessity of building a second railway line between Mongolia and China *Modern Technologies. System Analysis. Modeling* **60**:4 56–63 DOI: 10.26731/1813-9108.2018.4(60).56-63

[4] Borodin A V, Ivanova Yu A and Gric D B 2014 Influence of the radial load distribution on the life of the axle bearing of a freight car *The Journal Omsk Scientific Bulletin* **2** 35–37

[5] Nevmerzhiczkaya G V, Shlyushenkov A P 2002 Estimated reliability assessment of the axlebox bearing of the car according to the fatigue chipping criterion *Dynamics, strength and reliability of transport vehicles* (Bryansk: BSTU) 113-117

[6] Tao L, Wei S, Zhichao M, Junzhou H, Jianghui D, Liping W 2019 Dynamic investigation on railway vehicle considering the dynamic effect from the axle box bearings *Advances in mechanical engineering* **11**: 4 DOI: 10.1177/1687814019840503

[7] Hassan M, Bruni S 2019 Experimental and numerical investigation of the possibilities for the structural health monitoring of railway axles based on acceleration measurements *Structural health monitoring-an international journal* **18**:3 902-919 DOI: 10.1177/147592171878642

[8] Tsvik L B, Zapol`skij D V, Zen`kov E V and Eremeev V K 2015 Design profile railway wheels on a straight search in space radiuses hollow chamfer *Transport of the Urals* **3** 67–70

[9] Martua L, Ng AK, Sun G 2018 Prediction of Rail Rolling Contact Fatigue Crack Initiation Life via Three-Dimensional Finite Element Analysis *Int. Conf. on Intelligent Rail Transportation* (ICIRT, Singapore)

[10] Borodin A V, Kulinich E N and Ivanova Ju A 2010 Improvement of the roller box of the freight car *Journal of Transsib Railway Studies* **2** 15–20
[11] Borodin A V, Ivanova Ju A, Gric D B 2016 The axle box of a freight car with increased rigidity of the axle *The Journal Omsk Scientific Bulletin* **4** 5–8
[12] Petrov G and Tarmaev A 2018 Modeling of railway vehicles movement having deviations in the content of running parts *Proceedings of the International Conference: Aviamechanical Engineering and Transport* (*AVENT 2018*) (Atlantis Press: Series *Advances in Engineering Research*) **158** 410–415 DOI:10.2991/avent-18.2018.79.
[13] Mahutov N A 2005 Structural strength, life and man-made safety Part 1 (Novosibirsk: Science)
[14] Nevmerzhickaja G V 2006 Computer simulation of contact pressure distribution in a freight car bearing and assessment of its durability *Bulletin of Computer and Information Technologies* **2** 18-21
[15] Wang Q, Zhao Y and Wang H 2014 Experiments and characterization of the probabilistic fatigue lives and strength of D1 railway wheel steel *Journal of mechanical engineering* **50**:14 51-55 DOI: 10.3901/JME.201414.050
[16] Tomashevskij S B 2011 Wheelset – axle box: stress models *World of Transport and Transportation*] **1** 38–42
[17] Tomashevskij S B 2011 Clarification of the solution of contact problems on the example of calculating the details of railway transport *Transport of the Urals* **2** 66-70
[18] Tomashevskij S B 2011 The influence of elastic-plastic deformations on the results of solving contact problems of railway transport *Bulletin of Bryansk state technical university* **3** 17-
[19] Tomashevskij S B 2011 Numerical simulation of machine hardening by surface plastic deformation *Science and Technology in Transport* **2** 60-68
[20] Yao Ch, Zhiwei W, Bingyan Ch, Weihua Zh and Guanhua H 2019 An improved complementary ensemble empirical mode decomposition with adaptive noise and its application to rolling element bearing fault diagnosis *ISA transactions* DOI: 10.1016/j.isatra.2019.01.038
[21] Yang Z, Deng Xy, Li Zl 2018 Numerical modeling of dynamic frictional rolling contact with an explicit finite element method *Tribology international* **11** 214-231 DOI: 10.1016/j.triboint.2018.08.028
[22] Molodova M., Li Z., Núñez A., Dollevoet R. 2014 Validation of a finite element model for axle box acceleration at squats in the high frequency range *Comput Struct* **141** 84–93.
[23] State standard 15872-2014 2015 Rolling bearings. Axle box cylindrical roller bearings for rolling stock. Specification (Moscow: Standartinform)
[24] Classifier for defects and damage to rolling bearings 2007 (Moscow: Russian Railways)
[25] Parton V Z, Perlin P I 1981 Methods of the mathematical theory of elasticity (Moscow: Science)
[26] Bathe K-J and Wilson E L 1976 Numerical Methods in Finite Element Analysis (Prentice Hall)
[27] Duong Van Lam, A. A. Pykhalov 2018 Interpolation of geometry and inhomogeneity of material of deformable solids when constructing their 3d models with the finite elements method based on the computer tomograph scanning *Modern Technologies. System Analysis. Modeling* **60**:4 8 – 15. DOI: 10.26731/1813-9108.2017.3(55).8-16
[28] Vladimirov V S 1981 Equations of mathematical physics (Moscow: Science)

Assessment of contribution of human factor and factors of material-technical supply to safety risks due to poor repair and technology

O I Verevkina [1]

[1]Rostov State Transport University, Rostov-on-Don, Russian Federation

E-mail: ov18111966@mail.ru

Abstract. The paper discusses the method of assessment and analysis of functional risks within the infrastructure of the railway transport network. The basis of the method is laid down in [1-4], but the risk assessment subsystem was not described due to poor repair. This paper will try to fill this gap. A general diagram of traffic safety risks by factors was developed. The method of transforming the probability function of repair rules violation into the probability of traffic safety violation due to some repair is proposed. For this purpose, the problem of minimizing the disparity in non-linear equations determining scaled factors of traffic safety violation (TSV) due to repair is solved. The calculation results at the level of linear enterprise are presented.

1. Introduction

The paper presets the diagram and main features of the algorithm to assess actual and forecast contribution of a human factor (HF) and a material-technical supply factor (MTS) to safety violation risks of railway traffic due to low-quality repair and technology violation. The developed algorithm can be used as a subsystem for factor analysis of train safety risks at different levels of track economy and artificial structures.

2. Materials and methods

2.1 Overall risk assessment and indicator factorization framework

The general risk assessment diagram is presented in [1-4] under the name “hybrid method” since it includes the general logical-and-probabilistic method [5-7] and the statistics that are used to control the overall values.

2.2 Risk assessment subsystem according to MTS and HF factors

The general diagram of risk assessment subsystem for MTS and human factors is presented in Figure 1 alongside with the main stages of its implementation.

Unlike the assessment of functional risks by risk factors of technical condition (deviations of 3, 4 degree of the Urban Planning Code of the Russian Federation, rail deficiency, deficiency of metal parts of railroad switches, railway on wooden basis), in this case the factors leading to the violation of repair rules (RR) and repair technology (RT) are on the first place (item 2 of the diagram in Figure 1 and further, in the presence of data, feasible item 3). However, this tree gives the probability function of presence or absence of RR and RT violations, and in order to estimate the appearance of risk factors there is a need for item 3.a – assessment of actual probability of TSV due to RR and RT (per one risk factor of technical condition) and item 4 – such scaling of previously obtained probability functions (RR and RT violations), which leads to the risk factor as a result of repair. As a result of scaling, we obtain the probability function (item 5 of the diagram) of TSV risk factors due to repair.

Further (item 6 of the diagram) – construction of the event tree leading the risk factor to TSV (at the output), giving an opportunity to determine the probability of TSV for the 1st risk factor (item 7), which makes it possible to determine the probability of TSV in the end due to causes of violations of RR and RT (item 8) and to implement the Birnbaum schemes [5-7] to define the contribution into primary factors, and then to perform item 9-10 of the diagram.

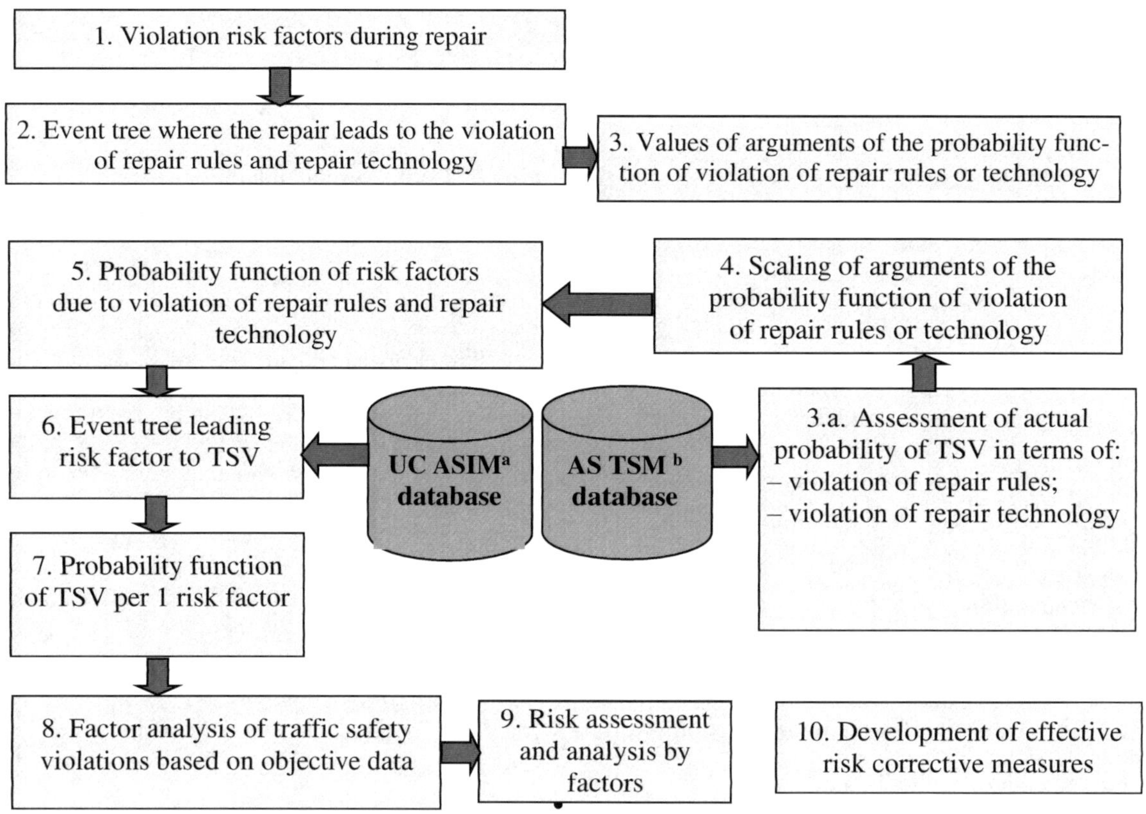

UC ASIMa – Unified corporate automated system of infrastructure management;
AS TSMb – Automated system of traffic safety management.

Figure 1. General diagram of factor risk analysis during repair

The diagram, on the basis of which scaling is performed, is given in Figure 2 for the assessment of risk for TSV according to deviations of 3, 4 degree of the Urban Planning Code of the Russian Federation.

Proportion of total violations (TSV)

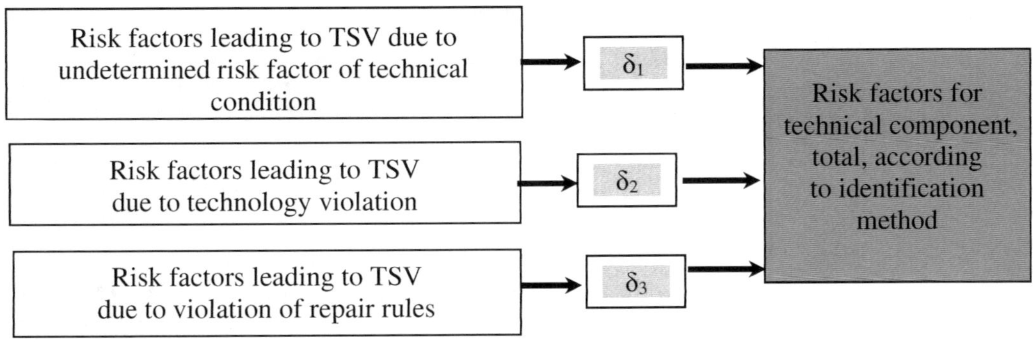

Figure 2. Scheme used to obtain conditions when scaling the values of the probability function arguments

Here, δ_1, δ_2, δ_3 – fractions in the quantitative assessment of the risk factor of j type, which we denote as λ.

$$\delta_1 + \delta_2 + \delta_3 = 1$$

Implementation of the diagram shown in Figure 1.

Item 1 of the diagram is implemented on the basis of [2] – method of risk factors identification during repair, factors are listed in Table 1.

Table 1. Types of factors

Group of factors	Factor	Reference number
Poor staff training and qualification	Understaffing of track servicemen	1
	Understaffing of section foremen	2
	Understaffing of track supervisors	3
	Understaffing of rail detector operators	4
	Turnover of servicemen	5
	Share of workers with the length of service in a position less than 1 year	6
MTS level	Lack of tools and labor saving devices	7
	Incomplete assembly of DBS, %	8
	Underperformance of material-technical resources supply plan, %	9
	Lack of equipment to measure track parameters and railroad switches	10

Item 2. The event tree is constructed according to [8], the horizontal line at the branching point – yes, the vertical line – not. The enlarged diagram of the event tree is shown in Figure 3.

Item 3. On the basis of the event tree, the probability function of no violations in repair is constructed by an expression that is built using the event tree (horizontal upper line – yes, vertical line – no [8]:

$$F(x_1, x_2, x_3, \ldots, x_8, x_9) = \prod_{k=1}^{9} x_i \quad (1)$$

Specific values of arguments x_i are generated by individual mathematical models [8-10].

The probability function of repair and technology violations: $P(x_1, x_2, x_3, \ldots, x_8, x_9) = 1 - \prod_{k=1}^{9} x_i$ (2)

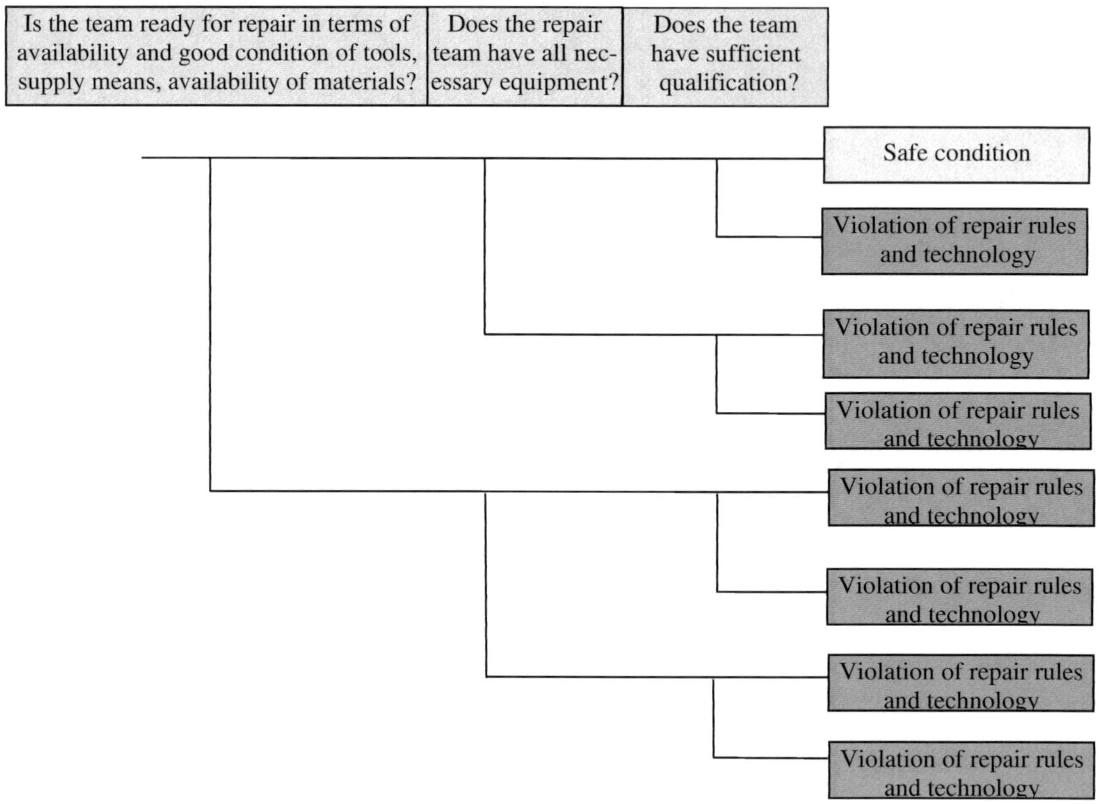

Figure 3. Enlarged event tree of violations during repair

For a specific situation at a linear enterprise, the number of repairs with the violation of repair technology or rules is calculated according to formulas:

$$Q = N \cdot P(q_1, q_2, q_3, \ldots, q_8, q_9) \quad (3)$$

$$P(q_1, q_2, q_3, \ldots, q_8, q_9) = 1 - \prod_{k=1}^{9} q_k$$

where $q_1, q_2, q_3, \ldots, q_8, q_9$ – specific values of arguments derived from material supply data and the state of the "human factor",

N – number of repairs;

x_i – variables ($0 \le x_i \le 1$) determining a condition of all repair components;

П – product sign;

I = 1 - 9 – whole index.

Here, in the time period of the task, there are clear values of the above variables, indicated respectively as $q_1, q_2, q_3, \ldots, q_8, q_9$

$$F(q_1, q_2, q_3, \ldots, q_8, q_9) = \prod_{k=1}^{9} q_k \,. \quad (4)$$

Item 4. The scaling of variables is based on the fact that the topology of the event tree leading to the risk factor is the same as in Figure 3, and therefore the probability function of no risk factors in repair is represented by the same expression, but the value of variables is different. Let us define them as follows:

$$q_1^*, q_2^*, q_3^*, \ldots q_8^*, q_9^* \quad (5)$$

and value of function: $$F(q_1^*, q_2^*, q_3^*, \dots q_8^*, q_9^*) = \prod_{k=1}^{9} q_k^* \quad (6)$$

The value of this function at such values of variables is known (usually from AS of TSM for infrastructure), but q_i^* is not known.

In factor analysis the problem is to determine the values (5) based on statistically known value Q = $(\delta_1+\delta_2) \cdot \lambda$ or

$$N \cdot P(q_1^*, q_2^*, q_3^*, \dots q_8^*, q_9^*) = (\delta_1 + \delta_2) * \lambda \quad (7)$$

where λ – current intensity of risk factor of type j = 1-7;
N – total number of repairs aimed at elimination and prevention of risk factors of j type;
$P(q_1^*, q_2^*, q_3^*, \dots q_8^*, q_9^*)$ – value of the probability function of risk factor due to violation of technology or repair rules.

$$P(q_1^*, q_2^*, q_3^*, \dots q_8^*, q_9^*) = 1 - \prod_{k=1}^{9} q_k^* \quad (8)$$

The mathematical formulation of this task is as follows:
to construct the conversion of numbers $q_1, q_2, q_3, \dots, q_8, q_9$ into numbers $q_1^*, q_2^*, q_3^*, \dots q_8^*, q_9^*$,
where: $F(q_1^*, q_2^*, \dots q_9^*) = C$, (9)
where C – constant obtained from calculations and fuzzy ratios expressed by terms:
Term 1: if $q_i = 1$, then $q_i^* = 1$ (if the state of the factor is complete, it does not contribute to the occurrence of risk factors). (10)
Term 2. The closer q_i to 1, the smaller the change in it when converted into q_i^*. (11)
Term 3: if $q_i = q_K$, then $q_i^* = q_k^*$ (equally imperfect factors contribute equally to risk occurrence). (12)
Term 4: if $q_i < q_K$, then $q_i^* < q_k^*$. (13)

Calculation of factors q_i^* is based on the solution of a problem of minimizing the difference:

$$\left(\prod_{k=1}^{9} q_k^* - C\right) \rightarrow \min. \quad (14)$$

In this case q_i^* (I = 1-n) should lie in the interval (δ, 1), δ > 0, and meet the above specified requirements (10)-(14).

After scaling, taking into account the implementation of items 5-6 and using the probability of TSV for 1 risk factor, we come to the conclusion that the probability of TSV due to repair is calculated according to the formula:

$$P_{TSV}(q_1^*, q_2^*, q_3^*, \dots q_8^*, q_9^*) = (\delta_1 + \delta_2) \cdot (\lambda/N) \cdot p \quad (15)$$

where p – TSV probability per 1 risk factor by technical component calculated using the event tree and the probability function, which is described in detail in [3-4].

The contribution of factor i according to Birnbaum [5-6] is calculated as follows:

$$P_{TSV}(\phi_i) = P_{TSV}(x_i = 1) - P_{TSV}(x_i = q_i^*) \quad (16)$$

Accordingly, the factor risk in a single economic equivalent is calculated by the formula:

$$R(\phi_i) = p_{TSV}(\phi_i) \cdot W \quad (17)$$

where W – given average damage from the type of TSV whose risk is analyzed;
$R(\phi_i)$ – contribution to the risk of TSV factor with number i.

IOP Conf. Series: Materials Science and Engineering **760** (2020) 012065 doi:10.1088/1757-899X/760/1/012065

With this approach, it is possible to obtain contributions from human and MTS factors by summing up all types of repairs in the main area of activities of a linear enterprise, regional and central management offices, i.e. in fact, in the railway network as a whole.

3. Results

Below is the example of calculation of contributions to the risk of MTS factors for one linear enterprise of the North Caucasus Office of Infrastructure and the influence of factors on contributions to risk.

The dynamics of risk factors (technical condition of track superstructure (TSS), MTS, staffing) in 2015 -2018 is presented in Table 2.

Table 2. Risk data

year	Deficiency of railway facility			Deviations in TSS maintenance				Deviations from maintenance standards of continuous welded rail		Staff						MTS factors			
	Number of flawed rails in a main trunk route	Number of defective rails in a main trunk route	Number of defective metal elements of railroad switches	Length of a route with unsatisfactory appraisal by points, km	Number of failures of a route of III, IV degrees per 100 km route	Number of repeated failures of a route of III, IV degrees per 100 km of a track	ength of track sections with excess operating tim	Number of places of temporary repair of continuous welded rail strings per 100 km of a main trunk route	Number of string pairs outside the optimal stress-free temperature	Understaffing of track servicemen	Understaffing of section foremen	Understaffing of track supervisors	Understaffing of rail detector operators	Turnover of servicemen, %	Share of workers with the length of service in a position less than 1 year, %	Lack of tools and labor saving devices	Incomplete assembly of DBS, %	Underperformance of material-technical resources supply plan, %	Lack of equipment to measure track parameters and railroad switches
2015	112	14	10	29	183	61	2.7	189	78	98	100	100	100	3.8	4	18.3	27	57	71
2016	157	52	12	28	280	93	2.7	237	196	100	100	100	100	0.0	0	18.3	29	57	71
2017	268	146	0	10	107	36	2.7	235	294	96	95	100	88	2.2	2	18.3	32.8	57	71
2018	298	52	22	4	165	22	28.3	417	1	98	98	100	85	2.0	2	37.2	100	65	95

*** DBS – distance-based stock*

There is an increase of indicators "Number of failures of a route of III, IV degrees per 100 km", "Number of places of temporary repair of continuous welded rail strings per 100 km of a main trunk route".

4. Discussion

There is positive dynamics of the factor within 2017-2018:

The following factors were improved:

- DBS from 62% to 100%,
- feasibility of the MTS plan,
- availability of route measuring tools and railroad switches.

- understaffing of track servicemen,
- understaffing of track servicemen,
- staffing of rail detector operators.

5. Risk calculation and risk dynamics.

One of the calculation steps is to calculate the risk matrix parameters according to [9, 10].

For this enterprise, the boundaries of risk levels are given in Table 3.

Table 3. Risk level limits (thousand rubles)

Category, color	Unacceptable	Unfavorable	Acceptable (Admissible)	Negligible
Category boundaries	73.01 < R	12.17 < R < 73.01	2.03 < R < 12.17	R < 2.03

The results of risk calculation and assessment by factors are presented in Table 4.

Table 4. Risk distribution by factors in 2017 and 2018 (thousand rubles/year)

Main factors	Poor staff training and qualification						MTS level				
Indicators defining the level of risk development	Understaffing of track servicemen	Understaffing of track servicemen	Understaffing of track servicemen	Understaffing of track servicemen	Understaffing of track servicemen	Understaffing of track servicemen	Understaffing of track servicemen	Understaffing of track servicemen	Understaffing of track servicemen	Understaffing of track servicemen	**Total risk level ***
2017	90	93	0.0	150	247	82	69	73	69	18	482
2018	32.6	32.4	0.0	64.4	92.4	28.1	28.6	29.0	28.6	9.2	212

** The total risk level was determined taking into account factors of technical condition, the calculation diagram of which is specified in [3-4].*

Due to measures taken, the total risk level has decreased by 2.5 times, but is in the category of "unacceptable".

6. Conclusion.

The calculations showed that the objective consideration of the impact of MTS completeness, staffing and staff turnover allowed adding the list of measures with measures positively affecting the risk dynamics towards reduction.

The calculations also show a significant non-linear degree of risk reduction with positive factors (improvement of conditions) of MTS and human factors. At the same time, even some deterioration of the condition of factors of "technical condition" and the increase of risks associated with it are more than compensated for through the improvement of MTS and human factor.

IOP Conf. Series: Materials Science and Engineering **760** (2020) 012065 doi:10.1088/1757-899X/760/1/012065

References

[1] Chernyak A A 1991 Combinatorial-graph method to analyze the reliability of complex systems with monotonic noise functions *Automation and telemetry* **4** 165-174.

[2] Chernyak A A, Chernyak Zh 1998 A Logical-and-probabilistic method to analyze the reliability of systems of higher dimension *Automatics and telemechanics*, **1** 171-180.

[3] Viktorova V S, Sverdlik Yu M, Stepanyantz A S 2010 Analysis of reliability of complex structure systems on multi-level models *Automation and telemetry* **7** 143-148.

[4] Quinlan J R 1986 Induction of decision trees *Mach.Learn* **l. I.** 81-106.

[5] Manwani N, Sastry P S 2012 Geometric decision tree *IEEE Trans. Syst. Man. Cybern.* **42** **1** 181-192.

[6] Moiseev A A 1991 Optimization of control at discrete control actions *Automatics and telemechanics*
9 123-132.

[7] Ayda-Zade K R 1989 Study and numerical solution of finite-difference approximations of distributed systems control tasks *Journal of computational mathematics and mathematical physics* **29** **3** 346-354.

[8] Ragimov A B 2015 Solution of problems of optimal control at piecewise-constant, piecewise-linear and piecewise-defined classes of functions of control effects *Control problems* **2** 13-23.

[9] Fedulov A S, Borisov V V 2016 Models of system dynamics based on odd cognitive maps *Control problems* **1**

[10] Karaev R A 2018 Cognitive tools for dynamic analysis of business strategy of enterprises. *Business Informatics* **1(43)** 7-16

Creation of a learning microprocessor system for protection of contact network feeders using adaptive parametric identification methods

M V Vostrikov[1], **K V Menaker**[1], **V A Ushakov**[2]

[1]Trans-Baikal Institute of Railway Transport, 11 Magistralnaya Str., Chita, 672040, Russia
[2]Irkutsk State Transport University, 15Chernyshevsky Str., Irkutsk, 664074, Russia

E-mail: aspirin1979@mail.ru, menkot@mail.ru

Abstract. Handling of high-weight trains, double trains, packet traffic schedule, recuperation on electric locomotives often leads to false protection relays, downtime of trains and increased wear of power switching and protective equipment of traction substations. The paper proposes methods to improve the selectivity of microprocessor relay protection of feeders of AC traction substations contact network by increasing the accuracy of extraction of the first current and voltage harmonic based on phase locked circuits, integration of learning subroutines by fixing the electric parameters and creating the sets of dynamic setpoints in accordance with the current train situation and adaptive parametric identification of dynamic assessment of current electric parameters of the traction network. It also analyzes structural diagrams of hardware and software of the proposed microprocessor protection system of contact network feeders.

1. Introduction

The main disadvantage of microprocessor protection devices of contact system feeders applied within the network of Russian railways is the probability of overlapping the areas of normal and emergency operation modes of reserve stages of remote protection related to short-term increase of the traction current and change of phase shift angle between current and voltage in normal operation mode when heavy trains take off, when switching the operation modes of electric locomotive engines, activation of recuperation modes, when trains get into protection coverage area and exit it in case of packed traffic schedule, as well as when handling heavy and double trains.

Some works of foreign scientists [1, 2, 4, 5, 6, 9, 10, 14, 15, 16] consider the methods to increase the selectivity of relay protection devices based on adaptive algorithms and learning elements. Among domestic studies we should mention the device of adaptive remote protection of standby stages of contact network feeders [7] and the device to monitor the state of high-voltage inputs [8]. The closest foreign analogues of adaptive systems are given in the following publications [11, 12, 13, 17].

The authors propose an upgraded microprocessor relay protection device containing, besides a set of standard units and algorithms of their operation, a control reference signals module based on phase locked frequency circuits, a learning routine and formation of a set of dynamic setpoints and adaptive parametric identification of transient operation modes of the traction power supply system.

2. Materials and methods

The object of the study includes standard protection and automation unit (PGU) of the 27.5 kV contact network feeder digital protection and automation device (DPA-27.5-CNF), which includes measuring modules of current and voltage sensors (MCVS), protection controller module (PCM), automation controller module (ACM), as well as an integrated module of control signal setting based on automatic frequency control circuits (AFC).

The subject of the study is the operation algorithm of the 27.5 kV contact network feeder digital protection and automation device (DPA-27.5-CNF) improved due to additional learning routines and the formation of a set of dynamic setpoints and adaptive parametric identification of transient operation modes of the traction power supply system.

The study was carried out using the methods of collecting and processing statistical information and experimental data, methods of system analysis and administration.

3. Creation of learning microprocessor system for the protection of contact network feeders

Data on the operation of DPA-27.5-CNF microprocessor protection made it possible to identify the main reasons for its actuation in different operation modes of the traction power supply system (TPSS). The highest percentage of actuations is recorded not due to emergency and forced operation modes of the TPSS, but as a result of many factors, including: overlapping the areas of normal and emergency operation modes of reserve stages of remote protection related to short-term increase of the traction current and change of phase shift angle between current and voltage in normal operation mode when heavy trains take off, when switching the operation modes of electric locomotive engines, activation of recuperation modes, when trains get into protection coverage area and exit it in case of packed traffic schedule, as well as when handling heavy and double trains.

Work [7] proposes an improved operation algorithm of the third stage of DPA-27.5-CNF directed remote protection, which includes an additional adaptive circuit for identification of complex resistance of the traction network. The test showed that the probability of overlapping the actuation zone of normal and emergency operation modes of standby stages during movement of heavy and double trains decreases only if the process is accompanied by gradual change of measured electrical parameters. The modes associated with short-term dynamic changes in current and resistance are not recognized.

The authors propose to use a more advanced algorithm, in which the above-mentioned disadvantages are eliminated by introducing an additional circuit of adaptive parametric identification of transient TPSS operation modes based on the analysis of the first current derivative and complex resistance.

Besides, it is possible to increase the selectivity of backup stages of remote protection of contact system feeders via the learning routine based on oscillograms of current, voltage, complex resistance of emergency modes in accordance with the existing train situation and the formation of the sets of dynamic setpoints.

Since the difference between current and voltage phases is a determining value when recognizing the emergency modes in the traction network, in the proposed system the accuracy of measurement of the initial phase of the first current and voltage harmonic is increased due to the subsystem for setting control reference signals based on automatic frequency control.

The essence of the proposed microprocessor protection system of contact system feeders using adaptive parametric identification methods is presented in the form of a flow diagram (Figure 1).

IOP Conf. Series: Materials Science and Engineering **760** (2020) 012066 doi:10.1088/1757-899X/760/1/012066

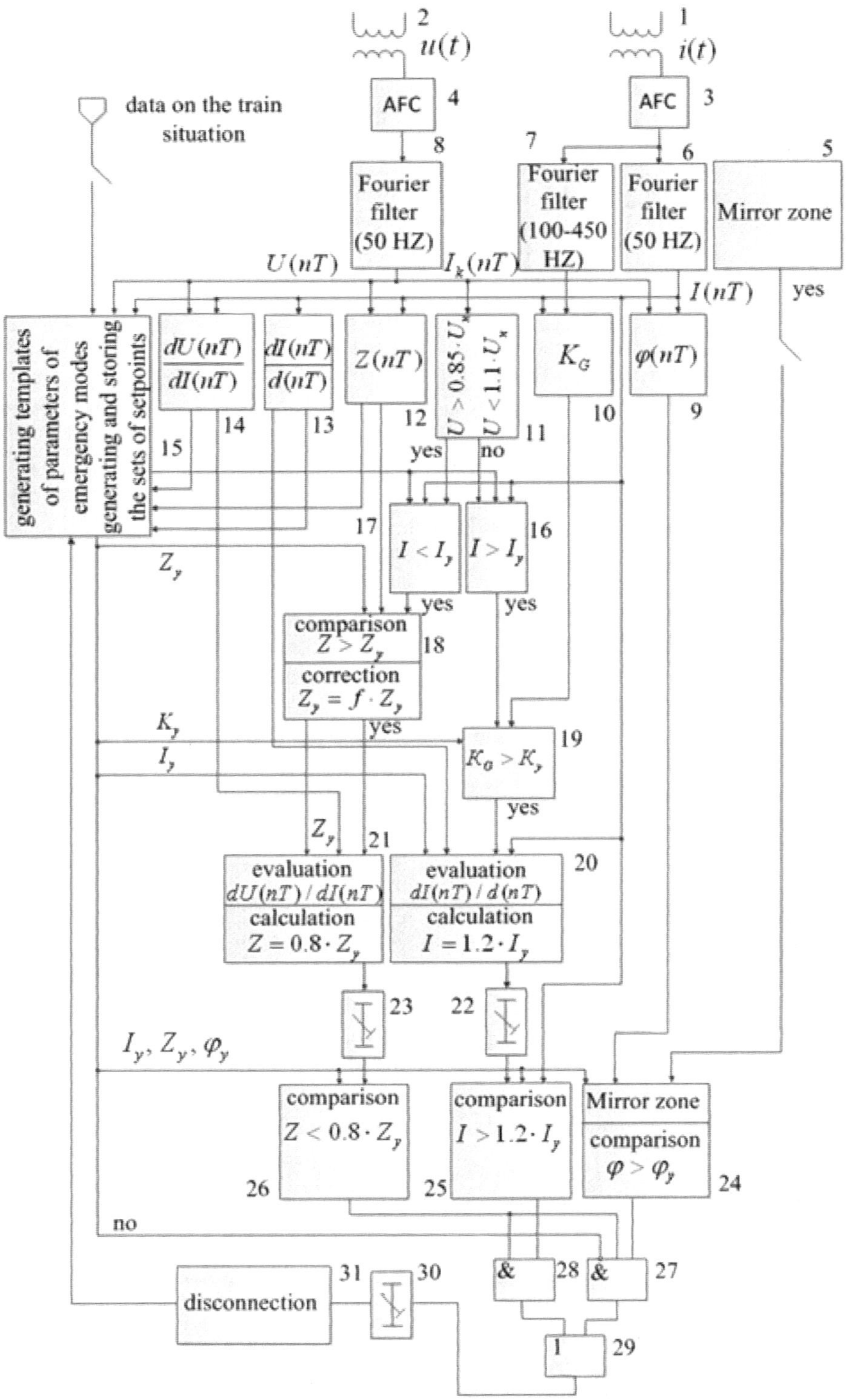

Figure 1. Flow diagram of hardware and software of microprocessor protection system of contact network feeders using adaptive parametric identification methods.

IOP Conf. Series: Materials Science and Engineering **760** (2020) 012066 doi:10.1088/1757-899X/760/1/012066

The measured signals from galvanic isolation and pre-scaling units of the input signals 1 and 2, after being pre-filtered, are to be phase evaluated by automatic frequency control circuits (AFC) 3 and 4 preceding the operations of program digital filtering of current and voltage 6 and 8. AFC circuits generate reference signals, front and rear edges of which coincide in phase with the measured signals of current and voltage at their transition through zero values [3]. The reference signals from the outputs of AFC circuits are supplied to microprocessor control device to determine current and voltage signal period and calculate the quantization interval by dividing by N duration of measured signals frequency expressed in clock frequency of a clock generator. The AFC circuits are primarily aimed to improve the accuracy of measurement of the initial phase of measured signals. The difference between phases of current and voltage is determining when recognizing the emergency modes in the traction network. Frequency filtering units 6, 7, 8, in the form of a software extract the main harmonic of current and voltage, as well as the harmonic signal in the frequency range from 100 to 450 Hz. On the basis of discrete values of current and voltage supplied from filters 6 and 8, the angle of phase shift between current and voltage of the main harmonic in subroutine 9 is calculated and further compared in subroutine 24 with the setpoint value.

Based on discrete values of a current signal coming from filters 6 and 7, current harmonics coefficient is calculated in subroutine 10. The calculated coefficient is supplied to subroutine 19, where it is compared with the value of the harmonic coefficient setpoint. Subroutines 12, 13, 14 calculate the complex resistance of the protected feeder and its first derivative, as well as the current derivative, based on data from filters 6 and 8.

Subroutine 11 is based on determination of voltage location in a preset limit characterizing normal operation mode of the traction network. If the voltage is within the preset setpoint range, the comparator units 16 and 17 compare the measured value of feeder current supplied from the filter 6 with the value of the setpoint current. If the measured current exceeds the setpoint value, then information is supplied from the output of the unit 16 to subroutine 19 for comparing the calculated harmonic coefficient with the setpoint value.

The need to analyze the harmonic coefficient is related to the nonlinear nature of the current signal. If harmonic coefficient exceeds the value of the setpoint, the current derivative coming from subroutine 13 is estimated and the forecast value of current time reduction exceeding the setpoint value by 20%. The “desensitization” of the current setpoint by 20% to the upper side is performed for the purpose of isolation from starting currents of electric locomotives. The analysis of the derivative current at the corresponding sign allows increasing the selectivity of protection at recognition of modes connected with short-term dynamic current shocks that occur when heavy trains take off, when switching the operation modes of electric locomotive engines, etc.

At the end of the forecast delay time implemented as subroutine 22, the current value with the setpoint value cut by 20% in subroutine 25 is re-checked. If the current exceeds the specified value and at the same time the complex resistance is reduced by 20% to the lower side of the setpoint value in subroutine 26, a command is generated to disconnect the feeder with arrangement of the set delay in subroutine 30.

If the current of the setpoint in the comparator 17 is not exceeded, the value of the complex resistance supplied from the subroutine 12 is compared. If the resistance exceeds the setpoint, a setpoint resistance correction signal is generated in block 18. The plurality of subroutines 11, 17, 18 form an adaptive resistance protection identification loop. Subroutine 21 evaluates the derivative of complex resistance and determines the forecast time of resistance reduction to a value less than the setpoint value by 20%. The “desensitization” of the resistance setpoint (by 20% to the lower side) is connected with the need to offset from starting currents of electric locomotives. The analysis of complex resistance derivative at the corresponding sign makes it possible to increase the selectivity of protection at recognition of modes that occur when heavy trains take off, when switching the operation modes of electric locomotive engines, etc.

At the end of the forecast delay time implemented as subroutine 22, the current value with the setpoint value cut by 20% in subroutine 25 is re-checked. If the complex resistance of the cut-off setpoint value is not exceeded and the current of the cut-off value is simultaneously exceeded by 20% to the higher side of the setpoint value in block 25, a command is generated to disconnect the feeder with arrangement of the set delay in subroutine 30. The plurality of subroutines 20, 21, 22, 23 form a loop of adaptive parametric protection identification by current derivative and complex resistance.

The command to disconnect the feeder is also supplied if the logic element 27 generates value "1". This event occurs provided three conditions are fulfilled: the argument of complex resistance of the traction network module exceeds the value of the setpoint when closing the program key "Mirror Zone", which increases the actuation zone due to addition of symmetrical actuation zones; absence of information from subroutine for generation of templates of electrical parameters of emergency modes and generation of sets of setpoints; effect of complex resistance of smaller value of the setpoint cut down by 20% in subroutine 26.

An important element of the proposed system, unlike [7], is the software-level subroutine 15 for generating templates of electrical parameters of emergency modes and generating the sets of setpoints. This subroutine forms an element of system learning. The essence of subroutine is to process information on the current train situation (traffic schedule, train coordinates, train weights, locomotive types, driving modes) received via external digital communication channel and to compare this information with voltage and current oscillograms, as well as change graphs of complex resistance, current derivative and impedance derivative. In process of operation of the protection device information on the change of electric parameters depending on train movement schedule, train weight, locomotive movement modes, individual features of track profile on a protected section is recorded. During processing of statistical information, templates of change of electric parameters are formed, on the basis of which sets of setpoints are developed. In case of frequent false actions of the protection device at repeated train and emergency situations, the user can analyze the information, find the reasons for actions, form sets of setpoints and templates of electrical parameters of emergency modes.

System training related to information accumulation contributes to significant increase of selectivity of remote protection device of contact system feeders and elimination of overlapping between actuation zone of normal and emergency modes of reserve operation stages. In case there is no information on external digital channel concerning current train situation, subroutine 15 performs the role of a device that stores the sets of electric parameters setpoints.

It shall be noted that a significant disadvantage of digital filters functioning as part of microprocessor protection units is the increased error of measurement of the initial phase of the first harmonic of current and voltage related to frequency deviation of the monitored signal from the nominal one. In case of signal deviation, the algorithms of digital filters change the number of samples both within integer and non-integral numbers. This fact actually leads to an increase in the error in determining the phase of the useful signal.

The authors have achieved a significant reduction of this error by introducing additional filtering modules into a standard protection unit and extracting the first harmonic based on AFC circuits (Figure 1, block 3 and 4).

The essence of operation of additional AFC modules is presented in the form of a flow diagram (Figure 2).

The module uses current transformer 1 and voltage transformer 3 as high-voltage inputs, the measuring outputs of which are connected to inputs of galvanic isolation and normalization units of current 2 and voltage 4. Signals from the output of measuring transformers of current and voltage are supplied to low frequencies filters 5 and 8 to exclude higher harmonics and reduce the error of the first harmonic extraction. In parallel, the signals from the measuring transformers are rectified by means of full-wave rectifiers 6 and 7.

From filter outputs the measured current and voltage signals are supplied to inputs of automatic frequency control circuits 9 and 10, at the output of which, when working together with pulse converters 12 and 15, the control signals are generated, the front and rear edges of which coincide in phase with the measured signals at their transition through zero values. The control signals control the operation of transistor switches of quasi-resonance inverters 13 and 14. Supply voltages proportional to amplitude values of current and voltage at the output of measuring transformers 2 and 4 are supplied to transistor switches of inverters 13 and 14 for generation of the first harmonics of measured signals.

In order to increase the reliability of operation, backup circuits bypassing the AFC are provided. In case of failure of PLL circuit, pulse converter or inverter, the control signals are supplied from microcontroller to electronic switches 11 and 16, which implement bypass circuits. From the outputs of inverters 13 and 14 the selected first harmonics of current and voltage are supplied to the inputs of ADC 17 and 18 and further from outputs of ADC to microcontroller PCM 19.

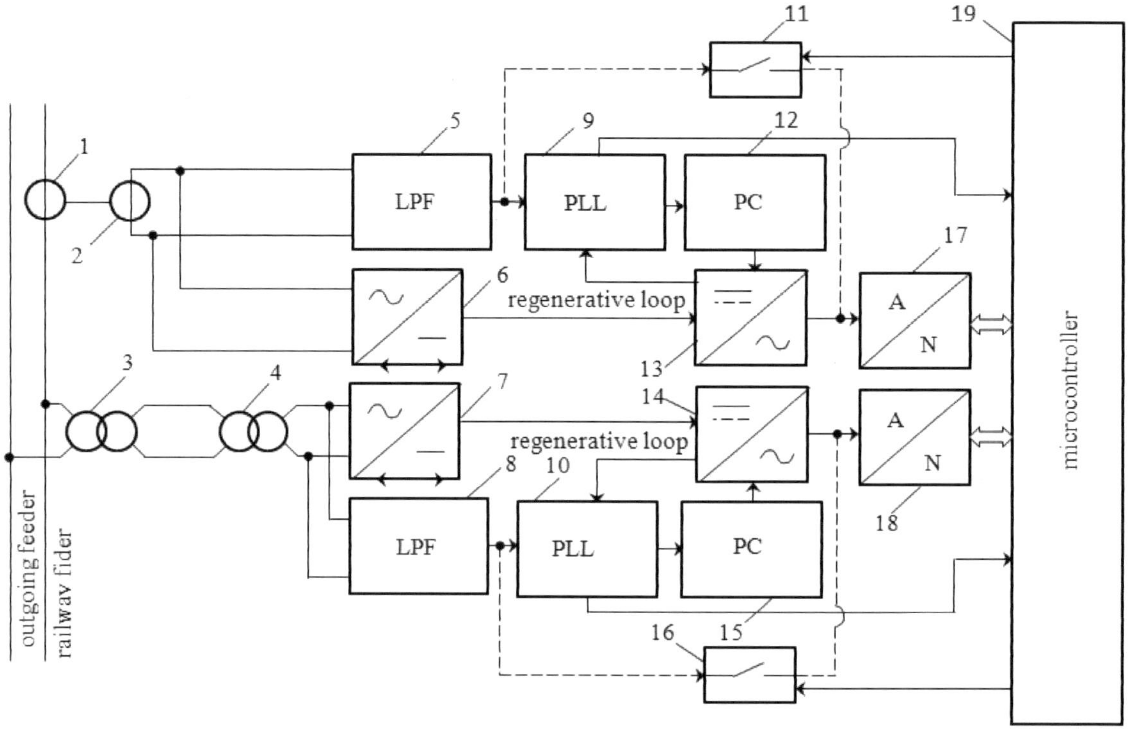

Figure 2. Flow diagram of additional modules for filtration and extraction of the first harmonic on the basis of AFC circuits as part of the standard DPA-27.5-CNF protection unit.

4. Conclusion

The simulation of the operation of the microprocessor protection system of contact network feeders using adaptive parametric identification methods in application packages confirmed the correctness and effectiveness of the proposed software and technical solutions related to modernization of the operation of existing standard DPA-27.5-CNF protection units. The results show that when using the proposed protection system, the number of false activations decreased in comparison with the operation of identical protection [7] by 3-5% at smooth change of electric parameters and by 15-17% in modes related to short-term dynamic changes of current and resistance.

In parallel, the error of measurement of initial phases of the first harmonic of current and voltage was estimated by standard digital filters and taking into account additional modules of filtration and extraction of the first harmonic on the basis of AFC circuits. In the second option, the measurement error decreased by 7-9%.

5. Acknowledgement

The study was carried out under financial support of the Trans-Baikal Institute of Railway Transport (Irkutsk State Transport University), Grant No. 46-gr dated 28.03.2018 “Creation and study of a useful model of an alternative device for filtering and extracting the first harmonic in microprocessor relay protection of contact network feeders", Grant No. 66-gr dated 15.04.2019 "Creation and study of a useful model of adaptive microprocessor relay protection of contact network feeders with alternative filter device".

References

[1] Halabi N El, Martín S, Borroy S, López D, García-Gracia M 2010 Improvement of resistive reach of distance protection through a power flow-based adaptive parameterization *Centre of Research of Energy Resources and Consumption*

[2] Avinash N S, Katti P K, Ghodekar J G 2012. A new adaptive technique for enhancement of zone-2 settings of distance relay *Energy and Power Engineering* **4.**

[3] Phadke A G, Thorp J S 2008 Synchronized Phasor Measurements and their Applications *Springer Science & Business Media* 248

[4] Hideyuki T, Yasutaka S 2011 Advanced line current differential relay with distance protection and adaptive fault locator *TOSHIBA Corporation*

[5] Kamel T S, Moustafa Hassan M A 2010 Adaptive neuro fuzzy inference system (ANFIS) for fault classification in the transmission lines *The Online Journal on Electronics and Electrical Engineering (OJEEE* **2 1** 164.

[6] Kamel T S, Moustafa Hassan M A 2010 On-line trained adaptive neuro-fuzzy inference system for distance relay of transmission line protection *The Online Journal on Electronics and Electrical Engineering (OJEEE)* **1 2** 73.

[7] Trifonov R S, Filippov S A, Yakovlev D A Russian Federation Patent No. 139536. MPK N 02N 7/26 (20.03.2014)

[8] G.M. Tsfasman Patent No. 2401434. MPK G01R 31/14 (28.04.2009)

[9] Srivani S G, Vittal K P Hybrid adaptive algorithm to fast distance relaying *International Journal of Recent Trends in Engineering* **1 1** 403

[10] Srivani S G, Vittal K P 2010 Integrated adaptive reach setting of distance relaying scheme in series compensated lines *International Journal on Electrical Engineering and Informatics* **2 4** 29

[11] Babu K V, Tripathy M and Singh A K 2011 Recent techniques used in transmission line protection: a review, *International Journal of Engineering, Science and Technology* 3 31-8.

[12] Shukla, Jaiswal P, Dubey R and Dohre S 2013 Techniques for transmission line failure detection and control: A review, *International Journal of Innovative Research and studies*, **2 3**, 1-10

[13] Rao M M and Deoghare P M, 2011 Novel Approach for Transmission Line Protection Using Wavelet Transform and Neural Network, *International Journal of Instrumentation, Control and Automation* **1 2** 16-21 ISSN : 2231-1890

[14] Wakhare K C and Wagh N B 2015 Review of Various Algorithms for Protection of Transmission Line, *International Journal of Engineering Research & Technology* **4 01** 346-351

[15] Chaithanya A and Rao A M 2014 PLL with Fuzzy Logic Controller Aided Fast Protection of Strong Power System, *International Journal of Advanced Research in Electrical, Electronics and Instrumentation Engineering* 3, 8 11501-11512

[16] Borkhade D 2014 Transmission Line Fault Detection Using Wavelet Transform, International *Journal on Recent and Innovation Trends in Computing and Communication* **2 10** 3138 – 3142

[17] Patel V and Christian A 2015 Wavelet Transform Application to Fault Classification on Transmission Line, *Indian Journal of Applied Research* **5 2** 208-212

AUTHOR INDEX

Abdurashitov, A.Y. 1
Akhverdiev, K S 10
Aleksandrov, A A 367
Alekseev, M E 20
Alekseeva, T L 20
Aleskovsky, S L 334
Alexandrov, A A 78
Alexeenko, V A 196
Almukhametova, E M 27
Antoshkin, S B 367
Arschinskii, M I 32
Arsentiev, G O 196
Arsentiev, M O 204
Arsentiev, O V 196
Arshinskiy, L V 42
Arshinsky, L V 47
Aslamov, A A 32
Aslamova, E A 47
Aslamova, V S 32, 47
Astashkov, N P 20
Astrakhantsev, L A 20
Averkina, E V 357
Bakanov, M V 367
Banshchikov, A V 57
Baranova, N N 441
Bezridnyj, E S 238
Bikbulatova, G G 63
Blinov, A A 319
Bocharov, I S 468
Bolshakov, R S 286
Butakova, L A 357
Butorin, D V 68, 78
Buyakova, N V 86
Bylina, E E 94
Byshlyago, A A 286
Cheremisin, V T 426
Cherepanov, A V 102, 230, 238
Chigir, A E 217
Christosenkov, S A 109
Chuiko, A D 109
Daneev, A V 57
Daneev, R A 367
Dimov, A V 286, 351
Dorofeev, A N 115
Dovgerd, G A 121
Dovgerd, I A 121
Dulskiy, E Y 127
Dulsky, E Y 134
Dunaev, M P 42
Egorova, M A 303
Eremeev, V K 468
Ermolenko, I Yu 142
Filippenko, N G 68, 149, 258
Filippov, V N 410
Galyautdinov, I I 155, 189
Gladkih, A M 155, 189
Golovnev, G E 426
Govorkov, A S 264
Gozbenko, V 161
Grigoriev, N P 435
Grishin, M V 441, 447, 461
Gryaznov, M V 115
Ignatenko, I V 169
Ignatenkov, A V 121
Ilyin, E S 180, 343
Ilyinykh, V. A. 174
Ivanov, P Y 127, 134
Kargapoltcev, S.K. 258
Karlina, A I 264
Karlina, Yu I 264
Khamnaeva, A A 127
Khudonogov, A M 127, 134
Khudonogov, I A 127, 134
Kim, V A 297
Klimov, N N 246
Klykov, M S 435
Kolmakov, O V 180
Kolmakov, V O 180, 343
Komissarov, A V 461
Komogorcev, M G 292
Konyuhov, V Yu 155, 189
Kopanev, M V 196, 204
Kovtun, M A 210, 319
Krasovskaya, O A 217
Kravchenko, D P 319
Kriklevskaya, L Y 223
Krivov, M V 47
Kryukov, A E 102
Kryukov, A V 86, 230, 238
Kuchera, L Y 204
Kuklina, O K 223
Kupreyeva, E N 326
Kurganov, V M 115
Kutsenko, S M 246
Kutsyi, A P 230, 250
Kutsyi, N N 250

Kuzmin, O V 418
Lagunova, E O 10
Larchenko, A G 78, 258
Lebedeva, O 161
Leonovich, D S 264
Leskovets, I V 270
Livshits, A V 68, 367
Livshits, A.V 258
Lobytsin, I O 127
Lyashko, S V 63
Lysikov, M G 121
Malanova, T V 250
Malovetskaya, E V 286
Manikovskij, A S 292
Martusov, A L 20
Melnikov, A A 297
Melnikov, A S 297
Menaker, K V 485
Mihailova, E A 223
Milovanova, E A 303
Mukutadze, M A 10
Nastasyak, O B 115
Nitezhuk, M S 42
Obuzdina, M V 309, 314
Olshansky, A M 121
Osipova, V.E 278
Parfenov, D Yu 343
Pervukhin, D V 115
Petrov, G I 410
Poltavskaya, Ju 161
Popov, I L 210, 319
Popov, S I 68
Pronina, L A 326
Pultyakov, A V 196
Pykhalov, A A 334
Ratushnyak, V S 180, 343
Rozenberg, E N 121
Rudykh, E A 32
Rusanov, V A 57
Rush, E A 309, 314
Ryabchyonok, N L 20
Sachkov, D I 351
Shafikov, A R 238
Shakirova, E V 357
Shchadova, E I 155, 189
Shtykin, E S 196
Sirina, N F 401
Sitdikov, R R 27
Sizykh, V N 367
Skopintseva, T A 94
Sludneva, L V 94
Solomin, A V 386
Solovyov, A L 395
Solovyov, L Yu 395
Stepanov, V V 292
Sychev, V.P. 1
Tarmaev, A A 410, 468
Tikhomirov, V A 20, 435
Toporkov, D M 210, 319
Tretyakov, E A 426
Trofimovich, P N 435
Trushnikov, V E 441, 447, 461
Tsvik, L B 468
Upyr, R U 351
Ushakov, V A 485
Valitov, R R 447
Verevkina, O I 477
Vetrov, A A 57
Vlasenko, S A 169
Vostrikov, M V 485
Yahina, A S 223
Yakovlev, D.A. 278
Yurkovskaya, E A 94
Yuryev, A V 343
Zakaryukin, V P 86
Zakharov, D V 418
Zakirov, A I 27
Zheleznyak, V N 142
Zhilycov, Y V 32
Zhuravlev, D A 264
Zubkov, V V 401
Zyong, Van Lam 334

Institute of Physics
Dirac House, Temple Back
Bristol BS1 6BE UK

ISSN: 1757-8981
ISBN 978-1-7138-3500-4

9 781713 835004

CURRAN ASSOCIATES INC.
proceedings.com